# Fundamentals of
# Mathematics

THIRD EDITION

# Fundamentals of Mathematics

**ARNOLD R. STEFFENSEN**
**L. MURPHY JOHNSON**
Northern Arizona University

HarperCollins*Publishers*

**To Barbara, Barbara, Becky, Cindy, and Pam**

Sponsoring Editor: Bill Poole
Development Editor: Pam Carlson
Project Editor: Randee Wire/Ann-Marie Buesing
Art Direction: Julie Anderson
Text and Cover Design: Lucy Lesiak Design/Lucy Lesiak
Cover Photo: Chicago Photographic Company ©
Photo Research: Judy Ladendorf
Director of Production: Jeanie Berke
Production Assistant: Linda Murray
Compositor: York Graphic Services
Printer and Binder: Courier Corporation

Fundamentals of Mathematics, Third Edition

ISBN 0-673-46285-4

90 91 92 93 9 8 7 6 5 4 3 2 1

# PREFACE

*Fundamentals of Mathematics, Third Edition,* is designed to give students a review of the basic skills of mathematics and to prepare students for beginning the study of algebra. Informal yet carefully worded explanations, detailed examples with accompanying practice exercises, pedagogical second color, abundant exercises, and comprehensive chapter reviews are hallmarks of the book. The text has been written for maximum instructor flexibility. Both core and peripheral topics can be selected to fit individual course needs. An annotated instructor's edition, testing manual, and test generator are provided for the instructor. Interactive tutorial software and instructional videotapes are also available.

## FEATURES

**STUDENT GUIDEPOSTS** Designed to help students locate important concepts as they study or review, student guideposts specify the major topics, rules, and procedures in each section. The guideposts are listed at the beginning of the section, then each is repeated as the corresponding material is discussed in the section.

**EXAMPLES** Approximately 775 carefully selected examples include detailed, step-by-step solutions and side annotations in color. Each example is headed by a brief descriptive title to help students focus on the concept being developed and to aid in review.

**PRACTICE EXERCISES** These parallel each example and keep students involved with the presentation by allowing them to check immediately their understanding of ideas. Answers immediately follow these exercises.

**CAUTIONS** This feature calls students' attention to common mistakes and special problems to avoid.

**HELPFUL HINTS** These have been supplied throughout the early portions of the text to emphasize useful memory devices or identify more important mechanical procedures.

**COLOR** Pedagogical color highlights important information throughout the book. Key definitions, rules, and procedures are set off in colored boxes for increased emphasis. Figures and graphs utilize color to clarify the concepts presented. Examples present important steps and helpful side comments in color.

**EXERCISES** As a key feature of the text, nearly 6000 exercises, including about 1000 applied problems, are provided. Two parallel exercise sets (A and B) and a collection of extension exercises (C) follow each section and offer a wealth of practice for students and flexibility for instructors. Exercises ranging from the routine to the more challenging, including application and calculator problems, are provided.

**Exercises A** This set of exercises includes space for working the problems, with answers immediately following the exercises. Some of these problems, identified by a colored circle ④, have their solutions worked out at the back of the book. Many of these solutions are to exercises that students frequently have difficulty solving.

**Exercises B** This set matches the exercises in set A problem for problem but is presented without work space or answers.

**Exercises C** This set is designed to give students an extra challenge. These problems extend the concepts of the section or demand more thought than exercises in sets A and B. Answers or hints are given for selected exercises in this set.

**REVIEW EXERCISES** To provide ample opportunities for review, the text features a variety of review exercises.

**For Review** exercises are located at the end of most A and B exercise sets. They not only encourage continuous review of previously covered material, but also often provide special review preparation for topics covered in the upcoming section.

**Chapter Review Exercises** and a practice **Chapter Test** conclude each chapter. The Chapter Review Exercises are divided into two parts. The problems in Part I are ordered and marked by section. Those in Part II are not referenced to the source section, and are thoroughly mixed to help students prepare for examinations. Answers to all review and test exercises are provided in the text.

**Final Review Exercises,** referenced to each chapter and with answers supplied, are located at the back of the book.

**CHAPTER REVIEWS** In addition to the Chapter Review Exercises and Chapter Tests, comprehensive chapter reviews also include **Key Words** and **Key Concepts**. Key Words, listed by section, have brief definitions. Key Concepts summarize the major points of each section.

**CALCULATORS** The use of a calculator is discouraged until Section 5.8, where a brief introduction is presented. After this point no special emphasis is placed on calculator use. A calculator can be a useful tool in working exercises that involve computing with decimals. Exercises of this type in the A and B exercise sets have not been labeled as calculator exercises since it is important for students to learn to judge for themselves when a calculator should or should not be used.

## INSTRUCTIONAL FLEXIBILITY

*Fundamentals of Mathematics, Third Edition,* offers proven flexibility for a variety of teaching situations such as individualized instruction, lab instruction, lecture classes, or a combination of methods.

Material in each section of the book is presented in a well-paced, easy-to-follow sequence. Students in a tutorial or lab instruction setting, aided by the student guideposts, can work through a section completely by reading the explanation, following the detailed steps in the examples, working the practice exercises, and then doing the exercises in set A.

The book can also serve as the basis for, or as a supplement to, classroom lectures. The straightforward presentation of material, numerous examples, practice exercises, and three sets of exercises offer the traditional lecture class an alternative approach within the convenient workbook format.

## NEW IN THIS EDITION

In a continuing effort to make this text even better suited to the needs of instructors and students, the following are some of the enhanced features of the new edition.

- The student guideposts are more visible.
- Explanations have been polished, reworded, and streamlined where appropriate. More figures and illustrations are used in discussions.
- Exercise sets have been reviewed for grading and balance of coverage. The number and variety of practical applications are increased. Additional challenging exercises have been incorporated into Exercises C.
- Additional For Review exercises review topics in preparation for the next section.
- Chapter Review Exercises are presented in two parts, one with sectional references and one without, to help students recognize problem types and better prepare for tests.

- Key Words given in the Chapter Review are expanded to include brief definitions.
- Greater emphasis has been given to the use of figures and illustrations in examples and exercises, particularly applications.
- Diagnostic Pretests have been added for Chapters 1 and 2 to help the instructor determine the extent of coverage necessary for a particular class.
- The proportion approach to percent has been given greater emphasis and is used with the percent equation in a side-by-side presentation in solving percent problems.
- A separate chapter on statistics expands the coverage of this important area.
- The development of operations with signed numbers has been rewritten to provide a more thorough discussion of the topic.
- Helpful Hints have been supplied throughout the text. They are designed to emphasize useful memory devices or identify more important mechanical procedures.

## SUPPLEMENTS

An expanded supplemental package is available for use with *Fundamentals of Mathematics, Third Edition*.

The ANNOTATED INSTRUCTOR'S EDITION provides instructors with immediate access to the answers to every exercise in the text; each answer is printed in color next to the corresponding text exercise.

The INSTRUCTOR'S TESTING MANUAL contains a series of **ready-to-duplicate tests,** including a Placement Test, six different but equivalent tests for each chapter (four open-response and two multiple-choice), and two final exams, all with answers supplied. Section-by-section **teaching tips** provide suggestions for content implementation that an instructor, tutor, or teaching assistant might find helpful.

HARPERCOLLINS TEST GENERATOR FOR MATHEMATICS Available in Apple, IBM, and Macintosh versions, the test generator enables instructors to select questions by objective, section, or chapter, or to use a ready-made test for each chapter. Instructors may generate tests in multiple-choice or open-response formats, scramble the order of questions while printing, and produce multiple versions of each test (up to 9 with Apple, up to 25 with IBM and Macintosh). The system features printed graphics and accurate mathematics symbols. It also features a preview option that allows instructors to view questions before printing, to regenerate variables, and to replace or skip questions if desired. The IBM version includes an editor that allows instructors to add their own problems to existing data disks.

VIDEOTAPES A new videotape series, ALGEBRA CONNECTION: *The Basic Mathematics Course,* has been developed to accompany *Fundamentals of Mathematics, Third Edition*. Produced by an Emmy Award–winning team in consultation with a task force of academicians from both two-year and four-year colleges, the tapes cover all objectives, topics, and problem-solving techniques within the text. In addition, each lesson is preceded by motivational ''launchers'' that connect classroom activity to real-world applications.

INTERACTIVE TUTORIAL SOFTWARE This innovative package is also available in Apple, IBM, and Macintosh versions. It offers interactive modular units, specifically linked to the text, for reinforcement of selected topics. The tutorial is self-paced and provides unlimited opportunities to review lessons and to practice problem solving. When students give a wrong answer, they can request to see the problem worked out. The program is menu-driven for ease of use, and on-screen help can be obtained at any time with a single keystroke. Students' scores are automatically recorded and can be printed for a permanent record.

## ACKNOWLEDGMENTS

We extend our sincere gratitude to the students and instructors who used the previous editions of this book and offered many suggestions for improvement. Special thanks go to the instructors at Northern Arizona University and Yavapai Community College. In particular, the assistance given over the years by James Kirk and Michael Ratliff is most appreciated. Also, we sincerely appreciate the support and encouragement of the Northern Arizona University administration, especially President Eugene M. Hughes. It is a pleasure and privilege to serve on the faculty of a university that recognizes quality teaching as its primary role.

We also express our thanks to the following instructors who responded to a questionnaire sent out by HarperCollins: Luther Henderson, Broward Community College; Susan L. Peterson, Rock Valley College; Frances Rosamond, National University; Andy Tomasulo, Erie Community College, North Campus; and Joseph Williams, Essex County College.

We are also indebted to the following reviewers for their countless beneficial suggestions at various stages of the book's revision:

Janet Bickham, *Del Mar College*

John E. Chavez, *Skyline College*

Doris Cox, *Sauk Valley College*

Gladys Cummings, *Saint Petersburg Junior College*

Patricia Deamer, *Skyline College*

Mary Lou Hart, *Brevard Community College*

Diana L. Hestwood, *Minneapolis Community College*

Winston Johnson, *Central Florida Community College*

Ginny Keen, *Western Michigan University*

Lucy Landesberg, *Nassau Community College*

John Pace, *Essex County College*

David Price, *Tarrant County Junior College*

C. Donald Rogers, *Erie Community College, North Campus*

Beverly Weatherwax, *Southwest Missouri State University*

Victoria J. Young, *Motlow State Community College*

We extend special appreciation to Joseph Mutter for the countless suggestions and support given over the years. Thanks also go to Diana Denlinger for typing this edition.

We thank our editors, Jack Pritchard, Bill Poole, Pam Carlson, Randee Wire, and Ann-Marie Buesing, whose support has been most appreciated.

Finally, we are indebted to our families and in particular our wives, Barbara and Barbara, whose encouragement over the years cannot be measured.

Arnold R. Steffensen
L. Murphy Johnson

# CONTENTS

Students in this course traditionally have a variety of backgrounds. For some, the pace of Chapters 1 and 2 will be appropriate and provide a thorough review in preparation for the more complex topics that follow. On the other hand, some students will be able to review operations on whole numbers at a faster rate or omit it altogether, leaving more time for the remaining sections. As a result, we have provided the following two Diagnostic Pretests for Chapters 1 and 2 to help you determine the extent of coverage necessary for your particular class. Since the questions in each chapter are grouped by section, you may discover that some topics will need to be discussed, while others can be omitted. Answers to the Diagnostic Pretests are given in the Instructor's Guide that accompanies this text and in the Annotated Instructor's Edition. Also, multiple-choice versions of the Pretests have been included in the Instructor's Guide for those who wish to use this format of testing.

## CHAPTER 1 DIAGNOSTIC PRETEST

**1.1**   **1.** Give the first five whole numbers.

     **2.** Consider the number line.

        **(a)** What number corresponds to $b$?

        **(b)** Place the appropriate symbol ($<$ or $>$) between $b$ and 4.

$$b \underline{\quad ? \quad} 4$$

        **(c)** Place the appropriate symbol ($<$ or $>$) between $a$ and $b$.

$$a \underline{\quad ? \quad} b$$

     **3.** Write $300{,}000 + 5000 + 200 + 6$ in standard notation.

     **4.** In the number 62,451,903, what digit represents the following places?

        **(a)** millions     **(b)** ten thousands     **(c)** tens

     **5.** Write 23,401 in expanded notation.

**1.2**   **6.** What law of addition is illustrated by $6 + 10 = 10 + 6$?

     **7.** What is the missing number? $2 + \underline{\quad ? \quad} = 2$

**1.3**   *Find the following sums.*

     **8.**    21     **9.**    648     **10.**    2049     **11.**    4235

        $\underline{+ 49}$         $\underline{+ 37}$        $\underline{+ 1167}$        217

                                                            $\underline{+ 6258}$

     **12.** What law of addition is illustrated by

$$(2 + 3) + 7 = 2 + (3 + 7)?$$

**1.4**   **13.** Round 43,584 to the nearest

        **(a)** ten     **(b)** hundred     **(c)** thousand.

1. _____

2. (a) _____

   (b) _____

   (c) _____

3. _____

4. (a) _____

   (b) _____

   (c) _____

5. _____

6. _____

7. _____

8. _____

9. _____

10. _____

11. _____

12. _____

13. (a) _____

   (b) _____

   (c) _____

**14.** Estimate $\quad$ 2459 by rounding to the nearest
$$\underline{+\quad 375}$$

(a) ten $\quad$ (b) hundred.

14. (a) _____

(b) _____

**15.** During a two-game championship, 8176 people attended the first game, and 9857 attended the second. First estimate the total attendance, and then find the exact attendance.

15. _____

**1.5** **16.** What is the missing number? $\quad 7 - \underline{\quad?\quad} = 2$

16. _____

**17.** What is the missing number? $\quad \underline{\quad?\quad} - 0 = 24$

17. _____

**1.6** *Find the following differences.*

| **18.** | 42 | **19.** | 643 | **20.** | 805 | **21.** | 9003 |
|---|---|---|---|---|---|---|---|
| | $-\ 35$ | | $-\ 97$ | | $-\ 486$ | | $-\ 157$ |

18. _____

19. _____

20. _____

21. _____

**1.7** **22.** Professor Perko has three classes with 32, 45, and 13 students, respectively. How many students does Professor Perko have?

22. _____

**23.** On the first day of football practice, 137 players reported. After one week, 48 players had been cut or quit the team. How many players were still on the team at the end of the week?

23. _____

**24.** Cari received two checks for her graduation, one for $250 and the other for $325. She used the money to buy a sweater for $68, a skirt for $45, and a book for $7. She kept $50 in cash and deposited the rest of the money in her savings account. How much did she deposit?

24. _____

## CHAPTER 2 DIAGNOSTIC PRETEST

**2.1**  **1.** Give the first three multiples of 5.

1. _____

**2.** What law of multiplication is illustrated by

$$5 \times 9 = 9 \times 5?$$

2. _____

**3.** What is the missing number?    $3 \times \underline{\quad ? \quad} = 3$

3. _____

**4.** What is the missing number?    $3 \times \underline{\quad ? \quad} = 0$

4. _____

**2.2**  *Find the following products.*

| **5.** 296 | **6.** 43 | **7.** 651 | **8.** 4351 |
|---|---|---|---|
| $\times\ 5$ | $\times\ 15$ | $\times\ 202$ | $\times\ 600$ |

5. _____

6. _____

7. _____

8. _____

**2.3**  **9.** What law of multiplication is illustrated by

$$(2 \times 3) \times 8 = 2 \times (3 \times 8)?$$

9. _____

**10.** Estimate the product $897 \times 603$ by rounding to the nearest hundred.

10. _____

**11.** Without actually finding the product, use estimation to determine whether the work shown appears to be correct.

```
    409
 × 817
   2863
   409
   3902
 397,153
```

11. _____

**2.4**  **12.** What is the missing number?    $15 \div 3 = \underline{\quad ? \quad}$

12. _____

**13.** What is the missing number?    $12 \div \underline{\quad ? \quad} = 6$

13. _____

**14.** What is the missing number?    $0 \div 7 = \underline{\quad ? \quad}$

14. _____

**15.** What number cannot be used as a divisor?

15. _____

**2.5**  *Find the following quotients and remainders.*

**16.** $7\overline{)168}$    **17.** $21\overline{)745}$    **18.** $12\overline{)3100}$

16. _____

17. _____

18. _____

**2.6**    **19.** Dr. Besnette has 3 children and he wishes to divide equally an inheritance of $7695 among them. How much will each child receive?

19. _____

**20.** Nancy purchases 7 books at $16 each and 3 posters at $4 each. How much change will she receive if she pays with seven 20-dollar bills?

20. _____

**2.7**    **21.** Is 17 a prime number? Explain.

21. _____

**22.** What is the only even prime number?

22. _____

**23.** Find the prime factorization of 150.

23. _____

**24.** Is 1357 divisible by 3?

24. _____

**25.** What property do numbers that are divisible by 10 have?

25. _____

**26.** Numbers that are divisible by 2 or that are multiples of 2 have a special name. What is that name?

26. _____

**2.8**    **27.** Is 16 a perfect square?

27. _____

**28.** What is the square of 25?

28. _____

**29.** What is the square root of 25?

29. _____

**30.** Evaluate $2^4$.

30. _____

**31.** Write $3 \cdot 3 \cdot 3 \cdot 3 \cdot 3 \cdot 3$ using an exponent.

31. _____

*Perform the indicated operations.*

**32.** $3 - 2 + 8 \div 4$

32. _____

**33.** $(2 + 3)^2 - 2^2 + 3^2$

33. _____

**34.** $5 \cdot 2^3$

34. _____

**35.** $\sqrt{9} \div 3$

35. _____

During the past few years we have taught fundamentals of mathematics to over 800 students and have heard the following comments more than just a few times. "I've always been afraid of math and have avoided it as much as possible." "I don't like math, but it's required to graduate." "I can't do percent problems!" If you have ever made a similar statement, now is the time to think positively, stop making negative comments, and start down the path toward success in mathematics. Don't worry about this course as a whole. The material in the text is presented in a way that lets you take one small step at a time. Here are some general and specific guidelines that are necessary and helpful.

## GENERAL GUIDELINES

1. Mastering mathematics requires motivation and dedication. A successful athlete does not become a champion without commitment to his or her goal. The same is true for the successful student of mathematics. Be prepared to work hard and spend time studying.

2. Mathematics is not learned simply by watching, listening, or reading; *it is learned by doing*. Use your pencil and practice. When your thoughts are organized and written in a neat and orderly fashion, you have taken a giant step toward success. Be complete and write out all details. The following are samples of two students' work on a word problem. Can you tell which one was the most successful in the course?

STUDENT A                    STUDENT F

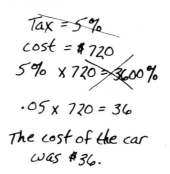

3. The use of a calculator is introduced in Section 5.8. Become familiar with the features of your calculator by consulting the owner's manual. When working problems that involve computing with decimals, a calculator can be a time-saving device. On the other hand, you should not be so dependent on a calculator that you use it to perform simple calculations that might be done mentally. For example, it would be ridiculous to use a calculator to find $8 \div 2$, while it would be helpful in finding $7343.84 \div 1.12$. It is important that you learn when to use and when not to use your calculator.

## SPECIFIC GUIDELINES

1. As you begin to study each section, glance through the material and obtain a preview of what is coming.

2. Return to the beginning of the section and start reading slowly. The STUDENT GUIDEPOSTS will help you locate important concepts as you progress or as you review.

3. Read through each EXAMPLE and make sure that you understand each step. The side comments in color will help you if something is not quite clear.

4. After reading each example, work the parallel PRACTICE EXERCISE. This will reinforce what you have just read and start the process of practice.

5. Periodically you will encounter a CAUTION. These warn you of common mistakes and special problems that should be avoided.

6. The HELPFUL HINTS throughout the text are designed to give you a better understanding of material or a more direct way of solving a problem.

7. After you have completed the material in the body of a section, you must test your mastery of the skills and practice, practice, practice! Begin with the exercises in set A. Answers to all of the problems are placed at the end of the set for easy reference. Some of the problems, identified by colored exercise numbers, have complete solutions at the back of the book. After trying these problems, refer to the step-by-step solutions if you have difficulty. You may want to practice more by doing the exercises in set B or to challenge yourself by trying the exercises in set C.

8. After you have completed all the sections in the chapter, read the CHAPTER RE- VIEW that contains key words and key concepts. The review exercises provide additional practice before you take the CHAPTER TEST. Answers to the tests are at the back of the book.

9. To aid you in studying for your final examination, we have concluded the book with a comprehensive set of FINAL REVIEW EXERCISES.

10. To review the definitions of important terms covered previously, refer to the GLOS- SARY that is also located at the back of the text.

If you follow these steps and work closely with your instructor, you will greatly improve your chance of success in mathematics.

Best of luck, and remember that you can do it!

# Adding and Subtracting Whole Numbers

## 1.1 WHOLE NUMBERS

### STUDENT GUIDEPOSTS

1 Numbers and the Number Line
2 Order and Inequality Symbols
3 Place-Value Number System
4 Word Names for Numbers

### 1 NUMBERS AND THE NUMBER LINE

The numbers that we learn about first are the **natural** or **counting numbers.** They are used to answer the question ''How many?'' The symbols or **numerals** representing these numbers are

$$1, 2, 3, 4, 5, 6, 7, 8, 9, 10, 11, \ldots .$$

The three dots are used to show that the pattern continues.

The **whole numbers** are formed by including 0 with the natural numbers. Thus,

$$0, 1, 2, 3, 4, 5, 6, 7, 8, 9, 10, 11$$

are the first twelve whole numbers.

Numbers are actually ideas that exist in our minds. It is helpful to ''picture'' these ideas using a **number line,** shown in Figure 1.1.

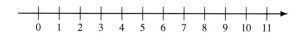

**Figure 1.1**

Every whole number corresponds to a point on a number line. The smallest whole number is 0. There is no largest whole number, since we can keep on extending the number line to the right. The arrowhead points to the direction in which the numbers get larger.

| EXAMPLE 1   Identifying Numbers on a Number Line | PRACTICE EXERCISE 1 |

What numbers are paired with *a, b,* and *c* in Figure 1.2?

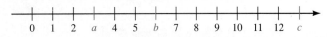

**Figure 1.2**

The point *a* is paired with the number 3, *b* with the number 6, and *c* with the number 13.

What numbers are paired with *x, y,* and *z* on the number line below?

Answer: *x* is paired with 5, *y* is paired with 1, and *z* is paired with 11.

## ❷ ORDER AND INEQUALITY SYMBOLS

The whole numbers occur in a natural **order.** For example, we know that 3 is less than 8, and that 8 is greater than 3. On a number line, 3 is to the left of 8, while 8 is to the right of 3.

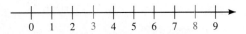

**Figure 1.3**

The symbol $<$ means "is less than." We write "3 is less than 8" as

$$3 < 8.$$

Similarly, the symbol $>$ means "is greater than." Thus, we write "8 is greater than 3" as

$$8 > 3.$$

The symbols $<$ and $>$ are called **inequality symbols.**

---

### Order on a Number Line

1. A number to the left of another number on a number line is *less* than the other number.
2. A number to the right of another number on a number line is *greater than* the other number.

---

## HINT

When the inequality symbols $<$ and $>$ are used, remember that the symbol points to the smaller of the two numbers. Also, when referring to a number line, it may help to remember that "left" and "less" both begin with the same letter.

| EXAMPLE 2   ORDERING NUMBERS | PRACTICE EXERCISE 2 |

Place the correct symbol ($<$ or $>$) between the numbers in the given pairs. Use part of a number line if necessary.

**(a)** 2____7

2 $<$ 7. Notice 2 is to the left of 7 on the number line in Figure 1.4.

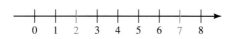

**Figure 1.4**

**(b)** 12____9

12 $>$ 9. Notice that 12 is to the right of 9 on the number line in Figure 1.5.

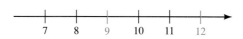

**Figure 1.5**

**(c)** 0____42

0 $<$ 42. See Figure 1.6.

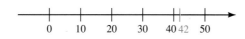

**Figure 1.6**

Place the correct symbol ($<$ or $>$) between the numbers in the given pairs.

**(a)** 3____10

**(b)** 15____5

**(c)** 0____106

Answers:  (a) $<$    (b) $>$    (c) $<$

## HINT

Notice that 13 $>$ 6, read "13 is greater than 6," and 6 $<$ 13, read "6 is less than 13," have the same meaning even though they are read differently.

## ❸ PLACE-VALUE NUMBER SYSTEM

The first ten whole numbers,

$$0, \ 1, \ 2, \ 3, \ 4, \ 5, \ 6, \ 7, \ 8, \ 9,$$

are called **digits** (from the Latin word *digitus* meaning finger). Any whole number can be written using a combination of these ten digits. For example,

245 is a three-digit number
3 is a one-digit number
75 is a two-digit number
4280 is a four-digit number.

The value of each digit in a number depends on its place in the number. In other words, our number system is a **place-value system.** For example, the four-digit number

$$2538,$$

written above in **standard notation,** is a short form of the **expanded notation**

$$2000 + 500 + 30 + 8,$$

as shown in the following diagram.

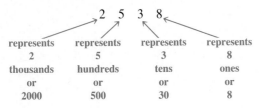

---

### EXAMPLE 3   WRITING IN EXPANDED NOTATION

Write each number in expanded notation.

(a)  532

$$5 \quad 3 \quad 2 = 500 + 30 + 2$$

represents 5 hundreds or 500
represents 3 tens or 30
represents 2 ones or 2

(b)  5032

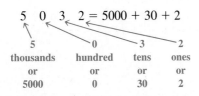

$$5 \quad 0 \quad 3 \quad 2 = 5000 + 30 + 2$$

5 thousands or 5000
0 hundred or 0
3 tens or 30
2 ones or 2

---

### PRACTICE EXERCISE 3

Write each number in expanded notation.

(a)  103

(b)  1030

Answers:  (a)  100 + 3
(b)  1000 + 30

---

Notice the difference between 5032 and 532 in Examples 3 **(b)** and 3 **(a).** In 5032, 0 represents 0 hundreds while the digit 5 represents 5 thousands. If 0 were not in the hundreds place in 5032, we would read the number as 532. It is clear from this example that 0 is an important **place-holder** in our system.

---

### EXAMPLE 4   WRITING IN STANDARD NOTATION

Write each number in standard notation.

(a)  300 + 20 + 8

$$300 + 20 + 8 = 328$$

(b)  5000 + 400 + 30 + 9

$$5000 + 400 + 30 + 9 = 5439$$

(c)  7000 + 20

$$7000 + 20 = 7020 \qquad \text{Zero in the hundreds place and ones place}$$

(d)  90,000 + 300 + 1

$$90,000 + 300 + 1 = 90,301$$

---

### PRACTICE EXERCISE 4

Write each number in standard notation.

(a)  900 + 20 + 7

(b)  8000 + 200 + 70 + 1

(c)  8000 + 70 + 1

(d)  10,000 + 300 + 5

Answers:  (a)  927   (b)  8271
(c)  8071   (d)  10,305

## ④ WORD NAMES FOR NUMBERS

When a number contains more than four digits, commas are used to separate the digits into groups of three, as in the number

$$2,431,897,502,113.$$

Notice that the commas are placed starting at the right. The last group of digits on the left can have one, two, or three digits. Each group of three has a name, as does each digit in the group. These names are given in the following chart.

| trillions | | | billions | | | millions | | | thousands | | | units | | |
|---|---|---|---|---|---|---|---|---|---|---|---|---|---|---|
| hundreds | tens | ones | hundreds | tens | ones | hundreds | tens | ones | hundreds | tens | ones | hundreds | tens | ones |
| 2 | , 4 | 3 | 1 | , 8 | 9 | 7 | , 5 | 0 | 2 | , 1 | 1 | 3 | | |

In words, the number above is

**two trillion, four hundred thirty-one billion, eight hundred ninety-seven million, five hundred two thousand, one hundred thirteen.**

The commas in the word name are in the same place as they are in the standard notation.

////////////////// **CAUTION** //////////////////

Do not use the word *and* when expressing whole numbers in words. For example, 423,123 is four hundred twenty-three thousand, one hundred twenty-three, *not* four hundred *and* twenty-three thousand, one hundred *and* twenty-three. The word *and* will be used later in Chapter 5 to refer to a decimal point in a number.

//////////////

---

**EXAMPLE 5　WRITING WORD NAMES FROM STANDARD NOTATION**

Write a word name for each number

**(a)** 39,205

$$39{,}205$$
$$\downarrow$$
thousand

The word name is thirty-nine thousand, two hundred five. The word *and* is not put between the words *hundred* and *five*.

**(b)** 27,532,141,269

**PRACTICE EXERCISE 5**

Write a word name for each number.

**(a)** 436,109

**(b)** 84,000,227

**(c)** 7,643,000,058

The word name is twenty-seven billion, five hundred thirty-two million, one hundred forty-one thousand, two hundred sixty-nine.

**(c)** 904,000,102,425,000

The word name is nine hundred four trillion, one hundred two million, four hundred twenty-five thousand.

---

**HINT**

Four-digit numbers are written without commas. For example, we write 4136 instead of 4,136. However, a number with five or more digits, such as 14,136, should be written with the comma.

---

**EXAMPLE 6  WRITING STANDARD NOTATION FROM WORD NAMES**

Write each number in standard notation.

**(a)** Four thousand, one hundred thirty-six.

4136

**(b)** Four hundred eighty-three billion, two hundred sixty million, five hundred twenty-one thousand, nine hundred ninety-six

| billions | millions | thousands | units |
|---|---|---|---|
| 4 8 3 , | 2 6 0 , | 5 2 1 , | 9 9 6 |

**(c)** Six trillion, three hundred twenty-four million, five hundred six

6,000,324,000,506

**PRACTICE EXERCISE 6**

Write each number in standard notation.

**(a)** Seven thousand, twenty-three

**(b)** Two hundred eighty-seven million, four hundred five thousand, nine hundred sixty-six

**(c)** Nine trillion, seven hundred fifty thousand, two hundred one

Answers: (a) 7023
(b) 287,405,966
(c) 9,000,000,750,201

---

The value of each digit in a number is determined by its place or position.

**EXAMPLE 7  GIVING THE VALUE OF A DIGIT IN A NUMBER**

Give the value of the digit 4 in each number.

**(a)** 304,152          4 thousands

**(b)** 23,241,309,117     4 ten millions

**(c)** 452,000,302,198    4 hundred billions

**(d)** 32,922,647        4 tens

**PRACTICE EXERCISE 7**

Give the value of the digit 7 in each number.

**(a)** 1,720,458

**(b)** 65,286,127,003

**(c)** 97,480,153

**(d)** 925,703

Answers: (a) 7 hundred thousands
(b) 7 thousands  (c) 7 millions
(d) 7 hundreds

| EXAMPLE 8  IDENTIFYING DIGITS BY NAME | PRACTICE EXERCISE 8 |

In the number 12,345,678 give the digit that represents the following places.

(a) Ten millions         1

(b) Hundred thousands    3

(c) Hundreds             6

(d) Ten thousands        4

(e) Ones                 8

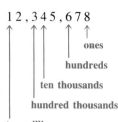

In the number 856,321,409 give the digit that represents the following places.

(a) Hundred millions

(b) Millions

(c) Ten thousands

(d) Thousands

(e) Tens

Answers: (a) 8   (b) 6   (c) 2
(d) 1   (e) 0

Extremely large numbers, in the billions or larger, are difficult to imagine. For example, traveling day and night at a rate of 55 miles per hour, it would take over 2000 years to travel one billion miles and over 2,000,000 years to travel one trillion miles.

## 1.1  EXERCISES A

**1.** What are the numbers 1, 2, 3, . . . called?

**2.** How are numbers often "pictured" in mathematics?

**3.** How do we read the symbols 7 > 2?

**4.** What is the largest three-digit number?

**5.** What is the smallest two-digit number?

**6.** What is the largest digit?

**7.** Is 10 a natural number?

**8.** Is 10 a digit?

**9.** What is the smallest whole number?

**10.** What is the largest whole number?

**11.** On the given number line, what number is paired with each point?
(a) $a$    (b) $b$    (c) $c$    (d) $d$

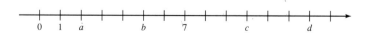

*Refer to the number line below to place the correct symbol (< or >) between the numbers in the following pairs.*

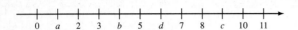

**12.** 2 _____ 3

**13.** 8 _____ 1

**14.** 0 _____ 10

**15.** 7 _____ 0

**16.** *a* _____ 2

**17.** 7 _____ *b*

**18.** 7 _____ *c*

**19.** *d* _____ *a*

**20.** *b* _____ *d*

*Write each number in expanded notation.*

**21.** 2479

**22.** 503

**23.** 207,519

**24.** 4,127,982

*Write each number in standard notation.*

**25.** 500 + 20 + 3

**26.** 10,000 + 600 + 20 + 1

**27.** 300 + 5

**28.** 400,000 + 30,000 + 9000 + 700 + 20 + 1

*Write a word name for each number.*

**29.** 4219

**30.** 107,586

**31.** 93,117

**32.** 13,219,475

*Write each number in standard notation.*

**33.** Six thousand, seven hundred five

**34.** Twenty-four thousand, one hundred fifty-nine

**35.** Three billion, four hundred twenty-seven million, one hundred ninety-three thousand, two hundred

**36.** Eight hundred twenty-five million, one

*Give the value of the digit 5 in each number.*

**37.** 325,076

**38.** 51,032,299

**39.** 3,279,005

**40.** 5,361,291,447

*In the number 63,572,189, what digit represents each of the following places?*

**41.** Millions

**42.** Ten thousands

**43.** Tens

**44.** Ten millions

**45.** Hundred thousands

**46.** Ones

*In Exercises 47–48, write the number in the statement in standard notation.*

**47.** The world population is approximately four billion, eight hundred forty-five million.

**48.** The U.S. Gross National Product in a recent year exceeded three trillion, six hundred twenty-five billion, one hundred fifty million dollars.

*In Exercises 49–50, write a word name for the number in the statement.*

**49.** One light-year, the distance light travels in one year, is approximately 5,879,195,000,000 miles.

**50.** It has been estimated that about 139,710,000 square kilometers of the earth's surface area is water.

---

ANSWERS:   1. natural or counting numbers   2. using a number line   3. 7 is greater than 2   4. 999   5. 10   6. 9   7. yes   8. no   9. 0   10. There is no largest whole number.   11. (a) 2 (b) 5 (c) 10 (d) 13   12. 2 < 3   13. 8 > 1   14. 0 < 10   15. 7 > 0   16. $a < 2$   17. $7 > b$   18. $7 < c$   19. $d > a$   20. $b < d$   21. 2000 + 400 + 70 + 9   22. 500 + 3   23. 200,000 + 7000 + 500 + 10 + 9   24. 4,000,000 + 100,000 + 20,000 + 7000 + 900 + 80 + 2   25. 523   26. 10,621   27. 305   28. 439,721   29. four thousand, two hundred nineteen   30. one hundred seven thousand, five hundred eighty-six   31. ninety-three thousand, one hundred seventeen   32. thirteen million, two hundred nineteen thousand, four hundred seventy-five   33. 6705   34. 24,159   35. 3,427,193,200   36. 825,000,001   37. 5 thousands   38. 5 ten millions   39. 5 ones   40. 5 billions   41. 3   42. 7   43. 8   44. 6   45. 5   46. 9   47. 4,845,000,000   48. $3,625,150,000,000   49. five trillion, eight hundred seventy-nine billion, one hundred ninety-five million   50. one hundred thirty-nine million, seven hundred ten thousand

## 1.1  EXERCISES B

**1.** What numbers are formed by including 0 with the natural numbers?

**2.** What is the symbol for a number called?

**3.** How do we read the symbols 3 < 4?

**4.** What is another name for the first ten whole numbers?

**5.** What is the largest four-digit number?

**6.** What is the smallest three-digit number?

**7.** Is 10 a counting number?

**8.** Is 10 a whole number?

**9.** What is the smallest natural number?

**10.** What is the largest natural number?

**11.** On the given number line, what number is paired with each point?    **(a)** $a$  **(b)** $b$  **(c)** $c$  **(d)** $d$

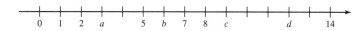

*Refer to the number line below to place the correct symbol (< or >) between the numbers in the following pairs.*

**12.** 4 _____ 11

**13.** 9 _____ 2

**14.** 5 _____ 0

**15.** 0 _____ 8

**16.** $b$ _____ 3

**17.** $a$ _____ 6

**18.** 2 _____ $c$

**19.** $c$ _____ $a$

**20.** $a$ _____ $d$

*Write each number in expanded notation.*

**21.** 409

**22.** 21,155

**23.** 603,114

**24.** 7,123,321

*Write each number in standard notation.*

**25.** 800 + 20 + 7       **26.** 5000 + 800 + 30 + 2     **27.** 30,000 + 500 + 40 + 3 **28.** 70,000 + 300

*Write a word name for each number.*

**29.** 5143          **30.** 203,195        **31.** 65,728        **32.** 21,205,414

*Write each number in standard notation.*

**33.** Eleven thousand, two hundred twenty-six      **34.** Ninety-five thousand, six hundred twenty-seven

**35.** Ten billion, two hundred five million, three hundred forty-seven      **36.** Seven hundred six million, fifteen

*Give the value of the digit 9 in each number.*

**37.** 193,040       **38.** 923,147        **39.** 27,925        **40.** 8,145,329,500

*In the number 520,193,678, what digit represents each of the following places?*

**41.** Ten millions       **42.** Ten thousands       **43.** Tens

**44.** Ones          **45.** Hundred thousands      **46.** Millions

*In Exercises 47–48, write the number in the statement in standard notation.*

**47.** The population of the greater Los Angeles area exceeds three million, eight hundred thousand.

**48.** There are over one hundred sixty-five million, five hundred thousand television sets in the United States.

*In Exercises 49–50, write a word name for the number in the statement.*

**49.** It has been estimated that motorists in the United States waste over 250,700,000 gallons of gasoline annually just warming up their automobiles.

**50.** A variety store has over 1,570,300 items in its inventory

## 1.1 EXERCISES C

*Assume that a and b represent whole numbers in Exercises 1–4.*

**1.** If $a < b$ is $2 + a < 2 + b$?      **2.** If $a < 5$ and $5 < b$ is $a < b$?

**3.** If $a < b$, could we also write $b > a$?      **4.** If $a < b$ and $b < a$ are both false, what could we conclude?

# 1.2 BASIC ADDITION FACTS

**STUDENT GUIDEPOSTS**

1. Adding Two Whole Numbers
2. Basic Addition Facts
3. Some Properties of Addition
4. Applications of Addition

## 1 ADDING TWO WHOLE NUMBERS

Addition of two whole numbers can be thought of as counting. The two numbers to be added are called **addends,** and the result is their **sum.** The symbol used in addition is the **plus sign, +.** As an example, suppose we add 2 coins and 5 coins. There are 2 coins in one stack and 5 coins in another stack, as in Figure 1.7. The addends are written

$$2 + 5.$$

The sum can be found by counting the total number of coins. When the two stacks are put together, the combined stack contains 7 coins, so

$$2 + 5 = 7.$$

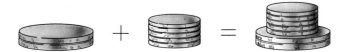

**Figure 1.7**

Addition problems are written either horizontally or vertically.

*Horizontal display*

2 + 5 = 7

(2: addend, 5: addend, 7: sum)

*Vertical display*

```
  2 ← addend
+ 5 ← addend
---
  7 ← sum
```

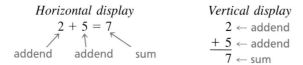

Addition can be shown using a number line. To add 2 and 5 on a number line, first move 2 units to the right of 0. From this point, move 5 more units to the right. The net result is a movement of 7 units from 0. Using the number line in Figure 1.8, we have shown that $2 + 5 = 7$.

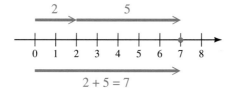

**Figure 1.8**

| EXAMPLE 1 ADDING TWO WHOLE NUMBERS | PRACTICE EXERCISE 1 |

Add.

**(a)** $4 + 2 = 6$   4 objects + 2 objects results in 6 objects

We could also show this using the number line in Figure 1.9.

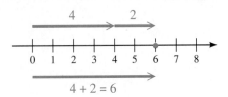

**Figure 1.9**

**(b)**
$$\begin{array}{r} 1 \\ + 8 \\ \hline 9 \end{array}$$   1 object + 8 objects results in 9 objects

The number line in Figure 1.10 shows the same problem.

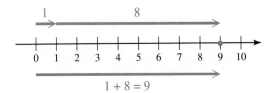

**Figure 1.10**

Add.

**(a)** $6 + 3$

**(b)**
$$\begin{array}{r} 9 \\ + 5 \\ \hline \end{array}$$

Answers: (a) 9   (b) 14

## ❷ BASIC ADDITION FACTS

Finding sums by counting or by using a number line takes time. However, by memorizing the **basic addition facts,** we can find the sum of any two one-digit numbers quickly. These facts, which serve as building blocks for larger sums, are summarized in the table below.

To find $7 + 3$ using the table, start at 7 in the left column and move straight across to the right until you are below the 3 in the top row. You have arrived at the number 10, which is $7 + 3$.

| + | 0 | 1 | 2 | 3 | 4 | 5 | 6 | 7 | 8 | 9 |
|---|---|---|---|---|---|---|---|---|---|---|
| **0** | 0 | 1 | 2 | 3 | 4 | 5 | 6 | 7 | 8 | 9 |
| **1** | 1 | 2 | 3 | 4 | 5 | 6 | 7 | 8 | 9 | 10 |
| **2** | 2 | 3 | 4 | 5 | 6 | 7 | 8 | 9 | 10 | 11 |
| **3** | 3 | 4 | 5 | 6 | 7 | 8 | 9 | 10 | 11 | 12 |
| **4** | 4 | 5 | 6 | 7 | 8 | 9 | 10 | 11 | 12 | 13 |
| **5** | 5 | 6 | 7 | 8 | 9 | 10 | 11 | 12 | 13 | 14 |
| **6** | 6 | 7 | 8 | 9 | 10 | 11 | 12 | 13 | 14 | 15 |
| **7** | 7 | 8 | 9 | 10 | 11 | 12 | 13 | 14 | 15 | 16 |
| **8** | 8 | 9 | 10 | 11 | 12 | 13 | 14 | 15 | 16 | 17 |
| **9** | 9 | 10 | 11 | 12 | 13 | 14 | 15 | 16 | 17 | 18 |

**Addition Table**

| EXAMPLE 2 USING THE ADDITION TABLE | PRACTICE EXERCISE 2 |

Use the Addition Table to find the sums.

**(a)** 8 + 4    Start at 8 in the left column and move across to below 4. The number is 12. Thus 8 + 4 = 12.

**(b)** 4 + 8    Start at 4 in the left column and move across to below 8. Thus 4 + 8 = 12.

**(c)** 0 + 4    Start at 0 in the left column and move across to below 4. Thus 0 + 4 = 4.

**(d)** 4 + 0    Start at 4 in the left column and move across to below 0. Thus 4 + 0 = 4.

Use the Addition Table to find the sums.

**(a)** 2 + 7

**(b)** 5 + 5

**(c)** 0 + 9

**(d)** 9 + 0

Answers: (a) **9**  (b) **10**  (c) **9**  (d) **9**

## ❸ SOME PROPERTIES OF ADDITION

Parts **(a)** and **(b)** of Example 2 show that changing the order of addition does not change the sum. We saw that

$$4 + 8 = 8 + 4.$$

This property of the number system is true in general.

### The Commutative Law of Addition

The **commutative law of addition** tells us that the order in which two (or more) numbers are added does not change the sum.

Parts **(c)** and **(d)** of Example 2 show that

$$0 + 4 = 4 \quad \text{and} \quad 4 + 0 = 4.$$

In general, if zero is added to any number, the sum is the *identical* number. For this reason, zero is sometimes called the **additive identity.**

It will be necessary to find the sum of three one-digit numbers in the next section. We can use a number line as shown in the next example.

| EXAMPLE 3 ADDING THREE ONE-DIGIT NUMBERS | PRACTICE EXERCISE 3 |

Add 3 + 4 + 2 using a number line.

First move 3 units to the right of 0. From this point (corresponding to 3) move 4 more units to the right arriving at 7. Finally, move two additional units to reach the point that corresponds to 9, as shown in Figure 1.11.

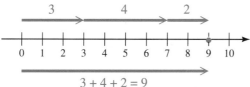

3 + 4 + 2 = 9

**Figure 1.11**

Add 1 + 5 + 3 using a number line.

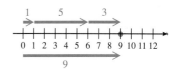

Answer: **9**

## ④ APPLICATIONS OF ADDITION

Many problems are solved by adding whole numbers as shown in the next example.

| EXAMPLE 4   AN APPLICATION OF ADDITION | PRACTICE EXERCISE 4 |

Figure 1.12 shows a sketch of a trail map in a wilderness area. (Note: **mi** is the abbreviation for miles.)

Refer to Figure 1.12

**(a)** A scout troop hiked from the Twin Buttes to Trout Lake and on to the Hangman's Tree. How far did they hike?

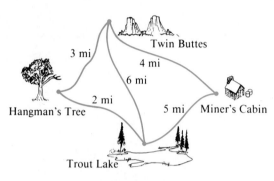

**Figure 1.12**

**(a)** If Joe hikes from the Miner's Cabin to Trout Lake and on to the Twin Buttes, how far has he hiked?

The distance hiked is the sum of 5 mi and 6 mi. Since

$$5 + 6 = 11,$$

Joe hiked 11 mi.

**(b)** Kelli hiked from Trout Lake to the Hangman's Tree, on to the Twin Buttes, and ended up at the Miner's Cabin. How far did she hike?

Since

$$2 + 3 + 4 = 9,$$

Kelli hiked a total of 9 mi.

**(b)** Ranger Dave hiked from the Miner's Cabin to the Twin Buttes, on to Trout Lake, and ended the hike at the Hangman's Tree. How far did he hike?

Answers:  (a)  8 mi    (b)  12 mi

## 1.2  EXERCISES A

*What addition problems are displayed below?*

**1.**
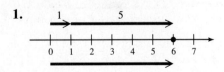

**2.**
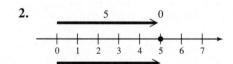

*You should be able to complete Exercises 3–52 in 90 seconds or less.*

| | | | | | | |
|---|---|---|---|---|---|---|
| **3.** 8<br>+ 9 | **4.** 0<br>+ 2 | **5.** 3<br>+ 7 | **6.** 6<br>+ 0 | **7.** 2<br>+ 4 | **8.** 7<br>+ 1 | **9.** 0<br>+ 0 |
| **10.** 2<br>+ 5 | **11.** 7<br>+ 7 | **12.** 8<br>+ 2 | **13.** 9<br>+ 0 | **14.** 6<br>+ 5 | **15.** 7<br>+ 8 | **16.** 3<br>+ 4 |
| **17.** 3<br>+ 0 | **18.** 2<br>+ 7 | **19.** 5<br>+ 9 | **20.** 1<br>+ 3 | **21.** 0<br>+ 4 | **22.** 5<br>+ 7 | **23.** 8<br>+ 1 |
| **24.** 1<br>+ 6 | **25.** 4<br>+ 0 | **26.** 6<br>+ 9 | **27.** 9<br>+ 2 | **28.** 8<br>+ 5 | **29.** 3<br>+ 5 | **30.** 7<br>+ 9 |
| **31.** 0<br>+ 3 | **32.** 7<br>+ 6 | **33.** 4<br>+ 7 | **34.** 9<br>+ 3 | **35.** 3<br>+ 1 | **36.** 8<br>+ 6 | **37.** 9<br>+ 9 |

**38.** $2 + 8 =$     **39.** $6 + 3 =$     **40.** $5 + 5 =$     **41.** $2 + 6 =$     **42.** $0 + 9 =$

**43.** $9 + 1 =$     **44.** $8 + 0 =$     **45.** $4 + 3 =$     **46.** $2 + 3 =$     **47.** $5 + 2 =$

**48.** $9 + 4 =$     **49.** $6 + 1 =$     **50.** $4 + 5 =$     **51.** $5 + 6 =$     **52.** $7 + 3 =$

*Add using a number line.*

**53.** $2 + 3 + 1$     **54.** $4 + 6 + 3$

**55.** $4 + 4 + 7$     **56.** $1 + 5 + 1$

*A hauling company has three trucks, one blue, one red, and one white. The blue truck can haul 2 tons of sand, the red one can haul 5 tons of sand, and the white one can haul 8 tons of sand. Use this information to solve Exercises 57–60.*

**57.** How many tons can be hauled using the blue truck and the red truck?

**58.** How many tons can be delivered using the white truck and the red truck?

**59.** How many tons of sand can be hauled by all three trucks?

**60.** If a contractor orders 10 tons of sand, which two trucks should be used for the delivery?

**FOR REVIEW**

**61.** Write eight trillion, two hundred twenty-five billion, five hundred million, four hundred eighty-six in standard notation.

**62.** Write 63,247 in expanded notation.

**63.** Write a word name for 8,275,111.

**64.** In the number 2,458,769, what digit represents each of the following?
   **(a)** ten thousands    **(b)** millions    **(c)** thousands    **(d)** tens

**65.** Refer to the given number line to place the correct symbol ($<$ or $>$) between the numbers in each pair.

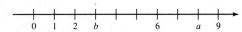

   **(a)** 2 _____ $a$    **(b)** 6 _____ $b$    **(c)** $a$ _____ 5    **(d)** $b$ _____ $a$

---

ANSWERS:   1. $1 + 5 = 6$   2. $5 + 0 = 5$   3. 17   4. 2   5. 10   6. 6   7. 6   8. 8   9. 0   10. 7   11. 14   12. 10   13. 9   14. 11   15. 15   16. 7   17. 3   18. 9   19. 14   20. 4   21. 4   22. 12   23. 9   24. 7   25. 4   26. 15   27. 11   28. 13   29. 8   30. 16   31. 3   32. 13   33. 11   34. 12   35. 4   36. 14   37. 18   38. 10   39. 9   40. 10   41. 8   42. 9   43. 10   44. 8   45. 7   46. 5   47. 7   48. 13   49. 7   50. 9   51. 11   52. 10

**53.**

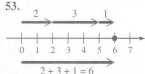

**54.**

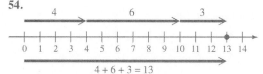

**55.**

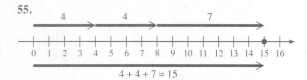

**56.**

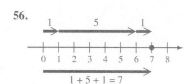

57. 7 tons   58. 13 tons   59. 15 tons   60. the blue truck and the white truck   61. 8,225,500,000,486   62. $60,000 + 3000 + 200 + 40 + 7$   63. eight million, two hundred seventy-five thousand, one hundred eleven   64. (a) 5 (b) 2 (c) 8 (d) 6   65. (a) $2 < a$ (b) $6 > b$ (c) $a > 5$ (d) $b < a$

## 1.2  EXERCISES B

*What addition problems are displayed below?*

**1.**

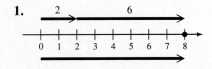

**2.**

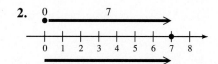

*You should be able to complete Exercises 3–52 in 90 seconds or less.*

| | | | | | | |
|---|---|---|---|---|---|---|
| **3.** 4<br>+ 6 | **4.** 3<br>+ 4 | **5.** 5<br>+ 2 | **6.** 7<br>+ 4 | **7.** 9<br>+ 3 | **8.** 8<br>+ 7 | **9.** 3<br>+ 8 |
| **10.** 9<br>+ 5 | **11.** 8<br>+ 2 | **12.** 6<br>+ 1 | **13.** 3<br>+ 5 | **14.** 2<br>+ 1 | **15.** 0<br>+ 0 | **16.** 4<br>+ 3 |
| **17.** 1<br>+ 7 | **18.** 0<br>+ 7 | **19.** 2<br>+ 5 | **20.** 3<br>+ 9 | **21.** 1<br>+ 1 | **22.** 0<br>+ 3 | **23.** 6<br>+ 4 |
| **24.** 2<br>+ 7 | **25.** 1<br>+ 8 | **26.** 0<br>+ 1 | **27.** 6<br>+ 2 | **28.** 9<br>+ 7 | **29.** 8<br>+ 5 | **30.** 6<br>+ 6 |
| **31.** 4<br>+ 5 | **32.** 0<br>+ 6 | **33.** 9<br>+ 1 | **34.** 2<br>+ 4 | **35.** 1<br>+ 3 | **36.** 0<br>+ 4 | **37.** 5<br>+ 5 |

**38.** 2 + 9 =       **39.** 1 + 3 =       **40.** 8 + 1 =       **41.** 9 + 2 =       **42.** 7 + 1 =

**43.** 3 + 6 =       **44.** 4 + 7 =       **45.** 3 + 7 =       **46.** 0 + 5 =       **47.** 9 + 4 =

**48.** 1 + 4 =       **49.** 4 + 9 =       **50.** 5 + 8 =       **51.** 6 + 7 =       **52.** 5 + 6 =

*Add using a number line.*

**53.** 4 + 5 + 3

**54.** 5 + 3 + 8

**55.** 5 + 5 + 6

**56.** 2 + 9 + 2

*Kim Doane has three boxes for storing cassette tapes, a tan one that holds 5 tapes, a gray one that holds 6 tapes, and a red one that holds 8 tapes. Use this information to solve Exercises 57–60.*

**57.** How many tapes can Kim store in the tan and gray cases?

**58.** How many tapes can be stored in the red case and the gray case?

**59.** How many tapes can Kim store in all three cases?

**60.** Kim plans to take 13 tapes along on a trip. Which two cases should she use?

**FOR REVIEW**

**61.** Write three hundred twenty-one million, one hundred five thousand, six hundred twenty-two in standard notation.

**62.** Write 136,403 in expanded notation.

**63.** Write a word name for 39,205,116.

**64.** In the number 6,258,403, what digit represents each of the following?
(a) millions    (b) ones    (c) thousands    (d) hundred thousands

**65.** Refer to the given number line to place the correct symbol ($<$ or $>$) between the numbers in each pair.

(a) $4$ ____ $y$    (b) $7$ ____ $x$    (c) $y$ ____ $z$    (d) $5$ ____ $z$

## 1.2  EXERCISES C

**1.** If $a$ and $b$ represent whole numbers, what law is illustrated by $a + b = b + a$?

**2.** If $a$ is a whole number, what does $a + 0$ equal?

## 1.3  ADDING WHOLE NUMBERS

**STUDENT GUIDEPOSTS**

❶ Adding a One-Digit Number and a Two-Digit Number

❷ Adding Two Two-Digit Numbers

❸ Adding Two Many-Digit Numbers

❹ Associative Law of Addition

❺ Checking Addition (Reverse-Order Method)

❶ **ADDING A ONE-DIGIT NUMBER AND A TWO-DIGIT NUMBER**

The last section gave the basic facts for adding two one-digit numbers. To add numbers with more digits, these facts are used together with knowledge of the place-value system. To illustrate, suppose we find the following sum.

$$\begin{array}{r} 45 \\ + \; 3 \\ \hline \end{array}$$

Write each addend in expanded notation.

$$45 = 40 + 5 = 4 \text{ tens} + 5 \text{ ones}$$
$$\underline{\phantom{4}3} = \underline{\phantom{40 + }3} = \underline{\phantom{4 \text{ tens} + }3 \text{ ones}}$$

Find the sum by adding the ones digits, 5 and 3, and placing the sum with the 4 tens.

$$\begin{array}{r} 4 \text{ tens} + 5 \text{ ones} \\ 3 \text{ ones} \\ \hline 4 \text{ tens} + 8 \text{ ones} = 48 \end{array}$$

The sum, 48, can be found more directly as shown below using the circled steps.

$$
\begin{array}{r}
45 \\
+\ \ 3 \\
\hline
48
\end{array}
$$

② Bring down tens digit →     48 ← ① Add ones digits

---

## HINT

To add 45 and 3 as shown above, remember that you will write only the sum, 48. The other remarks shown in color are used for explanation.

---

### ❷ ADDING TWO TWO-DIGIT NUMBERS

Next we find the sum of two two-digit numbers such as the following.

$$
\begin{array}{r}
35 \\
+\ 24 \\
\hline
\end{array}
$$

Remember that

$$35 = 3 \text{ tens} + 5 \text{ ones} \quad \text{and} \quad 24 = 2 \text{ tens} + 4 \text{ ones}.$$

Find the number of ones in the sum by adding 5 and 4. Then find the number of tens by adding 3 and 2.

$$
\begin{array}{r}
35 = 3 \text{ tens} + 5 \text{ ones} \\
+\ 24 = 2 \text{ tens} + 4 \text{ ones} \\
\hline
5 \text{ tens} + 9 \text{ ones} = 59
\end{array}
$$

Obviously, the sum can be found more quickly by simply adding the ones digits and then adding the tens digits.

$$
\begin{array}{r}
35 \\
+\ 24 \\
\hline
59
\end{array}
$$

② Add tens digits →    59 ← ① Add ones digits

    Try to find the following sum in the same way. Adding the ones digits and placing the sum directly below the ones digits in the addends gives

$$
\begin{array}{r}
27 \\
+\ 65 \\
\hline
12.
\end{array}
$$
    $7 + 5 = 12$

If the tens digits are added, the resulting sum looks strange and is obviously incorrect.

$$
\begin{array}{r}
27 \\
+\ 65 \\
\hline
812
\end{array}
$$
← This is not correct

What have we done wrong? For one thing, most of us would recognize that the sum of 27 and 65 could not possibly be as large a number as 812! In fact, we are forgetting what we know about place value. When we add the 2 and 6, we are adding 2 *tens* and 6 *tens* so that the sum is 8 *tens*, not 8 hundreds, as in 812. This is easier to see if we write the addends in expanded notation.

$$
\begin{array}{r}
27 = 20 + 7 = 2 \text{ tens} + \ 7 \text{ ones} \\
+\ 65 = 60 + 5 = 6 \text{ tens} + \ 5 \text{ ones} \\
\hline
8 \text{ tens} + 12 \text{ ones}
\end{array}
$$

Since

$$12 \text{ ones} = 10 \text{ ones} + 2 \text{ ones}$$
$$= 1 \text{ ten} + 2 \text{ ones}, \qquad \mathbf{1 \ ten = 10 \ ones}$$

the sum

$$8 \text{ tens} + \boxed{12 \text{ ones}} = 8 \text{ tens} + \boxed{1 \text{ ten}} + 2 \text{ ones}$$
$$= 8 \text{ tens} + 1 \text{ ten} + 2 \text{ ones}$$
$$= 9 \text{ tens} + 2 \text{ ones} = 92.$$

Thus,

$$\begin{array}{r} 27 \\ + \ 65 \\ \hline 92. \end{array}$$

The steps shown above are usually shortened by using a process called **carrying.**

$$\begin{array}{r} 27 \\ + \ 65 \\ \hline \end{array} \qquad \text{Add ones and think of} \begin{array}{r} 7 \\ + \ 5 \\ \hline 12 \end{array}$$

Write 2, the ones digit of this sum, below the ones digits of the addends. Write 1, the tens digit of the sum, above the tens digits of the addends.

$$\begin{array}{r} 1 \quad \leftarrow \text{Write tens digit in the tens column} \\ 27 \\ + \ 65 \\ \hline 2 \quad \leftarrow \text{Write ones digit here} \end{array}$$

The number 1 is called a **carry number.** Now add 1, 2, and 6 and write the sum, 9, below the tens digits.

$$\begin{array}{r} 1 \\ 27 \\ + \ 65 \\ \hline 92 \end{array}$$

---

## EXAMPLE 1  ADDING TWO TWO-DIGIT NUMBERS

Find each sum.

**(a)** $\begin{array}{r} 34 \\ + \ 49 \end{array}$

The steps to follow are listed below.

①  Add ones digits: $4 + 9 = 13$.

②  Write 3 under ones digits and carry 1.

③  Add 1, 3, and 4 to obtain 8.

$$\begin{array}{r} 1 \\ 34 \\ + \ 49 \\ \hline 83 \end{array}$$

**(b)** $\begin{array}{r} 69 \\ + \ 85 \end{array}$

Begin as before.

$$\begin{array}{r} 1 \\ 69 \\ + \ 85 \\ \hline 4 \end{array} \qquad 9 + 5 = 14; \text{ write 4 and carry 1}$$

---

## PRACTICE EXERCISE 1

Find each sum.

**(a)** $\begin{array}{r} 58 \\ + \ 25 \end{array}$

**(b)** $\begin{array}{r} 87 \\ + \ 65 \end{array}$

At this point, you must add 1, 6, and 8. Notice that the sum, 15, is really

$$15 \text{ tens} = 10 \text{ tens} + 5 \text{ tens}$$
$$= 1 \text{ hundred} + 5 \text{ tens}. \qquad 10 \text{ tens} = 1 \text{ hundred}$$

Write the 5 under the tens digits of the addends, and the 1 in the hundreds place of the sum.

$$\begin{array}{r} 1 \\ 69 \\ + \phantom{0}85 \\ \hline 154 \end{array}$$

Answers: (a) 83    (b) 152

## ③ ADDING TWO MANY-DIGIT NUMBERS

When there are more digits in the addends, continue the process as shown in the next example.

| EXAMPLE 2    ADDING TWO THREE-DIGIT NUMBERS |
|---|

**PRACTICE EXERCISE 2**

Find each sum.

**(a)**     $\begin{array}{r} 875 \\ + 398 \end{array}$

$$\begin{array}{r} 11 \\ 875 \\ + \phantom{0}398 \\ \hline 1273 \end{array}$$

The completed addition is shown above. Remember to use the following steps.

① Add 5 and 8, write 3 and carry 1.

② Add 1, 7, and 9, write 7 and carry 1.

③ Add 1, 8, and 3, getting 12. Since

$$12 \text{ hundreds} = 10 \text{ hundreds} + 2 \text{ hundreds}$$
$$= 1 \text{ thousand} + 2 \text{ hundreds} \qquad 10 \text{ hundreds} = 1 \text{ thousand}$$

write the 2 hundreds in the hundreds place and the 1 in the thousands place.

**(b)**     $\begin{array}{r} 1 \phantom{0} 1 \\ 5627 \\ + \phantom{0}7809 \\ \hline 13436 \end{array}$

Notice that there is no carry number for the hundreds digit, and that 13, the final sum of 1, 5, and 7, is placed in the thousands and ten thousands position.

Find each sum.

**(a)**     $\begin{array}{r} 963 \\ + 459 \end{array}$

**(b)**     $\begin{array}{r} 8246 \\ + 6195 \end{array}$

Answers: (a) 1422    (b) 14,441

## ④ ASSOCIATIVE LAW OF ADDITION

Addition is a **binary operation.** This means that only two numbers are added at a time (*bi* means two). Suppose we look again at the problem of adding three numbers. Find the following sum.

$$
\begin{array}{r}
3 \\
5 \\
+\ 2 \\
\end{array}
$$

Start by adding the top two numbers, 3 and 5. Then add their sum, 8, to 2, to obtain 10. The entire process is usually done mentally, but can be shown in the following way.

We could also calculate the sum by adding the bottom two numbers, 2 and 5, then adding their sum, 7, to 3 to obtain 10. Again, we can show the technique as follows.

Notice that both cases give the same result.

When the sum is written horizontally, as $3 + 5 + 2$, we use parentheses to show which numbers to add first. In the first case,

$$(3 + 5) + 2 = 8 + 2 = 10.$$

In the second case,

$$3 + (5 + 2) = 3 + 7 = 10.$$

Thus,

$$(3 + 5) + 2 = 3 + (5 + 2).$$

Regrouping the numbers does not change the sum. The same is true for any three whole numbers.

### The Associative Law of Addition

The **associative law of addition** tells us that the sum of more than two numbers is not changed by grouping. No matter how many numbers are added, they are always added two at a time.

---

### EXAMPLE 3   ADDING SEVERAL NUMBERS

Find the sum.        $9369 + 387 + 18 + 538$

$$
\begin{array}{r}
123 \\
9369 \\
387 \\
18 \\
+\ \ \ 538 \\
\hline
10312 \\
\end{array}
$$

### PRACTICE EXERCISE 3

Find the sum.

$$294 + 8451 + 12 + 1306$$

① Add 9, 7, 8, and 8; write 2 and carry 3.

② Add 3, 6, 8, 1, and 3; write 1 and carry 2.

③ Add 2, 3, 3, and 5; write 3 and carry 1.

④ Add 1 and 9; write 0 in thousands column and 1 in ten thousands column.

Answer: 10,063

## ❺ CHECKING ADDITION (REVERSE-ORDER METHOD)

We can check addition in one of two ways. One way is to repeat the addition. A second way, which avoids the risk of repeating any error, is to reverse the order of addition. That is, if we added from top to bottom the first time, we check by adding from bottom to top. This method, shown in Example 6, is called the **reverse-order check of addition.**

| EXAMPLE 4   REVERSE-ORDER CHECK | PRACTICE EXERCISE 4 |
|---|---|

Add and check.

**(a)** Add:
```
    3 ← ① 3 + 7 = 10      Check:    3    ② 15 + 3 = 18
    7                               7
 +  8  ② 10 + 8 = 18            +   8 ← ① 8 + 7 = 15
   18                              18
```

Add and check.

**(a)** Add:    6     Check:    6
    2          2
 + 9       + 9

**(b)** Add, and check in reverse order.
```
    21
    35
 +  76
```

The steps used to find the sum are numbered in order. Notice that both the ones column and the tens column are added from the top.

```
           1
④ 1 + 2 = 3  → 21 ← ① 1 + 5 = 6
⑤ 3 + 3 = 6  → 35
⑥ 6 + 7 = 13 → 76 ← ② 6 + 6 = 12
⑦ Write 13   →132 ← ③ Write 2 and carry 1
```

**(b)** Add:    36     Check:    36
    51        51
 + 89      + 89

The steps used to check are numbered in order.

```
⑥ 12+1 = 13  →  1
⑤ 10 + 2 = 12 → 21 ← ② 11 + 1 + 12
④ 7 + 3 = 10  → 35
                76 ← ① 6 + 5 = 11
⑦ Write 13    →132 ← ③ Write 2 and carry 1
```

Answers: (a) 17   (b) 176

## HINT

Sometimes when adding several numbers it helps to look for pairs that add to 10, 20, 30, and so on. This is illustrated below.

```
Add.    11
         7 ↘ 20 ↘
         4 ↗     30
         9 ↗  10 ↗  4
       + 3      4  ──
        ──         34
        34   Write only this sum
```

| EXAMPLE 5 AN APPLICATION OF ADDITION | PRACTICE EXERCISE 5 |

Farmer Adams owns 35 sheep and 48 cattle. How many sheep and cattle does he have on his farm?

Find the sum of 35 and 48.

$$\begin{array}{r} 35 \\ + 48 \\ \hline 83 \end{array}$$

Thus, Farmer Adams owns a total of 83 sheep and cattle.

Basketball player Andy Hurd scored 23 points in the first game of a tournament and 28 points in the second game. How many points did he score in the two games?

Answer: **51 points**

## 1.3 EXERCISES A

*Find the following sums and check using the reverse-order method.*

**1.**  $\begin{array}{r} 42 \\ + 5 \\ \hline \end{array}$
**2.**  $\begin{array}{r} 72 \\ + 6 \\ \hline \end{array}$
**3.** $84 + 3$
**4.**  $\begin{array}{r} 69 \\ + 30 \\ \hline \end{array}$
**5.**  $\begin{array}{r} 33 \\ + 22 \\ \hline \end{array}$

**6.**  $\begin{array}{r} 49 \\ + 35 \\ \hline \end{array}$
**7.**  $\begin{array}{r} 59 \\ + 76 \\ \hline \end{array}$
**8.** $43 + 79$
**9.**  $\begin{array}{r} 99 \\ + 56 \\ \hline \end{array}$
**10.**  $\begin{array}{r} 512 \\ + 65 \\ \hline \end{array}$

**11.**  $\begin{array}{r} 405 \\ + 26 \\ \hline \end{array}$
**12.** $727 + 73$
**13.** $173 + 469$
**14.**  $\begin{array}{r} 837 \\ + 175 \\ \hline \end{array}$
**15.**  $\begin{array}{r} 5824 \\ + 1736 \\ \hline \end{array}$

**16.**  $\begin{array}{r} 9999 \\ + 1111 \\ \hline \end{array}$
**17.**  $\begin{array}{r} 4002 \\ + 3009 \\ \hline \end{array}$
**18.** $3297 + 816$
**19.**  $\begin{array}{r} 7 \\ 3 \\ + 6 \\ \hline \end{array}$
**20.**  $\begin{array}{r} 3 \\ 5 \\ + 8 \\ \hline \end{array}$

**21.**  $\begin{array}{r} 1 \\ 9 \\ + 7 \\ \hline \end{array}$
**22.**  $\begin{array}{r} 31 \\ 46 \\ + 72 \\ \hline \end{array}$
**23.** $85 + 37 + 62$
**24.**  $\begin{array}{r} 456 \\ 767 \\ + 318 \\ \hline \end{array}$
**25.**  $\begin{array}{r} 3 \\ 25 \\ 14 \\ + 2 \\ \hline \end{array}$

**26.** $27 + 6 + 9 + 83$
**27.**  $\begin{array}{r} 3 \\ 79 \\ 5 \\ + 13 \\ \hline \end{array}$
**28.**  $\begin{array}{r} 615 \\ 203 \\ 970 \\ + 341 \\ \hline \end{array}$
**29.**  $\begin{array}{r} 3846 \\ 1092 \\ 637 \\ + 4425 \\ \hline \end{array}$
**30.**  $\begin{array}{r} 5298 \\ 6137 \\ 8056 \\ + 4321 \\ \hline \end{array}$

*Solve.*

**31.** Mary bought 8 pairs of shorts one day and 5 more pairs the next. How many pairs of shorts did she buy?

**32.** * Todd drove 135 miles on the first day of a trip and 467 miles on the second day. How many miles did he drive?

**33.** Attendance at the Friday-night performance of a play was 324. The next night, 275 attended. What was the total attendance these two nights?

**34.** Janet had a balance of $365 in her checking account before making deposits of $418 and $279. How much did she deposit, and what was her new balance?

---

ANSWERS:   1. 47   2. 78   3. 87   4. 99   5. 55   6. 84   7. 135   8. 122   9. 155   10. 577   11. 431   12. 800
13. 642   14. 1012   15. 7560   16. 11,110   17. 7011   18. 4113   19. 16   20. 16   21. 17   22. 149   23. 184
24. 1541   25. 44   26. 125   27. 100   28. 2129   29. 10,000   30. 23,812   31. 13 pairs of shorts   32. 602 mi   33. 599
34. $697; $1062

## 1.3  EXERCISES B

*Find the following sums and check using the reverse-order method.*

**1.**     32
    + 4

**2.**     91
    + 8

**3.** 52 + 7

**4.**     25
    + 72

**5.**     63
    + 18

**6.**     27
    + 48

**7.**     38
    + 83

**8.** 27 + 87

**9.**     213
    + 41

**10.**     392
    + 12

**11.**     519
    + 91

**12.** 989 + 11

**13.** 385 + 297

**14.**     487
    + 399

**15.**     7904
    + 2311

**16.**     6329
    + 4350

**17.**     6195
    + 4805

**18.** 4075 + 28

**19.**     8
    1
    + 5

**20.**     4
    5
    + 4

**21.**     7
    5
    + 2

**22.**     52
    29
    + 14

**23.** 47 + 38 + 91

**24.**     806
    359
    + 461

**25.**     16
    5
    70
    + 19

---

*The color indicates there is a complete solution in the back of the text. See *To the Student* for more details.

**26.** $4 + 27 + 9 + 41$    **27.**

$$\begin{array}{r} 52 \\ 8 \\ 63 \\ + \phantom{0}7 \\ \hline \end{array}$$

**28.**

$$\begin{array}{r} 225 \\ 407 \\ 311 \\ + 597 \\ \hline \end{array}$$

**29.**

$$\begin{array}{r} 8371 \\ 5 \\ 693 \\ + \phantom{00}12 \\ \hline \end{array}$$

**30.**

$$\begin{array}{r} 7896 \\ 4327 \\ 5131 \\ + 8454 \\ \hline \end{array}$$

*Solve.*

**31.** The first floor of the mathematics building has 18 classrooms and the second floor has 14 classrooms. How many classrooms are on the two floors?

**32.** A lab assistant poured 585 cubic centimeters of distilled water into a beaker followed by 428 cubic centimeters of alcohol. How much liquid was poured into the beaker?

**33.** Suppose you spent $47 for groceries, $18 for gas, and $85 for a jacket. What was the total amount that you spent?

**34.** The *odometer* on a car indicates how many miles the car has been driven. Prior to a trip, the odometer on Curt's car read 23,165 miles. He then drove 496 miles one day and 387 miles the next. How many miles did he drive those two days? What was the odometer reading at the end of the second day?

## 1.3  EXERCISES C

*Find the following sums.*

**1.**

$$\begin{array}{r} 30,120 \\ 1021 \\ 40,112 \\ \hline \end{array}$$

**2.**

$$\begin{array}{r} 99,876 \\ 29,432 \\ 67,132 \\ \hline \end{array}$$

**3.**

$$\begin{array}{r} 46,001 \\ 20,110 \\ 31,010 \\ 210 \\ \hline \end{array}$$

**4.**

$$\begin{array}{r} 76,432 \\ 49,612 \\ 58,668 \\ 24,688 \\ \hline \end{array}$$

[Answer: 209,400]

**5.** Sam Passamonte is an insurance salesman. For income tax purposes he keeps track of the number of miles that he drives each year. Last year, the miles driven each month were as follows: 2643 mi, 1298 mi, 3045 mi, 1127 mi, 2213 mi, 1778 mi, 896 mi, 743 mi, 512 mi, 919 mi, 1037 mi, and 1428 mi. How many miles did he drive last year?

**6.** If $a$, $b$, and $c$ represent whole numbers, what law is illustrated by $(a + b) + c = a + (b + c)$?

*Find the following sums.*

**7.**

$$\begin{array}{r} 6 \text{ ft} \quad 2 \text{ in} \\ + 5 \text{ ft} \quad 4 \text{ in} \\ \hline \end{array}$$

**8.**

$$\begin{array}{r} 6 \text{ ft} \quad 8 \text{ in} \\ + 5 \text{ ft} \quad 9 \text{ in} \\ \hline \end{array}$$

# 1.4 ROUNDING AND ESTIMATING WHOLE NUMBERS

STUDENT GUIDEPOSTS

  **1** Rounding a Whole Number       **2** Using Rounded Numbers to Estimate and Check Sums

## **1** ROUNDING A WHOLE NUMBER

In many situations we might use an approximate value of a whole number to obtain an *estimated* answer. For example, suppose we know that one clerk in a post office processed 12,780 letters while a second clerk processed 9125 letters during one working day. We could **estimate** that the total processed by the two was about 22,000 letters, if we *rounded* 12,780 to 13,000 and 9125 to 9000 and mentally added these rounded numbers.

Numbers may be rounded to the nearest ten, hundred, thousand, ten thousand, and so on. To **round** a number to the nearest ten, use the nearest multiple of ten (0, 10, 20, 30, 40, . . .) to approximate the given number.

For example, consider rounding 38 to the nearest ten. We may use part of a number line showing the multiples of ten on either side of the given number. The number 38 is between 30 and 40, as shown in Figure 1.13. Since 38 is closer to 40 than 30, we round up to 40. That is, 38 rounded to the nearest ten is 40.

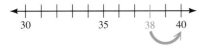

**Figure 1.13**                     **Figure 1.14**

To round 53 to the nearest ten, consider Figure 1.14. The number 53 is between 50 and 60. Since it is closer to 50, 53 rounded to the nearest ten is 50. What if we wanted to round 25 to the nearest ten? Since 25 is halfway between 20 and 30, we would round down to 20 or round up to 30. When a number is halfway between rounding numbers we will agree to round up. Thus, we will round 25 up to 30 as shown in Figure 1.15.

**Figure 1.15**

The following rule allows us to round numbers without using a number line.

| To Round a Whole Number to a Given Position |
| --- |
| **1.** Identify the digit in that position. |
| **2.** If the first digit to the right of that position is 0, 1, 2, 3, or 4, do not change the digit. |
| **3.** If the first digit to the right of that position is 5, 6, 7, 8, or 9, increase the digit by 1. |
| **4.** *In both cases* replace all digits to the right of that position with zeros. |

| EXAMPLE 1   ROUNDING TO THE NEAREST HUNDRED | PRACTICE EXERCISE 1 |
|---|---|

**(a)** Round 392 to the nearest hundred.

The digit 3 is in the hundreds position and 9 is the first digit to the right.

hundreds position
↓
392
↑
first digit to right

Increase 3 by 1 to 4, and replace 9 and 2 with zeros. Thus 400 is the desired rounded number.

**(b)** Round 6237 to the nearest hundred.

The digit 2 is in the hundreds position and 3 is the first digit to the right.

hundreds position
↓
6237
↑
first digit to right

Do not change the 2, and replace 3 and 7 with zeros. Thus 6200 is 6237 rounded to the nearest hundred.

**(c)** Round 2950 to the nearest hundred.

The 9 is in the hundreds position and 5 is the first digit to the right. Increasing 9 by 1 means that the thousands digit must be increased to 3 and the hundreds digit becomes 0. Thus 2950 to the nearest hundred is 3000.

**(a)** Round 541 to the nearest hundred.

**(b)** Round 1875 to the nearest hundred.

**(c)** Round 4950 to the nearest hundred.

Answers: (a) 500   (b) 1900
(c) 5000

| EXAMPLE 2   ROUNDING TO THE NEAREST THOUSAND | PRACTICE EXERCISE 2 |
|---|---|

**(a)** Round 7283 to the nearest thousand.

The 7 is in the thousands position. Since 2 is the first digit to the right, 7283 rounded to the nearest thousand is 7000.

**(b)** Round 26,501 to the nearest thousand.

The 6 is in the thousands position, and 5 is the first digit to the right. Increase 6 to 7 and the rounded number is 27,000.

**(a)** Round 4423 to the nearest thousand.

**(b)** Round 69,825 to the nearest thousand.

Answers: (a) 4000   (b) 70,000

| EXAMPLE 3   ROUNDING WHOLE NUMBERS | PRACTICE EXERCISE 3 |
|---|---|

**(a)** Round 4645 to the nearest ten.

There is a 4 in the tens position and 5 is the first digit to the right. Round 4645 to 4650.

**(b)** Round 4645 to the nearest hundred.

The 6 is in the hundreds position and 4 is the first digit to the right. Leave the 6 and write 4600 for the rounded number.

**(a)** Round 9745 to the nearest ten.

**(b)** Round 9745 to the nearest hundred.

Answers: (a) 9750   (b) 9700

//////////// **C A U T I O N** ////////////

In part **(b)** of Example 3, if 4645 had been rounded to tens before being rounded to hundreds, the final result would have been 4700 instead of 4600. This points out that this method of rounding *cannot* be done in steps; it must be done using the original number, not a rounded number.

////////////

## ② USING ROUNDED NUMBERS TO ESTIMATE AND CHECK SUMS

Rounded numbers are used to make quick *estimates* of sums. These estimates may also be used for a quick *check* for large errors in addition problems.

| EXAMPLE 4   ROUNDING TO THE NEAREST TEN |
| --- |

Estimate the sum by rounding to the nearest ten.

|  | Rounded to nearest ten |
| --- | --- |
| 18 | 20 |
| 82 | 80 |
| 65 | 70 |
| + 21 | + 20 |
|  | 190 |

   The estimated sum is 190. The exact sum, which is 186, can be checked by comparing it to the estimate.

| PRACTICE EXERCISE 4 |
| --- |

Estimate the sum by rounding to the nearest ten.

| 94 |
| --- |
| 48 |
| 25 |
| + 12 |

Answer:  The estimated sum is **180** (exact sum is 179).

| EXAMPLE 5   ROUNDING TO THE NEAREST HUNDRED |
| --- |

Estimate the sum by rounding to the nearest hundred.

|  | Rounded to nearest hundred |
| --- | --- |
| 425 | 400 |
| 607 | 600 |
| 587 | 600 |
| + 235 | + 200 |
|  | 1800 |

   The estimated sum is 1800. The exact sum, which is 1854, can be checked by comparing it to the estimate.

| PRACTICE EXERCISE 5 |
| --- |

Estimate the sum by rounding to the nearest hundred.

| 355 |
| --- |
| 739 |
| 462 |
| + 651 |

Answer:  The estimated sum is **2300** (exact sum is 2207).

////////////// **CAUTION** ///////////////

Using rounded numbers to estimate or check sums provides only a general technique that may not be totally accurate. It should be used with care. For example, in Practice Exercise 5, the estimated sum 2300 differs from the actual sum of 2207 by 93. In part **(a)** of Example 6 an error of 93 causes us to conclude that a mistake was made in addition. In one case, an error of 93 appears acceptable while in the other case it is not. As a result, remember that an *estimate* is an *approximation* not intended to replace actual calculations. Such approximations will help to provide us with some sense of accuracy in our work.

///////////

## EXAMPLE 6  ESTIMATING SUMS BY ROUNDING

Without finding the exact sum, estimate the sum by rounding and use the estimate to check the work shown.

|  |  | Rounded to ten |
|---|---|---|
| **(a)** | 382 | 380 |
|  | 37 | 40 |
|  | 409 | 410 |
|  | + 155 | + 160 |
|  | 983 | 990 |

With an estimated sum of 990, we would probably conclude that the given sum, 983, is correct.

|  |  | Rounded to hundred |
|---|---|---|
| **(b)** | 625 | 600 |
|  | 187 | 200 |
|  | 362 | 400 |
|  | + 717 | + 700 |
|  | 1691 | 1900 |

With an estimated sum of 1900, we would probably conclude that the given sum, 1691, is incorrect. The correct sum is 1891. Can you find where the error was made?

## PRACTICE EXERCISE 6

Estimate the sum by rounding and use your estimate to check the work shown.

**(a)** Check by rounding to the nearest ten.

```
   423
   165
    75
+  349
  1012
```

**(b)** Check by rounding to the nearest hundred.

```
   835
   293
   441
+  752
  2121
```

Answers: (a) The estimated sum is 1020; the sum appears to be correct. (b) The estimated sum is 2300; the sum appears to be incorrect.

Real-life situations often present problems similar to the one in the next example. By using rounded numbers, we can estimate or approximate a solution mentally.

## EXAMPLE 7  ESTIMATING IN AN APPLIED PROBLEM

Harvey has $589 in one savings account and $807 in a second. Without finding the exact total in the two accounts, estimate this total using rounded numbers.

## PRACTICE EXERCISE 7

There are 2985 acres on the Walter Ranch and 4120 acres on the Kirk Ranch. Without finding the exact number, estimate the total number of acres on the two ranches.

Harvey has $589 in one account. This is about $600, rounding to the nearest hundred. Similarly, he has about $800 in the second account. The total in the two accounts can be estimated by adding $600 and $800.

$$\begin{array}{r} 600 \\ + \ 800 \\ \hline 1400 \end{array}$$

Thus, Harvey has about $1400 in the two accounts.

**Answer: 7000 acres (using 3000 and 4000 as the two estimates)**

///////////// **CAUTION** /////////////

Do not estimate a sum by adding first and then rounding the answer. This would serve no useful purpose for checking or for getting a ''rough'' answer.

## 1.4   EXERCISES A

*Round to the nearest ten.*

**1.** 33          **2.** 68          **3.** 15          **4.** 625          **5.** 199

*Round to the nearest hundred.*

**6.** 327          **7.** 372          **8.** 68          **9.** 8550          **10.** 4977

*Round to the nearest thousand.*

**11.** 927          **12.** 9499          **13.** 9500          **14.** 23,450          **15.** 127,456

*Estimate the sum by rounding to the nearest ten. Do not find the exact sum.*

| **16** | **17.** | **18.** | **19.** |
|---|---|---|---|
| 23 | 65 | 602 | 1722 |
| 16 | 91 | 327 | 3345 |
| 54 | 19 | 455 | 2471 |
| + 79 | + 32 | + 281 | + 5128 |

*Estimate the sum by rounding to the nearest hundred. Do not find the exact sum.*

| **20** | **21.** | **22.** | **23.** |
|---|---|---|---|
| 329 | 103 | 602 | 1722 |
| 556 | 75 | 327 | 3345 |
| 113 | 969 | 455 | 2471 |
| + 925 | + 473 | + 281 | + 5128 |

*Estimate the sum by rounding to the nearest thousand. Do not find the exact sum.*

| 24. | 4321 | 25. | 6025 | 26 | 19,501 | 27. | 1722 |
|---|---|---|---|---|---|---|---|
| | 1530 | | 847 | | 19,499 | | 3345 |
| | 2905 | | 9327 | | 38,201 | | 2471 |
| | + 3040 | | + 1700 | | + 21,005 | | + 5128 |

*Without finding the exact sum, determine if the given sum seems correct or incorrect.*

| 28. | 72 | 29. | 357 | 30. | 245 | 31. | 2193 |
|---|---|---|---|---|---|---|---|
| | 16 | | 125 | | 61 | | 6407 |
| | 97 | | 845 | | 193 | | 3556 |
| | + 33 | | + 581 | | + 702 | | + 409 |
| | 218 | | 2308 | | 1201 | | 12,565 |

*Solve.*

**32.** According to a recent census, the population of New York City is 7,908,421. Round this number to the nearest **(a)** million **(b)** hundred thousand.

**33.** During the first few weeks of distribution, the movie *Beverly Hills Cop* grossed over $68,495,240. Round this number to the nearest **(a)** million **(b)** ten million.

**34** There are 7105 books on the first floor of a library, and 6898 books on the second floor. Without finding the exact total, estimate the number of books on the two floors.

**35.** Dr. Morgan purchased a new Buick for $14,958 and a new Honda for $11,141. Without finding the exact total, estimate the amount of money he spent buying these two automobiles.

---

ANSWERS: 1. 30 2. 70 3. 20 4. 630 5. 200 6. 300 7. 400 8. 100 9. 8600 10. 5000 11. 1000 12. 9000 13. 10,000 14. 23,000 15. 127,000 16. 170 17. 210 18. 1670 19. 12,670 20. 1900 21. 1700 22. 1700 23. 12,600 24. 12,000 25. 18,000 26. 98,000 27. 12,000 28. Rounding to nearest ten, the estimated sum is 220; given sum appears correct. 29. Rounding to nearest hundred, the estimated sum is 1900; given sum appears incorrect. 30. Rounding to nearest ten, the estimated sum is 1200; given sum appears correct. 31. Rounding to the nearest hundred, the estimated sum is 12,600; given sum appears correct. 32. (a) 8,000,000 (b) 7,900,000 33. (a) $68,000,000 (b) $70,000,000 34. 14,000 books 35. $26,000

## 1.4 EXERCISES B

*Round to the nearest ten.*

**1.** 47          **2.** 22          **3.** 548          **4.** 195          **5.** 699

*Round to the nearest hundred.*

**6.** 538          **7.** 662          **8.** 59          **9.** 6850          **10.** 2195

*Round to the nearest thousand.*

**11.** 4501          **12.** 4499          **13.** 2799          **14.** 8985          **15.** 135,525

*Estimate the sum by rounding to the nearest ten. Do not find the exact sum.*

| 16. | 17. | 18. | 19. |
|---|---|---|---|
| 34 | 75 | 503 | 2344 |
| 17 | 81 | 426 | 1665 |
| 65 | 18 | 545 | 2582 |
| + 88 | + 43 | + 397 | + 7138 |

*Estimate the sum by rounding to the nearest hundred. Do not find the exact sum.*

| 20. | 21. | 22. | 23. |
|---|---|---|---|
| 239 | 102 | 427 | 1856 |
| 665 | 875 | 581 | 2345 |
| 113 | 53 | 326 | 2281 |
| + 849 | + 271 | + 108 | + 6195 |

*Estimate the sum by rounding to the nearest thousand. Do not find the exact sum.*

| 24. | 25. | 26. | 27. |
|---|---|---|---|
| 6435 | 3175 | 39,501 | 4723 |
| 2505 | 955 | 39,499 | 2549 |
| 3916 | 7258 | 27,208 | 3449 |
| + 5050 | + 1900 | + 41,006 | + 6207 |

*Without finding the exact sum, determine if the given sum seems correct or incorrect.*

| 28. | 29. | 30. | 31. |
|---|---|---|---|
| 81 | 445 | 329 | 4293 |
| 19 | 165 | 83 | 5445 |
| 48 | 851 | 294 | 3795 |
| + 31 | + 558 | + 911 | + 603 |
| 179 | 1999 | 1417 | 14,136 |

*Solve.*

**32.** One year the profit made by the IBEX Corporation was $7,625,428. Round this number to the nearest **(a)** million **(b)** hundred thousand.

**33.** During the first few weeks of distribution, the movie *Raiders of the Lost Ark* grossed over $45,625,155. Round this number to the nearest **(a)** million **(b)** ten million.

**34.** Doug Christensen owns 5208 shares of one stock and 7112 shares of another. Without finding the exact total, estimate the number of shares of the two types of stock that he owns.

**35.** A jet is flying at an altitude of 7895 feet and then increases its altitude by 5120 feet. Without finding the exact altitude, estimate its new altitude.

## 1.4 EXERCISES C

*The Farm Investment Company has four properties in the state of Nebraska. The values of these properties are* $26,482,120, $5,842,580, $18,000,720, *and* $48,216,480. *Estimate the total value of these holdings by rounding to the indicated place.*

**1.** To the nearest thousand
[Answer: $98,542,000]

**2.** To the nearest hundred thousand

**3.** To the nearest million

**4.** To the nearest ten million
[Answer: $110,000,000]

**5.** Use your calculator to find the exact value of these four properties. Even with a calculator it is wise to know something about estimation, since errors can be made using a calculator by pressing the wrong button.

## 1.5 BASIC SUBTRACTION FACTS

**STUDENT GUIDEPOSTS**

**1** Subtracting Whole Numbers

**2** Related Addition and Subtraction Sentences

### 1 SUBTRACTING WHOLE NUMBERS

Subtraction of one whole number from another can be thought of as ''removing'' or ''taking away.'' For example, suppose we subtract 4 coins from 7 coins. If there are 7 coins in a stack and 4 are removed, 3 coins remain. See Figure 1.16.

} Remove 4 coins

7 coins {

} 3 coins remain

**Figure 1.16**

Subtraction is written with a **minus sign, −,** and can be written either horizontally or vertically.

*Horizontal display*

7 − 4 = 3

minuend     subtrahend     difference

*Vertical display*

$$
\begin{array}{r}
7 \leftarrow \text{minuend} \\
- 4 \leftarrow \text{subtrahend} \\
\hline
3 \leftarrow \text{difference}
\end{array}
$$

The number being taken away is called the **subtrahend.** The number it is taken from is the **minuend.** The number remaining is the **difference.**

minuend − subtrahend = difference

Subtraction can be shown on a number line. Recall that to find the sum 5 + 2 using a number line, we move right 5 units from 0 to the point associated with 5 and then move 2 more units to the right to the point 7, as shown in Figure 1.17.

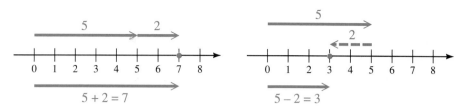

**Figure 1.17**                                    **Figure 1.18**

To find $5 - 2$ using a number line, we start the same way by moving 5 units to the right of 0 to the point 5. We then move 2 units to the *left* of 5 to the point 3. See Figure 1.18. Using a number line we have shown that $5 - 2 = 3$.

## ② RELATED ADDITION AND SUBTRACTION SENTENCES

Although we can subtract whole numbers using the ''take away'' method or the number line, it is better to relate subtraction to addition since we already know the basic addition facts. Each subtraction sentence is related to an addition sentence. For example, $4 + 5 = 9$ is the **related addition sentence** to $9 - 5 = 4$ and $9 - 5 = 4$ is the **related subtraction sentence** to $4 + 5 = 9$. Thus we can solve the subtraction problem,

$$9 - 5 = \square \qquad \text{by solving} \qquad \square + 5 = 9.$$

Since the addition sentence is true when 4 replaces $\square$, the subtraction sentence is true with 4 also.

| | |
|---|---|
| **EXAMPLE 1  USING THE RELATED ADDITION SENTENCE** | **PRACTICE EXERCISE 1** |

Subtract.

**(a)** $6 - 2 = 4$    4 is correct since $4 + 2 = 6$

**(b)** $8 - 6 = 2$    2 is correct since $2 + 6 = 8$

**(c)** $8 - 8 = 0$    0 is correct since $0 + 8 = 8$

**(d)** $9 - 1 = 8$    8 is correct since $8 + 1 = 9$

**(e)** $5 - 0 = 5$    5 is correct since $5 + 0 = 5$

Subtract.

**(a)** $5 - 1$

**(b)** $7 - 7$

**(c)** $9 - 4$

**(d)** $4 - 2$

**(e)** $8 - 0$

Answers:  (a) **4**    (b) **0**    (c) **5**
(d) **2**    (e) **8**

With practice the basic subtraction facts, which come from the basic addition facts, should be easy to remember. Thus we should be able to do basic subtraction quickly. Example 2 gives more practice with the vertical display.

| | |
|---|---|
| **EXAMPLE 2  BASIC SUBTRACTION USING A VERTICAL DISPLAY** | **PRACTICE EXERCISE 2** |

Subtract.

|      | *Subtraction* | *Related addition* |
|------|:---:|:---:|
| **(a)** | 9 | 2 |
|      | $-7$ | $+7$ |
|      | 2 | 9 |
| **(b)** | 5 | 4 |
|      | $-1$ | $+1$ |
|      | 4 | 5 |

Subtract.

**(a)**    8
         $-3$

**(b)**    7
         $-4$

**(c)**
$$\begin{array}{r} 4 \\ -4 \\ \hline 0 \end{array} \qquad \begin{array}{r} 0 \\ +4 \\ \hline 4 \end{array}$$

**(d)**
$$\begin{array}{r} 3 \\ -0 \\ \hline 3 \end{array} \qquad \begin{array}{r} 3 \\ +0 \\ \hline 3 \end{array}$$

**(c)**
$$\begin{array}{r} 3 \\ -1 \\ \hline \end{array}$$

**(d)**
$$\begin{array}{r} 9 \\ -9 \\ \hline \end{array}$$

Answers: (a) 5   (b) 3   (c) 2
(d) 0

## 1.5   EXERCISES A

*What subtraction problems are displayed below?*

**1.**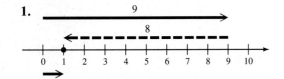

**2.**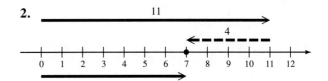

*You should be able to complete Exercises 3–37 in two minutes or less.*

**3.** $\begin{array}{r} 6 \\ -4 \\ \hline \end{array}$
  **4.** $\begin{array}{r} 5 \\ -1 \\ \hline \end{array}$
  **5.** $\begin{array}{r} 2 \\ -1 \\ \hline \end{array}$
  **6.** $\begin{array}{r} 9 \\ -6 \\ \hline \end{array}$
  **7.** $\begin{array}{r} 8 \\ -5 \\ \hline \end{array}$
  **8.** $\begin{array}{r} 7 \\ -0 \\ \hline \end{array}$
  **9.** $\begin{array}{r} 9 \\ -2 \\ \hline \end{array}$

**10.** $\begin{array}{r} 9 \\ -4 \\ \hline \end{array}$
  **11.** $\begin{array}{r} 8 \\ -2 \\ \hline \end{array}$
  **12.** $\begin{array}{r} 7 \\ -1 \\ \hline \end{array}$
  **13.** $\begin{array}{r} 7 \\ -4 \\ \hline \end{array}$
  **14.** $\begin{array}{r} 5 \\ -4 \\ \hline \end{array}$
  **15.** $\begin{array}{r} 6 \\ -2 \\ \hline \end{array}$
  **16.** $\begin{array}{r} 9 \\ -8 \\ \hline \end{array}$

**17.** $\begin{array}{r} 3 \\ -0 \\ \hline \end{array}$
  **18.** $\begin{array}{r} 1 \\ -1 \\ \hline \end{array}$
  **19.** $\begin{array}{r} 5 \\ -2 \\ \hline \end{array}$
  **20.** $\begin{array}{r} 8 \\ -4 \\ \hline \end{array}$
  **21.** $\begin{array}{r} 12 \\ -8 \\ \hline \end{array}$
  **22.** $\begin{array}{r} 10 \\ -4 \\ \hline \end{array}$
  **23.** $\begin{array}{r} 13 \\ -5 \\ \hline \end{array}$

**24.** $\begin{array}{r} 15 \\ -8 \\ \hline \end{array}$
  **25.** $\begin{array}{r} 10 \\ -2 \\ \hline \end{array}$
  **26.** $\begin{array}{r} 17 \\ -9 \\ \hline \end{array}$
  **27.** $\begin{array}{r} 11 \\ -5 \\ \hline \end{array}$
  **28.** $\begin{array}{r} 14 \\ -7 \\ \hline \end{array}$
  **29.** $\begin{array}{r} 10 \\ -1 \\ \hline \end{array}$
  **30.** $\begin{array}{r} 12 \\ -4 \\ \hline \end{array}$

**31.** $\begin{array}{r} 10 \\ -6 \\ \hline \end{array}$
  **32.** $\begin{array}{r} 18 \\ -9 \\ \hline \end{array}$
  **33.** $\begin{array}{r} 16 \\ -9 \\ \hline \end{array}$
  **34.** $\begin{array}{r} 13 \\ -6 \\ \hline \end{array}$
  **35.** $\begin{array}{r} 11 \\ -2 \\ \hline \end{array}$
  **36.** $\begin{array}{r} 15 \\ -7 \\ \hline \end{array}$
  **37.** $\begin{array}{r} 11 \\ -8 \\ \hline \end{array}$

**38.** $14 - 6 =$
  **39.** $10 - 5 =$
  **40.** $12 - 7 =$
  **41.** $11 - 6 =$
  **42.** $15 - 9 =$

**43.** $10 - 8 =$
  **44.** $11 - 3 =$
  **45.** $13 - 9 =$
  **46.** $12 - 5 =$
  **47.** $16 - 8 =$

**48.** $12 - 9 =$
  **49.** $17 - 8 =$
  **50.** $14 - 9 =$
  **51.** $12 - 3 =$
  **52.** $13 - 4 =$

*Solve.*

**53.** Kathy had 9 kittens and gave away 4 of them. How many kittens did she have left?

**54.** Greg has $14 in his wallet. If he buys a book for $6, how much money will he have left?

**FOR REVIEW**

**55.** Estimate the sum by rounding to the nearest   **(a)** ten   **(b)** hundred   **(c)** thousand.

```
  2351
  4932
+ 6145
```

*Without finding the exact sum, determine if the given sum seems correct or incorrect.*

**56.**
```
    19
    38
    52
+   87
   176
```

**57.**
```
   491
   650
   125
+  749
  2015
```

**58.**
```
  2395
   841
    89
+  153
  3478
```

ANSWERS:   1. $9 - 8 = 1$   2. $11 - 4 = 7$   3. 2   4. 4   5. 1   6. 3   7. 3   8. 7   9. 7   10. 5   11. 6   12. 6   13. 3   14. 1   15. 4   16. 1   17. 3   18. 0   19. 3   20. 4   21. 4   22. 6   23. 8   24. 7   25. 8   26. 8   27. 6   28. 7   29. 9   30. 8   31. 4   32. 9   33. 7   34. 7   35. 9   36. 8   37. 3   38. 8   39. 5   40. 5   41. 5   42. 6   43. 2   44. 8   45. 4   46. 7   47. 8   48. 3   49. 9   50. 5   51. 9   52. 9   53. 5 kittens   54. $8   55. (a) 13,430 (b) 13,400 (c) 13,000   **56. Rounded to the nearest ten, the estimated sum is 200; given sum appears incorrect.   57. Rounded to the nearest hundred, the estimated sum is 2000; given sum appears correct.   58. Rounded to the nearest hundred, the estimated sum is 3500; given sum appears correct.**

## 1.5   EXERCISES B

*What subtraction problems are displayed below?*

**1.**

**2.**

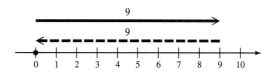

*You should be able to complete Exercises 3–37 in two minutes or less.*

| **3.** | **4.** | **5.** | **6.** | **7.** | **8.** | **9.** |
|---|---|---|---|---|---|---|
| 10<br>− 1 | 13<br>− 4 | 2<br>− 0 | 16<br>− 7 | 14<br>− 9 | 0<br>− 0 | 1<br>− 0 |

| **10.** | **11.** | **12.** | **13.** | **14.** | **15.** | **16.** |
|---|---|---|---|---|---|---|
| 16<br>− 8 | 6<br>− 1 | 6<br>− 2 | 4<br>− 2 | 9<br>− 0 | 8<br>− 5 | 12<br>− 9 |

| **17.** | **18.** | **19.** | **20.** | **21.** | **22.** | **23.** |
|---|---|---|---|---|---|---|
| 7<br>− 3 | 5<br>− 2 | 5<br>− 3 | 7<br>− 4 | 12<br>− 6 | 11<br>− 5 | 16<br>− 9 |

**24.**  11   **25.**  5   **26.**  9   **27.**  10   **28.**  14   **29.**  4   **30.**  10
    $-\ 6$       $-\ 1$       $-\ 8$        $-\ 5$        $-\ 7$       $-\ 0$         $-\ 6$

**31.**  6   **32.**  6   **33.**  9   **34.**  17   **35.**  11   **36.**  13   **37.**  10
    $-\ 4$       $-\ 3$       $-\ 1$        $-\ 8$        $-\ 8$        $-\ 6$          $-\ 7$

**38.** $9 - 4 =$        **39.** $7 - 6 =$        **40.** $15 - 8 =$        **41.** $12 - 3 =$        **42.** $17 - 9 =$

**43.** $3 - 1 =$        **44.** $7 - 7 =$        **45.** $6 - 6 =$        **46.** $8 - 3 =$        **47.** $7 - 0 =$

**48.** $14 - 5 =$        **49.** $12 - 4 =$        **50.** $2 - 2 =$        **51.** $18 - 9 =$        **52.** $13 - 9 =$

**53.** A delivery truck contains 17 cartons. At the first stop, 6 cartons are left off. How many cartons remain on the truck?

**54.** Wendy has 13 cassette tapes in a tan storage case and 7 cassette tapes in a blue case. How many more tapes are in the tan case than the blue case?

## FOR REVIEW

**55.** Estimate the sum by rounding to the nearest  **(a)** ten  **(b)** hundred  **(c)** thousand.

    6549
    2387
  $+\ 3501$

*Without finding the exact sum, determine if the given sum seems correct or incorrect.*

**56.**      38        **57.**      789        **58.**      4285
           45                      450                      91
           23                      235                      165
        $+\ 91$                 $+\ 549$                 $+\ 2035$
          197                     1823                     6576

## 1.5   EXERCISES C

*Let n represent a whole number. Find n by writing the related addition or subtraction sentence.*

**1.** $n - 5 = 9$        **2.** $n + 5 = 9$        **3.** $n - 1 = 6$        **4.** $n + 2 = 7$
[Answer: 14]                                                                   [Answer: 5]

**5.** $n - 0 = 3$        **6.** $n + 0 = 6$        **7.** $n - n = 0$        **8.** $n + 0 = n$
                                                   [Answer: $n$ can be
                                                   any whole number]

## 1.6 SUBTRACTING WHOLE NUMBERS

### **1** SUBTRACTING NUMBERS WITH MORE DIGITS

We now know basic facts of subtraction. To find differences between larger numbers, use the basic facts and the knowledge of place value. For example, find the following difference by writing the numbers in expanded form, then subtracting the ones digits, 8 and 5, and the tens digits, 3 and 2.

$$\begin{array}{rcl}
38 & = & 30 + 8 = 3 \text{ tens} + 8 \text{ ones} \\
- 25 & = & 20 + 5 = 2 \text{ tens} + 5 \text{ ones} \\
\hline
& & 1 \text{ ten} + 3 \text{ ones} = 13
\end{array}$$

The difference, 13, can be found more directly using the two steps shown below.

$$\begin{array}{r}
38 \\
- 25 \\
\hline
13
\end{array}$$

② Subtract tens digits → 13 ← ① Subtract ones digits

The same method works for numbers with more digits as shown in the next example.

| **EXAMPLE 1** SUBTRACTING LARGER NUMBERS | **PRACTICE EXERCISE 1** |
|---|---|
| Find the difference. $\begin{array}{r} 896 \\ - 572 \\ \hline \end{array}$ <br><br> Subtract the ones digits, the tens digits, and the hundreds digits. <br><br> $\begin{array}{rcl} 896 & = & 8 \text{ hundreds} + 9 \text{ tens} + 6 \text{ ones} \\ - 572 & = & 5 \text{ hundreds} + 7 \text{ tens} + 2 \text{ ones} \\ \hline 324 & & 3 \text{ hundreds} + 2 \text{ tens} + 4 \text{ ones} = 324 \end{array}$ | Find the difference. <br><br> $\begin{array}{r} 785 \\ - 643 \\ \hline \end{array}$ <br><br><br> Answer: 142 |

### **2** BORROWING IN SUBTRACTION PROBLEMS

You probably noticed that in Example 1 only subtraction of one-digit numbers was needed. We now look at problems which require subtraction of a one-digit number from a two-digit number. The notion of place value plays an important part in this process. Consider the following problem.

$$\begin{array}{r}
63 \\
- 48 \\
\hline
\end{array}$$

We quickly see that the ones digit 8 cannot be subtracted from the ones digit 3. The difficulty can be removed by ''borrowing'' ten ones from the tens column in 63. That is,

$$63 = \textbf{6 tens} + 3 \text{ ones}$$
$$= \textbf{5 tens} + \textbf{1 ten} + 3 \text{ ones} \qquad 6 \text{ tens} = 5 \text{ tens} + 1 \text{ ten}$$
$$= 5 \text{ tens} + \boxed{1 \text{ ten}} + 3 \text{ ones}$$
$$= 5 \text{ tens} + \boxed{10 \text{ ones}} + 3 \text{ ones} \qquad 1 \text{ ten} = 10 \text{ ones}$$
$$= 5 \text{ tens} + 13 \text{ ones}.$$

Thus,

$$
\begin{array}{r}
63 = 6 \text{ tens} + 3 \text{ ones} = 5 \text{ tens} + 13 \text{ ones} \\
- 48 = 4 \text{ tens} + 8 \text{ ones} = 4 \text{ tens} + \phantom{0}8 \text{ ones} \\
\hline
15 \qquad\qquad\qquad\qquad 1 \text{ ten} + \phantom{0}5 \text{ ones} = 15.
\end{array}
$$

We usually write the problem in shortened form as follows.

$$
\begin{array}{r}
\overset{5\ 13}{\cancel{63}} \\
- 48 \\
\hline
15
\end{array}
$$

Borrow 1 ten from 6 tens, change 6 to 5 and 3 to 13

Subtract corresponding columns

---

## H I N T

Borrowing in a subtraction problem is the reverse of carrying in an addition problem.

---

### EXAMPLE 2  BORROWING IN A SUBTRACTION PROBLEM

Find the difference.
$$
\begin{array}{r}
529 \\
- 365
\end{array}
$$

$$
\begin{array}{r}
\overset{4\ 12}{\cancel{529}} \\
- 365 \\
\hline
164
\end{array}
$$

Subtract 5 from 9 to obtain 4. Since $2 - 6$ is not a whole number, borrow 1 hundred from 5, change 5 to 4 and 2 to 12.

Subtract 6 from 12 and 3 from 4

### PRACTICE EXERCISE 2

Find the difference.

$$
\begin{array}{r}
638 \\
- 491
\end{array}
$$

Answer:  147

---

Sometimes we need to borrow more than once in the same problem. This is illustrated in the next example.

### EXAMPLE 3  REPEATED BORROWING

Find each difference.

**(a)**
$$
\begin{array}{r}
723 \\
- 475
\end{array}
$$

$$
\begin{array}{r}
\overset{1\ 13}{7\,\cancel{2}\,\cancel{3}} \\
- 4\ 7\ 5 \\
\hline
8
\end{array}
$$

Borrow 1 ten from 2, change 2 to 1 and 3 to 13

Subtract 5 from 13

### PRACTICE EXERCISE 3

Find each difference.

**(a)**
$$
\begin{array}{r}
425 \\
- 187
\end{array}
$$

11
6 1 13
7 2 3
− 4 7 5
    4 8    Since 1 − 7 is not a whole number, borrow 1 hundred
           from 7, change 7 to 6 and 1 to 11

           Subtract 7 from 11

11
6 1 13
7 2 3
− 4 7 5
2 4 8    Subtract 4 from 6

**(b)**    6312
         − 4795

**(b)**    7436
         − 2697

      0 12
6 3 1 2    Borrow 1 ten from 1, change 1 to 0 and 2 to 12
− 4 7 9 5
        7    Subtract 5 from 12

      10
    2 0 12
6 3 1 2    Since 0 − 9 is not a whole number, borrow 1 hundred
− 4 7 9 5       from 3, change 3 to 2 and 0 to 10
    1 7    Subtract 9 from 10

  12 10
  5 2 0 12
6 3 1 2    Since 2 − 7 is not a whole number, borrow 1 thousand
− 4 7 9 5       from 6, change 6 to 5 and 2 to 12
1 5 1 7    Subtract 7 from 12 and 4 from 5

Answers: (a) **238**   (b) **4739**

When there is a zero in a position from which we must borrow, we borrow across that position. This is shown in the next example.

| **EXAMPLE 4** BORROWING ACROSS A ZERO | **PRACTICE EXERCISE 4** |

Find each difference.

Find each difference.

**(a)**    605
         − 238

**(a)**    502
         − 267

    5 10
  6 0 5    We must borrow 1 ten, but since 605 has no tens,
− 2 3 8       we must first borrow 1 hundred (10 tens) from 6.
              Change 6 to 5 and 0 to 10.

         9
    5 10 15
  6 0 5    Now borrow 1 ten from 10, change 10 to 9 and 5 to 15
− 2 3 8

**(b)**    6007
         − 2659

    3 6 7    Subtract 8 from 15, 3 from 9, and 2 from 5

**(b)**    2003
         − 895

      1 10
    2 0 0 3    We must borrow 1 ten, but there are no tens, so we
  − 8 9 5          try to borrow 1 hundred. Since there are no
                   hundreds, we must first borrow 1 thousand.
                   Change 2 to 1 and 0 (the hundreds digit) to 10.

$$\begin{array}{r} {}^{9} \\ 1\ \cancel{10}\cancel{10} \\ \cancel{2}\ \cancel{0}\ \cancel{0}\ 3 \\ -\quad 8\ 9\ 5 \\ \hline \end{array}$$  Next, change 10 (hundreds) to 9 and 0 (tens) to 10

$$\begin{array}{r} {}^{9\ \ 9} \\ 1\ \cancel{10}\cancel{10}13 \\ \cancel{2}\ \cancel{0}\ \cancel{0}\ \cancel{3} \\ -\quad 8\ 9\ 5 \\ \hline \end{array}$$  Change 10 (tens) to 9 and 3 (ones) to 13

$$\begin{array}{r} {}^{9\ \ 9} \\ 1\ \cancel{10}\cancel{10}13 \\ \cancel{2}\ \cancel{0}\ \cancel{0}\ \cancel{3} \\ -\quad 8\ 9\ 5 \\ \hline 1\ 1\ 0\ 8 \end{array}$$  Subtract 5 from 13, 9 from 9, 8 from 9, and bring down the 1

**Answers:** (a) 235  (b) 3348

## CAUTION

From Section 1.2 we know that changing the order of addition does not change the sum. The same is *not* true for subtraction. For example,

$$8 - 5 = 3,$$

but $\qquad 5 - 8$  is not a whole number.

Using the symbol $\neq$, which is read "is not equal to," we write

$$8 - 5 \neq 5 - 8.$$

Similarly, in Section 1.3, we discovered that regrouping when adding does not change the sum. Again the same is *not* true for subtraction. For example,

$$(7 - 3) - 2 = 4 - 2 = 2,$$

but

$$7 - (3 - 2) = 7 - 1 = 6,$$

so that

$$(7 - 3) - 2 \neq 7 - (3 - 2).$$

Thus, subtraction is neither commutative nor associative.

## ❸ CHECKING SUBTRACTION BY ADDITION

Subtraction can be checked in one of two ways. If we check by repeating the subtraction step by step, chances are that any error made the first time will be repeated. Since it is better to check work using a different process, we use the related addition sentence.

$$7 - 3 = 4 \quad \text{is true since} \quad 4 + 3 = 7.$$

This example can be generalized.

## To Check a Subtraction Problem

Add the difference to the subtrahend. If the sum is the minuend, the problem is correct.

The format to follow when checking a subtraction problem is illustrated in the following example.

### EXAMPLE 5    CHECKING SUBTRACTION BY ADDITION

Check each subtraction problem.

(a)
$$
\begin{array}{r}
629 \\
-\ 342 \\
\hline
287
\end{array}
$$

These must be equal
$$
\begin{array}{r}
\rightarrow\ 629 \\
-\ 342 \\
\hline
287 \\
- - - - \\
\rightarrow\ \mathbf{629}
\end{array}
$$
Add these and place sum below dashed line

The dashed line can be thought of as the line for the addition problem
$$
\begin{array}{r}
342 \\
+\ 287 \\
\hline
629.
\end{array}
$$

(b)
$$
\begin{array}{r}
4030 \\
-\ 2376 \\
\hline
1664
\end{array}
$$

$$
\begin{array}{r}
4030 \\
-\ 2376 \\
\hline
1664 \\
- - - - \\
\mathbf{4040}
\end{array}
$$
Since these are not equal, the problem is incorrect. The difference should be 1654.

### PRACTICE EXERCISE 5

Check each subtraction problem.

(a)
$$
\begin{array}{r}
847 \\
-\ 269 \\
\hline
578
\end{array}
$$

(b)
$$
\begin{array}{r}
6701 \\
-\ 4985 \\
\hline
1726
\end{array}
$$

Answers: (a) Since $269 + 578 = 847$, the subtraction is correct. (b) Since $4985 + 1726 = 6711 \neq 6701$, the subtraction is incorrect.

### EXAMPLE 6    AN APPLICATION OF SUBTRACTION

Peggy has $883 in her checking account and writes a check for $209. First estimate the number of dollars left in her account, and then find the exact amount.

Since $883 is about $900 and $209 is about $200 (rounding to the nearest 100), estimate the number of dollars left in the account by subtracting 200 from 900.

$$
\left.
\begin{array}{r}
900 \\
-\ 200 \\
\hline
700
\end{array}
\right\} \text{mental work}
$$

Peggy has about $700 left in the account. To find the exact amount, subtract 209 from 883.

$$
\begin{array}{r}
883 \\
-\ 209 \\
\hline
674
\end{array}
$$

The exact amount left in her account is $674.

### PRACTICE EXERCISE 6

Phil Mortensen owned 3105 cattle and sold 1970 of them. First estimate the number of cattle he still owned, and then find the exact number.

Answer: Estimated number: 1000 cattle; exact number: 1135 cattle

//////////////// **CAUTION** ////////////////

Since subtraction is not commutative, we should be careful when solving a subtraction word problem. The numbers must be subtracted in the correct order.

////////////

## 1.6  EXERCISES  A

*Find the following differences.*

| | | | | | |
|---|---|---|---|---|---|
| **1.** 63<br>− 21 | **2.** 97<br>− 62 | **3.** 57<br>− 36 | **4.** 983<br>− 365 | **5.** 738<br>− 429 | **6.** 634<br>− 519 |

| | | | | | |
|---|---|---|---|---|---|
| **7.** 205<br>− 173 | **8.** 603<br>− 381 | **9.** 101<br>− 90 | **10.** 300<br>− 215 | **11.** 400<br>− 260 | **12.** 800<br>− 473 |

| | | | | | |
|---|---|---|---|---|---|
| **13.** 6973<br>− 4325 | **14.** 6281<br>− 3425 | **15.** 3005<br>− 1624 | **16.** 7002<br>− 1325 | **17.** 4315<br>− 2549 | **18.** 38,000<br>− 23,999 |

*Check the following subtraction problems.*

| | | | | |
|---|---|---|---|---|
| **19.** 96<br>− 23<br>73 | **20.** 425<br>− 132<br>293 | **21.** 325<br>− 279<br>156 | **22.** 4392<br>− 1782<br>2610 | **23.** 8002<br>− 4293<br>4709 |

*Find the difference and check your work.*

| | | | | |
|---|---|---|---|---|
| **24.** 423<br>− 105 | **25.** 700<br>− 268 | **26.** 302<br>− 78 | **27.** 7562<br>− 3157 | **28.** 6006<br>− 1991 |

*Solve.*

**29** The Kerns make a house payment of $389 per month. They are considering buying a new house for which the payment will be $605 per month. First estimate the increase in payment, and then find the exact increase.

**30.** A motor home will travel 402 miles on one 30-gallon tank of gasoline, while a Blazer can travel 595 miles on the same amount of gasoline. First estimate the difference in miles traveled, and then find the exact difference.

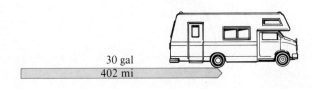

30 gal
402 mi

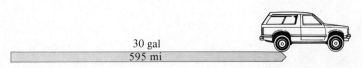

30 gal
595 mi

**31.** Suppose your bank balance was $614 when you wrote three checks for $105, $89, and $26.

   **(a)** What was the total of the checks written?

   **(b)** What was your new bank balance?

**32.** At the end of a three-day trip, the odometer on Walter's car read 36,297 miles. Suppose he drove 426 miles, 298 miles, and 512 miles on the three days.

   **(a)** What was the total number of miles driven on the three-day trip?

   **(b)** What was the odometer reading prior to the trip?

ANSWERS:  1. 42   2. 35   3. 21   4. 618   5. 309   6. 115   7. 32   8. 222   9. 11   10. 85   11. 140   12. 327   13. 2648   14. 2856   15. 1381   16. 5677   17. 1766   18. 14,001   19. correct   20. correct   21. incorrect (should be 46)   22. correct   23. incorrect (should be 3709)   24. 318   25. 432   26. 224   27. 4405   28. 4015   29. estimated increase: $200; actual increase: $216   30. estimated difference: 200 mi; actual difference: 193 mi   31. (a) $220   (b) $394   32. (a) 1236 mi   (b) 35,061 mi

## 1.6  EXERCISES B

*Find the following differences.*

| | | | | | | | | | | | |
|---|---|---|---|---|---|---|---|---|---|---|---|
| **1.** | 45 − 13 | **2.** | 43 − 20 | **3.** | 84 − 43 | **4.** | 581 − 124 | **5.** | 566 − 157 | **6.** | 511 − 208 |

| | | | | | | | | | | | |
|---|---|---|---|---|---|---|---|---|---|---|---|
| **7.** | 430 − 319 | **8.** | 270 − 115 | **9.** | 208 − 40 | **10.** | 600 − 103 | **11.** | 500 − 128 | **12.** | 900 − 303 |

| | | | | | | | | | | | |
|---|---|---|---|---|---|---|---|---|---|---|---|
| **13.** | 3728 − 1154 | **14.** | 5432 − 2734 | **15.** | 6005 − 251 | **16.** | 9030 − 675 | **17.** | 25,132 − 14,627 | **18.** | 42,135 − 11,329 |

*Check the following subtraction problems.*

| | | | | | | | | | |
|---|---|---|---|---|---|---|---|---|---|
| **19.** | 46 − 38 = 18 | **20.** | 407 − 229 = 178 | **21.** | 700 − 296 = 504 | **22.** | 3075 − 346 = 2729 | **23.** | 5040 − 2504 = 3546 |

*Find the difference and check your work.*

| | | | | | | | | | |
|---|---|---|---|---|---|---|---|---|---|
| **24.** | 208 − 129 | **25.** | 601 − 392 | **26.** | 7302 − 488 | **27.** | 4700 − 1308 | **28.** | 8000 − 6989 |

*Solve.*

**29.** A new car weighs 3105 pounds, while a new truck with a camper weighs 6970 pounds. First estimate the difference in weight, and then find the exact difference.

**30.** A new printing press can print 5890 brochures in one day, while an older model can print only 2995 brochures each day. First estimate the difference in the number of brochures per day, and then find the exact difference.

**31.** A new stereo system sells for $2650 plus sales tax in the amount of $106.

    **(a)** What is the cost of the system, including the sales tax?

    **(b)** If you make a down payment of $750, what amount remains to be paid?

**32.** A salesperson has a monthly expense account of $550. Suppose the amount already spent this month includes $125 for gasoline, $210 for food, and $86 for the telephone.

    **(a)** What is the total spent thus far this month?

    **(b)** What is the balance remaining in the expense account?

## 1.6  EXERCISES C

*In the following exercises do what is in the parentheses first.*

**1.** $(3 + 2) - 4$

**2.** $(8 + 6) - (9 + 1)$
[Answer: 4]

**3.** $(8 - 5) + 3$

**4.** $(6 - 2) - (9 - 8)$
[Answer: 3]

## 1.7  MORE APPLICATIONS

This section contains a variety of word problems that involve finding sums and differences. Follow these steps when solving word problems.

### To Solve a Word Problem

1. Read the problem several times to be certain you understand what is being asked. Draw a picture whenever possible.
2. Identify what is given and what must be found, then make a plan of attack.
3. Solve, using your plan and the given information.
4. Ask yourself if your answer is reasonable. If it is, check your work. If not, return to Step 1.

### HINT

Deciding if an answer is reasonable is important and can help you avoid making errors. For example, suppose the problem asks for the price of a new car. If you obtain an answer of $200, your answer is clearly unreasonable, and you should begin again.

### EXAMPLE 1  AN APPLICATION IN BANKING

Carlos has two passbook savings accounts, one with $2350 and the other with $1745. He also has three certificates of deposit with values of $5000, $2500, and $1250. What is the total of these cash assets?

    Total the amounts in the five different accounts at the right.

$$
\begin{array}{r}
2350 \\
1745 \\
5000 \\
2500 \\
+\ 1250 \\
\hline
12845
\end{array}
$$

This is a reasonable total, one which checks using the method of reverse-order addition. Thus, Carlos has $12,845 in cash assets.

### PRACTICE EXERCISE 1

At a recent concert, ticket prices were $7 for adults, $4 for students between 12 and 20 years of age, and $2 for children 12 years and younger. Becky is 13 years old and her sister Cindy is 9. If their parents take them to the concert, how much will the four of them pay?

Answer: $20

| EXAMPLE 2    AN APPLICATION IN SPORTS |
| --- |

The players' scores in a recent basketball game are shown in the table. What was the team score?

Total the numbers in the points column.

*First time*        *Second time*

| | | Player | Points |
| --- | --- | --- | --- |
| 21 | 21 | Hudson | 21 |
| 17 | 17 | Betton | 17 |
| 12 | 12 | Herman | 12 |
| 10 | 10 | Williams | 10 |
| 8 | 8 | Ingram | 8 |
| 7 | 7 | Hurd | 7 |
| 7 | 7 | Payne | 7 |
| 3 | 3 | Spencer | 3 |
| 3 | 3 | Rodriguez | 3 |
| 438  Incorrect | 88  Correct | | |

Since 438 is clearly unreasonable for the score of a team in a basketball game, reread the problem. Again it is clear that the nine numbers must be added. In adding a second time, the error is found. The actual sum is 88, which does check.

| PRACTICE EXERCISE 2 |
| --- |

When Rosa purchased a new car, the price she paid was posted on a window sticker like the one shown here. How much did Rosa pay for her car?

| | |
| --- | --- |
| Base price | $6500 |
| Power package | $1100 |
| Radio | $ 125 |
| Air conditioner | $ 700 |
| Special paint | $  95 |
| Tax and license | $ 436 |
| TOTAL | |

Answer: $8956

| EXAMPLE 3    READING A MAP |
| --- |

Figure 1.19 shows a sketch of a map of part of the western United States.

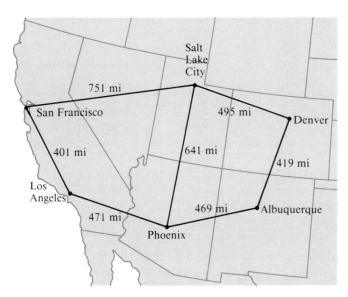

**Figure 1.19**

| PRACTICE EXERCISE 3 |
| --- |

The sketch below shows a map of the eastern United States.

**(a)** How far is it from Cleveland to Atlanta through Baltimore?

**(a)** How far is it from Denver to Phoenix through Albuquerque?

From Denver to Albuquerque is 419 mi, and from Albuquerque to Phoenix is 469 mi. Find the following sum.

$$
\begin{array}{r}
419 \\
+\ 469 \\
\hline
888
\end{array}
$$

Since this is reasonable, and does check, it is 888 mi from Denver to Phoenix through Albuquerque.

**(b)** Is it shorter to go from Los Angeles to Salt Lake City through San Francisco or through Phoenix?

The distance from Los Angeles to Salt Lake City through San Francisco is found by adding 401 and 751. The distance from Los Angeles to Salt Lake City through Phoenix is found by adding 471 and 641.

$$
\begin{array}{r}
401 \\
+\ 751 \\
\hline
1152
\end{array}
$$
Through San Francisco

$$
\begin{array}{r}
471 \\
+\ 641 \\
\hline
1112
\end{array}
$$
Through Phoenix

Since these sums appear reasonable, and do check, it is shorter to go by way of Phoenix.

**(c)** The Jacksons, who live in Los Angeles, plan a trip to Denver for their summer vacation. Going, they will take the southern route through Phoenix and Albuquerque. They plan to return through Salt Lake City and San Francisco. How far will they travel?

Add the six numbers 471, 469, 419, 495, 751, and 401. The total, 3006, seems reasonable and does check. Thus, the family will travel 3006 mi on their vacation.

$$
\begin{array}{r}
471 \\
469 \\
419 \\
495 \\
751 \\
+\ 401 \\
\hline
3006
\end{array}
$$

**(b)** Is it shorter to go from Baltimore to Nashville through Atlanta or through Cleveland and Louisville?

**(c)** The Crums, who live in Louisville, plan a trip to Baltimore. Going, they plan to take the direct route, and they plan to return home through Atlanta and Nashville. How far will they travel on their trip?

Answers: (a) **988 mi** (b) **It is shorter to go through Cleveland and Louisville (856 mi) than through Atlanta (887 mi).** (c) **1653 mi**

---

## EXAMPLE 4  AN APPLICATION IN GEOGRAPHY

The height of Mount Everest is twenty-nine thousand, twenty-eight feet, and the height of Mount McKinley is twenty thousand, three hundred twenty feet. How much higher is Mount Everest than Mount McKinley?

First visualize the problem as shown in Figure 1.20.

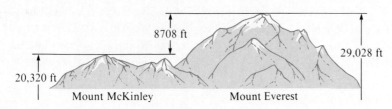

8708 ft

29,028 ft

20,320 ft

Mount McKinley        Mount Everest

**Figure 1.20**

## PRACTICE EXERCISE 4

The height of Pike's Peak is fourteen thousand, one hundred ten feet, and the height of Long's Peak is fourteen thousand, two hundred fifty-six feet. How much higher is Long's Peak than Pike's Peak?

Subtract 20,320 (the height of Mount McKinley) from 29,028 (the height of Mount Everest).

$$
\begin{array}{r}
{\scriptstyle 8\ 10} \\
29\cancel{0}28 \\
-\ 20320 \\
\hline
8708
\end{array}
$$

This difference is certainly reasonable, and it checks. Thus, Mount Everest is 8708 ft higher than Mount McKinley.

Answer: 146 feet

---

| **EXAMPLE 5   AN APPLICATION IN RECREATION** | **PRACTICE EXERCISE 5** |
|---|---|

On a two-week camping trip, a group of students took along 42 six-packs of soda. The first week they drank 18 six-packs, and the second week they drank another 21 six-packs. How many six-packs were left at the end of the trip?

   First find the total number of six-packs that were drunk during the two-week period.

$$
\begin{array}{r}
18 \\
+\ 21 \\
\hline
39
\end{array}
$$

Then, to find out how many six-packs were left, subtract this total from the number of six-packs taken along on the trip.

$$
\begin{array}{r}
42 \\
-\ 39 \\
\hline
3
\end{array}
$$

In view of what is given, 3 seems reasonable.

$$
\begin{array}{rl}
3 & \text{Number of six-packs left} \\
18 & \text{Number drunk first week} \\
+\ 21 & \text{Number drunk second week} \\
\hline
42 & \text{Number taken on trip}
\end{array}
$$

Thus, 3 checks and the group had 3 six-packs of soda left over.

Mike weighs 171 pounds and Arn weighs 185 pounds. Larry weighs 5 pounds less than Arn. What is the total weight of the three men?

Answer: 536 pounds

---

## 1.7   EXERCISES A

*Solve.*

1. A theater is divided into three sections. There are 285 seats in the section on the right, 430 seats in the center section, and 265 seats in the section on the left. How many people will the theater seat?

2. Sam, a salesman, drove 135 mi on Monday, 207 mi on Tuesday, 175 on Wednesday, 87 on Thursday, and 325 on Friday. How far did Sam drive that week?

**3.** The figure shown is called a **rectangle.** It has length 10 ft and width 4 ft. (Note: **ft** is the abbreviation for feet.) The **perimeter** of a rectangle is the distance around the figure, found by adding the measurements of all its sides. Find the perimeter of the rectangle shown.

10 ft

4 ft

**4.** Tana has a garden in the shape of the triangle shown in the figure. (Note: **m** is the abbreviation for meters.) If she plans to put a fence around the garden, how many meters of fencing will she need? [*Hint:* Find the perimeter of the triangle.]

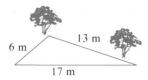

6 m    13 m    17 m

**5.** A bowler rolled games of 179, 203, and 188. What was her total for the three-game series?

**6.** Enrollments in the five colleges of a university are: Arts and Science, 3810; Business, 1608; Education, 1425; Engineering, 865; and Forestry, 239. What is the total university enrollment?

*Exercises 7–9 refer to the information given in the following table.*

Game Statistics

| Home Team | | | Visiting Team | | |
|---|---|---|---|---|---|
| Player | Points | Fouls | Player | Points | Fouls |
| Hanaki | 25 | 4 | Kirk | 18 | 3 |
| Bonnett | 19 | 3 | Mutter | 18 | 2 |
| Condon | 13 | 0 | Westmoreland | 15 | 4 |
| Bell | 7 | 5 | Anderson | 9 | 1 |
| Walter | 6 | 2 | Ratliff | 7 | 2 |
| Meyer | 3 | 1 | Romero | 1 | 3 |
| Shaff | 0 | 5 | Thoreson | 1 | 0 |

**7.** How many points were scored by the home team?

**8.** How many points were scored by the visiting team?

**9.** Which team had the most fouls?

**10.** Bob had $387 in his savings account. If he withdrew $175, how much was left in the account?

**11.** The Henderson well in western Arizona is 6483 ft deep. The Lupine well in southern Colorado is 3976 ft deep. How much deeper is the Henderson well?

**12.** Before leaving on a trip, Luisa noticed that the odometer reading on her car was 38,427. When she returned home from the trip, the reading was 41,003. How many miles did she drive?

**13** At the beginning of June, Mr. Hernandez had a balance of $625 in his checking account. During the month he made deposits of $300 and $235. He wrote checks for $23, $18, $113, $410, and $32. What was his balance at the end of the month?

**14.** Janet buys a radio for $80 and a portable TV for $140. If the tax on these two purchases totals $11, how much change will she receive if she pays with a check for $300?

*Exercises 15–16 refer to the map below.*

**15** What is the shortest route from Salt Lake City to Seattle? How much shorter is it than the next shortest route?

**16.** The Hardy family, who live in Portland, plan a trip to Cheyenne on their summer vacation. How much shorter is it to go through Butte than through Reno?

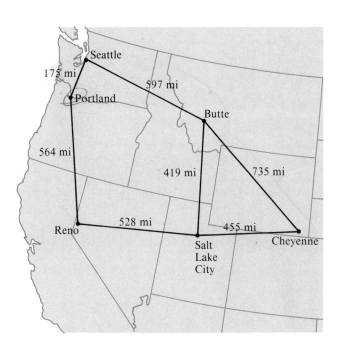

**17.** A bulk tanker contains 1478 gallons of gasoline. Three deliveries are made, one of 230 gallons, a second of 435 gallons, and the third of 170 gallons. How much gasoline remains in the tanker?

**18** Sales at Darrell's Men's Wear were $1245 on Friday. If this is $305 more than on Thursday, and $785 less than on Saturday, what were the sales on Thursday? On Saturday?

**19.** A truck weighs 4845 pounds when empty. When it is loaded with a camper, it weighs 7525 pounds. What is the weight of the camper?

**20.** The *margin of profit* when an item is sold is the list price of the item less the cost of the item. During a recent clearance sale, a store owner sold a bedroom set at cost, for $1150. If she had sold the set at its list price, $1595, what would the margin of profit have been?

4845 pounds

7525 pounds

**21.** On opening day of fall practice, 123 football players were present. If only 65 can eventually make the team, how many players must be cut from the squad?

**22** Mike has $25 and Jim has $43. Larry has $12 more than Jim, and Henri has $9 less than Mike and Jim together. How much money do the four men have together?

## FOR REVIEW

*Check the following subtraction problems.*

**23.**
$$
\begin{array}{r}
83 \\
- 47 \\
\hline
46
\end{array}
$$

**24.**
$$
\begin{array}{r}
649 \\
- 283 \\
\hline
526
\end{array}
$$

**25.**
$$
\begin{array}{r}
7003 \\
- 421 \\
\hline
6582
\end{array}
$$

**26.**
$$
\begin{array}{r}
8050 \\
- 2569 \\
\hline
6591
\end{array}
$$

---

ANSWERS:  1. 980   2. 929 mi   3. 28 ft   4. 36 m   5. 570   6. 7947   7. 73 points   8. 69 points   9. The home team had more (20) than the visiting team (15).   10. $212   11. 2507 ft   12. 2576 mi   13. $564   14. $69   15. The route through Butte is 251 mi shorter than the route through Reno and Portland.   16. 40 mi   17. 643 gallons   18. Thursday: $940; Saturday: $2030   19. 2680 pounds   20. $445   21. 58 players   22. $182   23. incorrect (should be 36)   24. incorrect (should be 366)   25. correct   26. incorrect (should be 5481)

## 1.7   EXERCISES B

*Solve.*

**1.** Professor Mutter teaches three classes. There are 42 students in his Math 151 class, 32 in his Math 136 class, and 38 in his Math 130 class. How many students does Professor Mutter have?

**2.** In the first five games of the season, running back Allan Clark picked up 137 yd, 215 yd, 120 yd, 154 yd, and 183 yd. What was his total yardage through the first five games?

**3.** The figure shown here is called a *square*. (Note: **cm** is the abbreviation for centimeters.) A **square** is a rectangle with four equal sides. Find the perimeter of the square shown.

11 cm

**4.** Dan's scorecard for playing the front nine holes of golf is shown here. How many shots did Dan take on the front nine?

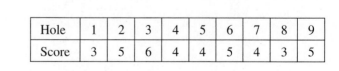

| Hole | 1 | 2 | 3 | 4 | 5 | 6 | 7 | 8 | 9 |
|------|---|---|---|---|---|---|---|---|---|
| Score | 3 | 5 | 6 | 4 | 4 | 5 | 4 | 3 | 5 |

**5.** Peggy has a garden in the shape shown below. If she plans to put a fence around the garden, how many feet of fencing will she need?

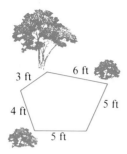

3 ft
6 ft
4 ft
5 ft
5 ft

**6.** On a five-day trip, the Watson family drove 135 mi, 380 mi, 427 mi, 265 mi, and 641 mi. What was the total mileage driven on the trip?

*Exercises 7–8 refer to the map of Lake Powell shown below.*

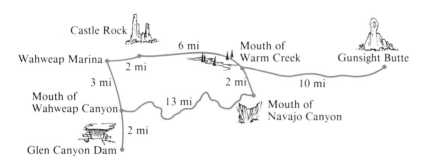

Castle Rock
Mouth of Warm Creek
6 mi
Gunsight Butte
Wahweap Marina
2 mi
3 mi
2 mi
10 mi
Mouth of Wahweap Canyon
13 mi
Mouth of Navajo Canyon
2 mi
Glen Canyon Dam

**7.** Which distance is greater, Castle Rock to Gunsight Butte, or Glen Canyon Dam to the mouth of the Navajo Canyon? How much greater?

**8.** What distance is traveled by a boat that leaves Wahweap Marina and makes a circular trip past the mouths of Wahweap Canyon, Navajo Canyon, and Warm Creek and returns to the marina by way of Castle Rock?

**9.** Sarah purchases 9 square yards of one type of cloth, 14 square yards of another, and 7 square yards of a third. How many square yards of material does she buy?

**10.** On a fishing trip, Mick caught 32 fish and Pete caught 17 fish. How many more fish did Mick catch than Pete?

**11.** Ms. Balushi has a subcompact car and a luxury sedan. She figures that over the period of one year, it will cost $735 to drive the luxury sedan to work while only $265 to drive the subcompact car. How much can she save by driving the subcompact car?

**12.** At the beginning of the month, Mr. Lopez had a balance of $735 in his checking account. If he made deposits of $400 and $375, and wrote checks for $87, $180, $112, $415, and $7 during the month, what was his balance at the end of the month?

**13.** Barb bought a coat for $74 and a purse for $17. If the tax on these two purchases totals $5, how much change will she receive if she pays with a check for $100?

**14.** Bridal Veil Falls is 1957 feet high. Niagara Falls is 193 feet high. How much higher is Bridal Veil Falls than Niagara Falls?

*Exercises 15–18 refer to the following backpacker's trail map.*

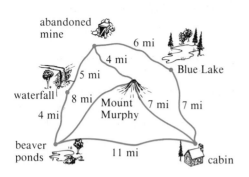

15. How far is it from the cabin to the abandoned mine by way of Mount Murphy?

16. If a hiker makes a complete loop around Mount Murphy, how far will she hike?

17. What is the shortest hike from Blue Lake to the beaver ponds?

18. What is the shortest hike from the waterfall to the cabin?

19. Lori bought a car for $3800. If she spent $625 on repairs and then sold it for $5100, how much money did she make on the deal?

20. Rob earns $724 a week. If he has $98 withheld for federal taxes, $43 withheld for Social Security taxes, and $140 withheld and deposited in his credit union, what is his net take-home pay each week?

21. Antwine bought a coat on sale. If the original price of the coat was $185 and it was discounted $40, how much change did he receive if he paid with two $100 bills?

22. At the start of a trip, the odometer on a car read 48,263 miles. At the end of the trip it read 51,191 miles. What distance was traveled on the trip?

## FOR REVIEW

*Check the following subtraction problems.*

23.
$$\begin{array}{r} 62 \\ -\ 39 \\ \hline 23 \end{array}$$

24.
$$\begin{array}{r} 828 \\ -\ 493 \\ \hline 435 \end{array}$$

25.
$$\begin{array}{r} 4002 \\ -\ 938 \\ \hline 3064 \end{array}$$

26.
$$\begin{array}{r} 2090 \\ -\ 1357 \\ \hline 1743 \end{array}$$

## 1.7  EXERCISES C

*Solve.*

1. Let $n$ represent the number of cars a dealer had available for sale one month. If she sold 60 during the month and had 120 left on the lot, find $n$.
[Answer: $n = 180$]

2. Let $n$ represent the number of boys in an eighth-grade class. If there are 35 total students including 17 girls, find $n$.

# CHAPTER 1 REVIEW

## KEY WORDS

1.1  The **natural** or **counting numbers** are 1, 2, 3, 4, . . . .

The **whole numbers** are the natural numbers together with zero: 0, 1, 2, 3, . . . .

A **numeral** is a symbol used to represent a number.

A **number line** is used to "picture" numbers.

The first ten whole numbers, 0, 1, 2, 3, 4, 5, 6, 7, 8, and 9, are also called **digits.**

Our number system is a **place-value system** because the value of each digit depends on its place or position in a numeral.

**1.2**  Two numbers to be added are called **addends** and the result is called their **sum.**

**1.3**  Addition is a **binary operation,** which means that only *two* numbers are added at a time.

**1.4**  Numbers can be **rounded** or approximated by numbers to the nearest ten, hundred, thousand, and so on.

**1.5**  In a subtraction problem, the number being taken away is called the **subtrahend,** the number it is taken from is the **minuend,** and the result is the **difference.** That is,

$$\text{minuend} - \text{subtrahend} = \text{difference}.$$

The **related addition sentence** to $7 - 3 = 4$, for example, is $4 + 3 = 7$.

Similarly, the **related subtraction sentence** to $4 + 3 = 7$, for example, is $7 - 3 = 4$ or $7 - 4 = 3$.

## KEY CONCEPTS

**1.1**  **1.** The whole numbers occur in a natural order. If one whole number is to the left of a second whole number on a number line, it is *less than* the second. The symbol $<$ represents ''less than,'' and $>$ represents ''greater than.''

**2.** The number 4379, written in standard notation, is a short form of expanded notation

$$4000 + 300 + 70 + 9.$$

**3.** Zero is used as a place-holder to show the difference between numbers such as 5032 and 532.

**4.** Write numbers such as 123 as ''one hundred twenty-three'' and *not* as ''one hundred *and* twenty-three.''

**1.2**  **1.** The *sum* of two numbers can be thought of as the number of objects in a combined collection.

**2.** Addition problems may be written either horizontally or vertically.

**3.** Changing the order of addition does not change the sum by the commutative law. For example, $2 + 9 = 9 + 2$.

**1.3**  **1.** The addition procedure can be shortened by using the process of carrying.

**2.** Regrouping sums does not change the answer by the associative law. For example, $(2 + 3) + 7 = 2 + (3 + 7)$.

**3.** Addition may be checked by the reverse-order method.

**1.4**  Rounded numbers can be used to estimate sums.

**1.5**  **1.** The *difference* of two numbers can be thought of as the number of objects which remain when objects are removed or taken away from a collection.

**2.** Subtraction can be defined by using the related addition sentence.

**1.6**  **1.** Borrowing in subtraction problems is the reverse of carrying in addition problems.

**2.** Subtraction can be checked by adding the difference to the subtrahend.

**1.7**  After you understand a word problem, make a plan of attack and identify what is given and what must be found. A picture or sketch may be helpful. Always make sure that any answer seems reasonable.

## REVIEW EXERCISES

*Part I*

**1.1**  **1.** What is the symbol for a number called?

**2.** Give the first five natural numbers.

**3.** Give the first five whole numbers.

**4.** What are the ten digits?

**5.** Why is the number system called a place-value system?

*Answer Exercises 6–11 using the given number line.*

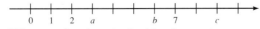

**6.** What number is paired with $a$?

**7.** What number is paired with $c$?

*Place the correct symbol (< or >) between the given numbers, remembering that a, b, and c are from the number line above.*

**8.** 2 _____ 7          **9.** *b* _____ *a*          **10.** *b* _____ *c*          **11.** *a* _____ *c*

**12.** Write 13,247,165 in expanded notation.

**13.** Write 400,000 + 3000 + 400 + 5 in standard notation.

**14.** Write a word name for 27,405,036.

*In the number 145,237,098 what digit represents each of the following places?*

**15.** Millions          **16.** Hundreds          **17.** Ten thousands          **18.** Ones

**1.2**  **19.** Consider the addition problem 7 + 3 = 10.
(a) What are 7 and 3 called?  (b) What is 10 called?  (c) What is the symbol + called?

**20.** Complete the statement using the commutative law of addition. 5 + 4 = _____

**21.** When a given number is added to 0, what is the sum?

**1.3**  *Find the following sums and check your work.*

| **22.** | **23.** | **24.** | **25.** | **26.** |
|---|---|---|---|---|
| 68 | 392 | 457 | 2507 | 5678 |
| + 97 | + 79 | + 168 | + 5926 | + 8765 |

| **27.** | **28.** | **29.** | **30.** | **31.** |
|---|---|---|---|---|
| 28 | 425 | 565 | 5654 | 10,137 |
| 31 | 384 | 137 | 327 | 4380 |
| + 54 | + 293 | 248 | + 4032 | 296 |
|  |  | + 309 |  | + 14 |

**32.** Why is the operation of addition called a binary operation?

**33.** What is the name of the law illustrated by (3 + 2) + 7 = 3 + (2 + 7)?

**1.4**  *Round to the nearest ten.*

**34.** 73          **35.** 285          **36.** 4176

*Round to the nearest hundred.*

**37.** 451          **38.** 83          **39.** 2550

*Round to the nearest thousand.*

**40.** 7350          **41.** 875          **42.** 12,627

**43.** Estimate the sum by rounding to the nearest ten.

1657
2349
+  836

**44.** Estimate the sum by rounding to the nearest hundred.

1657
2349
+  836

*Without finding the exact sum, determine if the given sum seems correct or incorrect.*

| **45.** | **46.** | **47.** |
|---|---|---|
| 25 | 328 | 3291 |
| 37 | 19 | 407 |
| 51 | + 476 | + 6930 |
| + 14 | 823 | 10,628 |
| 227 | | |

**48.** During a two-day holiday, 9942 people visited the Grand Canyon National Park on the first day and 8032 visited it the second. First estimate the total number of visitors during the two days, and then find the exact number.

**1.5** **49.** Consider the subtraction problem $54 - 23 = 31$. **(a)** What is the number 54 called? **(b)** What is the number 23 called? **(c)** What is the number 31 called? **(d)** What is the symbol "$-$" called?

**1.6** *Find the following differences and check your work.*

| **50.** | **51.** | **52.** | **53.** | **54.** |
|---|---|---|---|---|
| 85 | 63 | 238 | 654 | 406 |
| − 32 | − 47 | − 106 | − 235 | − 139 |

| **55.** | **56.** | **57.** | **58.** | **59.** |
|---|---|---|---|---|
| 352 | 605 | 6307 | 37,003 | 40,000 |
| − 87 | − 70 | − 2178 | − 10,526 | − 17,989 |

**1.7** **60.** A nursery has a stock consisting of 42 willows, 70 crabapples, 38 black pines, and 18 weeping birches. How many trees are in the inventory?

**61.** What is the perimeter of this figure?

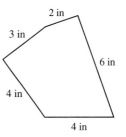

**62.** During one year, a magazine had 437,250 subscribers. The following year, an additional 45,593 families subscribed. How many subscribers did the magazine have that year?

**63.** It costs Ms. Spinelli $735 to drive her car to work during the year. If she carpools with two others, it only costs $278 a year. How much can she save by forming a carpool?

**64.** At the beginning of the month, Sid had a balance of $693 in his checking account. During the month he made deposits of $450 and $580. He wrote checks for $137, $320, $18, $5, and $79. What was his balance at the end of the month?

**65.** The Bonnetts bought a house for $55,280. After making several improvements which totaled $4375, they sold the house for $66,900. How much profit did they make on the sale?

*Part II*

*Find the following sums or differences and check your work.*

| **66.** | **67.** | **68.** | **69.** | **70.** |
|---|---|---|---|---|
| 307 | 347 | 2000 | 1478 | 35,001 |
| − 198 | 206 | − 743 | 296 | − 7296 |
| | + 493 | | + 3075 | |

**71.** Write 41,075 in expanded notation.

**72.** Write $10,000,000 + 16,000 + 200 + 5$ in standard notation.

**73.** What is the name of the law illustrated by $5 + 11 = 11 + 5$?

**74.** We know that $8 + 0 = 8$. What is 0 called in this sentence?

**75.** Round 4351 to the nearest **(a)** ten **(b)** hundred **(c)** thousand.

**76.** Estimate the following sum by rounding to the nearest hundred.

$$\begin{array}{r} 2641 \\ 375 \\ + 1529 \end{array}$$

**77.** Place the appropriate symbol ($<$ or $>$) between the given numbers. 9 _____ 0

**78.** Estimate the following difference by rounding to the nearest hundred.

$$\begin{array}{r} 6317 \\ - 4250 \end{array}$$

*Solve.*

**79.** Peter's monthly gross earnings are $2875. Suppose that $320 are taken out for federal income tax, $54 for state income tax, and $75 for an IRA.

  **(a)** What is the total taken out of his gross pay for these three items?

  **(b)** What is Peter's take-home pay?

**80.** It takes 130 kilowatt-hours of electricity to operate a stereo for a year, 510 kilowatt-hours to operate a TV for a year, and 4225 kilowatt-hours to operate a spa for a year.

  **(a)** How many kilowatt-hours are required to operate all three for a year?

  **(b)** How many more kilowatt-hours are required to operate the spa than the other two together?

**81.** The Shipps are retired and receive a Social Security payment of $491 each month. Next month they will receive an increase in benefits to $689 per month. First estimate the increase, and then find the actual increase.

**82.** During the month of November, a basketball player scored 22, 18, 11, 27, and 13 points in five games. First estimate the total points scored in these five games, and then find the exact number of points.

---

**ANSWERS:** 1. numeral 2. 1, 2, 3, 4, 5 3. 0, 1, 2, 3, 4 4. 0, 1, 2, 3, 4, 5, 6, 7, 8, 9 5. because the value of each digit is determined by its place or position in the numeral 6. 3 7. 9 8. $2 < 7$ 9. $b > a$ 10. $b < c$ 11. $a < c$ 12. $10,000,000 + 3,000,000 + 200,000 + 40,000 + 7000 + 100 + 60 + 5$ 13. 403,405 14. twenty-seven million, four hundred five thousand, thirty-six 15. 5 16. 0 17. 3 18. 8 19. (a) addends (b) sum (c) plus sign 20. $4 + 5$ 21. the given number 22. 165 23. 471 24. 625 25. 8433 26. 14,443 27. 113 28. 1102 29. 1259 30. 10,013 31. 14,827 32. because *two* numbers are added at a time 33. associative law of addition 34. 70 35. 290 36. 4180 37. 500 38. 100 39. 2600 40. 7000 41. 1000 42. 13,000 43. 4850 44. 4800 45. incorrect (should be 127) 46. correct 47. correct 48. estimated total: 18,000 people; exact total: 17,974 49. (a) minuend (b) subtrahend (c) difference (d) minus sign 50. 53 51. 16 52. 132 53. 419 54. 267 55. 265 56. 535 57. 4129 58. 26,477 59. 22,011 60. 168 61. 19 in 62. 483,843 subscribers 63. $457 64. $1164 65. $7245 66. 109 67. 1046 68. 1257 69. 4849 70. 27,705 71. $40,000 + 1000 + 70 + 5$ 72. 10,016,205 73. commutative law of addition 74. additive identity 75. (a) 4350 (b) 4400 (c) 4000 76. 4500 77. $9 > 0$ 78. 2000 79. (a) $449 (b) $2426 80. (a) 4865 kilowatt-hours (b) 3585 kilowatt-hours 81. estimated increase: $200; actual increase: $198 82. estimated total: 90 points; actual total: 91 points

**1.** Give the first 4 counting numbers.

1. _____

**2.** How do we read the symbol $<$?

2. _____

**3.** The statement $(2 + 3) + 7 = 2 + (3 + 7)$ illustrates what law of addition?

3. _____

**4.** Consider the number line.

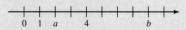

    **(a)** What number is paired with $a$?
    **(b)** What number is paired with $b$?
    **(c)** Place the appropriate symbol ($<$ or $>$) between $a$ and $b$.    $a$ ___?___ $b$.

4. (a) _____ (b) _____ (c) _____

**5.** Write 35,408,438 in expanded notation.

5. _____
_____

**6.** Write $300,000 + 80,000 + 400 + 7$ in standard notation.

6. _____

**7.** Write a word name for 1,659,288.

7. _____
_____

*Find the following sums.*

**8.**    51
     + 69

**9.**    427
     + 88

**10.**    5139
     + 4108

**11.**    6213
      107
       40
    + 6857

8. _____

9. _____

10. _____

11. _____

**12.** Estimate   8142
          650
       + 99

    **(a)** by rounding to the nearest ten.
    **(b)** by rounding to the nearest hundred.

12. (a) _____ (b) _____

*Find the following differences.*

**13.**   93
  − 56

**14.**   427
  − 78

**15.**   502
  − 295

**16.**   8205
  − 2136

13. _____

14. _____

15. _____

16. _____

*Solve.*

**17.** Juan purchased three books for $18, $35, and $19. How much did he pay for the three books?

17. _____

**18.** Mr. Hagood has a pickup and a subcompact car. He figures that over the period of one year, it will cost $887 to drive the pickup to work while only $403 to drive the subcompact car. First estimate his savings by driving the car, and then find the exact savings.

18. _____

**19.** A bulk tanker contains 1985 gallons of fuel. Three deliveries are made, one of 265 gallons, a second of 340 gallons, and the third of 125 gallons. How much fuel remains in the tanker?

19. _____

**20.** Find the perimeter of a rectangle of length 16 feet and width 7 feet.

20. _____

**21.** Abby had a balance of $397 in her checking account. She deposited two checks worth $265 and $147. She then wrote three checks for $35, $105, and $211. What was the balance in her account after these transactions?

21. _____

# CHAPTER 2

# Multiplying and Dividing Whole Numbers

## 2.1 BASIC MULTIPLICATION FACTS

**STUDENT GUIDEPOSTS**

1. Multiplication as Repeated Addition
2. Properties of Multiplication
3. The Multiplication Table
4. Multiples of a Number

### 1 MULTIPLICATION AS REPEATED ADDITION

In Chapter 1 we learned that subtraction was closely related to addition. Similarly, multiplication can be thought of as an extension of addition. For example, suppose we want to find the cost of four pieces of 5¢ candy. One way would be to find the sum

$$5 + 5 + 5 + 5 = 20.$$

Since repeated sums like this occur frequently, we should look for a shorter method for finding them. This shorter method is called **multiplication.** In our example, we are adding 5 to itself 4 times. In other words, we are finding

**4 times 5.**

The symbol ×, called the **multiplication sign** or **times sign,** means "times" in multiplication statements. The two numbers being multiplied are called **factors** (sometimes the **multiplier** and the **multiplicand**) and the result is called their **product.** Thus, in

$$4 \times 5 = 20,$$

4 and 5 are the factors (4 is the multiplier and the multiplicand is 5) whose product is 20. We can picture the product 4 × 5 as in Figure 2.1.

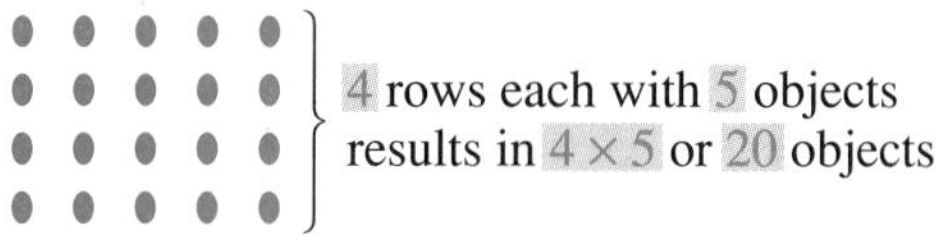
4 rows each with 5 objects
results in 4 × 5 or 20 objects

**Figure 2.1**

Remember that in 4 × 5, 4 tells us the number of 5's that must be added together to obtain 20. That is,

$$4 \times 5 = \underbrace{5 + 5 + 5 + 5}_{\text{four 5's}} = 20.$$

Multiplication problems are written either horizontally or vertically.

*Horizontal display*            *Vertical display*

$$4 \times 5 = 20$$

factors    product

$$\begin{array}{r} 5 \\ \times\ 4 \end{array}$$ factors

$$\overline{\phantom{0}20} \leftarrow \text{product}$$

Several other symbols or notations are used in multiplication. In place of the times symbol, $\times$, a raised dot is often used. For example,

$$4 \cdot 5 = 20 \quad \text{is the same as} \quad 4 \times 5 = 20.$$

Also, parentheses can be used around one or both of the factors with the multiplication signs $\times$ and $\cdot$ left out.

$$(4)(5) = 20 \quad \text{or} \quad 4(5) = 20 \quad \text{or} \quad (4)5 = 20$$

---

| **EXAMPLE 1** MULTIPLYING BY REPEATED ADDITION | **PRACTICE EXERCISE 1** |
|---|---|

Multiply.

**(a)** $3 \times 2 = \underbrace{2 + 2 + 2}_{\text{three 2's}} = 6$

$$\begin{array}{r} 2 \\ \times\ 3 \\ \hline 6 \end{array}$$

**(b)** $4 \times 6 = \underbrace{6 + 6 + 6 + 6}_{\text{four 6's}} = 24$

$$\begin{array}{r} 6 \\ \times\ 4 \\ \hline 24 \end{array}$$

**(c)** $6 \cdot 4 = \underbrace{4 + 4 + 4 + 4 + 4 + 4}_{\text{six 4's}} = 24$

$$\begin{array}{r} 4 \\ \times\ 6 \\ \hline 24 \end{array}$$

**(d)** $(5)(0) = \underbrace{0 + 0 + 0 + 0 + 0}_{\text{five 0's}} = 0$

$$\begin{array}{r} 0 \\ \times\ 5 \\ \hline 0 \end{array}$$

Multiply.

**(a)** $6 \times 3 =$
$3 + 3 + 3 + 3 + 3 + 3 = $ \_\_\_\_\_

**(b)** $5 \times 7 =$
$7 + 7 + 7 + 7 + 7 = $ \_\_\_\_\_

**(c)** $7 \cdot 5 =$
$5 + 5 + 5 + 5 + 5 + 5 + 5 = $ \_\_\_\_\_

**(d)** $(3)(0) = $ \_\_\_\_\_

Answers: (a) **18**   (b) **35**   (c) **35**
(d) **0**

---

## ② PROPERTIES OF MULTIPLICATION

Parts **(b)** and **(c)** of Example 1 suggest that changing the order of multiplication does not change the product, since $4 \times 6$ and $6 \times 4$ are both 24. From a visual standpoint, consider a page with 4 rows, each containing 6 objects, as shown in Figure 2.2(a). When the page is turned as in Figure 2.2(b), the result has 6 rows, each containing 4 objects. Since the number of objects remains the same, $4 \times 6 = 6 \times 4$. This demonstrates an important property of multiplication.

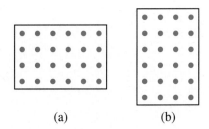

(a)            (b)

**Figure 2.2**

| The Commutative Law of Multiplication |
|---|

The **commutative law of multiplication** tells us that the order in which two (or more) numbers are multiplied does not change the product.

Also, part (**d**) of Example 1 shows that any whole number times zero is zero. This is sometimes called the **zero property of multiplication.**

| EXAMPLE 2   MULTIPLYING BY ONE |
|---|

Multiply.

(**a**) $7 \times 1 = \underbrace{1 + 1 + 1 + 1 + 1 + 1 + 1}_{\text{seven 1's}} = 7$

(**b**) $1 \cdot 7 = 7$
$\downarrow$
one 7

| PRACTICE EXERCISE 2 |
|---|

Multiply.

(**a**) $5 \times 1 =$
$1 + 1 + 1 + 1 + 1 = \underline{\qquad}$

(**b**) $1 \cdot 5 = \underline{\qquad}$

Answers:  (a)  5    (b)  5

We see from Example 2 that $7 \times 1 = 1 \times 7 = 7$. This is a special case of another important property of multiplication. If 1 is multiplied by any whole number, the product is the *identical* number. As a result, 1 is often called the **multiplicative identity.**

## ❸ THE MULTIPLICATION TABLE

Finding products by repeated addition takes time. By memorizing certain basic facts, we can find the product of any two one-digit numbers quickly. These facts are summarized in a Multiplication Table below. To find $8 \times 3$ using the table, start at 8 in the left column and move straight across to the right until you are below the 3 in the top row. You have arrived at the number 24 which is $8 \times 3$.

| × | 0 | 1 | 2 | 3 | 4 | 5 | 6 | 7 | 8 | 9 |
|---|---|---|---|---|---|---|---|---|---|---|
| 0 | 0 | 0 | 0 | 0 | 0 | 0 | 0 | 0 | 0 | 0 |
| 1 | 0 | 1 | 2 | 3 | 4 | 5 | 6 | 7 | 8 | 9 |
| 2 | 0 | 2 | 4 | 6 | 8 | 10 | 12 | 14 | 16 | 18 |
| 3 | 0 | 3 | 6 | 9 | 12 | 15 | 18 | 21 | 24 | 27 |
| 4 | 0 | 4 | 8 | 12 | 16 | 20 | 24 | 28 | 32 | 36 |
| 5 | 0 | 5 | 10 | 15 | 20 | 25 | 30 | 35 | 40 | 45 |
| 6 | 0 | 6 | 12 | 18 | 24 | 30 | 36 | 42 | 48 | 54 |
| 7 | 0 | 7 | 14 | 21 | 28 | 35 | 42 | 49 | 56 | 63 |
| 8 | 0 | 8 | 16 | 24 | 32 | 40 | 48 | 56 | 64 | 72 |
| 9 | 0 | 9 | 18 | 27 | 36 | 45 | 54 | 63 | 72 | 81 |

**Multiplication Table**

## ❹ MULTIPLES OF A NUMBER

A **multiple** of a number is the product of the number and a whole number. To find the multiples of a number, simply multiply it by each of the whole numbers, beginning with zero.

| EXAMPLE 3 | MULTIPLES OF A NUMBER |
|---|---|

**(a)** Multiples of 2

$0 \times 2 = 0$
$1 \times 2 = 2$
$2 \times 2 = 4$
$3 \times 2 = 6$
$4 \times 2 = 8$
$\vdots$

**(b)** Multiples of 5

$0 \times 5 = 0$
$1 \times 5 = 5$
$2 \times 5 = 10$
$3 \times 5 = 15$
$4 \times 5 = 20$
$\vdots$

**PRACTICE EXERCISE 3**

**(a)** List the multiples of 3.

**(b)** List the multiples of 10.

Answers: (a) 0, 3, 6, 9, 12, . . .
(b) 0, 10, 20, 30, 40, . . .

## HINT

In the Multiplication Table, the numbers in any row are multiples of the number at the left of that row. Also, the numbers in any column are multiples of the number at the top of that column.

## 2.1 EXERCISES A

**1.** Consider the multiplication problem $5 \times 8 = 40$. **(a)** What is 5 called? **(b)** What is 8 called? **(c)** What is 40 called?

**2.** The fact that $2 \times 7 = 7 \times 2$ illustrates what law of multiplication?

**3.** The number 1 times any number always gives what number for the product?

**4.** The numbers 0, 3, 6, 9, 12, and 15 are all multiples of what number?

*You should be able to complete Exercises 5–54 in 90 seconds or less.*

| | | | | | | | | | |
|---|---|---|---|---|---|---|---|---|---|
| **5.** $\begin{array}{r} 0 \\ \times\,0 \\ \hline \end{array}$ | **6.** $\begin{array}{r} 7 \\ \times\,1 \\ \hline \end{array}$ | **7.** $\begin{array}{r} 2 \\ \times\,4 \\ \hline \end{array}$ | **8.** $\begin{array}{r} 6 \\ \times\,0 \\ \hline \end{array}$ | **9.** $\begin{array}{r} 3 \\ \times\,7 \\ \hline \end{array}$ | **10.** $\begin{array}{r} 0 \\ \times\,2 \\ \hline \end{array}$ | **11.** $\begin{array}{r} 8 \\ \times\,9 \\ \hline \end{array}$ | **12.** $\begin{array}{r} 5 \\ \times\,4 \\ \hline \end{array}$ | **13.** $\begin{array}{r} 3 \\ \times\,8 \\ \hline \end{array}$ | **14.** $\begin{array}{r} 3 \\ \times\,6 \\ \hline \end{array}$ |
| **15.** $\begin{array}{r} 2 \\ \times\,5 \\ \hline \end{array}$ | **16.** $\begin{array}{r} 2 \\ \times\,9 \\ \hline \end{array}$ | **17.** $\begin{array}{r} 3 \\ \times\,3 \\ \hline \end{array}$ | **18.** $\begin{array}{r} 1 \\ \times\,2 \\ \hline \end{array}$ | **19.** $\begin{array}{r} 0 \\ \times\,7 \\ \hline \end{array}$ | **20.** $\begin{array}{r} 1 \\ \times\,5 \\ \hline \end{array}$ | **21.** $\begin{array}{r} 6 \\ \times\,7 \\ \hline \end{array}$ | **22.** $\begin{array}{r} 8 \\ \times\,3 \\ \hline \end{array}$ | **23.** $\begin{array}{r} 8 \\ \times\,1 \\ \hline \end{array}$ | **24.** $\begin{array}{r} 5 \\ \times\,7 \\ \hline \end{array}$ |
| **25.** $\begin{array}{r} 6 \\ \times\,8 \\ \hline \end{array}$ | **26.** $\begin{array}{r} 7 \\ \times\,2 \\ \hline \end{array}$ | **27.** $\begin{array}{r} 7 \\ \times\,9 \\ \hline \end{array}$ | **28.** $\begin{array}{r} 3 \\ \times\,5 \\ \hline \end{array}$ | **29.** $\begin{array}{r} 8 \\ \times\,5 \\ \hline \end{array}$ | **30.** $\begin{array}{r} 9 \\ \times\,2 \\ \hline \end{array}$ | **31.** $\begin{array}{r} 6 \\ \times\,9 \\ \hline \end{array}$ | **32.** $\begin{array}{r} 4 \\ \times\,0 \\ \hline \end{array}$ | **33.** $\begin{array}{r} 1 \\ \times\,6 \\ \hline \end{array}$ | **34.** $\begin{array}{r} 1 \\ \times\,7 \\ \hline \end{array}$ |
| **35.** $\begin{array}{r} 4 \\ \times\,7 \\ \hline \end{array}$ | **36.** $\begin{array}{r} 7 \\ \times\,6 \\ \hline \end{array}$ | **37.** $\begin{array}{r} 0 \\ \times\,3 \\ \hline \end{array}$ | **38.** $\begin{array}{r} 8 \\ \times\,4 \\ \hline \end{array}$ | **39.** $\begin{array}{r} 8 \\ \times\,7 \\ \hline \end{array}$ | **40.** $\begin{array}{r} 5 \\ \times\,0 \\ \hline \end{array}$ | **41.** $\begin{array}{r} 1 \\ \times\,1 \\ \hline \end{array}$ | **42.** $\begin{array}{r} 4 \\ \times\,9 \\ \hline \end{array}$ | **43.** $\begin{array}{r} 0 \\ \times\,5 \\ \hline \end{array}$ | **44.** $\begin{array}{r} 6 \\ \times\,2 \\ \hline \end{array}$ |

| **45.** $\begin{array}{r} 9 \\ \times 1 \\ \hline \end{array}$ | **46.** $\begin{array}{r} 2 \\ \times 0 \\ \hline \end{array}$ | **47.** $\begin{array}{r} 0 \\ \times 6 \\ \hline \end{array}$ | **48.** $\begin{array}{r} 9 \\ \times 5 \\ \hline \end{array}$ | **49.** $\begin{array}{r} 9 \\ \times 4 \\ \hline \end{array}$ | **50.** $\begin{array}{r} 3 \\ \times 2 \\ \hline \end{array}$ | **51.** $\begin{array}{r} 5 \\ \times 3 \\ \hline \end{array}$ | **52.** $\begin{array}{r} 1 \\ \times 8 \\ \hline \end{array}$ | **53.** $\begin{array}{r} 4 \\ \times 4 \\ \hline \end{array}$ | **54.** $\begin{array}{r} 9 \\ \times 8 \\ \hline \end{array}$ |

---

ANSWERS:  1. (a) multiplier or factor (b) multiplicand or factor (c) product  2. commutative law  3. that number
4. 3  5. 0  6. 7  7. 8  8. 0  9. 21  10. 0  11. 72  12. 20  13. 24  14. 18  15. 10  16. 18  17. 9  18. 2  19. 0
20. 5  21. 42  22. 24  23. 8  24. 35  25. 48  26. 14  27. 63  28. 15  29. 40  30. 18  31. 54  32. 0  33. 6
34. 7  35. 28  36. 42  37. 0  38. 32  39. 56  40. 0  41. 1  42. 36  43. 0  44. 12  45. 9  46. 0  47. 0  48. 45
49. 36  50. 6  51. 15  52. 8  53. 16  54. 72

## 2.1 EXERCISES B

**1.** Consider the multiplication problem $7 \times 9 = 63$. **(a)** What is 63 called? **(b)** What is 9 called? **(c)** What is 7 called?

**2.** Use the commutative law of multiplication to complete the following:  $9 \times 5 =$ _____.

**3.** The number 0 times any number always gives what number for the product?

**4.** The numbers 0, 7, 14, 21, 28, and 35 are all multiples of what number?

*You should be able to complete Exercises 5–54 in 90 seconds or less.*

| | | | | |
|---|---|---|---|---|
| **5.** $8 \cdot 1 =$ | **6.** $9 \cdot 2 =$ | **7.** $2 \cdot 0 =$ | **8.** $8 \cdot 8 =$ | **9.** $7 \cdot 9 =$ |
| **10.** $6 \cdot 1 =$ | **11.** $4 \cdot 2 =$ | **12.** $3 \cdot 5 =$ | **13.** $6 \cdot 7 =$ | **14.** $9 \cdot 9 =$ |
| **15.** $2 \cdot 6 =$ | **16.** $7 \cdot 7 =$ | **17.** $6 \cdot 9 =$ | **18.** $7 \cdot 2 =$ | **19.** $4 \cdot 1 =$ |
| **20.** $0 \cdot 2 =$ | **21.** $9 \cdot 8 =$ | **22.** $7 \cdot 1 =$ | **23.** $6 \cdot 8 =$ | **24.** $9 \cdot 5 =$ |
| **25.** $0 \cdot 0 =$ | **26.** $3 \cdot 4 =$ | **27.** $4 \cdot 0 =$ | **28.** $1 \cdot 8 =$ | **29.** $0 \cdot 6 =$ |
| **30.** $5 \cdot 0 =$ | **31.** $5 \cdot 5 =$ | **32.** $6 \cdot 3 =$ | **33.** $1 \cdot 2 =$ | **34.** $0 \cdot 7 =$ |
| **35.** $0 \cdot 8 =$ | **36.** $8 \cdot 9 =$ | **37.** $0 \cdot 5 =$ | **38.** $5 \cdot 7 =$ | **39.** $9 \cdot 3 =$ |
| **40.** $0 \cdot 4 =$ | **41.** $1 \cdot 7 =$ | **42.** $7 \cdot 4 =$ | **43.** $3 \cdot 2 =$ | **44.** $3 \cdot 0 =$ |
| **45.** $8 \cdot 6 =$ | **46.** $6 \cdot 5 =$ | **47.** $5 \cdot 8 =$ | **48.** $0 \cdot 3 =$ | **49.** $9 \cdot 7 =$ |
| **50.** $5 \cdot 9 =$ | **51.** $3 \cdot 7 =$ | **52.** $6 \cdot 4 =$ | **53.** $9 \cdot 0 =$ | **54.** $4 \cdot 8 =$ |

## 2.1 EXERCISES C

**1.** If $a$ and $b$ represent whole numbers, use the commutative law to complete.  $a \times b =$ _____

**2.** If $a$ represents a whole number, then **(a)** $a \times 1 =$ _____ **(b)** $a \times 0 =$ _____.

**3.** If your teacher assigns the multiples of 4 in 2.1 Exercises A, which problems should you work?

## 2.2 MULTIPLYING WHOLE NUMBERS

### 1 MULTIPLYING A TWO-DIGIT NUMBER BY A ONE-DIGIT NUMBER

The Multiplication Table in Section 2.1 includes all products of two one-digit numbers. We now consider the product of a one-digit number and a two-digit number, for example, $2 \times 43$. We know that

$$2 \times 43 = \underbrace{43 + 43}_{\text{two 43's}} = 86.$$

In expanded form,

$$2 \times 43 = 2 \times (40 + 3) = 2 \times (4 \text{ tens} + 3 \text{ ones})$$
$$= (4 \text{ tens} + 3 \text{ ones}) + (4 \text{ tens} + 3 \text{ ones})$$
$$= 4 \text{ tens} + 4 \text{ tens} + 3 \text{ ones} + 3 \text{ ones}$$
$$= 8 \text{ tens} + 6 \text{ ones} = 86.$$

There is a simpler way to find the product as shown in the following two steps.

$$\begin{array}{r} 43 \\ \times\ 2 \\ \hline 86 \end{array}$$

② Multiply 4, the tens digit in 43, by 2 and write the result in the tens column ——→ 86 ←—— ① Multiply 3, the ones digit in 43, by 2 and write the result in the ones column

---

**EXAMPLE 1** MULTIPLYING A TWO-DIGIT BY A ONE-DIGIT NUMBER

Find the product.
$$\begin{array}{r} 32 \\ \times\ 3 \end{array}$$

$$\begin{array}{r} 32 \\ \times\ 3 \\ \hline 96 \end{array}$$

② 9 is the product of 3 and the tens digit in 32 ——→ 96 ←—— ① 6 is the product of 3 and the ones digit in 32

**PRACTICE EXERCISE 1**

Find the product.
$$\begin{array}{r} 34 \\ \times\ 2 \end{array}$$

Answer: **68**

---

### 2 MULTIPLYING A MANY-DIGIT NUMBER BY A ONE-DIGIT NUMBER

The procedure shown above can be extended to some products of one-digit numbers and many-digit numbers.

| EXAMPLE 2 MULTIPLYING A THREE-DIGIT BY A ONE-DIGIT NUMBER | PRACTICE EXERCISE 2 |

Find the product.
$$\begin{array}{r} 423 \\ \times\ 2 \end{array}$$

③ Multiply 4, the hundreds digit in 423, by 2 and write the result in the hundreds column

$$\begin{array}{r} 423 \\ \times\ 2 \\ \hline 846 \end{array}$$

① Multiply 3, the ones digit in 423, by 2 and write the result in the ones column

② Multiply 2, the tens digit in 423, by 2 and write the result in the tens column

Find the product.

$$\begin{array}{r} 1233 \\ \times\ 2 \end{array}$$

Answer: 2466

In the examples above, each individual step resulted in a one-digit number. In some problems, such as $2 \times 73$, individual steps may give 2-digit numbers.

$$\begin{array}{r} 73 \\ \times 2 \\ \hline 146 \end{array}$$

② Multiply tens digit, 7, by 2 and write result here

① Multiply ones digit, 3, by 2 and write result here

| EXAMPLE 3    A ONE-DIGIT TIMES A THREE-DIGIT NUMBER | PRACTICE EXERCISE 3 |

Find the product.
$$\begin{array}{r} 723 \\ \times\ 3 \end{array}$$

$$\begin{array}{r} 723 \\ \times\ 3 \\ \hline 2169 \end{array}$$

③ $3 \times 7 = 21$

① $3 \times 3 = 9$

② $3 \times 2 = 6$

Find the product.

$$\begin{array}{r} 831 \\ \times\ 2 \end{array}$$

Answer: 1662

In Example 3, only the final step had two digits. We must now look at problems in which a two-digit number occurs sooner, for example, $2 \times 38$. Using expanded notation,

$$\begin{aligned} 2 \times 38 = 2(30 + 8) &= 2(3 \text{ tens} + 8 \text{ ones}) \\ &= (3 \text{ tens} + 8 \text{ ones}) + (3 \text{ tens} + 8 \text{ ones}) \\ &= 6 \text{ tens} + 16 \text{ ones} \\ &= 6 \text{ tens} + 10 \text{ ones} + 6 \text{ ones} \\ &= 6 \text{ tens} + 1 \text{ ten} + 6 \text{ ones} \qquad \textbf{10 ones = 1 ten} \\ &= 7 \text{ tens} + 6 \text{ ones} = 76. \end{aligned}$$

There is a simpler way to find this product. First, multiply 2 times the ones digit, 8, and obtain 16. Write 6 in the ones place and, to remember that 1 ten must still be recorded, write a small 1 above the tens digit in 38.

$$\begin{array}{r} {}^{1} \\ 38 \\ \times\ 2 \\ \hline 6 \end{array}$$

Next, multiplying 2 times 3, the tens digit, gives 6 which stands for 6 tens. These 6 tens added to the 1 ten obtained earlier gives 7 tens. Put the 7 in the tens place.

$$
\begin{array}{r}
1 \\
38 \\
\times\ 2 \\
\hline
76
\end{array}
$$

| EXAMPLE 4    MULTIPLYING USING REMINDER NUMBERS | PRACTICE EXERCISE 4 |
|---|---|

Find the products.

**(a)**
$$
\begin{array}{r}
96 \\
\times\ 7
\end{array}
$$

$$
\begin{array}{r}
4 \\
96 \\
\times\ 7 \\
\hline
672
\end{array}
$$

① $7 \times 6 = 42$; record 2, write 4 as a reminder
② $7 \times 9 = 63$; $63 + 4 = 67$; 67 tens = 60 tens + 7 tens = 6 hundreds + 7 tens
③ Record 7 in tens column and 6 in hundreds column

**(b)**
$$
\begin{array}{r}
2905 \\
\times\ 5
\end{array}
$$

$$
\begin{array}{r}
4\ \ 2 \\
2905 \\
\times\ 5 \\
\hline
14525
\end{array}
$$

① $5 \times 5 = 25$; record 5, write reminder 2
② $5 \times 0 = 0$; $0 + 2 = 2$; record 2
③ $5 \times 9 = 45$; record 5, write reminder 4
④ $5 \times 2 = 10$; $10 + 4 = 14$; record 14

*Practice Exercise 4:*

Find the products.

**(a)**
$$
\begin{array}{r}
75 \\
\times\ 8
\end{array}
$$

**(b)**
$$
\begin{array}{r}
3056 \\
\times\ 4
\end{array}
$$

Answers: (a) 600    (b) 12,224

## ❸ MULTIPLYING A NUMBER BY A MANY-DIGIT NUMBER

At the beginning of this section, we found the product $2 \times 43 = 86$. Since changing the order of multiplication does not change the product, according to the commutative law, we also know that $43 \times 2 = 86$. Suppose we attempt to multiply in that order. We first multiply the ones digit.

$$
\begin{array}{r}
2 \\
\times\ 43 \\
\hline
6
\end{array}
$$

Next we multiply 4 times 2. Remember that in 43, 4 represents 4 tens. Thus, we are actually multiplying (4 tens) $\times 2 = 8$ tens $= 80$. We show this by writing 80 below 6.

$$
\begin{array}{r}
2 \\
\times\ 43 \\
\hline
6 \\
80
\end{array}
$$

We draw a line below 80 and add the 8 tens and 6 ones.

$$
\begin{array}{r}
2 \\
\times\ 43 \\
\hline
6 \\
80 \\
\hline
86
\end{array}
$$

 First product
 Second product
 Sum of first and second products

Let us summarize what we have learned.

### To Multiply a Number by a Many-Digit Number

1. Find the first product in exactly the same way as when multiplying a many-digit number by a one-digit number.

2. To form the second product, place a 0 in the ones column, multiply the multiplicand by the tens digit in the multiplier. Place the result to the left of 0.

3. If the multiplier has three or more digits, a third product is formed by placing zeros in the ones and tens columns and multiplying the multiplicand by the appropriate digit. This process is continued for more digits.

4. Add all products to find the final product.

---

### EXAMPLE 5  MULTIPLYING BY A TWO-DIGIT NUMBER

Find the products.

**(a)**    78
       $\times$ 96

```
      7
      4
     78
   × 96
    468     ① First product: 6 × 78 = 468
   7020     ② Second product: Place 0, with 9 × 78 = 702
   7488     ③ Add first and second products
```

Notice that the first reminder number, 4, written when finding $6 \times 8$, was crossed out when the second reminder number, 7, was written.

**(b)**    438
       $\times$ 52

```
     438
   ×  52
    1876     ① 2 × 438 = 876
   21900     ② Place 0; 5 × 438 = 2190
   22776     ③ Add products
```

---

Multiplying a number by a 3-digit number uses the same approach.

### EXAMPLE 6  MULTIPLYING BY A THREE-DIGIT NUMBER

Find the products.

**(a)**    325
       $\times$ 618

The first two rows of this product are found the same way as in previous examples. The third row has two 0's to the right with $6 \times 325$ to the left.

---

### PRACTICE EXERCISE 5

Find the products.

**(a)**    69
       $\times$ 87

**(b)**    629
       $\times$ 43

Answers:  (a)  6003    (b)  27,047

---

### PRACTICE EXERCISE 6

Find the products.

**(a)**    437
       $\times$ 516

```
      13
      2̸4̸
     325
   × 618
    2600      First product: 8 × 325
  1̸1̸3250      Second product: 0 with 1 × 325
  195000      Third product: 00 with 6 × 325
  200850      The desired product is the sum of the three products
```

**(b)**    647
        × 305

**(b)**    835
        × 209

```
        12
        2̸3̸
       647
     × 305
      3235     5 × 647
      0000     0 with 0 × 647
    194100     00 with 3 × 647
    197335     Add products
```

Since the second row does not affect the sum, we can save time by omitting it from our work. Thus, we would show the product in the following way.

```
       647
     × 305
      3235
    194100     Keep 00 with 3 × 647
    197335
```

Answers: (a) 225,492
(b) 174,515

---

## HINT

With practice, and by paying close attention to the columns in the products, we may omit the extra zeros shown to the right in each row. For example, we might write the product in Example 6(a) in the following form.

```
       325
     × 618
      2600
       325
      1950
    200850
```

---

### ④ REVERSE-ORDER CHECK OF MULTIPLICATION

One method for checking a multiplication problem is based on the fact that changing the order of multiplication does not change the product. Suppose we found the following product.

```
        23
      × 47
       161
        92
      1081
```

To check our work, we interchange the multiplier and the multiplicand and find the new product.

$$
\begin{array}{r}
47 \\
\times\ 23 \\
\hline
141 \\
94\phantom{0} \\
\hline
1081
\end{array}
$$

Since we know that $47 \times 23$ is the same as $23 \times 47$, the answers should be equal. This check is the **reverse-order check of multiplication.**

---

| **EXAMPLE 7    USING THE REVERSE-ORDER CHECK** | **PRACTICE EXERCISE 7** |
|---|---|

Find the product of 329 and 54. Check your work using the reverse-order check.

Multiply:
$$
\begin{array}{r}
329 \\
\times\ 54 \\
\hline
1316 \\
1645\phantom{0} \\
\hline
17766
\end{array}
$$
Check:
$$
\begin{array}{r}
54 \\
\times\ 329 \\
\hline
486 \\
108\phantom{0} \\
162\phantom{00} \\
\hline
17766
\end{array}
$$
Factors are interchanged

The check on the right may look strange. We usually write the number with more digits above the one with fewer digits in a multiplication problem. However, this example shows that the multiplication method works either way.

Find the product of 457 and 35. Check your work using the reverse-order check.

$$
\begin{array}{r}
457 \\
\times\ 35 \\
\hline
\end{array}
\qquad
\text{Check:}
\qquad
\begin{array}{r}
35 \\
\times\ 457 \\
\hline
\end{array}
$$

Answer: 15,995

---

## 5 APPLICATIONS OF MULTIPLYING WHOLE NUMBERS

Some real-life applied problems involve multiplying whole numbers.

---

| **EXAMPLE 8    AN APPLICATION IN EDUCATION** | **PRACTICE EXERCISE 8** |
|---|---|

There are 32 mathematics classes with 25 students enrolled in each. What is the total number of students in these classes?

To solve this problem multiply 32 by 25.

$$
\begin{array}{r}
32 \\
\times\ 25 \\
\hline
160 \\
640\phantom{0} \\
\hline
800
\end{array}
$$

Thus, there are 800 students in the 32 classes.

There are 52 weeks in a year. How many weeks are there in 8 years?

Answer: 416 weeks

---

## 2.2 EXERCISES A

*Find the product.*

**1.**  $\begin{array}{r} 25 \\ \times\ 2 \\ \hline \end{array}$
**2.**  $\begin{array}{r} 22 \\ \times\ 3 \\ \hline \end{array}$
**3.**  $\begin{array}{r} 33 \\ \times\ 3 \\ \hline \end{array}$
**4.**  $\begin{array}{r} 36 \\ \times\ 2 \\ \hline \end{array}$
**5.**  $\begin{array}{r} 53 \\ \times\ 3 \\ \hline \end{array}$

**6.**  38
       × 2

**7.**  93
       × 3

**8.**  443
       × 2

**9.**  503
       × 3

**10.**  523
        × 3

**11.**  527
        × 3

**12.**  496
        × 7

**13.**  287
        × 8

**14.**  700
        × 3

**15.**  3213
        × 2

**16.**  5003
        × 3

**17.**  9325
        × 4

**18.**  8495
        × 7

**19.**  43
        × 12

**20.**  31
        × 23

**21.**  33
        × 22

**22.**  73
        × 45

**23.**  37
        × 40

**24.**  80
        × 80

**25.**  603
        × 42

**26.**  839
        × 42

**27.**  425
        × 312

**28.**  803
        × 601

**29.**  900
        × 900

**30.**  777
        × 888

*Find the product and check using the reverse-order check.*

**31.**  47
        × 19

**32.**  161
        × 37

**33.**  545
        × 139

*Solve.*

**34.** Playing tennis burns up 230 calories every hour. How many calories would be burned up by a player in a match taking 4 hours?

**35.** One bottle of soda holds 12 fluid ounces. How many fluid ounces of soda are in a case of 24 bottles?

**36.** A water tank contains 50,000 gallons of water. When opened, an outlet valve will release 40 gallons of water every minute.

(a) How many gallons of water will be released in 48 minutes?

(b) How many gallons of water remain in the tank after 48 minutes?

*Exercises 37–42 will help you prepare for the next section. Perform operations inside parentheses first.*

**37.** Find $5 + 3$.

**38.** Find $2 \times 5$.

**39.** Find $2 \times 3$.

**40.** Find $2 \times (5 + 3)$.

**41.** Find $(2 \times 5) + (2 \times 3)$.

**42.** Compare the results of Exercises 40 and 41. What do you discover?

---

ANSWERS:  1. 50  2. 66  3. 99  4. 72  5. 159  6. 76  7. 279  8. 886  9. 1509  10. 1569  11. 1581  12. 3472  13. 2296  14. 2100  15. 6426  16. 15,009  17. 37,300  18. 59,465  19. 516  20. 713  21. 726  22. 3285  23. 1480  24. 6400  25. 25,326  26. 35,238  27. 132,600  28. 482,603  29. 810,000  30. 689,976  31. 893  32. 5957  33. 75,755  34. 920 calories  35. 288 fluid ounces  36. (a) 1920 gallons (b) 48,080 gallons  37. 8  38. 10  39. 6  40. 16  41. 16  42. $2 \times (5 + 3) = 16$ and $(2 \times 5) + (2 \times 3) = 16$

## 2.2  EXERCISES B

*Find the products.*

| | | | | |
|---|---|---|---|---|
| **1.** 13 × 3 | **2.** 41 × 2 | **3.** 57 × 1 | **4.** 18 × 3 | **5.** 61 × 7 |
| **6.** 72 × 3 | **7.** 231 × 3 | **8.** 321 × 4 | **9.** 223 × 4 | **10.** 605 × 4 |
| **11.** 783 × 5 | **12.** 999 × 7 | **13.** 656 × 9 | **14.** 979 × 2 | **15.** 7013 × 2 |
| **16.** 6257 × 2 | **17.** 6987 × 1 | **18.** 6993 × 9 | **19.** 52 × 11 | **20.** 14 × 42 |
| **21.** 64 × 15 | **22.** 83 × 92 | **23.** 70 × 38 | **24.** 40 × 99 | **25.** 748 × 19 |
| **26.** 554 × 78 | **27.** 851 × 223 | **28.** 387 × 430 | **29.** 954 × 702 | **30.** 803 × 150 |

*Find the product and check using the reverse-order check.*

**31.**  83 × 48          **32.**  452 × 12          **33.**  893 × 452

*Solve.*

**34.** There are 365 days in a year. How many days are there in 5 years?

**35.** What is the cost of 15 stereos at $298 each?

**36.** The manufacturer of woodburning stoves has 247 stoves in stock. The company can make 12 stoves each day.

   **(a)** How many stoves can be made in 25 days?

   **(b)** Assuming no stoves are sold out of stock, how many stoves will be in stock 25 days from now?

*Exercises 37–42 will help you prepare for the next section. Perform operations inside parentheses first.*

**37.** Find $2 + 7$.

**38.** Find $4 \times 2$.

**39.** Find $4 \times 7$.

**40.** Find $4 \times (2 + 7)$.

**41.** Find $(4 \times 2) + (4 \times 7)$.

**42.** Compare the results of Exercises 40 and 41. What do you discover?

## 2.2 EXERCISES C

*Find the products.*

1.  4312
    $\times$ 648
    [Answer: 2,794,176]

2.  4006
    $\times$ 6004

3.  7592
    $\times$ 4362
    [Answer: 33,116,304]

4.  4932
    $\times$ 2715

5. A communications satellite travels 31,200 miles in one orbit of the earth. If the satellite makes 6 orbits per day, how many miles will it travel in 1 year? [Use 365 days for 1 year.]

## 2.3 PROPERTIES OF MULTIPLICATION AND SPECIAL PRODUCTS

### STUDENT GUIDEPOSTS

1. Associative Law of Multiplication
2. Distributive Laws
3. Multiplying by 10, 100, or 1000
4. Using Rounded Numbers to Estimate and Check Products

### 1 ASSOCIATIVE LAW OF MULTIPLICATION

Multiplication, like addition, is a binary operation (that is, we multiply two numbers at a time). To find the product of three or more numbers, we can multiply in steps. For example, consider the product $3 \times 2 \times 5$. We can first multiply $3 \times 2$, then multiply the result by 5.

$$(3 \times 2) \times 5 = 6 \times 5 = 30$$

Or we can multiply $2 \times 5$ first, then multiply the result by 3.

$$3 \times (2 \times 5) = 3 \times 10 = 30$$

The parentheses show which products are found first. The same result is obtained regardless of how the factors are grouped. That is,

$$(3 \times 2) \times 5 = 3 \times (2 \times 5),$$

which is an example of an important property of multiplication.

#### Associative Law of Multiplication

The **associative law of multiplication** tells us that the product of more than two numbers is not changed by grouping. No matter how many numbers are multiplied, they are always multiplied two at a time.

### HINT

Because of the commutative and associative laws of multiplication, when you multiply two or more numbers, the order of the numbers or the grouping of the numbers does not affect the product.

## ② DISTRIBUTIVE LAWS

An important relationship between multiplication and addition is shown in the following problem.

$$3 \times (4 + 5) = 3 \times 9 \qquad \text{Operate within parentheses first}$$
$$= 27 \qquad \text{Find the product of 3 and 9}$$

Next, find the following.

$$(3 \times 4) + (3 \times 5) = 12 + 15 \qquad \text{Operate within the parentheses first}$$
$$= 27 \qquad \text{Find the sum of 12 and 15}$$

As a result,

$$3 \times (4 + 5) = (3 \times 4) + (3 \times 5).$$

In a sense, the expression to the right of the equal sign is formed by "distributing" multiplication by 3 over the sum of 4 and 5. That is why this property is called the **distributive law** of multiplication over addition.

The multiplication procedure in Section 2.2 actually involves repeated use of the distributive law. Consider

$$3 \times 42 = 3 \times (40 + 2) \qquad \text{Expanded notation}$$
$$= (3 \times 40) + (3 \times 2) \qquad \text{Distributive law}$$
$$= 120 + 6 \qquad \text{Multiply}$$
$$= 126. \qquad \text{Add}$$

Notice that when the problem is written vertically, $3 \times 2$ is found in the first row and $3 \times 40$ in the second row.

$$
\begin{array}{r}
42 \\
\times 3 \\
\hline
6 \\
120 \\
\hline
126
\end{array}
\qquad
\begin{array}{l}
(3 \times 2) \\
+(3 \times 40) \\
\hline
(3 \times 40) + (3 \times 2)
\end{array}
$$

## ③ MULTIPLYING BY 10, 100, OR 1000

Multiplication problems involving the numbers 10, 100, and 1000 occur frequently. Recall that **multiples of 10** are numbers such as

$$10, \quad 20, \quad 30, \quad 40, \quad 50.$$

**Multiples of 100** include

$$100, \quad 200, \quad 300, \quad 400, \quad 500, \quad \text{and so on,}$$

while **multiples of 1000** include

$$1000, \quad 2000, \quad 3000, \quad 4000, \quad 5000, \quad \text{and so on.}$$

Products that involve multiples of 10, 100, or 1000 can be found quickly by using the following rule.

| To Multiply a Number: |
| --- |
| 1. by **10**, attach **one 0** at the end of the number, |
| 2. by **100**, attach **two 0's** at the end of the number, |
| 3. by **1000**, attach **three 0's** at the end of the number. |

---

**EXAMPLE 1  MULTIPLYING BY 10, 100, OR 1000**

Find each product.

**(a)** $10 \cdot 5 = 50$      Attach 0 to 5

**(b)** $10 \cdot 435 = 4350$      Attach 0 to 435

**(c)** $100 \cdot 4 = 400$      Attach two 0's to 4

**(d)** $100 \cdot 497 = 49{,}700$      Attach two 0's to 497

**(e)** $1000 \cdot 9 = 9000$      Attach three 0's to 9

**(f)** $1000 \cdot 138 = 138{,}000$      Attach three 0's to 138

**PRACTICE EXERCISE 1**

Find each product.

**(a)** $8 \cdot 10$

**(b)** $648 \cdot 10$

**(c)** $3 \cdot 100$

**(d)** $476 \cdot 100$

**(e)** $7 \cdot 1000$

**(f)** $929 \cdot 1000$

Answers: (a) 80   (b) 6480
(c) 300   (d) 47,600   (e) 7000
(f) 929,000

---

When multiplying two multiples of 10, 100, or 1000, similar shortcuts are possible.

---

**EXAMPLE 2 MULTIPLYING WITH MULTIPLES OF 10, 100, OR 1000**

Find each product.

**(a)**
$$\begin{array}{r} 40 \\ \times\ 70 \\ \hline 2800 \end{array}$$
Multiply $7 \cdot 4$ and attach two (the total number of zeros in the factors) zeros

**(b)**
$$\begin{array}{r} 700 \\ \times\ 90 \\ \hline 63000 \end{array}$$
Multiply $9 \cdot 7$ and attach three zeros $(2 + 1 = 3)$

In each case, the total number of zeros attached to the product is the sum of the zeros in the factors.

**PRACTICE EXERCISE 2**

Find each product.

**(a)**
$$\begin{array}{r} 400 \\ \times\ 700 \end{array}$$

**(b)**
$$\begin{array}{r} 9000 \\ \times\ 400 \end{array}$$

Answers: (a) 280,000
(b) 3,600,000

---

**④ USING ROUNDED NUMBERS TO ESTIMATE AND CHECK PRODUCTS**

In Section 2.2 we used the reverse-order method to check a multiplication problem. The obvious drawback of this technique is that it takes just as long to check as to find the original product. For this reason, it is helpful to have another method. Rounded approximations for the factors can be used for **estimating products** and discovering any major errors.

---

**EXAMPLE 3  CHECKING MULTIPLICATION BY ESTIMATING**

Find the product of 476 and 31; then check your work by estimating the product.

$$\begin{array}{ll}
\text{Multiply:} & \begin{array}{r} 476 \\ \times\ 31 \\ \hline 476 \\ 1428 \\ \hline 14756 \end{array}
\quad \text{Check:} \quad \begin{array}{r} 500 \\ \times\ 30 \\ \hline 15000 \end{array}
\quad \begin{array}{l} 476 \text{ rounds to } 500 \\ 31 \text{ rounds to } 30 \end{array}
\end{array}$$

Since 15,000 is "close" to 14,756, we would assume that no major error has been made.

**PRACTICE EXERCISE 3**

Find the product of 295 and 42; then check your work by estimating the product.

Answer: Rounding 295 to 300 and 42 to 40, the estimated product is $40 \cdot 300 = 12{,}000$ (the actual product is 12,390).

---
### **HINT**

When rounding numbers to find estimated products, round all four-digit numbers to the nearest thousand, all three-digit numbers to the nearest hundred, and all two-digit numbers to the nearest ten.

---

Rounded numbers can also be used to estimate solutions to applied problems.

| **EXAMPLE 4**   **ESTIMATING A PRODUCT IN AN APPLIED PROBLEM** | **PRACTICE EXERCISE 4** |
|---|---|

There are 4985 students in a school, and each spent $41 for athletic fees. Without finding the exact amount, estimate the total amount of money spent on athletic fees by these students.

    Since 4985 is about 5000 and $41 is about $40, we can estimate the total by multiplying 5000 by 40.

$$\left.\begin{array}{r} 5000 \\ \underline{40} \\ 200000 \end{array}\right\} \quad \text{Mental work}$$

Thus, about $200,000 was spent on these fees.

If one piece of pie has 205 calories, without finding the exact number, estimate the number of calories in 89 pieces.

**Answer: 18,000 calories (using 200 × 90)**

---

## 2.3  EXERCISES A

**1.** Use the associative law of multiplication to complete the following: $(3 \times 5) \times 7 =$ _____.

**2.** Use the distributive law to complete the following: $4 \times (5 + 7) =$ _____.

**3. (a)** Evaluate $7 \times (2 + 9)$. **(b)** Evaluate $(7 \times 2) + (7 \times 9)$. **(c)** Why are your answers to **(a)** and **(b)** the same?

**4. (a)** Evaluate $9 \times (3 \times 5)$. **(b)** Evaluate $(9 \times 3) \times 5$. **(c)** Why are your answers to **(a)** and **(b)** the same?

**5. (a)** Evaluate $8 \times 11$. **(b)** Evaluate $11 \times 8$. **(c)** Why are your answers to **(a)** and **(b)** the same?

*Find each product.*

| | | | | |
|---|---|---|---|---|
| **6.** $10 \cdot 9$ | **7.** $42 \cdot 10$ | **8.** $387 \cdot 10$ | **9.** $7 \cdot 100$ | **10.** $100 \cdot 77$ |
| **11.** $100 \cdot 427$ | **12.** $1000 \cdot 5$ | **13.** $1000 \cdot 42$ | **14.** $78 \cdot 1000$ | **15.** $1000 \cdot 529$ |
| **16.** $\begin{array}{r} 70 \\ \times 60 \end{array}$ | **17.** $\begin{array}{r} 800 \\ \times 400 \end{array}$ | **18.** $\begin{array}{r} 700 \\ \times 20 \end{array}$ | **19.** $\begin{array}{r} 6000 \\ \times 30 \end{array}$ | **20.** $\begin{array}{r} 7000 \\ \times 500 \end{array}$ |

*Check by estimating the product.*

| 21. | 78 | 22. | 409 | 23. | 813 |
|-----|----|-----|-----|-----|-----|
| | × 21 | | × 38 | | × 491 |
| | 78 | | 3272 | | 813 |
| | 156 | | 1227 | | 7317 |
| | 1638 | | 15542 | | 3252 |
| | | | | | 499183 |

*Solve.*

**24.** There are 29 volumes in a set of reference books, and each volume has 995 pages. Without finding the exact number, estimate the total number of pages in the set.

**25** A store had 103 customers on Saturday, and the average amount spent by each was $289. Without finding the exact amount, estimate the total receipts for Saturday.

**26.** Jeff's car can get 28 miles to a gallon of gasoline. Without finding the exact number of miles, estimate the distance Jeff can drive on 42 gallons of gasoline.

**27.** A block layer can lay 46 cinder blocks an hour. Estimate the number of blocks that he can lay in a 52-hour workweek. [Do not find the exact number.]

## FOR REVIEW

*Find the product and check your work using the reverse-order check.*

| 28. | 161 | 29. | 545 |
|-----|-----|-----|-----|
| | × 37 | | × 139 |

**ANSWERS:**   1. $3 \times (5 \times 7)$   2. $(4 \times 5) + (4 \times 7)$   3. (a) 77 (b) 77 (c) because of the distributive law   4. (a) 135 (b) 135 (c) because of the associative law   5. (a) 88 (b) 88 (c) because of the commutative law   6. 90   7. 420   8. 3870   9. 700   10. 7700   11. 42,700   12. 5000   13. 42,000   14. 78,000   15. 529,000   16. 4200   17. 320,000   18. 14,000   19. 180,000   20. 3,500,000   21. $20 \times 80 = 1600$; appears to be correct   22. $40 \times 400 = 16,000$; appears to be correct   23. $500 \times 800 = 400,000$; appears to be incorrect (correct product is 399,183)   24. 30,000 pages   25. $30,000   26. $30 \times 40 = 1200$ miles   27. $50 \times 50 = 2500$ blocks   28. 5957   29. 75,755

## 2.3  EXERCISES B

**1.** Use the associative law of multiplication to complete the following: $2 \times (4 \times 9) =$ _____.

**2.** Use the distributive law to complete the following: $(2 \times 5) + (2 \times 8) =$ _____.

**3.** (a) Evaluate $10 \times (2 + 3)$. (b) Evaluate $(10 \times 2) + (10 \times 3)$. (c) Why are your answers to (a) and (b) the same?

**4. (a)** Evaluate $3 \times (4 \times 10)$. **(b)** Evaluate $(3 \times 4) \times 10$. **(c)** Why are your answers to **(a)** and **(b)** the same?

**5. (a)** Evaluate $12 \times 4$. **(b)** Evaluate $4 \times 12$. **(c)** Why are your answers to **(a)** and **(b)** the same?

*Find each product.*

**6.** $10 \cdot 5$ 　　　**7.** $71 \cdot 10$ 　　　**8.** $593 \cdot 10$ 　　　**9.** $100 \cdot 8$ 　　　**10.** $43 \cdot 100$

**11.** $100 \cdot 645$ 　　**12.** $1000 \cdot 3$ 　　**13.** $1000 \cdot 57$ 　　**14.** $84 \cdot 1000$ 　　**15.** $259 \cdot 1000$

**16.** 
$$\begin{array}{r} 70 \\ \times\ 40 \\ \hline \end{array}$$
**17.** 
$$\begin{array}{r} 400 \\ \times\ 700 \\ \hline \end{array}$$
**18.** 
$$\begin{array}{r} 500 \\ \times\ 30 \\ \hline \end{array}$$
**19.** 
$$\begin{array}{r} 9000 \\ \times\ 40 \\ \hline \end{array}$$
**20.** 
$$\begin{array}{r} 8000 \\ \times\ 400 \\ \hline \end{array}$$

*Check by estimating the product.*

**21.** 
$$\begin{array}{r} 61 \\ \times\ 88 \\ \hline 488 \\ 488 \\ \hline 4368 \end{array}$$
**22.** 
$$\begin{array}{r} 678 \\ \times\ 51 \\ \hline 678 \\ 3390 \\ \hline 34578 \end{array}$$
**23.** 
$$\begin{array}{r} 290 \\ \times\ 603 \\ \hline 870 \\ 1740 \\ \hline 18270 \end{array}$$

*Solve.*

**24.** A man burns up 395 calories each day during his exercise period. Without finding the exact number, about how many calories will he burn up in 105 days?

**25.** A store has 7958 bottles of pills in its inventory. If the average number of pills per bottle is 98, approximately how many pills are in the store? [Do not find the exact number.]

**26.** Diane can type 85 words a minute. Estimate the number of words that she can type in 7 hours. [Do not find the actual number of words. Use the fact that 7 hours is 420 minutes.]

**27.** A merry-go-round in a carnival makes 11 revolutions a minute. Without finding the exact number, estimate the number of revolutions made in 103 minutes.

## FOR REVIEW

*Find the product and check your work using the reverse-order check.*

**28.** 
$$\begin{array}{r} 452 \\ \times\ 12 \\ \hline \end{array}$$
**29.** 
$$\begin{array}{r} 893 \\ \times\ 452 \\ \hline \end{array}$$

## 2.3　EXERCISES C

*Let a, b, and c represent any three whole numbers. Complete each law.*

**1.** Associative law: $a \times (b \times c) = $ _____.

**2.** Commutative law: $c \times b = $ _____.

**3.** Distributive law: $a \times (b + c) = $ _____.

## 2.4 BASIC DIVISION FACTS

### **1** DIVISION AS REPEATED SUBTRACTION

The final arithmetic operation we study is division. Recall that multiplication can be thought of as repeated addition. Similarly, division can be thought of as repeated subtraction.

The word *division* means "separation into parts." Consider dividing 15 by 3. We use ÷ as the **division symbol** and write

$$15 \div 3,$$

(read "15 divided by 3"). In the example, 15, the number being divided, is called the **dividend,** and 3, the dividing number, is the **divisor.** Suppose we have 15 objects and separate or divide them into 3 equal parts. As shown in Figure 2.3, each equal part has 5 objects in it.

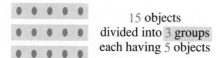

15 objects
divided into 3 groups
each having 5 objects

**Figure 2.3**

We write 15 ÷ 3 = 5, and call 5 the **quotient** of 15 and 3.

We could also look at the problem as repeated subtraction. That is, 15 ÷ 3 could be thought of as the number of times 3 can be subtracted from 15 until zero remains.

$$
\begin{array}{r}
15 \\
-\ 3 \\
\hline
12 \\
-\ 3 \\
\hline
9 \\
-\ 3 \\
\hline
6 \\
-\ 3 \\
\hline
3 \\
-\ 3 \\
\hline
0
\end{array}
$$

First subtraction of 3

Second subtraction of 3

Third subtraction of 3

Fourth subtraction of 3

Fifth subtraction of 3

We have subtracted 3 from 15 a total of 5 times. That is, 15 ÷ 3 = 5. The three ways to write a division problem are illustrated by

$$15 \div 3 = 5, \quad \frac{15}{3} = 5, \quad \text{and} \quad 3\overline{)15}.$$

### HINT

In general, division problems are written in one of the following three ways.

$$\text{dividend} \div \text{divisor} = \text{quotient} \qquad \frac{\text{dividend}}{\text{divisor}} = \text{quotient} \qquad \text{divisor}\overline{)\text{dividend}}^{\text{quotient}}$$

| | |
|---|---|
| **EXAMPLE 1** DIVIDING BY REPEATED SUBTRACTION | **PRACTICE EXERCISE 1** |

Find the quotient $20 \div 5$ by repeated subtraction.

Since 5 can be subtracted from 20 four times,

$$20 \div 5 = 4.$$

$$\begin{array}{r} 20 \\ -\ 5 \\ \hline 15 \\ -\ 5 \\ \hline 10 \\ -\ 5 \\ \hline 5 \\ -\ 5 \\ \hline 0 \end{array}$$

First subtraction

Second subtraction

Third subtraction

Fourth subtraction

Find the quotient by repeated subtraction.

$$\frac{30}{6}$$

Answer: 5; 6 can be subtracted 5 times

## ② RELATED MULTIPLICATION SENTENCE

Notice that

$$20 \div 5 = 4$$    Division sentence

and    $$5 \times 4 = 20.$$    Multiplication sentence

For a given division sentence, there is a **related multiplication sentence.**

### The Basic Principle of Division

For every division sentence

$$(\text{dividend}) \div (\text{divisor}) = (\text{quotient}),$$

there is a related multiplication sentence

$$(\text{divisor}) \times (\text{quotient}) = (\text{dividend}).$$

**HINT**

To obtain a related multiplication sentence for a division problem, move the divisor to the other side of the equal sign and multiply.

| | |
|---|---|
| **EXAMPLE 2** FINDING THE RELATED MULTIPLICATION SENTENCE | **PRACTICE EXERCISE 2** |

Give a related multiplication sentence for each division sentence.

**(a)** $14 \div 7 = 2$

Since 14 is the dividend, 7 the divisor, and 2 the quotient, a related multiplication sentence is $7 \times 2 = 14$.

Notice that by the commutative law, $2 \times 7 = 14$ is also a related multiplication sentence.

**(b)** $\dfrac{36}{4} = 9$

A related multiplication sentence is $4 \times 9 = 36$.

Give a related multiplication sentence for each division sentence.

**(a)** $25 \div 5 = 5$

**(b)** $6\overline{)42}$ with 7 above

Answers: (a) $5 \times 5 = 25$
(b) $6 \times 7 = 42$

Using the division principle we can learn the basic division facts. When we know them well enough, it will not be necessary to think of the related multiplication sentence.

| EXAMPLE 3   Using the Basic Division (Multiplication) Facts | PRACTICE EXERCISE 3 |

Divide.

**(a)** $18 \div 3 = 6$   This is true since $6 \times 3 = 18$

**(b)** $35 \div 7 = 5$   This is true since $5 \times 7 = 35$

**(c)** $8\overline{)64}^{\,8}$   This is true since $8 \times 8 = 64$

**(d)** $\dfrac{54}{9} = 6$   This is true since $6 \times 9 = 54$

Divide.

**(a)** $36 \div 9$

**(b)** $49 \div 7$

**(c)** $8\overline{)72}$

**(d)** $\dfrac{48}{6}$

Answers:   (a)  4   (b)  7   (c)  9
(d)  8

### ❸ DIVISION INVOLVING THE NUMBER 1

The basic principle of division is helpful for special division problems. In particular,

$$8 \div 8 = 1 \qquad \text{since} \qquad 8 \cdot 1 = 8$$

$$\frac{12}{12} = 1 \qquad \text{since} \qquad 12 \cdot 1 = 12$$

$$35\overline{)35}^{\,1} \qquad \text{since} \qquad 35 \cdot 1 = 35.$$

In general,

$$(\text{any nonzero whole number}) \div (\text{itself}\,) = 1.$$

Also,

$$5 \div 1 = 5 \qquad \text{since} \qquad 1 \cdot 5 = 5$$

$$\frac{9}{1} = 9 \qquad \text{since} \qquad 1 \cdot 9 = 9$$

$$1\overline{)13}^{\,13} \qquad \text{since} \qquad 1 \cdot 13 = 13.$$

These are examples of the following general rule.

$$(\text{any whole number}) \div 1 = (\text{that whole number})$$

### ❹ DIVISION INVOLVING THE NUMBER 0

When we divide 6 by 2, we see that

$$2\overline{)6}^{\,3} \qquad \text{means} \qquad 2 \times 3 = 6.$$

Suppose we try to divide 6 by 0 in the same manner. Let's use **?** for the quotient.

$$0\overline{)6}^{\,?} \qquad \text{means} \qquad 0 \times \mathbf{?} = 6.$$

Then **?** must represent what number? Suppose **?** is 6, then

$$\mathbf{?} \times 0 = 6 \times 0 = 0, \quad \text{not 6.}$$

What if **?** is 0? Then,

$$\mathbf{?} \times 0 = 0 \times 0 = 0, \quad \text{not } 6.$$

What if **?** is 2? Then,

$$\mathbf{?} \times 0 = 2 \times 0 = 0, \quad \text{not } 6.$$

In fact, what if **?** is any number? Then,

$$\mathbf{?} \times 0 = \text{(any number)} \times 0 = 0, \quad \text{not } 6.$$

Since there is no number we can use for **?** to make **?** $\times$ 0 equal to 6, we conclude that 0 cannot be used as a divisor of 6, or for that matter, as a divisor of any number.

### Division by Zero

**Division by 0 is undefined.** That is, we agree not to divide any number by 0.

Although division *by* zero is impossible, zero can be *divided by* any whole number (except 0). For example,

$$0 \div 3 = 0 \quad \text{since} \quad 3 \cdot 0 = 0$$

$$\frac{0}{12} = 0 \quad \text{since} \quad 12 \cdot 0 = 0$$

$$9\overline{)0}{}^{\,0} \quad \text{since} \quad 9 \cdot 0 = 0.$$

In general,

$$0 \div \text{(any nonzero whole number)} = 0.$$

The order of division, like that of subtraction, cannot be changed. A single example shows this. Since

$$8 \div 4 = 2 \quad \text{but} \quad 4 \div 8 \text{ is not a whole number,}$$

we see that $8 \div 4 \neq 4 \div 8$. Similarly,

$$(8 \div 4) \div 2 = 2 \div 2 = 1 \quad \text{but} \quad 8 \div (4 \div 2) = 8 \div 2 = 4.$$

Thus, $(8 \div 4) \div 2 \neq 8 \div (4 \div 2)$.

## HINT

Division is neither commutative nor associative. That is, changing the order of or grouping symbols in a division problem should never be done.

## 2.4 EXERCISES A

**1.** Consider the division problem $6 \div 2 = 3$. **(a)** What is 6 called? **(b)** What is 2 called? **(c)** What is 3 called?

**2.** In any division problem, the product of the divisor and the quotient is equal to what number?

**3.** If ☐ is any whole number, ☐ ÷ 1 = _____.

**4.** If ☐ is any whole number except 0, 0 ÷ ☐ = _____.

*You should be able to complete Exercises 5–49 in 90 seconds or less. Find each quotient.*

**5.** 18 ÷ 2 =   **6.** 28 ÷ 4 =   **7.** 30 ÷ 6 =   **8.** 16 ÷ 4 =   **9.** 8 ÷ 4 =

**10.** 5)‾25   **11.** 2)‾2   **12.** 6)‾18   **13.** 2)‾6   **14.** 8)‾72

**15.** 9 ÷ 1 =   **16.** 27 ÷ 9 =   **17.** 16 ÷ 8 =   **18.** 7 ÷ 7 =   **19.** 0 ÷ 8 =

**20.** $\dfrac{12}{6}$ =   **21.** $\dfrac{1}{1}$ =   **22.** $\dfrac{36}{9}$ =   **23.** $\dfrac{42}{6}$ =   **24.** $\dfrac{24}{4}$ =

**25.** 5 ÷ 5 =   **26.** 18 ÷ 9 =   **27.** 16 ÷ 2 =   **28.** 45 ÷ 5 =   **29.** 48 ÷ 6 =

**30.** 3)‾15   **31.** 2)‾12   **32.** 5)‾40   **33.** 5)‾0   **34.** 4)‾36

**35.** 6 ÷ 1 =   **36.** 24 ÷ 3 =   **37.** 64 ÷ 8 =   **38.** 63 ÷ 7 =   **39.** 20 ÷ 5 =

**40.** $\dfrac{0}{1}$ =   **41.** $\dfrac{48}{8}$ =   **42.** $\dfrac{54}{6}$ =   **43.** $\dfrac{0}{2}$ =   **44.** $\dfrac{8}{8}$ =

**45.** 24 ÷ 8 =   **46.** 6 ÷ 3 =   **47.** 36 ÷ 6 =   **48.** 21 ÷ 3 =   **49.** 35 ÷ 5 =

## FOR REVIEW

*Find the products.*

**50.**  500
$\underline{\times\ 30}$

**51.**  600
$\underline{\times\ 900}$

**52.** 4000
$\underline{\times\ 20}$

*Check by estimating the product.*

**53.**   92
$\underline{\times\ 19}$
 828
  $\underline{\ 92\ }$
1748

**54.**   743
$\underline{\times\ 102}$
1486
 $\underline{743\ \ }$
8916

---

**ANSWERS:** 1. (a) dividend (b) divisor (c) quotient 2. dividend 3. ☐ 4. 0 5. 9 6. 7 7. 5 8. 4 9. 2 10. 5 11. 1 12. 3 13. 3 14. 9 15. 9 16. 3 17. 2 18. 1 19. 0 20. 2 21. 1 22. 4 23. 7 24. 6 25. 1 26. 2 27. 8 28. 9 29. 8 30. 5 31. 6 32. 8 33. 0 34. 9 35. 6 36. 8 37. 8 38. 9 39. 4 40. 0 41. 6 42. 9 43. 0 44. 1 45. 3 46. 2 47. 6 48. 7 49. 7 50. 15,000 51. 540,000 52. 80,000 53. 90 × 20 = 1800; appears to be correct 54. 700 × 100 = 70,000; appears to be incorrect (correct product is 75,786)

## 2.4 EXERCISES B

**1.** Consider the division problem $54 \div 9 = 6$. **(a)** What is 6 called? **(b)** What is 9 called? **(c)** What is 54 called?

**2.** In any division problem, the dividend is equal to the product of the quotient and what number?

**3.** Any nonzero whole number divided by itself is equal to what number?

**4.** What whole number can never be used as a divisor?

*You should be able to complete Exercises 5–49 in 90 seconds or less. Find each quotient.*

**5.** $\dfrac{20}{4} =$     **6.** $\dfrac{48}{6} =$     **7.** $\dfrac{12}{4} =$     **8.** $\dfrac{24}{4} =$     **9.** $\dfrac{28}{7} =$

**10.** $16 \div 4 =$     **11.** $56 \div 7 =$     **12.** $6 \div 2 =$     **13.** $4 \div 2 =$     **14.** $7 \div 7 =$

**15.** $\dfrac{0}{5} =$     **16.** $\dfrac{12}{3} =$     **17.** $\dfrac{63}{7} =$     **18.** $\dfrac{4}{4} =$     **19.** $\dfrac{0}{2} =$

**20.** $5\overline{)15}$     **21.** $2\overline{)16}$     **22.** $9\overline{)63}$     **23.** $9\overline{)36}$     **24.** $7\overline{)49}$

**25.** $\dfrac{27}{3} =$     **26.** $\dfrac{35}{5} =$     **27.** $\dfrac{30}{5} =$     **28.** $\dfrac{9}{3} =$     **29.** $\dfrac{21}{3} =$

**30.** $6 \div 3 =$     **31.** $18 \div 3 =$     **32.** $42 \div 7 =$     **33.** $24 \div 8 =$     **34.** $0 \div 9 =$

**35.** $\dfrac{32}{4} =$     **36.** $\dfrac{1}{1} =$     **37.** $\dfrac{0}{7} =$     **38.** $\dfrac{18}{9} =$     **39.** $\dfrac{35}{7} =$

**40.** $1\overline{)0}$     **41.** $9\overline{)72}$     **42.** $1\overline{)6}$     **43.** $3\overline{)15}$     **44.** $1\overline{)3}$

**45.** $\dfrac{9}{1} =$     **46.** $\dfrac{56}{8} =$     **47.** $\dfrac{25}{5} =$     **48.** $\dfrac{9}{9} =$     **49.** $\dfrac{18}{2} =$

## FOR REVIEW

*Find the products.*

**50.**  $\begin{array}{r} 700 \\ \times\ 70 \\ \hline \end{array}$     **51.**  $\begin{array}{r} 800 \\ \times\ 800 \\ \hline \end{array}$     **52.**  $\begin{array}{r} 1000 \\ \times\ 500 \\ \hline \end{array}$

*Check by estimating the product.*

**53.**  $\begin{array}{r} 28 \\ \times\ 69 \\ \hline 252 \\ 128\phantom{0} \\ \hline 1532 \end{array}$     **54.**  $\begin{array}{r} 802 \\ \times\ 489 \\ \hline 7218 \\ 6416\phantom{0} \\ 3208\phantom{00} \\ \hline 392{,}178 \end{array}$

## 2.4 EXERCISES C

*Let n represent a whole number. Find n by writing the related multiplication or division sentence.*

**1.** $n \div 5 = 6$      **2.** $n \times 8 = 24$      **3.** $n \div 7 = 5$      **4.** $n \times 6 = 42$
[Answer: 30]                                        [Answer: 35]

*Assume that a, b, and c represent whole numbers in Exercises 5–10. Complete each statement.*

**5.** $a \div a =$ \_\_\_\_\_      **6.** $a \div 1 =$ \_\_\_\_\_      **7.** $0 \div a =$ \_\_\_\_
                                                                 (Assume $a$ is not 0.)

**8.** Are $a \div b$ and $b \div a$ equal?

**9.** Are $(a \div b) \div c$ and $a \div (b \div c)$ equal?

**10.** What does $a \div 0$ equal?

## 2.5 DIVIDING WHOLE NUMBERS

### STUDENT GUIDEPOSTS

1 Dividing by a One-Digit Number      4 Dividing by a Three-Digit Number
2 Remainder in a Division Problem      5 Checking Division
3 Dividing by a Two-Digit Number

### 1 DIVIDING BY A ONE-DIGIT NUMBER

The basic division facts allow us to divide special two-digit numbers by one-digit numbers. We now consider more general problems like the one below. If we try to find the quotient

$$86 \div 2$$

by repeated subtraction, we will have to subtract 2 a great number of times. We can shorten the process by subtracting many 2's at once, as follows.

```
    86
  - 20     Ten 2's subtracted
    66
  - 20     Ten more 2's subtracted
    46
  - 20     Ten more 2's subtracted
    26
  - 20     Ten more 2's subtracted
     6
  -  6     Subtract three 2's
     0
```

The total number of 2's subtracted is

$$10 + 10 + 10 + 10 + 3 = 43.$$

As a result, we see that

$$86 \div 2 = 43 \quad \text{or} \quad 2\overline{)86}.^{43}$$

In the previous display, we see that if the tens digit, 8, is divided by 2, we obtain the tens digit, 4, of the quotient. Similarly, dividing the ones digit, 6, by 2, we obtain the ones digit, 3, of the quotient.

$$8 \div 2 = 4 \longrightarrow \quad \underline{43} \quad \longleftarrow 3 = 6 \div 2$$
$$2\overline{)86}$$

When dividing the tens digit and the ones digit, we use basic division facts.

---

| **EXAMPLE 1  DIVIDING BY A ONE-DIGIT NUMBER** | **PRACTICE EXERCISE 1** |
|---|---|

Find the quotient.    $3\overline{)96}$

$$\begin{array}{r} \longmapsto 3 = 9 \div 3 \quad \text{(tens digit)} \\ \underline{32} \longleftarrow 2 = 6 \div 3 \quad \text{(ones digit)} \\ 3\overline{)96} \end{array}$$

Divide tens digit first, then divide ones digit.

Find the quotient.

$$3\overline{)690}$$

Answer: 230

---

In Example 1, the quotient in each step was a basic division fact. We next consider problems in which this is not the case. For example, suppose we find the quotient

$$3\overline{)48}.$$

First we use repeated subtraction.

$$\begin{array}{r} 48 \\ -\ 30 \\ \hline 18 \\ -\ 18 \\ \hline 0 \end{array} \quad \begin{array}{l} \text{Subtract ten 3's} \\[18pt] \text{Subtract six 3's} \end{array}$$

Adding 10 and 6 gives the quotient, 16. Had we written the problem as before, we would have

$$\begin{array}{r} 16 \\ 3\overline{)48}. \end{array}$$

Since $4 \div 3$ is not a basic division fact, we must find this quotient in several steps.

① Find the largest number less than 4 which is divisible by 3. This number is 3, and $3 \div 3 = 1$. Place 1 above the 4.

$$\begin{array}{r} 1 \\ 3\overline{)48} \end{array}$$

② Multiply 1 times 3 and place the product below 4.

③ Draw a line below 3 and subtract.

$$\begin{array}{r} 1 \\ 3\overline{)48} \\ \underline{3} \\ 1 \end{array}$$

④ Bring down the eight.

⑤ Divide 18 by 3 and place the quotient, 6, to the right of 1.

$$\begin{array}{r} 16 \\ 3\overline{)48} \\ \underline{3} \\ 18 \end{array}$$

⑥ Multiply 6 times 3 and place the result below 18.

⑦ Draw a line below 18 and subtract.

$$\begin{array}{r} 16 \\ 3\overline{)48} \\ \underline{3} \\ 18 \\ \underline{18} \\ 0 \end{array}$$

In the second step, the 1 in the quotient stands for 1 ten, so the product $1 \times 3$ represents 3 tens or 30. Compare this to the result of repeated subtraction.

## ② REMAINDER IN A DIVISION PROBLEM

When we divided 48 by 3 above, the final difference, 0, is called the **remainder.** In general, the remainder can be any number less than the divisor.

| EXAMPLE 2 DIVISION WITH A NONZERO REMAINDER | PRACTICE EXERCISE 2 |
|---|---|

Find the quotient. Follow the numbered steps. $4\overline{)537}$

⑤ $12 \div 4 = 3$

① $4 \div 4 = 1 \longrightarrow$ $\quad 134 \longleftarrow$ ⑨ $16 \div 4 = 4$
$\quad\quad\quad\quad\quad\quad\quad 4\overline{)537}$
② $1 \times 4 = 4 \longrightarrow$ $\quad 4$
③ $5 - 4 = 1 \longrightarrow$ $\quad 13 \longleftarrow$ ④ Bring down 3
⑥ $3 \times 4 = 12 \longrightarrow$ $\quad 12$
⑦ $13 - 12 = 1 \longrightarrow$ $\quad 17 \longleftarrow$ ⑧ Bring down 7
⑩ $4 \times 4 = 16 \longrightarrow$ $\quad 16$
$\quad\quad\quad\quad\quad\quad\quad\quad 1 \longleftarrow$ ⑪ $17 - 16 = 1$ (remainder)

Find the quotient.

$3\overline{)473}$

Answer: 157 R2

## HINT

If you obtain a remainder that is greater than or equal to the divisor, you have made a mistake. In the final division step, the divisor went into the dividend more times. Remember that your remainder must always be less than the divisor.

In many cases the first digit of the dividend is less than the divisor.

| EXAMPLE 3 DIVIDING INTO THE FIRST TWO DIGITS | PRACTICE EXERCISE 3 |
|---|---|

Find the quotient. $3\overline{)263}$

Since there is no number less than or equal to 2 which is divisible by 3, use the first *two* digits of the dividend in the first step. Find the largest number less than or equal to 26 which is divisible by 3. That number, 24, when divided by 3 gives a quotient of 8. Place 8 above 6 and continue.

① $24 \div 3 = 8 \longrightarrow$ $\quad 87 \longleftarrow$ ⑤ $21 \div 3 = 7$
$\quad\quad\quad\quad\quad\quad\quad 3\overline{)263}$
② $8 \times 3 = 24 \longrightarrow$ $\quad 24$
③ $26 - 24 = 2 \longrightarrow$ $\quad 23 \longleftarrow$ ④ Bring down 3
⑥ $7 \times 3 = 21 \longrightarrow$ $\quad 21$
$\quad\quad\quad\quad\quad\quad\quad\quad 2 \longleftarrow$ ⑦ $23 - 21 = 2$ (remainder)

Find the quotient.

$6\overline{)656}$

Answer: 109 R2

It is easy to extend the process to dividends with more than three digits.

| **EXAMPLE 4    DIVIDING A FOUR-DIGIT NUMBER** | **PRACTICE EXERCISE 4** |

Find the quotient.    $5\overline{)3825}$

$$\boxed{5}\; 30 \div 5 = 6$$
$$\boxed{1}\; 35 \div 5 = 7 \underline{\qquad\qquad}\;\;\Big\downarrow\Big\downarrow\Big\downarrow \underline{\qquad} \boxed{9}\; 25 \div 5 = 5$$

$$
\begin{array}{r}
765 \\
5\overline{)3825}
\end{array}
$$

$\boxed{2}\; 7 \times 5 = 35 \longrightarrow \quad 35$

$\boxed{3}\; 38 - 35 = 3 \longrightarrow \quad 32 \longleftarrow \boxed{4}$ Bring down 2

$\boxed{6}\; 6 \times 5 = 30 \longrightarrow \quad 30$

$\boxed{7}\; 32 - 30 = 2 \longrightarrow \quad 25 \longleftarrow \boxed{8}$ Bring down 5

$\boxed{10}\; 5 \times 5 = 25 \longrightarrow \quad 25$

$\qquad\qquad\qquad\qquad\qquad\quad 0 \longleftarrow \boxed{11}\; 25 - 25 = 0$

Find the quotient.

$$2\overline{)4056}$$

Answer: 2028

With practice you will learn a shortcut illustrated in the next example.

| **EXAMPLE 5    A SHORTCUT OCCURRING IN SOME PROBLEMS** | **PRACTICE EXERCISE 5** |

Find the quotient.    $7\overline{)3528}$

$$
\begin{array}{r}
504 \\
7\overline{)3528} \\
\underline{35\phantom{00}} \\
028 \\
\underline{28} \\
0
\end{array}
$$

Since 7 cannot divide 2, bring down 8 and write 0 above 2

Find the quotient.

$$8\overline{)4824}$$

Answer: 603

## CAUTION

If you forget to write the digit 0 in a quotient, such as in Example 5, your answer will be wrong. Remember that 0 is an important placeholder in our number system.

## ③ DIVIDING BY A TWO-DIGIT NUMBER

Finding quotients when the divisor has more than one digit is a bit more challenging. Consider

$$25\overline{)825}.$$

The quotient can be found directly using what we have learned about dividing by one-digit numbers. We first decide which of the first few digits of the dividend (825) make a number greater than or equal to the divisor (25). In our example, that number is 82. The largest number less than 82 which is divisible by 25 is 75. The quotient, 3, is placed above the last digit in 82.

$$
\begin{array}{r}
3\phantom{0} \\
25\overline{)82\,5}
\end{array}
\qquad 82 \geq 25 \text{ and } 75 \div 25 = 3
$$

Next multiply $3 \times 25$, place the result below 82 and subtract. The difference is 7. Bring down 5.

$$
\begin{array}{r}
3\phantom{0} \\
25\overline{)825} \\
75\phantom{0} \\
\hline
75
\end{array}
$$

Finally divide 75 by 25. The quotient is 3 which is placed above 5. Multiply $3 \times 25$, place the product below 75, and subtract to obtain 0.

$$
\begin{array}{r}
33 \\
25\,\overline{)825} \\
75\phantom{0} \\
\hline
75 \\
75 \\
\hline
0
\end{array}
$$

Dividing by a two-digit number sometimes involves a bit of guesswork and luck as we try to find each digit in the quotient. For example, suppose we want to find the quotient $960 \div 42$. We might start by looking at the first digits in 960 and 42 and asking

What number less than or equal to 9 can be divided by 4?

Since $8 \le 9$, and $8 \div 4 = 2$, we try 2. This technique, while not foolproof, gives us a starting point, as shown in the next example.

---

### EXAMPLE 6   DIVIDING BY A TWO-DIGIT NUMBER

Find the quotient.   $32\overline{)739}$

The number less than or equal to 7 which is divisible by 3 is 6, and since $6 \div 3 = 2$, begin by placing 2 above 3 in 739.

$$
\begin{array}{l}
\text{①}\ 6 \div 3 = 2 \longrightarrow \quad 2 \\
\phantom{\text{①}\ 6 \div 3 = 2 \longrightarrow }32\overline{)739} \\
\text{②}\ 2 \times 32 = 64 \longrightarrow \quad 64 \\
\text{③}\ 73 - 64 = 9 \longrightarrow \quad 99 \longleftarrow \text{④ Bring down 9}
\end{array}
$$

To divide 99 by 32, find the largest number less than or equal to 9 which is divisible by 3. That number is 9 and $9 \div 3 = 3$. Thus, place 3 above 9 in 739 and proceed as before.

$$
\begin{array}{r}
23 \\
3\,2\overline{)739} \\
64 \\
\hline
99 \\
3 \times 32 = 96 \longrightarrow \quad 96 \\
\hline
3 \longleftarrow 99 - 96 = 3\ \text{(remainder)}
\end{array}
$$

This method can be applied to dividends with more digits.

---

### EXAMPLE 7   DIVIDING BY A TWO-DIGIT NUMBER

Find the quotient.   $67\overline{)5294}$

The number less than or equal to 52 which is divisible by 6 is 48, so we try the quotient, 8.

$$
\begin{array}{r}
8\phantom{00} \\
67\overline{)5294} \\
8 \times 67 = 536 \longrightarrow \quad 536
\end{array}
$$

---

### PRACTICE EXERCISE 6

Find the quotient.

$$41\overline{)861}$$

Answer: 21

---

### PRACTICE EXERCISE 7

Find the quotient.

$$46\overline{)2181}$$

But $536 > 529$ so we cannot subtract. (As we said earlier, this method is not foolproof.) We try a smaller quotient, 7.

$$\begin{array}{r} 7 \\ 67\overline{)5294} \\ 7 \times 67 = 469 \longrightarrow \phantom{67)}469 \\ \hline \phantom{67)}604 \end{array}$$

Now we try 60 divided by 6. But since this quotient is the two-digit number 10, we "back off" and use the largest one-digit number, 9.

$$\begin{array}{r} 79 \\ 67\overline{)5294} \\ \phantom{67)}469 \\ \hline \phantom{67)}604 \\ 9 \times 67 = 603 \longrightarrow \phantom{67)}603 \\ \hline \phantom{67)}1 \end{array}$$

Answer: 47 R19

## ④ DIVIDING BY A THREE-DIGIT NUMBER

An **algorithm** is a procedure or method used to perform a certain task. The algorithm for finding quotients that we have shown involves "trial-and-error." Sometimes estimating can make "trial-and-error" faster. The number of trials that you make before finding the desired quotient will decrease with practice.

This algorithm can be used to find quotients involving three-digit divisors.

| **EXAMPLE 8    DIVIDING BY A THREE-DIGIT NUMBER** |
| --- |

Find the quotient.    $562\overline{)116527}$

$$\begin{array}{r} \phantom{562)}207 \\ 562\overline{)116527} \\ \phantom{562)}1124 \\ \hline \phantom{562)}412 \\ \phantom{562)}000 \\ \hline \phantom{562)}4127 \\ \phantom{562)}3934 \\ \hline \phantom{562)}193 \end{array}$$

- 10 ÷ 5 = 2
- 412 cannot be divided by 562
- 35 ÷ 5 = 7

The work can be shortened by bringing down the next digit, 7, when we realize that 412 cannot be divided by 562. But be sure to place 0 in the quotient.

$$\begin{array}{r} 207 \\ 562\overline{)116527} \\ \phantom{562)}1124 \\ \hline \phantom{562)}4127 \\ \phantom{562)}3934 \\ \hline \phantom{562)}193 \end{array}$$

Remember this 0

| **PRACTICE EXERCISE 8** |
| --- |

Find the quotient.

$$451\overline{)93808}$$

Answer: 208

---

## HINT

Some students may question the necessity of learning the division algorithms since division problems can be solved more quickly with a calculator. Although this is true, the techniques here are needed for understanding similar algorithms found in algebra.

---

## 5 CHECKING DIVISION

As with other arithmetic operations, there should be a way to check if a quotient is correct. Consider the following division problem.

$$
\text{divisor} \longrightarrow 4\overline{)25} \quad \begin{array}{l} \longleftarrow \text{quotient} \\ \longleftarrow \text{dividend} \end{array}
$$
$$
\begin{array}{r} \underline{24} \\ 1 \longleftarrow \text{remainder} \end{array}
$$

Observe that

$$(6 \times 4) + 1 = 25$$
$$24 \quad + 1 = 25.$$

This relationship is true in general and provides a check for division.

### Check for Division

In any division problem

$$(\text{quotient} \times \text{divisor}) + \text{remainder} = \text{dividend}.$$

Suppose we check the division problem given in Example 7. Since the quotient is 79, the divisor is 67, the remainder is 1, and the dividend is 5294, we have the following:

$$
\begin{array}{ccc} (\text{quotient} & \times \text{divisor}) & + \text{remainder} \\ \downarrow & \downarrow & \downarrow \end{array}
$$
$$(79 \times 67) + 1 = (5293) + 1$$
$$= 5294 \longleftarrow \text{dividend}$$

Thus, the work was correct.

The final example in this section illustrates how rounded numbers can be used to find approximate solutions to applied division problems.

---

### EXAMPLE 9  ESTIMATING IN A DIVISION APPLICATION

Manuel wants to divide 117 pieces of candy among 9 children. First find the approximate number each will receive; and then find the exact number.

Since 117 is about 120 and 9 is about 10, we can mentally divide 120 by 10 to obtain approximately 12 pieces for each. To find the exact number, divide 117 by 9. The quotient is 13, so each child receives exactly 13 pieces.

---

### PRACTICE EXERCISE 9

There are 1092 weeks in 21 years. First find the approximate number of weeks in one year; then find the exact number.

**Answer:** Approximate number is 50 (using $1000 \div 20$); exact number is 52 (using $1092 \div 21$)

## 2.5 EXERCISES A

*Find each quotient.*

**1.** 2)46          **2.** 3)93          **3.** 2)446          **4.** 4)408          **5.** 2)468

**6.** 2)58          **7.** 5)90          **8.** 3)438          **9.** 6)966          **10.** 3)225

**11.** 6)546          **12.** 5)905          **13.** 5)545          **14.** 6)654          **15.** 4)4076

**16** 2)6054          **17.** 4)2836          **18.** 5)4535          **19.** 32)672          **20.** 63)756

**21** 23)2231          **22.** 28)1316          **23** 473)13244          **24.** 623)35511          **25.** 755)31710

*Find each quotient and remainder. Check your answer.*

**26.** 7)67          **27.** 6)714          **28.** 42)737          **29** 83)7553          **30.** 466)10736

*Solve.*

**31.** A man buys 9 suits for $1782. First find the approximate cost of each suit, and then find the exact cost.

**32** There are 4011 calories in 21 candy bars. First find the approximate number of calories in each bar, and then find the exact number.

---

ANSWERS:   1. 23   2. 31   3. 223   4. 102   5. 234   6. 29   7. 18   8. 146   9. 161   10. 75   11. 91   12. 181
13. 109   14. 109   15. 1019   16. 3027   17. 709   18. 907   19. 21   20. 12   21. 97   22. 47   23. 28   24. 57   25. 42
26. Q: 9; R: 4 [(9 × 7) + 4 = 63 + 4 = 67]   27. Q: 119; R: 0 [(119 × 6) + 0 = 714 + 0 = 714]   28. Q: 17; R: 23
[(17 × 42) + 23 = 714 + 23 = 737]   29. Q: 91; R: 0 [(91 × 83) + 0 = 7553 + 0 = 7553]   30. Q: 23; R: 18
[(23 × 466) + 18 = 10,718 + 18 = 10,736]   31. approximate cost: $180 (using 1800 ÷ 10); exact cost: $198 (using
1782 ÷ 9)   32. approximate number: 200 calories (using 4000 ÷ 20); exact number: 191 calories (using 4011 ÷ 21)

## 2.5 EXERCISES B

*Find each quotient.*

**1.** 2)64          **2.** 6)60          **3.** 3)639          **4.** 3)609          **5.** 2)96

**6.** 3)84          **7.** 6)84          **8.** 4)712          **9.** 2)986          **10.** 4)384

**11.** $2\overline{)134}$     **12.** $3\overline{)327}$     **13.** $4\overline{)208}$     **14.** $3\overline{)3645}$     **15.** $6\overline{)6066}$

**16.** $3\overline{)6309}$     **17.** $8\overline{)4064}$     **18.** $9\overline{)5418}$     **19.** $45\overline{)720}$     **20.** $43\overline{)3698}$

**21.** $52\overline{)3796}$     **22.** $67\overline{)3484}$     **23.** $175\overline{)8575}$     **24.** $89\overline{)31773}$     **25.** $347\overline{)92996}$

*Find each quotient and remainder. Check your answer.*

**26.** $9\overline{)93}$     **27.** $23\overline{)1081}$     **28.** $67\overline{)5728}$     **29.** $253\overline{)14173}$     **30.** $342\overline{)68742}$

*Solve.*

**31.** A man buys 11 radios at a total cost of $396. First find the approximate cost of 1 radio, then find the exact cost.

**32.** A man's heart beats 2016 times during an exercise period of 21 minutes. First find the approximate number of beats per minute, and then find the actual number (which in this case is also really an approximation).

## 2.5 EXERCISES C

*Find the quotient and remainder.*

**1.** $7\overline{)18114}$

**2.** $27\overline{)177696}$
[Answer: Q: 6581; R: 9]

**3.** $208\overline{)106029}$

## 2.6 MORE APPLICATIONS

### STUDENT GUIDEPOSTS

1. Total Value Relationship
2. Area of a Rectangle
3. Business Applications
4. Calculating Miles Per Gallon
5. Unit Pricing

Many word problems involve multiplication and division of whole numbers. As with addition and subtraction word problems, you should make sure that your answer is reasonable.

### 1 TOTAL VALUE RELATIONSHIP

A basic principle used in many applied problems involves finding a *total value*. The total value of a quantity made up of several parts is the value per part multiplied by the number of parts.

$$\textbf{(total value)} = \textbf{(value per part)} \cdot \textbf{(number of parts)}.$$

For example, if you have 35 dimes, the total value of these coins is given by:

(total value of 35 dimes) = (value of 1 dime) · (number of dimes)

$$= \quad (10¢) \quad \cdot \quad (35)$$
$$= \quad 350¢.$$

We use this principle in the next example.

| **EXAMPLE 1**  TOTAL VALUE IN A BUSINESS PROBLEM | **PRACTICE EXERCISE 1** |
|---|---|

During a sale on men's sweaters, a store owner sold 43 sweaters at $18 a sweater. How much money did he take in on the sweater sale?

For 1 sweater sold, he receives $18 \cdot 1 = \$18$.

For 2 sweaters sold, he receives $18 \cdot 2 = \$36$.

For 3 sweaters sold, he receives $18 \cdot 3 = \$54$.

For 43 sweaters sold,

(total value of 43 sweaters sold) = (value per sweater) · (number of sweaters)

$\qquad$ \$774 $\qquad$ = $\qquad$ \$18 $\qquad$ · $\qquad$ 43.

That is, we multiply 18 by 43 to obtain $774. Since $774 seems reasonable (40 sweaters at $20 each would bring in $800; so 774 appears to check), the store owner took in $774 during the sale.

Lucy had a job babysitting which paid $3 per hour. How much was she paid for sitting a total of 14 hours?

Answer: $42

## 2 AREA OF A RECTANGLE

The formula for finding the area of a rectangle has many applications involving multiplication.

### Area of a Rectangle

The area of a rectangle is given by the formula

**Area = (length) · (width).**

The rectangle in Figure 2.4 has area given by

$$Area = (length) \cdot (width)$$
$$= 3 \cdot 2$$
$$= 6.$$

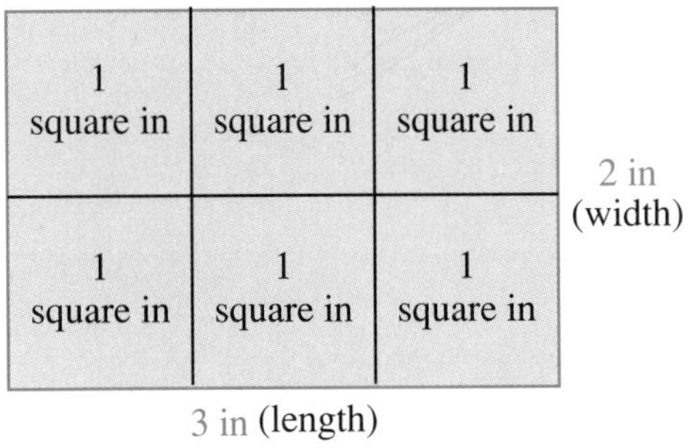

Figure 2.4

Since the rectangle contains 6 squares which have 1 inch sides, the measure of the area of the rectangle is given in square inches (sometimes abbreviated **sq in**). Thus, the rectangle has area 6 sq in. Had the sides of the rectangle been 3 ft and 2 ft, the area would have been 6 sq ft.

| EXAMPLE 2 Area and Perimeter of a Rectangle | PRACTICE EXERCISE 2 |

A room is 4 yd wide and 5 yd long (**yd** is the abbreviation for yards).

**(a)** Find the floor area of the room.

In many word problems, it is helpful to make a sketch which describes the situation. A sketch of the floor of the room is shown in Figure 2.5. The area of the floor is

$$\text{Area} = (\text{length}) \cdot (\text{width})$$

$$= 5 \cdot 4 = 20 \text{ sq yd.}$$

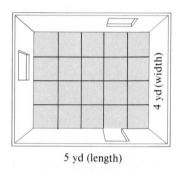

4 yd (width)

5 yd (length)

**Figure 2.5**

**(b)** Find the perimeter of the room.

The perimeter is the distance around the room found by adding the lengths of its sides. Thus, the perimeter is

$$5 + 5 + 4 + 4 = 18 \text{ yd.}$$

**(c)** If carpet costs $16 a square yard, how much would it cost to carpet the room?

One square yard of carpet costs $16. Then 20 sq yd at $16 a yard would cost

$$20 \cdot 16 = \$320.$$

The floor in a recreation hall is 20 yd long and 14 yd wide.

**(a)** Find the area of the floor.

**(b)** Find the perimeter of the hall.

**(c)** If carpeting costs $13 a square yard, how much would it cost to carpet the hall?

Answers: **(a)** 280 sq yd   **(b)** 68 yd
**(c)** $3640

---

## ③ BUSINESS APPLICATIONS

Many applications of multiplication can be found in business. Forms such as invoices, shipping orders, packing slips, and purchase orders use multiplication. A sample purchase order is shown in the next example.

| EXAMPLE 3  AN APPLICATION IN BUSINESS | PRACTICE EXERCISE 3 |

Complete the purchase order in Figure 2.6.

Complete the purchase order.

### NORTHERN PENNSYLVANIA UNIVERSITY
### General Stores

Customer's Order No. ___11937-Q___    Date ___March 20___    19_89_
Name ___Mathematics Department___
Address ___Box 5717    MAT 115___

| Sold By A.S. | Cash | C.O.D. | Charge | On Acct. ✔ | Mdse. Retd. | Paid Out | |
|---|---|---|---|---|---|---|---|

| Quantity | Description | Unit Price | Amount |
|---|---|---|---|
| 50 reams | Multi-purpose Paper | $2 | |
| 6 cans | Duplicating Fluid | $3 | |
| 3 | Staplers | $4 | |
| 8 pkg | Writing Pads | $3 | |
| 1 | Paper Trimmer | $42 | |
| | | | |
| | | | |
| | | | |
| | | TOTAL | |

Rec'd. by _____

**Figure 2.6**

| Quantity | Description | Unit price | Amount |
|---|---|---|---|
| 12 | Shirts | $14 | (a) |
| 36 | Pair socks | $ 4 | (b) |
| 50 | Pair slacks | $24 | (c) |
| 25 | Belts | $13 | (d) |
| | TOTAL | | (e) |

| Quantity | Description | Unit Price | Amount |
|---|---|---|---|
| 50 reams | Multi-purpose Paper | $2 | $100 |
| 6 cans | Duplicating Fluid | $3 | 18 |
| 3 | Staplers | $4 | 12 |
| 8 pkg | Writing Pads | $3 | 24 |
| 1 | Paper Trimmer | $42 | 42 |
| | | TOTAL | $196 |

**Figure 2.7**

To complete this form, compute the numbers that go into the Amount column. The price of each ream of Multi-purpose Paper is $2 so the cost of 50 reams is $50 \times \$2$, which is $100. Enter this cost in the Amount column. Compute and enter the remaining costs the same way. See Figure 2.7.

Find the total cost of all items by adding the separate costs in the Amount column ($100 + 18 + 12 + 24 + 42$). Enter this sum, $196, next to TOTAL as shown.

Answers: (a) **$168**    (b) **$144**
(c) **$1200**    (d) **$325**    (e) **$1837**

**HINT**

Read an applied problem carefully and be aware that a given problem might involve more than one arithmetic operation. In Example 3, for instance, to complete the purchase order we first find five products and then find their sum.

## ❹ CALCULATING MILES PER GALLON

In the next example we provide a formula for finding the number of miles that a vehicle can be driven using one gallon of gasoline.

| EXAMPLE 4   CALCULATING MILES PER GALLON | PRACTICE EXERCISE 4 |
|---|---|

In order to check the gas mileage on his car, Mike noted that the odometer reading was 18,432 when he filled the tank with gas. At 18,696 he refilled the tank with 11 gallons of gas. What was his mileage per gallon?

The miles per gallon (mpg) is found by:

$$\text{mpg} = \frac{\text{number of miles driven}}{\text{number of gallons used}}$$

To find the number of miles driven, subtract odometer readings.

$$\begin{array}{r} 18696 \\ -\ 18432 \\ \hline 264 \end{array} \quad \text{Miles driven}$$

Thus,

$$\text{mpg} = \frac{264}{11} = 24.$$

Since 24 mpg seems reasonable and does check ($24 \times 11 = 264$ and $18{,}432 + 264 = 18{,}696$), Mike is getting 24 miles per gallon of gas.

On a recent trip, Sharon traveled from El Paso, Texas, to Cheyenne, Wyoming, a distance of 750 miles, and returned to El Paso. If she used a total of 75 gallons of gas on the trip, how many miles per gallon did she get?

Answer: 20 mpg

The next two examples provide other types of applications involving arithmetic operations.

| EXAMPLE 5   MAKING PAYMENTS ON A PURCHASE | PRACTICE EXERCISE 5 |
|---|---|

Toni agreed to purchase a used car from her parents for $1100, by paying $80 down and the rest in equal monthly payments for one year. How much were her monthly payments if she is not being charged any interest?

Because she paid $80 down, the difference $1100 − $80 = $1020 is the amount she has to pay over 12 months (1 year = 12 months). To find the amount paid each month, divide 1020 by 12.

$$\begin{array}{r} 85 \\ 12\overline{)1020} \\ \underline{96} \\ 60 \\ \underline{60} \\ 0 \end{array}$$

Check:  
$$\begin{array}{r} 85 \\ \times\ 12 \\ \hline 170 \\ \underline{85} \\ 1020 \end{array} \quad \text{and} \quad \begin{array}{r} 1020 \\ +\ 80 \\ \hline 1100 \end{array}$$

Since this amount seems reasonable and does check, her monthly payment is $85.

Lennie purchased four new tires for his car. If each tire cost $75 and the tax on the purchase of the four was $12, what was the total cost of the tires? If the dealer agreed to let him pay for the tires in 3 equal monthly payments, how much does he pay each month?

Answer: $312; $104

| EXAMPLE 6   DISTRIBUTING PRIZE MONEY | PRACTICE EXERCISE 6 |
|---|---|

A seven-man basketball squad won a local tournament that had a $500 prize for the winning team. The seven players and the coach agree to divide the money equally (in dollars) among themselves with any remainder going to the coach. How much does the coach receive?

$$\begin{array}{r} 62 \\ 8\overline{)500} \\ 48 \\ \hline 20 \\ 16 \\ \hline 4 \end{array}$$  Remainder

Check:  $(62 \times 8) + 4 = 496 + 4$
$= 500$

Each of the seven players should receive $62 with the coach receiving $62 plus the $4 left over. Thus, the coach receives $66. (Does the answer seem reasonable?)

The twelve members of a fraternity bought a raffle ticket and won the $1000 prize. They agree to split the prize among themselves equally (in dollars) with any remainder going into their party fund. How much does each receive and how much is deposited in the fund?

Answer: $83 to each member and $4 to the fund

## ⑤ UNIT PRICING

The wise shopper has learned to compare prices by determining *unit prices*. For example, if a can of beans weighs 15 oz (**oz** stands for ounces) and costs 45¢, the unit price of the beans is

$$\frac{45}{15} = 3¢ \text{ per ounce.}$$

If a different brand of beans sells for 52¢ but weighs 13 oz, its unit price is

$$\frac{52}{13} = 4¢ \text{ per ounce.}$$

Assuming equal quality of the products, the first can is clearly the better buy. Unit pricing gives us excellent examples of division problems.

## 2.6   EXERCISES A

*Solve.*

1. A compact car has a 12-gallon gas tank and can travel 29 miles on one gallon of gas. How far can it travel on a full tank?

2. If a can of cola contains 86 calories, how many calories are in a six-pack of cola?

3. There are 365 days in a year.
   (a) How many days are there in 5 years?
   (b) How many hours are there in a year? [1 day = 24 hours]
   (c) How many hours are there in 5 years?

4. A hardware store sells 35 snow blowers at $329 each and 40 chain saws at $289 each.
   (a) How much money does the store take in on the sale of snow blowers?
   (b) How much money does the store take in on the sale of chain saws?
   (c) How much money does the store take in on the sale of both snow blowers and chain saws?

**5.** Suppose that 225 lift tickets are sold per day at the Sugar Loaf Ski Area.

 **(a)** How many tickets are sold in a 14-day period?

 **(b)** If each ticket costs $7, how much money is taken in during a 14-day period?

**6.** A rectangular vegetable garden is 14 m wide and 26 m long (**m** stands for meter).

 **(a)** What is the area of the garden?

 **(b)** What is the perimeter of the garden?

**7.** A woman buys 3 suits at $175 each, 5 blouses at $21 each, and 6 scarves at $8 each. How much has she spent?

**8** Barbara wants to put a rectangular mirror measuring 8 ft by 5 ft on the wall of her living room. If a mirror costs $3 a square foot, how much will the mirror cost?

**9.** A rectangular room measures 9 yd long and 5 yd wide. How much will it cost to carpet the room over a foam pad if carpet costs $20 a sq yd, padding costs $2 a sq yd, and installation costs are included in the price of carpet and padding?

**10.** A sketch of the floor plan of a family room is shown in the figure. Mr. Kirk plans to cover the floor with vinyl floor covering costing $9 a sq yd. How much will it cost to complete the job? [*Hint:* Divide the floor plan into two rectangular areas.]

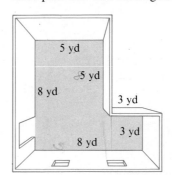

**11.** Complete the following purchase order.

| Quantity | Description | Unit Price | Amount |
|---|---|---|---|
| 3 | Back Packs | $90 | **(a)** _____ |
| 1 | Tent | 140 | **(b)** _____ |
| 4 | Sleeping Bags | 110 | **(c)** _____ |
| 4 | Ground Pads | 15 | **(d)** _____ |
| 1 | Pack Stove | 20 | **(e)** _____ |
| 8 | Complete Meals | 7 | **(f)** _____ |
| | TOTAL | | **(g)** _____ |

**12.** Sam and Carolyn have a home mortgage which requires a monthly payment of $520 for 30 years.

    **(a)** How many monthly payments must they make over the 30-year period?

    **(b)** How much will they have paid in total at the end of 30 years?

**13** A domed football stadium is divided into 20 sections of seats. Suppose that 5 sections each seat 400 spectators, 8 sections each seat 1200 spectators, and the remaining sections each seat 950 spectators.

    **(a)** What is the seating capacity of the dome?

    **(b)** If every seat is sold for the championship game, and the price of each ticket is $5, how much money is taken in for the game?

**14.** Karen can type 85 words per minute. How long will it take her to type 21,760 words?

**15.** At the start of a trip, Mr. Creighten's odometer read 24,273. When he returned from the trip, it read 25,561. If he used 56 gallons of gas, what was his gas mileage?

**16.** A certain brand of toothpaste sells for 80¢. How many tubes can be purchased for $6, and how much change is left? [*Hint:* Use 600¢ for $6.]

**17** Which is the better buy, a 12-lb bag of dog food selling for $3.72 (372¢) or a 9-lb bag of dog food selling for $2.88 (288¢)?

**18.** Eleven people pool their money and buy a lottery ticket. The prize is $1500 and they agree, if they win, to divide the money equally (in dollars) and draw straws for any amount left over after the split. What is the most that a winner could receive?

**19.** A Toyota get 26 mpg. How many gallons would it take to make a 2444-mile trip?

**20** The enrollment at a university is 12,100 students. If the faculty numbers 550, what is the number of students per teacher?

**21.** Dr. Yard buys 225 shares of stock for a total cost of $31,050. What was the cost of each share?

**22.** The gate receipts at a concert were $87,850. If tickets sold for $7 each, how many were in attendance?

**23.** Assuming equal quality, which is the better buy, a 43-oz box of soap for $1.29 (129¢) or a 78-oz box of soap for $1.56 (156¢)?

## FOR REVIEW

*Find each quotient and check your answer.*

**24.** $8\overline{)1880}$

**25.** $37\overline{)7750}$

**26.** $259\overline{)107744}$

*Exercise 27 will help you prepare for the next section.*

**27.** List the multiples of 2 between 1 and 15.

ANSWERS: 1. 348 mi  2. 516 calories  3. (a) 1825 days (b) 8760 hours (c) 43,800 hours  4. (a) $11,515
(b) $11,560 (c) $23,075  5. (a) 3150 tickets (b) $22,050  6. (a) 364 sq m (b) 80 m  7. $678  8. $120  9. $990
10. $441  11. (a) $270 (b) $140 (c) $440 (d) $60 (e) $20 (f) $56 (g) $986  12. (a) 360 payments (b) $187,200
13. (a) 18,250 seats (b) $91,250  14. 256 minutes  15. 23 mpg  16. 7 tubes; 40¢ in change  17. The 12-lb bag at 31¢
per lb is a better buy than the 9-lb bag at 32¢ per lb.  18. $140  19. 94 gallons  20. 22  21. $138  22. 12,550
23. At 2¢ per oz, the 78-oz box is a better buy than the 43-oz box at 3¢ per oz.  24. Q: 235; R: 0. [(235 × 8) + 0 =
1880 + 0 = 1880]  25. Q: 209; R: 17. [(209 × 37) + 17 = 7733 + 17 = 7750]  26. Q: 416; R: 0. [(416 × 259) + 0 =
107,744 + 0 = 107,744]  27. 2, 4, 6, 8, 10, 12, 14

## 2.6 EXERCISES B

*Solve.*

**1.** A boat has a 45-gallon gas tank. If the boat can travel 4 miles on one gallon of gas, how far can it go on a full tank of gas?

**2.** During a sale on CB radios, Chuck sold 18 radios at $78 each. How much money did he take in?

**3.** There are 365 days in a year.
  **(a)** How many days are there in 7 years?
  **(b)** How many hours are there in 7 years?
  **(c)** How many minutes are there in 7 years?

**4.** A room is 12 ft wide and 15 ft long.
  **(a)** Find the floor area of the room.
  **(b)** Find the perimeter of the room.
  **(c)** If carpet costs $2 a sq ft, how much would it cost to carpet the room?

**5.** Suppose that 120 tickets are sold per day to see *The Mummy* at The Tourist Trap.
  **(a)** How many tickets are sold in a 30-day period?
  **(b)** If each ticket costs $4, how much money is taken in during a 30-day period?

**6.** A wall is 20 ft long and 8 ft high.
  **(a)** What is the area of the wall?
  **(b)** What is the perimeter of the wall?

**7.** Betsy buys 3 sweaters at $39 each, 2 skirts at $42 each, and 7 pairs of earrings at $4 a pair. How much has she spent?

**8.** The flat roof of a mobile home measures 60 ft long and 15 ft wide. If it costs $2 a square foot to put sealer on the roof, how much will the seal job cost?

**9.** A rectangular patio measures 30 ft long by 12 ft wide. How much will it cost to put outdoor carpeting on the patio if carpet costs $3 a square foot and installation costs are $1 a square foot?

**10.** A sketch of the floor plan of the first floor in the Andrews' home is shown in the figure. How much will it cost to lay carpeting, costing $20 a square yard, over padding, costing $3 a square yard, over the entire floor?

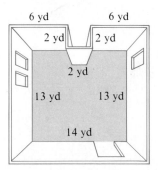

**11.** Complete the following purchase order.

| Quantity | Description | Unit Price | Amount |
|---|---|---|---|
| 120 | Dress Shirts | $17 | **(a)** _____ |
| 85 | Ties | 8 | **(b)** _____ |
| 36 | Suits | 170 | **(c)** _____ |
| 24 | Belts | 13 | **(d)** _____ |
| | | TOTAL | **(e)** _____ |

**12.** The Orlandos have a house mortgage which requires a monthly payment of $480 for 25 years.

(a) How many monthly payments must they make over the 25-year period?

(b) How much will they have paid in total at the end of 25 years?

**13.** A basketball arena is divided into 16 sections of seats. Suppose that 8 sections each seat 300 spectators, 4 sections will each seat 420 spectators, and the remaining sections will each seat 565 spectators.

(a) What is the seating capacity of the arena?

(b) If every seat is sold for $4 each, how much money will the athletic department take in?

**14.** Max used 1470 bricks to cover a wall. If there are 35 rows of bricks, how many bricks are in each row?

**15.** When Anita filled the gas tank on her car before leaving on a trip, she observed that the odometer reading was 27,154. When she returned from the trip, she refilled the tank and noted that the odometer reading was then 27,518. If she purchased 14 gallons of gas, what was her mileage per gallon?

**16.** A certain brand of cereal sells for 95¢. How many boxes can be purchased for $12, and how much change is left?

**17.** Dyane purchased a used car from a friend for $1500 by paying $150 down and the rest in equal monthly payments over a period of 18 months (no interest was charged). How much were her monthly payments?

**18.** Which is a better buy, a 9-lb bag of kitty litter selling for 99¢ or a 13-lb bag selling for $1.30 (130¢)?

**19.** The receipts at a play were $15,300. If tickets sold for $6 each, how many people were in attendance?

**20.** The enrollment at a community college is 18,200 students. If the faculty number 728, what is the number of students per teacher?

**21.** Mr. Wilkins divides $135,000 among his daughters. If each receives $45,000, how many daughters does he have?

**22.** Dr. Johnson, a chemist, pours 384 milliliters of acid into 4 flasks in such a way that each flask has the same amount. How much acid is in each flask?

**23.** Assuming equal quality, which is a better buy, a 5-lb bag of dog food for $2.00 (200¢), or an 8-lb bag of dog food for $3.04 (304¢)?

**FOR REVIEW**

*Find each quotient and check your answer.*

**24.** $7\overline{)2282}$

**25.** $23\overline{)1499}$

**26.** $157\overline{)32105}$

*Exercise 27 will help you prepare for the next section.*

**27.** List the multiples of 3 between 1 and 20.

## 2.6 EXERCISES C

**1.** Let $n$ represent the number of people at a concert. If tickets were sold for $11 each and $2816 was collected, find $n$.

**2.** Let $n$ represent the number of shirts that Sherman bought. The shirts cost $16 each. If he paid a total of $68 for the shirts plus a $20 pair of pants, find $n$. [Answer: $n = 3$]

## 2.7 DIVISIBILITY AND PRIMES

### STUDENT GUIDEPOSTS

**1** Divisors of a Whole Number
**2** Prime and Composite Numbers
**3** Finding the Prime Factorization of a Number
**4** Divisibility Tests

### **1** DIVISORS OF A WHOLE NUMBER

In this section we consider several topics related to multiplication and division of whole numbers. Recall that a **multiple** of a whole number is the product of the number and some other whole number. For example, some multiples of 3 are

$$0 = 3 \times 0, \qquad 3 = 3 \times 1, \qquad 6 = 3 \times 2, \qquad 9 = 3 \times 3, \qquad 12 = 3 \times 4.$$

Another way of saying the same thing (if we exclude zero) is that one number is a **divisor** or **factor** of a second if the second is a multiple of the first. For example,

$$6 = 3 \times 2.$$

6 is a multiple of both 3 and 2      3 and 2 are both divisors or factors of 6

| EXAMPLE 1   FINDING MULTIPLES OF A NUMBER | PRACTICE EXERCISE 1 |
|---|---|

Find the first ten multiples of 6.

Multiply 6 by each of the first ten whole numbers.

| | |
|---|---|
| $0 \times 6 = 0$ | $5 \times 6 = 30$ |
| $1 \times 6 = 6$ | $6 \times 6 = 36$ |
| $2 \times 6 = 12$ | $7 \times 6 = 42$ |
| $3 \times 6 = 18$ | $8 \times 6 = 48$ |
| $4 \times 6 = 24$ | $9 \times 6 = 54$ |

Find the first ten multiples of 5.

Answer: 0, 5, 10, 15, 20, 25, 30, 35, 40, 45

| **EXAMPLE 2**   **FINDING DIVISORS OF A NUMBER** | **PRACTICE EXERCISE 2** |

Show that 1, 2, 3, 4, 6, 8, 12, and 24 are all divisors of 24.

Since 24 is a multiple of each of these numbers, as shown below,

$$1 \times 24 = 24 \qquad 2 \times 12 = 24 \qquad 3 \times 8 = 24 \qquad 4 \times 6 = 24$$

each is a divisor of 24.

Show that 1, 2, 5, 10, 25, and 50 are all divisors of 50.

Answer: Since $1 \times 50 = 50$, $2 \times 25 = 50$, and $5 \times 10 = 50$, 50 is a multiple of each number, which means each of the numbers is a divisor of 50.

---

In Example 2, notice that 1, 2, 3, 4, 6, 8, 12, and 24 all divide 24 evenly; that is, the remainder is zero. This is true in general.

### Divisor of a Number

One number is a divisor of another if division results in a remainder of zero.

### HINT

To find the divisors of a number such as 15, we would need to divide 15 by each of the numbers less or equal to 15: 1, 2, 3, 4, 5, 6, 7, 8, 9, 10, 11, 12, 13, 14, and 15, and look at the remainders. This process would show that the only divisors of 15 are 1, 3, 5, and 15, since dividing 15 by these numbers, and only these numbers, results in a remainder of 0. Actually, there is no need to try 8, 9, 10, 11, 12, 13, and 14 since they are greater than half of 15.

## ② PRIME AND COMPOSITE NUMBERS

Certain natural numbers have only two divisors, 1 and the number itself. For example, the only divisors of 7 are 1 and 7.

### Prime Numbers

A natural number that has *exactly two different* divisors, 1 and itself, is called a **prime number.**

| **EXAMPLE 3**   **RECOGNIZING PRIME NUMBERS** | **PRACTICE EXERCISE 3** |

Are the following numbers prime?

**(a)** 1   is *not* prime since 1 can only be written as $1 = 1 \times 1$. Thus, 1 does not have two *different* divisors.

**(b)** 2   is prime. The only divisors of 2 are 1 and 2 ($2 = 1 \times 2$). Thus, 2 is the first prime number.

**(c)** 4   is *not* prime since $4 = 2 \times 2$.

Are the following numbers prime?

**(a)** 3

**(b)** 5

**(c)** 9

Answers: (a) **3** is prime.   (b) **5** is prime.   (c) **9** is not prime since $9 = 3 \times 3$.

### Composite Numbers

A natural number greater than 1 that is not a prime is a **composite number.**

---

## H I N T

Every composite number can be written as a product of two natural numbers *other than* the number itself and 1.

---

### EXAMPLE 4   RECOGNIZING COMPOSITE NUMBERS

Show that the following numbers are composite.

**(a)**  4 is composite since $4 = 2 \times 2$.

**(b)**  6 is composite since $6 = 2 \times 3$.

**(c)**  8 is composite since $8 = 2 \times 4$.

**(d)**  9 is composite since $9 = 3 \times 3$.

### PRACTICE EXERCISE 4

Show that the following numbers are composite.

**(a)**  10

**(b)**  12

**(c)**  18

**(d)**  25

Answers:  (a)  $10 = 2 \times 5$
(b)  $12 = 3 \times 4$    (c)  $18 = 2 \times 9$
(d)  $25 = 5 \times 5$

---

## H I N T

The number 1 is neither prime nor composite. It is the only natural number with this property.

---

### ③ FINDING THE PRIME FACTORIZATION OF A NUMBER

Every composite number can be written as a product of primes. In Example 4, 4 was written as $2 \times 2$, 6 as $2 \times 3$, and 9 as $3 \times 3$. Although 8 was written as $2 \times 4$, we could also write 8 as $2 \times 2 \times 2$. In all cases, the composite number has been written as a product of prime factors, called the **prime factorization** of the number.

### Prime Factorization of a Composite Number

Each composite number can be written as a product of primes, and except for the order in which the primes are written, this factorization is the only one possible.

Although we know that a composite number can be expressed as a product of primes, we do not yet have a good way to find these prime factors. The best way is to try to divide the number by each of the primes

$$2, \quad 3, \quad 5, \quad 7, \quad 11, \quad 13, \quad 17, \quad 19, \quad 23, \ldots,$$

starting with 2 and continuing in order of size as long as necessary. When a remainder is zero, the prime is a factor of the number. The process is repeated using the new quotient as the dividend. For example, to find the prime factors of 60, start by dividing 60 by 2.

$$60 \div 2 = 30, \quad \text{so} \quad 60 = 2 \cdot 30$$

Now try possible prime divisors of 30, starting again with 2.

$$30 \div 2 = 15, \qquad \text{so} \qquad 60 = 2 \cdot 30 = 2 \cdot 2 \cdot 15$$

Next look for prime divisors of 15. Since the remainder when 15 is divided by 2 is 1, not zero, move on to the next prime, 3.

$$15 \div 3 = 5 \qquad \text{so} \qquad 60 = 2 \cdot 2 \cdot 15 = 2 \cdot 2 \cdot 3 \cdot 5$$

Since all factors listed (2, 2, 3, and 5) are prime, we say the prime factorization of 60 is $2 \cdot 2 \cdot 3 \cdot 5$.

One way to display the prime factorization of 60 shown above is to use a *factor tree* like the following.

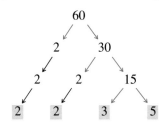

Divide 60 by 2, write down 2 and the quotient, 30

Divide 30 by 2; on the next row, write both 2's and the new quotient, 15

2 does not divide 15, so try the next larger prime, 3

These are the prime factors

This process can be continued for larger numbers. Multiplying the numbers across each row of the factor tree gives the original number, with the bottom row giving the prime factors.

Another technique that can be used to obtain the same prime factorization of 60 is shown below.

$$
\begin{array}{r}
2 \,\lfloor\,60 \\
2 \,\lfloor\,30 \qquad \leftarrow 30 = 60 \div 2 \\
3 \,\lfloor\,15 \qquad \leftarrow 15 = 30 \div 2 \\
5 \qquad \leftarrow 5 = 15 \div 3
\end{array}
$$

The prime factorization of 60 is $2 \cdot 2 \cdot 3 \cdot 5$.

The information given in color in the above display is used for explanation and would not be written if you use this technique.

---

**EXAMPLE 5   FINDING PRIME FACTORIZATIONS**

Find the prime factors of each number.

**(a)** 28

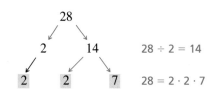

$28 \div 2 = 14$

$28 = 2 \cdot 2 \cdot 7$

**(b)** 182

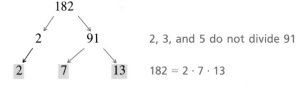

2, 3, and 5 do not divide 91

$182 = 2 \cdot 7 \cdot 13$

**PRACTICE EXERCISE 5**

Find the prime factors of each number.

**(a)** 20

**(b)** 120

**(c)** 825

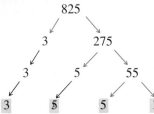

3 is the smallest prime that divides 825

$825 = 3 \cdot 5 \cdot 5 \cdot 11$

Alternatively, we find the same prime factorization using the following.

$$
\begin{array}{r|l}
3 & 825 \\ \hline
5 & 275 \\ \hline
5 & 55 \\ \hline
& 11
\end{array}
$$

← $275 = 825 \div 3$
← $55 = 275 \div 5$
← $11 = 55 \div 5$

Again we obtain $825 = 3 \cdot 5 \cdot 5 \cdot 11$.

**(d)** 3185

$$
\begin{array}{r|l}
5 & 3185 \\ \hline
7 & 637 \\ \hline
7 & 91 \\ \hline
& 13
\end{array}
$$

Thus, $3185 = 5 \cdot 7 \cdot 7 \cdot 13$.

**(c)** 490

**(d)** 3003

Answers: (a) $20 = 2 \cdot 2 \cdot 5$
(b) $120 = 2 \cdot 2 \cdot 2 \cdot 3 \cdot 5$
(c) $490 = 2 \cdot 5 \cdot 7 \cdot 7$
(d) $3003 = 3 \cdot 7 \cdot 11 \cdot 13$

##  DIVISIBILITY TESTS

To find out whether a given number (prime or composite) is a divisor of another without actually dividing, we can use simple tests called *divisibility tests*. We shall give tests for the numbers

2,  3,  5,  and  10.

| Divisibility Test for 2 |
| --- |
| A number is divisible by 2 if its ones digit is 0, 2, 4, 6, or 8. |

Numbers divisible by 2 are called **even numbers.** The first ten even whole numbers are 0, 2, 4, 6, 8, 10, 12, 14, 16, and 18. Numbers that are not even, such as 1, 3, 5, 7, 9, 11, and 13, are called **odd numbers.**

| **EXAMPLE 6** TESTING DIVISIBILITY BY 2 |
| --- |

Are the following numbers divisible by 2?

**(a)** 235 is not divisible by 2 since the ones digit is 5.

**(b)** 4284 is divisible by 2 since the ones digit is 4.

| PRACTICE EXERCISE 6 |
| --- |

Are the following numbers divisible by 2?

**(a)** 350

**(b)** 3273

Answers: (a) Yes; its ones digit is 0.   (b) No; its ones digit is 3 and not 0, 2, 4, 6, or 8.

## Divisibility Test for 3

A number is divisible by 3 if the sum of its digits is divisible by 3.

Since the sum of the digits of a number is smaller than the original number, it is easier to find out if the sum is divisible by 3.

### EXAMPLE 7    TESTING DIVISIBILITY BY 3

Are the following numbers divisible by 3?

**(a)** 24 is divisible by 3 since $2 + 4 = 6$ and 6 is divisible by 3.

**(b)** 3241 is not divisible by 3 since $3 + 2 + 4 + 1 = 10$ and 10 is not divisible by 3.

### PRACTICE EXERCISE 7

Are the following numbers divisible by 3?

**(a)** 73

**(b)** 6594

Answers: (a) No; $7 + 3 = 10$, and 10 is not divisible by 3.  (b) Yes; $6 + 5 + 9 + 4 = 24$, and 24 is divisible by 3.

## Divisibility Test for 5

A number is divisible by 5 if its ones digit is 0 or 5.

### EXAMPLE 8    TESTING DIVISIBILITY BY 5

Are the following numbers divisible by 5?

**(a)** 320 is divisible by 5 since the ones digit is 0.

**(b)** 4304 is not divisible by 5 since the ones digit is 4.

### PRACTICE EXERCISE 8

Are the following numbers divisible by 5?

**(a)** 625

**(b)** 3948

Answers: (a) Yes; its ones digit is 5.  (b) No; its ones digit, 8, is neither 0 nor 5.

### H I N T

For a number to be divisible by 6, it must be divisible by both 2 and 3, since $6 = 2 \times 3$. Thus, a divisibility test for 6 is a combination of the tests for 2 and 3.

## Divisibility Test for 10

A number is divisible by 10 if its ones digit is 0.

| EXAMPLE 9   TESTING DIVISIBILITY BY 10 | PRACTICE EXERCISE 9 |
|---|---|

Are the following numbers divisible by 10?

**(a)** 240 is divisible by 10 since its ones digit is 0.

**(b)** 325,361 is not divisible by 10 since its ones digit is 1, not 0.

Are the following numbers divisible by 10?

**(a)** 8395

**(b)** 23,486,280

Answers: (a) No; its ones digit is 5 not 0.   (b) Yes; its ones digit is 0.

## 2.7   EXERCISES A

*Answer true or false. If the answer is false, explain why.*

**1.** Since $15 = 3 \times 5$, 15 is a multiple of both 3 and 5.

**2.** A natural number with exactly two different divisors, 1 and itself, is called a composite number.

**3.** The only even prime number is 2.

*Find all the divisors of each number.*

**4.** 18                **5.** 28                **6.** 56

*Find the first ten multiples of each number.*

**7.** 4                **8.** 7                **9.** 100

*Find the prime factorization of each number.*

**10.** 55               **11.** 65               **12.** 78

**13** 140               **14.** 169              **15.** 252

**16.** 294              **17.** 429              **18.** 693

**19.** 1300             **20.** 3675             **21.** 5100

*Determine whether each number is prime or composite.*

**22.** 13               **23.** 27               **24.** 31

**25.** 49               **26.** 51               **27.** 97

*Use the following numbers to answer Exercises 28–31.*

120, 126, 240, 147, 130, 104, 70, 135, 72, 110, 88, 4125

**28.** Which numbers are divisible by 2?

**29.** Which numbers are divisible by 3?

**30.** Which numbers are divisible by 5?

**31.** Which numbers are divisible by 10?

## FOR REVIEW

*Solve.*

**32.** In a particular organization, there are 5 times as many women as men. If there are 465 women, how many men are there?

**33.** Burford purchased a TV from his roommate for $614. He paid $50 down and the rest in equal monthly payments over a period of one year. If no interest was charged, what were his monthly payments?

**34.** A hall measures 2 yd wide by 10 yd long, and a living room measures 12 yd long by 7 yd wide. If it costs $20 a square yard for carpeting, how much will it cost to carpet the hall and living room?

**35.** The Connors have a second mortgage which requires a monthly payment of $200 for 12 years.

  **(a)** How many monthly payments must they make over the 12-year period?

  **(b)** How much will they have paid in total at the end of 12 years?

*Exercises 36–38 will help you prepare for the next section. Find each product.*

**36.** $2 \cdot 2 \cdot 2 \cdot 2 \cdot 2$

**37.** $3 \cdot 3 \cdot 3 \cdot 3$

**38.** $5 \cdot 5$

ANSWERS: 1. true  2. false (prime number)  3. true  4. 1, 2, 3, 6, 9, 18  5. 1, 2, 4, 7, 14, 28  6. 1, 2, 4, 7, 8, 14, 28, 56  7. 0, 4, 8, 12, 16, 20, 24, 28, 32, 36  8. 0, 7, 14, 21, 28, 35, 42, 49, 56, 63  9. 0, 100, 200, 300, 400, 500, 600, 700, 800, 900  10. $55 = 5 \cdot 11$  11. $65 = 5 \cdot 13$  12. $78 = 2 \cdot 3 \cdot 13$  13. $140 = 2 \cdot 2 \cdot 5 \cdot 7$  14. $169 = 13 \cdot 13$  15. $252 = 2 \cdot 2 \cdot 3 \cdot 3 \cdot 7$  16. $294 = 2 \cdot 3 \cdot 7 \cdot 7$  17. $429 = 3 \cdot 11 \cdot 13$  18. $693 = 3 \cdot 3 \cdot 7 \cdot 11$  19. $1300 = 2 \cdot 2 \cdot 5 \cdot 5 \cdot 13$  20. $3675 = 3 \cdot 5 \cdot 5 \cdot 7 \cdot 7$  21. $5100 = 2 \cdot 2 \cdot 3 \cdot 5 \cdot 5 \cdot 17$  22. prime  23. composite  24. prime  25. composite  26. composite  27. prime  28. 120, 126, 240, 130, 104, 70, 72, 110, 88  29. 120, 126, 240, 147, 135, 72, 4125  30. 120, 240, 130, 70, 135, 110, 4125  31. 120, 240, 130, 70, 110  32. 93 men  33. $47 per month  34. $2080  35. (a) 144 (b) $28,800  36. 32  37. 81  38. 25

## 2.7 EXERCISES B

*Answer true or false. If the answer is false, explain why.*

**1.** Since $12 = 2 \times 6$, 2 and 6 are divisors of 12.

**2.** The multiples of 2 are odd numbers.

**3.** The only natural number which is neither prime nor composite is 1.

*Find all divisors of each number.*

**4.** 21

**5.** 36

**6.** 54

*Find the first ten multiples of each number.*

**7.** 3                       **8.** 6                       **9.** 1000

*Find the prime factorization of each number.*

**10.** 66                     **11.** 82                     **12.** 96

**13.** 135                    **14.** 273                    **15.** 420

**16.** 630                    **17.** 945                    **18.** 1750

**19.** 1050                   **20.** 5733                   **21.** 5610

*Determine whether each number is prime or composite.*

**22.** 17                     **23.** 26                     **24.** 31

**25.** 55                     **26.** 63                     **27.** 213

*Use the following numbers to answer Exercises 28–31.*

815, 150, 156, 42, 784, 125, 343, 90, 210, 66, 56, 3280

**28.** Which numbers are divisible by 2?          **29.** Which numbers are divisible by 3?

**30.** Which numbers are divisible by 5?          **31.** Which numbers are divisible by 10?

## FOR REVIEW

*Solve.*

**32.** A map has a scale of 35 miles to the inch. If two cities are actually 280 miles apart, how far apart are they on the map?

**33.** A man has only $20 bills in his wallet. If he buys 3 pairs of shoes at $42 each, how many bills must he use? How much change will he receive?

**34.** An auditorium is divided into three sections. There are 20 rows, each with 12 seats, in the first section. The second section consists of 30 rows, each with 18 seats. The third section has the same number of seats as the first section. How many seats are in the auditorium?

**35.** A student pays $120 a month for rent and $75 a month for food during the 9-month school year.

(a) How much does she spend on rent?

(b) How much does she spend on food?

(c) How much does she spend on rent and food together?

*Exercises 36–38 will help you prepare for the next section. Find each product.*

**36.** $3 \cdot 3 \cdot 3$          **37.** $2 \cdot 2 \cdot 2 \cdot 2 \cdot 2 \cdot 2 \cdot 2$          **38.** $5 \cdot 5 \cdot 5 \cdot 5$

## 2.7  EXERCISES C

*Find the prime factorization of each number.*

**1.** 13,650

**2.** 17,199

**3.** 39,270
[Answer: $2 \cdot 3 \cdot 5 \cdot 7 \cdot 11 \cdot 17$]

## 2.8  POWERS, ROOTS, AND ORDER OF OPERATIONS

=== STUDENT GUIDEPOSTS ===

**1** Powers of a Number

**3** Order of Operations

**2** Perfect Squares and Square Roots

**4** Evaluating Using Parentheses

### **1** POWERS OF A NUMBER

Often we are asked to multiply the same number by itself several times. For example, the number 64 expressed as a product of primes is

$$64 = 2 \cdot 2 \cdot 2 \cdot 2 \cdot 2 \cdot 2.$$

Such expanded products can be abbreviated by using *exponents*. We write

$$2^6 \quad \text{for} \quad \underbrace{2 \cdot 2 \cdot 2 \cdot 2 \cdot 2 \cdot 2.}_{\text{six 2's}}$$

base    exponent

The expression $2^6$, read ''2 to the sixth,'' is called the **sixth power** of 2; 6 is an **exponent,** and 2 is the **base.** An exponent tells how many times a base is used as a factor. Suppose we find the first five powers of 2.

The **first power** of 2 is $2^1$ or 2.
The **second power** of 2, also read ''two **squared,**'' is $2^2 = 2 \cdot 2 = 4$.
The **third power** of 2, also read ''two **cubed,**'' is $2^3 = 2 \cdot 2 \cdot 2 = 8$.
The **fourth power** of 2 is $2^4 = 2 \cdot 2 \cdot 2 \cdot 2 = 16$.
The **fifth power** of 2 is $2^5 = 2 \cdot 2 \cdot 2 \cdot 2 \cdot 2 = 32$.

---

**EXAMPLE 1    EVALUATING NUMBERS HAVING EXPONENTS**

Find the value by writing without exponents.

**(a)** $5^3$    We have $5^3 = 5 \cdot 5 \cdot 5 = 125$.

**(b)** $4^5$    We have $4^5 = 4 \cdot 4 \cdot 4 \cdot 4 \cdot 4 = 1024$.

**PRACTICE EXERCISE 1**

Find the value by writing without exponents.

**(a)** $3^7$

**(b)** $7^3$

Answers:  (a)  **2187**   (b)  **343**

---

**EXAMPLE 2    WRITING NUMBERS USING EXPONENTS**

Write each product using exponents.

**(a)** $6 \cdot 6 \cdot 6 \cdot 6$    Since there are four 6's, the base is 6 and the exponent is 4. Thus, $6 \cdot 6 \cdot 6 \cdot 6 = 6^4$.

**(b)** $10 \cdot 10 \cdot 10$    There are three 10's, so we have $10^3$.

**PRACTICE EXERCISE 2**

Write each product using exponents.

**(a)** $5 \cdot 5 \cdot 5 \cdot 5 \cdot 5 \cdot 5$

**(b)** $100 \cdot 100 \cdot 100 \cdot 100$

Answers:  (a)  $5^6$   (b)  $100^4$

| EXAMPLE 3 PRIME FACTORIZATION USING EXPONENTS | PRACTICE EXERCISE 3 |

Write the prime factorization for each number using exponents.

**(a)** 100     Using a factor tree, we have $100 = 2 \cdot 2 \cdot 5 \cdot 5$. Using exponents, $2 \cdot 2$ becomes $2^2$ and $5 \cdot 5$ becomes $5^2$. Thus, $100 = 2^2 \cdot 5^2$.

**(b)** 540     Using a factor tree, we have $540 = 2 \cdot 2 \cdot 3 \cdot 3 \cdot 3 \cdot 5$. Using exponents, $540 = 2^2 \cdot 3^3 \cdot 5$.

Write the prime factorization for each number using exponents.

**(a)** 1323

**(b)** 2200

Answers: (a) $3^3 \cdot 7^2$
(b) $2^3 \cdot 5^2 \cdot 11$

| EXAMPLE 4 EVALUATING PRODUCTS HAVING EXPONENTS | PRACTICE EXERCISE 4 |

Find the value of each product.

**(a)** $2^5 \cdot 5^2 = \underbrace{2 \cdot 2 \cdot 2 \cdot 2 \cdot 2} \cdot \underbrace{5 \cdot 5}$
$= \quad\quad 32 \quad\quad \cdot \quad 25$
$= 800$

**(b)** $2 \cdot 3^4 \cdot 7^2 = \underbrace{2} \cdot \underbrace{3 \cdot 3 \cdot 3 \cdot 3} \cdot \underbrace{7 \cdot 7}$
$= 2 \cdot \quad\quad 81 \quad\quad \cdot 49$
$= 7938$

Find the value of each product.

**(a)** $3^3 \cdot 5^3$

**(b)** $2^3 \cdot 3 \cdot 7^2 \cdot 13$

Answers: (a) 3375    (b) 15,288

## ❷ PERFECT SQUARES AND SQUARE ROOTS

The opposite of finding powers is taking *roots*. For example, we know that the square of 3 is 9, $3^2 = 9$. We call 3 a *square root* of 9, and 9 is called a *perfect square*. That is,

$$3^2 = 9.$$

     9 is the square of 3, making 9 a perfect square

3 is a square root of 9

---

### Perfect Squares and Square Roots

When a whole number is squared, the result is a **perfect square.** The number which is being squared is a **square root** of the perfect square.

---

| EXAMPLE 5 FINDING SQUARE ROOTS | PRACTICE EXERCISE 5 |

Find the square root of each number.

**(a)** 25     Since $25 = 5^2$, 5 is a square root of 25.

**(b)** 64     Since $64 = 8^2$, 8 is a square root of 64.

**(c)** 0     Since $0 = 0^2$, 0 is a square root of 0.

Find the square root of each number.

**(a)** 16

**(b)** 81

**(c)** 1

Answers: (a) 4    (b) 9    (c) 1

---

### H I N T

Remember that to find the square root of a number, you must ask:

     "What number squared is equal to the given number?"

The symbol $\sqrt{\ \ }$ , called a **radical,** means "square root." For example, $\sqrt{25}$ is read "the square root of 25." Similarly, $\sqrt{64}$ is read "the square root of 64." By Example 5 we see that

$$\sqrt{25} = 5 \quad \text{and} \quad \sqrt{64} = 8.$$

---

## C A U T I O N

Do not confuse the terms *square* and *square root*. For example,

**4 is a *square root* of 16**
$$\downarrow$$
$$16 \ = \ 4^2.$$
$$\downarrow$$
**16 is the *square* of 4**

---

| EXAMPLE 6   FINDING SQUARE ROOTS | PRACTICE EXERCISE 6 |

Find each square root.

**(a)** $\sqrt{9}$     Since $9 = 3^2$, $\sqrt{9} = 3$.

**(b)** $\sqrt{100}$     Since $100 = 10^2$, $\sqrt{100} = 10$.

Find each square root.

**(a)** $\sqrt{36}$

**(b)** $\sqrt{144}$

Answers: **(a)** 6 (not $6^2$)    **(b)** 12

---

## C A U T I O N

A common mistake is to say that $\sqrt{9} = 3^2$. Notice that $3^2 = 9$ so that $\sqrt{9}$ is just 3, *not* $3^2$.

---

The first sixteen perfect squares and their whole number square roots are listed in the following table.

| Perfect square | Whole number square root |
|---|---|
| 0 | $\sqrt{0} = 0$ |
| 1 | $\sqrt{1} = 1$ |
| 4 | $\sqrt{4} = 2$ |
| 9 | $\sqrt{9} = 3$ |
| 16 | $\sqrt{16} = 4$ |
| 25 | $\sqrt{25} = 5$ |
| 36 | $\sqrt{36} = 6$ |
| 49 | $\sqrt{49} = 7$ |
| 64 | $\sqrt{64} = 8$ |
| 81 | $\sqrt{81} = 9$ |
| 100 | $\sqrt{100} = 10$ |
| 121 | $\sqrt{121} = 11$ |
| 144 | $\sqrt{144} = 12$ |
| 169 | $\sqrt{169} = 13$ |
| 196 | $\sqrt{196} = 14$ |
| 225 | $\sqrt{225} = 15$ |

## ③ ORDER OF OPERATIONS

Some problems involve several operations. For example, when we checked a division problem such as

$$6 \overline{\smash{)}25} \\ \phantom{6)}\frac{\phantom{2}4}{\phantom{0}}$$

$$\begin{array}{r} 4 \\ 6\overline{\smash{)}25} \\ \underline{24} \\ 1 \end{array}$$

we wrote $(6 \times 4) + 1 = 24 + 1 = 25$. The parentheses show that 6 and 4 are multiplied first, and then the product is added to 1. If we had omitted the parentheses and written

$$6 \times 4 + 1,$$

there could be some confusion about the *order* of operations. Should we multiply $6 \times 4$ first and then add 1, or should we add $4 + 1$ first, and then multiply the sum by 6 to obtain $6 \cdot 5 = 30$? The parentheses were necessary to avoid this confusion. Another way to eliminate this problem is to agree on a definite **order of operations** when addition, subtraction, multiplication, division, finding powers, and taking roots occur in a problem.

| Order of Operations |
| --- |
| 1. Find all powers and square roots, in any order, first. |
| 2. Multiply and divide, in order, from left to right. |
| 3. Add and subtract, in order, from left to right. |

---

**EXAMPLE 7  USING ORDER OF OPERATIONS**

Perform the indicated operations.

**(a)** $2 + 3 \cdot 4$

$$2 + 3 \cdot 4 = 2 + 12 \qquad \text{Multiply first}$$
$$= 14 \qquad \text{Then add}$$

**(b)** $5 \cdot 6 - 12 \div 4$

$$5 \cdot 6 - 12 \div 4 = 30 - 3 \qquad \text{Multiply and divide first}$$
$$= 27 \qquad \text{Then subtract}$$

**(c)** $\sqrt{64} - 2^3$

$$\sqrt{64} - 2^3 = 8 - 8 \qquad \text{Take root and cube first}$$
$$= 0 \qquad \text{Then subtract}$$

**(d)** $4^2 \div 2 + \sqrt{9} - 1$

$$4^2 \div 2 + \sqrt{9} - 1 = 16 \div 2 + 3 - 1 \qquad \text{Find power and root first}$$
$$= 8 + 3 - 1 \qquad \text{Divide next}$$
$$= 11 - 1 \qquad \text{Add next}$$
$$= 10 \qquad \text{Subtract last}$$

---

**PRACTICE EXERCISE 7**

Perform the indicated operations.

**(a)** $5 \cdot 7 - 10$

**(b)** $18 \div 3 + 2 \cdot 4$

**(c)** $5^2 - \sqrt{36}$

**(d)** $\sqrt{81} + 6^2 \div 9 - 2$

Answers: (a) **25**  (b) **14**  (c) **19** (d) **11**

**4 EVALUATING USING PARENTHESES**

When parentheses are used, they affect the order of operations.

| Evaluating Using Parentheses |
| --- |
| Always operate within parentheses first. Then follow the standard order of operations. |

| EXAMPLE 8   OPERATIONS INVOLVING PARENTHESES | PRACTICE EXERCISE 8 |
| --- | --- |

Perform the indicated operations.

**(a)** $(5 - 1) \cdot 3 + 7$

$$(5 - 1) \cdot 3 + 7 = 4 \cdot 3 + 7 \qquad \text{Subtract inside parentheses first}$$
$$= 12 + 7 \qquad \text{Multiply before adding}$$
$$= 19$$

**(b)** $\sqrt{9}(6 \div 2) + 5^2$

$$\sqrt{9}(6 \div 2) + 5^2 = \sqrt{9}(3) + 5^2 \qquad \text{Divide inside the parentheses first}$$
$$= 3 \cdot 3 + 25 \qquad \text{Take root and square next}$$
$$= 9 + 25 \qquad \text{Multiply before adding}$$
$$= 34$$

**(c)** $(3 \cdot 5)^2$

$$(3 \cdot 5)^2 = (15)^2 \qquad \text{Multiply inside parentheses first}$$
$$= 225 \qquad \text{Then square}$$

**(d)** $3 \cdot 5^2$

$$3 \cdot 5^2 = 3 \cdot 25 \qquad \text{Square first}$$
$$= 75 \qquad \text{Then multiply}$$

**(e)** $(2 + 3)^3$

$$(2 + 3)^3 = (5)^3 \qquad \text{Add inside parentheses first}$$
$$= 125 \qquad \text{Then cube}$$

**(f)** $2^3 + 3^3$

$$2^3 + 3^3 = 8 + 27 \qquad \text{Cube first}$$
$$= 35 \qquad \text{Then add}$$

Perform the indicated operations.

**(a)** $2(6 + 1) - 3$

**(b)** $3^2 - \sqrt{4}(10 \div 5)$

**(c)** $(2 \cdot 7)^2$

**(d)** $2 \cdot 7^2$

**(e)** $(1 + 4)^3$

**(f)** $1^3 + 4^3$

Answers: **(a) 11    (b) 5    (c) 196    (d) 98    (e) 125    (f) 65**

///////////// CAUTION ////////////

Parts **(c)** and **(d)** in Example 8 show that

$$3 \cdot 5^2 \neq (3 \cdot 5)^2,$$

and parts **(e)** and **(f)** show that

$$(2 + 3)^3 \neq 2^3 + 3^3.$$

A common mistake when working with exponents is to assume that expressions like those above are equal. Remember to evaluate within parentheses before finding a power.

/////////////

## 2.8 EXERCISES A

**1.** The number $3^4$ is given. **(a)** What is the number 4 called? **(b)** What is the number 3 called? **(c)** What power of 3 is this?

**2.** What is the second power of a number called?

**3.** When parentheses are not used, which operation is performed first, addition or multiplication?

*Find the value by writing without exponents.*

**4.** $2^8$                    **5.** $10^4$                    **6.** $8^3$

*Write each product using exponents.*

**7.** $3 \cdot 3 \cdot 3 \cdot 3$          **8.** $9 \cdot 9 \cdot 9 \cdot 9 \cdot 9$          **9.** $4 \cdot 4 \cdot 4 \cdot 4 \cdot 4 \cdot 4$

*Find the prime factorization of each number using exponents.*

**10.** 252                **11.** 2646                **12** 4400

*Evaluate each product without using exponents.*

**13.** $2 \cdot 3^3 \cdot 11$          **14.** $3^2 \cdot 5^2 \cdot 7^2$          **15.** $3 \cdot 5^3 \cdot 13^2$

*Find the square root of each number.*

**16.** 49                **17.** 100                **18.** 121

*Find the square of each number.*

**19.** 49                **20.** 100                **21.** 121

*Find each square root.*

**22.** $\sqrt{16}$          **23.** $\sqrt{1}$          **24.** $\sqrt{169}$

*Perform the indicated operations.*

**25.** $3 + 9 - 5$          **26.** $14 \div 7 - 1$          **27.** $4 - 8 \div 2 + 1$

**28.** $7 + 2 \cdot 3 - 4$          **29.** $36 \div 6 - 3 \cdot 2$          **30.** $5 + 2^2$

**31.** $\sqrt{25} - 2^2$          **32** $5^2 - \sqrt{49}$          **33.** $3^2 + 5 - \sqrt{4} + 1$

**34** $\sqrt{25} + 15 - 2^2 \cdot 5$          **35.** $2^5 - \sqrt{100} + 7$          **36.** $2^3 \div \sqrt{16} + 3$

**37.** $(9 - 2) \cdot 3 + 4$          **38.** $(2 + 3) \div 5 - 0$          **39.** $6 + 2(5 - 3)$

**40.** $(5 + 1) \cdot (4 - 2) - 3$          **41** $\sqrt{4}(9 \div 3) + 2^3$          **42.** $2(\sqrt{4} \div 2) - 2$

**43** $(3 \cdot 7)^2$

**44** $3 \cdot 7^2$

**45.** $3^2 \cdot 7^2$

**46** $(3 + 7)^2$

**47** $3^2 + 7^2$

**48.** $2(3 + 7)^2$

## FOR REVIEW

**49.** Find all the divisors of 68.

**50.** Is 85 prime or composite?

**51.** Is 38 divisible by 2?

**52.** Is 38 divisible by 3?

**53.** Is 425 divisible by 5?

**54.** Is 425 divisible by 10?

**55.** Find the prime factorization of 1650.

---

ANSWERS:  1. (a) exponent (b) base (c) fourth  2. the square of the number  3. multiplication  4. 256  5. 10,000  6. 512  7. $3^4$  8. $9^5$  9. $4^6$  10. $2^2 \cdot 3^2 \cdot 7$  11. $2 \cdot 3^3 \cdot 7^2$  12. $2^4 \cdot 5^2 \cdot 11$  13. 594  14. 11,025  15. 63,375  16. 7  17. 10  18. 11  19. 2401  20. 10,000  21. 14,641  22. 4  23. 1  24. 13  25. 7  26. 1  27. 1  28. 9  29. 0  30. 9  31. 1  32. 18  33. 13  34. 0  35. 29  36. 5  37. 25  38. 1  39. 10  40. 9  41. 14  42. 0  43. 441  44. 147  45. 441  46. 100  47. 58  48. 200  49. 1, 2, 4, 17, 34, 68  50. composite ($85 = 5 \cdot 17$)  51. yes  52. no  53. yes  54. no  55. $2 \cdot 3 \cdot 5 \cdot 5 \cdot 11$

## 2.8  EXERCISES B

**1.** The number $5^4$ is given. **(a)** What is the number 5 called? **(b)** What is the number 4 called? **(c)** What power of 5 is this?

**2.** What is the third power of a number called?

**3.** When parentheses are not used, which operation is performed first, subtraction or division?

*Find the value by writing without exponents.*

**4.** $3^5$

**5.** $5^3$

**6.** $10^5$

*Write each product using exponents.*

**7.** $2 \cdot 2 \cdot 2$

**8.** $5 \cdot 5 \cdot 5 \cdot 5 \cdot 5 \cdot 5 \cdot 5$

**9.** $8 \cdot 8 \cdot 8 \cdot 8$

*Find the prime factorization of each number using exponents.*

**10.** 270

**11.** 1008

**12.** 11,880

*Evaluate each product without using exponents.*

**13.** $2^2 \cdot 3 \cdot 13$

**14.** $3^2 \cdot 5^3 \cdot 7$

**15.** $3 \cdot 7^2 \cdot 13^2$

*Find the square root of each number.*

**16.** 36

**17.** 81

**18.** 144

*Find the square of each number.*

**19.** 36          **20.** 81          **21.** 144

*Find each square root.*

**22.** $\sqrt{64}$          **23.** $\sqrt{169}$          **24.** $\sqrt{225}$

*Perform the indicated operations.*

**25.** $5 + 8 - 3$     **26.** $18 \div 9 - 1$     **27.** $6 - 12 \div 4 + 7$

**28.** $9 + 3 \cdot 5 - 7$     **29.** $49 \div 7 - 2 \cdot 2$     **30.** $\sqrt{25} + 75$

**31.** $\sqrt{81} - 2^3$     **32.** $6^2 - \sqrt{121}$     **33.** $4^2 + 3 - \sqrt{9} + 2$

**34.** $\sqrt{49} + 5 - 2^2 \cdot 3$     **35.** $2^6 - \sqrt{144} + 5$     **36.** $3^2 \div \sqrt{9} + 1$

**37.** $(8 - 4) \cdot 2 + 6$     **38.** $(3 + 1) \div 2 - 0$     **39.** $5 + 2(7 - 3)$

**40.** $(2 + 3) \cdot (6 - 1) - 2$     **41.** $\sqrt{25}(6 \div 2) + 3^2$     **42.** $3(\sqrt{16} \div 2) - 6$

**43.** $(2 \cdot 11)^2$     **44.** $2 \cdot 11^2$     **45.** $2^2 \cdot 11^2$

**46.** $(2 + 6)^2$     **47.** $2^2 + 6^2$     **48.** $3(2 + 6)^2$

## FOR REVIEW

**49.** Find all divisors of 80.       **50.** Is 83 prime or composite?

**51.** Is 147 divisible by 2?       **52.** Is 147 divisible by 3?

**53.** Is 147 divisible by 5?       **54.** Is 2350 divisible by 10?

**55.** Find the prime factorization of 4875.

## 2.8 EXERCISES C

*Perform the indicated operations.*

**1.** $\sqrt{64}(6 \div 2 - 1) - 8 \div 4$       **2.** $9 \div 3 \cdot 5 - 8 \div 2 + \sqrt{4}$
[Answer: 13]

**3.** $\sqrt{1 + 8} \cdot \sqrt{6 - 2} + (5 - 2)^2 \div 3$       **4.** $(4^2 - 3^2) \div 7 + (\sqrt{25} - \sqrt{4})$
[Answer: 9]

# CHAPTER 2 REVIEW

## KEY WORDS

**2.1** In a **multiplication** problem, such as $2 \times 4 = 8$, $\times$ is called the **multiplication sign** or **times sign,** 2 and 4 are **factors,** 2 is the **multiplier,** 4 is the **multiplicand,** and 8 is the **product.**

A **multiple** of a number is the product of the number and a whole number.

**2.4** In a division problem, such as $12 \div 3 = 4$, $\div$ is called the **division symbol,** 12 is the **dividend,** 3 is the **divisor,** and 4 is the **quotient.**

The final difference in a division problem is called the **remainder,** a number that is less than the divisor.

An **algorithm** is a method or procedure used to perform a certain task.

**2.7** One number is a **divisor** of a second if the second is a multiple of the first.

A **prime number** is a natural number that has exactly two different divisors, 1 and itself.

A **composite number** is a natural number greater than 1 that is not prime.

An **even number** is divisible by 2. Remember that 0 is an even number.

A number that is not even is called an **odd number.**

**2.8** In the expression $3^5$, 5 is called the **exponent,** 3 is called the **base,** and $3^5$ is called the **fifth power** of 3.

The second power of a number is called the **square** of the number.

The third power of a number is called the **cube** of the number.

When a whole number is squared, the result is called a **perfect square** and the number being squared is a **square root** of the perfect square. The symbol $\sqrt{\phantom{x}}$ is called a **radical** and is used to indicate square roots.

## KEY CONCEPTS

**2.1** When two whole numbers are multiplied, the multiplier shows the number of times the multiplicand is to be added. For example,

$$4 \times 3 = 3 + 3 + 3 + 3.$$
four 3's in the sum

2. By the commutative law, changing the order of multiplication does not change the product. For example,

$$2 \times 5 = 5 \times 2.$$

3. The number 1 is the multiplicative identity since

$$1 \times (\text{any number}) = \text{that number.}$$

4. Any whole number times 0 is 0.

5. Any whole number times 1 is that whole number.

**2.2** 1. When multiplying a many-digit number by a one-digit number, write down the ones digit and add the tens digit to the next product.

2. When multiplying by a many-digit number, remember to use zeros as placeholders or else to move each new product over one more place to the left.

3. To check a multiplication problem, interchange the multiplier and the multiplicand and find the new product (the reverse-order check).

**2.3** 1. By the associative law, regrouping the factors in a multiplication problem does not change the product. For example,

$$(2 \times 3) \times 5 = 2 \times (3 \times 5).$$

2. The distributive law illustrates an important property relating multiplication and addition. For example,

$$2(3 + 5) = (2 \times 3) + (2 \times 5).$$

3. To multiply a number by 10, attach one 0 at the end of the number. To multiply by 100, attach two 0's, and to multiply by 1000, attach three 0's.

4. Products can be estimated by rounding the multiplier and multiplicand and finding the product of the results.

**2.4** 1. A related multiplication sentence for

$$10 \div 2 = 5 \quad \text{is} \quad 2 \times 5 = 10.$$

2. Any whole number, except 0, divided by itself is 1.

**3.** Any whole number divided by 1 is that whole number.

**4.** Division by zero is undefined.

**5.** Zero divided by any whole number, except 0, is 0.

**2.5** A division problem can be checked by using (quotient × divisor) + remainder = dividend.

**2.6** **1.** (total value) = (value per part) · (number of parts)

**2.** The area of a rectangle is given by the formula

$$\text{Area} = (\text{length}) \cdot (\text{width}).$$

**3.** Miles per gallon (mpg) is found by

$$\text{mpg} = \frac{\text{number of miles driven}}{\text{number of gallons used}}.$$

**2.7** Each composite number can be written as a product of primes called the prime factorization of the number. A factor tree is useful in finding these primes.

**2.8** **1.** Do not confuse the terms square and square root. For example,

4 is the **square root** of 16

16 is the **square** of 4.

**2.** Operations should be performed in the following order.
**(a)** Operate within parentheses.
**(b)** Evaluate powers and roots.
**(c)** Multiply and divide from left to right.
**(d)** Add and subtract from left to right.

## REVIEW EXERCISES

*Part I*

**2.1** **1.** Consider the multiplication problem $7 \times 9 = 63$. **(a)** What is 7 called? **(b)** What is 9 called? **(c)** What is 63 called?

**2.** The fact that $4 \times 3 = 3 \times 4$ illustrates what law of multiplication?

**3.** Any number multiplied by 1 always gives what number for the product?

**4.** The numbers 0, 4, 8, 12, 16, and 20 are the first six multiples of what number?

**2.2** *Find the following products and check using the reverse-order check.*

| | | | | |
|---|---|---|---|---|
| **5.** 347 <br> × 5 | **6.** 6007 <br> × 6 | **7.** 4398 <br> × 7 | **8.** 41 <br> × 23 | **9.** 79 <br> × 65 |
| **10.** 723 <br> × 31 | **11.** 479 <br> × 628 | **12.** 848 <br> × 207 | **13.** 4039 <br> × 705 | **14.** 6493 <br> × 3000 |

**2.3** **15.** The fact that $(4 \times 7) \times 9 = 4 \times (7 \times 9)$ illustrates what law of multiplication?

**16.** The fact that $3 \times (4 + 1) = (3 \times 4) + (3 \times 1)$ illustrates what law of multiplication and addition?

*Check by estimating the product.*

| | |
|---|---|
| **17.** 86 <br> × 31 <br> 86 <br> 258 <br> 2666 | **18.** 307 <br> × 513 <br> 921 <br> 307 <br> 1835 <br> 187491 |

**2.4** **19.** Consider the division problem $12 \div 4 = 3$.

(a) What is 12 called?

(b) What is 4 called?

(c) What is 3 called?

**20.** What is the related multiplication sentence for $12 \div 4 = 3$?

*Complete the following.*

**21.** $5 \div 5 =$ \_\_\_\_\_

**22.** $5 \div 1 =$ \_\_\_\_\_

**23.** $0 \div 5 =$ \_\_\_\_\_

**24.** $5 \div 0 =$ \_\_\_\_\_

**2.5** *Find each quotient and check your work.*

**25.** $4\overline{)292}$

**26.** $7\overline{)6531}$

**27.** $6\overline{)2793}$

**28.** $62\overline{)806}$

**29.** $47\overline{)1252}$

**30.** $83\overline{)10375}$

**2.6** *Solve.*

**31.** Mr. Kwan buys a new car for $8420. If he pays $500 down and the balance in 36 equal monthly payments, how much will he pay each month? (Do not worry about interest at this time.)

**32.** A nine-man softball team wins a tournament and receives a prize of $1000. They agree to split the prize money evenly (in dollars) with the remainder to be given to the captain. How much does the captain receive?

**33.** Tickets for a rock concert, each selling for $8, went on sale at 9:00 A.M. At 1:00 the tickets were all sold. There were 3200 tickets sold the first hour, 2950 the second, 2400 the third, and 1500 the fourth.

(a) How many tickets were sold?

(b) What were the total receipts?

**34.** Dick wants to carpet his dining and living rooms with carpet selling for $18 a square yard. A sketch of the floor plan of the rooms is shown in the figure. How much will the carpet cost? [*Hint:* Draw a line dividing the area into two rectangles.]

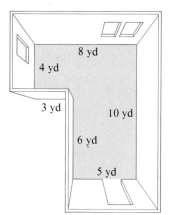

**35.** Assuming equal quality, which is the better buy, a 12-oz box of cat food for 84¢ or a 13-oz box of food for 78¢?

**2.7** **36.** The numbers 2 and 5 are divisors or factors of 10 since $2 \times 5 = 10$. What is 10 relative to 2 and 5?

**37.** What do we call a natural number that has exactly two different divisors, 1 and itself?

**38.** What is a number called if it is greater than 1 and not prime?

**39.** What is the only natural number which is neither prime nor composite?

**40.** What do we call a natural number that is divisible by 2?

**41.** How many even prime numbers are there?

**42.** Find all divisors of 105.         **43.** Find the first five multiples of 13.

*Find the prime factorization of each number.*

**44.** 195              **45.** 112              **46.** 490

**47.** 1089             **48.** 2730             **49.** 3400

*Use the numbers* 140, 273, 105, *and* 150 *to answer Exercises 50–53.*

**50.** Which numbers are divisible by 2?        **51.** Which numbers are divisible by 3?

**52.** Which numbers are divisible by 5?        **53.** Which numbers are divisible by 10?

2.8 **54.** Is 50 a perfect square?     **55.** What is the square of 9?     **56.** What is the square root of 9?

**57.** Evaluate $6^3$ without using an exponent.        **58.** Write $5 \cdot 5 \cdot 5 \cdot 5 \cdot 5 \cdot 5$ using an exponent.

**59.** Evaluate $2^3 \cdot 11^2$ without using exponents.      **60.** Find the prime factorization of 11,000 using exponents.

*Find each square root.*

**61.** $\sqrt{81}$              **62.** $\sqrt{144}$

*Perform the indicated operations.*

**63.** $2 - 1 + 3$         **64.** $3 - 8 \div 4 + 2$         **65.** $\sqrt{36} - 2^2$

**66.** $(7 - 3) \cdot 2 + 1$         **67.** $(5 \cdot 4)^2$         **68.** $5 \cdot 4^2$

**Part II**

*Solve.*

**69.** A collection of 308 people are to be divided into subgroups of 11. First find the approximate number of people in each subgroup, and then find the exact number.

**70.** An average of 10,990 people live in each of 11 counties. First estimate the number of people living in these counties; then find the actual number based on these figures.

*Perform the indicated operations.*

**71.**  623
   $\times$ 8

**72.**    52
   $\times$ 43

**73.**  $43\overline{)1204}$

**74.**  $27\overline{)1515}$

**75.** $28 \div 1$

**76.** $0 \div 28$

**77.** $28 \div 28$

**78.** $8 - 5 + 6 - 2$

**79.** $\sqrt{81} - 2^2$

**80.** $(4 - 2) \cdot 6 - 3$

**81.** $3^2 \cdot 2^3$

**82.** $(3 \cdot 2)^4$

**83.** What is the square of 16?

**84.** What is the square root of 16?

**85.** Is 4311 divisible by 3?

**86.** Find the prime factorization of 3234.

**87.** Check by estimating the product.
    91
 $\times$ 48
  728
  364
 4368

**88.** Estimate the following product by rounding to the nearest hundred.
   708
 $\times$ 489

*Solve.*

**89.** The total enrollment in the four classes (freshmen, sophomores, juniors, and seniors) at State University is 8227. If there are 2436 freshmen, 2107 sophomores, and 1939 juniors, how many seniors are there in attendance?

**90.** On Monday, the hens at the Chicken Farm laid 4942 eggs, and on Tuesday they laid 4802 eggs. How many 12-egg cartons will be needed to hold the eggs laid on these two days?

---

ANSWERS:   **1.** (a) multiplier (or factor) (b) multiplicand (or factor) (c) product   **2.** commutative law   **3.** that number   **4.** 4   **5.** 1735   **6.** 36,042   **7.** 30,786   **8.** 943   **9.** 5135   **10.** 22,413   **11.** 300,812   **12.** 175,536   **13.** 2,847,495   **14.** 19,479,000   **15.** associative law   **16.** distributive law   **17.** $90 \cdot 30 = 2700$; appears to be correct   **18.** $300 \cdot 500 = 150,000$; appears to be incorrect (correct product is 157,491)   **19.** (a) dividend (b) divisor (c) quotient   **20.** $4 \times 3 = 12$   **21.** 1   **22.** 5   **23.** 0   **24.** undefined   **25.** Q: 73; R: 0   **26.** Q: 933; R: 0   **27.** Q: 465; R: 3   **28.** Q: 13; R: 0   **29.** Q: 26; R: 30   **30.** Q: 125; R: 0   **31.** $220   **32.** $112   **33.** (a) 10,050 tickets (b) $80,400   **34.** $1116   **35.** The 13-oz box for 78¢ is a better buy at 6¢ per ounce than the 12-oz box for 84¢ at 7¢ per ounce.   **36.** multiple   **37.** prime   **38.** composite   **39.** 1   **40.** even   **41.** one (only the number 2)   **42.** 1, 3, 5, 7, 15, 21, 35, 105   **43.** 0, 13, 26, 39, 52   **44.** $195 = 3 \cdot 5 \cdot 13$   **45.** $112 = 2 \cdot 2 \cdot 2 \cdot 2 \cdot 7$   **46.** $490 = 2 \cdot 5 \cdot 7 \cdot 7$   **47.** $1089 = 3 \cdot 3 \cdot 11 \cdot 11$   **48.** $2730 = 2 \cdot 3 \cdot 5 \cdot 7 \cdot 13$   **49.** $3400 = 2 \cdot 2 \cdot 2 \cdot 5 \cdot 5 \cdot 17$   **50.** 140, 150   **51.** 273, 105, 150   **52.** 140, 105, 150   **53.** 140, 150   **54.** no   **55.** 81   **56.** 3   **57.** $6 \cdot 6 \cdot 6 = 216$   **58.** $5^6$   **59.** 968   **60.** $2^3 \cdot 5^3 \cdot 11$   **61.** 9   **62.** 12   **63.** 4   **64.** 3   **65.** 2   **66.** 9   **67.** 400   **68.** 80   **69.** approximate number: 30 people; exact number: 28 people   **70.** estimated number: 110,000 people; exact number: 120,890 people   **71.** 4984   **72.** 2236   **73.** Q: 28; R: 0   **74.** Q: 56; R: 3   **75.** 28   **76.** 0   **77.** 1   **78.** 7   **79.** 5   **80.** 9   **81.** 72   **82.** 1296   **83.** 256   **84.** 4   **85.** yes   **86.** $2 \cdot 3 \cdot 7 \cdot 7 \cdot 11$   **87.** $90 \cdot 50 = 4500$; appears to be correct   **88.** 350,000   **89.** 1745 seniors   **90.** 812 cartons

**1.** What law of multiplication is illustrated by the following?

$$2 \times 9 = 9 \times 2$$

1. _____

**2.** What do we call a natural number greater than 1 that is not a prime?

2. _____

**3.** When is a number divisible by 2?

3. _____

*Find the products.*

**4.**  329
 $\times\ 4$

**5.**  4037
 $\times\ 209$

**6.**  9695
 $\times\ 400$

4. _____

5. _____

6. _____

**7.** Estimate the product $6103 \times 296$ by using rounded numbers.

7. _____

*Complete the following.*

**8.** $15 \div 15 = \underline{\ \ ?\ \ }$

8. _____

**9.** $0 \div 15 = \underline{\ \ ?\ \ }$

9. _____

**10.** $15 \div 1 = \underline{\ \ ?\ \ }$

10. _____

**11.** $15 \div 0 = \underline{\ \ ?\ \ }$

11. _____

*Find the quotients.*

**12.** $5\overline{)1180}$

**13.** $23\overline{)3556}$

**14.** $32\overline{)14668}$

12. _____

13. _____

14. _____

**15.** Find the prime factorization of 350.

15. _____

**16.** Wanda buys 3 coats for $130 each, 6 blouses for $27 each, and 5 pairs of earrings for $14 each. How much has she spent?

16. _____

**17.** Dick wants to put carpeting costing $30 a square yard on the floor in his bedroom. If the room measures 5 yd long and 4 yd wide, how much will the carpeting cost?

17. _____

**18.** What is the square root of 25?

18. _____

**19.** What is the square of 25?

19. _____

*Find each square root.*

**20.** $\sqrt{64}$

20. _____

**21.** $\sqrt{25}$

21. _____

*Perform the indicated operations.*

**22.** $4 + 2 - 1$

22. _____

**23.** $(8 - 2) \div 3 + 7$

23. _____

**24.** $2 \cdot 3^3$

24. _____

**25.** $(2 \cdot 3)^3$

25. _____

**26.** $\sqrt{49} - 2^2$

26. _____

**27.** A car dealer has 51 cars on a lot, and each is valued at $11,150. Without finding the exact amount, estimate the total value of the cars on the lot.

27. _____

## CHAPTER 3

# Multiplying and Dividing Fractions

## 3.1 FRACTIONS

━━━━━━━━━━━━━━ **STUDENT GUIDEPOSTS** ━━━━━━━━━━━━━━

| | |
|---|---|
| ❶ Fractions | ❸ Types of Fractions |
| ❷ Proper and Improper Fractions | |

### ❶ FRACTIONS — *Rational # (same thing)*

In the first two chapters we looked at whole numbers. Whole numbers are used to name "whole" or entire quantities such as 3 (whole) miles, 15 (whole) dollars, or 1 (whole) day. At times, however, we may be interested in parts of a single quantity such as part of a mile, part of a dollar, or part of a day. For this, we use *fractions*.

Suppose we have a bar like the one in Figure 3.1.

**Figure 3.1**

This bar may be thought of as a piece of wood, a bar of butter, a bar of candy, or any similar quantity. If the bar is divided into two equal parts, each of the parts is called a **half** of the whole bar. See Figure 3.2. This quotient is a new number, called *one half,* and is written

$$\frac{1}{2}.$$

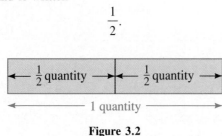

**Figure 3.2**

If we divide the bar into three equal parts as in Figure 3.3, each part is a **third** of the whole. The quotient

$$\frac{1}{3}$$

is a new number called *one third*. If the bar were a bar of candy to be split equally among 3 children, each child would receive $\frac{1}{3}$ bar. If there were 2 girls and 1 boy, the girls would receive 2 of the 3 equal parts, that is, *two thirds* of the bar. The shaded part in Figure 3.3 corresponds to $\frac{2}{3}$. The number $\frac{2}{3}$, read "2 over 3" or "2 divided by 3," is a quotient of two whole numbers, 2 and 3.

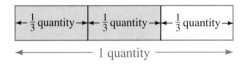

**Figure 3.3**

Any quotient of two whole numbers is called a **fraction.** The top number (the dividend) is called the **numerator** of the fraction, while the bottom number (the divisor) is called the **denominator.** In our example with the candy bar, we would have the following.

numerator $\longrightarrow$ 2 $\longleftarrow$ Number of parts the girls received
denominator $\longrightarrow$ 3 $\longleftarrow$ Bar divided into 3 equal parts

---

## EXAMPLE 1    FRACTION OF A WHOLE

A pie is to be divided into equal parts and served to 5 boys and 1 girl. Suppose the pie is represented by a circle, as in Figure 3.4, which has been divided into 6 equal parts $(5 + 1 = 6)$. Each part is a **sixth** of the pie, and each child receives $\frac{1}{6}$ of the pie. Since there are 5 boys, they will receive

numerator $\longrightarrow$ 5 $\longleftarrow$ Number of parts the boys receive
denominator $\longrightarrow$ 6 $\longleftarrow$ Number of equal parts of pie

of the pie. The shaded part of the pie corresponds to this fraction, while the unshaded part, $\frac{1}{6}$, corresponds to the part received by the girl.

**Figure 3.4**

## PRACTICE EXERCISE 1

A man divided his land as shown in the figure. If his daughter got the shaded portion, what fraction of the land did she get?

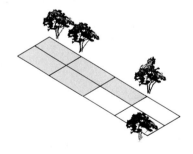

Answer: $\frac{5}{8}$

---

## EXAMPLE 2    FRACTION OF A WHOLE

The square in Figure 3.5 is divided into equal parts. What fraction of the square is shaded?

There are 4 equal parts with 3 of them shaded. Thus, the shaded fraction of the square is *three fourths,* written $\frac{3}{4}$.

## PRACTICE EXERCISE 2

The shaded part of the lot shown on the next page was used for a building. What fraction of the lot was used?

**Figure 3.5**

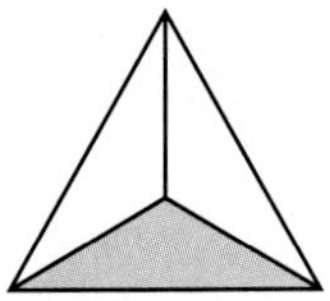

Answer: $\frac{1}{3}$

## ② PROPER AND IMPROPER FRACTIONS

All the fractions that we have considered thus far have been *proper fractions*. A **proper fraction** is one whose numerator is less than its denominator. For example,

$$\frac{1}{2}, \quad \frac{1}{3}, \quad \frac{2}{3}, \quad \frac{5}{6}, \quad \frac{3}{4}, \quad \frac{32}{67}, \quad \frac{101}{102}$$

are all proper fractions.

Some fractions are not proper fractions. For example, consider the bar in Figure 3.6 which is divided into five equal parts. All five parts are shaded. In this case, the fraction of the bar which is shaded is *five fifths,* or $\frac{5}{5}$. But since the whole bar is shaded,

$$1 = \frac{5}{5}. \qquad \text{One whole = five fifths}$$

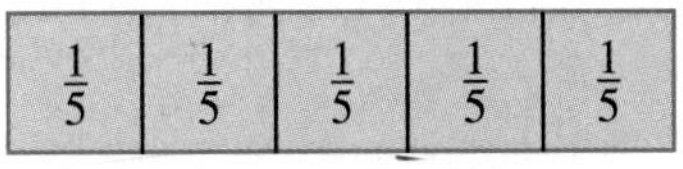

| $\frac{1}{5}$ | $\frac{1}{5}$ | $\frac{1}{5}$ | $\frac{1}{5}$ | $\frac{1}{5}$ |
|---|---|---|---|---|

**Figure 3.6**

This agrees with what we have learned about division of whole numbers. Recall that any whole number except zero divided by itself is 1.

$$\frac{2}{2} = 1, \qquad \frac{3}{3} = 1, \qquad \frac{4}{4} = 1, \qquad \frac{25}{25} = 1$$

The fractions above are special types of improper fractions. An **improper fraction** is one whose numerator is greater than or equal to its denominator. Other examples of improper fractions are:

$$\frac{3}{2}, \quad \frac{5}{4}, \quad \frac{7}{3}, \quad \frac{12}{12}, \quad \frac{9}{3}, \quad \frac{8}{7}.$$

Consider the fraction $\frac{4}{3}$, read *four thirds*. Let us use as a unit the bar divided into 3 equal parts in Figure 3.7.

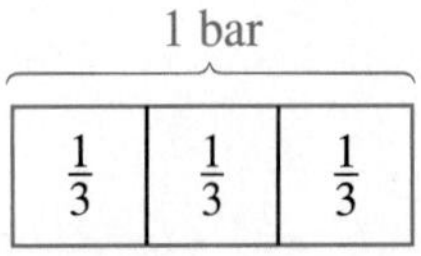

1 bar

| $\frac{1}{3}$ | $\frac{1}{3}$ | $\frac{1}{3}$ |
|---|---|---|

**Figure 3.7**

The question is, what fraction of the bar in Figure 3.7 is shaded in Figure 3.8? Four of the thirds are shaded, so that the shaded region represents $\frac{4}{3}$.

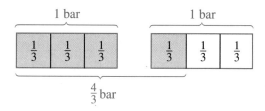

$$\frac{4}{3}\begin{array}{l}\longleftarrow \text{ Number of parts shaded}\\ \longleftarrow \text{ Number of parts in } \textit{one} \text{ unit}\end{array}$$

**Figure 3.8**

---

## CAUTION

Notice that the unit in Figure 3.7 is *one* bar divided into *thirds*. Thus the fraction shaded is

$$\frac{4}{3} \quad \text{not} \quad \frac{4}{6}.$$

---

| **EXAMPLE 3 IMPROPER FRACTION** | **PRACTICE EXERCISE 3** |
| --- | --- |

Consider the bar in Figure 3.9 as one whole unit.

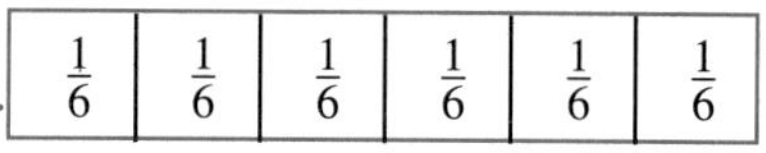

Because of → | $\frac{1}{6}$ | $\frac{1}{6}$ | $\frac{1}{6}$ | $\frac{1}{6}$ | $\frac{1}{6}$ | $\frac{1}{6}$ | 6 is the Denominator

**Figure 3.9**

What fraction of the bar shown in Figure 3.9 is shaded in Figure 3.10? What fraction is left unshaded?

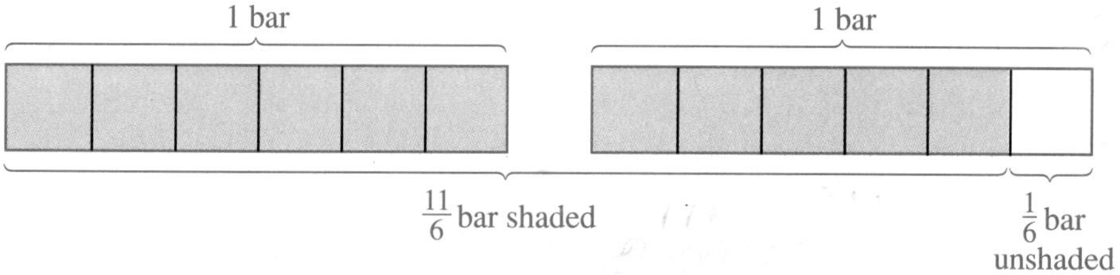

$\frac{11}{6}$ bar shaded          $\frac{1}{6}$ bar unshaded

**Figure 3.10**

Since each bar has been divided into 6 equal parts and 11 of the parts are shaded, the shaded portion corresponds to $\frac{11}{6}$. The unshaded part represents $\frac{1}{6}$.

Shaded: $\quad \dfrac{11}{6}\begin{array}{l}\longleftarrow \text{ Number of parts shaded}\\ \longleftarrow \text{ Number of parts in } \textit{one} \text{ bar}\end{array}$

Unshaded: $\quad \dfrac{1}{6}\begin{array}{l}\longleftarrow \text{ Number of parts not shaded}\\ \longleftarrow \text{ Number of parts in } \textit{one} \text{ bar}\end{array}$

**PRACTICE EXERCISE 3**

What fraction of the figure is shaded below?

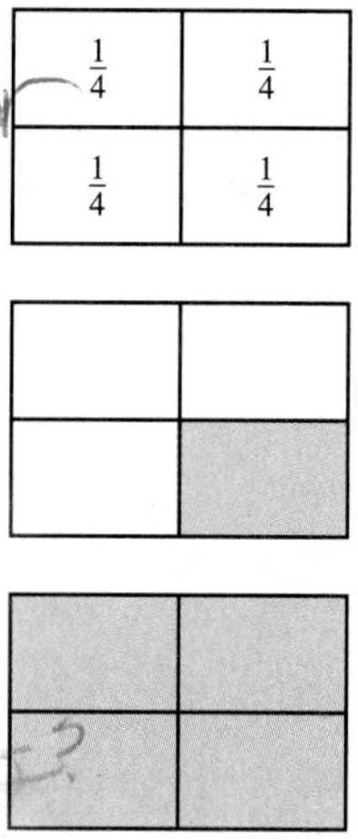

| $\frac{1}{4}$ | $\frac{1}{4}$ |
| --- | --- |
| $\frac{1}{4}$ | $\frac{1}{4}$ |

Answer: $\frac{5}{4}$

We have seen that the whole number 1 can be written as a fraction such as $\frac{3}{3}$. Other whole numbers can also be written as fractions. For example,

$$\frac{4}{2} = 2, \qquad \frac{6}{2} = 3, \qquad \frac{8}{2} = 4.$$

In each case, if the numerator is divided by the denominator, the result is a whole number. The whole number 0 can also be represented by a fraction. For example,

$$\frac{0}{5} \quad \text{corresponds to 0.}$$

We can interpret $\frac{0}{5}$ as dividing an object into 5 equal parts and taking none of them.

## ❸ TYPES OF FRACTIONS

We now summarize what we have learned about fractions below.

| Types of Fractions |
| --- |
| Proper fractions |
| $\frac{4}{9}$      Numerator less than denominator |
| $\frac{0}{9} = 0$      Whole number zero |
| Improper fractions |
| $\frac{9}{4}$      Numerator greater than denominator |
| $\frac{4}{4} = 1$      Numerator equals denominator; whole number one |
| $\frac{12}{4} = 3$      Whole number 3 |

---

### EXAMPLE 4   FRACTION OF A NUMBER OF OBJECTS

A dozen bottles of soda were taken to a picnic. If seven bottles were drunk, what fractional part was left?

Note that 7 of the 12 bottles are empty and 5 are full ($12 - 7 = 5$). Thus, $\frac{5}{12}$ of the original amount was left over.

### PRACTICE EXERCISE 4

Wendy Reed made two dozen cookies. Her son Paul ate 13 of them before dinner. What fraction of the cookies was left?

Answer: $\frac{11}{24}$

---

### EXAMPLE 5   FRACTION OF A NUMBER OF PEOPLE

During a basketball game, 3 of the 10 players on a team fouled out. What fractional part of the team fouled out?

Of the 10 players, 3 fouled out, or $\frac{3}{10}$ of the team.

### PRACTICE EXERCISE 5

A basketball player made 9 of 11 shots during a game. What fraction of her shots did she make?

Answer: $\frac{9}{11}$

## 3.1  EXERCISES A

*Give the numerator and the denominator of each fraction.*

**1.** $\dfrac{5}{7}$          **2.** $\dfrac{11}{4}$          **3.** $\dfrac{0}{8}$          **4.** $\dfrac{12}{12}$

*What fractional part of each figure is shaded?*

**5.** 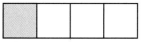      **6.**       **7.**

**8.**       **9.**       **10.**

**11.**       **12.** 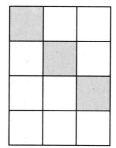      **13.**

**14** The bar

| $\frac{1}{4}$ | $\frac{1}{4}$ | $\frac{1}{4}$ | $\frac{1}{4}$ |
|---|---|---|---|

represents one whole unit. What fraction of the unit is shaded below? Consider the whole figure. There is one answer, and it is an improper fraction.

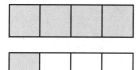

**15.** The figure

| $\frac{1}{6}$ | $\frac{1}{6}$ | $\frac{1}{6}$ | $\frac{1}{6}$ | $\frac{1}{6}$ | $\frac{1}{6}$ |
|---|---|---|---|---|---|

represents one whole unit. What fraction of the unit is shaded below? Consider the whole figure. There is one answer, and it is an improper fraction.

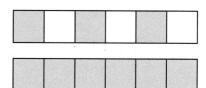

**16.** The circle

represents one whole unit. What fraction of the unit is shaded below?

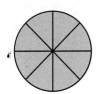 

**17.** The triangle

represents one whole unit. What fraction of the unit is shaded below?

**18.** The numerator of the fraction $\frac{4}{7}$ stands for how many of the equal parts of a whole?

**19.** A class is divided into 8 equal groups and 3 of the groups are chosen for a special project.

  **(a)** What fractional part of the class is chosen?

  **(b)** What fractional part is not chosen?

**20.** A baseball player gets 4 hits in 9 times at bat during a double header. The number of hits is what fraction of his total at bats?

**21.** Two pies are each cut into 6 pieces, and each of 7 people eats one piece.

  **(a)** What fractional part of one pie was eaten?

  **(b)** What fractional part of one pie was not eaten?

**22.** A student received $300 from home. She spent $30 for books, $75 for a bicycle, $40 for utilities, and put the rest into savings. What fractional part of the $300 went for **(a)** books? **(b)** the bicycle? **(c)** utilities? **(d)** savings?

*Express each fraction as a whole number.*

**23.** $\dfrac{12}{3}$

**24.** $\dfrac{0}{5}$

**25.** $\dfrac{43}{43}$

**26.** $\dfrac{26}{13}$

---

ANSWERS:   1. numerator: 5; denominator: 7   2. numerator: 11; denominator: 4   3. numerator: 0; denominator: 8
4. numerator: 12; denominator: 12   5. $\frac{1}{4}$   6. $\frac{1}{3}$   7. $\frac{2}{5}$   8. $\frac{2}{4}$   9. $\frac{3}{8}$   10. $\frac{3}{3}$   11. $\frac{1}{2}$   12. $\frac{3}{12}$   13. $\frac{5}{16}$   14. $\frac{5}{4}$   15. $\frac{9}{6}$
16. $\frac{11}{8}$   17. $\frac{5}{3}$   18. 4   19. (a) $\frac{3}{8}$ (b) $\frac{5}{8}$   20. $\frac{4}{9}$   21. (a) $\frac{7}{6}$ (b) $\frac{5}{6}$   22. (a) $\frac{30}{300}$ (b) $\frac{75}{300}$ (c) $\frac{40}{300}$ (d) $\frac{155}{300}$   23. 4   24. 0
25. 1   26. 2

## 3.1  EXERCISES B

*Give the numerator and the denominator of each fraction.*

**1.** $\dfrac{9}{17}$    **2.** $\dfrac{25}{3}$    **3.** $\dfrac{7}{1}$    **4.** $\dfrac{23}{23}$

*What fractional part of each figure is shaded?*

**5.**     **6.**     **7.**

**8.** 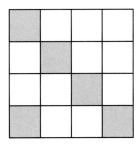    **9.** 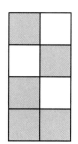    **10.**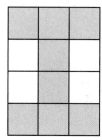

**11.**     **12.**     **13.**

**14.** The square

represents one whole unit. What fraction of the unit is shaded below? Consider the whole figure. There is one answer, and it is an improper fraction.

**15.** The bar

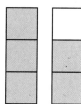

represents one whole unit. What fraction of the unit is shaded below? Consider the whole figure. There is one answer, and it is an improper fraction.

**16.** The circle

represents one whole unit. What fraction of the unit is shaded below?

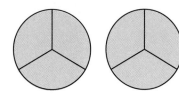

**17.** The bar

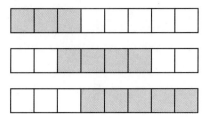

represents one whole unit. What fraction of the unit is shaded below?

**18.** The numerator of the fraction $\frac{11}{7}$ stands for how many of the equal parts of a whole?

**19.** A committee of 13 women is divided into two sub-committees, one with 7 members which will study a particular project, and the other which will search for a guest speaker. What fraction of the committee has been chosen to find the speaker?

**20.** A basketball player made 7 free throws in 11 attempts during the first half of a game. The number of free throws made is what fraction of his total attempts?

**21.** Three pies are each cut into 6 pieces. Each of 11 people eat one piece.

**(a)** What fractional part of one pie was eaten?

**(b)** What fractional part of one pie was not eaten?

**22.** Maria Lopez had $500 in her checking account. She spent $100 for food, $125 for clothes, $50 for a night out, and put the rest in savings. What fractional part of the $500 went for **(a)** food? **(b)** clothes? **(c)** the night out? **(d)** savings?

*Express each fraction as a whole number.*

**23.** $\frac{20}{4}$       **24.** $\frac{18}{1}$       **25.** $\frac{0}{101}$       **26.** $\frac{500}{500}$

## 3.1 EXERCISES C

*Let a and b represent natural numbers with a > b. Tell which of the following fractions are proper and which are improper.*

**1.** $\frac{a}{b}$       **2.** $\frac{b}{a}$       **3.** $\frac{a+b}{a-b}$       **4.** $\frac{2a}{a}$

[Answer: proper]       [Answer: improper]

## 3.2 RENAMING FRACTIONS

**STUDENT GUIDEPOSTS**

1 Equal Fractions
2 Building and Reducing Fractions
3 Reducing to Lowest Terms by Factoring into Primes
4 Dividing Out Common, Nonprime Factors

### 1 EQUAL FRACTIONS

Every fraction has many names. This can be shown by the three bars in Figure 3.11. In each, exactly the same part of the bar is shaded. In **(a),** the amount shaded is

$$\frac{1}{3} \quad \text{of the bar.}$$

In **(b),** $\quad \frac{2}{6} \quad$ of the bar is shaded,

and in **(c),** $\quad \frac{4}{12} \quad$ of the bar is shaded.

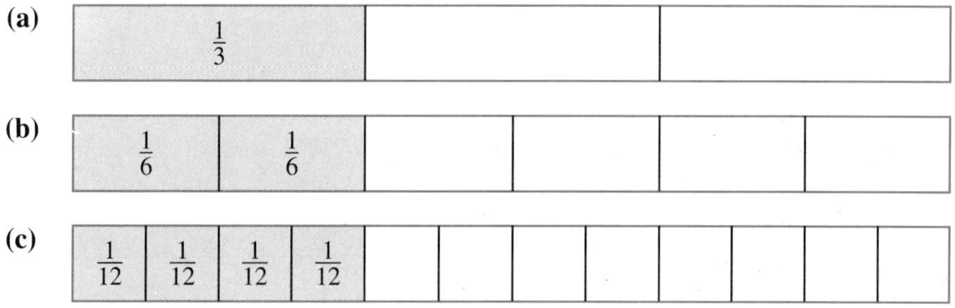

**(a)**

| $\frac{1}{3}$ | | |
|---|---|---|

**(b)**

| $\frac{1}{6}$ | $\frac{1}{6}$ | | | | |
|---|---|---|---|---|---|

**(c)**

| $\frac{1}{12}$ | $\frac{1}{12}$ | $\frac{1}{12}$ | $\frac{1}{12}$ | | | | | | | | |
|---|---|---|---|---|---|---|---|---|---|---|---|

**Figure 3.11**

Since the same part of the bar is shaded in all three cases,

$$\frac{1}{3}, \quad \frac{2}{6}, \quad \text{and} \quad \frac{4}{12}$$

are all names for the same fraction. Fractions which are names for the same number are **equal fractions.** Thus,

$$\frac{1}{3} = \frac{2}{6} = \frac{4}{12}.$$

Knowing that $\qquad \frac{2}{6} = \frac{4}{12}$

leads us to a test for equality of fractions. Multiply as indicated.

$$2 \cdot 12 = 24 \qquad \frac{2}{6} \bowtie \frac{4}{12} \qquad 6 \cdot 4 = 24$$

This gives the equality $\quad 2 \cdot 12 = 6 \cdot 4.$

These products are called **cross products.**

*Cross Multiplying*

| Equal Fractions |
| --- |
| Two fractions are equal when the cross products are equal. |

We can use this to test for equality of any two fractions.

| EXAMPLE 1 TESTING FOR EQUALITY |
| --- |

Are the two fractions equal?

**(a)** $\dfrac{3}{4}$ and $\dfrac{6}{8}$

$3 \cdot 8 = 24$    $\dfrac{3}{4} \diagdown\!\!\!\!\diagup \dfrac{6}{8}$    $4 \cdot 6 = 24$     Thus, $\dfrac{3}{4} = \dfrac{6}{8}$.

**(b)** $\dfrac{2}{9}$ and $\dfrac{6}{27}$     Since $2 \cdot 27 = 54$ and $9 \cdot 6 = 54$,

$$\frac{2}{9} = \frac{6}{27}.$$

**(c)** $\dfrac{5}{9}$ and $\dfrac{4}{7}$     Since $5 \cdot 7 = 35$ and $9 \cdot 4 = 36$,

$$\frac{5}{9} \neq \frac{4}{7}.$$

| PRACTICE EXERCISE 1 |
| --- |

Are the two fractions equal?

**(a)** $\dfrac{9}{12}$ and $\dfrac{15}{20}$

**(b)** $\dfrac{20}{50}$ and $\dfrac{16}{40}$

**(c)** $\dfrac{9}{11}$ and $\dfrac{5}{6}$

Answers: (a) yes   (b) yes (c) no

## ② BUILDING AND REDUCING FRACTIONS

You probably noticed that equal fractions are related in another way besides having equal cross products. As an example, multiplying both the numerator and denominator of $\frac{1}{3}$ by 2 gives $\frac{2}{6}$.

$$\frac{1 \times 2}{3 \times 2} = \frac{2}{6}$$

Similarly, multiplying both the numerator and denominator of $\frac{1}{3}$ by 4 gives $\frac{4}{12}$.

$$\frac{1 \times 4}{3 \times 4} = \frac{4}{12}$$

These are examples of *building fractions*.

| Building Fractions |
| --- |
| If both the numerator and denominator of a fraction are *multiplied* by the same whole number (except 0), an **equal fraction** is formed. |

| EXAMPLE 2 BUILDING FRACTIONS |
| --- |

**(a)** Find a fraction with denominator 20 that is equal to $\frac{1}{4}$.

We are looking for a fraction of the form $\dfrac{?}{20}$ such that

$$\frac{1}{4} = \frac{?}{20}.$$

| PRACTICE EXERCISE 2 |
| --- |

**(a)** Find a fraction with denominator 16 that is equal to $\frac{5}{2}$.

To build an equal fraction, the numerator and denominator of $\frac{1}{4}$ must be multiplied by the *same* number. To get 20 for a denominator, that number must be 5, since $4 \times 5 = 20$.

$$\frac{1}{4} = \frac{1 \times 5}{4 \times 5} = \frac{5}{20}$$

**(b)** Find a fraction with numerator 18 that is equal to $\frac{3}{2}$.

Find $\dfrac{18}{?}$ such that

$$\frac{3}{2} = \frac{18}{?}.$$

What number multiplied by 3 gives 18? That number is 6.

$$\frac{3}{2} = \frac{3 \times 6}{2 \times 6} = \frac{18}{12}$$

**(b)** Find a fraction with numerator 30 that is equal to $\frac{5}{8}$.

Answers:  (a) $\frac{40}{16}$   (b) $\frac{30}{48}$

---

The reverse of building fractions is *reducing fractions*. Starting with the fraction $\frac{5}{20}$ and dividing both the numerator and the denominator by 5 gives

$$\frac{5 \div 5}{20 \div 5} = \frac{1}{4},$$

which is equal to $\frac{5}{20}$. This is an example of the following.

| Reducing Fractions |
|---|
| If both the numerator and denominator of a fraction are *divided* by the same whole number (except 0), an **equal fraction** is formed. |

| EXAMPLE 3    REDUCING FRACTIONS |
|---|

**(a)** Find a fraction with denominator 3 that is equal to $\frac{14}{21}$.

We are looking for a fraction of the form $\dfrac{?}{3}$ such that

$$\frac{14}{21} = \frac{?}{3}.$$

To get an equal fraction, the numerator and denominator of $\frac{14}{21}$ must be divided by the *same* number. To get 3 for a denominator, that number must be 7, since $21 \div 7 = 3$.

$$\frac{14}{21} = \frac{14 \div 7}{21 \div 7} = \frac{2}{3}$$

**(b)** Find a fraction with numerator 2 that is equal to $\frac{22}{11}$.

Find $\dfrac{2}{?}$ such that

$$\frac{22}{11} = \frac{2}{?}.$$

| PRACTICE EXERCISE 3 |
|---|

**(a)** Find a fraction with denominator 5 that is equal to $\frac{49}{35}$.

**(b)** Find a fraction with numerator 6 that is equal to $\frac{72}{12}$.

Divide numerator and denominator by 11, because $22 \div 11 = 2$.

$$\frac{22}{11} = \frac{22 \div 11}{11 \div 11} = \frac{2}{1}$$

Notice that this fraction is really the whole number 2.

## ③ REDUCING TO LOWEST TERMS BY FACTORING INTO PRIMES

We have seen that every fraction has many names. For example, the fraction $\frac{1}{4}$ can be expressed as

$$\frac{1}{4}, \quad \frac{2}{8}, \quad \frac{3}{12}, \quad \frac{4}{16}, \quad \frac{5}{20}, \quad \text{and so on.}$$

In this list, every fraction to the right of $\frac{1}{4}$ has been *built up* from $\frac{1}{4}$ by multiplying numerator and denominator by some number. On the other hand, $\frac{1}{4}$ can be obtained by *reducing* each fraction to the right of it, by dividing numerator and denominator by some number.

Of all the possible names for the fraction $\frac{1}{4}$, 1 and 4 are the *smallest* numbers we can use in the name. In a sense, $\frac{1}{4}$ is the best name for the number since it cannot be reduced.

### Reduced to Lowest Terms

When 1 is the only nonzero whole number divisor of both the numerator and denominator of a fraction, the fraction is **reduced to lowest terms.**

Notice that

$\frac{2}{8}$ is not in lowest terms since 2 is a divisor of 2 and 8,

$\frac{3}{12}$ is not in lowest terms since 3 is a divisor of 3 and 12.

However,

$\frac{1}{4}$ is in lowest terms since 1 is the only divisor of 1 and 4.

Reducing a fraction to lowest terms is simplified by using prime numbers. Recall from Section 2.7 that a whole number greater than 1 is *prime* if its only divisors are itself and 1. Also, recall that every whole number greater than 1 is either prime or can be expressed as a product of primes.

### To Reduce a Fraction to Lowest Terms

1. Factor the numerator and denominator into primes.
2. Divide out all prime factors common to both the numerator and denominator.
3. Multiply remaining primes, first in the numerator and then in the denominator.

| EXAMPLE 4   REDUCING TO LOWEST TERMS | PRACTICE EXERCISE 4 |

Reduce $\frac{4}{6}$ to lowest terms.

Factor the numerator and denominator into primes.

$$\frac{4}{6} = \frac{2 \cdot 2}{3 \cdot 2}$$

Since 2 is a prime factor common to both the numerator and denominator, divide it out of both, giving an *equal fraction*.

$$\frac{(2 \cdot 2) \div 2}{(3 \cdot 2) \div 2} = \frac{2 \cdot (2 \div 2)}{3 \cdot (2 \div 2)}$$

$$= \frac{2 \cdot 1}{3 \cdot 1} = \frac{2}{3} \qquad \text{A number divided by itself equals 1}$$

A shorthand way of writing this division is by simply crossing out the common factors.

$$\frac{4}{6} = \frac{2 \cdot \overset{1}{\cancel{2}}}{3 \cdot \underset{1}{\cancel{2}}} = \frac{2 \cdot 1}{3 \cdot 1} = \frac{2}{3}$$

**Reduce $\frac{10}{25}$ to lowest terms.**

Answer: $\frac{2}{5}$

---

## $\boxed{\text{H I N T}}$

Notice that we inserted small "ones" in our work in Example 4. It is important to get in the habit of writing them so that you will not make mistakes when reducing fractions such as the following:

$$\frac{3}{15} = \frac{\overset{1}{\cancel{3}}}{5 \cdot \underset{1}{\cancel{3}}} = \frac{1}{5}.$$

Failure to write the 1s could lead you to the answer $\frac{0}{5}$ or 0, which is wrong.

---

The process of dividing out common factors by crossing them out, as shown in Example 4 and in the following examples, is sometimes called *canceling factors*. However, we will use the words *divide out*.

| EXAMPLE 5   DIVIDING OUT COMMON PRIME FACTORS | PRACTICE EXERCISE 5 |

Reduce to lowest terms by factoring into primes.

**(a)** $\dfrac{12}{42} = \dfrac{2 \cdot 2 \cdot 3}{2 \cdot 3 \cdot 7}$     Factor numerator and denominator into primes

$$= \frac{\overset{1}{\cancel{2}} \cdot 2 \cdot \overset{1}{\cancel{3}}}{\underset{1}{\cancel{2}} \cdot \underset{1}{\cancel{3}} \cdot 7} \qquad \text{Divide out common prime factors 2 and 3}$$

$$= \frac{2}{7}$$

Reduce to lowest terms by factoring into primes.

**(a)** $\dfrac{98}{63}$

**(b)** $\dfrac{5}{20} = \dfrac{5}{2 \cdot 2 \cdot 5}$     Factor numerator and denominator into primes

$\quad\quad = \dfrac{\overset{1}{\cancel{5}}}{2 \cdot 2 \cdot \underset{1}{\cancel{5}}}$     Divide out common prime factor 5

$\quad\quad = \dfrac{1}{4}$     Multiply remaining factors in the denominator

Remember to write the small 1s to avoid getting the wrong answer $\frac{0}{4}$ or 0.

**(c)** $\dfrac{210}{273} = \dfrac{2 \cdot \overset{1}{\cancel{3}} \cdot 5 \cdot \overset{1}{\cancel{7}}}{\underset{1}{\cancel{3}} \cdot \underset{1}{\cancel{7}} \cdot 13} = \dfrac{2 \cdot 5}{13} = \dfrac{10}{13}$

**(b)** $\dfrac{7}{35}$

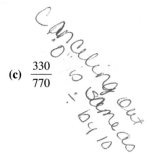

**(c)** $\dfrac{330}{770}$

**Answers:** (a) $\frac{14}{9}$ (b) $\frac{1}{5}$ (c) $\frac{3}{7}$

## 4 DIVIDING OUT COMMON, NONPRIME FACTORS

In Example 5 **(a)** we could have noticed that 6 was a factor of both the numerator and denominator and written

$$\dfrac{12}{42} = \dfrac{2 \cdot 6}{7 \cdot 6} = \dfrac{2 \cdot \overset{1}{\cancel{6}}}{7 \cdot \underset{1}{\cancel{6}}} = \dfrac{2}{7}.$$

*must Be A form of 1*

This technique does not require factoring into primes and can sometimes save a few steps.

---

| EXAMPLE 6 DIVIDING OUT COMMON, NONPRIME FACTORS | PRACTICE EXERCISE 6 |
|---|---|

Reduce to lowest terms.

**(a)** $\dfrac{4}{20} = \dfrac{\overset{1}{\cancel{4}}}{5 \cdot \underset{1}{\cancel{4}}} = \dfrac{1}{5}$

**(b)** $\dfrac{16}{40} = \dfrac{2 \cdot \overset{1}{\cancel{8}}}{5 \cdot \underset{1}{\cancel{8}}} = \dfrac{2}{5}$

**(c)** $\dfrac{36}{120} = \dfrac{3 \cdot \overset{1}{\cancel{12}}}{10 \cdot \underset{1}{\cancel{12}}} = \dfrac{3}{10}$

Reduce to lowest terms.

**(a)** $\dfrac{5}{40}$

**(b)** $\dfrac{18}{63}$

**(c)** $\dfrac{20}{150}$

**Answers:** (a) $\frac{1}{8}$ (b) $\frac{2}{7}$ (c) $\frac{2}{15}$

When reducing fractions by dividing out common, nonprime factors, be sure to reduce the fraction to lowest terms. For example, in Example 6 **(b),** had we only divided out the common factor 4,

$$\frac{16}{40} = \frac{4 \cdot \overset{1}{\cancel{4}}}{10 \cdot \underset{1}{\cancel{4}}} = \frac{4}{10},$$

we would not be finished since the fraction is not yet in lowest terms.

////////|

Before continuing, you should review the divisibility tests for 2, 3, 5, and 10 in Chapter 2. Reducing fractions can be simplified by using these tests.

| EXAMPLE 7    APPLICATION IN EDUCATION | PRACTICE EXERCISE 7 |
|---|---|

On a math test, Marvin got 60 correct out of 80 problems. On his history test, he got 75 correct out of 100 problems. What fractional part of each test did he get correct? On which test did he perform better?

On the math test, 60 correct out of 80 problems results in $\frac{60}{80}$ of the problems being correct. On the history test, he had $\frac{75}{100}$ correct. But

$$60 \cdot 100 = 6000 = 80 \cdot 75.$$

Thus,
$$\frac{60}{80} = \frac{75}{100}$$

since the cross-products are equal. Also, note that each fraction reduces to $\frac{3}{4}$. Thus, Marvin performed the same on the two tests.

Nina had 21 correct out of 30 on a math placement test and 70 out of 100 on an English test. What fractional part of each test did she get correct? On which test did she perform better?

Answer: Math, $\frac{7}{10}$; English, $\frac{7}{10}$; she scored the same on both tests.

## 3.2 EXERCISES A

*The shaded portions of the bars illustrate that what fractions are equal?*

**1.**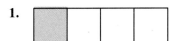

**2.**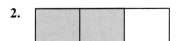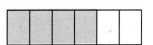

*Decide which fractions are equal by using the cross-product rule.*

**3.** $\dfrac{9}{12}$ and $\dfrac{3}{4}$

**4.** $\dfrac{2}{5}$ and $\dfrac{3}{7}$

**5.** $\dfrac{6}{16}$ and $\dfrac{3}{8}$

**6.** $\dfrac{6}{8}$ and $\dfrac{3}{2}$

**7.** $\dfrac{12}{32}$ and $\dfrac{4}{11}$

**8** $\dfrac{8}{9}$ and $\dfrac{96}{108}$

*In each of the following, find the numerator of the new fraction having the given denominator.*

**9.** $\dfrac{3}{4} = \dfrac{?}{12}$

**10.** $\dfrac{3}{2} = \dfrac{?}{6}$

**11.** $\dfrac{1}{2} = \dfrac{?}{22}$

**12.** $\dfrac{5}{8} = \dfrac{?}{16}$

**13** $\dfrac{9}{18} = \dfrac{?}{6}$

**14.** $\dfrac{5}{20} = \dfrac{?}{4}$

**15.** $\dfrac{20}{18} = \dfrac{?}{9}$

**16.** $\dfrac{8}{12} = \dfrac{?}{3}$

**17.** $\dfrac{6}{18} = \dfrac{?}{3}$

*In each of the following, find the denominator of the new fraction having the given numerator.*

**18.** $\dfrac{2}{7} = \dfrac{4}{?}$

**19.** $\dfrac{1}{9} = \dfrac{3}{?}$

**20.** $\dfrac{3}{5} = \dfrac{9}{?}$

**21.** $\dfrac{12}{20} = \dfrac{3}{?}$

**22.** $\dfrac{9}{2} = \dfrac{27}{?}$

**23** $\dfrac{36}{14} = \dfrac{18}{?}$

*Reduce each fraction to lowest terms.*

**24.** $\dfrac{4}{10}$

**25.** $\dfrac{15}{10}$

**26.** $\dfrac{6}{18}$

**27.** $\dfrac{16}{8}$

**28.** $\dfrac{39}{52}$

**29.** $\dfrac{12}{30}$

**30.** $\dfrac{121}{22}$

**31** $\dfrac{210}{105}$

**32.** $\dfrac{42}{18}$

*Solve.*

**33.** During the season, a starting ball player made 144 free throws out of 162 attempts. A reserve player went to the line only 9 times but made 8 shots. Which player made the greater fractional part of his free-throw attempts?

**34.** During May, one scientific equipment dealer sold 21 microscopes from his inventory of 77 microscopes. A second dealer sold 18 microscopes from his inventory of 66 microscopes. Which dealer sold the greater fractional part of his stock?

## FOR REVIEW

*What fractional part of each figure is shaded?*

**35.**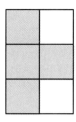

**36.**

*Express each fraction as a whole number.*

**37.** $\dfrac{17}{17}$ **38.** $\dfrac{0}{99}$ **39.** $\dfrac{100}{5}$ **40.** $\dfrac{84}{12}$

---

**ANSWERS:** 1. $\frac{1}{4} = \frac{2}{8} = \frac{4}{16}$ 2. $\frac{2}{3} = \frac{4}{6} = \frac{6}{9}$ 3. $9 \cdot 4 = 12 \cdot 3$, so $\frac{9}{12} = \frac{3}{4}$ 4. $2 \cdot 7 \neq 5 \cdot 3$; so $\frac{2}{5} \neq \frac{3}{7}$ 5. $6 \cdot 8 = 16 \cdot 3$; so $\frac{6}{16} = \frac{3}{8}$ 6. $6 \cdot 2 \neq 8 \cdot 3$; so $\frac{6}{8} \neq \frac{3}{2}$ 7. $12 \cdot 11 \neq 32 \cdot 4$; so $\frac{12}{32} \neq \frac{4}{11}$ 8. $8 \cdot 108 = 9 \cdot 96$; so $\frac{8}{9} = \frac{96}{108}$ 9. 9 10. 9 11. 11 12. 10 13. 3 14. 1 15. 10 16. 2 17. 1 18. 14 19. 27 20. 15 21. 5 22. 6 23. 7 24. $\frac{2}{5}$ 25. $\frac{3}{2}$ 26. $\frac{1}{3}$ 27. $\frac{2}{1}$ or 2 28. $\frac{3}{4}$ 29. $\frac{2}{5}$ 30. $\frac{11}{2}$ 31. $\frac{2}{1}$ or 2 32. $\frac{7}{3}$ 33. The starter made $\frac{144}{162} = \frac{8}{9}$ of his attempts, and the reserve made $\frac{8}{9}$ of his shots. Thus, each made the same fractional part of his attempts. 34. $\frac{21}{77}$ and $\frac{18}{66}$ are the same fractional parts (both equal $\frac{3}{11}$ reduced to lowest terms). 35. $\frac{4}{6}$ $\left(\text{or } \frac{2}{3} \text{ reduced}\right)$ 36. $\frac{4}{8}$ $\left(\text{or } \frac{1}{2} \text{ reduced}\right)$ 37. 1 38. 0 39. 20 40. 7

---

## 3.2 EXERCISES B

*The shaded portions of the bars illustrate that what fractions are equal?*

**1.**

**2.**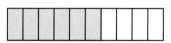

*Decide which fractions are equal by using the cross-product rule.*

**3.** $\dfrac{7}{8}$ and $\dfrac{21}{24}$ **4.** $\dfrac{5}{12}$ and $\dfrac{6}{13}$ **5.** $\dfrac{9}{6}$ and $\dfrac{18}{12}$

**6.** $\dfrac{13}{15}$ and $\dfrac{27}{30}$ **7.** $\dfrac{11}{12}$ and $\dfrac{132}{144}$ **8.** $\dfrac{4}{5}$ and $\dfrac{400}{500}$

*In each of the following, find the numerator of the new fraction having the given denominator.*

**9.** $\dfrac{3}{8} = \dfrac{?}{24}$

**10.** $\dfrac{9}{7} = \dfrac{?}{35}$

**11.** $\dfrac{1}{11} = \dfrac{?}{55}$

**12.** $\dfrac{?}{5} = \dfrac{8}{20}$

**13.** $\dfrac{4}{12} = \dfrac{?}{3}$

**14.** $\dfrac{21}{14} = \dfrac{?}{2}$

**15.** $\dfrac{25}{75} = \dfrac{?}{3}$

**16.** $\dfrac{15}{12} = \dfrac{?}{4}$

**17.** $\dfrac{100}{500} = \dfrac{?}{5}$

*In each of the following, find the denominator of the new fraction having the given numerator.*

**18.** $\dfrac{6}{?} = \dfrac{12}{10}$

**19.** $\dfrac{4}{?} = \dfrac{24}{18}$

**20.** $\dfrac{28}{4} = \dfrac{7}{?}$

**21.** $\dfrac{25}{15} = \dfrac{5}{?}$

**22.** $\dfrac{36}{24} = \dfrac{6}{?}$

**23.** $\dfrac{100}{20} = \dfrac{5}{?}$

*Reduce each fraction to lowest terms.*

**24.** $\dfrac{9}{18}$

**25.** $\dfrac{24}{16}$

**26.** $\dfrac{20}{4}$

**27.** $\dfrac{36}{48}$

**28.** $\dfrac{24}{60}$

**29.** $\dfrac{14}{126}$

**30.** $\dfrac{55}{33}$

**31.** $\dfrac{42}{6}$

**32.** $\dfrac{96}{528}$

*Solve.*

**33.** In one shipment of tires 4 of the 200 were defective. The next shipment of 700 tires contained 14 which were defective. Which shipment had the greater fraction of defective tires?

**34.** In June a poll showed that 25 of 30 people surveyed favored the new freeway. The next month 55 of 66 people favored the freeway. During which month did the greater fraction favor the new freeway?

## FOR REVIEW

*What fractional part of each figure is shaded?*

**35.**

**36.**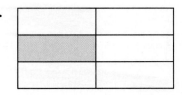

*Express each fraction as a whole number.*

**37.** $\dfrac{217}{217}$

**38.** $\dfrac{0}{581}$

**39.** $\dfrac{666}{18}$

**40.** $\dfrac{2000}{200}$

## 3.2 EXERCISES C

*Let a and b represent natural numbers. Reduce to lowest terms. [Hint: $a^2 = a \cdot a$]*

**1.** $\dfrac{3^3 4^2}{3^2 4}$

[Answer: 12]

**2.** $\dfrac{5^4 a^2}{5^2 a}$

**3.** $\dfrac{27 a^3}{81 a^7}$

$\left[\text{Answer: } \frac{1}{3a^4}\right]$

**4.** $\dfrac{a^2 b^3}{ab}$

**5.** $\dfrac{5 a^3 b^4}{10 a^2 b^6}$

**6.** $\dfrac{42 a^4 b^2}{14 a b^3}$

$\left[\text{Answer: } \frac{3a^3}{b}\right]$

## 3.3 MULTIPLYING FRACTIONS

### STUDENT GUIDEPOSTS

**1** Meaning of Multiplication of Fractions

**2** Rule for Multiplying Fractions

**3** Properties of Multiplication

**4** Powers and Roots of Fractions

### **1** MEANING OF MULTIPLICATION OF FRACTIONS

The simplest arithmetic operation on fractions is multiplication. Notice that the word *of* after a fraction means ''times.''

What is one half of 8?

Obviously, half of 8 is 4. Let us multiply $\frac{1}{2}$ times 8 as follows.

$$\frac{1}{2} \cdot 8 = \frac{1}{2} \cdot \frac{8}{1} \qquad \text{Write 8 as the fraction } \tfrac{8}{1}$$

$$= \frac{1 \cdot 8}{2 \cdot 1} \qquad \text{Multiply numerators}$$
$$\qquad\qquad \text{Multiply denominators}$$

$$= \frac{8}{2} \qquad \text{Divide result}$$

$$= 4$$

The final answer is 4, which agrees with our intuitive notion of $\frac{1}{2}$ of 8.

### **HINT**

Two facts are useful:

**1.** The word *of* means ''times'' when it follows a fraction.

**2.** The product of two fractions $= \dfrac{\text{product of their numerators}}{\text{product of their denominators}}$.

Suppose we look at a second example and find the product of $\frac{1}{2}$ and $\frac{1}{3}$. The shaded part of the bar in Figure 3.12 corresponds to the fraction $\frac{1}{3}$. If we multiply $\frac{1}{2} \cdot \frac{1}{3}$, we recognize this as

$$\frac{1}{2} \text{ of } \frac{1}{3}.$$

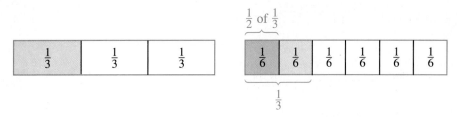

**Figure 3.12**                                 **Figure 3.13**

Just as we found $\frac{1}{2}$ of 8 (which we recognized as 4), suppose we find $\frac{1}{2}$ of $\frac{1}{3}$ by taking one half of the shaded part in Figure 3.12. Figure 3.13 shows that

$$\frac{1}{2} \text{ of } \frac{1}{3} \text{ is equal to } \frac{1}{6}.$$

Finding the product of $\frac{1}{2}$ and $\frac{1}{3}$ by multiplying numerators, multiplying denominators, and dividing the results gives the same answer.

$$\frac{1}{2} \cdot \frac{1}{3} = \frac{1 \cdot 1}{2 \cdot 3} = \frac{1}{6}$$

| EXAMPLE 1   MULTIPLYING FRACTIONS | PRACTICE EXERCISE 1 |
|---|---|

Find each product.

**(a)** $\frac{1}{3} \cdot \frac{1}{7} = \frac{1 \cdot 1}{3 \cdot 7} = \frac{1}{21}$   Multiply numerators, multiply denominators and divide the products

**(b)** $\frac{2}{3} \cdot \frac{5}{9} = \frac{2 \cdot 5}{3 \cdot 9} = \frac{10}{27}$

Find each product.

**(a)** $\frac{1}{6} \cdot \frac{1}{8}$

**(b)** $\frac{3}{8} \cdot \frac{7}{5}$

Answers:   (a) $\frac{1}{48}$   (b) $\frac{21}{40}$

## ❷ RULE FOR MULTIPLYING FRACTIONS

In Example 1, we multiplied fractions by first multiplying the numerators, then multiplying the denominators, and then dividing the numerators' product by the denominators' product. For example,

$$\frac{3}{8} \cdot \frac{4}{5} = \frac{3 \cdot 4}{8 \cdot 5} = \frac{12}{40}.$$

We know from Section 3.2 that to reduce $\frac{12}{40}$ to lowest terms, we need to factor both the numerator and the denominator into primes and divide out common prime factors.

$$\frac{12}{40} = \frac{\overset{1}{\cancel{2}} \cdot \overset{1}{\cancel{2}} \cdot 3}{\underset{1}{\cancel{2}} \cdot \underset{1}{\cancel{2}} \cdot 2 \cdot 5} = \frac{3}{10}$$

Multiplying and then reducing fractions can be time consuming. To reduce the number of steps, and thus have less chance of making an error, use the following rule.

| To Multiply Two (or More) Fractions |
| :--- |

1. Factor all numerators and denominators into primes.
2. Place all numerator factors over all denominator factors.
3. Divide out all prime factors common to both the numerator and denominator.
4. Multiply the remaining primes, first in the numerator, and then in the denominator.

If we return to the original example of $\frac{3}{8} \cdot \frac{4}{5}$ and use these rules, we get

$$\frac{3}{8} \cdot \frac{4}{5} = \frac{3}{2 \cdot 2 \cdot 2} \cdot \frac{2 \cdot 2}{5}$$

$$= \frac{3 \cdot \overset{1}{\cancel{2}} \cdot \overset{1}{\cancel{2}}}{\underset{1}{\cancel{2}} \cdot \underset{1}{\cancel{2}} \cdot 2 \cdot 5}$$

$$= \frac{3}{10}.$$

This gives us the same answer we obtained above, but has fewer steps. Notice that we could have factored out the common nonprime factor of 4 to shorten our work even further.

$$\frac{3}{8} \cdot \frac{4}{5} = \frac{3 \cdot \overset{1}{\cancel{4}}}{\underset{2}{\cancel{8}} \cdot 5}$$

$$= \frac{3 \cdot 1}{2 \cdot 5}$$

$$= \frac{3}{10}$$

| EXAMPLE 2   MULTIPLYING FRACTIONS | PRACTICE EXERCISE 2 |
| :--- | :--- |

Find each product.

Find each product.

**(a)** $\dfrac{2}{9} \cdot \dfrac{3}{4} = \dfrac{2}{3 \cdot 3} \cdot \dfrac{3}{2 \cdot 2}$    Factor into primes

$$= \frac{\overset{1}{\cancel{2}} \cdot \overset{1}{\cancel{3}}}{3 \cdot \underset{1}{\cancel{3}} \cdot \underset{1}{\cancel{2}} \cdot 2}$$    Put all numerator factors over all denominator factors and divide out common prime factors

$$= \frac{1}{3 \cdot 2}$$    Remember that the numerator is 1, not 0

$$= \frac{1}{6}$$

**(a)** $\dfrac{3}{7} \cdot \dfrac{5}{6}$

**(b)** $\dfrac{2}{7} \cdot 14 = \dfrac{2}{7} \cdot \dfrac{14}{1}$    Write 14 as the fraction $\frac{14}{1}$

$$= \frac{2}{7} \cdot \frac{2 \cdot 7}{1}$$    Factor

**(b)** $\dfrac{8}{13} \cdot 39$

$$= \frac{2 \cdot 2 \cdot \overset{1}{\cancel{7}}}{\underset{1}{\cancel{7}} \cdot 1} \qquad \text{Indicate product and divide out 7}$$

$$= \frac{4}{1} = 4$$

**(c)** $\dfrac{2}{15} \cdot \dfrac{3}{5} = \dfrac{2}{3 \cdot 5} \cdot \dfrac{3}{5} = \dfrac{2 \cdot \overset{1}{\cancel{3}}}{\underset{1}{\cancel{3}} \cdot 5 \cdot 5} = \dfrac{2}{25}$

By recognizing that 15 divided by 3 is 5, we can shorten our work as follows.

$$\frac{2}{\underset{5}{\cancel{15}}} \cdot \frac{\overset{1}{\cancel{3}}}{5} = \frac{2}{5 \cdot 5} = \frac{2}{25}$$

**(c)** $\dfrac{7}{18} \cdot \dfrac{9}{2}$

Answers: (a) $\frac{5}{14}$ (b) 24 (c) $\frac{7}{4}$

## ③ PROPERTIES OF MULTIPLICATION

Multiplication of fractions has the same properties as multiplication of whole numbers. The commutative law of multiplication states that changing the order of a product does not change the result. For example,

$$\frac{2}{3} \cdot \frac{7}{8} = \frac{7}{8} \cdot \frac{2}{3}.$$

The associative law of multiplication states that regrouping products does not change the result. For example,

$$\left(\frac{2}{3} \cdot \frac{7}{8}\right) \cdot \frac{3}{4} = \frac{2}{3} \cdot \left(\frac{7}{8} \cdot \frac{3}{4}\right).$$

The number 1, written as $\frac{1}{1}$, is the multiplicative identity, since we can multiply any fraction by 1 and obtain that same fraction. For example,

$$1 \cdot \frac{2}{3} = \frac{2}{3} \cdot 1 = \frac{2}{3}.$$

The associative law allows us to multiply three fractions without using parentheses.

---

| **EXAMPLE 3  MULTIPLYING THREE FRACTIONS** |
|---|

Find the product of $\frac{4}{9}$, $\frac{3}{14}$, and $\frac{7}{6}$.

$$\frac{4}{9} \cdot \frac{3}{14} \cdot \frac{7}{6} = \frac{2 \cdot 2}{3 \cdot 3} \cdot \frac{3}{2 \cdot 7} \cdot \frac{7}{2 \cdot 3} \qquad \text{Factor all numerators and denominators into primes}$$

$$= \frac{\cancel{2} \cdot \cancel{2} \cdot \cancel{3} \cdot \cancel{7}}{\cancel{3} \cdot 3 \cdot \cancel{2} \cdot \cancel{7} \cdot \cancel{2} \cdot 3} \qquad \text{Indicate products and divide out common prime factors}$$

$$= \frac{1}{3 \cdot 3} \qquad \text{Remember that the numerator is 1}$$

$$= \frac{1}{9}$$

| **PRACTICE EXERCISE 3** |
|---|

Find the product of $\frac{7}{5}$, $\frac{10}{21}$, and $\frac{3}{8}$.

Answer: $\frac{1}{4}$

## ④ POWERS AND ROOTS OF FRACTIONS

In Section 2.8 we studied powers and roots of whole numbers. We can find powers and roots of fractions in a similar manner.

---

| **EXAMPLE 4** **POWERS OF FRACTIONS** |
|---|

Find the value of each power.

$\longrightarrow$ two $\frac{2}{3}$'s

**(a)** $\left(\dfrac{2}{3}\right)^2 = \overbrace{\dfrac{2}{3}\cdot\dfrac{2}{3}}^{} = \dfrac{2\cdot 2}{3\cdot 3} = \dfrac{4}{9}$

**(b)** $\left(\dfrac{5}{4}\right)^2 = \dfrac{5}{4}\cdot\dfrac{5}{4} = \dfrac{5\cdot 5}{4\cdot 4} = \dfrac{25}{16}$

| **PRACTICE EXERCISE 4** |
|---|

Find the value of each power.

**(a)** $\left(\dfrac{5}{7}\right)^2$

**(b)** $\left(\dfrac{8}{3}\right)^2$

Answers: (a) $\frac{25}{49}$    (b) $\frac{64}{9}$

---

| **EXAMPLE 5** **ROOTS OF FRACTIONS** |
|---|

Find the value of each radical expression.

**(a)** $\sqrt{\dfrac{4}{9}}$      Since $\dfrac{2}{3}\cdot\dfrac{2}{3} = \dfrac{4}{9}$, $\sqrt{\dfrac{4}{9}} = \dfrac{2}{3}$.      $\frac{2}{3}$ **is a square root of** $\frac{4}{9}$

**(b)** $\sqrt{\dfrac{121}{25}}$      Since $\dfrac{11}{5}\cdot\dfrac{11}{5} = \dfrac{121}{25}$, $\sqrt{\dfrac{121}{25}} = \dfrac{11}{5}$.

| **PRACTICE EXERCISE 5** |
|---|

Find the value of each radical expression.

**(a)** $\sqrt{\dfrac{25}{16}}$

**(b)** $\sqrt{\dfrac{144}{49}}$

Answers: (a) $\frac{5}{4}$    (b) $\frac{12}{7}$

---

Many word problems involve finding products of fractions.

---

| **EXAMPLE 6** **APPLICATION TO AREA** |
|---|

A farmer has a plot of land in the form of a rectangle $\frac{1}{3}$ mi wide and $\frac{7}{8}$ mi long. What is the area of the parcel of land?

First, make a sketch of the information as in Figure 3.14. Recall that the area of a rectangle is the length times the width.

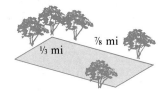

**Figure 3.14**

$$\text{Area} = \frac{7}{8}\cdot\frac{1}{3} = \frac{7}{2\cdot 2\cdot 2}\cdot\frac{1}{3}$$

$$= \frac{7\cdot 1}{2\cdot 2\cdot 2\cdot 3} = \frac{7}{24}$$

| **PRACTICE EXERCISE 6** |
|---|

A picture is $\frac{3}{4}$ yd wide and $\frac{10}{7}$ yd long. What is the area of the picture?

Thus, the area of the land is $\frac{7}{24}$ square miles. Figure 3.15 shows how this answer can be interpreted.

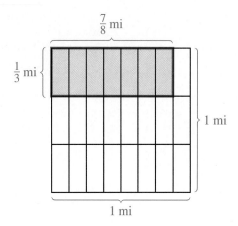

**Figure 3.15**

Notice that 1 square mile has been divided into 24 equal parts and that 7 of these parts, or $\frac{7}{24}$ of one square mile, corresponds to the area of the parcel of land.

Answer: $\frac{15}{14}$ sq yd

---

| **EXAMPLE 7** APPLICATION TO MAP READING | **PRACTICE EXERCISE 7** |

On a map, 1 inch represents 32 miles. How many miles does $\frac{3}{4}$ of an inch represent?

   Find $\frac{3}{4}$ of 32 miles. Used in this way *of* means *multiply*.

$$\frac{3}{4} \text{ of } 32$$

$$\downarrow \downarrow \downarrow$$

$$\frac{3}{4} \cdot 32 = \frac{3}{4} \cdot \frac{32}{1} = \frac{3}{2 \cdot 2} \cdot \frac{2 \cdot 2 \cdot 2 \cdot 2 \cdot 2}{1}$$

$$= \frac{3 \cdot \overset{1}{\cancel{2}} \cdot \overset{1}{\cancel{2}} \cdot 2 \cdot 2 \cdot 2}{\underset{1}{\cancel{2}} \cdot \underset{1}{\cancel{2}} \cdot 1}$$

$$= \frac{3 \cdot 2 \cdot 2 \cdot 2}{1}$$

$$= \frac{24}{1} = 24$$

Thus, $\frac{3}{4}$ of an inch represents 24 mi. If we notice that 4 divides into 32, we can shorten our work.

$$\frac{3}{4} \cdot \frac{32}{1} = \frac{3}{\underset{1}{\cancel{4}}} \cdot \frac{\overset{8}{\cancel{32}}}{1} = \frac{24}{1} = 24$$

If a large loaf of bread weighs 3 pounds, how much would $\frac{2}{3}$ of a loaf weigh?

Answer: 2 pounds

| EXAMPLE 8   APPLICATION TO LABOR | PRACTICE EXERCISE 8 |
|---|---|

Maria makes \$35 for working one full day. How much should she receive for working $\frac{3}{5}$ of a day?

  Calculate $\frac{3}{5}$ of \$35.

Frank is paid \$12 per hour for construction work. How much should he be paid if he works $\frac{7}{3}$ hours?

$$\frac{3}{5} \text{ of } 35$$
$$\downarrow \downarrow \downarrow$$
$$\frac{3}{5} \cdot 35 = \frac{3}{5} \cdot \frac{35}{1} = \frac{3}{\overset{5}{\cancel{5}}} \cdot \frac{\overset{7}{\cancel{35}}}{\underset{1}{1}}$$

$$= \frac{3 \cdot 7}{1} = \frac{21}{1} = 21$$

Thus, she should receive \$21 for $\frac{3}{5}$ of a day's work.

Answer: \$28

# 3.3   EXERCISES A

*Answer true or false. If the answer is false, explain why.*

**1.** The phrase "$\frac{3}{4}$ of a number" translates to $\frac{3}{4} \cdot$ (that number).

**2.** The product of two fractions can be found by multiplying all numerators and dividing the result by the product of all denominators.

**3.** When multiplying fractions, the final product should always be reduced to lowest terms.

**4.** The fact that $\left(\frac{2}{3} \cdot \frac{3}{4}\right) \cdot \frac{2}{5} = \frac{2}{3} \cdot \left(\frac{3}{4} \cdot \frac{2}{5}\right)$ illustrates the commutative law.

**5.** The fact that $1 \cdot \frac{7}{8} = \frac{7}{8}$ illustrates the associative law.

**6.** The fact that $\frac{2}{3} \cdot \frac{8}{9} = \frac{8}{9} \cdot \frac{2}{3}$ illustrates the commutative law.

*Find each product.*

**7.** $\frac{2}{9} \cdot \frac{1}{3}$

**8.** $\frac{4}{3} \cdot \frac{3}{10}$

**9.** $\frac{1}{4} \cdot \frac{8}{9}$

**10.** $\frac{6}{7} \cdot \frac{14}{3}$

**11.** $\frac{8}{9} \cdot \frac{3}{4}$

**12.** $\frac{3}{5} \cdot 40$

**13.** $7 \cdot \frac{3}{21}$

**14.** $\frac{6}{5} \cdot \frac{1}{3}$

**15.** $\frac{6}{35} \cdot \frac{20}{12}$

**16.** $\frac{21}{6} \cdot \frac{4}{7}$

**17.** $\frac{37}{19} \cdot 0$

**18.** $\frac{7}{5} \cdot 10$

**19.** $\dfrac{7}{8} \cdot \dfrac{4}{35}$

**20** $\dfrac{18}{84} \cdot \dfrac{36}{27}$

**21.** $\dfrac{35}{9} \cdot \dfrac{3}{28}$

**22.** $\dfrac{2}{3} \cdot \dfrac{10}{9} \cdot \dfrac{6}{5}$

**23.** $\dfrac{3}{8} \cdot \dfrac{24}{9} \cdot 7$

**24.** $\dfrac{160}{169} \cdot \dfrac{13}{8} \cdot \dfrac{26}{5}$

*Find the value of each power.*

**25.** $\left(\dfrac{3}{4}\right)^2$

**26.** $\left(\dfrac{1}{9}\right)^2$

**27.** $\left(\dfrac{7}{3}\right)^2$

*Find the value of each radical.*

**28.** $\sqrt{\dfrac{9}{16}}$

**29.** $\sqrt{\dfrac{1}{49}}$

**30.** $\sqrt{\dfrac{144}{25}}$

*Solve.*

**31** Claude owns a plot of land in the shape of a rectangle $\frac{4}{7}$ km wide by $\frac{7}{8}$ km long (**km** is the abbreviation for *kilometer*). What is the area of his plot?

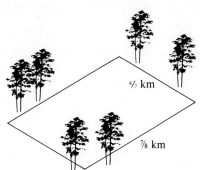

**32.** On a map, 1 inch represents 27 miles. How many miles does $\frac{4}{3}$ inch represent?

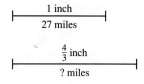

**33.** Max earned $225 for working one five-day week. How much would he earn working three days? [*Hint:* Three days is $\frac{3}{5}$ of one five-day week.]

**34.** Ramona took out a loan for two thirds the cost of a remodeling job. If the job cost $1230, how much did she borrow?

## FOR REVIEW

*Decide which fractions are equal by using cross products.*

**35.** $\dfrac{4}{5}$ and $\dfrac{5}{6}$

**36.** $\dfrac{4}{16}$ and $\dfrac{3}{12}$

**37.** $\dfrac{9}{5}$ and $\dfrac{7}{4}$

*Reduce each fraction to lowest terms.*

**38.** $\dfrac{12}{18}$

**39.** $\dfrac{65}{91}$

**40.** $\dfrac{273}{182}$

*Find the missing term in each fraction.*

**41.** $\dfrac{16}{6} = \dfrac{?}{3}$

**42.** $\dfrac{30}{5} = \dfrac{12}{?}$

**43.** $\dfrac{9}{42} = \dfrac{?}{14}$

---

ANSWERS: 1. true  2. true  3. true  4. false (associative law)  5. false (multiplicative identity)  6. true  7. $\frac{2}{27}$
8. $\frac{2}{5}$  9. $\frac{2}{9}$  10. 4  11. $\frac{2}{3}$  12. 24  13. 1  14. $\frac{2}{5}$  15. $\frac{2}{7}$  16. 2  17. 0  18. 14  19. $\frac{1}{10}$  20. $\frac{2}{7}$  21. $\frac{5}{12}$  22. $\frac{8}{9}$  23. 7
24. 8  25. $\frac{9}{16}$  26. $\frac{1}{81}$  27. $\frac{49}{9}$  28. $\frac{3}{4}$  29. $\frac{1}{7}$  30. $\frac{12}{5}$  31. $\frac{1}{2}$ sq km  32. 36 mi  33. \$135  34. \$820  35. $4 \cdot 6 \neq$
$5 \cdot 5$; so $\frac{4}{5} \neq \frac{5}{6}$  36. $4 \cdot 12 = 16 \cdot 3$; so $\frac{4}{16} = \frac{3}{12}$  37. $9 \cdot 4 \neq 5 \cdot 7$; so $\frac{9}{5} \neq \frac{7}{4}$  38. $\frac{2}{3}$  39. $\frac{5}{7}$  40. $\frac{3}{2}$  41. 8  42. 2  43. 3

## 3.3 EXERCISES B

*Answer true or false. If the answer is false, explain why.*

1. To reduce the number of steps in multiplying fractions, divide out common factors before multiplying numerators and multiplying denominators.

2. When common factors are divided out before multiplying numerators and denominators, the resulting fraction will be reduced to lowest terms.

3. To square a fraction, square the numerator and square the denominator.

4. The fact that $\frac{1}{2} \cdot \frac{3}{7} = \frac{3}{7} \cdot \frac{1}{2}$ illustrates the associative law.

5. The fact that $\left(\frac{2}{9} \cdot \frac{5}{13}\right) \cdot \frac{4}{7} = \frac{2}{9} \cdot \left(\frac{5}{13} \cdot \frac{4}{7}\right)$ illustrates the commutative law.

6. Any fraction multiplied by 1 gives that fraction as the product.

*Find each product.*

**7.** $\dfrac{7}{8} \cdot \dfrac{1}{2}$

**8.** $\dfrac{9}{4} \cdot \dfrac{4}{3}$

**9.** $\dfrac{6}{5} \cdot \dfrac{10}{3}$

**10.** $\dfrac{2}{7} \cdot \dfrac{21}{8}$

**11.** $\dfrac{3}{8} \cdot 40$

**12.** $9 \cdot \dfrac{17}{18}$

**13.** $\dfrac{20}{35} \cdot \dfrac{7}{4}$

**14.** $\dfrac{2}{3} \cdot \dfrac{5}{6}$

**15.** $\dfrac{62}{9} \cdot 0$

**16.** $\dfrac{7}{8} \cdot \dfrac{24}{42}$

**17.** $\dfrac{27}{26} \cdot \dfrac{13}{9}$

**18.** $10 \cdot \dfrac{9}{2}$

**19.** $\dfrac{35}{40} \cdot \dfrac{8}{7}$

**20.** $\dfrac{99}{13} \cdot \dfrac{39}{11}$

**21.** $\dfrac{32}{25} \cdot \dfrac{15}{16}$

**22.** $\dfrac{1}{3} \cdot \dfrac{20}{9} \cdot \dfrac{6}{5}$

**23.** $6 \cdot \dfrac{3}{7} \cdot \dfrac{28}{12}$

**24.** $\dfrac{169}{5} \cdot \dfrac{25}{13} \cdot \dfrac{1}{13}$

*Find the value of each power.*

**25.** $\left(\dfrac{1}{7}\right)^2$

**26.** $\left(\dfrac{5}{8}\right)^2$

**27.** $\left(\dfrac{10}{3}\right)^2$

*Find the value of each radical.*

**28.** $\sqrt{\dfrac{25}{9}}$

**29.** $\sqrt{\dfrac{1}{100}}$

**30.** $\sqrt{\dfrac{169}{121}}$

*Solve.*

**31.** A rectangular field is $\frac{3}{7}$ mi wide and $\frac{7}{9}$ mi long. What is the area of the field?

**32.** On a map, 1 inch represents 12 miles. How many miles are represented by $\frac{5}{6}$ inches?

**33.** Tuition for one semester amounts to $2352. If Jack paid two thirds of this amount at early registration, how much is left to be paid?

**34.** Susan's annual salary is $28,500. What is her monthly salary?

## FOR REVIEW

*Decide which fractions are equal by using cross products.*

**35.** $\dfrac{7}{9}$ and $\dfrac{9}{11}$

**36.** $\dfrac{20}{15}$ and $\dfrac{8}{6}$

**37.** $\dfrac{3}{12}$ and $\dfrac{2}{6}$

*Reduce each fraction to lowest terms.*

**38.** $\dfrac{24}{15}$

**39.** $\dfrac{81}{90}$

**40.** $\dfrac{175}{350}$

*Find the missing term in each fraction.*

**41.** $\dfrac{20}{8} = \dfrac{?}{2}$

**42.** $\dfrac{4}{1} = \dfrac{12}{?}$

**43.** $\dfrac{22}{?} = \dfrac{88}{44}$

## 3.3 EXERCISES C

*Let a and b represent natural numbers. Find the products.*

**1.** $\dfrac{9}{a} \cdot \dfrac{a^2}{36}$

**2.** $\dfrac{7a^2}{6b^3} \cdot \dfrac{36ab}{35a^2}$

$\left[\text{Answer: } \frac{6a}{5b^2}\right]$

**3.** $\dfrac{2^2 a^4}{3^3 b^4} \cdot \dfrac{3a^2 b^3}{4ab}$

## **3.4** DIVIDING FRACTIONS

━━━━━ STUDENT GUIDEPOSTS ━━━━━

1 Reciprocal of a Fraction          3 Related Division Sentence
2 Method of Dividing Fractions

### 1 RECIPROCAL OF A FRACTION

Division of fractions is closely related to multiplication. In fact, a division problem can be changed to an equivalent multiplication problem using a *reciprocal*.

> **Reciprocal of a Fraction**
>
> Two fractions are **reciprocals** if their product is 1.

Consider the fractions $\frac{3}{4}$ and $\frac{4}{3}$.

$$\frac{3}{4} \cdot \frac{4}{3} = \frac{\overset{1}{\cancel{3}} \cdot \overset{1}{\cancel{4}}}{\underset{1}{\cancel{4}} \cdot \underset{1}{\cancel{3}}} = \frac{1}{1} = 1$$

Thus $\frac{3}{4}$ and $\frac{4}{3}$ are reciprocals of each other. Notice that $\frac{4}{3}$ is obtained when we interchange the numerator and denominator of $\frac{3}{4}$. This is true of all reciprocals and gives an easy way to find the reciprocal of a number.

> **Finding Reciprocals**
>
> To find the reciprocal of a fraction, interchange the numerator and denominator. Thus the reciprocal of $\frac{a}{b}$ is $\frac{b}{a}$.

---

| **EXAMPLE 1**  FINDING RECIPROCALS | **PRACTICE EXERCISE 1** |
|---|---|
| Find the reciprocal of each number. | Find the reciprocal of each number. |
| **(a)** $\frac{3}{8}$    The reciprocal of $\frac{3}{8}$ is $\frac{8}{3}$. | **(a)** $\frac{13}{11}$ |
| **(b)** $\frac{1}{3}$    The reciprocal of $\frac{1}{3}$ is $\frac{3}{1}$ or 3. | **(b)** $\frac{1}{20}$ |
| | **(c)** 9 |
| **(c)** 5    Since 5 can be written as $\frac{5}{1}$, its reciprocal is $\frac{1}{5}$. | **(d)** 1 |
| **(d)** 0    0 has no reciprocal since 0 is $\frac{0}{1}$ and $\frac{1}{0}$ is undefined. | Answers: (a) $\frac{11}{13}$  (b) 20  (c) $\frac{1}{9}$ (d) 1 |

---

## ② METHOD OF DIVIDING FRACTIONS

The fact that the product of reciprocals is 1 gives us a way to convert a division problem into a multiplication problem. For example, consider

$$\frac{2}{3} \div \frac{4}{9} \qquad \text{or} \qquad \frac{\frac{2}{3}}{\frac{4}{9}}.$$

Recall from Section 3.2 that when the numerator and denominator of a fraction are multiplied by the same nonzero whole number, an equal fraction is formed. The same holds true when both numerator and denominator are multiplied by the same nonzero fraction. Suppose we multiply numerator and denominator of

$$\frac{\frac{2}{3}}{\frac{4}{9}}$$

by $\frac{9}{4}$, the reciprocal of the denominator, $\frac{4}{9}$.

$$\frac{\frac{2}{3} \cdot \frac{9}{4}}{\frac{4}{9} \cdot \frac{9}{4}} = \frac{\frac{2}{3} \cdot \frac{9}{4}}{\overset{1}{\cancel{4}} \cdot \overset{1}{\cancel{9}}} = \frac{\frac{2}{3} \cdot \frac{9}{4}}{1}$$

Just as

$$\frac{5}{1} = 5, \qquad \frac{10}{1} = 10, \qquad \text{and} \qquad \frac{127}{1} = 127,$$

so, also, does

$$\frac{\frac{2}{3} \cdot \frac{9}{4}}{1} = \frac{2}{3} \cdot \frac{9}{4}.$$

We have just shown that the division problem

$$\frac{2}{3} \div \frac{4}{9}$$

can be converted to the multiplication problem

$$\frac{2}{3} \cdot \frac{9}{4}.$$

This is true in general.

| To Divide Two Fractions |
| --- |
| 1. Replace the divisor by its reciprocal and change the division sign to multiplication. |
| 2. Proceed as in multiplication. |

| **EXAMPLE 2**   DIVIDING FRACTIONS | **PRACTICE EXERCISE 2** |

Find each quotient.

**(a)** $\dfrac{7}{12} \div \dfrac{14}{9} = \dfrac{7}{12} \cdot \dfrac{9}{14}$   Replace divisor by its reciprocal and multiply

$= \dfrac{7}{2 \cdot 2 \cdot 3} \cdot \dfrac{3 \cdot 3}{2 \cdot 7}$   Factor

$= \dfrac{7 \cdot 3 \cdot 3}{2 \cdot 2 \cdot 3 \cdot 2 \cdot 7}$   Divide out common factors

$= \dfrac{3}{2 \cdot 2 \cdot 2} = \dfrac{3}{8}$

**(b)** $\dfrac{20}{3} \div 5 = \dfrac{20}{3} \div \dfrac{5}{1}$   5 can be written as $\frac{5}{1}$

$= \dfrac{20}{3} \cdot \dfrac{1}{5}$   The reciprocal of $\frac{5}{1}$ is $\frac{1}{5}$

$= \dfrac{2 \cdot 2 \cdot 5}{3} \cdot \dfrac{1}{5}$

$= \dfrac{2 \cdot 2 \cdot 5 \cdot 1}{3 \cdot 5} = \dfrac{2 \cdot 2 \cdot 1}{3} = \dfrac{4}{3}$

Find each quotient.

**(a)** $\dfrac{8}{15} \div \dfrac{2}{5}$

**(b)** $\dfrac{14}{9} \div 7$

---

Before considering applied problems, remember that $10 \div 2$ can be interpreted as the number of 2's in 10. There are five 2's in 10. Similarly,

$$\frac{2}{5} \div \frac{1}{10}$$

means "How many $\frac{1}{10}$ths are in $\frac{2}{5}$?" Consider Figure 3.16.

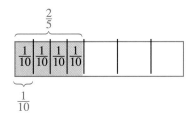

**Figure 3.16**

By counting we see that there are four $\frac{1}{10}$ths in $\frac{2}{5}$. But notice that

$$\frac{2}{5} \div \frac{1}{10} = \frac{2}{5} \cdot \frac{10}{1} = \frac{2 \cdot 10}{5 \cdot 1}$$

$$= \frac{2 \cdot 2 \cdot 5}{5 \cdot 1} = 4.$$

Thus, using the division rule we find the number of $\frac{1}{10}$ths in $\frac{2}{5}$ without relying on a figure.

| | |
|---|---|
| **EXAMPLE 3** APPLICATION OF DIVISION | **PRACTICE EXERCISE 3** |

A piece of string 12 m long is to be cut into pieces each of length $\frac{3}{4}$ m. How many pieces can be cut?

A butcher wants to cut a roast weighing 14 lb into steaks that each weigh $\frac{2}{3}$ lb. How many steaks will he get?

Find the number of $\frac{3}{4}$'s in 12. That is, find $12 \div \frac{3}{4}$.

$$12 \div \frac{3}{4} = \frac{12}{1} \cdot \frac{4}{3} = \frac{3 \cdot 4}{1} \cdot \frac{4}{3} = \frac{\overset{1}{\cancel{3}} \cdot 4 \cdot 4}{1 \cdot \underset{1}{\cancel{3}}} = \frac{16}{1} = 16$$

Thus, 16 pieces of length $\frac{3}{4}$ m can be cut.

Answer: 21 steaks

---

## $\boxed{\textbf{H I N T}}$

Notice that 12 was not factored completely to primes in Example 3. Since the only factor (besides 1) in the denominator was 3, there was no need to factor 12 beyond $3 \cdot 4$. With practice, you can take shortcuts such as this.

---

## ③ RELATED DIVISION SENTENCE

Many division problems are more easily solved by first considering a multiplication sentence and then changing it to the **related division sentence.** For example, suppose we know that $\frac{1}{2}$ of a trip is 220 miles, and want to know how long the whole trip is. If we know that

$$\frac{1}{2} \text{ of the trip is 220,}$$

we can write

$$\begin{array}{ccccc} \frac{1}{2} & \text{of} & \square & \text{is} & 220 \\ \downarrow & \downarrow & \downarrow & \downarrow & \downarrow \end{array}$$
$$\frac{1}{2} \cdot \square = 220. \quad \text{Multiplication sentence}$$

We can find out what $\square$ is by writing the related division sentence.

$$\square = \frac{220}{\frac{1}{2}} \quad \text{Related division sentence}$$

$$\frac{220}{\frac{1}{2}} = 220 \cdot \frac{2}{1} = \frac{220}{1} \cdot \frac{2}{1} = \frac{440}{1} = 440$$

Thus $\qquad \frac{1}{2} \cdot 440 = 220 \quad$ Multiplication sentence

and $\qquad \dfrac{220}{\frac{1}{2}} = 440 \quad$ Related division sentence

are two different forms of the same sentence. We use this information in the next examples.

| EXAMPLE 4   USING THE RELATED DIVISION SENTENCE | PRACTICE EXERCISE 4 |

Find the missing number.

**(a)** $\frac{2}{3}$ of a number is 42.

$$\frac{2}{3} \text{ of } \square \text{ is } 42$$
$$\downarrow\downarrow\downarrow \; \downarrow\;\downarrow$$
$$\frac{2}{3} \cdot \square = 42 \qquad \text{Multiplication sentence}$$

$$\square = \frac{42}{\frac{2}{3}} \qquad \text{Related division sentence}$$

$$= \frac{42}{1} \cdot \frac{3}{2}$$

$$= \frac{21 \cdot \overset{1}{\cancel{2}} \cdot 3}{1 \cdot \underset{1}{\cancel{2}}} = 63$$

**(b)** $\frac{5}{9}$ of the weight is 20 lb.

$$\frac{5}{9} \text{ of } \square \text{ is } 20$$
$$\downarrow\downarrow\downarrow \; \downarrow\;\downarrow$$
$$\frac{5}{9} \cdot \square = 20 \qquad \text{Multiplication sentence}$$

$$\square = \frac{20}{\frac{5}{9}} \qquad \text{Related division sentence}$$

$$= \frac{20}{1} \cdot \frac{9}{5}$$

$$= \frac{4 \cdot \overset{1}{\cancel{5}} \cdot 9}{1 \cdot \underset{1}{\cancel{5}}} = 36 \text{ lb}$$

Find the missing number.

**(a)** $\frac{9}{2}$ of a number is 30.

**(b)** $\frac{7}{8}$ of the length is 350 cm.

Answers: (a) $\frac{20}{3}$   (b) 400 cm

| EXAMPLE 5   APPLICATION OF DIVISION | PRACTICE EXERCISE 5 |

After driving 160 miles, the Carlsons had completed $\frac{5}{7}$ of their trip. What was the total length of their trip?

    We know that $\frac{5}{7}$ *of* the trip *is* 160 miles. Remember *of* becomes *times* and *is* becomes *equals* in the multiplication sentence.

$$\frac{5}{7} \text{ of } \square \text{ is } 160$$
$$\downarrow\downarrow\downarrow \; \downarrow\;\;\downarrow$$
$$\frac{5}{7} \cdot \square = 160 \qquad \text{Multiplication sentence}$$

$$\square = \frac{160}{\frac{5}{7}} \qquad \text{Related division sentence}$$

After working 16 hours, Carla had completed $\frac{8}{11}$ of a job. How long will it take do the total job?

$$\frac{8}{11} \text{ of the job is } 16$$
$$\downarrow \; \downarrow \;\; \downarrow \;\; \downarrow\;\downarrow$$
$$\frac{8}{11} \cdot \quad \square \quad = 16$$

$$= \frac{160}{1} \cdot \frac{7}{5}$$

$$= \frac{\overset{1}{\cancel{5}} \cdot 32 \cdot 7}{1 \cdot \underset{1}{\cancel{5}}} = 224$$

Thus, the total length of the trip was 224 miles.

Answer: 22 hr

## 3.4 EXERCISES A

*Find the reciprocal of each number.*

**1.** $\dfrac{2}{3}$     **2.** $\dfrac{4}{5}$     **3.** $\dfrac{1}{7}$     **4.** 3     **5.** 0

*Find each quotient.*

**6.** $\dfrac{1}{3} \div \dfrac{5}{6}$     **7.** $\dfrac{2}{3} \div \dfrac{1}{9}$     **8.** $\dfrac{2}{3} \div \dfrac{4}{9}$     **9.** $\dfrac{3}{7} \div \dfrac{9}{28}$

**10.** $7 \div \dfrac{14}{3}$     **11.** $\dfrac{5}{4} \div \dfrac{1}{2}$     **12.** $\dfrac{20}{9} \div \dfrac{10}{15}$     **13.** $\dfrac{8}{11} \div 4$

**14** $\dfrac{13}{15} \div \dfrac{39}{5}$     **15.** $\dfrac{2}{7} \div \dfrac{2}{7}$     **16.** $\dfrac{4}{7} \div 44$     **17.** $\dfrac{20}{27} \div \dfrac{35}{36}$

*Find the missing number.*

**18.** $\dfrac{3}{8} \cdot \square = 18$     **19.** $\dfrac{11}{2} \cdot \square = 33$     **20.** $\square \cdot \dfrac{7}{3} = 35$

**21.** $\dfrac{10}{7}$ of $\square$ is 20     **22** $\dfrac{10}{9}$ of $\square$ is 25     **23.** $\dfrac{1}{8}$ of $\square$ is $\dfrac{2}{9}$

**24.** $\dfrac{2}{11}$ of $\square$ is 10     **25.** $\dfrac{8}{1}$ of $\square$ is 56     **26.** $\dfrac{10}{9}$ of $\square$ is $\dfrac{8}{3}$

*Solve.*

**27.** A piece of rope 10 m long is to be cut into pieces each of length $\frac{2}{3}$ m. How many pieces can be cut?

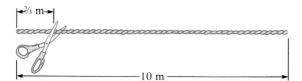

**28** After driving 365 miles, the Johnsons had completed $\frac{5}{6}$ of their trip. What was the total length of the trip?

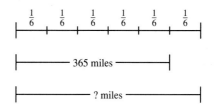

**29.** A piece of wire $\frac{7}{8}$ m long is to be cut into 14 equal pieces. What is the length of each piece?

**30.** How many pills, each containing $\frac{9}{2}$ grain of an anti-biotic, must be taken to make an 18-grain dosage?

## FOR REVIEW

**31.** The value of a house is $75,000. For insurance purposes, the value of the contents of the house are estimated to be $\frac{1}{5}$ of the value of the house. What is the estimated value of the contents? [*Hint:* $\frac{1}{5}$ *of* value of house = value of contents. What does *of* translate to?]

**32.** Pam receives an inheritance of $8000. If she invests $\frac{3}{16}$ of this amount in bonds and $\frac{5}{8}$ of it in a mutual fund, how much does she invest in each category?

*Find each product.*

**33.** $\frac{4}{11} \cdot \frac{121}{32}$

**34.** $13 \cdot \frac{7}{169}$

**35.** $\frac{2}{7} \cdot \frac{28}{3} \cdot \frac{9}{4}$

---

ANSWERS: 1. $\frac{3}{2}$  2. $\frac{5}{4}$  3. 7  4. $\frac{1}{3}$  5. 0 has no reciprocal  6. $\frac{2}{5}$  7. 6  8. $\frac{3}{2}$  9. $\frac{4}{3}$  10. $\frac{3}{2}$  11. $\frac{5}{2}$  12. $\frac{10}{3}$  13. $\frac{2}{11}$
14. $\frac{1}{9}$  15. 1  16. $\frac{1}{77}$  17. $\frac{16}{21}$  18. 48  19. 6  20. 15  21. 14  22. $\frac{45}{2}$  23. $\frac{16}{9}$  24. 55  25. 7  26. $\frac{12}{5}$  27. **15 pieces**
28. **438 mi**  29. $\frac{1}{16}$ m  30. **4 pills**  31. $15,000  32. $1500 in bonds; $5000 in the mutual fund  33. $\frac{11}{8}$  34. $\frac{7}{13}$
35. **6**

## 3.4 EXERCISES B

*Find the reciprocal of each number.*

**1.** $\frac{4}{7}$

**2.** $\frac{11}{3}$

**3.** $\frac{1}{15}$

**4.** 6

**5.** $\frac{100}{99}$

*Find each quotient.*

**6.** $\frac{1}{5} \div \frac{2}{15}$

**7.** $\frac{3}{4} \div \frac{1}{8}$

**8.** $\frac{2}{3} \div \frac{1}{6}$

**9.** $\frac{4}{9} \div \frac{1}{3}$

**10.** $9 \div \dfrac{18}{7}$  **11.** $\dfrac{7}{4} \div \dfrac{1}{2}$  **12.** $\dfrac{5}{7} \div \dfrac{15}{28}$  **13.** $\dfrac{8}{13} \div 4$

**14.** $\dfrac{13}{17} \div \dfrac{39}{34}$  **15.** $\dfrac{4}{3} \div \dfrac{4}{3}$  **16.** $\dfrac{6}{11} \div 36$  **17.** $\dfrac{20}{16} \div \dfrac{10}{8}$

*Find the missing number.*

**18.** $\dfrac{2}{9} \cdot \square = 12$  **19.** $\dfrac{13}{7} \cdot \square = 39$  **20.** $\square \cdot \dfrac{3}{20} = 27$

**21.** $\dfrac{5}{16}$ of $\square$ is 15  **22.** $\dfrac{8}{7}$ of $\square$ is 20  **23.** $\dfrac{1}{6}$ of $\square$ is $\dfrac{7}{13}$

**24.** $\dfrac{4}{15}$ of $\square$ is 30  **25.** $\dfrac{6}{1}$ of $\square$ is 24  **26.** $\dfrac{15}{7}$ of $\square$ is $\dfrac{5}{14}$

*Solve.*

**27.** Beth needs $\frac{3}{8}$ of a yard of material to make a pillow. How many pillows can she make from 18 yards of material?

**28.** A tank holds 220 gallons of fuel when it is $\frac{4}{5}$ full. What is the capacity of the tank?

**29.** If five people divide a $\frac{3}{4}$-pound box of candy, how much should each receive?

**30.** When Joey hammers a nail into a board, he sinks it $\frac{2}{9}$ of an inch with each blow. How many times will he have to hit a 4-inch nail to drive it completely into the board?

## FOR REVIEW

**31.** The day before spring break, only $\frac{3}{5}$ of the students come to Professor Condon's class. If she has 60 students enrolled, how many are in attendance?

**32.** A container contains $\frac{2}{3}$ of a gallon of milk when full. How much milk is in the container when it is $\frac{1}{4}$ full?

*Find each product.*

**33.** $\dfrac{5}{13} \cdot \dfrac{52}{55}$  **34.** $\dfrac{19}{144} \cdot 12$  **35.** $\dfrac{3}{4} \cdot \dfrac{42}{6} \cdot \dfrac{5}{14}$

## 3.4 EXERCISES C

*Let a represent a nonzero fraction. Find a, by writing the related division sentence. The letter a is used in place of the symbol $\square$ in the examples.*

**1.** $5 \cdot a = 7$
$\left[\text{Answer: } \frac{7}{5}\right]$

**2.** $\dfrac{2}{3} \cdot a = \dfrac{14}{9}$

**3.** $\dfrac{a}{\frac{3}{4}} = \dfrac{1}{3}$
$\left[\text{Answer: } \frac{1}{4}\right]$

## 3.5 MIXED NUMBERS

### 1 MIXED NUMBERS

Many of the answers to word problems are improper fractions which may look a bit strange. For example, $\frac{9}{2}$ miles is not as familiar to us as $4\frac{1}{2}$ miles. For this reason, improper fractions are often written as *mixed numbers*.

A **mixed number** is the sum of a whole number and a proper fraction. The improper fraction $\frac{7}{6}$ can be represented as in Figure 3.17.

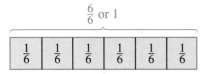  

**Figure 3.17**

Since all 6 parts of the first bar and 1 additional part of the second are shaded,

$$\frac{7}{6} \quad \text{is} \quad 1 + \frac{1}{6}. \qquad \text{1 bar} + \tfrac{1}{6} \text{ of another bar}$$

To write $1 + \dfrac{1}{6}$ as a mixed number, omit the plus sign and write

$$1\frac{1}{6}, \text{ read ``one and one sixth.''}$$

### 2 CHANGING AN IMPROPER FRACTION TO A MIXED NUMBER

To change an improper fraction to a mixed number, think of the improper fraction as a division problem. For example, $\frac{7}{6}$ is

$$7 \div 6.$$

Perform the division.

$$\text{divisor} \longrightarrow 6\overline{)7} \begin{array}{l} 1 \longleftarrow \text{quotient} \\ \phantom{6)7} \longleftarrow \text{dividend} \\ \underline{6} \\ 1 \longleftarrow \text{remainder} \end{array}$$

Recall that this division problem can be checked as follows:

$$7 = 6 \cdot 1 + 1 \qquad (\text{dividend}) = (\text{divisor}) \cdot (\text{quotient}) + (\text{remainder})$$

Dividing through by 6, $\dfrac{7}{6} = \dfrac{\cancel{6} \cdot 1}{\cancel{6}} + \dfrac{1}{6} = 1 + \dfrac{1}{6} = 1\dfrac{1}{6}.$

This is a way to obtain a mixed number from an improper fraction.

| To Change an Improper Fraction to a Mixed Number |
| --- |

1. Reduce the fraction to lowest terms (if necessary).
2. Divide the numerator by the denominator to obtain the quotient and remainder.
3. The mixed number is

$$\text{quotient} + \frac{\text{remainder}}{\text{divisor}}.$$

| EXAMPLE 1 CHANGING IMPROPER FRACTIONS TO MIXED NUMBERS | PRACTICE EXERCISE 1 |
| --- | --- |

Change each improper fraction to a mixed number.

**(a)** $\dfrac{15}{4}$

$$\text{divisor} \longrightarrow 4\overline{)15} \begin{array}{l} 3 \longleftarrow \text{quotient} \\ \phantom{4)1}\longleftarrow \text{dividend} \\ \underline{12} \\ 3 \longleftarrow \text{remainder} \end{array}$$

Thus, the mixed number is $3\frac{3}{4}$, read "three and three fourths."

**(b)** $\dfrac{121}{35}$

$$\text{divisor} \longrightarrow 35\overline{)121} \begin{array}{l} 3 \longleftarrow \text{quotient} \\ \phantom{35)12}\longleftarrow \text{dividend} \\ \underline{105} \\ 16 \longleftarrow \text{remainder} \end{array}$$

Thus, $\dfrac{121}{35} = 3\dfrac{16}{35}$.

**(c)** $\dfrac{27}{6}$

In this case, first reduce $\frac{27}{6}$ to lowest terms.

$$\frac{27}{6} = \frac{\overset{1}{\cancel{3}} \cdot 9}{\underset{1}{\cancel{3}} \cdot 2} = \frac{9}{2} \qquad 2\overline{)9}\begin{array}{l}4\\ \underline{8}\\1\end{array}$$

Thus, $\dfrac{27}{6} = \dfrac{9}{2} = 4\dfrac{1}{2}$.

Change each improper fraction to a mixed number.

**(a)** $\dfrac{23}{7}$

**(b)** $\dfrac{114}{23}$

**(c)** $\dfrac{35}{15}$

Answers: (a) $3\frac{2}{7}$  (b) $4\frac{22}{23}$
(c) $2\frac{1}{3}$

## ③ CHANGING A MIXED NUMBER TO AN IMPROPER FRACTION

We now want to change a mixed number into an improper fraction. From Figure 3.18, we can see that the mixed number $3\frac{1}{2}$ is the same as the improper fraction $\frac{7}{2}$.

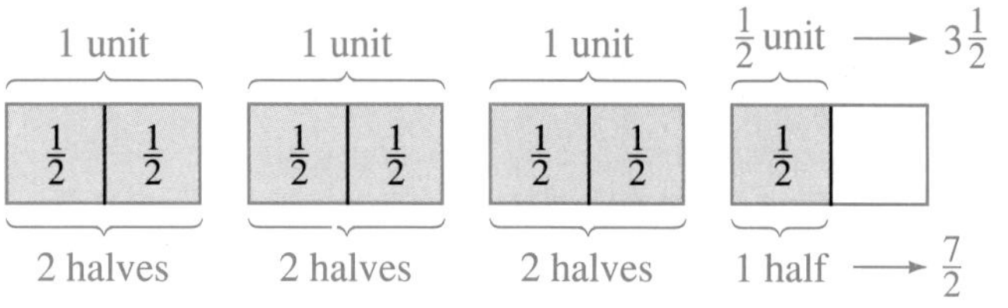

**Figure 3.18**

Each whole unit has 2 halves and there are 3 whole units plus 1 half unit.

$$2 \cdot 3 + 1 = 7.$$

Putting this result over the denominator, 2, gives the improper fraction $\frac{7}{2}$. The same result can be obtained by the following procedure.

$$3\,\frac{1}{2}\ \text{over} = \frac{2 \cdot 3 + 1}{2} = \frac{7}{2}$$

| To Change a Mixed Number to an Improper Fraction |
| --- |

1. Multiply the denominator by the whole number.
2. Add the result to the numerator of the fraction.
3. This sum, over the denominator, is the improper fraction.

---

**EXAMPLE 2   CHANGING MIXED NUMBERS TO IMPROPER FRACTIONS**

Change each mixed number to an improper fraction.

**(a)** $3\frac{1}{4}$

$$3\,\frac{1}{4}\ \text{over} = \frac{4 \cdot 3 + 1}{4} = \frac{12 + 1}{4} = \frac{13}{4}$$

**(b)** $17\frac{2}{3}$

$$17\,\frac{2}{3} = \frac{3 \cdot 17 + 2}{3} = \frac{51 + 2}{3} = \frac{53}{3}$$

**PRACTICE EXERCISE 2**

Change each mixed number to an improper fraction.

**(a)** $7\frac{2}{5}$

**(b)** $22\frac{1}{4}$

Answers: (a) $\frac{37}{5}$   (b) $\frac{89}{4}$

---

## 3.5 EXERCISES A

*Change each improper fraction to a mixed number.*

**1.** $\frac{7}{5}$

**2.** $\frac{17}{3}$

**3.** $\frac{23}{8}$

**4.** $\dfrac{45}{6}$

**5.** $\dfrac{71}{10}$

**6.** $\dfrac{19}{2}$

**7.** $\dfrac{135}{4}$

**8** $\dfrac{257}{9}$

**9.** $\dfrac{142}{11}$

*Change each mixed number to an improper fraction.*

**10.** $3\dfrac{1}{3}$

**11.** $4\dfrac{3}{5}$

**12.** $7\dfrac{1}{8}$

**13.** $10\dfrac{3}{4}$

**14.** $9\dfrac{7}{11}$

**15.** $21\dfrac{2}{3}$

**16.** $5\dfrac{7}{8}$

**17** $32\dfrac{7}{10}$

**18.** $66\dfrac{2}{3}$

## FOR REVIEW

*Solve.*

**19.** A tank holds 560 gallons of water when it is $\frac{5}{8}$ full. What is the capacity of the tank?

**20.** How many steaks each weighing $\frac{3}{8}$ pound can be cut from a 15-pound roast?

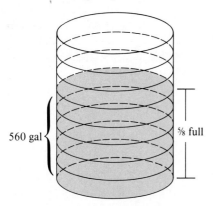

560 gal   ⅝ full

*Find each quotient.*

**21.** $\dfrac{10}{21} \div \dfrac{20}{14}$

**22.** $\dfrac{5}{13} \div 10$

**23.** $38 \div \dfrac{19}{2}$

---

ANSWERS:  1. $1\frac{2}{5}$  2. $5\frac{2}{3}$  3. $2\frac{7}{8}$  4. $7\frac{1}{2}$  5. $7\frac{1}{10}$  6. $9\frac{1}{2}$  7. $33\frac{3}{4}$  8. $28\frac{5}{9}$  9. $12\frac{10}{11}$  10. $\frac{10}{3}$  11. $\frac{23}{5}$  12. $\frac{57}{8}$  13. $\frac{43}{4}$  14. $\frac{106}{11}$  15. $\frac{65}{3}$  16. $\frac{47}{8}$  17. $\frac{327}{10}$  18. $\frac{200}{3}$  19. 896 gallons  20. 40 steaks  21. $\frac{1}{3}$  22. $\frac{1}{26}$  23. 4

## 3.5 EXERCISES B

*Change each improper fraction to a mixed number.*

**1.** $\dfrac{8}{5}$

**2.** $\dfrac{19}{3}$

**3.** $\dfrac{27}{8}$

**4.** $\dfrac{51}{6}$

**5.** $\dfrac{86}{10}$

**6.** $\dfrac{17}{2}$

**7.** $\dfrac{173}{4}$

**8.** $\dfrac{269}{9}$

**9.** $\dfrac{312}{11}$

*Change each mixed number to an improper fraction.*

**10.** $3\dfrac{2}{3}$

**11.** $5\dfrac{2}{5}$

**12.** $6\dfrac{3}{8}$

**13.** $12\dfrac{1}{4}$

**14.** $11\dfrac{7}{11}$

**15.** $26\dfrac{1}{3}$

**16.** $3\dfrac{7}{8}$

**17.** $4\dfrac{7}{10}$

**18.** $76\dfrac{2}{3}$

### FOR REVIEW

*Solve.*

**19.** A wire 20 inches long is to be cut into pieces each of length $\dfrac{5}{8}$ inch. How many pieces can be cut?

**20.** After hiking 26 miles, the campers had completed $\dfrac{2}{3}$ of a hike. What was the total length of the hike?

*Find each quotient.*

**21.** $\dfrac{8}{15} \div \dfrac{22}{6}$

**22.** $\dfrac{9}{11} \div 18$

**23.** $108 \div \dfrac{12}{7}$

## 3.5 EXERCISES C

*Change each improper fraction to a mixed number.*

**1.** $\dfrac{1055}{7}$

**2.** $\dfrac{2615}{23}$

$\left[\text{Answer: } 113\dfrac{16}{23}\right]$

*Change each mixed number to an improper fraction.*

**3.** $201\dfrac{3}{17}$

**4.** $463\dfrac{51}{103}$

$\left[\text{Answer: } \dfrac{47,740}{103}\right]$

# 3.6 MULTIPLYING AND DIVIDING MIXED NUMBERS

## STUDENT GUIDEPOSTS

**1** Method for Multiplying and Dividing Mixed Numbers

**2** Applications of Mixed Numbers

## 1 METHOD FOR MULTIPLYING AND DIVIDING MIXED NUMBERS

Mixed numbers can be multiplied and divided just like proper and improper fractions. Change any mixed numbers to improper fractions and proceed as in Sections 3.3 and 3.4.

### To Multiply or Divide Mixed Numbers

1. Change any mixed numbers to improper fractions.
2. Proceed as when multiplying or dividing fractions.

---

| EXAMPLE 1  MULTIPLYING MIXED NUMBERS | PRACTICE EXERCISE 1 |
|---|---|

Multipy.

Multiply.

**(a)** $2\frac{1}{2} \cdot 3\frac{1}{4}$

**(a)** $1\frac{1}{5} \cdot 5\frac{2}{7}$

First change to improper fractions.

$$2\frac{1}{2} = \frac{2 \cdot 2 + 1}{2} = \frac{5}{2} \quad \text{and} \quad 3\frac{1}{4} = \frac{4 \cdot 3 + 1}{4} = \frac{13}{4}$$

$$2\frac{1}{2} \cdot 3\frac{1}{4} = \frac{5}{2} \cdot \frac{13}{4} = \frac{65}{8} = 8\frac{1}{8} \quad \text{Change improper fraction to a mixed number}$$

**(b)** $12 \cdot 10\frac{2}{9}$

**(b)** $7\frac{3}{8} \cdot 4 = \frac{8 \cdot 7 + 3}{8} \cdot \frac{4}{1} = \frac{59}{8} \cdot \frac{4}{1} = \frac{59 \cdot \overset{1}{\cancel{4}}}{2 \cdot \underset{1}{\cancel{4}} \cdot 1}$

$$= \frac{59}{2} = 29\frac{1}{2}$$

**(c)** $\left(5\frac{1}{4}\right)^2$

**(c)** $\left(7\frac{2}{3}\right)^2$

Find $\left(5\frac{1}{4}\right)\left(5\frac{1}{4}\right)$.

$$5\frac{1}{4} = \frac{4 \cdot 5 + 1}{4} = \frac{21}{4} \quad \text{Change to improper fraction}$$

$$\left(5\frac{1}{4}\right)^2 = \left(\frac{21}{4}\right)^2 = \left(\frac{21}{4}\right)\left(\frac{21}{4}\right)$$

$$= \frac{21 \cdot 21}{4 \cdot 4} = \frac{441}{16} = 27\frac{9}{16}$$

**Answers:** (a) $6\frac{12}{35}$  (b) $122\frac{2}{3}$  (c) $58\frac{7}{9}$

//////////////// **CAUTION** ////////////////

It is tempting to try to multiply

$$2\frac{1}{2} \cdot 3\frac{1}{4}$$

by multiplying the whole numbers ($2 \cdot 3 = 6$) and then multiplying the fractions $\left(\frac{1}{2} \cdot \frac{1}{4} = \frac{1}{8}\right)$. That would make the product $6\frac{1}{8}$. But from Example 1(a), this product is $8\frac{1}{8}$. Thus, the parts of mixed numbers *should not* be multiplied or divided separately.

////////////

---

### EXAMPLE 2   DIVIDING MIXED NUMBERS

Divide.

**(a)** $4\frac{2}{3} \div 3\frac{1}{2}$

First change to improper fractions.

$$4\frac{2}{3} = \frac{3 \cdot 4 + 2}{3} = \frac{14}{3} \quad \text{and} \quad 3\frac{1}{2} = \frac{2 \cdot 3 + 1}{2} = \frac{7}{2}$$

$$4\frac{2}{3} \div 3\frac{1}{2} = \frac{14}{3} \div \frac{7}{2} = \frac{14}{3} \cdot \frac{2}{7} \qquad \text{Multiply by } \tfrac{2}{7}, \text{ the reciprocal of } \tfrac{7}{2}$$

$$= \frac{2 \cdot 7}{3} \cdot \frac{2}{7}$$

$$= \frac{2 \cdot \overset{1}{7} \cdot 2}{3 \cdot \underset{1}{7}} = \frac{4}{3} = 1\frac{1}{3}$$

**(b)** $5\frac{3}{5} \div 2 = \frac{28}{5} \div 2$

$$= \frac{28}{5} \cdot \frac{1}{2} = \frac{\overset{1}{2} \cdot 14 \cdot 1}{5 \cdot \underset{1}{2}} = \frac{14}{5} = 2\frac{4}{5}$$

**(c)** $7 \div 8\frac{2}{5} = 7 \div \frac{42}{5} = 7 \cdot \frac{5}{42} = \frac{\overset{1}{7} \cdot 5}{\underset{1}{7} \cdot 6} = \frac{5}{6}$

### PRACTICE EXERCISE 2

Divide.

**(a)** $2\frac{4}{5} \div 1\frac{3}{4}$

**(b)** $10\frac{5}{8} \div 5$

**(c)** $4 \div 7\frac{1}{3}$

Answers:  (a) $1\frac{3}{5}$   (b) $2\frac{1}{8}$   (c) $\frac{6}{11}$

---

## ② APPLICATIONS OF MIXED NUMBERS

Many applied problems can be expressed better with mixed numbers than with improper fractions. For example, we would normally say that Michael can hike $4\frac{1}{2}$ miles and not that he can hike $\frac{9}{2}$ miles.

| EXAMPLE 3   APPLICATION TO RECREATION | PRACTICE EXERCISE 3 |
|---|---|

Michael can hike $4\frac{1}{2}$ miles in one hour. At the same rate, how far can he hike in 6 hours?

One bag of fruit weighs $8\frac{3}{4}$ pounds. How much would 12 of these bags weigh?

Since he can go $4\frac{1}{2}$ miles in one hour, he can go 6 times that distance in 6 hours.

$$6 \cdot 4\frac{1}{2} = 6 \cdot \frac{9}{2} \quad \text{Change to an improper fraction}$$

$$= \frac{6}{1} \cdot \frac{9}{2}$$

$$= \frac{\overset{1}{\cancel{2}} \cdot 3 \cdot 9}{\underset{1}{\cancel{2}}} = 27$$

Thus, Michael can go 27 miles in 6 hours.

Answer:  105 pounds

| EXAMPLE 4   APPLICATION TO SEWING | PRACTICE EXERCISE 4 |
|---|---|

A uniform requires $2\frac{5}{8}$ yards of material. How many uniforms can be made from $57\frac{3}{4}$ yards of material?

How many $2\frac{2}{5}$-lb packages of fish can be obtained from 132 lb of fish?

Since each uniform requires $2\frac{5}{8}$ yards, then $2\frac{5}{8}$ times the number of uniforms must be $57\frac{3}{4}$.

$$2\frac{5}{8} \cdot \square = 57\frac{3}{4}$$

$$\square = 57\frac{3}{4} \div 2\frac{5}{8} \quad \text{Related division sentence}$$

$$= \frac{231}{4} \div \frac{21}{8}$$

$$= \frac{231}{4} \cdot \frac{8}{21}$$

$$= \frac{11 \cdot \overset{1}{\cancel{21}} \cdot \overset{1}{\cancel{4}} \cdot 2}{\underset{1}{\cancel{4}} \cdot \underset{1}{\cancel{21}}} = 22$$

Thus, 22 uniforms can be made.

Answer:  55 packages

## 3.6  EXERCISES A

*In the following exercises, give all answers as mixed numbers.*

*Multiply.*

**1.** $3\frac{1}{5} \cdot 4\frac{2}{3}$

**2.** $5\frac{4}{7} \cdot 2\frac{2}{13}$

**3.** $12\frac{2}{9} \cdot 4\frac{1}{5}$

**4.** $6\dfrac{4}{5} \cdot 11$

**5.** $\left(3\dfrac{1}{2}\right)^2$

**6.** $\left(2\dfrac{2}{5}\right)^2$

**7.** $3 \cdot 1\dfrac{1}{2} \cdot 2\dfrac{1}{3}$

**8** $2\dfrac{3}{5} \cdot 6\dfrac{1}{4} \cdot 12$

**9.** $11\dfrac{1}{2} \cdot 5\dfrac{1}{4} \cdot \dfrac{2}{7}$

*Divide.*

**10.** $3\dfrac{1}{6} \div 2\dfrac{1}{2}$

**11.** $8\dfrac{5}{6} \div 4\dfrac{1}{9}$

**12.** $3 \div 7\dfrac{4}{5}$

**13.** $16\dfrac{1}{2} \div 8$

**14.** $\dfrac{7}{8} \div 5\dfrac{1}{4}$

**15.** $10\dfrac{1}{5} \div \dfrac{3}{10}$

**16.** $9\dfrac{2}{7} \div 1\dfrac{2}{3}$

**17.** $3\dfrac{1}{8} \div 30$

**18.** $80 \div 4\dfrac{2}{7}$

*Solve.*

**19.** A wheel on a child's car is turning at a rate of $105\dfrac{1}{2}$ revolutions per minute. If the car is driven for $14$ minutes, how many revolutions does the wheel make?

**20.** Cindy is paid $1\dfrac{3}{4}$ dollars per hour for babysitting. If she sits for a period of $4\dfrac{1}{2}$ hours, how much is she paid?

**21.** A car traveled 420 miles on $12\dfrac{4}{5}$ gallons of gas. How many miles per gallon did it get?

**22** A pompon girl's uniform requires $3\dfrac{1}{8}$ yards of material. How many uniforms can be made from 25 yards of material?

**23.** The weight of one cubic foot of water is $62\dfrac{1}{2}$ lb. How much does $3\dfrac{1}{3}$ cubic feet of water weigh?

**24.** How many $5\dfrac{1}{2}$-gram samples can be obtained from 176 grams of a chemical?

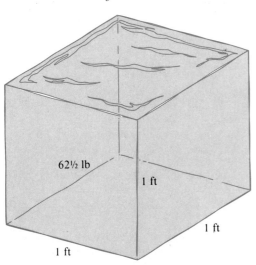

62½ lb

1 ft

1 ft

1 ft

1 cubic foot of water

**FOR REVIEW**

*Change each improper fraction to a mixed number.*

**25.** $\dfrac{9}{7}$

**26.** $\dfrac{18}{5}$

**27.** $\dfrac{123}{15}$

*Change each mixed number to an improper fraction.*

**28.** $5\dfrac{2}{11}$

**29.** $10\dfrac{6}{13}$

**30.** $81\dfrac{3}{20}$

---

ANSWERS:  1. $14\frac{14}{15}$  2. 12  3. $51\frac{1}{3}$  4. $74\frac{4}{5}$  5. $12\frac{1}{4}$  6. $5\frac{19}{25}$  7. $10\frac{1}{2}$  8. 195  9. $17\frac{1}{4}$  10. $1\frac{4}{15}$  11. $2\frac{11}{74}$  12. $\frac{5}{13}$
13. $2\frac{1}{16}$  14. $\frac{1}{6}$  15. 34  16. $5\frac{4}{7}$  17. $\frac{5}{48}$  18. $18\frac{2}{3}$  19. 1477 revolutions  20. $7\frac{7}{8}$ dollars  21. $32\frac{13}{16}$ mpg  22. 8 uniforms  23. $208\frac{1}{3}$ lb  24. 32 samples  25. $1\frac{2}{7}$  26. $3\frac{3}{5}$  27. $8\frac{1}{5}$  28. $\frac{57}{11}$  29. $\frac{136}{13}$  30. $\frac{1623}{20}$

## 3.6 EXERCISES B

*In the following exercises, give all answers as mixed numbers.*

*Multiply.*

**1.** $1\dfrac{3}{4} \cdot 2\dfrac{1}{7}$

**2.** $6\dfrac{1}{8} \cdot 3\dfrac{3}{14}$

**3.** $2\dfrac{2}{5} \cdot 8\dfrac{1}{3}$

**4.** $5\dfrac{2}{9} \cdot 12$

**5.** $\left(2\dfrac{1}{3}\right)^2$

**6.** $\left(6\dfrac{1}{2}\right)^2$

**7.** $4 \cdot 2\dfrac{1}{3} \cdot 5\dfrac{1}{4}$

**8.** $1\dfrac{1}{10} \cdot 5\dfrac{5}{7} \cdot 14$

**9.** $\dfrac{2}{11} \cdot 7\dfrac{1}{3} \cdot 10\dfrac{1}{4}$

*Divide.*

**10.** $5\dfrac{5}{6} \div 1\dfrac{1}{2}$

**11.** $7\dfrac{4}{5} \div 3\dfrac{1}{4}$

**12.** $5 \div 7\dfrac{1}{4}$

**13.** $13\dfrac{1}{2} \div 5$

**14.** $\dfrac{4}{7} \div 2\dfrac{2}{9}$

**15.** $9\dfrac{3}{5} \div \dfrac{4}{15}$

**16.** $8\dfrac{1}{15} \div 7\dfrac{1}{5}$

**17.** $6\dfrac{3}{4} \div 60$

**18.** $20 \div 5\dfrac{5}{6}$

*Solve.*

**19.** A motor rotates at $221\dfrac{1}{2}$ revolutions per second. How many revolutions will it make in one minute?

**20.** Bill worked for $5\dfrac{1}{5}$ hours as a babysitter. How much was he paid if he received $1\dfrac{1}{4}$ dollars per hour?

**21.** A truck traveled 378 miles on $15\dfrac{2}{5}$ gallons of fuel. How many miles per gallon did it get?

**22.** How many pieces of steel rod $3\dfrac{1}{3}$ feet long can be cut from a rod which is 110 feet long?

**23.** The area of a room is its length times its width. What is the area of a room which is $8\frac{3}{4}$ yards long and $6\frac{2}{5}$ yards wide?

**24.** There are 26 pieces of candy of the same size in a box. The total weight of the candy is $5\frac{1}{5}$ pounds. How much does each piece of candy weigh?

## FOR REVIEW

*Change each improper fraction to a mixed number.*

**25.** $\dfrac{11}{7}$

**26.** $\dfrac{21}{4}$

**27.** $\dfrac{218}{12}$

*Change each mixed number to an improper fraction.*

**28.** $4\dfrac{1}{12}$

**29.** $11\dfrac{3}{5}$

**30.** $101\dfrac{1}{8}$

## 3.6 EXERCISES C

*Perform the indicated operations.*

**1.** $36\dfrac{2}{13} \cdot 27\dfrac{5}{8}$

**2.** $105\dfrac{7}{25} \div 62\dfrac{8}{15}$

$\left[\text{Answer: } 1\frac{229}{335}\right]$

## 3.7 MORE APPLICATIONS

This section considers a variety of applications of fractions and mixed numbers. Having different types of problems in the same section will give practice at deciding which operation should be used.

---

| EXAMPLE 1 APPLICATION TO CONSTRUCTION | PRACTICE EXERCISE 1 |

A rectangular room is $12\frac{1}{2}$ feet long and $10\frac{1}{5}$ feet wide. How many square feet of tile are needed to cover the floor of the room? If tile costs $1\frac{1}{5}$ dollars for one square foot, how much will it cost to tile the floor?

Since the area of the floor is the length of the room times its width, this problem requires multiplication. See Figure 3.19.

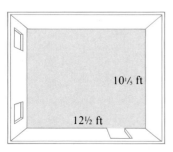

$$\text{Area} = \text{length} \cdot \text{width}$$

$$= 12\frac{1}{2} \cdot 10\frac{1}{5}$$

$$= \frac{25}{2} \cdot \frac{51}{5}$$

**Figure 3.19**

A rectangular room is $6\frac{2}{3}$ yards long and $4\frac{1}{4}$ yards wide. What is the area of the room? If carpet costs $16\frac{1}{2}$ dollars per square yard installed, how much will it cost to carpet the room?

$$= \frac{\overset{5}{\cancel{25}}}{2} \cdot \frac{51}{\underset{1}{\cancel{5}}}$$

$$= \frac{255}{2} = 127\frac{1}{2}$$

Thus, $127\frac{1}{2}$ square feet of tile are required. The total cost is found by multiplying the number of square feet by the cost for each square foot.

$$127\frac{1}{2} \cdot 1\frac{1}{5} = \frac{255}{2} \cdot \frac{6}{5}$$

$$= \frac{51 \cdot \overset{1}{\cancel{6}} \cdot \overset{1}{\cancel{2}} \cdot 3}{\underset{1}{\cancel{2}} \cdot \underset{1}{\cancel{5}}}$$

$$= 153$$

The cost to tile the floor is 153 dollars.

Answer: $28\frac{1}{3}$ square yards, $467\frac{1}{2}$ dollars

---

| EXAMPLE 2 HOUSEHOLD APPLICATION |
|---|

Howard bought $27\frac{1}{3}$ pounds of nuts. If he put $18\frac{2}{3}$ pounds in the freezer, what fraction of the nuts did he freeze?

To find a fraction of the nuts, divide the number of pounds frozen by the total number of pounds of nuts.

$$18\frac{2}{3} \div 27\frac{1}{3} = \frac{56}{3} \div \frac{82}{3} \quad \text{Change to improper fractions}$$

$$= \frac{56}{3} \cdot \frac{3}{82}$$

$$= \frac{28 \cdot \overset{1}{\cancel{2}} \cdot \overset{1}{\cancel{3}}}{\underset{1}{\cancel{3}} \cdot \underset{1}{\cancel{2}} \cdot 41} = \frac{28}{41}$$

The fraction of the nuts which were frozen is $\frac{28}{41}$.

---

| PRACTICE EXERCISE 2 |
|---|

After working for several hours, Andrew had processed $42\frac{1}{2}$ pounds of an 80-pound shipment of cheese. What fraction of the cheese had he processed?

Answer: $\frac{17}{32}$

---

| EXAMPLE 3 MARKETING APPLICATION |
|---|

A fruit market sells grapes in bags. If $3\frac{1}{3}$ pounds of grapes are put into each bag, how many bags can be prepared from a shipment of 120 pounds of grapes?

If we knew the number of bags we could multiply $3\frac{1}{3}$ times that number to obtain 120 pounds.

$$3\frac{1}{3} \cdot \square = 120$$

$$\square = 120 \div 3\frac{1}{3} \quad \text{Related division sentence}$$

$$= 120 \div \frac{10}{3}$$

$$= \frac{120}{1} \cdot \frac{3}{10}$$

---

| PRACTICE EXERCISE 3 |
|---|

When a fuel tank is $\frac{2}{5}$ full, it contains 320 gallons. How many gallons of fuel will the tank hold?

$$\frac{2}{5} \text{ of } \square \text{ is } 320$$
$$\downarrow \ \downarrow \ \downarrow \ \downarrow \ \downarrow$$
$$\frac{2}{5} \cdot \square = 320$$

$$= \frac{12 \cdot \overset{1}{\cancel{10}} \cdot 3}{1 \cdot \underset{1}{\cancel{10}}} = 36$$

Thus, 36 bags can be prepared for sale.

---

| **EXAMPLE 4** **REVOLUTIONS OF A RECORD** | **PRACTICE EXERCISE 4** |

A record revolves at a rate of $33\frac{1}{3}$ revolutions per minute. If it takes $15\frac{1}{2}$ minutes to play one song, how many revolutions does the record make during the song?

A flywheel turns at the rate of $102\frac{1}{2}$ revolutions per second. How many revolutions will it make in $5\frac{2}{5}$ seconds?

   Look at a simpler case first: If the song lasted 2 minutes, the record would make $2 \cdot 33\frac{1}{3}$ revolutions. Since the song takes $15\frac{1}{2}$ minutes, the record makes $15\frac{1}{2} \cdot 33\frac{1}{3}$ revolutions.

$$15\frac{1}{2} \cdot 33\frac{1}{3} = \frac{2 \cdot 15 + 1}{2} \cdot \frac{3 \cdot 33 + 1}{3}$$

$$= \frac{31}{2} \cdot \frac{100}{3}$$

$$= \frac{31 \cdot \overset{1}{\cancel{2}} \cdot 50}{\underset{1}{\cancel{2}} \cdot 3}$$

$$= \frac{1550}{3} = 516\frac{2}{3}$$

During the song, the record makes $516\frac{2}{3}$ revolutions.

---

   Mixed numbers give a better idea of the size of a number than improper fractions. For example, it is easier to understand

$$516\frac{2}{3} \text{ revolutions } \quad \text{than} \quad \frac{1550}{3} \text{ revolutions.}$$

## ⚠ CAUTION ⚠

It is easy to confuse the mixed number notation with multiplication. Often, when no symbol is used, multiplication is assumed. This is not true with mixed numbers.

$$516\frac{2}{3} \quad \text{is} \quad 516 + \frac{2}{3} \quad \textit{not} \quad 516 \cdot \frac{2}{3}$$

To show this,

$$516 + \frac{2}{3} = 516\frac{2}{3} = \frac{1550}{3},$$

but

$$516 \cdot \frac{2}{3} = \frac{\overset{172}{\cancel{516}} \cdot 2}{\underset{1}{\cancel{3}}} = 172 \cdot 2 = 344.$$

## 3.7 EXERCISES A

*Solve.*

**1.** A tank holds 240 gallons when it is $\frac{2}{3}$ full. How many gallons does it hold when full?

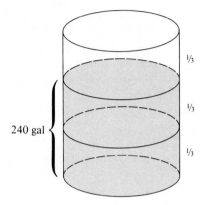

**2.** A tank holds 240 gallons when it is full. How many gallons does it contain when it is $\frac{2}{3}$ full?

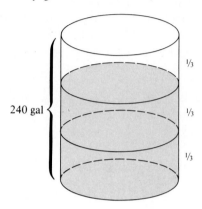

**3.** If $\frac{5}{6}$ of a class consists of boys and 42 students are in the class, how many students are boys? How many are girls?

**4.** Walter ate $2\frac{1}{5}$ pounds of a $3\frac{1}{2}$-pound box of candy the first day. What fraction of the candy had he eaten?

**5.** Twila can run $7\frac{1}{2}$ miles in one hour. At the same rate how far can she run in 2 hours?

**6.** Sam added $2\frac{1}{2}$ cups of flour to a cake recipe. He later found out that this was $1\frac{1}{2}$ times the amount that he should have added. How much should he have used?

**7.** A recipe calls for $1\frac{1}{4}$ cups of sugar. How much sugar should be used if the recipe is doubled? How much if $\frac{1}{3}$ the recipe is to be used?

**8.** A man's shirt requires $2\frac{3}{5}$ yards of material. How many shirts can be made from 52 yards of material?

**9.** A domed stadium has rectangular floor dimensions of 120 yards by 70 yards. If $\frac{9}{10}$ of the floor area is to be covered with synthetic turf, how many square yards of turf must be laid? If the turf costs $15 a square yard installed, how much will the project cost?

**10** A family has an income of $18,000 a year. If $\frac{1}{5}$ the income is spent for food, $\frac{1}{4}$ for housing, $\frac{1}{10}$ for clothing, and $\frac{3}{8}$ for taxes, how much is spent for each?

**11.** Brenda worked $10\frac{2}{3}$ hours and received $136. What was her hourly rate of pay?

**12.** A tank holds 560 gallons of water when it is full. When a leak was discovered, $\frac{5}{8}$ of the water had drained out. How much water was left in the tank?

**13** A record makes 45 revolutions per minute. How long will it take to make $321\frac{1}{2}$ revolutions?

**14.** A record makes $33\frac{1}{3}$ revolutions per minute. How many revolutions are made in $6\frac{3}{5}$ minutes?

## FOR REVIEW

*Perform the indicated operations.*

**15.** $2\frac{1}{9} \cdot 3\frac{3}{5}$

**16.** $5\frac{3}{7} \div 4\frac{2}{5}$

**17.** $\left(6\frac{2}{3}\right)^2$

**18.** $3 \cdot 5\frac{1}{7} \cdot 4\frac{2}{3}$

**19.** $9\frac{3}{5} \div 10$

**20.** $16 \div 7\frac{1}{3}$

---

ANSWERS: 1. 360 gallons  2. 160 gallons  3. 35 boys, 7 girls  4. $\frac{22}{35}$  5. 15 mi  6. $1\frac{2}{3}$ cups  7. $2\frac{1}{2}$ cups, $\frac{5}{12}$ cup  8. 20 shirts  9. 7560 sq yd of turf at a cost of $113,400  10. food: $3600; housing: $4500; clothing: $1800; taxes: $6750  11. $12\frac{3}{4}$ dollars per hour  12. 210 gallons  13. $7\frac{13}{90}$ minutes  14. 220 revolutions  15. $7\frac{3}{5}$  16. $1\frac{18}{77}$  17. $44\frac{4}{9}$  18. 72  19. $\frac{24}{25}$  20. $2\frac{2}{11}$

---

## 3.7  EXERCISES B

*Solve.*

**1.** A container is $\frac{3}{10}$ full and has 15 quarts in it. How much will it hold when full?

**2.** A container holds 15 quarts when it is full. How many quarts does it contain when it is $\frac{3}{10}$ full?

**3.** A bag contains 120 marbles which are either red or blue. If $\frac{3}{5}$ of the marbles are blue, how many of each color are in the bag?

**4.** There were $7\frac{1}{3}$ gallons of punch prepared for a party. If $6\frac{2}{5}$ gallons were drunk, what fraction of the punch was consumed?

**5.** Hank rides his bike at a rate of $12\frac{2}{3}$ miles per hour. How far can he go in $6\frac{1}{2}$ hours?

**6.** A chemist added $5\frac{1}{4}$ liters of alcohol to a solution. She later discovered that this was $2\frac{1}{2}$ times as much as she should have added. How much should have been added?

**7.** To make a party punch, $2\frac{3}{4}$ quarts of ice cream are required. How much ice cream should be added if three times the punch recipe is to be made? How much if $\frac{1}{5}$ the recipe is to be made?

**8.** A connecting wire must be $8\frac{1}{4}$ inches long. How many of these wires can be cut from a roll of wire 198 inches long?

9. A large living room is 12 yards wide and 16 yards long. If $\frac{5}{8}$ of the floor area is to be carpeted, how many square yards are required? If carpet costs $24 per square yard installed, how much will the job cost?

10. Gloria Moore has $2400 for college expenses. If she spends $\frac{1}{4}$ of the money for clothing, $\frac{1}{8}$ for books, $\frac{1}{6}$ for travel, and $\frac{2}{5}$ for tuition, how much is spent for each?

11. Nick worked $72\frac{4}{5}$ hours and received $673\frac{2}{5}$. What was his hourly rate of pay?

12. A water truck holds 250 gallons when it is full. After $\frac{3}{5}$ of the tank is drained out, how much is left in the tank?

13. A record makes $33\frac{1}{3}$ revolutions each minute. How long will it take to make $137\frac{1}{2}$ revolutions?

14. A pulley on a machine is turning at a rate of $214\frac{1}{2}$ revolutions per minute. How many revolutions will it make in $10\frac{2}{3}$ minutes?

## FOR REVIEW

*Perform the indicated operation.*

15. $5\frac{2}{3} \cdot 9\frac{3}{5}$

16. $1\frac{3}{8} \div 9\frac{3}{7}$

17. $\left(1\frac{1}{10}\right)^2$

18. $\frac{5}{8} \cdot 9\frac{3}{5} \cdot 4$

19. $12\frac{2}{3} \div 20$

20. $42 \div 11\frac{1}{3}$

## 3.7 EXERCISES C

*Solve. Let a represent a natural number.*

1. If $a$ is multiplied by $\frac{3}{4}$ the result is $\frac{1}{12}$. Find $a$.

2. If $a$ is divided by $\frac{5}{6}$ the result is $\frac{3}{25}$. Find $a$.

$\left[\text{Answer: } \frac{1}{10}\right]$

# CHAPTER 3 REVIEW

## KEY WORDS

3.1 A **fraction** is the quotient of two whole numbers.

The **numerator** is the top number in a fraction.

The **denominator** is the bottom number in a fraction.

A **proper fraction** is one whose numerator is less than its denominator.

An **improper fraction** is one whose numerator is greater than or equal to its denominator.

3.2 Fractions that are names for the same number are **equal fractions.**

A fraction is **reduced to lowest terms** when 1 is the only nonzero whole number which is a divisor of both the numerator and denominator.

3.3 The **commutative law of multiplication** states that changing the order of a product does not change the result.

The **associative law of multiplication** states that regrouping products does not change the result.

The **multiplicative identity** tells us we can multiply any number by 1 and always obtain that same number.

**3.4** Two fractions are **reciprocals** if their product is 1.

**3.5** A **mixed number** is the sum of a whole number and a proper fraction.

## KEY CONCEPTS

**3.1**  **1.** Any fraction like $\frac{0}{9}$ is equal to 0.

**2.** Any fraction like $\frac{4}{4}$ is equal to 1.

**3.2**  **1.** Fractions are equal when their cross products are equal.

**2.** When the numerator and denominator of a fraction are multiplied or divided by the same whole number (except 0), an equal fraction is formed.

**3.** To reduce a fraction to lowest terms, factor the numerator and denominator into primes and divide out all common factors.

**3.3**  **1.** When multiplying fractions, divide out common factors of the numerator and denominator first before multiplying the remaining factors.

**2.** The word *of* translates to "times" when it follows a fraction in a word problem.

**3.** Fractions satisfy the commutative and associative laws of multiplication, and 1 is the multiplicative identity.

**3.4**  **1.** The reciprocal of a fraction is found by interchanging the numerator and the denominator of the fraction.

**2.** To divide fractions, multiply the dividend by the reciprocal of the divisor.

**3.** Zero is the only number that does not have a reciprocal.

**4.** To solve some applied problems, we first write a multiplication sentence then change to the related division sentence.

**3.5**  **1.** A mixed number like $3\frac{1}{2}$ is $3 + \frac{1}{2}$ $\left(\text{not } 3 \text{ times } \frac{1}{2}\right)$.

**2.** Improper fractions can be changed to mixed numbers by dividing.

**3.** A mixed number such as $5\frac{2}{7}$ can be changed to an improper fraction as follows:

$$5\frac{2}{7} = \frac{7 \cdot 5 + 2}{7} = \frac{37}{7}$$

**3.6**  **1.** Change mixed numbers to improper fractions in order to multiply or divide.

**2.** You should not multiply whole number parts and fraction parts separately when multiplying mixed numbers.

**3.7** In solving applied problems determine which of the following is required.

**(a)** Finding a fractional part

**(b)** Multiplying fractions

**(c)** Dividing fractions

## REVIEW EXERCIS

*Part I*

**3.1**  **1.** What is the denominator of the fraction $\frac{2}{9}$?

**2.** Is the fraction $\frac{21}{2}$ a proper or an improper fraction?

**3.** Is $\frac{7}{7}$ a proper or an improper fraction?

**4.** The fraction $\frac{0}{11}$ is the same as what number?

*What fractional part of each figure is shaded?*

**5.**

**6.**

**7.**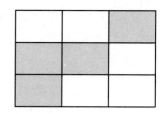

**8.** Let the figure

be one whole unit. What fraction of the unit is shaded below?

**9.** A ball player made 11 freethrows in 14 attempts during a game. The number of shots she made is what fraction of the shots that she attempted?

*Express each fraction as a whole number.*

**10.** $\dfrac{9}{9}$

**11.** $\dfrac{31}{0}$

**12.** $\dfrac{0}{15}$

**3.2** *Determine which fractions are equal by using the cross-product rule.*

**13.** $\dfrac{8}{11}$ and $\dfrac{5}{7}$

**14.** $\dfrac{3}{12}$ and $\dfrac{5}{20}$

*In each of the following, find the missing term of the fraction.*

**15.** $\dfrac{5}{8} = \dfrac{?}{24}$

**16.** $\dfrac{3}{9} = \dfrac{1}{?}$

**17.** $\dfrac{4}{?} = \dfrac{32}{8}$

*Reduce each fraction to lowest terms.*

**18.** $\dfrac{20}{30}$

**19.** $\dfrac{180}{144}$

**20.** $\dfrac{42}{28}$

**3.3** **21.** The word *of* translates to what operation when it follows a fraction in a problem?

**22.** The fact that $\frac{1}{3} \cdot \frac{7}{8} = \frac{7}{8} \cdot \frac{1}{3}$ illustrates which of the properties of multiplication?

*Find each product.*

**23.** $\dfrac{2}{3} \cdot \dfrac{18}{7}$

**24.** $5 \cdot \dfrac{3}{25}$

**25.** $\dfrac{6}{5} \cdot \dfrac{4}{7} \cdot \dfrac{5}{12}$

*Find the value of each power or root.*

**26.** $\left(\dfrac{7}{6}\right)^2$

**27.** $\sqrt{\dfrac{121}{25}}$

*Solve.*

**28.** There are 65 members of a photography club, and four-fifths of the members own a Pantex camera. How many members do not own this kind of camera?

**3.4**  **29.** What is the reciprocal of $\frac{13}{7}$?

**30.** What is the only number that does not have a reciprocal?

*Find each quotient.*

**31.** $\dfrac{2}{9} \div \dfrac{4}{27}$          **32.** $12 \div \dfrac{3}{4}$          **33.** $\dfrac{7}{3} \div 21$

*Find the missing number.*

**34.** $\dfrac{5}{7} \cdot \square = 25$          **35.** $\dfrac{2}{9}$ of $\square$ is $\dfrac{4}{25}$

*Solve.*

**36.** When a tank is $\frac{1}{6}$ full, it contains 130 gallons of water. What is the capacity of the tank?

**3.5**  **37.** What do we call the sum of a whole number and a proper fraction?          **38.** What is the understood operation between 3 and $\frac{7}{8}$ in the mixed number $3\frac{7}{8}$?

*Change to a mixed number.*

**39.** $\dfrac{127}{4}$          **40.** $\dfrac{243}{11}$          **41.** $\dfrac{97}{2}$

*Change to an improper fraction.*

**42.** $3\dfrac{9}{13}$          **43.** $21\dfrac{1}{5}$          **44.** $16\dfrac{2}{3}$

**3.6**  *Perform the indicated operations.*

**45.** $2\dfrac{3}{4} \cdot 5\dfrac{1}{8}$          **46.** $7\dfrac{2}{3} \div 4\dfrac{3}{5}$          **47.** $4\dfrac{2}{11} \cdot 8\dfrac{1}{4}$

**48.** $5 \cdot 7\dfrac{2}{3} \cdot 10\dfrac{1}{5}$          **49.** $18 \div 12\dfrac{2}{5}$          **50.** $22\dfrac{1}{2} \div 15$

*Solve.*

**51.** A wheel turns at $32\frac{1}{2}$ revolutions per second. How many revolutions will it make in $10\frac{1}{5}$ seconds?          **52.** A wheel turns at $32\frac{1}{2}$ revolutions per second. How long will it take to make 120 revolutions?

**3.3–**  *Solve.*
**3.7**

**53.** On a map 1 inch represents 35 miles. How many miles does $1\frac{3}{5}$ in represent?          **54.** A rectangular room is $2\frac{1}{8}$ yards wide and $4\frac{1}{3}$ yards long. How much will it cost to carpet the room using carpet which sells for $13\frac{1}{2}$ dollars per sq yd?

**55.** A container holds 186 gallons of fuel when it is $\frac{2}{3}$ full. What is the capacity of the container?

**56.** A container holds 186 gallons of fuel when full. If $\frac{2}{3}$ of a full tank is drained off, how much fuel remains?

**57.** A merry-go-round makes $8\frac{3}{4}$ revolutions per minute. How many revolutions will be made during a $3\frac{3}{5}$ minute ride?

**58.** An Oldsmobile Omega traveled 480 miles on $13\frac{3}{4}$ gallons of gas. How many miles per gallon did it get?

**Part II**

*Find the missing number.*

**59.** $\frac{3}{4} \cdot \square = 12$

**60.** $\frac{1}{7}$ of $\square$ is $\frac{5}{21}$

*Perform the indicated operation.*

**61.** $\left(\frac{2}{7}\right)^2$

**62.** $\sqrt{\frac{144}{4}}$

**63.** $\frac{3}{5} \cdot \frac{2}{9} \cdot \frac{5}{4}$

**64.** $\frac{16}{5} \div \frac{4}{25}$

**65.** $6\frac{2}{3} \cdot 8\frac{1}{2}$

**66.** $12\frac{3}{4} \div 4\frac{1}{2}$

*Find the missing term of each fraction.*

**67.** $\frac{5}{12} = \frac{?}{36}$

**68.** $\frac{8}{?} = \frac{24}{15}$

*Solve.*

**69.** A Buick traveled 234 miles on $10\frac{2}{5}$ gallons of gas. How many miles per gallon did it get?

**70.** On a map 1 inch represents 50 miles. How many miles are represented by $5\frac{3}{4}$ inches?

*Express each fraction as a whole number.*

**71.** $\frac{45}{45}$

**72.** $\frac{0}{21}$

*Determine which fractions are equal by using the cross-product rule.*

**73.** $\frac{5}{30}$ and $\frac{2}{12}$

**74.** $\frac{11}{9}$ and $\frac{21}{18}$

---

ANSWERS: 1. 9  2. improper  3. improper  4. 0  5. $\frac{2}{4}$ or $\frac{1}{2}$  6. $\frac{1}{3}$  7. $\frac{4}{9}$  8. $\frac{3}{2}$  9. $\frac{11}{14}$  10. 1  11. undefined  12. 0  13. $\frac{8}{11} \neq \frac{5}{7}$  14. $\frac{3}{12} = \frac{5}{20}$  15. 15  16. 3  17. 1  18. $\frac{2}{3}$  19. $\frac{5}{4}$  20. $\frac{3}{2}$  21. multiplication  22. commutative law  23. $\frac{12}{7}$ or $1\frac{5}{7}$  24. $\frac{3}{5}$  25. $\frac{2}{7}$  26. $\frac{49}{36}$  27. $\frac{11}{5}$  28. 13 members  29. $\frac{7}{13}$  30. 0  31. $\frac{3}{2}$  32. 16  33. $\frac{1}{9}$  34. 35  35. $\frac{18}{25}$  36. 780 gallons  37. mixed number  38. addition  39. $31\frac{3}{4}$  40. $22\frac{1}{11}$  41. $48\frac{1}{2}$  42. $\frac{48}{13}$  43. $\frac{106}{5}$  44. $\frac{50}{3}$  45. $14\frac{3}{32}$  46. $1\frac{2}{3}$  47. $34\frac{1}{2}$  48. 391  49. $1\frac{14}{31}$  50. $1\frac{1}{2}$  51. $331\frac{1}{2}$ revolutions  52. $3\frac{9}{13}$ seconds  53. 56 mi  54. $124\frac{5}{16}$ dollars  55. 279 gallons  56. 62 gallons  57. $31\frac{1}{2}$ revolutions  58. $34\frac{10}{11}$ mpg  59. 16  60. $\frac{5}{3}$  61. $\frac{4}{49}$  62. 6  63. $\frac{1}{6}$  64. 20  65. $56\frac{2}{3}$  66. $2\frac{5}{6}$  67. 15  68. 5  69. $22\frac{1}{2}$ mpg  70. $287\frac{1}{2}$ mi  71. 1  72. 0  73. $\frac{5}{30} = \frac{2}{12}$  74. $\frac{11}{9} \neq \frac{21}{18}$

1. What kind of fraction has the numerator less than the denominator?

1. _____

2. Which law is illustrated by $\frac{2}{3} \cdot \left(\frac{3}{4} \cdot \frac{7}{8}\right) = \left(\frac{2}{3} \cdot \frac{3}{4}\right) \cdot \frac{7}{8}$?

2. _____

3. What fractional part of the figure is shaded?

3. _____

4. Determine if $\frac{9}{2}$ and $\frac{36}{8}$ are equal by using the cross-product rule.

4. _____

*Find the missing term.*

5. $\frac{3}{5} = \frac{?}{35}$

5. _____

6. $\frac{9}{11} = \frac{27}{?}$

6. _____

7. Reduce $\frac{105}{45}$ to lowest terms.

7. _____

*Perform the indicated operations.*

8. $\frac{9}{5} \cdot \frac{20}{39}$

8. _____

9. $\frac{3}{8} \div \frac{21}{2}$

9. _____

10. $\frac{5}{8} \cdot \frac{12}{5} \cdot \frac{2}{11}$

10. _____

11. $\frac{9}{7} \div \frac{3}{35}$

11. _____

12. $\left(\frac{3}{8}\right)^2$

12. _____

13. $\sqrt{\frac{49}{144}}$

13. _____

**14.** Find the missing number.

$\frac{4}{5}$ of $\square$ is 22

14. _____

**15.** Change $\frac{48}{20}$ to a mixed number.

15. _____

**16.** Change $7\frac{2}{11}$ to an improper fraction.

16. _____

*Perform the indicated operations and give the answer as a mixed number.*

**17.** $5\frac{1}{5} \cdot 4\frac{1}{4}$

17. _____

**18.** $7\frac{1}{3} \div 9\frac{2}{9}$

18. _____

**19.** $7 \cdot 4\frac{3}{8} \cdot 10\frac{2}{7}$

19. _____

**20.** $6 \div 2\frac{4}{5}$

20. _____

*Solve.*

**21.** A container holds 18 gallons when it is $\frac{8}{9}$ full. How many gallons does it contain when it is full?

21. _____

**22.** A record makes $33\frac{1}{3}$ revolutions per minute. How many revolutions does it make in $16\frac{1}{2}$ minutes?

22. _____

# Adding and Subtracting Fractions

## 4.1 ADDING AND SUBTRACTING LIKE FRACTIONS

### STUDENT GUIDEPOSTS

1. Like and Unlike Fractions
2. Adding Like Fractions
3. Subtracting Like Fractions
4. Combination Problems and Applications

### 1 LIKE AND UNLIKE FRACTIONS

Fractions which have the same denominator are called **like fractions.** For example, the fractions

$$\frac{1}{6}, \quad \frac{5}{6}, \quad \frac{7}{6}, \quad \frac{9}{6}, \quad \frac{11}{6}$$

are like fractions. Fractions with different denominators are called **unlike fractions.** Thus,

$$\frac{2}{3}, \quad \frac{1}{5}, \quad \frac{3}{2}, \quad \frac{1}{4}, \quad \frac{7}{8}$$

are all unlike fractions.

### 2 ADDING LIKE FRACTIONS

We begin the study of addition of fractions with like fractions. If we want to add

$$\frac{1}{8} + \frac{5}{8},$$

the sum becomes obvious when the problem is rewritten as

1 eighth + 5 eighths.

Just as

1 book + 5 books = 6 books,
1 eighth + 5 eighths = 6 eighths.

That is,

$$\frac{1}{8} + \frac{5}{8} = \frac{6}{8}.$$

We simply add numerators (1 + 5) and place the result over the common denominator, 8. The sum, $\frac{6}{8}$, should then be reduced to lowest terms, $\frac{3}{4}$. The same is true in general.

| **To Add Like Fractions** |
| --- |

1. Add their numerators.
2. Place this sum over the common denominator.
3. Reduce the fraction to lowest terms, if possible.

| **EXAMPLE 1  ADDING LIKE FRACTIONS** | **PRACTICE EXERCISE 1** |
| --- | --- |

Find each sum.

**(a)** $\dfrac{3}{4} + \dfrac{7}{4} = \dfrac{3+7}{4}$   Place sum of numerators over denominator, 4

$= \dfrac{10}{4}$   Add

$= \dfrac{\cancel{2}\cdot 5}{\cancel{2}\cdot 2} = \dfrac{5}{2}$   Reduce to lowest terms

**(b)** $\dfrac{7}{30} + \dfrac{1}{30} + \dfrac{12}{30} = \dfrac{7+1+12}{30}$   Add numerators

$= \dfrac{20}{30}$

$= \dfrac{\cancel{10}\cdot 2}{\cancel{10}\cdot 3} = \dfrac{2}{3}$   Reduce

In **(b)**, we took a shortcut and did not factor numerator and denominator to primes when reducing $\frac{20}{30}$, because it was obvious that 10 was a common factor which could be divided out.

Find each sum.

**(a)** $\dfrac{2}{5} + \dfrac{1}{5}$

**(b)** $\dfrac{11}{42} + \dfrac{17}{42} + \dfrac{5}{42}$

Answers:  (a) $\frac{3}{5}$   (b) $\frac{11}{14}$

### ③ SUBTRACTING LIKE FRACTIONS

It is as easy to subtract like fractions as to add them. For example, to find

$$\frac{7}{8} - \frac{3}{8},$$

we think of

**7 eighths − 3 eighths = 4 eighths**

and subtract as we would if we were subtracting

**7 books − 3 books = 4 books.**

Thus,

$$\frac{7}{8} - \frac{3}{8} = \frac{4}{8}, \quad \text{which reduces to} \quad \frac{1}{2}.$$

### To Subtract Like Fractions

1. Subtract their numerators.
2. Place this difference over the common denominator.
3. Reduce the fraction to lowest terms, if possible.

---

**EXAMPLE 2    SUBTRACTING LIKE FRACTIONS**

Find each difference.

**(a)** $\dfrac{5}{6} - \dfrac{1}{6} = \dfrac{5-1}{6}$    Place difference of numerators over common denominator, 6

$\qquad\qquad = \dfrac{4}{6}$    Subtract

$\qquad\qquad = \dfrac{\cancel{2} \cdot 2}{\cancel{2} \cdot 3} = \dfrac{2}{3}$    Reduce to lowest terms

**(b)** $\dfrac{7}{11} - \dfrac{7}{11} = \dfrac{7-7}{11}$    Subtract numerators and place difference over common denominator, 11

$\qquad\qquad = \dfrac{0}{11} = 0$

**PRACTICE EXERCISE 2**

Find each difference.

**(a)** $\dfrac{7}{17} - \dfrac{2}{17}$

**(b)** $\dfrac{9}{15} - \dfrac{4}{15}$

Answers:  (a) $\frac{5}{17}$    (b) $\frac{1}{3}$

---

### ⚠ CAUTION

Never add or subtract fractions by <u>adding</u> or <u>subtracting</u> denominators. For example,

$$\frac{2}{5} + \frac{1}{5} \quad \textit{is not the same as} \quad \frac{2+1}{5+5} = \frac{3}{10},$$

since

$$\frac{2}{5} + \frac{1}{5} = \frac{3}{5}.$$

Also,

$$\frac{2}{5} - \frac{1}{5} \quad \textit{is not the same as} \quad \frac{2-1}{5-5},$$

which is undefined.

---

### ❹ COMBINATION PROBLEMS AND APPLICATIONS

We can find combinations of sums and differences of like fractions by adding or subtracting in order from left to right. An example will help to make this clear.

---

### EXAMPLE 3  COMBINATION PROBLEM

Perform the indicated operations.

$$\frac{4}{13} + \frac{11}{13} - \frac{3}{13} = \frac{4 + 11 - 3}{13} \qquad \text{Combine numerators as indicated}$$

$$= \frac{15 - 3}{13} \qquad \text{Add first from left}$$

$$= \frac{12}{13} \qquad \text{Subtract}$$

---

### PRACTICE EXERCISE 3

Perform the indicated operations.

$$\frac{7}{23} + \frac{9}{23} - \frac{4}{23}$$

Answer: $\frac{12}{23}$

---

### EXAMPLE 4  APPLICATION TO RECREATION

On Monday Murphy swam $\frac{9}{20}$ of a mile and on Tuesday he swam $\frac{7}{20}$ of a mile. What is the total distance that he swam on the two days?
   The total distance is the sum of the two distances $\frac{9}{20}$ and $\frac{7}{20}$.

$$\frac{9}{20} + \frac{7}{20} = \frac{9 + 7}{20} = \frac{16}{20} = \frac{4}{5}$$

Thus, he swam a total of $\frac{4}{5}$ of a mile.

---

### PRACTICE EXERCISE 4

Joe typed for $\frac{5}{4}$ hours one day and $\frac{15}{4}$ hours the next day. How many total hours did he type during the two days?

Answer: 5 hours

---

### EXAMPLE 5  APPLICATION TO ENTERTAINMENT

A recipe for a party drink calls for $\frac{3}{10}$ of a gallon of 7-Up, $\frac{7}{10}$ of a gallon of lemonade, and $\frac{1}{10}$ of a gallon of orange sherbet.

**(a)** How many gallons of drink can be made from the recipe?
   Add:

$$\frac{3}{10} + \frac{7}{10} + \frac{1}{10} = \frac{3 + 7 + 1}{10} = \frac{11}{10}.$$

Thus, the recipe will make $\frac{11}{10}$ or $1\frac{1}{10}$ gallons of drink.

**(b)** If the recipe is doubled, how much of each ingredient is needed and how much drink will be made?

Multiply the amount of each ingredient by 2.

$$2 \cdot \frac{3}{10} = \frac{2}{1} \cdot \frac{3}{10} = \frac{6}{10} = \frac{3}{5} \qquad \text{Amount of 7-Up}$$

$$2 \cdot \frac{7}{10} = \frac{2}{1} \cdot \frac{7}{10} = \frac{14}{10} = \frac{7}{5} \qquad \text{Amount of lemonade}$$

$$2 \cdot \frac{1}{10} = \frac{2}{1} \cdot \frac{1}{10} = \frac{2}{10} = \frac{1}{5} \qquad \text{Amount of orange sherbet}$$

---

### PRACTICE EXERCISE 5

A mixture of nuts calls for $\frac{1}{8}$ pound of almonds, $\frac{7}{8}$ pound of peanuts, and $\frac{5}{8}$ pound of walnuts.

**(a)** How many pounds are in the final mixture?

**(b)** If four times the recipe is to be made, how many pounds of each nut is required, and how many pounds are in the final mixture?

Adding,

$$\frac{3}{5} + \frac{7}{5} + \frac{1}{5} = \frac{3 + 7 + 1}{5} = \frac{11}{5}.$$

Thus, if the recipe is doubled, there will be $\frac{11}{5}$ or $2\frac{1}{5}$ gallons of drink which is reasonable since $2 \cdot \frac{11}{10} = \frac{2 \cdot 11}{2 \cdot 5} = \frac{11}{5}$.

---

**EXAMPLE 6**  APPLICATION TO WORK

If it takes $\frac{11}{24}$ of a day to do a job and Mike has been working for $\frac{7}{24}$ of a day, how much longer will it take him to complete the job?

Subtract the time worked from the total time required to complete the job.

$$\frac{11}{24} - \frac{7}{24} = \frac{4}{24} = \frac{1}{6}$$

Thus, it will take Mike $\frac{1}{6}$ of a day to finish the job.

**PRACTICE EXERCISE 6**

A contractor estimated that it would take about $\frac{24}{5}$ weeks to do a job. After work has gone on for $\frac{18}{5}$ weeks, how many weeks are left to complete the job?

Answer: $\frac{6}{5}$ weeks

---

## 4.1  EXERCISES A

*Find the sum.*

**1.** $\frac{2}{7} + \frac{3}{7}$

**2.** $\frac{1}{4} + \frac{5}{4}$

**3.** $\frac{3}{12} + \frac{11}{12}$

**4.** $\frac{1}{6} + \frac{5}{6}$

**5.** $\frac{3}{2} + \frac{9}{2}$

**6.** $\frac{7}{16} + \frac{13}{16}$

**7.** $\frac{3}{8} + \frac{5}{8} + \frac{7}{8}$

**8** $\frac{2}{35} + \frac{18}{35} + \frac{25}{35}$

**9.** $\frac{17}{24} + \frac{23}{24} + \frac{8}{24}$

*Find the difference.*

**10.** $\frac{7}{6} - \frac{5}{6}$

**11.** $\frac{9}{11} - \frac{3}{11}$

**12.** $\frac{5}{9} - \frac{2}{9}$

**13.** $\frac{23}{25} - \frac{8}{25}$

**14** $\frac{28}{35} - \frac{18}{35}$

**15.** $\frac{73}{81} - \frac{10}{81}$

*Perform the indicated operations.*

**16.** $\frac{2}{13} + \frac{9}{13} - \frac{5}{13}$

**17.** $\frac{9}{25} - \frac{3}{25} + \frac{4}{25}$

**18.** $\frac{22}{12} - \frac{10}{12} - \frac{6}{12}$

**19.** $\frac{8}{3} + \frac{14}{3} - \frac{1}{3}$

**20** $\frac{7}{10} - \frac{1}{10} + \frac{3}{10}$

**21.** $\frac{9}{5} - \frac{6}{5} - \frac{3}{5}$

*Solve.*

**22.** Jim hiked from Skunk Creek to Hidden Valley, a distance of $\frac{11}{15}$ mi, and from there to Indian Cave, a distance of $\frac{8}{15}$ mi. How far did he hike?

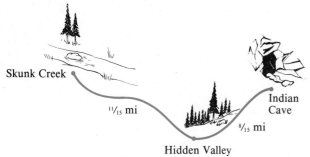

Skunk Creek

$^{11}/_{15}$ mi

Indian Cave

$^{8}/_{15}$ mi

Hidden Valley

**23** Becky had 3 days to do a job. If she did $\frac{5}{12}$ of it the first day and $\frac{3}{12}$ of it the second, how much must she do the third day to complete the job? [*Hint:* 1 job can be written as $\frac{12}{12}$ job.]

**24.** To get the exact color he wanted, Bjorn mixed $\frac{1}{12}$ of a gallon of white paint with $\frac{7}{12}$ of a gallon of yellow and $\frac{8}{12}$ of a gallon of red paint. How much paint did he get?

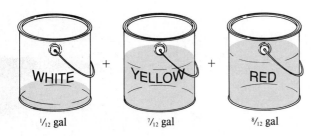

WHITE    +    YELLOW    +    RED

$^{1}/_{12}$ gal         $^{7}/_{12}$ gal         $^{8}/_{12}$ gal

**25.** If Bjorn in Exercise 24 decides to triple the amount used, how much of each color will he need, and how much paint will result?

---

ANSWERS: 1. $\frac{5}{7}$ 2. $\frac{3}{2}$ 3. $\frac{7}{6}$ 4. 1 5. 6 6. $\frac{5}{4}$ 7. $\frac{15}{8}$ 8. $\frac{9}{7}$ 9. 2 10. $\frac{1}{3}$ 11. $\frac{6}{11}$ 12. $\frac{1}{3}$ 13. $\frac{3}{5}$ 14. $\frac{2}{7}$ 15. $\frac{7}{9}$
16. $\frac{6}{13}$ 17. $\frac{2}{5}$ 18. $\frac{1}{2}$ 19. 7 20. $\frac{9}{10}$ 21. 0 22. $1\frac{4}{15}$ mi 23. $\frac{1}{3}$ of the job $\left(\frac{5}{12} + \frac{3}{12} = \frac{8}{12}, \text{ then } \frac{12}{12} - \frac{8}{12} = \frac{4}{12} = \frac{1}{3}\right)$
24. $\frac{4}{3}$ gallons 25. $\frac{3}{12} = \frac{1}{4}$ gallon of white; $\frac{21}{12} = \frac{7}{4}$ gallons of yellow; and $\frac{24}{12} = 2$ gallons of red add up to $\frac{48}{12} = 4$ gallons of paint.

## 4.1 EXERCISES B

*Find the sum.*

**1.** $\frac{3}{5} + \frac{4}{5}$

**2.** $\frac{5}{8} + \frac{3}{8}$

**3.** $\frac{2}{15} + \frac{11}{15}$

**4.** $\frac{2}{9} + \frac{4}{9}$

**5.** $\frac{7}{3} + \frac{11}{3}$

**6.** $\frac{9}{20} + \frac{3}{20}$

**7.** $\frac{1}{4} + \frac{3}{4} + \frac{5}{4}$

**8.** $\frac{5}{48} + \frac{9}{48} + \frac{1}{48}$

**9.** $\frac{7}{60} + \frac{51}{60} + \frac{2}{60}$

*Find the difference.*

**10.** $\frac{4}{9} - \frac{1}{9}$

**11.** $\frac{8}{13} - \frac{6}{13}$

**12.** $\frac{11}{15} - \frac{2}{15}$

**13.** $\dfrac{17}{12} - \dfrac{11}{12}$

**14.** $\dfrac{42}{55} - \dfrac{17}{55}$

**15.** $\dfrac{72}{105} - \dfrac{37}{105}$

*Perform the indicated operations.*

**16.** $\dfrac{7}{17} + \dfrac{5}{17} - \dfrac{6}{17}$

**17.** $\dfrac{10}{30} - \dfrac{7}{30} + \dfrac{2}{30}$

**18.** $\dfrac{14}{6} - \dfrac{5}{6} - \dfrac{1}{6}$

**19.** $\dfrac{7}{15} + \dfrac{17}{15} - \dfrac{11}{15}$

**20.** $\dfrac{23}{28} - \dfrac{20}{28} + \dfrac{3}{28}$

**21.** $\dfrac{8}{12} - \dfrac{2}{12} - \dfrac{5}{12}$

*Solve.*

**22.** There was $\frac{5}{24}$ of a gallon of water in a container. How many gallons were in the container after $\frac{7}{24}$ gallon was added?

**23.** It takes $\frac{25}{36}$ of an hour to do a job. Beth worked $\frac{7}{36}$ of an hour one day and $\frac{11}{36}$ of an hour the next day. How many hours would she need to work the third day to complete the job?

**24.** Nancy has $\frac{1}{9}$ gallon of white paint, $\frac{7}{9}$ gallon of blue paint, and $\frac{5}{9}$ gallon of green paint. How much paint would she have if she mixed them all together?

**25.** If Nancy in Exercise 24 decided that she only wants $\frac{1}{4}$ of the amount of each color, how much of each would she mix and how much paint would result?

## 4.1 EXERCISES C

*Let a represent a natural number. Find the sum or difference.*

**1.** $\dfrac{16}{a} + \dfrac{11}{a}$

$\left[\text{Answer: } \dfrac{27}{a}\right]$

**2.** $\dfrac{16}{a} - \dfrac{11}{a}$

**3.** Burford added $\frac{1}{4}$ and $\frac{3}{4}$ as follows.

$$\frac{1}{4} + \frac{3}{4} = \frac{1+3}{4+4} = \frac{4}{8} = \frac{1}{2}$$

What is wrong with his work?

## 4.2 LEAST COMMON MULTIPLES

=========== STUDENT GUIDEPOSTS ===========

**1** Listing Method for Finding the LCM

**3** Special Algorithm for Finding the LCM

**2** Prime Factorization Method for Finding the LCM

**4** LCM of Denominators

## ❶ LISTING METHOD FOR FINDING THE LCM

Recall from Section 2.1 that the multiples of a number are obtained by multiplying that number by each whole number. Consider the numbers 6 and 8. The nonzero multiples of 6 are

$$6, 12, 18, \textbf{24}, 30, 36, 42, \textbf{48}, 54, 60, 66, \textbf{72}, \ldots,$$

and the nonzero multiples of 8 are

$$8, 16, \textbf{24}, 32, 40, \textbf{48}, 56, 64, \textbf{72}, \ldots.$$

Any number which is common to both lists of nonzero multiples is called a common multiple. The common multiples of 6 and 8 are

$$24, 48, 72, \ldots.$$

The smallest of the common multiples of two (or more) counting numbers is called the **least common multiple,** or **LCM,** of the numbers. Since 24 is the smallest number in the list of common multiples of 6 and 8, we say that 24 is the LCM of 6 and 8. This method of finding the LCM of two or more numbers is called the **listing method.** When using the listing method, it is a good idea to use a chart to organize your work as shown below.

| Multiples of 6 | 6 | 12 | 18 | **24** | 30 | 36 | 42 |
|---|---|---|---|---|---|---|---|
| Multiples of 8 | 8 | 16 | **24** | | | | |

| EXAMPLE 1 · FINDING THE LCM USING THE LISTING METHOD |
|---|

Find the LCM of the given numbers.

**(a)** 12 and 30

| Multiples of 12: | 12, | 24, | 36, | 48, | **60**, | 72, | 84 |
|---|---|---|---|---|---|---|---|
| Multiples of 30: | 30, | **60** | | | | | |

We stop writing multiples of 30 once we find a multiple that is also in the first list. Since the smallest number common to both lists is 60, we say the LCM of 12 and 30 is 60.

**(b)** 6, 10, and 15

| Multiples of 6: | 6, | 12, | 18, | 24, | **30**, | 36, | 42 |
|---|---|---|---|---|---|---|---|
| Multiples of 10: | 10, | 20, | **30**, | 40, | 50, | 60 | |
| Multiples of 15: | 15, | **30** | | | | | |

Since 30 is in the first two lists, we can stop after 30 in the third list. The LCM of 6, 10, and 15 is 30.

## ❷ PRIME FACTORIZATION METHOD FOR FINDING THE LCM

Recall that every counting number greater than 1 is either prime or can be factored into primes. Any multiple (including the LCM) of two or more natural numbers must have as factors (divisors) all primes which are factors of each of the natural numbers. This leads to another efficient method for finding LCMs.

Suppose we factor both 6 and 8 into primes.

$$6 = 2 \cdot 3$$
$$8 = 2 \cdot 2 \cdot 2 = 2^3$$

The LCM of 6 and 8 must have $2 \cdot 3$ as a factor since the LCM is a multiple of 6. It must also have $2 \cdot 2 \cdot 2$ or $2^3$ as a factor since it must be a multiple of 8. The smallest number which fits these restrictions is

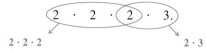

$$2 \cdot 2 \cdot 2 \cdot 3.$$
$$2 \cdot 2 \cdot 2 \qquad 2 \cdot 3$$

Thus, the LCM is $2 \cdot 2 \cdot 2 \cdot 3 = 2^3 \cdot 3 = 24$. This illustrates a general method for finding LCMs.

---

### To Find the LCM Using Prime Factors

1. Factor each number into primes using exponents.
2. If there are no common prime factors, the LCM is the product of all the prime factors.
3. If there are common prime factors, the LCM is the product of the highest power of each prime factor.

---

To use the rule above to find the LCM of 6 and 8, first factor each number into primes using exponents.

$$6 = 2 \cdot 3$$
$$8 = 2 \cdot 2 \cdot 2 = 2^3$$

Since these two numbers have a common prime factor of 2, we use the highest power of each prime factor. We highlight where each prime occurs the greatest number of times.

$$6 = 2 \cdot \mathbf{3} \qquad \text{One 2, one 3}$$
$$8 = \mathbf{2 \cdot 2 \cdot 2} = 2^3 \qquad \text{Three 2s}$$

Now we can find the LCM by taking the product of $2^3$ and 3.

$$\text{LCM} = \mathbf{2 \cdot 2 \cdot 2 \cdot 3} = 2^3 \cdot 3 = 24$$

---

| EXAMPLE 2    FINDING THE LCM USING PRIME FACTORS | PRACTICE EXERCISE 2 |
|---|---|

Find the LCM of the numbers in Example 1 by using prime factors.

**(a)** 12 and 30

$$12 = \mathbf{2 \cdot 2 \cdot 3} = 2^2 \cdot 3 \qquad \text{Two 2's, one 3}$$
$$30 = 2 \cdot 3 \cdot \mathbf{5} \qquad \text{One 2, one 3, one 5}$$

The LCM must be the product of two 2's, one 3, and one 5. Thus, the LCM of 12 and 30 is $2 \cdot 2 \cdot 3 \cdot 5 = 60$.

Find the LCM using prime factors.

**(a)** 14 and 18

**(b)** 6, 10, and 15

$$6 = \mathbf{2} \cdot \mathbf{3} \qquad \text{One 2, one 3}$$
$$10 = 2 \cdot \mathbf{5} \qquad \text{One 2, one 5}$$
$$15 = 3 \cdot 5 \qquad \text{One 3, one 5}$$

Thus, the LCM of 6, 10, and 15 is $2 \cdot 3 \cdot 5 = 30$.

**(b)** 5, 8, and 20

Answers: (a) 126    (b) 40

Compare the results of Example 2 with those of Example 1.

| EXAMPLE 3  FINDING THE **LCM** USING PRIME FACTORS |
|---|

Find the LCM using prime factors.

**(a)** 7 and 28

$$7 = \mathbf{7} \qquad\qquad\qquad \text{One 7}$$
$$28 = \mathbf{2} \cdot \mathbf{2} \cdot 7 = 2^2 \cdot 7 \qquad \text{Two 2's, one 7}$$

The LCM $= 2 \cdot 2 \cdot 7 = 2^2 \cdot 7 = 28$. Note that the LCM of 7 and 28 is one of the two numbers itself. This will happen when one number is a multiple of the other (28 is a multiple of 7).

**(b)** 6, 10, and 25

$$6 = \mathbf{2} \cdot \mathbf{3} \qquad\qquad \text{One 2, one 3}$$
$$10 = 2 \cdot 5 \qquad\qquad \text{One 2, one 5}$$
$$25 = \mathbf{5} \cdot \mathbf{5} = 5^2 \qquad \text{Two 5's}$$

The LCM is $2 \cdot 3 \cdot 5 \cdot 5 = 2 \cdot 3 \cdot 5^2 = 150$.

| PRACTICE EXERCISE 3 |
|---|

Find the LCM using prime factors.

**(a)** 13 and 39

**(b)** 8, 12, and 30

Answers: (a) 39    (b) 120

| EXAMPLE 4  APPLICATION OF **LCM** |
|---|

A man wishes to apply ceramic tiles to a portion of a wall. He plans to use three types of tiles, one 5 inches long, another 6 inches long, and a third 14 inches long. What is the shortest length of wall space he can cover if the first row contains only 14-in tiles, the second row contains only 6-in tiles, and the third contains only 5-in tiles? [Assume that the tiles are not cut and are laid with no space between them.]

The length of the first row is a multiple of 14 inches, the length of the second row is a multiple of 6 inches, and the length of the third row is a multiple of 5 inches. Thus, the shortest distance that can be covered by these three lengths of tile is the LCM of 5, 6, and 14.

$$5 = \mathbf{5} \qquad\qquad \text{One 5}$$
$$6 = \mathbf{2} \cdot \mathbf{3} \qquad\quad \text{One 2, one 3}$$
$$14 = 2 \cdot \mathbf{7} \qquad\quad \text{One 2, one 7}$$

The LCM is $2 \cdot 3 \cdot 5 \cdot 7 = 210$. The shortest distance than can be covered is 210 inches by using 15 of the 14-in tiles ($210 \div 14 = 15$) in the bottom row, 35 of the 6-in tiles ($210 \div 6 = 35$) in the next row, and 42 of the 5-in tiles ($210 \div 5 = 42$) in the third row.

| PRACTICE EXERCISE 4 |
|---|

Three types of pipe are to be laid in the same ditch. One pipe is 3 feet long, another 4 feet long, and the third is 10 feet long. What is the shortest distance from the start to where the three will have a seam at the same point?

Answer: 60 feet

### ❸ SPECIAL ALGORITHM FOR FINDING THE LCM (OPTIONAL)

Although the prime factorization method is probably best for use in later mathematics courses, the following method works well for whole numbers. To explain the procedure we will find the LCM of 36 and 45 using this method.

The basic idea is to divide these numbers by the prime numbers to see which primes are factors of each. It is best to start with the lowest prime and work up. Remember the first few prime numbers are

$$2, \ 3, \ 5, \ 7, \ 11.$$

Try to divide 36 and 45 by 2.

| 2 | 36 | 45 |
|---|----|----|
|   | 18 | 45 |

Two divides 36 but not 45. We put the quotient $18 = 36 \div 2$ below 36 and bring down 45 unchanged. Two will also divide 18.

| 2 | 36 | 45 |
|---|----|----|
| 2 | 18 | 45 |
|   | 9  | 45 |

Since $18 \div 2 = 9$, we put 9 and 45 on the next line. Now 2 will not divide 9 or 45 so we try the next prime, 3.

| 2 | 36 | 45 |
|---|----|----|
| 2 | 18 | 45 |
| 3 | 9  | 45 |
|   | 3  | 15 |

Since $9 \div 3 = 3$ and $45 \div 3 = 15$, the next line has a 3 and a 15. Divide by 3 again.

| 2 | 36 | 45 |
|---|----|----|
| 2 | 18 | 45 |
| 3 | 9  | 45 |
| 3 | 3  | 15 |
|   | 1  | 5  |

We complete the process by dividing by 5.

| 2 | 36 | 45 |
|---|----|----|
| 2 | 18 | 45 |
| 3 | 9  | 45 |
| 3 | 3  | 15 |
| 5 | 1  | 5  |
|   | 1  | 1  |

When the bottom row is all ones we stop the process and the LCM is the product of the prime factors in the left column since each divided one of the two numbers. Thus the LCM of 36 and 45 is

$$LCM = 2 \cdot 2 \cdot 3 \cdot 3 \cdot 5 = 180.$$

In the next example we find the LCM of three numbers using this procedure.

---

**EXAMPLE 5** FINDING THE **LCM** BY SPECIAL ALGORITHM

Find the LCM of 7, 9, and 12.

| 2 | 7 | 9 | 12 | Divide by 2 |
|---|---|---|----|-------------|
| 2 | 7 | 9 | 6  | Divide by 2 again |
| 3 | 7 | 9 | 3  | Divide by 3 |
| 3 | 7 | 3 | 1  | Divide by 3 again |
| 7 | 7 | 1 | 1  | Divide by 7 |
|   | 1 | 1 | 1  | |

Since the last row is all ones, the process stops and the

$$LCM = 2 \cdot 2 \cdot 3 \cdot 3 \cdot 7 = 252.$$

---

**PRACTICE EXERCISE 5**

Find the LCM of 6, 11, and 14.

Answer: 462

---

## ❹ LCM OF DENOMINATORS

We conclude this section by finding the LCM of the denominators of a pair of fractions and expressing each as an equal fraction having the LCM as denominator. This will prepare you for addition and subtraction of fractions in Section 4.3.

---

**EXAMPLE 6** FINDING THE **LCM** OF DENOMINATORS

Find the LCM of the denominators of $\frac{2}{15}$ and $\frac{3}{20}$ and express each as an equal fraction with the LCM as denominator.

We must find the LCM of 15 and 20.

$$15 = \boxed{3 \cdot 5} \qquad \text{One 3, one 5}$$
$$20 = \boxed{2 \cdot 2} \cdot 5 = 2^2 \cdot 5 \qquad \text{Two 2s, one 5}$$

Thus the LCM $= 2 \cdot 2 \cdot 3 \cdot 5 = 2^2 \cdot 3 \cdot 5 = 60$. To express $\frac{2}{15}$ with denominator 60, we must multiply numerator and denominator by $2^2$, the missing factors which make the denominator equal $2^2 \cdot 3 \cdot 5$ or 60.

$$\frac{2}{15} = \frac{2 \cdot 2^2}{15 \cdot 2^2} = \frac{8}{60}$$

Now multiply numerator and denominator of $\frac{3}{20}$ by 3, the missing factor which makes the denominator equal $2^2 \cdot 3 \cdot 5$ or 60.

$$\frac{3}{20} = \frac{3 \cdot 3}{20 \cdot 3} = \frac{9}{60}$$

With the LCM $= 60$, then $\frac{2}{15} = \frac{8}{60}$ and $\frac{3}{20} = \frac{9}{60}$.

---

**PRACTICE EXERCISE 6**

Find the LCM of the denominators of $\frac{3}{14}$ and $\frac{7}{18}$ and express each as an equal fraction with the LCM as denominator.

Answer: LCM = 126; $\frac{3}{14} = \frac{27}{126}$; $\frac{7}{18} = \frac{49}{126}$

## 4.2  EXERCISES A

*Use the listing method to find the LCM.*

**1.** 24 and 18

**2.** 15 and 9

**3.** 20 and 30

**4.** 3 and 8

**5.** 10, 12, and 20

**6.** 8, 9, and 12

*Use prime factors to find the LCM.*

**7.** 36 and 15

**8.** 30 and 28

**9.** 13 and 9

**10.** 2, 5, and 7

**11.** 121 and 22

**12.** 4, 6, and 9

**13.** 52 and 66

**14** 18, 24, and 30

**15.** 22, 55, and 121

*Use any method to find the LCM.*

**16.** 20 and 45

**17.** 18 and 28

**18.** 25 and 30

**19.** 10, 15, and 27

**20.** 12, 14, and 28

**21.** 9, 20, and 25

*Find the LCM of the denominators and express each fraction as an equal fraction with the LCM for a denominator.*

**22.** $\dfrac{1}{8}$ and $\dfrac{7}{12}$

**23.** $\dfrac{3}{4}$ and $\dfrac{5}{6}$

**24.** $\dfrac{5}{28}$ and $\dfrac{11}{42}$

**25.** $\dfrac{18}{25}$, $\dfrac{1}{15}$, and $\dfrac{1}{3}$

**26.** $\dfrac{3}{8}$, $\dfrac{2}{9}$, and $\dfrac{5}{12}$

**27.** $\dfrac{3}{10}$, $\dfrac{7}{100}$, and $\dfrac{19}{1000}$

*Solve.*

**28.** Mercury, Venus, and Earth revolve around the sun once every 3, 7, and 12 months, respectively. If the three planets are now in the same straight line, what is the smallest number of months that must pass before they line up again?

**29** A blocklayer has three lengths of blocks: 8 inches, 9 inches, and 14 inches. He plans to lay a wall of each type of block so that the three walls are the same length. Neglecting the mortar seams, what is the shortest length of wall possible?

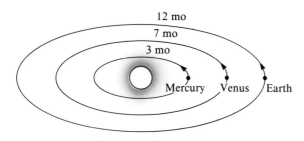

12 mo
7 mo
3 mo
Mercury  Venus  Earth

## FOR REVIEW

*Perform the indicated operations.*

**30.** $\dfrac{7}{12} + \dfrac{2}{12}$

**31.** $\dfrac{13}{11} - \dfrac{6}{11}$

**32.** $\dfrac{4}{7} + \dfrac{1}{7} + \dfrac{2}{7}$

**33.** $\dfrac{11}{6} - \dfrac{7}{6}$

**34.** $\dfrac{18}{15} - \dfrac{13}{15} + \dfrac{1}{15}$

**35.** $\dfrac{6}{3} - \dfrac{2}{3} - \dfrac{1}{3}$

**36.** Dennis walked $\dfrac{13}{4}$ miles due north. He then walked $\dfrac{7}{4}$ miles due south. Finally he walked $\dfrac{10}{4}$ miles due north. **(a)** How far north is he from his starting point?  **(b)** How far did he walk?

---

ANSWERS:  1. 72  2. 45  3. 60  4. 24  5. 60  6. 72  7. 180  8. 420  9. 117  10. 70  11. 242  12. 36
13. 1716  14. 360  15. 1210  16. 180  17. 252  18. 150  19. 270  20. 84  21. 900  22. LCM = 24; $\frac{1}{8} = \frac{3}{24}$; $\frac{7}{12} = \frac{14}{24}$
23. LCM = 12; $\frac{3}{4} = \frac{9}{12}$; $\frac{5}{6} = \frac{10}{12}$  24. LCM = 84; $\frac{5}{28} = \frac{15}{84}$, $\frac{11}{42} = \frac{22}{84}$  25. LCM = 75; $\frac{18}{25} = \frac{54}{75}$, $\frac{1}{15} = \frac{5}{75}$, $\frac{1}{3} = \frac{25}{75}$
26. LCM = 72; $\frac{3}{8} = \frac{27}{72}$, $\frac{2}{9} = \frac{16}{72}$, $\frac{5}{12} = \frac{30}{72}$  27. LCM = 1000; $\frac{3}{10} = \frac{300}{1000}$, $\frac{7}{100} = \frac{70}{1000}$, $\frac{19}{1000} = \frac{19}{1000}$  28. 84 months  29. 504 in
30. $\frac{3}{4}$  31. $\frac{7}{11}$  32. 1  33. $\frac{2}{3}$  34. $\frac{2}{5}$  35. 1  36. (a) 4 mi (b) $\frac{15}{2}$ mi

## 4.2 EXERCISES B

*Use the listing method to find the LCM.*

**1.** 15 and 6

**2.** 30 and 40

**3.** 45 and 30

**4.** 5 and 16

**5.** 5, 8, and 12

**6.** 7, 21, and 18

*Use prime factors to find the LCM.*

**7.** 18 and 20

**8.** 30 and 21

**9.** 13 and 12

**10.** 3, 8, and 9

**11.** 169 and 26

**12.** 9, 15, and 22

**13.** 45 and 42

**14.** 20, 26, and 34

**15.** 49, 35, and 65

*Use any method to find the LCM.*

**16.** 10 and 55

**17.** 15 and 42

**18.** 40 and 75

**19.** 9, 12, and 20

**20.** 8, 10, and 36

**21.** 12, 20, and 42

*Find the LCM of the denominators and express each fraction as an equal fraction with the LCM for a denominator.*

**22.** $\dfrac{3}{10}$ and $\dfrac{4}{15}$

**23.** $\dfrac{5}{14}$ and $\dfrac{1}{4}$

**24.** $\dfrac{6}{55}$ and $\dfrac{5}{44}$

**25.** $\dfrac{2}{3}, \dfrac{5}{18},$ and $\dfrac{7}{30}$

**26.** $\dfrac{7}{27}, \dfrac{3}{4},$ and $\dfrac{1}{24}$

**27.** $\dfrac{7}{20}, \dfrac{9}{200}, \dfrac{11}{2000}$

*Solve.*

**28.** Three satellites in the same plane revolve around the earth once every 6, 12, and 14 hours. If the three satellites are in the same straight line, what is the least number of hours that must pass before they line up again?

**29.** A bricklayer has three lengths of bricks: 4 inches, 5 inches, and 6 inches. He plans to lay a wall of each type of brick so that the three walls are the same length. Neglecting mortar seams, what is the shortest length of wall possible?

## FOR REVIEW

*Perform the indicated operations.*

**30.** $\dfrac{3}{11} + \dfrac{1}{11}$

**31.** $\dfrac{13}{9} - \dfrac{7}{9}$

**32.** $\dfrac{21}{48} + \dfrac{17}{48} + \dfrac{5}{48}$

**33.** $\dfrac{7}{20} - \dfrac{2}{20}$

**34.** $\dfrac{8}{5} - \dfrac{1}{5} + \dfrac{3}{5}$

**35.** $\dfrac{13}{11} - \dfrac{2}{11} - \dfrac{7}{11}$

**36.** Jane has 3 days to do a job. She did $\frac{7}{15}$ of the job on the first day and $\frac{2}{15}$ on the second. **(a)** How much had she done in two days? **(b)** How much must she do the third day to complete the job?

## 4.2  EXERCISES C

*Find the LCM.*

**1.** 126 and 735     [Answer: 4410]

**2.** 60, 198, and 495

# 4.3  ADDING AND SUBTRACTING UNLIKE FRACTIONS

━━━━━━━━━━ STUDENT GUIDEPOSTS ━━━━━━━━━━

**1** Least Common Denominator (LCD)     **2** Method for Adding and Subtracting Unlike Fractions

## 1 LEAST COMMON DENOMINATOR (LCD)

In Section 4.2 we found the least common multiple (LCM) of the denominators of fractions. The LCM of the denominators of two or more fractions is called the **least common denominator (LCD)** of the fractions. The first step in adding or subtracting unlike fractions is to find the LCD and change unlike fractions to like fractions with the LCD as denominator.

## 2 METHOD FOR ADDING AND SUBTRACTING UNLIKE FRACTIONS

Suppose we use the problem

$$\frac{1}{3} + \frac{3}{4}$$

to illustrate the addition method. You can probably tell by inspection that the least common denominator (LCD) is 12. As we did at the end of Section 4.2, change each fraction to an equal fraction with the LCD = 12 as denominator.

$$\frac{1}{3} = \frac{1 \cdot 4}{3 \cdot 4} = \frac{4}{12}$$ Multiply numerator and denominator by 4, the missing factor which makes the denominator 12.

$$\frac{3}{4} = \frac{3 \cdot 3}{4 \cdot 3} = \frac{9}{12}$$ Multiply numerator and denominator by 3, the missing factor which makes the denominator 12.

Thus, $\frac{1}{3} + \frac{3}{4} = \frac{4}{12} + \frac{9}{12} = \frac{4 + 9}{12} = \frac{13}{12}$.

---

### To Add or Subtract Unlike Fractions

1. Rewrite the sum or difference with each denominator written as a prime or factored into primes.
2. Find the LCD (the LCM of all denominators).
3. Multiply the numerator and denominator of each fraction by all factors present in the LCD but missing in the denominator of the particular fraction.
4. Add or subtract the resulting like fractions.

---

### EXAMPLE 1   ADDING AND SUBTRACTING FRACTIONS

Perform the indicated operation.

**(a)** $\dfrac{2}{3} + \dfrac{1}{4} = \dfrac{2}{3} + \dfrac{1}{2 \cdot 2}$   Rewrite denominators as primes; LCD is $2 \cdot 2 \cdot 3$

$\qquad = \dfrac{2 \cdot 2 \cdot 2}{2 \cdot 2 \cdot 3} + \dfrac{1 \cdot 3}{2 \cdot 2 \cdot 3}$   Multiply by missing factors so denominators equal $2 \cdot 2 \cdot 3$

$\qquad = \dfrac{8}{12} + \dfrac{3}{12}$   Simplify and add

$\qquad = \dfrac{8 + 3}{12}$

$\qquad = \dfrac{11}{12}$

**(b)** $\dfrac{5}{9} + \dfrac{1}{6} = \dfrac{5}{3 \cdot 3} + \dfrac{1}{2 \cdot 3}$   Rewrite denominators as primes; LCD is $2 \cdot 3 \cdot 3$

$\qquad = \dfrac{2 \cdot 5}{2 \cdot 3 \cdot 3} + \dfrac{1 \cdot 3}{2 \cdot 2 \cdot 3}$   Multiply by missing factors so denominators equal $2 \cdot 3 \cdot 3$

$\qquad = \dfrac{10}{18} + \dfrac{3}{18}$   Simplify and add

$\qquad = \dfrac{10 + 3}{18}$

$\qquad = \dfrac{13}{18}$

### PRACTICE EXERCISE 1

Perform the indicated operations.

**(a)** $\dfrac{2}{5} + \dfrac{1}{6}$

**(b)** $\dfrac{1}{4} + \dfrac{3}{10}$

With practice, we may be able to take some shortcuts. For example, if we know that 18 is the LCD of 9 and 6, we can skip the factoring step and add as follows.

**(c)** $7 - \dfrac{5}{11}$

$$\frac{5}{9} + \frac{1}{6} = \frac{5 \cdot 2}{9 \cdot 2} + \frac{1 \cdot 3}{6 \cdot 3}$$

Multiply by missing factors so denominators equal 18

$$= \frac{10}{18} + \frac{3}{18} = \frac{13}{18}$$

**(c)** $2 - \dfrac{3}{7} = \dfrac{2}{1} - \dfrac{3}{7}$     LCD = 7

$$= \frac{2 \cdot 7}{1 \cdot 7} - \frac{3}{7}$$     Multiply by missing factor

$$= \frac{14 - 3}{7}$$     Simplify and subtract

$$= \frac{11}{7}$$

Answers: (a) $\frac{17}{30}$   (b) $\frac{11}{20}$   (c) $\frac{72}{11}$

Sometimes fractions are added vertically. For example, $\frac{2}{3}$ and $\frac{1}{4}$ could be added as follows.

$$\frac{2}{3} = \frac{2}{3} = \frac{2 \cdot 2 \cdot 2}{2 \cdot 2 \cdot 3} = \frac{8}{12}$$     The LCD = 2 · 2 · 3 = 12

$$+ \frac{1}{4} = \frac{1}{2 \cdot 2} = \frac{1 \cdot 3}{2 \cdot 2 \cdot 3} = \frac{3}{12}$$

$$\frac{11}{12}$$     8 + 3 = 11

Compare this with Example 1(**a**).

---

## EXAMPLE 2   ADDING AND SUBTRACTING FRACTIONS

Perform the indicated operation.

**(a)** $\dfrac{7}{12} - \dfrac{2}{9} = \dfrac{7}{2 \cdot 2 \cdot 3} - \dfrac{2}{3 \cdot 3}$

Rewrite denominators

LCD = 2 · 2 · 3 · 3 = 36

$$= \frac{7 \cdot 3}{2 \cdot 2 \cdot 3 \cdot 3} - \frac{2 \cdot 2 \cdot 2}{2 \cdot 2 \cdot 3 \cdot 3}$$

Multiply by missing factors so denominators equal 2 · 2 · 3 · 3

$$= \frac{21 - 8}{36}$$     Simplify and subtract

$$= \frac{13}{36}$$

---

### PRACTICE EXERCISE 2

Perform the indicated operations.

**(a)** $\dfrac{11}{15} - \dfrac{3}{20}$

**(b)** $\dfrac{2}{3} - \dfrac{1}{6} = \dfrac{2}{3} - \dfrac{1}{2 \cdot 3}$    LCD $= 2 \cdot 3 = 6$

$\qquad = \dfrac{2 \cdot 2}{2 \cdot 3} - \dfrac{1}{2 \cdot 3}$    Multiply by missing factor

$\qquad = \dfrac{4 - 1}{6}$    Simplify and subtract

$\qquad = \dfrac{3}{6}$

$\qquad = \dfrac{1 \cdot \cancel{3}}{2 \cdot \cancel{3}}$    Reduce to lowest terms

$\qquad = \dfrac{1}{2}$

**(c)** $\dfrac{7}{9} + 3 = \dfrac{7}{3 \cdot 3} + \dfrac{3}{1}$    LCD $= 3 \cdot 3 = 9$

$\qquad = \dfrac{7}{3 \cdot 3} + \dfrac{3 \cdot 3 \cdot 3}{1 \cdot 3 \cdot 3}$    Multiply by missing factors

$\qquad = \dfrac{7}{9} + \dfrac{27}{9}$    Simplify and add

$\qquad = \dfrac{7 + 27}{9}$

$\qquad = \dfrac{34}{9}$

**(b)** $\dfrac{7}{9} - \dfrac{5}{18}$

**(c)** $\dfrac{4}{11} + 1$

Answers: (a) $\frac{7}{12}$   (b) $\frac{1}{2}$   (c) $\frac{15}{11}$

As with addition, fractions can also be subtracted vertically. For example, $\frac{7}{12} - \frac{2}{9}$ can be written as follows.

$$\dfrac{7}{12} = \dfrac{7}{2 \cdot 2 \cdot 3} = \dfrac{7 \cdot 3}{2 \cdot 2 \cdot 3 \cdot 3} = \dfrac{21}{36} \qquad \text{The LCD} = 2 \cdot 2 \cdot 3 \cdot 3 = 36$$

$$-\dfrac{2}{9} = -\dfrac{2}{3 \cdot 3} = -\dfrac{2 \cdot 2 \cdot 2}{2 \cdot 2 \cdot 3 \cdot 3} = -\dfrac{8}{36}$$

$$\dfrac{13}{36} \qquad 21 - 8 = 13$$

Compare this with Example 2**(a)**.

Sums and differences of more than two fractions can also be found.

| **EXAMPLE 3   OPERATIONS ON THREE FRACTIONS** | **PRACTICE EXERCISE 3** |

Perform the indicated operations.

**(a)** $\dfrac{1}{3} + \dfrac{1}{4} + \dfrac{5}{6} = \dfrac{1}{3} + \dfrac{1}{2 \cdot 2} + \dfrac{5}{2 \cdot 3}$    Factor denominators: LCD $= 2 \cdot 2 \cdot 3 = 12$

$\qquad = \dfrac{2 \cdot 2 \cdot 1}{2 \cdot 2 \cdot 3} + \dfrac{1 \cdot 3}{2 \cdot 2 \cdot 3} + \dfrac{2 \cdot 5}{2 \cdot 2 \cdot 3}$    Supply missing factors

$\qquad = \dfrac{4 + 3 + 10}{2 \cdot 2 \cdot 3} = \dfrac{17}{12}$

Perform the indicated operations.

**(a)** $\dfrac{1}{5} + \dfrac{3}{10} + \dfrac{5}{14}$

**(b)** $\dfrac{13}{15} - \dfrac{1}{5} - \dfrac{1}{2} = \dfrac{13}{3 \cdot 5} - \dfrac{1}{5} - \dfrac{1}{2}$     $\text{LCD} = 2 \cdot 3 \cdot 5 = 30$

$= \dfrac{2 \cdot 13}{2 \cdot 3 \cdot 5} - \dfrac{2 \cdot 3 \cdot 1}{2 \cdot 3 \cdot 5} - \dfrac{1 \cdot 3 \cdot 5}{2 \cdot 3 \cdot 5}$

$= \dfrac{26 - 6 - 15}{2 \cdot 3 \cdot 5}$

$= \dfrac{5}{2 \cdot 3 \cdot 5} = \dfrac{\cancel{5}}{2 \cdot 3 \cdot \cancel{5}} = \dfrac{1}{6}$     Reduce to lowest terms

**(b)** $\dfrac{19}{20} - \dfrac{1}{4} - \dfrac{2}{3}$

Answers:  (a) $\frac{6}{7}$   (b) $\frac{1}{30}$

---

| EXAMPLE 4   APPLICATION OF ADDITION | PRACTICE EXERCISE 4 |

In making three different kinds of pastries, Mr. Chandler used $\frac{3}{4}$ cup of flour for the first batch, $\frac{7}{8}$ cup for the second, and $\frac{5}{3}$ cups for the third. How much flour did he use?

Add the three fractions.

$\dfrac{3}{4} + \dfrac{7}{8} + \dfrac{5}{3} = \dfrac{3}{2 \cdot 2} + \dfrac{7}{2 \cdot 2 \cdot 2} + \dfrac{5}{3}$     $\text{LCD} = 2 \cdot 2 \cdot 2 \cdot 3 = 24$

$= \dfrac{3 \cdot 2 \cdot 3}{2 \cdot 2 \cdot 2 \cdot 3} + \dfrac{7 \cdot 3}{2 \cdot 2 \cdot 2 \cdot 3} + \dfrac{2 \cdot 2 \cdot 2 \cdot 5}{2 \cdot 2 \cdot 2 \cdot 3}$

$= \dfrac{18 + 21 + 40}{24} = \dfrac{79}{24} = 3\dfrac{7}{24}$

Thus, Mr. Chandler used $3\frac{7}{24}$ cups of flour.

Wilma ran $\frac{7}{8}$ mi, $\frac{3}{4}$ mi, and $\frac{9}{10}$ mi during one training session. How many miles did she run?

Answer: $\frac{101}{40}$ or $2\frac{21}{40}$ mi

---

**CAUTION**

Never add or subtract fractions by adding or subtracting numerators and denominators. For example,

$\dfrac{2}{5} + \dfrac{1}{4}$     *is not the same as*     $\dfrac{2 + 1}{5 + 4} = \dfrac{3}{9} = \dfrac{1}{3},$

since

$\dfrac{2}{5} + \dfrac{1}{4} = \dfrac{2 \cdot 4}{5 \cdot 4} + \dfrac{1 \cdot 5}{4 \cdot 5} = \dfrac{8 + 5}{20} = \dfrac{13}{20}.$

---

## 4.3 EXERCISES A

*Add.*

**1.** $\dfrac{2}{3} + \dfrac{1}{2}$

**2.** $\dfrac{11}{12} + \dfrac{5}{6}$

**3.** $\dfrac{3}{10} + \dfrac{5}{12}$

**4.** $\dfrac{7}{10} + \dfrac{2}{3}$

**5.** $\dfrac{7}{12} + \dfrac{1}{18}$    **6.** $\dfrac{11}{21} + \dfrac{2}{35}$    **7.** $\dfrac{3}{28} + \dfrac{13}{70}$    **8** $4 + \dfrac{4}{5}$

**9.** $\dfrac{7}{8} + 3$    **10.** $\begin{array}{r} \dfrac{5}{12} \\[2mm] + \dfrac{3}{4} \\ \hline \end{array}$    **11** $\begin{array}{r} \dfrac{1}{6} \\[2mm] + \dfrac{3}{5} \\ \hline \end{array}$    **12.** $\begin{array}{r} \dfrac{7}{15} \\[2mm] + \dfrac{4}{25} \\ \hline \end{array}$

*Subtract.*

**13.** $\dfrac{3}{8} - \dfrac{1}{4}$    **14.** $\dfrac{3}{4} - \dfrac{2}{3}$    **15.** $\dfrac{7}{11} - \dfrac{2}{7}$    **16.** $\dfrac{7}{15} - \dfrac{13}{35}$

**17** $4 - \dfrac{4}{5}$    **18.** $1 - \dfrac{8}{9}$    **19.** $\dfrac{5}{24} - \dfrac{1}{8}$    **20.** $\dfrac{17}{12} - \dfrac{9}{16}$

**21.** $\dfrac{19}{15} - 1$    **22.** $\begin{array}{r} \dfrac{4}{7} \\[2mm] - \dfrac{1}{14} \\ \hline \end{array}$    **23** $\begin{array}{r} \dfrac{8}{15} \\[2mm] - \dfrac{3}{20} \\ \hline \end{array}$    **24.** $\begin{array}{r} \dfrac{29}{12} \\[2mm] - 2 \\ \hline \end{array}$

*Perform the indicated operations.*

**25.** $\dfrac{2}{3} + \dfrac{3}{4} + \dfrac{1}{6}$    **26.** $\dfrac{3}{5} + \dfrac{1}{3} + \dfrac{7}{10}$    **27.** $\dfrac{14}{15} - \dfrac{2}{5} - \dfrac{1}{3}$

**28.** $\dfrac{7}{20} - \dfrac{1}{4} + \dfrac{3}{8}$    **29** $\dfrac{8}{15} + \dfrac{1}{12} - \dfrac{5}{20}$    **30.** $\dfrac{7}{3} - 2 + \dfrac{1}{7}$

*Solve.*

**31.** To obtain the right shade of paint for his living room, Alphonse mixed $\frac{7}{8}$ of a gallon of white paint with $\frac{2}{3}$ of a gallon of yellow and $\frac{3}{5}$ of a gallon of blue. How much paint did he have?

**32.** Suppose Alphonse in Exercise 31 used 2 gallons of the paint that he mixed to paint his living room. How much paint did he have left?

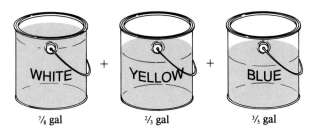

WHITE    +    YELLOW    +    BLUE

⅞ gal           ⅔ gal           ⅗ gal

## FOR REVIEW

*Find the LCM.*

**33.** 27 and 35

**34.** 10 and 45

**35.** 12, 14, and 21

*The following exercises review material from Section 3.6. We review multiplication and division of mixed numbers to give a contrast with addition and subtraction of mixed numbers in the next section.*

**36.** $3\frac{2}{3} \cdot 6\frac{1}{11}$

**37.** $5\frac{1}{2} \div 8\frac{3}{4}$

**38.** $10 \div 7\frac{6}{7}$

---

ANSWERS:   1. $\frac{7}{6}$  2. $\frac{7}{4}$  3. $\frac{43}{60}$  4. $\frac{41}{30}$  5. $\frac{23}{36}$  6. $\frac{61}{105}$  7. $\frac{41}{140}$  8. $\frac{24}{5}$  9. $\frac{31}{8}$  10. $\frac{7}{6}$  11. $\frac{23}{30}$  12. $\frac{47}{75}$  13. $\frac{1}{8}$  14. $\frac{1}{12}$
15. $\frac{27}{77}$  16. $\frac{2}{21}$  17. $\frac{16}{5}$  18. $\frac{1}{9}$  19. $\frac{1}{12}$  20. $\frac{41}{48}$  21. $\frac{4}{15}$  22. $\frac{1}{2}$  23. $\frac{23}{60}$  24. $\frac{5}{12}$  25. $\frac{19}{12}$  26. $\frac{49}{30}$  27. $\frac{1}{5}$  28. $\frac{19}{40}$  29. $\frac{11}{30}$
30. $\frac{10}{21}$  31. $\frac{257}{120}$ gallons  32. $\frac{17}{120}$ of a gallon  33. 945  34. 90  35. 84  36. $22\frac{1}{3}$  37. $\frac{22}{35}$  38. $1\frac{3}{11}$

## 4.3    EXERCISES B

*Add.*

**1.** $\frac{1}{3} + \frac{1}{4}$

**2.** $\frac{5}{12} + \frac{1}{6}$

**3.** $\frac{1}{10} + \frac{7}{12}$

**4.** $\frac{2}{7} + \frac{3}{8}$

**5.** $\frac{9}{16} + \frac{5}{6}$

**6.** $\frac{9}{22} + \frac{4}{33}$

**7.** $\frac{21}{80} + \frac{11}{24}$

**8.** $\frac{3}{10} + 7$

**9.** $3 + \frac{9}{11}$

**10.** $\begin{array}{r} \frac{7}{10} \\ + \frac{2}{5} \\ \hline \end{array}$

**11.** $\begin{array}{r} \frac{5}{8} \\ + \frac{5}{12} \\ \hline \end{array}$

**12.** $\begin{array}{r} \frac{8}{35} \\ + \frac{11}{21} \\ \hline \end{array}$

*Subtract.*

**13.** $\dfrac{5}{8} - \dfrac{1}{3}$
**14.** $\dfrac{8}{11} - \dfrac{2}{7}$
**15.** $\dfrac{11}{15} - \dfrac{6}{35}$
**16.** $\dfrac{13}{14} - \dfrac{2}{21}$

**17.** $5 - \dfrac{3}{7}$
**18.** $7 - \dfrac{3}{8}$
**19.** $\dfrac{1}{7} - \dfrac{1}{13}$
**20.** $\dfrac{7}{24} - \dfrac{9}{40}$

**21.** $5 - \dfrac{34}{7}$
**22.** $\begin{array}{r} \dfrac{7}{10} \\ -\ \dfrac{2}{5} \\ \hline \end{array}$
**23.** $\begin{array}{r} \dfrac{13}{22} \\ -\ \dfrac{5}{33} \\ \hline \end{array}$
**24.** $\begin{array}{r} 4 \\ -\ \dfrac{17}{5} \\ \hline \end{array}$

*Perform the indicated operations.*

**25.** $\dfrac{1}{4} + \dfrac{1}{5} + \dfrac{3}{10}$
**26.** $\dfrac{3}{8} + \dfrac{1}{6} + \dfrac{5}{12}$
**27.** $\dfrac{20}{21} - \dfrac{2}{3} - \dfrac{2}{7}$

**28.** $\dfrac{13}{30} - \dfrac{2}{5} + \dfrac{7}{10}$
**29.** $\dfrac{7}{16} + \dfrac{9}{24} - \dfrac{7}{10}$
**30.** $\dfrac{11}{2} - 5 - \dfrac{3}{11}$

**31.** There were three candidates from the Western Party in an election. One got $\frac{1}{5}$ of the votes, another $\frac{1}{12}$ of the votes, and the third $\frac{3}{8}$ of the votes. What total fraction of the votes did the Western Party receive?

**32.** Two candidates from the Northern Party received the remaining votes in the election in Exercise 31. What fraction of the votes did the Northern Party get?

## FOR REVIEW

*Find the LCM.*

**33.** 22 and 77
**34.** 75 and 95
**35.** 6, 10, and 15

*The following exercises review material from Section 3.6. We review multiplication and division of mixed numbers to give a contrast with addition and subtraction of mixed numbers in the next section.*

**36.** $4\dfrac{1}{2} \cdot 10\dfrac{3}{8}$
**37.** $7\dfrac{4}{5} \div 3\dfrac{6}{7}$
**38.** $8\dfrac{4}{5} \div 4$

## 4.3 EXERCISES C

*Let a and b represent natural numbers. Perform the indicated operations.*

**1.** $\dfrac{3}{a} + \dfrac{2}{b}$
**2.** $\dfrac{5}{ab} - \dfrac{2}{b}$   $\left[\text{Answer: } \frac{5 - 2a}{ab}\right]$

# 4.4 ADDING AND SUBTRACTING MIXED NUMBERS

## 1 ADDING MIXED NUMBERS

In Chapter 3 we defined a mixed number as the sum of a whole number and a proper fraction. For example,

$$6\frac{3}{4} = 6 + \frac{3}{4}.$$

In Section 3.6 we changed mixed numbers to improper fractions before multiplying or dividing. This same procedure works for addition, but most problems are more easily done by adding the whole number parts and then adding the fraction parts.

---

**EXAMPLE 1    ADDING MIXED NUMBERS**

Add $3\frac{5}{8}$ and $2\frac{1}{6}$.

$$3\frac{5}{8} = 3\frac{5}{2 \cdot 2 \cdot 2} = 3\frac{5 \cdot 3}{2 \cdot 2 \cdot 2 \cdot 3} = 3\frac{15}{24}$$

$$+ 2\frac{1}{6} = 2\frac{1}{2 \cdot 3} = 2\frac{1 \cdot 2 \cdot 2}{2 \cdot 2 \cdot 2 \cdot 3} = 2\frac{4}{24}$$

$$5\frac{19}{24}$$

add whole numbers    add fractions

---

**PRACTICE EXERCISE 1**

Add $7\frac{1}{10}$ and $3\frac{4}{15}$.

Answer: $10\frac{11}{30}$

---

**EXAMPLE 2    FRACTIONS ADDING TO AN IMPROPER FRACTION**

Add $425\frac{3}{8}$ and $211\frac{2}{3}$.

$$425\frac{3}{8} = 425\frac{3}{2 \cdot 2 \cdot 2} = 425\frac{3 \cdot 3}{2 \cdot 2 \cdot 2 \cdot 3} = 425\frac{9}{24}$$

$$+ 211\frac{2}{3} = + 211\frac{2}{3} = + 211\frac{2 \cdot 2 \cdot 2 \cdot 2}{2 \cdot 2 \cdot 2 \cdot 3} = + 211\frac{16}{24}$$

$$636\frac{25}{24}$$

add whole numbers    add fractions

---

**PRACTICE EXERCISE 2**

Add $324\frac{2}{5}$ and $447\frac{3}{4}$.

Since $\frac{25}{24}$ is an improper fraction, change it to the mixed number $1\frac{1}{24}$ and rewrite the answer as follows.

$$636\frac{25}{24} = 636 + 1\frac{1}{24} \qquad \frac{25}{24} \text{ is } 1\frac{1}{24}$$

$$= \boxed{636 + 1} + \frac{1}{24} \qquad \text{Rewrite mixed number as a sum}$$

$$= \boxed{637} + \frac{1}{24} \qquad \text{Add whole numbers}$$

$$= 637\frac{1}{24} \qquad \text{Final answer}$$

Answer: $772\frac{3}{20}$

Use the same method to add three or more mixed numbers.

---

| **EXAMPLE 3**    **ADDING THREE MIXED NUMBERS** | **PRACTICE EXERCISE 3** |
|---|---|

Add $10\frac{1}{3}$, $5\frac{5}{6}$, and $3\frac{3}{4}$.

Add $8\frac{1}{4}$, $3\frac{5}{12}$, and $1\frac{5}{6}$.

$$10\frac{1}{3} = 10\frac{1 \cdot 4}{3 \cdot 4} = 10\frac{4}{12} \qquad \text{LCD is 12}$$

$$5\frac{5}{6} = 5\frac{5 \cdot 2}{6 \cdot 2} = 5\frac{10}{12}$$

$$+ \; 3\frac{3}{4} = 3\frac{3 \cdot 3}{4 \cdot 3} = 3\frac{9}{12}$$

$$18\frac{23}{12}$$

add whole numbers    add fractions

Change $\frac{23}{12}$ to $1\frac{11}{12}$ and add to 18.

$$18 + 1\frac{11}{12} = 18 + 1 + \frac{11}{12}$$

$$= 19\frac{11}{12}$$

Answer: $13\frac{1}{2}$

---

## ② SUBTRACTING MIXED NUMBERS

As with addition, mixed numbers can be subtracted by subtracting the whole numbers and the fractions separately.

| EXAMPLE 4    SUBTRACTING MIXED NUMBERS | PRACTICE EXERCISE 4 |

Subtract: $48\frac{3}{4} - 27\frac{1}{3}$.

$$48\frac{3}{4} = \quad 48\frac{3}{2 \cdot 2} = \quad 48\frac{3 \cdot 3}{2 \cdot 2 \cdot 3} = \quad 48\frac{9}{12}$$

$$- 27\frac{1}{3} = -27\frac{1}{3} \quad = -27\frac{2 \cdot 2 \cdot 1}{2 \cdot 2 \cdot 3} = -27\frac{4}{12}$$

$$21\frac{5}{12}$$

subtract whole numbers    subtract fractions

Subtract: $121\frac{5}{8} - 102\frac{1}{5}$.

Answer: $19\frac{17}{40}$

## **HINT**

An extra step is sometimes needed for this kind of subtraction. Sometimes the fraction being subtracted from is smaller than the fraction being subtracted. When this occurs, borrow.

| EXAMPLE 5    BORROWING WHEN SUBTRACTING MIXED NUMBERS | PRACTICE EXERCISE 5 |

Subtract.

**(a)**
$$57\frac{1}{4}$$
$$- 33\frac{1}{2}$$

When subtracting the fractions, we try

$$\frac{1}{4} - \frac{1}{2} = \frac{1}{4} - \frac{2}{4} = \frac{1-2}{4},$$

which we do not know how to do. Therefore, we must borrow 1 $\left(\text{in the form } \frac{4}{4}\right)$ from 57 and add it to $\frac{1}{4}$ $\left(\text{making } \frac{5}{4}\right)$ before we can subtract.

$$56\frac{\frac{1}{4} + \frac{4}{4}}{}$$

$$57\frac{1}{4} = \quad 57\frac{1}{4} = \quad 56\frac{5}{4} = \quad 56\frac{5}{4}$$

$$- 33\frac{1}{2} = -33\frac{1}{2} = -33\frac{1}{2} = -33\frac{2}{4}$$

$$23\frac{3}{4}$$

Subtract.

**(a)**    $92\frac{1}{6}$
$- 54\frac{2}{3}$

**(b)**    $8\frac{9}{10}$
$- 6$

**(b)** $\quad 6\dfrac{3}{4} = \quad 6\dfrac{3}{4}$

$\dfrac{-\ 2\quad = -\ 2\dfrac{0}{4}}{\qquad\quad 4\dfrac{3}{4}}$  $\qquad 2 = 2 + 0 = 2 + \dfrac{0}{4}$

$\dfrac{3}{4} - \dfrac{0}{4} = \dfrac{3-0}{4} = \dfrac{3}{4}$

**(c)** $\quad 6\ =\ 5\dfrac{4}{4}$

$\dfrac{-\ 2\dfrac{3}{4} = -\ 2\dfrac{3}{4}}{\qquad 3\dfrac{1}{4}}$  Compare this with **(b)** above

**(c)** $\qquad 15$

$\dfrac{-\ 10\frac{7}{8}}{}$

Answers: (a) $37\frac{1}{2}$   (b) $2\frac{9}{10}$
(c) $4\frac{1}{8}$

As was true with whole numbers, subtraction of mixed numbers can be checked by addition.

These must be equal
$$\left.\begin{array}{c} \rightarrow 6 \\ -\ 2\dfrac{3}{4} \leftarrow \\ \overline{\quad 3\dfrac{1}{4}} \leftarrow \\ ----\ \\ \rightarrow 5\dfrac{4}{4} = 6 \leftarrow \end{array}\right.$$
Add these and place the sum below the dashed line

## ③ APPLICATIONS OF MIXED NUMBERS

Many applied problems involve the addition or subtraction of mixed numbers.

---

### EXAMPLE 6   APPLICATION OF ADDITION

Mike weighs $158\frac{1}{2}$ lb, Den weighs $138\frac{3}{4}$ lb, and Murph weighs $172\frac{1}{8}$ lb. Find the combined weight of these men.

We must add the three weights.

$$158\frac{1}{2} = 158\frac{4}{8} \qquad \text{LCD is 8}$$

$$138\frac{3}{4} = 138\frac{6}{8}$$

$$\dfrac{172\frac{1}{8} = 172\frac{1}{8}}{468\frac{11}{8} = 468 + 1 + \frac{3}{8} = 469\frac{3}{8}}$$

Thus, the combined weight of the men is $469\frac{3}{8}$ lb.

### PRACTICE EXERCISE 6

On a deep sea fishing trip Reva caught three fish weighing $122\frac{2}{5}$ lb, $97\frac{7}{10}$ lb, and $171\frac{4}{5}$ lb. What was the total weight of her fish?

Answer: $391\frac{9}{10}$ lb

| **EXAMPLE 7**   **APPLICATION OF SUBTRACTION** | **PRACTICE EXERCISE 7** |

It took Barb $2\frac{3}{4}$ days to sew a set of draperies. Had she hired a professional seamstress, it would have taken $1\frac{1}{5}$ days. How much time could she have saved by using the professional?

The time she could have saved is the difference between the two times.

$$2\frac{3}{4} = \quad 2\frac{3 \cdot 5}{4 \cdot 5} = \quad 2\frac{15}{20} \qquad \text{LCD is 20}$$
$$-1\frac{1}{5} = -1\frac{1 \cdot 4}{5 \cdot 4} = -1\frac{4}{20}$$
$$\overline{\qquad\qquad\qquad\qquad\qquad 1\frac{11}{20}}$$

Thus, she could have saved $1\frac{11}{20}$ days.

Randy had a job to do that required $6\frac{3}{8}$ days to finish. After he had worked for $4\frac{3}{4}$ days, how many days were left to complete the job?

Answer: $1\frac{5}{8}$

# 4.4 EXERCISES A

*Add.*

**1.**  $5\frac{3}{8}$
$+ 2\frac{1}{2}$

**2.**  $6\frac{1}{6}$
$+ 5\frac{1}{5}$

**3.**  $3\frac{2}{5}$
$+ 4\frac{5}{6}$

**4.**  $8\frac{1}{12}$
$+ 7\frac{3}{4}$

**5.**  $2\frac{2}{7}$
$+ 4\frac{5}{21}$

**6.**  $3\frac{3}{20}$
$+ 2\frac{5}{24}$

**7.**  $6\frac{7}{33}$
$+ 10\frac{3}{11}$

**8**  $15\frac{7}{8}$
$+ 4$

**9.**  $8$
$+ 7\frac{3}{10}$

**10**  $215\frac{7}{8}$
$+ 147\frac{1}{2}$

**11.**  $685\frac{2}{11}$
$+ 296\frac{2}{3}$

**12.**  $45\frac{5}{12}$
$+ 88\frac{15}{16}$

*Subtract.*

**13.**  $9\frac{2}{9}$
$- 3\frac{1}{3}$

**14.**  $11\frac{4}{5}$
$- 5\frac{2}{3}$

**15.**  $6\frac{3}{8}$
$- 5\frac{3}{4}$

**16.**  $10\frac{2}{5}$
$- 1\frac{7}{20}$

17.  $4\frac{8}{9}$
     $-\ 3\frac{11}{15}$

18.  $8\frac{1}{4}$
     $-\ 3\frac{2}{3}$

19.  $4\frac{11}{35}$
     $-\ 2\frac{4}{21}$

20.  $17\frac{6}{11}$
     $-\ \ \ 8$

21.  $15$
     $-\ 4\frac{9}{16}$

22.  $485\frac{9}{10}$
     $-\ 316\frac{2}{5}$

23.  $925\frac{1}{7}$
     $-495\frac{11}{14}$

24.  $211\frac{3}{22}$
     $-\ 201\frac{7}{33}$

*Add.*

25.  $10\frac{2}{3}$
     $4\frac{1}{5}$
     $+\ 7\frac{2}{15}$

26.  $7\frac{1}{5}$
     $3\frac{2}{3}$
     $+\ 1\frac{1}{15}$

27.  $8\frac{3}{4}$
     $2\frac{1}{12}$
     $+\ 4\frac{5}{6}$

28.  $5\frac{8}{9}$
     $2\frac{1}{3}$
     $+\ 1\frac{5}{12}$

*Solve.*

29.  Juan took a hike over three different trails. If the three trails were $5\frac{1}{3}$ mi, $8\frac{3}{4}$ mi, and $7\frac{7}{8}$ mi long, what was the total distance that he hiked?

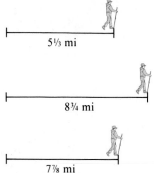

5⅓ mi

8¾ mi

7⅞ mi

30.  It took Arn $7\frac{1}{8}$ days to type a manuscript. Had he not been broke, he could have hired a typist who could do the job in $1\frac{2}{3}$ days. How many days could have been saved?

31.  Mr. Horn owns $25\frac{1}{4}$ acres of land in Colorado, $160\frac{2}{3}$ acres in Utah, and $185\frac{1}{6}$ acres in Florida. What is the total number of acres that he owns?

32.  It took $3\frac{5}{8}$ hours for Burford to do a job. With help he could have done the job in $2\frac{3}{4}$ hours. How much time could he have saved had he had help?

## FOR REVIEW

*Perform the indicated operations.*

**33.** $\dfrac{4}{5} + \dfrac{1}{15} + \dfrac{7}{30}$

**34.** $\dfrac{4}{7} + \dfrac{11}{14} - \dfrac{8}{21}$

**35.** $\dfrac{9}{5} - \dfrac{17}{15} - \dfrac{16}{25}$

**36.** $\dfrac{1}{10} + \dfrac{4}{35} - \dfrac{2}{21}$

*The following exercises review material from Section 1.1 to help prepare you for the next section. Place the appropriate sign, $<$ or $>$, between the whole numbers.*

**37.** 16 _____ 4

**38.** 2 _____ 45

**39.** 0 _____ 6

**40.** 8 _____ 7

---

ANSWERS:  1. $7\frac{7}{8}$  2. $11\frac{11}{30}$  3. $8\frac{7}{30}$  4. $15\frac{5}{6}$  5. $6\frac{11}{21}$  6. $5\frac{43}{120}$  7. $16\frac{16}{33}$  8. $19\frac{7}{8}$  9. $15\frac{3}{10}$  10. $363\frac{3}{8}$  11. $981\frac{28}{33}$  12. $134\frac{17}{48}$  13. $5\frac{8}{9}$  14. $6\frac{2}{15}$  15. $\frac{5}{8}$  16. $9\frac{1}{20}$  17. $1\frac{7}{45}$  18. $4\frac{7}{12}$  19. $2\frac{13}{105}$  20. $9\frac{6}{11}$  21. $10\frac{7}{16}$  22. $169\frac{1}{2}$  23. $429\frac{5}{14}$  24. $9\frac{61}{66}$  25. 22  26. $11\frac{14}{15}$  27. $15\frac{2}{3}$  28. $9\frac{23}{36}$  29. $21\frac{23}{24}$ mi  30. $5\frac{11}{24}$ days  31. $371\frac{1}{12}$ acres  32. $\frac{7}{8}$ hr  33. $1\frac{1}{10}$  34. $\frac{41}{42}$  35. $\frac{2}{75}$  36. $\frac{5}{42}$  37. $>$  38. $<$  39. $<$  40. $>$

## 4.4  EXERCISES B

*Add.*

**1.** $\quad 2\dfrac{3}{10}$

$\quad +\ 3\dfrac{2}{5}$

**2.** $\quad 7\dfrac{1}{8}$

$\quad +\ 9\dfrac{1}{6}$

**3.** $\quad 4\dfrac{2}{11}$

$\quad +\ 3\dfrac{3}{4}$

**4.** $\quad 8\dfrac{4}{15}$

$\quad +\ 1\dfrac{4}{5}$

**5.** $\quad 4\dfrac{7}{8}$

$\quad +\ 2\dfrac{7}{10}$

**6.** $\quad 8\dfrac{5}{12}$

$\quad +\ 10\dfrac{7}{15}$

**7.** $\quad 6\dfrac{3}{13}$

$\quad +\ 9\dfrac{7}{26}$

**8.** $\quad 19\dfrac{4}{21}$

$\quad +\ 10$

**9.** 17

$\quad +\ 1\dfrac{5}{8}$

**10.** $\quad 106\dfrac{4}{5}$

$\quad +\ 317\dfrac{4}{15}$

**11.** $\quad 1060\dfrac{1}{7}$

$\quad +\ 1121\dfrac{2}{3}$

**12.** $\quad 21\dfrac{7}{10}$

$\quad +\ 19\dfrac{19}{20}$

*Subtract.*

**13.** $\quad 8\dfrac{2}{5}$

$\quad -\ 6\dfrac{8}{15}$

**14.** $\quad 15\dfrac{5}{7}$

$\quad -\ 12\dfrac{2}{3}$

**15.** $\quad 7\dfrac{3}{10}$

$\quad -\ 4\dfrac{4}{5}$

**16.** $\quad 3\dfrac{1}{6}$

$\quad -\ 1\dfrac{7}{18}$

**17.**  $3\frac{3}{10}$

$-\ 2\frac{4}{15}$

**18.**  $8\frac{1}{8}$

$-\ 4\frac{3}{5}$

**19.**  $12\frac{11}{15}$

$-\ 11\frac{5}{6}$

**20.**  $9\frac{7}{11}$

$-\ 4$

**21.**  $11$

$-\ 7\frac{3}{14}$

**22.**  $325\frac{4}{5}$

$-\ 170\frac{1}{4}$

**23.**  $428\frac{1}{5}$

$-\ 147\frac{7}{8}$

**24.**  $901\frac{1}{15}$

$-\ 899\frac{20}{21}$

*Add.*

**25.**  $1\frac{2}{11}$

$2\frac{3}{22}$

$+\ 4\frac{1}{2}$

**26.**  $4\frac{2}{3}$

$6\frac{5}{6}$

$+\ 7\frac{1}{12}$

**27.**  $4\frac{3}{4}$

$4\frac{3}{8}$

$+\ 4\frac{3}{16}$

**28.**  $9\frac{5}{6}$

$7\frac{1}{4}$

$+\ 1\frac{2}{3}$

*Solve.*

**29.** On Monday Debbie worked out for $2\frac{1}{4}$ hours, on Wednesday $3\frac{2}{5}$ hours, and on Friday $3\frac{1}{3}$ hours. What is the sum of her hours for these three days?

**30.** Harry's paper for a history class has to be 16 pages long. How many more pages must he write if he has done $9\frac{2}{5}$ pages?

**31.** On a trip the Williamses filled their gas tank three times. They required $16\frac{2}{5}$ gallons, $21\frac{7}{10}$ gallons, and $19\frac{1}{5}$ gallons. How many gallons of gasoline did they buy on the trip?

**32.** A contractor estimates that it will take $17\frac{1}{2}$ days to complete a job. How many days of work are left after $9\frac{3}{4}$ days have been completed?

## FOR REVIEW

*Perform the indicated operations.*

**33.**  $\frac{1}{6} + \frac{1}{8} + \frac{1}{10}$

**34.**  $\frac{7}{12} + \frac{3}{4} - \frac{5}{8}$

**35.**  $\frac{10}{7} - \frac{7}{10} - \frac{3}{14}$

**36.**  $\frac{9}{11} + \frac{5}{33} - \frac{7}{22}$

*The following exercises review material from Section 1.1 to help prepare you for the next section. Place the appropriate sign, < or >, between the whole numbers.*

**37.** 1 _____ 9          **38.** 4 _____ 0          **39.** 16 _____ 18          **40.** 25 _____ 15

## 4.4  EXERCISES C

*Perform the indicated operations.*

**1.** $209\frac{4}{5} - 122\frac{4}{15} + 65\frac{1}{3}$

**2.** $489\frac{3}{4} - 280\frac{5}{8} - 2\frac{1}{2}$   $\left[\text{Answer: } 206\frac{5}{8}\right]$

## 4.5  ORDER OF OPERATIONS AND COMPARING FRACTIONS

=== STUDENT GUIDEPOSTS ===

**1** Rule for the Order of Operations   **2** Comparing Fractions

### **1** RULE FOR THE ORDER OF OPERATIONS

The order of operations on whole numbers that we studied in Section 2.8 applies to fractions as well, and is reviewed below.

| Order of Operations |
|---|
| Operations should be performed in the following order. |
| 1. Operate within parentheses. |
| 2. Find all powers and roots in any order. |
| 3. Multiply and divide, in order, from left to right. |
| 4. Add and subtract, in order, from left to right. |

---

**EXAMPLE 1  ORDER OF OPERATIONS**

Evaluate each expression.

**(a)** $\left(\dfrac{5}{6} - \dfrac{1}{6}\right) \cdot \dfrac{3}{4} = \dfrac{4}{6} \cdot \dfrac{3}{4}$   Operate inside parentheses first

$= \dfrac{2}{3} \cdot \dfrac{3}{4}$   Reduce fraction

$= \dfrac{\overset{1}{2} \cdot \overset{1}{3}}{\underset{1}{3} \cdot \underset{1}{2} \cdot 2}$   Multiply

$= \dfrac{1}{2}$

**PRACTICE EXERCISE 1**

Evaluate each expression.

**(a)** $\dfrac{4}{15} \cdot \left(\dfrac{3}{8} + \dfrac{1}{8}\right)$

**(b)** $\dfrac{5}{12} + \dfrac{3}{8} \div \dfrac{3}{4} = \dfrac{5}{12} + \dfrac{3}{8} \cdot \dfrac{4}{3}$     Divide first

$$= \dfrac{5}{12} + \dfrac{\overset{1}{\cancel{3}} \cdot \overset{1}{\cancel{4}}}{2 \cdot \underset{1}{\cancel{4}} \cdot \underset{1}{\cancel{3}}}$$

$$= \dfrac{5}{12} + \dfrac{1}{2} = \dfrac{5}{12} + \dfrac{6}{12}$$     Then add, the LCD is 12

$$= \dfrac{11}{12}$$

**(c)** $\dfrac{5}{7} \cdot \dfrac{7}{2} - \dfrac{1}{11} \div \dfrac{3}{22} = \dfrac{5 \cdot 7}{7 \cdot 2} - \dfrac{1}{11} \cdot \dfrac{22}{3}$     Multiply and divide first

$$= \dfrac{5 \cdot \overset{1}{\cancel{7}}}{\underset{1}{\cancel{7}} \cdot 2} - \dfrac{1 \cdot 2 \cdot \overset{1}{\cancel{11}}}{\underset{1}{\cancel{11}} \cdot 3}$$

$$= \dfrac{5}{2} - \dfrac{2}{3}$$

$$= \dfrac{15}{6} - \dfrac{4}{6}$$     LCD is 6

$$= \dfrac{11}{6} = 1\dfrac{5}{6}$$     Subtract last

---

**(b)** $\dfrac{5}{9} \div \dfrac{5}{3} - \dfrac{1}{10}$

**(c)** $\dfrac{9}{20} \div \dfrac{3}{10} - \dfrac{4}{11} \cdot \dfrac{11}{20}$

**Answers:** (a) $\frac{2}{15}$   (b) $\frac{7}{30}$
(c) $1\frac{3}{10}$

---

## EXAMPLE 2   ORDER OF OPERATIONS

Evaluate each expression.

**(a)** $\left(\dfrac{9}{11} - \dfrac{3}{11}\right) \cdot \left(\dfrac{1}{2}\right)^2 = \dfrac{6}{11} \cdot \left(\dfrac{1}{2}\right)^2$     Evaluate inside parentheses first

$$= \dfrac{6}{11} \cdot \dfrac{1}{4}$$     Find square next

$$= \dfrac{\overset{1}{\cancel{2}} \cdot 3 \cdot 1}{11 \cdot \underset{1}{\cancel{2}} \cdot 2}$$     Multiply

$$= \dfrac{3}{22}$$

**(b)** $\sqrt{\dfrac{4}{9}} - \dfrac{1}{3} \cdot \dfrac{1}{2} = \dfrac{2}{3} - \dfrac{1}{3} \cdot \dfrac{1}{2}$     Find square root first

$$= \dfrac{2}{3} - \dfrac{1}{6}$$     Multiply

$$= \dfrac{4}{6} - \dfrac{1}{6}$$     Now subtract

$$= \dfrac{3}{6} = \dfrac{1}{2}$$     Reduce

---

## PRACTICE EXERCISE 2

Evaluate each expression.

**(a)** $\left(\dfrac{2}{3}\right)^2 \div \left(\dfrac{4}{7} + \dfrac{5}{14}\right)$

**(b)** $\dfrac{3}{5} - \sqrt{\dfrac{16}{25}} \cdot \dfrac{1}{2}$

**(c)** $\left(\dfrac{3}{4}\right)^2 + \dfrac{5}{3} \div \dfrac{4}{3} - \dfrac{2}{5} \cdot \dfrac{5}{4} = \dfrac{9}{16} + \dfrac{5}{3} \div \dfrac{4}{3} - \dfrac{2}{5} \cdot \dfrac{5}{4}$    Square first

$= \dfrac{9}{16} + \dfrac{5}{3} \cdot \dfrac{3}{4} - \dfrac{2 \cdot 5}{5 \cdot 4}$    Divide and multiply

$= \dfrac{9}{16} + \dfrac{5 \cdot \overset{1}{\cancel{3}}}{\underset{1}{\cancel{3}} \cdot 4} - \dfrac{\overset{1}{\cancel{2}} \cdot \overset{1}{\cancel{5}}}{\underset{1}{\cancel{5}} \cdot \underset{1}{\cancel{2}} \cdot 2}$

$= \dfrac{9}{16} + \dfrac{5}{4} - \dfrac{1}{2}$

$= \dfrac{9}{16} + \dfrac{20}{16} - \dfrac{8}{16}$    Now add and subtract

$= \dfrac{21}{16} = 1\dfrac{5}{16}$

**(c)** $\dfrac{4}{15} - \dfrac{3}{5} \cdot \dfrac{2}{15} + \left(\dfrac{1}{5}\right)^2$

Answers:  (a) $\frac{56}{117}$   (b) $\frac{1}{5}$   (c) $\frac{17}{75}$

## ❷ COMPARING FRACTIONS

Recall that one whole number is greater than another if the first is to the right of the second on a number line. Deciding which of two fractions is larger is just as simple, when the denominators are the same. Consider the fractions

$$\dfrac{2}{3} \quad \text{and} \quad \dfrac{3}{5}.$$

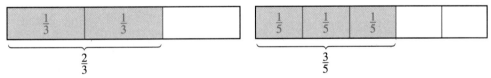

**Figure 4.1**

Compare the bars in Figure 4.1. It is not obvious which fraction is larger. Now find the LCD of these fractions.

$$\dfrac{2}{3} = \dfrac{2 \cdot 5}{3 \cdot 5} = \dfrac{10}{15} \quad \text{and} \quad \dfrac{3}{5} = \dfrac{3 \cdot 3}{3 \cdot 5} = \dfrac{9}{15}$$

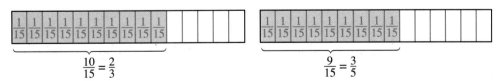

**Figure 4.2**

Comparing the bars in Figure 4.2 makes it clear that the shaded portion $\frac{10}{15}$ is larger than the shaded portion $\frac{9}{15}$. Thus

$$\dfrac{10}{15} > \dfrac{9}{15} \quad \text{and} \quad \dfrac{2}{3} > \dfrac{3}{5}.$$

But it is not necessary to compare fractional parts of bars since certainly

**10** fifteenths is more than **9** fifteenths.

When two fractions have the same denominator, the larger fraction has the larger numerator.

### To Tell Which of Two Fractions Is Larger

1. Find a common denominator and convert each fraction to an equal fraction having that denominator.
2. The larger fraction has the larger numerator.

This rule allows us to compare fractions using what we know about the order of whole numbers.

---

### EXAMPLE 3   COMPARING FRACTIONS

Which of the two fractions is larger?

**(a)** $\dfrac{5}{6}$ or $\dfrac{3}{4}$

$$\frac{5}{6} = \frac{5}{2 \cdot 3} = \frac{2 \cdot 5}{2 \cdot 2 \cdot 3} = \frac{10}{12}$$

$$\frac{3}{4} = \frac{3}{2 \cdot 2} = \frac{3 \cdot 3}{2 \cdot 2 \cdot 3} = \frac{9}{12}$$

The LCD of 4 and 6 is $2 \cdot 2 \cdot 3 = 12$

Since $10 > 9$, $\dfrac{10}{12} > \dfrac{9}{12}$. Thus, $\dfrac{5}{6} > \dfrac{3}{4}$.

**(b)** $\dfrac{3}{20}$ or $\dfrac{1}{5}$

$$\frac{3}{20} = \frac{3}{2 \cdot 2 \cdot 5} = \frac{3}{30}$$

$$\frac{1}{5} = \frac{2 \cdot 2 \cdot 1}{2 \cdot 2 \cdot 5} = \frac{4}{20}$$

The LCD of 5 and 20 is $2 \cdot 2 \cdot 5 = 20$

Since $4 > 3$, $\dfrac{4}{20} > \dfrac{3}{20}$. Thus $\dfrac{1}{5} > \dfrac{3}{20}$.

### PRACTICE EXERCISE 3

Which of the two fractions is larger?

**(a)** $\dfrac{7}{12}$ or $\dfrac{6}{11}$

**(b)** $\dfrac{1}{7}$ or $\dfrac{2}{17}$

Answers:  (a) $\frac{7}{12} > \frac{6}{11}$   (b) $\frac{1}{7} > \frac{2}{17}$

---

### EXAMPLE 4   APPLICATION TO RECREATION

A map of trails joining Blue Ridge Hill and the blackberry patch is shown in Figure 4.3. Is it farther from Blue Ridge Hill to the blackberry patch by way of Luna Pond or by way of White Rock Creek? How much farther?

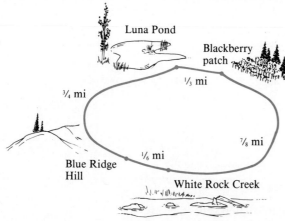

Luna Pond

Blackberry patch

¾ mi

⅓ mi

⅞ mi

⅙ mi

Blue Ridge Hill

White Rock Creek

**Figure 4.3**

### PRACTICE EXERCISE 4

In June, Chicago had two rainstorms, one which dropped $\frac{3}{8}$ inch and the other $1\frac{1}{4}$ inches of precipitation. During July, there were three storms with $\frac{1}{3}$, $\frac{4}{5}$, and $\frac{1}{8}$ inch of rain. During which of these months did Chicago have more precipitation? How much more?

The distance from Blue Ridge Hill to the blackberry patch by way of Luna Pond can be found by adding $\frac{3}{4}$ and $\frac{1}{3}$.

$$\frac{3}{4} + \frac{1}{3} = \frac{3 \cdot 3}{4 \cdot 3} + \frac{1 \cdot 4}{3 \cdot 4} = \frac{9}{12} + \frac{4}{12} = \frac{13}{12}$$

The distance by way of White Rock Creek is found by adding $\frac{1}{6}$ and $\frac{7}{8}$.

$$\frac{1}{6} + \frac{7}{8} = \frac{1}{2 \cdot 3} + \frac{7}{2 \cdot 2 \cdot 2} = \frac{2 \cdot 2 \cdot 1}{2 \cdot 2 \cdot 2 \cdot 3} + \frac{7 \cdot 3}{2 \cdot 2 \cdot 2 \cdot 3}$$
$$= \frac{4 + 21}{24} = \frac{25}{24}$$

Write $\frac{13}{12}$ with the same denominator as $\frac{25}{24}$.

$$\frac{13}{12} = \frac{13 \cdot 2}{12 \cdot 2} = \frac{26}{24}$$

Since $26 > 25$, $\frac{26}{24} > \frac{25}{24}$. Thus, $\frac{13}{12} > \frac{25}{24}$ so it is farther by way of Luna Pond.

To find out how much farther it is, subtract $\frac{25}{24}$ from $\frac{13}{12}$. Using the fractions above,

$$\frac{13}{12} - \frac{25}{24} = \frac{26}{24} - \frac{25}{24} = \frac{1}{24}.$$

Thus, it is $\frac{1}{24}$ of a mile farther.

Answer: $\frac{11}{30}$ inch more in June

## 4.5 EXERCISES A

*Evaluate each expression.*

**1.** $\left(\dfrac{5}{9} + \dfrac{4}{9}\right) \cdot \dfrac{11}{15}$

**2.** $\dfrac{4}{5} \div \dfrac{2}{3} - \dfrac{2}{5}$

**3.** $\dfrac{4}{3} \cdot \left(\dfrac{7}{8} - \dfrac{1}{4}\right)$

**4.** $\dfrac{2}{3} \div \dfrac{1}{9} \cdot \dfrac{1}{4}$

**5.** $\dfrac{15}{16} - \dfrac{3}{4} \cdot \dfrac{1}{2}$

**6.** $\left(\dfrac{4}{7} - \dfrac{3}{14}\right) \div \dfrac{4}{21}$

**7.** $\sqrt{\dfrac{1}{4}} - \left(\dfrac{1}{4}\right)^2$

**8** $\left(\dfrac{3}{5}\right)^2 - \dfrac{1}{5} + \dfrac{6}{25}$

**9.** $\dfrac{2}{3} \cdot \dfrac{6}{11} - 2 \div \dfrac{11}{2}$

**10** $\sqrt{\dfrac{1}{9}} + \dfrac{6}{7} \div \dfrac{3}{14} - 3 \cdot \dfrac{1}{9}$

**11.** $\left(\dfrac{4}{5} - \dfrac{1}{10}\right)^2 \div \dfrac{7}{10}$

**12.** $\dfrac{5}{6} \div \dfrac{2}{3} - \dfrac{1}{4} \div \dfrac{1}{2} - \left(\dfrac{1}{2}\right)^2$

*Which of the two fractions is larger?*

**13.** $\dfrac{4}{15}$ or $\dfrac{1}{4}$

**14.** $\dfrac{29}{21}$ or $\dfrac{25}{18}$

**15.** $\dfrac{4}{6}$ or $\dfrac{27}{42}$

**16.** $\dfrac{35}{11}$ or $\dfrac{41}{13}$

**17** $\dfrac{17}{30}$ or $\dfrac{21}{40}$

**18.** $\dfrac{19}{8}$ or $\dfrac{81}{36}$

*Answer true or false.*

**19.** $\dfrac{1}{4} < \dfrac{1}{3}$

**20.** $\dfrac{3}{8} > \dfrac{4}{7}$

**21.** $\dfrac{25}{3} < \dfrac{35}{4}$

**22.** $3\dfrac{2}{5} > 3\dfrac{5}{12}$

**23.** $\dfrac{1}{2} \div \dfrac{1}{4} > \dfrac{1}{3} \div \dfrac{1}{9}$

**24.** $\dfrac{1}{2} < \dfrac{1}{3} < \dfrac{1}{4}$

*Solve.*

**25.** Fahrenheit temperature can be found by multiplying Celsius temperature by $\dfrac{9}{5}$ and adding 32°. If Celsius temperature is 25°, what is the Fahrenheit temperature?

**26** Carl did $\dfrac{1}{4}$ of a job on Monday morning and $\dfrac{7}{40}$ in the afternoon. He did $\dfrac{1}{4}$ of the job on Tuesday morning and $\dfrac{1}{5}$ on Tuesday afternoon. On which day did he do more of the job? How much more?

## FOR REVIEW

*Perform the indicated operations.*

**27.** $\begin{array}{r} 3\dfrac{2}{3} \\ + \, 2\dfrac{1}{6} \\ \hline \end{array}$

**28.** $\begin{array}{r} 4\dfrac{2}{7} \\ - \, 3\dfrac{5}{14} \\ \hline \end{array}$

**29.** $\begin{array}{r} 416\dfrac{1}{9} \\ + \, 201\dfrac{2}{3} \\ \hline \end{array}$

**30.** $\begin{array}{r} 8\dfrac{3}{10} \\ 2\dfrac{4}{5} \\ + \, 5\dfrac{2}{5} \\ \hline \end{array}$

ANSWERS: 1. $\frac{11}{15}$  2. $\frac{4}{5}$  3. $\frac{5}{6}$  4. $1\frac{1}{2}$  5. $\frac{9}{16}$  6. $1\frac{7}{8}$  7. $\frac{7}{16}$  8. $\frac{2}{5}$  9. 0  10. 4  11. $\frac{7}{10}$  12. $\frac{1}{2}$  13. $\frac{4}{15} > \frac{1}{4}$ 14. $\frac{29}{21} < \frac{25}{18}$  15. $\frac{4}{6} > \frac{27}{42}$  16. $\frac{35}{11} > \frac{41}{13}$  17. $\frac{17}{30} > \frac{21}{40}$  18. $\frac{19}{8} > \frac{81}{36}$  19. true  20. false  21. true  22. false  23. false 24. false  25. 77°  26. Carl did $\frac{1}{40}$ more of the job on Tuesday  27. $5\frac{5}{6}$  28. $\frac{13}{14}$  29. $617\frac{7}{9}$  30. $16\frac{1}{2}$

## 4.5  EXERCISES B

*Evaluate each expression.*

**1.** $\left(\dfrac{9}{7} - \dfrac{3}{7}\right) \cdot \dfrac{7}{12}$

**2.** $\dfrac{2}{9} \cdot \dfrac{3}{4} - \dfrac{1}{6}$

**3.** $\left(\dfrac{8}{7} - \dfrac{9}{14}\right) \div \dfrac{3}{4}$

**4.** $\dfrac{3}{5} \cdot \dfrac{5}{6} \div \dfrac{1}{4}$

**5.** $\dfrac{9}{11} - \dfrac{2}{7} \div \dfrac{11}{7}$

**6.** $\dfrac{4}{21} \cdot \left(\dfrac{9}{14} - \dfrac{1}{7}\right)$

**7.** $\left(\dfrac{2}{3}\right)^2 - \sqrt{\dfrac{4}{81}}$

**8.** $\dfrac{5}{8} - \dfrac{1}{4} - \left(\dfrac{1}{2}\right)^2$

**9.** $\dfrac{15}{16} \div \dfrac{3}{8} - 4 \cdot \dfrac{1}{12}$

**10.** $\dfrac{14}{20} \cdot \dfrac{8}{7} - \sqrt{\dfrac{1}{25}} + \dfrac{1}{3} \div \dfrac{5}{3}$

**11.** $\dfrac{12}{21} \div \left(\dfrac{7}{3} - \dfrac{5}{3}\right)^2$

**12.** $\dfrac{6}{11} \div \dfrac{3}{22} - \left(\dfrac{1}{4}\right)^2 - \dfrac{5}{2} \div \dfrac{10}{3}$

*Which of the two fractions is larger?*

**13.** $\dfrac{2}{5}$ or $\dfrac{3}{7}$

**14.** $\dfrac{19}{18}$ or $\dfrac{37}{36}$

**15.** $\dfrac{22}{33}$ or $\dfrac{14}{22}$

**16.** $\dfrac{49}{10}$ or $\dfrac{64}{15}$

**17.** $\dfrac{31}{50}$ or $\dfrac{13}{20}$

**18.** $\dfrac{25}{6}$ or $\dfrac{109}{28}$

*Answer true or false.*

**19.** $\dfrac{2}{5} < \dfrac{4}{9}$

**20.** $\dfrac{7}{11} > \dfrac{13}{20}$

**21.** $\dfrac{71}{8} > \dfrac{53}{6}$

**22.** $5\dfrac{9}{13} < 5\dfrac{17}{26}$

**23.** $\dfrac{3}{8} \cdot \dfrac{4}{9} < \dfrac{2}{3} \div 5$

**24.** $\dfrac{1}{2} > \dfrac{1}{3} > \dfrac{1}{4}$

*Solve.*

**25.** Celsius temperature can be found by subtracting 32° from Fahrenheit temperature and then multiplying by $\frac{5}{9}$. If Fahrenheit temperature is 95°, what is the Celsius temperature?

**26.** Dorci hiked $5\frac{3}{16}$ miles on Thursday morning and $4\frac{1}{8}$ miles on Thursday afternoon. She went $4\frac{1}{6}$ miles on Friday morning and $5\frac{1}{8}$ miles on Friday afternoon. On which day did she cover more miles? How much more?

**FOR REVIEW**

*Perform the indicated operations.*

**27.** $5\dfrac{3}{8}$
  $+ 7\dfrac{1}{4}$

**28.** $9\dfrac{2}{5}$
  $- 8\dfrac{9}{10}$

**29.** $925\dfrac{3}{10}$
  $- 416\dfrac{3}{8}$

**30.** $4\dfrac{1}{2}$
  $2\dfrac{3}{8}$
  $+ 9\dfrac{5}{24}$

## 4.5 EXERCISES C

*Evaluate each expression.*

**1.** $\left(\dfrac{1}{3} - \dfrac{1}{4}\right)^2 \div \dfrac{5}{12} + \sqrt{\dfrac{6}{25} - \dfrac{2}{25}} - \dfrac{3}{5} \cdot \dfrac{1}{6}$

**2.** $\left[\left(\dfrac{4}{7} - \dfrac{3}{14}\right) \div \dfrac{1}{21}\right] - \sqrt{\left(\dfrac{2}{7} + \dfrac{1}{7}\right)^2}$

$\left[\text{Answer: } 7\frac{1}{14}\right]$

## 4.6 MORE APPLICATIONS

This section is designed to give practice with a variety of applications. Read each problem carefully to determine which operations are required and in what order.

---

### EXAMPLE 1   CONSTRUCTION APPLICATION

Vince is building a cabin in the mountains. To minimize transportation cost he wants to buy boards from which he can cut pieces that are either all 10 inches in length, all 12 inches, or all 16 inches. What is the shortest board that he can buy and have no waste?

To prevent waste the board must be a length which is a multiple of each of the numbers 10, 12, and 16. The shortest such board will be of length the least common multiple (LCM).

$10 = 2 \cdot 5$      Factor 10
$12 = 2 \cdot 2 \cdot 3 = 2^2 \cdot 3$      Factor 12
$16 = 2 \cdot 2 \cdot 2 \cdot 2 = 2^4$      Factor 16

The LCM is $2 \cdot 2 \cdot 2 \cdot 2 \cdot 3 \cdot 5 = 2^4 \cdot 3 \cdot 5 = 240$. Thus the shortest board that will work is 240 inches or 20 ft long.

### PRACTICE EXERCISE 1

An artist makes three kinds of geometric decorations out of wire. He uses 8 inches of wire for one kind, 14 inches of wire for another, and 21 inches for the third. What is the shortest length of wire from which he can cut pieces all of which are either 8 inches, 14 inches, or 21 inches, with no waste?

Answer:  168 inches, the LCM of 8, 14, and 21

---

### EXAMPLE 2   APPLICATION TO AGRICULTURE

A grain dealer needed to fill a silo with wheat in three days. The first day deliveries filled $\frac{3}{8}$ of the silo. The second day an additional $\frac{1}{3}$ of the silo was filled. See Figure 4.4.

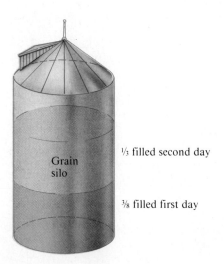

⅓ filled second day

Grain silo

⅜ filled first day

**Figure 4.4**

### PRACTICE EXERCISE 2

A family business is owned by a father and his two children. The son owns $\frac{5}{14}$ of the business and the daughter owns $\frac{2}{7}$ of it.

**(a)** How much of the silo was filled after two days?

The fraction in the silo is the sum of the two days' deliveries

$$\frac{3}{8} + \frac{1}{3} = \frac{3 \cdot 3}{8 \cdot 3} + \frac{1 \cdot 8}{3 \cdot 8}$$    LCD = 8 · 3 since 8 and 3 have no common factors

$$= \frac{9}{24} + \frac{8}{24}$$

$$= \frac{9 + 8}{24} = \frac{17}{24}$$

Thus, $\frac{17}{24}$ of the silo was filled.

**(b)** What fraction of the silo must be filled the third day to complete the job?

First note that the fraction for a full silo is 1 and then subtract $\frac{17}{24}$ from 1.

$$1 - \frac{17}{24} = \frac{24}{24} - \frac{17}{24}$$    LCD = 24

$$= \frac{24 - 17}{24} = \frac{7}{24}$$

Thus, $\frac{7}{24}$ of the silo must be filled the third day.

**(a)** How much do the children own?

**(b)** How much does the father own?

Answers:  (a) $\frac{9}{14}$   (b) $\frac{5}{14}$

---

### EXAMPLE 3    APPLICATION TO WORK

Carrie worked $6\frac{1}{2}$ hours on Monday, $5\frac{2}{3}$ hours on Tuesday, and $2\frac{1}{4}$ hours on Wednesday. How many hours did she work during the three days?

To find her total hours, add the three mixed numbers.

$$6\frac{1}{2} = 6\frac{6}{12}$$    LCD = 12

$$5\frac{2}{3} = 5\frac{8}{12}$$

$$+ 2\frac{1}{4} = 2\frac{3}{12}$$

$$13\frac{17}{12} = 13 + 1 + \frac{5}{12} = 14\frac{5}{12}$$

Carrie worked $14\frac{5}{12}$ hours.

### PRACTICE EXERCISE 3

On Thursday, Dennis ran $3\frac{2}{3}$ miles, on Friday he ran $4\frac{1}{2}$ miles, and on Saturday, $\frac{3}{4}$ mile. What is the total distance that he ran on the three days?

Answer: $8\frac{11}{12}$ mi

---

## 4.6  EXERCISES A

*Solve.*

**1.** Pam did $\frac{5}{8}$ of a job and Marsha did $\frac{1}{4}$ of the job. How much of the job was left to do?

**2.** It takes $6\frac{3}{4}$ hours to roast a turkey. If it has been in the oven for $2\frac{1}{4}$ hours, how long will it be before it is done?

**3.** A grain dealer received shipments of $7\frac{3}{8}$ tons, $9\frac{1}{6}$ tons, and $5\frac{1}{4}$ tons of wheat. How much wheat did she receive in the three shipments?

**4.** If the dealer in Exercise 3 needed a total of 30 tons of wheat, how much more must be received?

**5.** For a wiring job, an electronic technician needs lengths of 9 cm, 12 cm, and 15 cm. What is the shortest piece of wire from which he can cut pieces of the same length with no waste?

**6.** In June, Pittsburgh had two rainstorms, one which dropped $\frac{5}{16}$ inch and the other $2\frac{1}{2}$ inches of precipitation. During August, there were three storms with $\frac{1}{4}$, $\frac{7}{8}$, and $1\frac{5}{12}$ inches of rain. During which month did Pittsburgh have more precipitation? How much more?

**7.** Celsius temperature can be found by subtracting 32° from Fahrenheit temperature and then multiplying by $\frac{5}{9}$. If Fahrenheit temperature is 50°, what is the Celsius temperature?

**8.** Three sides of a four-sided pen are $120\frac{1}{4}$ ft, $135\frac{2}{3}$ ft, and $160\frac{1}{2}$ ft. If the perimeter of the pen is $515\frac{4}{15}$ ft, what is the length of the fourth side?

**9** Lon worked $7\frac{3}{8}$ days, $6\frac{2}{5}$ days, $5\frac{4}{5}$ days, $8\frac{1}{4}$ days, and $7\frac{1}{2}$ days to complete a project. What was his total number of days worked?

**10** What was Lon paid for working the days in Exercise 9 if he received $40 per day?

**11.** On Friday Carlos walked $1\frac{1}{5}$ mi and ran $\frac{5}{8}$ mi. On Saturday, he walked $\frac{3}{4}$ mi and ran $1\frac{1}{6}$ mi. On which day did he travel a greater distance? How much greater?

**12** The stock of the Mina Corporation opened at $\$121\frac{3}{8}$. It increased $\$1\frac{3}{4}$ during the first hour of trading, but fell $\$3\frac{1}{8}$ before noon. It then increased $\$2\frac{7}{8}$ before closing time. What was the closing value of the stock?

## FOR REVIEW

*Evaluate each expression.*

**13.** $\left(\frac{7}{9} - \frac{1}{3}\right) \div \frac{1}{3} \cdot \frac{1}{4}$

**14.** $\sqrt{\frac{25}{16} - \left(\frac{3}{4}\right)^2}$

**15.** $\frac{2}{5} \div \frac{3}{10} - \frac{4}{9} \cdot \frac{3}{2}$

*Answer true or false.*

**16.** $\frac{3}{8} < \frac{8}{24}$

**17.** $\frac{22}{7} < \frac{10}{3}$

**18.** $\frac{4}{5} \cdot \frac{1}{2} > \frac{5}{6} \div \frac{10}{3}$

---

ANSWERS: 1. $\frac{1}{8}$  2. $4\frac{1}{2}$ hours  3. $21\frac{19}{24}$ tons  4. $8\frac{5}{24}$ tons  5. 180 cm  6. $\frac{13}{48}$ inches more in June  7. 10°  8. $99\frac{3}{20}$ ft  9. $35\frac{13}{40}$ days  10. $1413  11. $\frac{11}{120}$ mi farther on Saturday  12. $\$122\frac{7}{8}$  13. $\frac{1}{3}$  14. $\frac{11}{16}$  15. $\frac{2}{3}$  16. false  17. true  18. true

## 4.6 EXERCISES B

*Solve.*

**1.** In the morning Paul completed $\frac{5}{12}$ of his project for the day. By the middle of the afternoon he had done only another $\frac{1}{6}$ of the project. How much of the job was left to do?

**2.** Victoria must wait $5\frac{5}{6}$ hours for the results of her test. If she has been waiting for $4\frac{1}{3}$ hours, how much longer must she wait?

**3.** A contractor for a construction job ordered three loads of gravel. One weighed $2\frac{1}{4}$ tons, one $5\frac{5}{6}$ tons, and the other $4\frac{5}{8}$ tons. What total weight of gravel did he receive?

**4.** If the contractor in Exercise 3 really needed 20 tons of gravel, how much more must he order?

**5.** To construct a brick wall, three lengths of bricks are used, 6 inches, 14 inches, and 16 inches. What is the shortest length of wall that can be constructed if only one length of brick is used in any one row?

**6.** During the first week of the spring thaw, two measurements of lake level were made. On the first, the lake level had increased $2\frac{3}{4}$ ft from the previous measurement. On the second, the level had increased another $3\frac{5}{12}$ ft. Three measurements the second week showed increases of $1\frac{5}{6}$ ft, $2\frac{2}{3}$ ft, and $2\frac{1}{4}$ ft. During which week did the level increase more? How much more?

**7.** Fahrenheit temperature can be found by multiplying Celsius temperature by $1\frac{4}{5}$ and adding 32°. If Celsius temperature is 15°, what is the Fahrenheit temperature?

**8.** Jo must have $216\frac{1}{2}$ points to win a contest. If for three projects she has received $25\frac{1}{4}$ points, $17\frac{1}{6}$ points, and $42\frac{3}{8}$ points, how many more points does she need to win?

**9.** Fran worked for $32\frac{3}{4}$ hours, $28\frac{5}{6}$ hours, $42\frac{1}{6}$ hours, and $19\frac{3}{4}$ hours over a four-week period. What was her total number of hours worked?

**10.** Fran in Exercise 9 is paid $10 per hour. How much should she be paid for the four weeks' work?

**11.** On Wednesday a grocery distributor received two shipments of grapes, $395\frac{3}{8}$ lb and $215\frac{3}{4}$ lb. On Thursday two more shipments of $198\frac{1}{2}$ lb and $416\frac{1}{8}$ lb were received. On which day did he receive more grapes? How many pounds more?

**12.** The opening quote for stock of the Nina Corporation was $62\frac{1}{8}$. During the morning it dropped $4\frac{3}{4}$ but increased again $6\frac{5}{8}$ before noon. It then dropped $1\frac{1}{8}$ before closing time. What was the closing value of the stock?

## FOR REVIEW

*Evaluate each expression.*

**13.** $\left(\frac{5}{8}-\frac{1}{4}\right)\cdot\frac{1}{9}\div\frac{3}{4}$

**14.** $\left(\frac{3}{5}\right)^2-\sqrt{\frac{1}{25}}$

**15.** $\frac{3}{8}\cdot\frac{2}{9}-\frac{3}{10}\div4\frac{1}{2}$

*Answer true or false.*

**16.** $\frac{7}{8}>\frac{17}{20}$

**17.** $\frac{35}{6}<\frac{43}{7}$

**18.** $\frac{9}{2}\div\frac{3}{8}>\frac{10}{3}\cdot\frac{15}{4}$

## 4.6 EXERCISES C

*Let a represent a natural number.*

**1.** If one shipment was $2\frac{1}{2}a$ units and a second shipment was $4\frac{2}{3}a$ units, what was the total of the two shipments?

$\left[\text{Answer: } 7\frac{1}{6}a \text{ units}\right]$

**2.** Lonna has $10\frac{3}{4}a$ units and Maria has $7\frac{2}{5}a$ units. How many more units does Lonna have?

## CHAPTER 4 REVIEW

### KEY WORDS

**4.1** **Like fractions** have the same denominator.
**Unlike fractions** have different denominators.

**4.2** A whole number that is in the lists of nonzero multiples of two or more numbers is called a **common multiple.**

The **least common multiple (LCM)** of two or more counting numbers is the smallest of all common multiples of the numbers.

**4.3** The **least common denominator (LCD)** is the least common multiple of the denominators of fractions in an addition or subtraction problem.

### KEY CONCEPTS

**4.1** **1.** Like fractions are added or subtracted by adding or subtracting the numerators and placing the result over the common denominator.

**2.** *Never* add or subtract fractions by adding or subtracting denominators.

**4.2** Three methods of finding LCM's are the listing method, finding prime factors, and the special optional algorithm.

**4.3** Fractions to be added or subtracted must be changed to equal fractions with the same LCD. You should *not* find the LCD when multiplying or dividing fractions.

**4.4** The method used for adding or subtracting mixed numbers is to add or subtract the whole number parts and the proper fraction parts separately. Do *not* try to multiply or divide mixed fractions using this method.

**4.5** **1.** Operations should be performed in the following order.
    **a.** Operate within parentheses.
    **b.** Evaluate powers and roots.
    **c.** Multiply and divide from left to right.
    **d.** Add and subtract from left to right.

**2.** To tell which of two fractions is larger, change them to equal fractions with the same LCD and compare numerators.

### REVIEW EXERCISES

*Part I*

**4.1** *Perform the indicated operations.*

**1.** $\dfrac{3}{22} + \dfrac{9}{22}$

**2.** $\dfrac{17}{18} - \dfrac{1}{18}$

**3.** $\dfrac{4}{25} + \dfrac{1}{25}$

**4.** $\dfrac{8}{35} - \dfrac{3}{35}$

**5.** $\dfrac{7}{10} + \dfrac{11}{10} - \dfrac{3}{10}$

**6.** $\dfrac{9}{17} - \dfrac{3}{17} - \dfrac{6}{17}$

**4.2**  *Find the LCM of the given numbers.*

**7.** 40 and 8

**8.** 12 and 30

**9.** 15 and 35

**10.** 9 and 30

**11.** 120 and 14

**12.** 12, 14, and 18

**4.3**  *Perform the indicated operations.*

**13.** $\dfrac{5}{12} + \dfrac{3}{18}$

**14.** $3 + \dfrac{7}{9}$

**15.** $\dfrac{4}{7} - \dfrac{1}{6}$

**16.** $3 - \dfrac{4}{3}$

**17.** $\dfrac{12}{15} + \dfrac{2}{3} - \dfrac{1}{5}$

**18.** $\dfrac{5}{3} - 1 + \dfrac{2}{7}$

**4.4**  *Perform the indicated operations.*

**19.**  $\begin{array}{r} 2\dfrac{1}{8} \\ + 7\dfrac{3}{4} \\ \hline \end{array}$

**20.**  $\begin{array}{r} 5\dfrac{2}{7} \\ - 3\dfrac{2}{3} \\ \hline \end{array}$

**21.**  $\begin{array}{r} 6\dfrac{2}{5} \\ + 8\dfrac{7}{8} \\ \hline \end{array}$

**22.**  $\begin{array}{r} 9\dfrac{3}{7} \\ - 4\dfrac{5}{6} \\ \hline \end{array}$

**23.**  $\begin{array}{r} 215\dfrac{1}{4} \\ + 375\dfrac{2}{11} \\ \hline \end{array}$

**24.**  $\begin{array}{r} 627\dfrac{1}{2} \\ - 421\dfrac{5}{8} \\ \hline \end{array}$

**4.5**  *Evaluate each expression.*

**25.** $\left(\dfrac{2}{5} - \dfrac{1}{15}\right) \div \dfrac{5}{3}$

**26.** $\sqrt{\dfrac{4}{9}} + \left(\dfrac{1}{3}\right)^2$

**27.** $\dfrac{1}{8} \cdot \dfrac{2}{3} - \dfrac{1}{12}$

**28.** $\dfrac{1}{2} \cdot \left(\dfrac{7}{9} + \dfrac{1}{3}\right) \div \dfrac{5}{6}$

**29.** $\left(\dfrac{3}{8} - \dfrac{1}{4}\right)^2 \cdot \dfrac{8}{9}$

**30.** $\dfrac{1}{6} \div \dfrac{5}{9} - \left(\dfrac{1}{5}\right)^2$

*Which of the two fractions is larger?*

**31.** $\dfrac{4}{17}$ or $\dfrac{1}{4}$

**32.** $\dfrac{8}{11}$ or $\dfrac{5}{7}$

**33.** $\dfrac{38}{9}$ or $\dfrac{47}{11}$

*Answer true or false.*

**34.** $\dfrac{7}{12} < \dfrac{10}{18}$

**35.** $\dfrac{25}{3} > \dfrac{41}{5}$

**36.** $7\dfrac{19}{21} < 7\dfrac{49}{51}$

**4.6** *Solve.*

**37.** A contractor needs steel rods which measure 2 feet, 6 feet, and 9 feet. He will cut only one length from each rod. What is the shortest rod that he can order and be able to cut each length with no waste?

**38.** Carla worked for $6\frac{1}{2}$ hours, $8\frac{3}{4}$ hours, $9\frac{1}{6}$ hours, and $7\frac{1}{4}$ hours the first four days of the week. How many total hours did she work in the four days?

**Part II**

**39.** It takes $3\frac{11}{15}$ weeks to complete a building project, and the contractor has been working for $2\frac{2}{5}$ weeks. How much longer will it be before the job is finished?

**40.** Robert ran $7\frac{3}{4}$ miles and $6\frac{2}{3}$ miles during one week. The next week he ran $5\frac{1}{2}$ miles, $3\frac{1}{3}$ miles, and $5\frac{1}{4}$ miles. During which week did he run farther? How much farther?

*Find the LCM of the given numbers.*

**41.** 16 and 22

**42.** 35 and 45

**43.** 6, 14, and 24

*Place the appropriate sign, $<$ or $>$, between the two fractions.*

**44.** $\dfrac{5}{6}$  $\dfrac{9}{11}$

**45.** $\dfrac{16}{7}$  $\dfrac{23}{9}$

**46.** $\dfrac{41}{37}$  $\dfrac{62}{55}$

*Perform the indicated operations.*

**47.**
$$3\frac{1}{2}$$
$$+\,7\frac{2}{3}$$

**48.**
$$16\frac{1}{9}$$
$$-\,12\frac{17}{18}$$

**49.**
$$6\frac{2}{5}$$
$$9\frac{1}{10}$$
$$+\,12\frac{14}{15}$$

**50.** $\left(\dfrac{2}{3} - \dfrac{1}{9}\right) \div \dfrac{11}{18}$

**51.** $\dfrac{6}{11} + \dfrac{13}{22} - \dfrac{5}{33}$

**52.** $\sqrt{\dfrac{25}{64}} - \left(\dfrac{1}{4}\right)^2$

ANSWERS: 1. $\frac{6}{11}$  2. $\frac{8}{9}$  3. $\frac{1}{5}$  4. $\frac{1}{7}$  5. $\frac{3}{2}$ or $1\frac{1}{2}$  6. 0  7. 40  8. 60  9. 105  10. 90  11. 840  12. 252  13. $\frac{7}{12}$  14. $\frac{34}{9}$ or $3\frac{7}{9}$  15. $\frac{17}{42}$  16. $\frac{5}{3}$ or $1\frac{2}{3}$  17. $\frac{19}{15}$ or $1\frac{4}{15}$  18. $\frac{20}{21}$  19. $9\frac{7}{8}$  20. $1\frac{13}{21}$  21. $15\frac{11}{40}$  22. $4\frac{25}{42}$  23. $590\frac{19}{44}$  24. $205\frac{7}{8}$  25. $\frac{1}{5}$  26. $\frac{7}{9}$  27. 0  28. $\frac{2}{3}$  29. $\frac{1}{72}$  30. $\frac{13}{50}$  31. $\frac{1}{4}$ is larger  32. $\frac{8}{11}$ is larger  33. $\frac{47}{11}$ is larger  34. false  35. true  36. true  37. 18 ft  38. $31\frac{2}{3}$ hours  39. $1\frac{1}{3}$ weeks  40. $\frac{1}{3}$ mi more the first week  41. 176  42. 315  43. 168  44. $>$  45. $<$  46. $<$  47. $11\frac{1}{6}$  48. $3\frac{1}{6}$  49. $28\frac{13}{30}$  50. $\frac{10}{11}$  51. $\frac{65}{66}$  52. $\frac{9}{16}$

*Find the LCM of the given numbers.*

**1.** 14 and 15

1. _____

**2.** 8, 12, and 20

2. _____

*Perform the indicated operations.*

**3.** $\dfrac{5}{21} + \dfrac{2}{21}$

3. _____

**4.** $\dfrac{14}{15} - \dfrac{4}{15}$

4. _____

**5.** $\dfrac{7}{12} + \dfrac{3}{8}$

5. _____

**6.** $\dfrac{9}{20} - \dfrac{1}{6}$

6. _____

**7.** $\dfrac{12}{7} - \dfrac{13}{14} + \dfrac{1}{2}$

7. _____

**8.** $\dfrac{7}{8} - \dfrac{5}{12} - \dfrac{1}{6}$

8. _____

**9.** $\quad 2\dfrac{5}{9}$
$\quad + 7\dfrac{1}{3}$

9. _____

**10.** $\quad 42\dfrac{7}{15}$
$\quad - 33\dfrac{3}{5}$

10. _____

**11.** $\quad 3\dfrac{3}{4}$
$\quad\; 8\dfrac{1}{3}$
$\quad + 5\dfrac{7}{12}$

11. _____

**12.** 
$$4\frac{3}{8}$$
$$2\frac{3}{4}$$
$$+\ 1\frac{1}{2}$$

12. _____

*Evaluate each expression.*

**13.** $\sqrt{\dfrac{4}{9}} - \dfrac{2}{3} \cdot \dfrac{15}{20} + 4 \div \dfrac{1}{2}$

13. _____

**14.** $\left(\dfrac{8}{9} - \dfrac{2}{3}\right)^2 \div \dfrac{4}{9} - \dfrac{1}{2} \cdot \dfrac{1}{5}$

14. _____

*Answer true or false.*

**15.** $\dfrac{17}{20} < \dfrac{27}{30}$

15. _____

**16.** $5\dfrac{12}{13} > 5\dfrac{43}{40}$

16. _____

*Solve.*

**17.** During one hour, Maria made three sales of material: $7\frac{1}{2}$ yd, $3\frac{1}{4}$ yd, and $2\frac{7}{8}$ yd. What was the total yardage sold?

17. _____

**18.** A repair job is estimated to require $6\frac{5}{6}$ hours to complete. If Hank has been working for $2\frac{11}{12}$ hours, how much longer will he have to work to finish the job?

18. _____

# Decimals

## 5.1 DECIMAL NOTATION

### 1 DECIMAL FRACTIONS AND DECIMALS

As we saw in Chapter 3, a fraction has many names. For example,

$$\frac{1}{2}, \quad \frac{2}{4}, \quad \frac{3}{6}, \quad \frac{4}{8}, \quad \frac{5}{10}, \quad \frac{25}{50}, \quad \frac{50}{100}$$

are all names for the fraction *one half*. A fraction which has a denominator equal to a power of 10 (10, 100, 1000, 10,000, and so forth) is called a **decimal fraction.** The following are decimal fractions.

$\frac{5}{10}$ is read "**five tenths.**"

$\frac{3}{100}$ is read "**three hundredths.**"

$\frac{47}{1000}$ is read "**forty-seven thousandths.**"

$\frac{121}{10,000}$ is read "**one hundred twenty-one ten thousandths.**"

Decimal fractions are usually referred to simply as **decimals.** Each decimal has two forms, a *fractional form,*

$$\frac{5}{10}, \quad \frac{3}{100}, \quad \frac{47}{1000}, \quad \frac{121}{10,000},$$

and a *decimal form* or *decimal notation,*

$$0.5, \quad 0.03, \quad 0.047, \quad \text{and} \quad 0.0121.$$

233

Recall that the number system is a place-value system. The decimal form of a number is part of this system, as shown in the following revised place-value chart first seen in Chapter 1. Notice that the value of any position is one tenth the value of the position to its immediate left.

| Thousands | Hundreds | Tens | Ones | Tenths | Hundredths | Thousandths | Ten thousandths | Hundred thousandths | Millionths |
|---|---|---|---|---|---|---|---|---|---|
| 1000 | 100 | 10 | 1 | $\frac{1}{10}$ | $\frac{1}{100}$ | $\frac{1}{1000}$ | $\frac{1}{10,000}$ | $\frac{1}{100,000}$ | $\frac{1}{1,000,000}$ |

When a number is written in decimal form, a period, called a **decimal point,** separates the ones digit from the tenths digit. It serves as a reference point: positions to the left of the decimal point end in *s* while those to the right end in *ths*. Consider the number 352.1783 shown below.

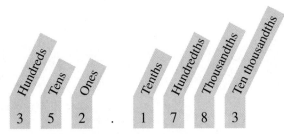

The decimal point separates the ones digit from the tenths digit.

This number is read    **three hundred fifty-two *and* one thousand seven hundred eighty-three ten thousandths.**

The three points to remember about word names for decimals are given below.

### To Read a Decimal

1. The part of the number to the left of the decimal point is read just like a whole number.
2. The decimal point is read "and."
3. The part of the number to the right of the decimal point is read like a whole number but is followed by the name of the place filled by the farthest right digit.

### EXAMPLE 1    WORD NAMES FOR DECIMALS

Give the word name for each decimal.

**(a)** .5 or 0.5 is read "five tenths."
         ↑
    tenths place

Both .5 and 0.5 are correct ways to write five tenths. The second form is preferred since it calls attention to the decimal point.

### PRACTICE EXERCISE 1

Give the word name for each decimal.

**(a)** 0.3

**(b)** 3.5 is read "three and five tenths."

**(c)** 0.03 is read "three hundredths."
↑
hundredths place

**(d)** 4.03 is read "four and three hundredths."

**(e)** 0.008 is read "eight thousandths."
↑
thousandths place

**(f)** 21.008 is read "twenty-one and eight thousandths."

**(g)** 4.107 is read "four and one hundred seven thousandths."

**(b)** 5.9

**(c)** 0.08

**(d)** 2.01

**(e)** 0.005

**(f)** 57.005

**(g)** 6.242

Answers: (a) three tenths
(b) five and nine tenths    (c) eight
hundredths    (d) two and one
hundredth    (e) five thousandths
(f) fifty-seven and five thousandths
(g) six and two hundred forty-two
thousandths

If there are no digits (or possibly only zeros) to the right of the decimal point, the decimal is actually a whole number. In such cases, the decimal point is usually omitted. For example,

$$47, \quad 47., \quad \text{and} \quad 47.0$$

are all names for the whole number forty-seven.

| EXAMPLE 2    WRITING NUMBERS IN DECIMAL NOTATION |
| --- |

Write each number in decimal notation.

**(a)** Three  and  nine tenths is written 3.9.
      ↓      ↓      ↓
      3      .      9

**(b)** Thirteen  and  three hundredths is written 13.03.
       ↓      ↓      ↓
       13     .      03

**(c)** Five thousand four and twenty-seven thousandths is written 5004.027.

**(d)** Forty-two thousandths is written 0.042.

| PRACTICE EXERCISE 2 |
| --- |

Write each number in decimal notation.

**(a)** Six and four tenths

**(b)** Thirty-five and two hundredths

**(c)** Eight thousand forty-five and six hundred eight thousandths

**(d)** Ninety-three ten thousandths

Answers: (a) 6.4    (b) 35.02
(c) 8045.608    (d) 0.0093

| **H I N T** |
| --- |

A less formal method for reading decimals is commonly used. For example,

3.42       is often read    "three *point* four two,"
10.001    is often read    "one zero *point* zero zero one,"
6254.398  is often read    "six two five four *point* three nine eight."

We simply read each digit in order from left to right and say "point" when we reach the decimal point.

## 2 WRITING DECIMALS IN EXPANDED NOTATION

When whole numbers were first studied, we discussed expanded notation. For example,

$$2375 \quad \text{is} \quad 2000 + 300 + 70 + 5$$

in expanded notation. This notation can be extended to all decimals, as shown below for the decimal 23.456.

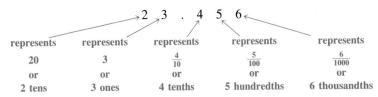

| represents | represents | represents | represents | represents |
|---|---|---|---|---|
| 20 | 3 | $\frac{4}{10}$ | $\frac{5}{100}$ | $\frac{6}{1000}$ |
| or | or | or | or | or |
| 2 tens | 3 ones | 4 tenths | 5 hundredths | 6 thousandths |

Thus, we can write 23.456 in expanded notation as

$$20 + 3 + \frac{4}{10} + \frac{5}{100} + \frac{6}{1000}.$$

---

| EXAMPLE 3 WRITING DECIMALS IN EXPANDED NOTATION | PRACTICE EXERCISE 3 |
|---|---|

Write each decimal in expanded notation.

**(a)** 432.78

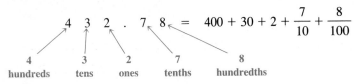

$$= \quad 400 + 30 + 2 + \frac{7}{10} + \frac{8}{100}$$

4 — hundreds    3 — tens    2 — ones    7 — tenths    8 — hundredths

**(b)** $25.6794 = 20 + 5 + \frac{6}{10} + \frac{7}{100} + \frac{9}{1000} + \frac{4}{10,000}$

**(c)** $403.002 = 400 + 3 + \frac{2}{1000}$

**(d)** $3050.7020 = 3000 + 50 + \frac{7}{10} + \frac{2}{1000}$

The final zero to the right does not affect the value of the decimal and could be omitted.

Write each decimal in expanded notation.

**(a)** 258.65

**(b)** 43.2963

**(c)** 605.007

**(d)** 1060.05030

Answers: (a) $200 + 50 + 8 + \frac{6}{10} + \frac{5}{100}$ (b) $40 + 3 + \frac{2}{10} + \frac{9}{100} + \frac{6}{1000} + \frac{3}{10,000}$ (c) $600 + 5 + \frac{7}{1000}$ (d) $1000 + 60 + \frac{5}{100} + \frac{3}{10,000}$

---

## 3 USING DECIMALS FOR AMOUNTS OF MONEY

Perhaps the most common use of decimal notation is for amounts of money. For example, we could read

    **$25.76** as **twenty-five** *and* **seventy-six hundredths dollars.**

However, since one hundredth of a dollar is one *cent* (1¢) we usually say

    **twenty-five dollars** *and* **seventy-six cents.**

Notice that the word *and* is still identified with the decimal point. We must put the decimal point in the right place. For instance,

$$\$25 \quad \text{and} \quad \$0.25$$

do not represent the same amount: \$25 or \$25.00 is twenty-five dollars while \$0.25 is twenty-five hundredths of a dollar or twenty-five cents (25¢).

| EXAMPLE 4 WRITING WORD NAMES FOR MONEY |
| --- |

Write a word name for each amount of money.

**(a)** \$4.76     Four and seventy-six hundredths dollars *or* four dollars and seventy-six cents

**(b)** \$1327.03     One thousand three hundred twenty-seven and three hundredths dollars *or* one thousand three hundred twenty-seven dollars and three cents

**(c)** \$41.00     Forty-one and zero hundredths dollars *or* forty-one dollars and no cents *or simply* forty-one dollars

| EXAMPLE 5 WRITING DECIMALS FOR AMOUNTS OF MONEY |
| --- |

Write a decimal for each amount of money.

**(a)** Three hundred twenty-seven and twelve hundredths dollars is written \$327.12.

**(b)** Two dollars and thirty-seven cents is written \$2.37.

**(c)** Seventy-six and no hundredths dollars is written \$76.00.

| PRACTICE EXERCISE 4 |
| --- |

Write a word name for each amount of money.

**(a)** \$3.98

**(b)** \$2650.07

**(c)** \$75.00

Answers: (a) Three and ninety-eight hundredths dollars *or* three dollars and ninety-eight cents (b) Two thousand six hundred fifty and seven hundredths dollars *or* two thousand six hundred fifty dollars and seven cents (c) Seventy-five and zero hundredths dollars *or* seventy-five dollars and no cents *or simply* seventy-five dollars

| PRACTICE EXERCISE 5 |
| --- |

Write a decimal for each amount of money.

**(a)** Six hundred fifty-four and twenty-three hundredths dollars

**(b)** Ten dollars and ninety-five cents

**(c)** Eighty-four and no hundredths dollars

Answers: (a) \$654.23   (b) \$10.95 (c) \$84.00

Writing checks requires writing word names for decimals. It has become customary to write fractions for the cents portion of an amount of money. For example, \$3.48 is written

$$\text{three and } \frac{48}{100} \text{ dollars}$$

instead of

three and forty-eight hundredths dollars.

This is done on the sample check in Figure 5.1.

**Figure 5.1**

# 5.1 EXERCISES A

**1.** What are fractions called that have a denominator equal to a power of 10?

**2.** What do we call the period used in decimal notation?

*Give the* formal *word name for each decimal.*

**3.** 0.7

**4.** 2.05

**5.** 43.29

**6.** 127.562

**7.** 302.008

**8.** 0.0009

*Write each number in decimal notation.*

**9.** four hundredths

**10.** two hundred twenty-eight and nine tenths

**11.** twenty-three and seven hundred forty-eight thousandths

**12.** eight hundred one and one hundred eight thousandths

**13.** three and one ten thousandth

**14.** four hundred ninety-seven and six thousand forty-three ten thousandths

*Write each decimal in expanded notation.*

**15.** 41.8

**16.** 30.07

**17.** 2050.703

**18.** 40.1005

*Write a word name for the dollar amount using dollars and cents.*

**19.** $1.47

**20.** $23.04

**21.** $162.00

**22.** $10.40

*Write a word name as it would appear on a check.*

**23.** $427.68

**24.** $20.00

**25.** $3.02

## FOR REVIEW

*The following exercises will help you prepare for the next section. Write each improper fraction as a mixed number.*

**26.** $\dfrac{9}{4}$

**27.** $\dfrac{42}{11}$

**28.** $\dfrac{103}{100}$

*Write each mixed number as an improper fraction.*

**29.** $2\dfrac{1}{5}$

**30.** $5\dfrac{7}{10}$

**31.** $7\dfrac{23}{100}$

ANSWERS: 1. decimal fractions or decimals 2. decimal point 3. seven tenths 4. two and five hundredths 5. forty-three and twenty-nine hundredths 6. one hundred twenty-seven and five hundred sixty-two thousandths 7. three hundred two and eight thousandths 8. nine ten thousandths 9. 0.04 10. 228.9 11. 23.748 12. 801.108 13. 3.0001 14. 497.6043 15. $40 + 1 + \frac{8}{10}$ 16. $30 + \frac{7}{100}$ 17. $2000 + 50 + \frac{7}{10} + \frac{3}{1000}$ 18. $40 + \frac{1}{10} + \frac{5}{10,000}$ 19. one dollar and forty-seven cents 20. twenty-three dollars and four cents 21. one hundred sixty-two dollars [and zero (or no) cents] 22. ten dollars and forty cents 23. Four hundred twenty-seven and $\frac{68}{100}$ dollars 24. Twenty and $\frac{00}{100}$ dollars 25. Three and $\frac{02}{100}$ dollars 26. $2\frac{1}{4}$ 27. $3\frac{9}{11}$ 28. $1\frac{3}{100}$ 29. $\frac{11}{5}$ 30. $\frac{57}{10}$ 31. $\frac{723}{100}$

## 5.1 EXERCISES B

**1.** Decimal fractions have denominators that are powers of what number?

**2.** The decimal point in decimal notation separates which two digits?

*Give the* formal *word name for each decimal.*

**3.** 0.9

**4.** 3.06

**5.** 52.39

**6.** 147.053

**7.** 602.007

**8.** 0.0005

*Write each number in decimal notation.*

**9.** seven hundredths

**10.** three hundred thirty-five and eight tenths

**11.** forty-two and eight hundred sixty-five thousandths

**12.** seven and two ten thousandths

**13.** five hundred one and one hundred five thousandths

**14.** three hundred six and one thousand forty-two ten thousandths

*Write each decimal in expanded notation.*

**15.** 32.5          **16.** 60.03          **17.** 51.006          **18.** 80.2006

*Write a word name for the dollar amount using dollars and cents.*

**19.** $1.58          **20.** $47.09          **21.** $20.06          **22.** $535.00

*Write a word name as it would appear on a check.*

**23.** $6.05               **24.** $70.00               **25.** $395.49

## FOR REVIEW

*The following exercises will help you prepare for the next section. Write each improper fraction as a mixed number.*

**26.** $\dfrac{13}{4}$          **27.** $\dfrac{28}{13}$          **28.** $\dfrac{201}{100}$

*Write each mixed number as an improper fraction.*

**29.** $3\dfrac{1}{4}$          **30.** $4\dfrac{7}{10}$          **31.** $8\dfrac{31}{100}$

## 5.1 EXERCISES C

**1.** The speed of light is approximately 273,213.6 feet per second. Give the formal word name for this number.

**2.** Write a word name for $10,428.64 as it would appear on a check.

## 5.2 CONVERSIONS BETWEEN FRACTIONS AND DECIMALS

### STUDENT GUIDEPOSTS

**1** Changing Decimals to Fractions

**2** Changing Fractions to Decimals

**3** Summary List of Equivalent Fractions and Decimals

## ❶ CHANGING DECIMALS TO FRACTIONS

Recall from Section 5.1 that any decimal fraction has two forms, a decimal form and a fractional form. We now consider how to change from one form to the other.

Converting decimals to fractions is simply a matter of thinking of the decimal in words and writing the fraction, as in the following table.

| Decimal | Word Name | Fraction |
|---------|-----------|----------|
| 0.7 | seven tenths | $\dfrac{7}{10}$ |
| 0.23 | twenty-three hundredths | $\dfrac{23}{100}$ |
| 0.423 | four hundred twenty-three thousandths | $\dfrac{423}{1000}$ |
| 3.1 | three and one tenth | $3\dfrac{1}{10}$ |

The resulting fraction (which may be part of a mixed number) should be reduced to lowest terms.

---

### To Change a Decimal to a Fraction   (First Method)

1. Say the word name for the decimal.
2. Write the word name as a fraction or a mixed number.
3. Reduce the fraction to lowest terms.

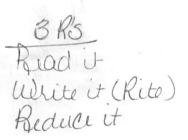

---

### EXAMPLE 1   DECIMALS TO FRACTIONS (FIRST METHOD)

Change each decimal to a fraction or mixed number.

**(a)** 0.6 is read "six tenths." Thus,

$$0.6 = \frac{6}{10} \qquad \text{Six tenths is } \tfrac{6}{10}$$

$$= \frac{\cancel{2} \cdot 3}{\cancel{2} \cdot 5} = \frac{3}{5}. \qquad \text{Reduce } \tfrac{6}{10} \text{ to lowest terms}$$

**(b)** 0.35 is read "thirty-five hundredths."

$$0.35 = \frac{35}{100}$$

$$= \frac{\cancel{5} \cdot 7}{\cancel{5} \cdot 20} = \frac{7}{20} \qquad \text{Reduce to lowest terms}$$

**(c)** 0.348 is read "three hundred forty-eight thousandths."

$$0.348 = \frac{348}{1000}$$

$$= \frac{\cancel{4} \cdot 87}{\cancel{4} \cdot 250} = \frac{87}{250}$$

### PRACTICE EXERCISE 1

Change each decimal to a fraction or mixed number.

**(a)** 0.2

**(b)** 0.65

**(c)** 0.804

**(d)** 4.8 is read "four and eight tenths."

$$4.8 = 4 + \frac{8}{10}$$

$$= 4\frac{8}{10}$$

$$= 4\frac{\cancel{2} \cdot 4}{\cancel{2} \cdot 5}$$

$$= 4\frac{4}{5}$$

**(d)** 53.625

Answers: (a) $\frac{1}{5}$  (b) $\frac{13}{20}$  (c) $\frac{201}{250}$
(d) $53\frac{5}{8}$

Another method of conversion allows us to write any decimal as a proper or improper fraction. Consider 0.38, for example. Since *two* digits follow the decimal point, put *two* zeros after 1 to form the denominator of the fraction. When the decimal point is removed, the resulting number is the numerator of the desired fraction. Sometimes the fraction must be reduced.

$$0.38 = \frac{38}{100} = \frac{\cancel{2} \cdot 19}{\cancel{2} \cdot 50} = \frac{19}{50} \qquad \text{Reduced}$$

↑ two digits after the decimal point        ↑ two zeros after 1

---

### To Change a Decimal to a Fraction (Second Method)

1. Count the digits following the decimal point; put that number of zeros after 1 in the denominator of the fraction.

2. Remove the decimal point; put the resulting whole number in the numerator of the fraction.

3. Reduce the fraction to lowest terms.

---

### EXAMPLE 2  DECIMALS TO FRACTIONS (SECOND METHOD)

Change each decimal to a fraction or mixed number using the second method of conversion.

**(a)** 0.6

$$\frac{6}{10} \xleftarrow{\hspace{1cm}} \text{Remove the decimal point and put 6 in the numerator}$$

Since one digit follows the decimal point in 0.6, 1 followed by *one* zero, or 10, is the denominator

$$= \frac{\cancel{2} \cdot 3}{\cancel{2} \cdot 5} = \frac{3}{5} \qquad \text{Reduce to lowest terms}$$

**(b)** 0.35

$$\frac{35}{100} \xleftarrow{\hspace{1cm}} \text{Remove decimal point}$$

1 followed by two zeros (0.35 two digits)

$$= \frac{\cancel{5} \cdot 7}{\cancel{5} \cdot 20} = \frac{7}{20} \qquad \text{Reduce to lowest terms}$$

### PRACTICE EXERCISE 2

Change each decimal to a fraction or mixed number using the second method of conversion.

**(a)** 0.2

**(b)** 0.65

**(c)**  0.348

$$\frac{348}{1000} \longleftarrow \text{Remove decimal point}$$
$$\longleftarrow \text{1 followed by three zeros (0.348 three digits)}$$

$$= \frac{\cancel{4} \cdot 87}{\cancel{4} \cdot 250} = \frac{87}{250} \qquad \text{Reduce}$$

**(d)**  4.8

$$\frac{48}{10} \longleftarrow \text{Remove decimal point}$$
$$\longleftarrow \text{1 followed by one zero}$$

$$= \frac{\cancel{2} \cdot 24}{\cancel{2} \cdot 5} \qquad \text{Reduce}$$

$$= \frac{24}{5} \quad \text{or} \quad 4\frac{4}{5}$$

**(c)**  0.804

**(d)**  53.625

**Answers:** (a) $\frac{1}{5}$  (b) $\frac{13}{20}$  (c) $\frac{201}{250}$
(d) $53\frac{5}{8}$

## ② CHANGING FRACTIONS TO DECIMALS

Next we change fractions to decimals. Remember that one meaning for a fraction is division. For example,

$$\frac{2}{3} \quad \text{means} \quad 2 \div 3 \quad \text{or} \quad 3\overline{)2}.$$

In Chapter 2 we learned how to divide one whole number (dividend) by a smaller whole number (divisor). The same method can be used when the divisor is greater than the dividend, as in the fractions $\frac{2}{3}$ or $\frac{1}{4}$. The result is a decimal. Place a decimal point to the right of the dividend and attach zeros as necessary. Place a decimal point in the quotient directly above the decimal point in the dividend. Divide as if there were no decimals present. The following example illustrates.

To change $\frac{1}{4}$ to a decimal, divide 1 by 4.

.← ③ Place decimal point in quotient above decimal point in dividend
$4\overline{)1.0}$ ← ② Attach one zero
↑
① Place decimal point after 1

Ignore the decimal points and divide 10 by 4. Then, attach a second 0 to the dividend, bring it down, and divide 4 into 20.

$$\begin{array}{r} .25 \\ 4\overline{)1.00} \\ \underline{8}\downarrow \\ 20 \\ \underline{20} \\ 0 \end{array} \qquad 20 \div 4 = 5$$

Stop when the remainder is 0. The quotient is the desired decimal.

$$\frac{1}{4} = 0.25$$

We can check that the method is correct by reversing the steps and changing 0.25 to a fraction.

$$0.25 \quad \text{is} \quad \frac{25}{100} \quad \text{which reduces to} \quad \frac{1 \cdot \cancel{25}}{4 \cdot \cancel{25}} = \frac{1}{4}.$$

Suppose we change $\frac{2}{3}$ to a decimal.

.  ← ② Place decimal point in quotient
3)2.0 ← ① Attach one zero after decimal point

We divide 20 by 3, then attach another zero, bring it down, and again divide 20 by 3.

$$
\begin{array}{r}
.66 \\
3\overline{)2.00} \\
1\,8\!\downarrow \\
\hline
20 \\
18 \\
\hline
2
\end{array}
$$

If we attach another zero and bring it down, we would again divide 20 by 3. By now, it is clear that this pattern will continue. Therefore, $\frac{2}{3}$ can be represented by

$$0.666 \cdots$$

The three dots show that the pattern of 6's continues. We often omit the dots and place a bar over the repeating digit (or digits) to shorten the notation.

$$0.\overline{6} \quad \text{means} \quad 0.666 \cdots$$

The two examples above illustrate the following important property of fractions.

### Terminating and Repeating Decimals

Every fraction can be written as a decimal.

1. If the division process ends with a remainder of zero, the decimal is a **terminating decimal.**

2. If the division process does not end, there is a digit or block of digits which repeats, and the decimal is a **repeating decimal.**

Thus, 0.25 is a terminating decimal and $0.\overline{6}$ or $0.666 \cdots$ is a repeating decimal.

### To Change a Fraction to a Decimal

1. Use the numerator as the dividend and the denominator as the divisor in a division problem.

2. Place decimal points after the numerator (dividend) and in the same position in the quotient.

3. Attach one zero at a time to the numerator (now the dividend) and divide as whole numbers.

4. Stop when a remainder of 0 is obtained or when a repeating pattern is found.

| **EXAMPLE 3** CHANGING FRACTIONS TO DECIMALS | **PRACTICE EXERCISE 3** |
|---|---|

Change each fraction to a decimal.

Change each fraction to a decimal.

**(a)** $\frac{1}{2}$    Put in decimal points and attach a zero, then divide 10 by 2.

**(a)** $\frac{1}{5}$

$$
\begin{array}{r}
.5 \\
2\overline{)1.0} \\
\underline{1\ 0} \\
0
\end{array}
$$

Since the remainder is 0, division stops. Thus, $\frac{1}{2}$ is the terminating decimal 0.5 $\left(\text{remember that } 0.5 = \frac{5}{10} = \frac{1 \cdot \cancel{5}}{2 \cdot \cancel{5}} = \frac{1}{2}\right)$.

**(b)** $\frac{3}{8}$

**(b)** $\frac{7}{8}$

$$
\begin{array}{r}
.375 \\
8\overline{)3.000} \\
\underline{2\ 4} \\
60 \\
\underline{56} \\
40 \\
\underline{40} \\
0
\end{array}
$$

Thus, $\frac{3}{8}$ is the terminating decimal 0.375 $\left(\text{remember that } 0.375 = \frac{375}{1000} = \frac{3 \cdot \cancel{125}}{8 \cdot \cancel{125}} = \frac{3}{8}\right)$.

**(c)** $\frac{1}{3}$

**(c)** $\frac{2}{9}$

$$
\begin{array}{r}
.33 \\
3\overline{)1.000} \\
\underline{9} \\
10 \\
\underline{9} \\
10
\end{array}
$$

Obviously, the pattern of 3's will continue. Thus, $\frac{1}{3}$ is the repeating decimal

$$0.\overline{3} \quad \text{or} \quad 0.333 \cdots$$

**(d)** $\frac{7}{33}$

**(d)** $\frac{12}{33}$

$$
\begin{array}{r}
.2121 \\
33\overline{)7.0000} \\
\underline{6\ 6} \\
40 \\
\underline{33} \\
70 \\
\underline{66} \\
40 \\
\underline{33} \\
7
\end{array}
$$

Again, we see a pattern: the block of two digits, 21, will continue to repeat. Thus,

$$\frac{7}{33} = 0.2121 \cdots = 0.\overline{21}.$$

Notice that the bar is placed over the *two* repeating digits, 2 and 1.

Improper fractions can be changed to decimals using the same procedure.

| EXAMPLE 4 CHANGING IMPROPER FRACTIONS TO DECIMALS |
| --- |

Change each fraction to a decimal.

**(a)** $\dfrac{7}{4}$

$$\overset{\cdot}{4)7.0} \quad \text{Place decimal points} \atop \text{Attach a zero}$$

This time we divide 7 by 4 and place 1 in the quotient above 7 before the decimal point.

$$
\begin{array}{r}
1.75 \\
4)\overline{7.00} \\
\underline{4}\phantom{.00} \\
3\,0 \\
\underline{2\,8} \\
20 \\
\underline{20} \\
0
\end{array}
$$

Place decimal points and attach a zero

Attach and bring down another zero

Thus, $\frac{7}{4}$ is equal to 1.75.

**(b)** $\dfrac{14}{3}$

$$
\begin{array}{r}
4.66 \\
3)\overline{14.00} \\
\underline{12}\phantom{.00} \\
2\,0 \\
\underline{1\,8} \\
20 \\
\underline{18} \\
2
\end{array}
$$

The pattern of 6's will repeat. Thus, $\frac{14}{3} = 4.666 \cdots = 4.\overline{6}$.

| PRACTICE EXERCISE 4 |
| --- |

Change each fraction to a decimal.

**(a)** $\dfrac{11}{4}$

**(b)** $\dfrac{11}{3}$

## ❸ SUMMARY LIST OF EQUIVALENT FRACTIONS AND DECIMALS

You have probably noticed that we changed terminating decimals to fractions but did not change repeating decimals to fractions. The method for making such conversions is beyond this level of work. Certain basic conversions occur often

enough that they should be memorized. These are included in the following list.
Memorize these now, before doing the exercises.

$$\frac{1}{2} = 0.5 \qquad \frac{1}{8} = 0.125 \qquad \frac{3}{4} = 0.75$$

$$\frac{1}{3} = 0.\overline{3} \qquad \frac{1}{9} = 0.\overline{1} \qquad \frac{3}{8} = 0.375$$

$$\frac{1}{4} = 0.25 \qquad \frac{1}{10} = 0.1 \qquad \frac{5}{8} = 0.625$$

$$\frac{1}{5} = 0.2 \qquad \frac{2}{3} = 0.\overline{6} \qquad \frac{7}{8} = 0.875$$

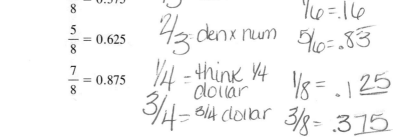

*[Handwritten margin notes:]*

Memorize

$\frac{2}{3} = 0.\overline{6}$

$\frac{2 \times 2}{2 \cdot 5}$

1/2 - think of 1/2 dollar
1/3 = den. x num
2/3 = den x num
1/4 - think 1/4 dollar
3/4 = 3/4 dollar

1/5 - 4/5 = .2 x ea. num.
1/6 = .1\overline{6}
5/6 = .8\overline{3}
1/8 = .125
3/8 = .375
5/8 = .625
7/8 = .875
(8 in to 10,30,50,70)

1/9 - 8/9 = .\overline{1}  .\overline{2}  .\overline{3}  .\overline{4}  etc.

## 5.2  EXERCISES A

*Change each decimal to a fraction (or mixed number).*

**1.** 1.4    **2.** 3.25    **3.** 21.9    **4.** 4.05

**5.** 3.002    **6.** 0.015    **7** 0.1302    **8.** 493.72

*Change each decimal to a fraction or mixed number using the second method as shown in Example 2.*

**9.** 0.3    **10.** 0.65    **11.** 2.7

**12.** 0.305    **13.** 1.29    **14.** 2.005

*Change each fraction to a decimal.*

**15.** $\frac{3}{4}$    **16.** $\frac{5}{6}$    **17.** $\frac{4}{9}$    **18.** $\frac{7}{11}$

**19.** $\frac{1}{16}$    **20.** $\frac{7}{3}$    **21.** $\frac{11}{4}$    **22** $\frac{19}{16}$

*Write each of the following as a fraction, using the table in the text that you memorized.*

**23.** $0.\overline{3}$    **24.** 0.375    **25.** 0.125    **26.** 1.5    **27.** $0.\overline{6}$

**28.** 0.875    **29.** 0.2    **30.** 0.625    **31.** $0.\overline{1}$    **32.** 0.25

**33.** 0.1    **34.** 1.25    **35.** 0.5    **36.** 0.75    **37.** $1.\overline{3}$

## FOR REVIEW

**38.** Give the formal word name for 401.003.

**39.** Write twenty-three and one hundred seven thousandths in decimal notation.

**40.** Write a word name for $34.85   **(a)** as it would appear on a check;   **(b)** using dollars and cents.

*The following exercises will help you prepare for the next section. Round each number to the nearest (a) ten, (b) hundred, and (c) thousand.*

**41.** 1365
**42.** 12,501
**43.** 7494

---

ANSWERS:   1. $1\frac{2}{5}$ or $\frac{7}{5}$   2. $3\frac{1}{4}$ or $\frac{13}{4}$   3. $21\frac{9}{10}$ or $\frac{219}{10}$   4. $4\frac{1}{20}$ or $\frac{81}{20}$   5. $3\frac{1}{500}$ or $\frac{1501}{500}$   6. $\frac{3}{200}$   7. $\frac{651}{5000}$   8. $493\frac{18}{25}$ or $\frac{12,343}{25}$   9. $\frac{3}{10}$   10. $\frac{13}{20}$   11. $2\frac{7}{10}$ or $\frac{27}{10}$   12. $\frac{61}{200}$   13. $1\frac{29}{100}$ or $\frac{129}{100}$   14. $2\frac{1}{200}$ or $\frac{401}{200}$   15. 0.75   16. $0.8\overline{3}$ (Place the bar over 3 only; 8 does not repeat.)   17. $0.\overline{4}$   18. $0.\overline{63}$   19. 0.0625   20. $2.\overline{3}$   21. 2.75   22. 1.1875   23. $\frac{1}{3}$   24. $\frac{3}{8}$   25. $\frac{1}{8}$   26. $\frac{3}{2}$   27. $\frac{2}{3}$   28. $\frac{7}{8}$   29. $\frac{1}{5}$   30. $\frac{5}{4}$   31. $\frac{1}{9}$   32. $\frac{1}{4}$   33. $\frac{1}{10}$   34. $\frac{5}{4}$   35. $\frac{1}{2}$   36. $\frac{3}{4}$   37. $\frac{4}{3}$   38. four hundred one and three thousandths   39. 23.107   40. (a) Thirty-four and $\frac{85}{100}$ dollars (b) thirty four dollars and eighty-five cents   41. (a) 1370 (b) 1400 (c) 1000   42. (a) 12,500 (b) 12,500 (c) 13,000   43. (a) 7490 (b) 7500 (c) 7000

## 5.2   EXERCISES B

*Change each decimal to a fraction (or mixed number).*

**1.** 2.4
**2.** 5.75
**3.** 22.8
**4.** 7.03

**5.** 4.002
**6.** 0.035
**7.** 0.3212
**8.** 525.25

*Change each decimal to a fraction or mixed number using the second method as shown in Example 2.*

**9.** 0.4
**10.** 0.85
**11.** 2.8

**12.** 0.205
**13.** 1.39
**14.** 3.007

*Change each fraction to a decimal.*

**15.** $\dfrac{1}{4}$
**16.** $\dfrac{3}{20}$
**17.** $\dfrac{7}{9}$
**18.** $\dfrac{3}{11}$

**19.** $\dfrac{3}{16}$
**20.** $\dfrac{19}{5}$
**21.** $\dfrac{13}{4}$
**22.** $\dfrac{10}{3}$

*Write each of the following as a decimal, using the table in the text that you memorized.*

**23.** $\dfrac{2}{3}$
**24.** $\dfrac{1}{10}$
**25.** $\dfrac{3}{2}$
**26.** $\dfrac{1}{5}$
**27.** $\dfrac{7}{8}$

**28.** $\dfrac{5}{8}$    **29.** $\dfrac{1}{2}$    **30.** $\dfrac{4}{3}$    **31.** $\dfrac{3}{4}$    **32.** $\dfrac{3}{8}$

**33.** $\dfrac{1}{8}$    **34.** $\dfrac{1}{3}$    **35.** $\dfrac{1}{4}$    **36.** $\dfrac{1}{9}$    **37.** $\dfrac{5}{4}$

## FOR REVIEW

**38.** Give a formal word name for 639.0007.

**39.** Write one hundred fifteen and two hundred sixty-nine ten thousandths in decimal notation.

**40.** Write a word name for $82.89  **(a)** as it would appear on a check;  **(b)** using dollars and cents.

*The following exercises will help you prepare for the next section. Round each number to the nearest (a) ten, (b) hundred, and (c) thousand.*

**41.** 2526                **42.** 17,495                **43.** 6351

## 5.2 EXERCISES C

**1.** Change 236.525 to a mixed number.

**2.** Change 1000.0001 to a mixed number.

**3.** Change $\frac{1}{7}$ to a decimal.
[Answer: $0.\overline{142857}$]

**4.** Change $\frac{7}{13}$ to a decimal.

## 5.3  ROUNDING AND ESTIMATING DECIMALS

=== STUDENT GUIDEPOSTS ===

**1** Finding the Number of Decimal Places

**2** Rounding Decimals

**3** Estimating Decimal Computations

### 1 FINDING THE NUMBER OF DECIMAL PLACES

In Section 1.4 we learned how to round whole numbers to the nearest ten, hundred, thousand, and so forth. We now see how to round decimals to the nearest one, tenth, hundredth, and so forth, using the notion of *decimal places*.

### Number of Decimal Places

If a number is written in decimal notation, the number of digits to the right of the decimal point is called the number of **decimal places** in the number.

| **EXAMPLE 1**   **FINDING THE NUMBER OF DECIMAL PLACES** |
|---|

Give the number of decimal places in each number.

**(a)** 3. 15 has *two* decimal places.

       ↑

   *two* digits to the right of the decimal point

**(b)** 40. 397 has *three* decimal places.

      ↑

    *three* digits

**(c)** 12 or 12. has zero or no decimal places.

**(d)** 0.600 has three decimal places.

**(e)** 2.6060 has four decimal places.

| **PRACTICE EXERCISE 1** |
|---|

Give the number of decimal places in each number.

**(a)** 5.2

**(b)** 37.029

**(c)** 75

**(d)** 0.350

**(e)** 6.0003

Answers: (a) one (b) three (c) zero (d) three (e) four

//////////////// **CAUTION** ///////////////

Compare the three decimals 0.6, 0.60, and 0.600, having one, two, and three decimal places, respectively. Although these numbers have the same value, by using zeros to the right of the digit 6, we are emphasizing a greater degree of precision or accuracy. If the numbers were used for measurements, 0.600 would indicate accuracy to the nearest thousandth, 0.60 to the nearest hundredth, and 0.6 to the nearest tenth.

//////////

## ❷ ROUNDING DECIMALS

The following table summarizes the relationship between the number of decimal places and the position corresponding to this number, and is helpful when rounding decimals.

| Decimal Places | Last Position on the Right | Example |
|---|---|---|
| 0 | ones | 5 or 5. |
| 1 | tenths | 5.7 |
| 2 | hundredths | 5.76 |
| 3 | thousandths | 5.764 |
| 4 | ten thousandths | 5.7648 |

| To Round a Decimal to a Particular Number of Places |
|---|

1. Locate the position (ones, tenths, hundredths, etc.) corresponding to this number of places.

2. If the first digit to the right of this position is less than 5, drop it and all digits to the right of it. Do not change the digit in the place to which you are rounding.

3. If the first digit to the right of this position is greater than or equal to 5, increase the digit in the desired place by one and drop all digits to the right of it.

Other methods of rounding can be used. However, the method given here is perhaps the most useful and certainly the simplest. Notice that rounding decimals is much like rounding whole numbers.

---

| **EXAMPLE 2  ROUNDING TO ONE DECIMAL PLACE** | **PRACTICE EXERCISE 2** |

Round 35.24 to one decimal place (to the nearest tenth).

We are rounding to the tenths position.

$$35.\boxed{2}4$$

↑
round to this position

Since 4, the first digit to the right of 2, is less than 5, we drop 4 and do not change 2. Thus, rounded to one decimal place, 35.24 becomes 35.2.

**Practice Exercise 2:** Round 68.32 to one decimal place (to the nearest tenth).

Answer: 68.3

---

We often round to obtain approximate values for numbers. For example, 35.24 is approximately equal to 35.2. The symbols $\approx$ and $\doteq$ stand for "is approximately equal to." Thus, from Example 2,

$$35.24 \approx 35.2 \quad \text{or} \quad 35.24 \doteq 35.2.$$

---

| **EXAMPLE 3  ROUNDING TO TWO DECIMAL PLACES** | **PRACTICE EXERCISE 3** |

Round 2.3482 to two decimal places (to the nearest hundredth).

Round to the hundredths position.

first digit to the right of rounding position

$$2.3\boxed{4}82$$

↑
round to this position

Since the first digit to the right of the hundredths position is 8, which is greater than 5, increase 4 to 5 and drop 8 and 2. Thus,

$$2.3482 \approx 2.35, \quad \text{rounded to two decimal places.}$$

**Practice Exercise 3:** Round 6.5473 to two decimal places (to the nearest hundredth).

Answer: 6.55

---

| **EXAMPLE 4  ROUNDING TO THE NEAREST TENTH** | **PRACTICE EXERCISE 4** |

Round 53.85 to the nearest tenth (to one decimal place).

first digit is 5, so increase 8 to 9 and drop 5

$$53.\boxed{8}5$$

↑
round to this position

Thus, $53.85 \approx 53.9$, rounded to the nearest tenth.

**Practice Exercise 4:** Round 3.125 to the nearest hundredth (to two decimal places).

Answer: 3.13

---

Increases in a particular digit may need to be carried over to the next digit. This is shown in the next example.

| **EXAMPLE 5   ROUNDING ACROSS A DIGIT** | **PRACTICE EXERCISE 5** |

Round 1.3895 to three decimal places (to the nearest thousandth).

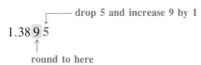

When 9 is increased by 1, the result is 10, so replace 9 by 0 and increase 8 to 9. Thus,

$$1.3895 \approx 1.390.$$

Do not give 1.39 for the answer; 1.39 has only two decimal places. The final zero in 1.390 must be present.

Round 41.4975 to two decimal places (to the nearest hundredth).

Answer: **41.50**

---

| **CAUTION** |

When rounding, do not round off one digit at a time from the right. For example, to round 2.6149 to the nearest hundredth, do not round to the nearest thousandth,

$$2.615$$

and then to the nearest hundredth,

$$2.62.$$

This is incorrect, since using our method,

$$2.6149 \approx 2.61. \qquad \text{Rounded to the nearest hundredth}$$

---

| **EXAMPLE 6   ROUNDING TO THE NEAREST ONE** | **PRACTICE EXERCISE 6** |

Round 21.654 to the nearest one (0 decimal places).

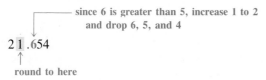

Thus, $21.654 \approx 22$, rounded to the nearest one (or whole number).

Round 16.539 to the nearest one (0 decimal places).

Answer: **17**

---

Decimals that do not terminate, but have repeating blocks of digits such as

$$0.333 \cdots \quad \text{or} \quad 0.\overline{3} \quad \text{and} \quad 0.212121 \cdots \quad \text{or} \quad 0.\overline{21},$$

can also be rounded. The procedure is identical to that for terminating decimals.

| **EXAMPLE 7   ROUNDING REPEATING DECIMALS** | **PRACTICE EXERCISE 7** |

**(a)** Round $0.\overline{3}$ to the nearest tenth.

$$3 < 5, \text{ so drop remaining digits}$$

$$0.\overline{3} = 0.\,3\,33\cdots$$

round to here

Thus, $0.\overline{3} \approx 0.3$.

**(b)** Round $0.\overline{6}$ to the nearest thousandth.

$$6 > 5, \text{ so increase thousandths digit from 6 to 7 and drop remaining digits}$$

$$0.\overline{6} = 0.66\,6\,6\cdots$$

round to here

Thus, $0.\overline{6} \approx 0.667$.

**(c)** Round $0.\overline{28}$ to the nearest tenth.

$$8 > 5 \text{ so increase tenths digit from 2 to 3 and drop remaining digits}$$

$$0.\overline{28} = 0.282828\cdots$$

round to here

Thus, $0.\overline{28} \approx 0.3$.

**(d)** Round $0.\overline{28}$ to the nearest hundredth.

$$2 < 5 \text{ so drop remaining digits}$$

$$0.\overline{28} = 0.282828\cdots$$

round to here

Thus, $0.\overline{28} \approx 0.28$.

**(a)** Round $0.\overline{2}$ to the nearest tenth.

**(b)** Round $0.\overline{8}$ to the nearest hundredth.

**(c)** Round $0.\overline{51}$ to the nearest hundredth.

**(d)** Round $0.\overline{51}$ to the nearest thousandth.

Answers: (a) 0.2   (b) 0.89   (c) 0.52   (d) 0.515

Calculations involving amounts of money must often be rounded to the nearest dollar (one) or to the nearest cent (hundredths). For example, suppose we obtain

$$\$57.432$$

as an answer. Since our smallest unit of money is the cent, or one hundredth of a dollar, we usually round such figures to the nearest cent and write

$$\$57.43 \quad \text{instead of} \quad \$57.432.$$

At times (when completing an income tax form, for example), we may round an amount of money to the nearest dollar. For example, to the nearest dollar,

$$\$57.43 \approx \$57.00 \text{ or } \$57,$$

and

$$\$112.50 \approx \$113.00 \quad \text{or} \quad \$113.$$

## ③ ESTIMATING DECIMAL COMPUTATIONS

In the sections that follow we will consider sums, differences, products, and quotients of decimals. Often in real-life practical situations, an estimated result might be all that is needed. For example, if we were deciding whether to buy 50 pounds of ground beef at $1.97 per pound, we might want to know the approximate cost. Since $1.97 is about $2.00, the cost is approximately

$$50 \times \$2 = \$100.$$

Similarly, if we have borrowed $8897.50 and plan to pay it back in 30 equal monthly payments, the approximate monthly payment is

$$\$9000 \div 30 = \$300.$$

In the first situation, we rounded $1.97 to the nearest dollar, and in the second, we rounded to the nearest thousand dollars. Such estimates are useful in daily activities and can often be made mentally.

| **EXAMPLE 8**  ESTIMATING IN A CONSUMER PROBLEM | **PRACTICE EXERCISE 8** |
|---|---|

The Thoresons plan to carpet their living room using carpeting that costs $14.95 a sq yd. If the room is rectangular with length 8.1 yd and width 5.8 yd, what would be the approximate cost of the job?

The approximate area of the room is

$8 \times 6 = 48$ sq yd    Rounding 8.1 and 5.8 to the nearest one

Since each square yard of carpeting costs approximately $15 (rounding to the nearest one), the total cost is about

$$15 \times 48 = \$720.$$

The Randolphs borrowed $8095.50 and plan to pay it back in 30 equal monthly payments. Estimate their monthly payment.

Answer: $270 (using $8100 \div 30$)

---

### ┌─────┐ HINT └─────┐

In any applied problem involving decimals, try to estimate the solution by rounding to the nearest whole numbers and use this estimation to help you decide whether the answer you obtain seems reasonable and accurate.

## 5.3 EXERCISES A

*Give the number of decimal places in each number.*

**1.** 32.1          **2.** 4.23          **3.** 6.453          **4.** 42.3965

**5.** 13          **6.** 42.80          **7.** 4.280          **8.** 0.4280

*Round each number to one decimal place (to the nearest tenth).*

**9.** 3.21          **10.** 4.67          **11.** 3.256

**12.** 5.4249          **13.** 32.75          **14.** 151.951

*Round each number to two decimal places (to the nearest hundredth).*

**15.** 7.321  **16.** 3.487  **17.** 4.3209

**18.** 5.4249  **19.** 34.0085  **20.** 34.0995

*Round each number to three decimal places (to the nearest thousandth).*

**21.** 3.4579  **22.** 7.2571  **23.** 3.4906

**24.** 5.4249  **25.** 45.0085  **26.** 9.0307

*Round each number to the nearest one.*

**27.** 24.1  **28.** 63.5129  **29.** 5.4249

**30.** 9.003  **31.** 19.84  **32.** 39.099

*Round each repeating decimal to the nearest **(a)** tenth, **(b)** hundredth, and **(c)** thousandth.*

**33.** $0.\overline{4}$  **34** $0.\overline{35}$  **35.** $0.\overline{123}$
  **(a)**      **(a)**      **(a)**
  **(b)**      **(b)**      **(b)**
  **(c)**      **(c)**      **(c)**

**36.** $0.\overline{7}$  **37.** $0.\overline{93}$  **38.** $0.\overline{455}$
  **(a)**      **(a)**      **(a)**
  **(b)**      **(b)**      **(b)**
  **(c)**      **(c)**      **(c)**

*Round each amount of money to the nearest cent.*

**39.** $4.237  **40.** $6.235  **41.** $7.231  **42.** $8.105

*Round each amount of money to the nearest dollar.*

**43.** $37.29  **44.** $13.49  **45.** $28.50  **46.** $16.99

*Solve.*

**47** A bricklayer will use 290 bricks at a cost of 9.5¢ per brick. What is the approximate total cost of the bricks?

**48.** What is the approximate cost of 95 pounds of beans selling for 41¢ per pound?

## FOR REVIEW

**49.** Change 7.08 to a mixed number.

**50.** Change $\frac{8}{9}$ to a decimal.

---

ANSWERS:  1. 1  2. 2  3. 3  4. 4  5. 0 or none  6. 2  7. 3  8. 4  9. 3.2  10. 4.7  11. 3.3  12. 5.4  13. 32.8  14. 152.0  15. 7.32  16. 3.49  17. 4.32  18. 5.42  19. 34.01  20. 34.10  21. 3.458  22. 7.257  23. 3.491  24. 5.425  25. 45.009  26. 9.031  27. 24  28. 64  29. 5  30. 9  31. 20  32. 39  33. (a) 0.4 (b) 0.44 (c) 0.444  34. (a) 0.4 (b) 0.35 (c) 0.354  35. (a) 0.1 (b) 0.12 (c) 0.123  36. (a) 0.8 (b) 0.78 (c) 0.778  37. (a) 0.9 (b) 0.94 (c) 0.939  38. (a) 0.5 (b) 0.46 (c) 0.455  39. $4.24  40. $6.24  41. $7.23  42. $8.11  43. $37  44. $13  45. $29  46. $17  47. $30 [3000¢]  48. $40 [using $100 \times 40 = 4000$¢]  49. $7\frac{2}{25}$  50. $0.\overline{8}$

## 5.3 EXERCISES B

*Give the number of decimal places in each number.*

**1.** 63.1      **2.** 7.45      **3.** 8.425      **4.** 51.405

**5.** 15      **6.** 71.90      **7.** 7.190      **8.** 0.7190

*Round each number to one decimal place (to the nearest tenth).*

**9.** 4.31      **10.** 5.87      **11.** 6.357

**12.** 2.6249      **13.** 41.06      **14.** 0.964

*Round each number to two decimal places (to the nearest hundredth).*

**15.** 9.432      **16.** 6.278      **17.** 12.4208

**18.** 13.5091      **19.** 3.0995      **20.** 0.0355

*Round each number to three decimal places (to the nearest thousandth).*

**21.** 4.3569      **22.** 2.3581      **23.** 6.3585

**24.** 76.0095      **25.** 8.0606      **26.** 7.4008

*Round each number to the nearest one.*

**27.** 26.1      **28.** 32.507      **29.** 7.499

**30.** 6.005      **31.** 29.78      **32.** 8.0099

*Round each repeating decimal to the nearest (**a**) tenth, (**b**) hundredth, and (**c**) thousandth.*

**33.** $0.\overline{9}$      **34.** $0.\overline{17}$      **35.** $0.\overline{234}$

**36.** $0.\overline{1}$      **37.** $0.\overline{84}$      **38.** $0.\overline{655}$

*Round each amount of money to the nearest cent.*

**39.** $7.382      **40.** $4.535      **41.** $9.448      **42.** $10.205

*Round each amount of money to the nearest dollar.*

**43.** $62.38      **44.** $41.48      **45.** $74.50      **46.** $32.98

*Solve.*

**47.** What is the approximate cost of 32 pounds of coffee selling for $2.95 per pound?

**48.** A clock radio sells for $39.95. If Bill has $243.65 in his checking account, approximately how many radios could he buy?

## FOR REVIEW

**49.** Change 2.016 to a mixed number.

**50.** Change $\frac{5}{33}$ to a decimal.

## 5.3 EXERCISES C

**1.** Change $\frac{4}{7}$ to a decimal and round the result to the nearest **(a)** tenth, **(b)** hundredth, and **(c)** thousandth.

**2.** What is the approximate cost of 305 books each selling for $9.95?

**3.** John and Sue borrowed $6123.50 and plan to pay it back in 30 equal monthly payments. What will be their approximate monthly payment?
[Answer: $200 (using $6000 ÷ 30)]

**4.** A paneling job requires 592 sheets of wood paneling at a cost of $20.75 per sheet. What is the approximate cost of the paneling?

**5.** The Wards plan to carpet their living room using carpet costing $19.50 a square yard. If the room is rectangular in shape as shown to the right, what would be the approximate cost of the job?
[Answer: $1080 (using $9 \times 6 \times \$20$)]

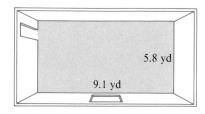
5.8 yd
9.1 yd

## 5.4 ADDING AND SUBTRACTING DECIMALS

### STUDENT GUIDEPOSTS

**1** Adding Two or More Decimals
**2** Applications Adding Decimals
**3** Subtracting Two Decimals
**4** Applications Subtracting Decimals

### **1** ADDING TWO OR MORE DECIMALS

When we studied the basic operations on fractions, we began with multiplication and division, and then went on to the more difficult operations of addition and subtraction. For numbers in decimal notation, addition and subtraction are easier. For example, suppose we want to find the following sum.

$$
\begin{array}{r}
2.35 \\
+ \ 1.21 \\
\end{array}
$$

We might convert each decimal to fractional form and then add the fractions. But if we add the columns in the original problem and move the decimal point straight down, we obtain the same sum much more easily.

$$
\begin{array}{r}
2.35 \\
1.21 \\
\hline
3.56 \\
\end{array}
$$

This illustrates a general method.

### To Add Two or More Decimals

1. Arrange the numbers in columns so that the decimal points line up.

2. Add in columns from right to left as if adding whole numbers. (Additional zeros may be placed to the right of the decimal points if needed.)

3. Place the decimal point in the sum in line with the other decimal points.

---

### EXAMPLE 1 ADDING DECIMALS

Find each sum.

**(a)** $21.3 + 4.8$

$$
\begin{array}{r}
1\phantom{0.0} \\
21.3 \\
+\ \ 4.8 \\
\hline
26.1 \\
\end{array}
$$

① Arrange the numbers in columns so that the decimal points line up

② Carry 1 after adding 8 and 3 just as in addition of whole numbers

③ Place the decimal point in line with others

**(b)** $37 + 2.5 + 10.03 + 425.008$

$$
\begin{array}{r}
37. \\
2.5 \\
10.03 \\
+\ 425.008 \\
\hline
474.538 \\
\end{array}
\quad \text{or} \quad
\begin{array}{r}
37.000 \\
2.500 \\
10.030 \\
+\ 425.008 \\
\hline
474.538 \\
\end{array}
$$

Writing 0's to the right of the decimal points may help you keep the decimal points lined up vertically

---

### PRACTICE EXERCISE 1

Find each sum.

**(a)** $154.5 + 60.7$

**(b)** $3.8 + 68 + 40.07 + 653.009$

Answers: (a) 215.2 (b) 764.879

---

## ② APPLICATIONS ADDING DECIMALS

You can see that adding decimals is much like adding whole numbers. Following are several examples of word problems that involve adding decimals. Make certain that your answers are reasonable.

---

### EXAMPLE 2 ADDING DECIMALS IN A MILEAGE PROBLEM

Before leaving on a trip, Mike's odometer read 43,271.9. If he drove 827.6 miles on the trip, what did it read when he returned?

The return reading must be the beginning reading plus the number of miles traveled.

$$
\begin{array}{r}
43271.9 \\
+\ \ \ \ 827.6 \\
\hline
44099.5 \\
\end{array}
$$

Thus, when he returned, his odometer read 44,099.5.

---

### PRACTICE EXERCISE 2

On a two-day trip, Steve drove 489.3 miles the first day and 501.8 miles the second. How far did he drive on the trip?

Answer: 991.1 miles

| EXAMPLE 3 | ADDING DECIMALS ON A TELEPHONE BILL |

A good example of adding decimals can be found in a long-distance telephone bill. A typical bill is shown in Figure 5.2.

```
602 555-0801   FEB.25, 1991                    Mountain Bell

ITEMIZED CALLS

NO.    DATE    TIME    TO PLACE         TO AREA-NO.    MIN    AMOUNT
 1     115     747P    MONTGOMERY,AL    205 999-4801    2     $1.46
 2     123     1035A   NEWARK,NJ        503 888-3213    3      2.56
 3     2 1     345P    LOS ANGELES,CA   301 777-4243    1      0.87
 4     219     832A    BROOKLYN,NY      212 423-2131    5      3.27

                       TOTAL OF ITEMIZED CALLS EXCLUDING TAX $8.16
```

**Figure 5.2**

The total due on long-distance calls is $8.16, the sum of the decimals in the last column.

| PRACTICE EXERCISE 3 |

Part of a monthly telephone bill is given below. Find the total bill for long-distance calls.

```
        Southern Telephone

   MIN        AMOUNT

    4         $2.63
    7          8.48
   12         11.73
              _____
  TOTAL
```

Answer: **$22.84**

| EXAMPLE 4 | ADDING DECIMALS IN A DISTANCE PROBLEM |

Consider the map in Figure 5.3. Starting in West Yorkville, Margot hiked the complete circular trail past Devon Falls, Mount Blanc, Sanditon, Brunswick, East Yorkville, and back to West Yorkville. How long was her hike?

| PRACTICE EXERCISE 4 |

Use the map in Figure 5.3. Henry decided to hike along the river from Sanditon to Devon Falls and then on to West Yorkville where his wife would meet him with their car. How long was his hike?

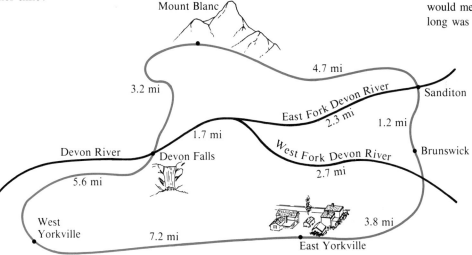

**Figure 5.3**

Add the distances on the map.

$$
\begin{array}{r}
5.6 \\
3.2 \\
4.7 \\
1.2 \\
3.8 \\
+\ 7.2 \\
\hline
25.7
\end{array}
$$

Thus, Margot hiked a distance of 25.7 miles.

Answer: **9.6 mi**

| EXAMPLE 5 ADDING DECIMALS ON A DEPOSIT SLIP | PRACTICE EXERCISE 5 |

A typical bank checking account deposit slip is shown in Figure 5.4. The total deposit is found by adding all amounts in the right-hand column. Notice that a vertical line separates the dollars column from the cents column. It is used instead of a decimal point because it has been found that people are more likely to arrange the numbers correctly using this method.

Part of a bank checking account deposit slip is shown below. Find the total deposit.

| | DEPOSIT | |
|---|---|---|
| | Dollars | Cents |
| CASH | 3025 | 51 |
| CHECKS | 431 | 68 |
| | 15 | 06 |
| | 283 | 93 |
| | 4 | 58 |
| TOTAL | | |

**First Federal Bank**
New Orleans, LA

Date *August 2* 19 *91*   DEPOSIT

Account No. **894-667-0523**

Sarah Williams
120 Charles St.
New Orleans, LA

| | Dollars | Cents |
|---|---|---|
| CASH | 13 | 27 |
| CHECKS (List Seperately) | 127 | 40 |
| | 312 | 80 |
| | 4 | 95 |
| | | |
| | | |
| TOTAL DEPOSIT | 458 | 42 |

**Figure 5.4**

Answer: $3760.76

## ❸ SUBTRACTING TWO DECIMALS

Subtraction of decimals is similar to addition. For example, suppose we want to find the following difference.

$$\begin{array}{r} 2.35 \\ -\ 1.21 \end{array}$$

By subtracting in columns, we obtain the difference.

$$\begin{array}{r} 2.35 \\ -\ 1.21 \\ \hline 1.14 \end{array}$$

The decimal point is moved straight down just as in an addition problem.

| To Subtract Two Decimals |
|---|

1. Arrange the numbers in columns so that the decimal points line up.
2. Subtract in columns from right to left as if subtracting whole numbers. (Additional zeros may be placed to the right of the decimal points if needed.)
3. Place the decimal point in the difference in line with the other decimal points.

| **EXAMPLE 6** SUBTRACTING DECIMALS | **PRACTICE EXERCISE 6** |
|---|---|

Find each difference.

**(a)** $43.7 - 12.5$

Arrange the numbers in columns so that the decimal points line up, then subtract.

$$
\begin{array}{r}
43.7 \\
-\ 12.5 \\
\hline
31.2
\end{array}
$$

↑
place decimal point in line with others

**(b)** $121.37 - 48.2$

$$
\begin{array}{r}
\overset{1\ 11}{1\,\cancel{2}\,\cancel{1}.37} \\
-\ \ \ 4\,8.20 \leftarrow \text{supply extra zero} \\
\hline
7\,3.17
\end{array}
$$

$7 = 7 - 0$
$1 = 3 - 2$
$3 = 11 - 8$
$7 = 11 - 4$

Notice that we had to borrow 1 ten from the 2 tens (or 20) in order to subtract 8.

**(c)** $48.52 - 19.67$

$$
\begin{array}{r}
\overset{17\ \ \ 14}{\overset{3\ 7\ \ 4\ 12}{\cancel{4}\,\cancel{8}.\cancel{5}\,\cancel{2}}} \\
-\ 1\,9.67 \\
\hline
2\,8.85
\end{array}
$$

$5 = 12 - 7$ (borrow 1 tenth from .5)
$8 = 14 - 6$ (borrow 1 from 8)
$8 = 17 - 9$ (borrow 1 ten from 40)
$2 = 3 - 1$

Borrowing is done as if we were subtracting whole numbers. Of course, with practice, the small numbers and the "crossing out" can be eliminated. The problem would then look like

$$
\begin{array}{r}
48.52 \\
-\ 19.67 \\
\hline
28.85.
\end{array}
$$

Find each difference.

**(a)** $75.8 - 23.6$

**(b)** $433.41 - 84.3$

**(c)** $94.31 - 36.58$

Answers: (a) **52.2**   (b) **349.11**
(c) **57.73**

**HINT**

Computationally, the only real difference between adding and subtracting decimals and adding and subtracting whole numbers is placing the decimal point. This is easy to do once the numbers have been written with the decimal points lined up.

## ❹ APPLICATIONS SUBTRACTING DECIMALS

Many word problems involve subtracting decimals. As before, use common sense and estimation to be sure that your answers are reasonable.

---

| **EXAMPLE 7** SUBTRACTING DECIMALS IN A DISTANCE PROBLEM | **PRACTICE EXERCISE 7** |

It is 13.6 miles from Wetumpka to Slapout and 32.1 miles from Wetumpka to Pine Level. How much farther is it from Wetumpka to Pine Level than to Slapout? See Figure 5.5.

   Subtract 13.6 from 32.1.

$$\begin{array}{r} 32.1 \\ -\ 13.6 \\ \hline 18.5 \end{array}$$

It is 18.5 miles farther to Pine Level.

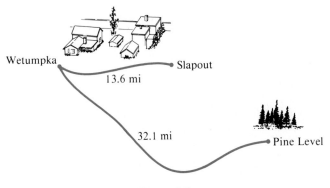

**Figure 5.5**

Donna noticed that the odometer on her car read 35,126.8 miles when she filled her gas tank in Oklahoma City. She then drove to Houston where her odometer reading was 35,576.1. How far is it from Oklahoma City to Houston?

Answer: 449.3 miles

---

| **EXAMPLE 8** SUBTRACTING DECIMALS IN A TAX PROBLEM | **PRACTICE EXERCISE 8** |

Mr. Wong had a total income of $27,085.50 last year. He had deductions of $427.56, $2037.15, $402.03, and $1327.48. If taxable income is total income minus deductions, what was Mr. Wong's taxable income?

   First find the total of all deductions.

$$\begin{array}{r} \$\ \ 427.56 \\ 2037.15 \\ 402.03 \\ +\ \ 1327.48 \\ \hline \$4194.22 \end{array}$$

Next subtract this total from his total income.

$$\begin{array}{r} \$27085.50 \\ -\ \ \ \ 4194.22 \\ \hline \$22891.28 \end{array}$$

Thus, Mr. Wong had a taxable income of $22,891.28.

Dr. Weston bought a shirt for $22.95 and a pair of pants for $34.98. Tax on the two purchases was $1.15 and $1.75, respectively. He paid for the items with a $100 bill and received a $20 bill, a $10 bill, a $5 bill, four $1 bills, one dime, one nickel, and two pennies in change. Was the change correct?

Answer: Yes; the change should be $39.17.

Example 8 illustrates that more than one operation is often used when solving a word problem. This is also shown in the next example.

| EXAMPLE 9   OPERATIONS IN A CHECKING ACCOUNT |
| --- |

To keep track of the balance in a checking account, most people use a check register similar to the one in Figure 5.6.

The ''Balance Forward'' of $608.27 is copied from the preceding page in the register. The first entry is a check written for $85.23. This amount is entered in the ''Amount of Check'' column and also in the working column under $608.27. The balance in the account, $523.04, is found by subtracting. Next, a deposit of $728.30 is entered in both the ''Amount of Deposit'' column and in the working column. Adding gives the new balance of $1251.34. The next check for $435.75 is entered as before, and the amount subtracted in the working column yields a new balance of $815.59. Thus, keeping a check register involves repeated additions and subtractions of decimals.

| Check No. | Date | Checks Issued to or Deposit Description | Amount of Check | Amount of Deposit | | Balance Forward |
| --- | --- | --- | --- | --- | --- | --- |
| | | | | | | 608 \| 27 |
| 301 | 6/22 | Marvin's Market | 85 \| 23 | | Check or Deposit | 85 \| 23 |
| | | | | | Balance | 523 \| 04 |
| | 6/25 | Paycheck Deposited | | 728 \| 30 | Check or Deposit | 728 \| 30 |
| | | | | | Balance | 1251 \| 34 |
| 302 | 7/1 | Home Mortgage for July | 435 \| 75 | | Check or Deposit | 435 \| 75 |
| | | | | | Balance | 815 \| 59 |

**Figure 5.6**

**PRACTICE EXERCISE 9**

On April 1, Cindi Franks had $958.12 in her checking account. During the month, she made two deposits, one of $327.50 and a second of $693.73. She wrote three checks in the amounts $483.16, $103.09, and $723.49. What was her balance at the end of the month?

Answer: $669.61

---

| 5.4   **EXERCISES A** |
| --- |

*Find each sum.*

**1.** 41.2 + 33.9

**2.** 4.37 + 2.09

**3.** 22.3 + 2.23 + 223

**4.** 0.003 + 2.107 + 135.1

**5.** 6.035 + 10.09 + 100.9

**6.** 4.3 + 5.01 + 0.005 + 41.376

**7.**   6.05
   21.3
\+  4.712

**8.**   32.079
   1.428
   403.6
\+  10.05

**9.**   105.30
   6.4193
   10.03
\+ 4217.508

*Arrange the following numbers in columns and find their sum.*

**10.** $132.57, $41.03, $1.79, $10.00

**11.** $4237.88, $1.23, $14.40, $103.05, $15.00

*Parts of monthly telephone bills are shown in Exercises 12–13. Find the total bills for the long-distance calls.*

**12.**

| Pacific Telephone | |
|---|---|
| MIN | AMOUNT |
| 3 | $1.57 |
| 5 | 6.23 |
| 1 | 0.37 |
| TOTAL | |

**13.**

| New England Telephone | |
|---|---|
| MIN | AMOUNT |
| 10 | $13.25 |
| 2 | 4.40 |
| 5 | 7.23 |
| 1 | 0.87 |
| TOTAL | |

*Exercises 14–15 refer to the following map.*

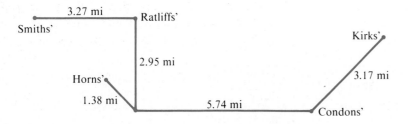

**14.** How far is it by road from the Smiths' to the Horns'?

**15.** How far is it by road from the Kirks' to the Smiths'?

*Parts of bank checking account deposit slips are shown in Exercises 16–17. Find the total deposits.*

**16.**

| DEPOSIT | | |
|---|---|---|
| | Dollars | Cents |
| CASH | 47 | 23 |
| CHECKS | 185 | 60 |
| | 13 | 07 |
| | 478 | 39 |
| | 6 | 50 |
| | | |
| TOTAL | | |

**17.**

| DEPOSIT | | |
|---|---|---|
| | Dollars | Cents |
| CASH | 1030 | 47 |
| CHECKS | 427 | 33 |
| | 48 | 05 |
| | 11 | 27 |
| | 4 | 38 |
| | 677 | 07 |
| TOTAL | | |

*Solve.*

**18.** Find the perimeter of the figure. (Perimeter is the distance around.)

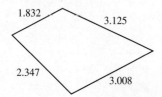

**19.** Find the sum of the following numbers: two thousand one hundred thirty-six and forty-one hundredths; five hundred three and two hundred seven thousandths; nine thousand one and one thousand nine ten thousandths. Give the answer in decimal notation and in words.

*Find each difference.*

**20.** 41.3 − 22.5

**21.** 6.03 − 4.2

**22.** 42.0 − 7.5

**23.** 101.3 − 1.013

**24.** 4.00 − 3.75

**25.** 12.056 − 3.009

**26.**  67.284
     − 13.498

**27.**  28.4
     −  3.45

**28.**    7
     − 1.009

*Solve.*

**29** The Marshall family had a total income of $31,255.75 last year. If they had deductions of $1237.40, $725.36, $2347.01, and $444.83, how much taxable income did they have?

**30.** Becky received a check for $125.00. If she spent $27.45 for a pair of shoes, $45.30 for a dress, and $18.78 for books, how much did she have left?

**31.** Complete the missing items in the following check register.

| Check No. | Date | Checks Issued to or Deposit Description | Amount of Check | Amount of Deposit | | Balance Forward 325 40 | |
|---|---|---|---|---|---|---|---|
| 420 | 3/2 | Harry's Hair Hut | 25 50 | | Check or Deposit | | (a) |
| | | | | | Balance | | (b) |
| 421 | 3/3 | Credit Union March Car Payment | 165 35 | | Check or Deposit | | (c) |
| | | | | | Balance | | (d) |
| | 3/4 | Gift from Uncle Claude | | 225 00 | Check or Deposit | | (e) |
| | | | | | Balance | | (f) |

**32.** When Nicky was sick, he ran a temperature of 102.3° Fahrenheit. If normal body temperature is 98.6° Fahrenheit, how many degrees above normal was his temperature?

**33.** When Den filled the gas tank on his car, the odometer read 32,478.8. The next time he filled it, the odometer read 32,920.3. How far had he driven?

**34.** Lucy bought a record for $8.65 and paid with a $10 bill. She received one dime, one quarter, and one $1 bill in change. Was this the correct change?

**35.** Subtract three hundred twenty-seven and eight hundredths from two thousand seven and seven hundred three thousandths. Give the difference in decimal notation and in words.

**36** When Jerri was sick in the hospital, she received four antibiotic injections of 2.65 milligrams, 2.75 milligrams, 3.5 milligrams, and 4.0 milligrams.

(a) What was the total of the four injections?

(b) If all injections were drawn from the same bottle that contained 30.5 milligrams of the antibiotic, how much remained in the bottle?

**37** In the figure below, find the length of $x$ and $y$.

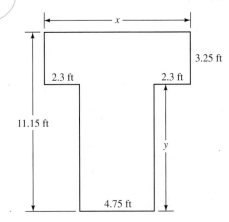

38. The table to the right shows the amount of rainfall for three months during the summer. If the normal rainfall for this three-month period is 9.6 inches, was the rainfall above or below average and by how much?

| Month | Rainfall (in inches) |
|--------|----------------------|
| June | 2.1 |
| July | 4.3 |
| August | 3.7 |

## FOR REVIEW

*Give the number of decimal places in each number.*

**39.** 42.37      **40.** 103.301      **41.** 21.0506      **42.** 1139

**43.** Round 3.1594 to the nearest (a) tenth, (b) hundredth, (c) thousandth.

*Use the following newspaper advertisements to answer Exercises 44–47.*

AM/FM Stereo 3-In-One Compact Unit

• AM/FM stereo receiver
• Front loading cassette deck
• 6 ½ full range speaker system

$139⁸⁸

19″ Color TV
Quick Start Picture Tube

• 105 channel cable ready
• 100% solid state
• Infrared remote control

$283³³

**44.** Estimate the cost of one TV and one stereo. Then find the exact cost.

**45.** About how much more is the price of the TV than the price of the stereo? What is the exact difference in price?

**46.** About how many stereos could be purchased for $600?

**47.** Estimate the total cost of purchasing 5 television sets.

ANSWERS: 1. 75.1  2. 6.46  3. 247.53  4. 137.21  5. 117.025  6. 50.691  7. 32.062  8. 447.157  9. 4339.2573  10. $185.39  11. $4371.56  12. $8.17  13. $25.75  14. 7.6 mi  15. 15.13 mi  16. $730.79  17. $2138.57  18. 10.312  19. 11,640.7179; eleven thousand six hundred forty and seven thousand one hundred seventy-nine ten thousandths  20. 18.8  21. 1.83  22. 34.5  23. 100.287  24. 0.25  25. 9.047  26. 53.786  27. 24.95  28. 5.991  29. $26,501.15  30. $33.47  31. (a) $25.50 (b) $299.90 (c) $165.35 (d) $134.55 (e) $225.00 (f) $359.55  32. 3.7°  33. 441.5 mi  34. yes  35. 1680.623; one thousand six hundred eighty and six hundred twenty-three thousandths  36. (a) 12.9 milligrams (b) 17.6 milligrams  37. $x = 9.35$ ft; $y = 7.9$ ft  38. above the average by 0.5 inches  39. 2  40. 3  41. 4  42. 0  43. (a) 3.2 (b) 3.16 (c) 3.159  44. $420 [using 280 + 140]; $423.21  45. $140 [using 280 − 140]; $143.45  46. 4 [using 600 ÷ 150]  47. $1500 [using 5 × 300]

## 5.4  EXERCISES B

*Find each sum.*

**1.** 6.21 + 70.3

**2.** 2.59 + 1.07

**3.** 46.3 + 4.63 + 0.463

**4.** 4.032 + 40.32 + 403.2

**5.** 1.017 + 40.08 + 200.6

**6.** 2.5 + 6.02 + 0.009 + 23.765

**7.**
```
    9.01
   37.005
 +  2.3
```

**8.**
```
   65.028
    1.351
  609.5
 +  20.07
```

**9.**
```
  121.73
    2.407
   52.1
 + 1035.113
```

*Arrange the following numbers in columns and find their sum.*

**10.** $20.00, $55.95, $1.32, $163.55

**11.** $135.62, $827.03, $2.06, $11,427.21, $25.33

*Parts of monthly telephone bills are shown in Exercises 12–13. Find the total bills for the long-distance calls.*

**12.**
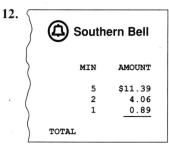

| Southern Bell | |
|---|---|
| MIN | AMOUNT |
| 5 | $11.39 |
| 2 | 4.06 |
| 1 | 0.89 |
| TOTAL | |

**13.**
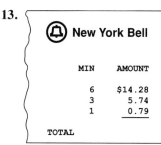

| New York Bell | |
|---|---|
| MIN | AMOUNT |
| 6 | $14.28 |
| 3 | 5.74 |
| 1 | 0.79 |
| TOTAL | |

*Exercises 14–15 refer to the following map.*

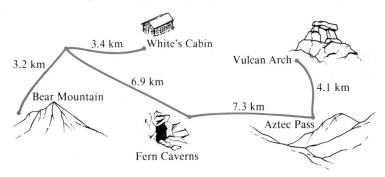

3.4 km  White's Cabin

3.2 km

Vulcan Arch

6.9 km

4.1 km

Bear Mountain

7.3 km

Aztec Pass

Fern Caverns

**14.** Following the trails, how far is it from Vulcan Arch to White's Cabin?

**15.** Following the trails, how far is it from Aztec Pass to Bear Mountain?

*Parts of bank checking account deposit slips are shown in Exercises 16–17. Find the total deposits.*

**16.**

| | DEPOSIT | |
|---|---|---|
| | Dollars | Cents |
| CASH | 62 | 09 |
| CHECKS | 477 | 51 |
| | 1035 | 20 |
| | 6 | 05 |
| | | |
| TOTAL | | |

**17.**

| | DEPOSIT | |
|---|---|---|
| | Dollars | Cents |
| CASH | 148 | 75 |
| CHECKS | 10 | 96 |
| | 1069 | 47 |
| | 285 | 32 |
| | | |
| TOTAL | | |

**18.** Find the perimeter of the figure.

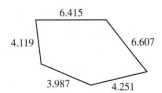

6.415

4.119

6.607

3.987

4.251

**19.** Find the sum of the following numbers: three thousand and twenty-one hundredths; four hundred two and six thousandths; fifty-six and two hundred forty-nine ten thousandths. Give the answer in decimal notation and in words.

*Find each difference.*

**20.** $42.1 - 31.5$

**21.** $7.05 - 6.2$

**22.** $51.0 - 6.3$

**23.** $202.5 - 2.025$

**24.** $3.00 - 1.25$

**25.** $15.055 - 2.009$

**26.** 97.317
   − 18.529

**27.** 38.6
   − 9.75

**28.** 8
   − 1.003

*Solve.*

**29.** The Lopez family had a total income of $42,621.35 last year. If they had deductions of $2427.51, $425.30, $612.12, and $1230.07, how much taxable income did they have?

**30.** Michelle plans a trip of 1285.5 miles. If she drove 195.6 mi on Monday, 283.2 mi on Tuesday, and 641.7 mi on Wednesday, how far must she drive on Thursday to complete the trip?

**31.** Complete the missing items in the check register shown below.

| Check No. | Date | Checks Issued to or Deposit Description | Amount of Check | Amount of Deposit | | Balance Forward 614 30 | |
|---|---|---|---|---|---|---|---|
| 315 | 10/2 | Graydon's Grocery Store | 32 47 | | Check or Deposit | | (a) |
| | | | | | Balance | | (b) |
| | 10/2 | Deposit Week's Wages | | 176 19 | Check or Deposit | | (c) |
| | | | | | Balance | | (d) |
| 316 | 10/5 | County Farm Insurance | 110 59 | | Check or Deposit | | (e) |
| | | | | | Balance | | (f) |

**32.** When Ralph had the flu he ran a temperature of 103.1° Fahrenheit. If normal body temperature is 98.6° Fahrenheit, how many degrees above normal was his temperature?

**33.** When Laurie filled the gas tank on her truck, the odometer read 19,627.2. The next time she filled it, the odometer read 19,810.1. How far had she driven?

**34.** Jack bought a stereo for $289.75, and paid with three $100 bills and three quarters. He received one $10 bill, one $1 bill, and one quarter in change. Was this the correct change?

**35.** Subtract two hundred fifty-seven and three thousandths from four thousand six and one tenth. Give the difference in decimal notation and in words.

**36.** The times recorded for the members of the ASU 400-meter relay team were 10.5 sec, 10.8 sec, 11.2 sec, and 10.3 sec.

   **(a)** What was the combined time for the event?

   **(b)** The team's best time prior to this race was 42.9 sec. Did the team better its time; and if so, by how much?

**37.** In the figure below find the length of *x* and *y*.

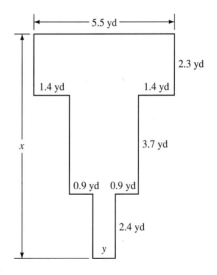

**38.** The table to the right shows the number of gallons of gasoline used by a delivery truck during a four-month period. During the same period last year a total of 642.3 gallons were used. How many gallons more or less were used this year than last?

| Month | Number of gallons |
|---|---|
| January | 132.5 |
| February | 147.8 |
| March | 201.3 |
| April | 187.6 |

## FOR REVIEW

*Give the number of decimal places in each number.*

**39.** 26.3                **40.** 4.005                **41.** 6.0109                **42.** 247

**43.** Round 6.2584 to the nearest **(a)** tenth, **(b)** hundredth, **(c)** thousandth.

*Use the following newspaper advertisements to answer Exercises 44–47.*

14 Day Programmable Remote VHS Videocassette Recorder with Stereo Sound

• Automatic scan tuner
• Shuttle Search for easy location of desired program segment

**$597²⁷**

Microwave Oven with 600 watts of cooking power

• Microwave cookbook included

New Low Price

**$178³²**

**44.** Estimate the cost of one video recorder and one microwave oven. Then find the exact cost.

**45.** About how much more is the recorder than the oven? What is the exact difference in price?

**46.** About how many video recorders could be purchased for $3000?

**47.** Estimate the total cost of purchasing 6 microwave ovens.

## 5.4  EXERCISES C

*Solve.*

**1.** The average of three numbers is their sum divided by 3. Marvin owns three cats weighing 9.1 lb, 3.3 lb, and 11.6 lb. What is the average weight of his cats?

**2.** Find the length of $x$ in the figure at right if both rectangles in the figure are squares (all four sides are equal).

[Answer: 5.05 cm]

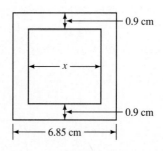

## 5.5  MULTIPLYING DECIMALS

### **1** BASIC MULTIPLICATION

Multiplying decimals is very much like multiplying whole numbers once we decide what to do with the decimal points. Suppose we multiply

$$0.7 \times 0.21$$

as fractions. We know

$$0.7 = \frac{7}{10} \quad \text{and} \quad 0.21 = \frac{21}{100}.$$

Then

$$
\begin{aligned}
0.7 \times 0.21 &= \frac{7}{10} \times \frac{21}{100} \\
&= \frac{7 \times 21}{10 \times 100} = \frac{147}{1000} \\
&= 0.147.
\end{aligned}
$$

Now write the problem vertically. Ignoring the decimal points, multiply as if multiplying whole numbers.

$$
\begin{array}{r}
0.21 \\
\times\ 0.7 \\
\hline
147
\end{array}
$$
    Decimal point not yet placed

Placing the decimal point in the product depends on the number of decimal places in the numbers multiplied. The factor 0.21 has two decimal places and 0.7 has one, so the product has three decimal places, the sum of the decimal places in the factors.

$$
\begin{array}{r}
0.\,21 \\
\times\ 0.\,7 \\
\hline
0.\,147
\end{array}
$$
    2 decimal places
    1 decimal place
    3 decimal places (2 + 1 = 3)

| **To Multiply Two Decimals** |
| --- |

1. Ignore the decimal points and multiply as if the numbers were whole numbers.

2. Find the sum of the decimal places in the two factors.

3. Place the decimal point so that the product has the same number of decimal places as the sum in Step **2.** If the whole number product obtained in Step **1** has zeros on the end, they must be counted when placing the decimal point.

| EXAMPLE 1 MULTIPLYING DECIMALS | PRACTICE EXERCISE 1 |
|---|---|

Find each product.

**(a)** $0.6 \times 1.7$

$$
\begin{array}{r}
1.\;7 \\
\times\; 0.\;6 \\
\hline
1.\;02
\end{array}
$$
↑
place decimal point here

1 decimal place
1 decimal place
2 decimal places in product (1 + 1 = 2)

**(b)** $0.2 \times 0.3$

$$
\begin{array}{r}
0.3 \\
\times\; 0.2 \\
\hline
0.06
\end{array}
$$
↑
place decimal point here after inserting
a 0 to the left of the 6.

1 decimal place
1 decimal place
2 decimal places (1 + 1 = 2)

Notice that the product is 0.06 and *not* 0.6 or 0.60.

**(c)** $0.4 \times 1.35$

$$
\begin{array}{r}
1.\;35 \\
\times\; 0.\;4 \\
\hline
0.\;540
\end{array}
$$

2 decimal places
1 decimal place
3 decimal places in product (2 + 1 = 3)

↑ ↑
count the zero for placing decimal point
place decimal point here

**(d)** $1.3 \times 6.002$

$$
\begin{array}{r}
6.\;002 \\
\times\; 1.\;3 \\
\hline
18006 \\
6002\phantom{0} \\
\hline
7.8026
\end{array}
$$
↑
place decimal point here

3 decimal places
1 decimal place

4 decimal places in product (3 + 1 = 4)

**(e)** $42 \times 11.03$

$$
\begin{array}{r}
11.\;03 \\
\times\; 42 \\
\hline
2206 \\
4412\phantom{0} \\
\hline
463.26
\end{array}
$$

2 decimal places
0 decimal places

2 decimal places in product (2 + 0 = 2)

---

**PRACTICE EXERCISE 1**

Find each product.

**(a)** $0.4 \times 2.8$

**(b)** $0.3 \times 0.3$

**(c)** $0.6 \times 2.45$

**(d)** $1.2 \times 4.007$

**(e)** $10.05 \times 23$

Answers: (a) **1.12** (b) **0.09**
(c) **1.470 or 1.47** (d) **4.8084**
(e) **231.15**

**HINT**

Incorrect placing of the decimal point in the answer to a problem is a common error. Estimating can give a quick check. For example, suppose we had to multiply

$$3.2 \times 21.037.$$

We could round 3.2 to 3 and 21.037 to 21. Since $3 \times 21 = 63$, we would expect the product to be somewhere near 63.

$$
\begin{array}{r}
21.037 \\
\times\ 3.2 \\
\hline
42074 \\
63101\phantom{0} \\
\hline
67.3084
\end{array}
$$

— decimal point *must* be placed here according to the estimated product

---

**EXAMPLE 2   PLACING THE DECIMAL POINT BY ESTIMATION**

Find $2.07 \times 495.3$ and place the decimal point by estimating the product.

estimate

$$
\begin{array}{r}
495.3 \longrightarrow 500 \\
\times\ 2.07 \longrightarrow \times\ 2 \\
\hline
34671 \qquad\quad 1000 \leftarrow \text{estimated product} \\
99060\phantom{0} \\
\hline
1025.271
\end{array}
$$

— since estimated product is 1000, decimal point must be placed here

Notice that counting decimal places would give the same placement. The work shown in color should be done mentally.

**PRACTICE EXERCISE 2**

Find $29.1 \times 195.07$ and place the decimal point by estimating the product.

Answer: **5676.537 [using 30 × 200 = 6000, the decimal point must be placed as shown]**

---

## ② MULTIPLYING DECIMALS BY A POWER OF 10

Multiplying decimals by a power of 10 (10, 100, 1000, 10,000, and so forth), can be simplified by merely moving the decimal point. For example, consider the two products below.

$$
\begin{array}{cc}
\begin{array}{r}
12.428 \\
\times\ 10 \\
\hline
124.280
\end{array}
&
\begin{array}{r}
12.428 \\
\times\ 1000 \\
\hline
12,428.000
\end{array}
\end{array}
$$

place decimal point here     place decimal point here

When we multiplied 12.428 by 10, the product 124.28 could be obtained by moving the decimal point one place to the right in 12.428. Similarly, the product of 12.428 and 1000 could be obtained by moving the decimal point three places to the right. In general, we have the following rule.

### To Multiply a Decimal by a Power of 10

Move the decimal point to the *right* the same number of decimal places as the number of zeros in the power of 10.

---

### EXAMPLE 3   MULTIPLYING BY POWERS OF 10

Find each product.

**(a)** $2.135 \times 10$

Since there is one zero in 10, move the decimal point in 2.135 one place to the right.

$$2.135 \times 10 = 21.35$$
1 zero    right 1 place

**(b)** $2.135 \times 1000$

Move the decimal point 3 places to the right.

$$2.135 \times 1000 = 2135.$$
3 zeros    right 3 places

**(c)** $2.135 \times 100,000$

Move the decimal point 5 places to the right.

$$2.135 \times 100,000 = 213500. = 213,500$$
5 zeros    right 5 places

Notice that we needed to attach two zeros when moving the decimal point.

### PRACTICE EXERCISE 3

Find each product.

**(a)**  $10 \times 48.215$

**(b)**  $100 \times 48.215$

**(c)**  $10,000 \times 48.215$

Answers:  (a)  **482.15**    (b)  **4821.5**
(c)  **482,150**

---

### ③ CONVERTING DOLLARS TO CENTS

A familiar application of multiplying by 100 involves converting dollars to cents. For example,

$1.98 converted to cents is 198¢.

Multiplying 1.98 by 100 gives 198.

### To Convert Dollars to Cents

Discard the $ sign, move the decimal point two places to the right (multiply by 100), and attach a ¢ sign to the right of the result.

---

### EXAMPLE 4   CONVERTING DOLLARS TO CENTS

Convert from dollars to cents.

**(a)** $45.35   Discard the $ sign, move the decimal point two places to the right, and attach the ¢ symbol. The result is 4535¢.

**(b)** $0.25 = 25¢

### PRACTICE EXERCISE 4

Convert from dollars to cents.

**(a)** $295.39

**(b)** $0.30

Answers:  (a)  **29,539¢**    (b)  **30¢**

## ❹ MULTIPLYING DECIMALS BY 0.1, 0.01, AND 0.001

When a decimal is multiplied by 0.1, 0.01, or 0.001, the product can also be found by moving the decimal point, this time to the left.

> ### To Multiply a Decimal by 0.1, 0.01, or 0.001
> Move the decimal point to the *left* the same number of decimal places as decimal places in 0.1, 0.01, or 0.001.

| EXAMPLE 5    MULTIPLYING BY 0.1, 0.01, AND 0.001 | PRACTICE EXERCISE 5 |
| --- | --- |

Find the product.

**(a)** $0.1 \times 2.135$

Since 0.1 has 1 decimal place, the decimal point in 2.135 is moved 1 place to the left.

$$0.1 \times 2.135 = 0.2135$$

1 decimal place    left 1 place

| 2.135 | 3 decimal places |
| --- | --- |
| 0.1 | 1 decimal place |
| 0.2135 | 3 + 1 = 4 decimal places |

The product to the right shows the accuracy of this method.

**(b)** $0.01 \times 2.135$

Move the decimal point 2 places to the left.

$$0.01 \times 2.135 = 0.02135$$

2 decimal places    left 2 places

Notice that an extra zero was attached.

**(c)** $0.001 \times 2.135$

Move the decimal point 3 places to the left.

$$0.001 \times 2.135 = 0.002135$$

3 decimal places        left 3 places

This time two zeros were supplied.

### PRACTICE EXERCISE 5

Find each product.

**(a)** $0.1 \times 48.215$

**(b)** $0.01 \times 48.215$

**(c)** $0.001 \times 48.215$

Answers: (a) **4.8215**    (b) **0.48215**
(c) **0.048215**

## ❺ CONVERTING CENTS TO DOLLARS

Multiplying by 0.01 converts from cents to dollars. For example,

198¢ converted to dollars is $1.98.

Multiplying 198 by 0.01 gives 1.98.

> ### To Convert Cents to Dollars
> Discard the ¢ sign, move the decimal point two places to the left (multiply by 0.01), and attach a $ sign at the left of the result.

| EXAMPLE 6  CONVERTING CENTS TO DOLLARS |
|---|

Convert from cents to dollars.

**(a)** 7955¢   Discard the ¢ sign, move the decimal point two places to the left (remember that 7955 is the same as 7955.), and attach a $ sign. The result is $79.55.

**(b)** 8¢ = $0.08

Notice that the extra zeros are supplied in the result.

| PRACTICE EXERCISE 6 |
|---|

Convert from cents to dollars.

**(a)** 12,995¢

**(b)** 15¢

Answers:  (a)  **$129.95**   (b)  **$0.15**

## ⑥ APPLICATIONS MULTIPLYING DECIMALS

Many applied problems are solved by multiplying decimals.

| EXAMPLE 7  MULTIPLYING DECIMALS IN A CONSUMER PROBLEM |
|---|

Bob bought 8 books at a cost of $2.95 each. How much did he spend?
  Multiply $2.95 by 8.

$$\begin{array}{r} \$2.95 \\ \times\ 8 \\ \hline \$23.60 \end{array}$$
$2.95   2 decimal places
× 8   0 decimal places
$23.60   2 decimal places (2 + 0 = 2)

Thus, Bob spent $23.60 for the books. Since 8 books at $3 each would cost $24 [24 = 8 × 3], our work appears correct.

| PRACTICE EXERCISE 7 |
|---|

Shawn Herman makes a car payment of $324.30 per month for 36 months. How much will he pay altogether?

Answer:  **$11,674.80**

| EXAMPLE 8  CALCULATING WEEKLY EARNINGS |
|---|

Alberta is paid $7.23 per hour. Her hourly time card for one week is shown in Figure 5.7. How much was she paid that week?
  To find out how much she was paid, we need to know the total number of hours that she worked.

| Day | Hours worked |
|---|---|
| Mon | 5.7 |
| Tues | 3.25 |
| Wed | 4.9 |
| Thur | 6.5 |
| Fri | 7.8 |

**Figure 5.7**

$$\begin{array}{r} 5.7 \\ 3.25 \\ 4.9 \\ 6.5 \\ +\ 7.8 \\ \hline 28.15 \end{array}$$

Since she worked 28.15 hours at $7.23 per hour, multiply.

$$\begin{array}{r} 28.15 \\ \times\ 7.23 \\ \hline 8445 \\ 5630\ \\ 19705\ \ \\ \hline 203.5245 \end{array}$$

Rounded to the nearest cent, Alberta earned $203.52. Since 30 hours at $7 an hour would result in $210, our work appears to be correct.

| PRACTICE EXERCISE 8 |
|---|

A used car salesman earns a salary of 0.04 times his total sales. During one week, he sold three cars for $2450.75, $3675.40, and $4425.35, respectively. How much did he earn that week?

Answer:  **$422.06**

| EXAMPLE 9   MULTIPLYING DECIMALS IN A PHYSICS PROBLEM | PRACTICE EXERCISE 9 |

The air pressure at sea level is 14.7 pounds per square inch of surface area. The surface area of the body of a man is approximately 2225 square inches. To the nearest pound, what is the total air pressure on his body at sea level?

Multiply the pressure per square inch by the total number of square inches of surface area.

$$
\begin{array}{r}
2225 \\
\times\ 14.7 \\
\hline
15575 \\
8900\phantom{0} \\
2225\phantom{00} \\
\hline
32707.5
\end{array}
$$   0 + 1 = 1 decimal place

Thus, the total pressure on the man's body (to the nearest pound) is 32,708 pounds, a result that may surprise you.

Rounded to the nearest cent, what is the cost in dollars and cents of 22.5 gallons of gasoline at 131.9¢ per gallon?

Answer: $29.68

## 5.5   EXERCISES A

*Find the products. Check placement of the decimal point by estimating the product.*

**1.** 14.03
  × 0.5

**2.** 6.58
  × 1.2

**3.** 12.05
  × 2.3

**4.** 3.004
  × 1.7

**5.** 5.107
  × 8.4

**6.** 4.0108
  × 2.6

**7.** 12.411
  × 5.4

**8.** 12.28
  × 13

*Without making actual computations, use an estimate to decide which of the given answers is correct.*

**9.** 4.1 × 39.047     **(a)** 1.60093     **(b)** 16.0093     **(c)** 160.093     **(d)** 1600.93

**10.** 2.15 × 403.7     **(a)** 8.67955     **(b)** 86.7955     **(c)** 867.955     **(d)** 8679.55

**11.** 32.3 × 48.1     **(a)** 1553.63     **(b)** 15,536.3     **(c)** 155,363.0     **(d)** 1,553,630.0

**12.** 203.475 × 9.407   **(a)** 191.409     **(b)** 1914.09     **(c)** 19,140.9     **(d)** 191,409

*Find the products.*

**13.** 15.237 × 10

**14.** 15.237 × 100

**15.** 15.237 × 10,000

**16.** 15.237 × 0.1

**17.** 15.237 × 0.01

**18.** 15.237 × 0.001

*Convert from dollars to cents.*

**19.** $5.45          **20.** $21.98          **21.** $0.79

*Convert from cents to dollars.*

**22.** 250¢          **23.** 7936¢          **24.** 4¢

*Solve.*

**25.** Joe bought 5 paperback books for $2.75 each. How much did he spend?

**26.** Carlos is paid $8.19 per hour. His timecard for one week is shown in the figure. How much was he paid that week?

| Day | Hours worked |
|-----|--------------|
| Mon | 6.3 |
| Tues | 4.7 |
| Wed | 8.2 |
| Thur | 6.9 |
| Fri | 7.4 |

**27.** The EPA-estimated miles per gallon for a midsized car is 21.8 mpg. The capacity of its gas tank is 19.4 gallons. To the nearest mile, what is the driving range of the car?

**28.** The air pressure at sea level is 14.7 pounds per square inch of surface area. The surface area of a student's body is approximately 2085 square inches. To the nearest pound, what is the total air pressure on his body at sea level?

**29.** Pedro buys 12.5 pounds of peaches at $0.48 per pound and 9.6 pounds of hamburger at $1.37 per pound. How much change will he receive if he pays with a check for $25.00?

**30.** Find the area of a rectangle, to the nearest hundredth of a square centimeter, if its length is 14.38 cm and its width is 9.41 cm.

**31** Barbara plans to cover a rectangular wall with vinyl covering costing $0.67 a sq ft. If the wall is 13.5 ft long and 7.6 ft high, how much will the project cost?

**32.** Steve Hellmann's contract specifies that he be paid $14.38 per hour for a 40-hour work week. For all hours over 40, he must be paid time and a half (1.5 times the normal hourly rate). Last week Mr. Hellmann worked 51 hours. How much was he paid?

**33.** It costs $17.50 a day plus $0.18 per mile to rent a car at Cheapo Car Rentals. How much would it cost to rent the car for two days if it is driven a total of 240 miles?

**34.** Dave worked 41 hours one week and was paid $7.87 an hour. What was his approximate pay for the week? His exact pay?

**35** Herman wishes to buy four new tires for his car. He is interested in the tires shown in the advertisement to the right. What would be the approximate cost of the tires? The exact cost?

All Season Radials
with White Walls

EVERYDAY
LOW PRICES    $38^{75}

## FOR REVIEW

**36.** Mr. and Mrs. Bonnett stayed at the Frontier Hotel in Las Vegas on a 3-day/2-night package. Their room cost $139.95, and they charged $37.28 in food and beverage to their room account. If Mr. Bonnett paid the bill with two $100 bills, how much change did he receive?

**37.** Find the perimeter of the garden sketched in the figure below.

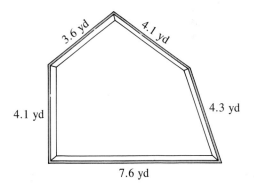

3.6 yd    4.1 yd

4.1 yd    4.3 yd

7.6 yd

**38.** On a five-day trip, a family traveled 2450.3 miles. On the first day they traveled 233.2 mi, on the second 537.3 mi, on the third 641.4 mi, and on the fourth 596.3 mi.

**(a)** How far did they travel on the fifth day?

**(b)** What is the answer to **(a)** rounded to the nearest mile?

**(c)** What is the answer to **(a)** if all distances are rounded to the nearest mile before any calculation?

**(d)** Compare the answers to **(b)** and **(c)**.

ANSWERS:  1. 7.015  2. 7.896  3. 27.715  4. 5.1068  5. 42.8988  6. 10.42808  7. 67.0194  8. 159.64  9. (c)
10. (c)  11. (a)  12. (b)  13. 152.37  14. 1523.7  15. 152,370  16. 1.5237  17. 0.15237  18. 0.015237  19. 545¢
20. 2198¢  21. 79¢  22. $2.50  23. $79.36  24. $0.04  25. $13.75  26. $274.37  27. 423 mi  28. 30,650 pounds
29. $5.85  30. 135.32 sq cm  31. $68.74  32. $812.47  33. $78.20  34. $320 [using 40 × 8]; $322.67  35. $160
[using 4 × 40]; $155.00  36. $22.77  37. 23.7 yd  38. (a) 442.1 mi (b) 442 mi (c) 443 mi (d) Since answers differ, it
is better not to round until the last step (unless you are estimating an answer).

## 5.5  EXERCISES B

*Find the products. Check placement of the decimal point by estimating the product.*

**1.** 21.02
  × 0.7

**2.**  2.59
  × 1.5

**3.**  14.06
  × 5.4

**4.**  2.001
  × 1.2

**5.**  3.109
  × 5.6

**6.**  3.0109
  × 4.3

**7.**  13.611
  × 3.2

**8.**  14.56
  × 18

*Without making actual computations, use an estimate to decide which of the given answers is correct.*

**9.** 6.2 × 51.075    **(a)** 3.16665    **(b)** 31.6665    **(c)** 316.665    **(d)** 3166.65

**10.** 4.25 × 291.8    **(a)** 1.24015    **(b)** 12.4015    **(c)** 124.015    **(d)** 1240.15

**11.** 41.5 × 62.3    **(a)** 2585.45    **(b)** 258.545    **(c)** 25.8545    **(d)** 2.58545

**12.** 195.625 × 3.508    **(a)** 68.62525    **(b)** 686.2525    **(c)** 6862.525    **(d)** 68625.25

*Find the products.*

**13.** 341.59 × 10

**14.** 341.59 × 1000

**15.** 341.59 × 100

**16.** 341.59 × 0.1

**17.** 341.59 × 0.001

**18.** 341.59 × 0.01

*Convert from dollars to cents.*

**19.** $8.23

**20.** $45.15

**21.** $0.67

*Convert from cents to dollars.*

**22.** 705¢

**23.** 2009¢

**24.** 12¢

*Solve.*

**25.** Dyane bought 6 greeting cards at $1.25 each. How much did she spend?

**26.** Peter is paid $11.53 per hour. His hourly time card for one week is shown below. How much was he paid that week (to the nearest cent)?

| Day | Hours worked |
| --- | --- |
| Mon | 7.2 |
| Tues | 6.5 |
| Wed | 8.1 |
| Thur | 6.9 |
| Fri | 5.4 |

**27.** What is the driving range (to the nearest mile) of a car with EPA-estimated miles per gallon of 26.2 and a gas tank with capacity of 14.5 gallons?

**28.** The air pressure at sea level is 14.7 pounds per square inch of surface area. If the surface area of a young child is 1125 square inches, to the nearest pound, what is the total air pressure on his body at sea level?

**29.** Rosa buys 11.3 pounds of bananas at $0.35 per pound and 8.2 pounds of ground meat for $1.09 per pound. How much change will she receive if she pays with a $20 bill?

**30.** Find the area of a rectangle, to the nearest tenth of a square foot, if its length is 6.34 ft and its width is 2.59 ft.

**31.** What is the price of a mirror measuring 5.2 ft long and 3.7 ft wide if the cost per square foot is $0.85?

**32.** Ms. Morgan is paid $12.35 per hour for a 40-hour work week. For all hours over 40 she is paid time-and-a-half as overtime. If she worked 53 hours one week, how much was she paid?

**33.** It costs $18.25 a day plus $0.22 per mile to rent a car at Local Car Rentals. How much would it cost to rent the car for three days if it is driven 395 miles?

**34.** It costs $12.05 per hour to operate a particular machine. What is the approximate cost of running the machine for 48 hours? The exact cost?

**35.** Mr. and Mrs. Washington decide to purchase a vacuum cleaner, advertised at right, for each of their five children. What would be the approximate cost of the cleaners? The exact cost?

$59⁸⁸

Upright Vacuum Cleaner with
12″ Beater Bar Brush

## FOR REVIEW

**36.** The Bennings spent three nights at the Desert Oasis Hotel in Palm Springs. The price of the room was $105.50 per night, and they charged $62.39 in food and beverages to their room. If Mr. Benning paid the bill with four $100 bills, how much change did he receive?

**37.** Find the perimeter of the lot sketched in the figure.

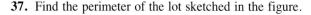
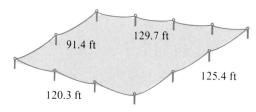

129.7 ft

91.4 ft

125.4 ft

120.3 ft

**38.** On a four-day vacation, a family spent a total of $1123.06. They spent $418.19 on the first day, $176.43 on the second day, and $212.05 on the third day.

 (a) How much did they spend on the fourth day?

 (b) What is the answer to (a) rounded to the nearest dollar?

 (c) What is the answer to (a) if all amounts are rounded to the nearest dollar before making any calculations?

 (d) Compare the answers to (b) and (c).

## 5.5 EXERCISES C

*Solve.*

**1.** The cost of heating fuel is 98.9¢ per gallon. To the nearest cent, how much will 38.5 gallons cost?
 [Answer: $38.08]

**2.** Terry McGinnis plans to carpet her family room with carpeting which costs $18.95 a square yard. If the room is rectangular in shape, 7.2 yards long and 4.9 yards wide, what would be the approximate cost of the job? The exact cost?

**3.** The Hagoods must pay $8.35 in property tax for every $1000 of assessed value on their home. How much tax do they pay if the assessed value of their home is $102,000?

**4.** Find the area of the region below by dividing the region into two rectangles.

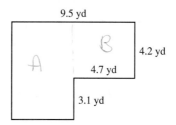

## 5.6 DIVIDING DECIMALS

### STUDENT GUIDEPOSTS

**1** Dividing a Decimal by a Whole Number

**2** Dividing a Decimal by a Decimal

**3** Finding Quotients Rounded to a Given Decimal Place

**4** Dividing Decimals by a Power of 10

**5** Dividing Decimals by 0.1, 0.01, and 0.001

**6** Quotients With a Repeating Block of Digits

**7** Finding Quotients of Two Whole Numbers

**8** Applications Dividing Decimals

## 1 DIVIDING A DECIMAL BY A WHOLE NUMBER

Dividing decimals is much like dividing whole numbers, except for placing the decimal point. Converting fractions to decimals as was done in Section 5.2 is a special case of this division process.

Suppose we divide

$$2.7 \div 3$$

using fractions. We know

$$2.7 = 2\frac{7}{10} = \frac{27}{10}.$$

Then

$$2.7 \div 3 = \frac{27}{10} \div \frac{3}{1}$$

$$= \frac{27}{10} \times \frac{1}{3}$$

$$= \frac{9 \cdot \cancel{3} \cdot 1}{10 \cdot \cancel{3}}$$

$$= \frac{9}{10}$$

$$= 0.9.$$

Thus, 2.7 divided by 3 must be 0.9. Now, suppose we write the quotient as follows.

$$3\overline{)2.7}$$

Ignoring the decimal point and dividing as we divided whole numbers gives the following result.

$$
\begin{array}{r}
9 \\
3\overline{)2.7} \\
\underline{2\ 7} \\
0
\end{array}
$$
Decimal point not yet placed

Moving the decimal point in the dividend (2.7) straight up results in the desired quotient, 0.9.

$$
\begin{array}{r}
.9 \\
3\overline{)2.\!\uparrow 7} \\
\underline{2\ 7} \\
0
\end{array}
$$

This illustrates a general procedure.

### To Divide a Decimal by a Whole Number

1. Place a decimal point directly above the decimal point in the dividend.
2. Divide as if both numbers were whole numbers.

| EXAMPLE 1 DIVIDING DECIMALS BY A WHOLE NUMBER | PRACTICE EXERCISE 1 |

Find the quotients.

**(a)** $1.12 \div 7$

Place a decimal point in the quotient directly above the decimal point in the dividend, then divide as if the dividend were a whole number.

$$
\begin{array}{r}
.16 \\
7\overline{)1.12} \\
7\phantom{.12} \\
\hline
42 \\
42 \\
\hline
0
\end{array}
$$

The quotient is 0.16.

**(b)** $11.34 \div 21$

$$
\begin{array}{r}
.54 \\
21\overline{)11.34} \\
10\ 5\phantom{4} \\
\hline
84 \\
84 \\
\hline
0
\end{array}
$$
Place decimal point
Divide as if whole numbers

The quotient is 0.54.

**(c)** $113.95 \div 43$

$$
\begin{array}{r}
2.65 \\
43\overline{)113.95} \\
86\phantom{.95} \\
\hline
27\ 9\phantom{5} \\
25\ 8\phantom{5} \\
\hline
2\ 15 \\
2\ 15 \\
\hline
0
\end{array}
$$
Place decimal point and divide

The quotient is 2.65.

---

Find the quotients.

**(a)** $1.08 \div 4$

**(b)** $15.04 \div 32$

**(c)** $149.15 \div 19$

Answers: **(a)** 0.27   **(b)** 0.47
**(c)** 7.85

---

## ❷ DIVIDING A DECIMAL BY A DECIMAL

Division problems with decimals for both the dividend and the divisor can be changed into problems with whole number divisors. For example, suppose we write the division problem

$$0.35\overline{)1.498}$$

as a fraction.

$$\frac{1.498}{0.35}$$

Multiplying numerator and denominator by 100

$$\frac{1.498 \times 100}{0.35 \times 100} = \frac{149.8}{35}$$

would give a fraction which has a whole number in the denominator. That is, to divide

$$1.498 \text{ by } 0.35$$

we could just as well divide

$$149.8 \text{ by } 35.$$

To do this, zeros must be supplied to the right of 8 as shown.

```
        4.28
  35)149.80      Attach a zero to continue the process
     140
      9 8
      7 0
      2 80
      2 80
         0
```

---

### To Divide a Decimal by a Decimal

1. Move the decimal point to the right in the divisor (multiply by a power of 10) to obtain a whole number.

2. Move the decimal point in the dividend the same number of places to the right (multiply by the same power of 10). Extra zeros may need to be attached.

3. Divide by the whole number divisor as before.

---

For example, suppose we consider $1.498 \div 0.35$ again.

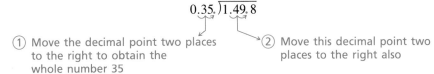

① Move the decimal point two places to the right to obtain the whole number 35

② Move this decimal point two places to the right also

---

| **EXAMPLE 2   DIVIDING DECIMALS** | **PRACTICE EXERCISE 2** |

Find the quotients.

**(a)** $4.465 \div 1.9$

```
  1.9, )4.4, 65      ① Move decimal point one place to the right
                        in both divisor and dividend

       2. 35
  19)44↑.65          ② Place decimal point directly above and
     38                 divide by the whole number 19
      6 6
      5 7
        95
        95
         0
```

Find the quotients.

**(a)** $7.266 \div 2.1$

**(b)** 2.7168 ÷ 0.48

```
              5.66
  0.48.)2.71.68
         2 40
         31 6
         28 8
          2 88
          2 88
             0
```

**(b)** 2.7195 ÷ 0.35

**(c)** 42 ÷ 1.4    The whole number 42 can be thought of as the decimal 42. or 42.0.

```
            3 0.
  1.4.)42.0.
        42
         0 0
```

**(c)** 84 ÷ 1.2

Answers: (a) **3.46**   (b) **7.77**
(c) **70**

## ❸ FINDING QUOTIENTS ROUNDED TO A GIVEN DECIMAL PLACE

In Example 2, each quotient was a terminating decimal which had very few decimal places. When this is not the case, we often round answers to a given number of decimal places.

| EXAMPLE 3   FINDING QUOTIENTS TO A GIVEN DECIMAL PLACE |
| --- |

**(a)** Find 3.257 ÷ 2.7, rounded to the nearest tenth.

```
            1.20
  2.7.)3.2.57
        2 7
         5 5
         5 4
           17
```

We could continue dividing, but since the first digit to the right of the tenths position is 0, the quotient is approximately 1.2, rounded to the nearest tenth. That is,

$$3.257 \div 2.7 \approx 1.2.$$

**(b)** Find 2.551 ÷ 0.16, rounded to the nearest hundredth.

```
             15.943
  0.16.)2.55.100
         1 6
          95
          80
          15 1
          14 4
             70
             64
             60
             48
             12
```

PRACTICE EXERCISE 3

**(a)** Find 2.107 ÷ 3.1 rounded to the nearest tenth.

**(b)** Find 3.205 ÷ 0.22 rounded to the nearest hundredth.

Since the first digit to the right of the hundredths position is 3, the quotient is approximately 15.94, rounded to the nearest hundredth. That is,

$$2.551 \div 0.16 \approx 15.94.$$

Answers: (a) 0.7    (b) 14.57

---

## H I N T

In Section 5.5 we placed decimals and checked decimal products by estimating the answer. The same technique can be used when dividing decimals.

---

| **EXAMPLE 4**   ESTIMATING TO LOCATE DECIMAL PLACE | **PRACTICE EXERCISE 4** |
| --- | --- |

Find 156.3 ÷ 12 and place the decimal point by estimating the answer.
Since the quotient is about 150 ÷ 10, the result should be about 15.

```
        13.025        decimal point must be placed here
   12)156.300         according to estimated quotient
       12
       36
       36
        30
        24
        60
        60
         0
```

Find 17.064 ÷ 8 and place the decimal point by estimating the quotient.

Answer: 2.133 [Since 17.064 is about 16 and 16 ÷ 8 is 2, the quotient must be about 2.]

---

## 4 DIVIDING DECIMALS BY A POWER OF 10

The quotient in Example 4 was approximated by dividing 150 by 10. Dividing a decimal by 10 is the same as multiplying the decimal by 0.1 since

$$150 \div 10 = 150 \times \frac{1}{10} = 150 \times 0.1 \,.$$

As a result, from Section 5.5, dividing by 10 (multiplying by 0.1) is simply a matter of moving the decimal point one place to the left. Similar rules apply when dividing by any power of 10 such as 10, 100, 1000, or 10,000.

### To Divide a Decimal by a Power of 10

Move the decimal point to the *left* the same number of decimal places as the number of zeros in the power of 10.

| **EXAMPLE 5** **DIVIDING DECIMALS BY A POWER OF 10** | **PRACTICE EXERCISE 5** |
|---|---|

Find each quotient.

**(a)** $43.21 \div 100$

    Since there are 2 zeros in 100, move the decimal point in 43.21 two places to the left.

$$43.21 \div 100 = 0.4321$$

          2 zeros    left 2 places

**(b)** $43.21 \div 1000$     Move the decimal point 3 places to the left.

$$43.21 \div 1000 = 0.04321$$

          3 zeros    left 3 places

Notice that an extra zero was needed when moving the decimal point.

**(c)** $43.21 \div 100,000$

    Move the decimal point 5 places to the left.

$$43.21 \div 100,000 = 0.0004321$$

Find each quotient.

**(a)** $652.5 \div 100$

**(b)** $652.5 \div 1000$

**(c)** $652.5 \div 100,000$

Answers: (a) **6.525**   (b) **0.6525** (c) **0.006525**

## 5 DIVIDING DECIMALS BY 0.1, 0.01, AND 0.001

Suppose we divide 1.235 by 0.1. Dividing by 0.1 is the same as multiplying by 10.

$$1.235 \div 0.1 = 1.235 \div \frac{1}{10} = 1.235 \times \frac{10}{1} = 1.235 \times 10$$

As a result, from Section 5.5, dividing by 0.1 (multiplying by 10) is the same as moving the decimal point one place to the right. Similar rules apply when dividing by 0.01 or 0.001. This is summarized in the following rule.

> ### To Divide a Decimal by 0.1, 0.01, or 0.001
>
> Move the decimal point to the *right* the same number of decimal places as decimal places in 0.1, 0.01, or 0.001.

| **EXAMPLE 6** **DIVIDING DECIMALS BY 0.1, 0.01, OR 0.001** | **PRACTICE EXERCISE 6** |
|---|---|

Find each quotient.

**(a)** $54.32 \div 0.1$
    Since 0.1 has one decimal place, move the decimal point in 54.32 one place to the right.

$$54.32 \div 0.1 = 543.2$$

          right 1 place

Find each quotient.

**(a)** $439.7 \div 0.1$

**(b)** 54.32 ÷ 0.001

Move the decimal point three places to the right since 0.001 has three decimal places.

$$54.32 \div 0.001 = 54\underset{\text{right 3 places}}{320.}$$

Notice that a zero must be attached to obtain the quotient 54,320.

**(b)** 439.7 ÷ 0.01

Answers: (a) 4397    (b) 43,970

---

## H I N T

The difference between multiplying a decimal by a power of 10 and dividing by a power of 10 is simply reversing the direction we move the decimal point. This is easy to remember since multiplication and division are *opposite* processes. The same remarks apply to multiplying and dividing by 0.1, 0.01, or 0.001.

---

## 6 QUOTIENTS WITH A REPEATING BLOCK OF DIGITS

Some division problems result in quotients which have a repeating block of digits, as mentioned in Section 5.2. When this repetition is discovered, stop the process and place a bar over the repeating digits.

| EXAMPLE 7    A QUOTIENT WITH A REPEATING BLOCK OF DIGITS | PRACTICE EXERCISE 7 |

Find 2.19 ÷ 0.99.

$$
\begin{array}{r}
2.2121 \\
0.99\overline{)2.19.0000} \\
\underline{1\ 98} \\
21\ 0 \\
\underline{19\ 8} \\
1\ 20 \\
\underline{99} \\
210 \\
\underline{198} \\
120 \\
\underline{99} \\
21
\end{array}
$$

We can see that the digits 21 will continue to repeat after the decimal point. Thus,

$$2.19 \div 0.99 = 2.\overline{21}.$$

Find 1.74 ÷ 0.33.

Answer: 5.$\overline{27}$

---

## 7 FINDING QUOTIENTS OF TWO WHOLE NUMBERS

When one whole number is divided by another, the quotient is sometimes given as a decimal rounded to a particular degree of accuracy.

---

### EXAMPLE 8  DIVIDING WHOLE NUMBERS

Find $431 \div 29$, rounded to the nearest hundredth.

$$
\begin{array}{r}
14.862 \\
29\overline{)431.000} \quad \text{Add three zeros} \\
\underline{29} \\
141 \\
\underline{116} \\
25\ 0 \\
\underline{23\ 2} \\
1\ 80 \\
\underline{1\ 74} \\
60 \\
\underline{58} \\
2
\end{array}
$$

Since the first digit to the right of the hundredths position is 2, the quotient is approximately 14.86, rounded to the nearest hundredth.

---

### PRACTICE EXERCISE 8

Find $243 \div 21$, rounded to the nearest tenth.

Answer: 11.6

---

## ⑧ APPLICATIONS DIVIDING DECIMALS

Many applied problems are solved by dividing one decimal by another. As always, estimate the answer and ask yourself if the actual answer seems reasonable.

---

### EXAMPLE 9  CALCULATING MONTHLY PAYMENTS

How many months will it take to pay off a loan of $894.18 if the monthly payments are $42.58?

Divide the total to be paid ($894.18) by the amount paid each month ($42.58).

$$
\begin{array}{r}
21. \\
42.58\overline{)894.18} \\
\underline{851\ 6} \\
42\ 58 \\
\underline{42\ 58} \\
0
\end{array}
$$

Thus, it will take 21 months to pay off the loan. Since $900 \div 40 = 22.5$, this answer seems reasonable.

---

### PRACTICE EXERCISE 9

Fern took out a loan for $7752.96 which is to be paid off in 36 equal monthly payments. How much is each payment?

Answer: $215.36

---

### EXAMPLE 10  FINDING A GRADE POINT AVERAGE

Finding the grade point average (GPA) of a student is an important application of decimals. Calculate a GPA using a 4-point scale, that is, a perfect GPA (every grade an A) is 4.00. Each letter grade is assigned a value: A = 4, B = 3, C = 2, D = 1, and F = 0. The first step is to find the *course value,* that is, the product of the credits and the grade value, as shown in Figure 5.8.

---

### PRACTICE EXERCISE 10

Burford earned a C in a 4-credit English course, a D in a 3-credit biology course, a B in a 2-credit physical education course, and an F in a 4-credit history course. Calculate Burford's GPA (rounded to the nearest hundredth) on a 4-point scale.

| Course | Credits | Grade | Value | Course Value |
|--------|---------|-------|-------|--------------|
| Math | 4 | B | 3 | $4 \cdot 3 = 12$ |
| English | 3 | A | 4 | $3 \cdot 4 = 12$ |
| Music | 2 | A | 4 | $2 \cdot 4 = \phantom{0}8$ |
| History | 3 | C | 2 | $3 \cdot 2 = \phantom{0}6$ |
| Physics | 3 | D | 1 | $3 \cdot 1 = \phantom{0}\underline{3}$ |
| Totals | 15 | | | 41 |

**Figure 5.8**

$$\text{GPA} = \frac{\text{sum of course values}}{\text{total credits}} = \frac{41}{15} = 2.733$$

The student's GPA, rounded to the nearest hundredth, is 2.73.

Answer: **1.31**

## 5.6 EXERCISES A

*Find the quotients.*

**1.** $6\overline{)5.4}$     **2.** $8\overline{)18.4}$     **3.** $12\overline{)43.2}$     **4.** $25\overline{)171.5}$

**5.** $1.4\overline{)1.12}$     **6.** $1.4\overline{)0.112}$     **7.** $0.27\overline{)0.378}$     **8** $0.31\overline{)2.232}$

**9.** $4.2\overline{)8.988}$     **10.** $0.015\overline{)0.0945}$     **11.** $1.6\overline{)56}$     **12.** $25\overline{)69}$

*Find the quotient, rounded to the nearest tenth.*

**13.** $1.9\overline{)2.36}$     **14.** $48\overline{)315}$     **15.** $1.02\overline{)3.185}$

*Find the quotient, rounded to the nearest hundredth.*

**16.** $1.9\overline{)2.36}$     **17** $48\overline{)315}$     **18.** $1.02\overline{)3.185}$

*Without making actual calculations, use an estimate to decide which of the given answers is correct.*

**19.** $243.41 \div 19$    **(a)** 1.28111    **(b)** 12.8111    **(c)** 128.111    **(d)** 1281.11

**20.** $625.03 \div 25.1$    **(a)** 24.9016    **(b)** 2.49016    **(c)** 2490.16    **(d)** 24,901.6

**21.** $357.103 \div 62.8$    **(a)** 0.56864    **(b)** 5.6864    **(c)** 56.864    **(d)** 568.64

**22.** $403.229 \div 6.417$    **(a)** 0.062838    **(b)** 0.62838    **(c)** 6.2838    **(d)** 62.838

*Find the quotients.*

**23.** $32.71 \div 10$

**24.** $41.38 \div 100$

**25.** $5.329 \div 10,000$

**26.** $621.43 \div 100,000$

**27.** $8.319 \div 1000$

**28.** $0.0327 \div 0.1$

**29.** $426.5 \div 0.001$

**30.** $0.627 \div 0.1$

**31.** $4.398 \div 0.01$

*Find the quotient. (Each quotient is a repeating decimal.)*

**32** $1.8\overline{)21.2}$

**33.** $1.05\overline{)2.135}$

**34.** $0.36\overline{)2.58}$

*Solve.*

**35.** How many months will it take to pay off a loan of $1176.30 if the monthly payments are $65.35?

**36.** The Rappaports used 260 stepping stones to build a patio. If the total cost of the stones was $107.15, how much did each stone cost (to the nearest tenth of a cent)?

**37** The assessed value of a parcel of property can be found by dividing the amount of tax by the tax rate. Margaret's property was taxed at a rate of 0.147, and she paid a property tax of $382.40. Find the assessed value of her property.

**38.** The balance owed on three rooms of furniture amounts to $3157.42. If Tana wishes to pay off the account in 24 equal monthly payments, how much should each payment be?

**39.** The following table shows Sam's grades during one semester. Complete the table and calculate his GPA, correct to the nearest hundredth.

| Course | Credits | Grade | Value | Course Value |
|--------|---------|-------|-------|--------------|
| History | 3 | D | 1 | |
| English | 3 | A | 4 | |
| Algebra | 5 | B | 3 | |
| Chemistry | 3 | F | 0 | |
| Art | 2 | C | 2 | |

**40.** Which is the better buy, 9 apples for $1.62 or 13 apples for $2.21?

## FOR REVIEW

*Find the products.*

**41.**   $3.02$
     $\times\ 0.5$

**42.**   $7.38$
     $\times\ 1.2$

**43.**   $7.011$
     $\times\ 2.01$

**44.** $691.35 \times 100$        **45.** $691.35 \times 0.1$        **46.** $691.35 \times 0.01$

**47.** Convert $12.45 to cents.        **48.** Convert 17¢ to dollars.

**49.** Marvin bought 8 complete backpacking meals at a cost of $6.78 each. If tax on the purchases amounted to $2.17, and he paid with three $20 bills, how much change did he receive?

**50.** It costs $98.00 a week plus 16.5¢ ($0.165) per mile to rent a car at Horn's Rentals. How much would it cost to rent the car for one week if it is driven a total of 683 miles?

**51.** A blocklayer earns $13.78 per hour for a 40-hour work week and time and a half for each hour of overtime. If Pete worked 52 hours during the week laying blocks, how much was he paid?

**52.** Suppose all we knew was that Pete earned the total amount in the answer to Exercise 51 by working 52 hours. What would we say was his hourly rate?

*The following exercises will help you prepare for the next section. Place the correct symbol ($<$ or $>$) between each pair of numbers.*

**53.** 0    7        **54.** $\dfrac{3}{2}$    $\dfrac{5}{4}$        **55.** 6    $\dfrac{25}{4}$

*Perform the indicated operations.*

**56.** $2 + 8 \div 4 - 1$        **57.** $\left(\dfrac{1}{2} + \dfrac{1}{4}\right) \cdot 2 - 1$        **58.** $\sqrt{9} \div 4 + \dfrac{1}{4}$

---

ANSWERS: 1. 0.9  2. 2.3  3. 3.6  4. 6.86  5. 0.8  6. 0.08  7. 1.4  8. 7.2  9. 2.14  10. 6.3  11. 35  12. 2.76
13. 1.2  14. 6.6  15. 3.1  16. 1.24  17. 6.56  18. 3.12  19. (b)  20. (a)  21. (b)  22. (d)  23. 3.271  24. 0.4138
25. 0.0005329  26. 0.0062143  27. 0.008319  28. 0.327  29. 426,500  30. 6.27  31. 439.8  32. $11.\overline{7}$  33. $2.0\overline{3}$
34. $7.\overline{16}$  35. 18 months  36. 41.2¢ or $0.412  37. $2601.36  38. $131.56  39. GPA = 2.13  40. 13 apples at 17¢
each is better than 9 apples at 18¢ each  41. 1.51  42. 8.856  43. 14.09211  44. 69,135  45. 69.135  46. 6.9135
47. 1245¢  48. $0.17  49. $3.59  50. $210.70  51. $799.24  52. $15.37 (He earned $799.24 for 52 hours of work, so
divide $799.24 by 52.)  53. $<$  54. $>$  55. $<$  56. 3  57. $\frac{1}{2}$  58. 1

## 5.6  EXERCISES B

*Find the quotients.*

**1.** $9\overline{)23.4}$        **2.** $5\overline{)0.45}$        **3.** $14\overline{)37.8}$        **4.** $35\overline{)89.6}$

**5.** $1.7\overline{)7.31}$        **6.** $1.7\overline{)0.731}$        **7.** $0.41\overline{)0.8487}$        **8.** $0.52\overline{)3.536}$

**9.** $5.1\overline{)116.79}$        **10.** $0.022\overline{)0.1056}$        **11.** $1.3\overline{)74.1}$        **12.** $32\overline{)86}$

*Find the quotient, rounded to the nearest tenth.*

**13.** $1.8\overline{)2.47}$        **14.** $41\overline{)365}$        **15.** $1.05\overline{)4.172}$

*Find the quotient, rounded to the nearest hundredth.*

**16.** $1.8\overline{)2.47}$        **17.** $41\overline{)365}$        **18.** $1.05\overline{)4.172}$

*Without making actual calculations, use an estimate to decide which of the given answers is correct.*

**19.** $321.7 \div 21$      **(a)** 1.53190      **(b)** 15.3190      **(c)** 153.190      **(d)** 1531.90

**20.** $489.02 \div 92.2$      **(a)** 530.390      **(b)** 53.0390      **(c)** 5.30390      **(d)** 0.0530390

**21.** $449.207 \div 49.3$      **(a)** 0.091117      **(b)** 0.91117      **(c)** 9.1117      **(d)** 91.117

**22.** $702.116 \div 9.328$      **(a)** 0.0752697      **(b)** 0.752697      **(c)** 7.52697      **(d)** 75.2697

*Find the quotients.*

**23.** $64.85 \div 100$

**24.** $31.65 \div 10$

**25.** $1.357 \div 1000$

**26.** $435.07 \div 10,000$

**27.** $921.05 \div 100,000$

**28.** $0.0569 \div 0.1$

**29.** $327.5 \div 0.001$

**30.** $0.0659 \div 0.01$

**31.** $2.678 \div 0.001$

*Find the quotient. (Each quotient is a repeating decimal.)*

**32.** $4.5\overline{)1.45}$

**33.** $33\overline{)189}$

**34.** $2.16\overline{)11.16}$

*Solve.*

**35.** How many months will it take to pay off a loan of $4054.68 if the monthly payments are $112.63?

**36.** The Giffords used 320 tiles to build a patio. If the total cost of the tiles was $45.75, how much did each tile cost (to the nearest tenth of a cent)?

**37.** The assessed value of a lot can be found by dividing the amount of tax by the tax rate. If Denise's lot was taxed at a rate of 0.151, and she paid a property tax of $468.10, find the assessed value of her property.

**38.** If Johnny used 112.6 gallons of gas on a trip of 2780 miles, to the nearest tenth, what was his mpg?

**39.** The following table shows Wes's grades during the fall term. Complete the table and calculate his GPA, correct to the nearest tenth.

| Course | Credits | Grade | Value | Course Value |
|--------|---------|-------|-------|--------------|
| English | 3 | B | 3 | |
| Chemistry | 4 | C | 2 | |
| French | 3 | A | 4 | |
| Geometry | 3 | B | 3 | |
| History | 2 | F | 0 | |

**40.** Which is the better buy, a 32-ounce bottle of ketchup for $1.60 or a 48-ounce bottle for $2.88?

**FOR REVIEW**

*Find the products*

**41.** $\begin{array}{r} 6.08 \\ \times\ 0.4 \\ \hline \end{array}$

**42.** $\begin{array}{r} 4.41 \\ \times\ 2.3 \\ \hline \end{array}$

**43.** $\begin{array}{r} 2.022 \\ \times\ 3.01 \\ \hline \end{array}$

**44.** $21.683 \times 10$

**45.** $21.683 \times 0.01$

**46.** $21.683 \times 1000$

**47.** Convert $3.86 to cents.

**48.** Convert 6439¢ to dollars.

**49.** Susan bought 7 TV dinners at a cost of $2.79 each. If tax on the purchases amounted to $0.98, and she paid with three $10 bills, how much change did she receive?

**50.** It costs $158.00 a week plus 18.5¢ ($0.185) per mile to rent a motor home at Wannabegger Rentals. How much would it cost to rent a motor home for three weeks if it is driven a total of 1800 miles?

**51.** Jess earns $12.35 per hour for a 40-hour work week, and time and a half for each hour of overtime. How much should Jess be paid this week if he worked a total of 54 hours?

**52.** Suppose all we knew was that Jess earned the total amount in the answer to Exercise 51 by working 54 hours. What would we say was his hourly rate?

*The following exercises will help you prepare for the next section. Place the correct symbol ($<$ or $>$) between each pair of numbers.*

**53.** $\dfrac{1}{2}$    $0$

**54.** $\dfrac{9}{8}$    $\dfrac{5}{4}$

**55.** $3$    $\dfrac{44}{15}$

*Perform the indicated operations.*

**56.** $3 + 9 \times 2 - 5$

**57.** $\left(\dfrac{3}{8} - \dfrac{1}{4}\right) \cdot 4 - \dfrac{1}{2}$

**58.** $5^2 - \sqrt{16} + 2$

## 5.6 EXERCISES C

*Solve.*

**1.** On his two-week summer vacation, Sam Passamonte drove 715.6 miles the first week and 527.3 miles the second week. He purchased gas three times during the vacation, the last time when he arrived home. If he left home with a full tank and bought 17.2 gallons, 21.3 gallons, and 19.7 gallons, to the nearest tenth, how many miles did he get to the gallon?
[Answer: 21.4 mpg]

**2.** A water tank is filled with water. The tank, when empty, weighs 187.6 pounds and, when full, weighs 839.4 pounds. If one cubic foot of water weighs 62.5 pounds, how many cubic feet of water does the tank hold? Give answer rounded to the nearest tenth of a cubic foot.

## 5.7 ORDER OF OPERATIONS AND COMPARING DECIMALS

### STUDENT GUIDEPOSTS

**1** Operations Involving Both Fractions and Decimals

**2** Order of Operations for Decimals

**3** Arranging Decimals in Order of Size

**4** Comparing Unit Prices

# ① OPERATIONS INVOLVING BOTH FRACTIONS AND DECIMALS

In Chapters 3 and 4 we considered operations on fractions, and in this chapter we considered the same operations on decimals. At times, fractions and decimals may occur in the same problem.

| EXAMPLE 1 MULTIPLYING A FRACTION AND A DECIMAL | PRACTICE EXERCISE 1 |

Find the product.    $\frac{3}{4} \times 4.28$

We could change $\frac{3}{4}$ to decimal form, 0.75, and then multiply.

$$
\begin{array}{r}
4.28 \\
\times\ 0.75 \\
\hline
2140 \\
2996 \\
\hline
3.2100
\end{array}
$$

Or, we could change 4.28 to fractional form,

$$4.28 = 4\frac{28}{100} = \frac{100 \cdot 4 + 28}{100} = \frac{428}{100},$$

and then multiply.

$$\frac{3}{4} \times \frac{428}{100} = \frac{3}{4} \times \frac{4 \cdot 107}{100}$$

$$= \frac{3 \cdot \cancel{4} \cdot 107}{\cancel{4} \cdot 100} = \frac{321}{100} = 3\frac{21}{100}$$

Find the quotient.

$$4.9 \div \frac{7}{8}$$

Answer: $5\frac{3}{5}$ or 5.6

In general, problems which involve both decimals and fractions may be solved either by changing all numbers to decimals or by changing all numbers to fractions. When fractions are changed to decimals we must often make approximations.

| EXAMPLE 2 ADDING A FRACTION AND A DECIMAL | PRACTICE EXERCISE 2 |

Find the sum.    $\frac{1}{3} + 5.75$

Since $\frac{1}{3} = 0.\overline{3} = 0.333 \ldots$ , we can only *approximate* the sum using decimals. As a result, it might be better to change 5.75 to fractional form.

$$5.75 = 5\frac{75}{100} = 5\frac{3}{4} = \frac{4 \cdot 5 + 3}{4} = \frac{23}{4}$$

Find the difference.

$$2.25 - \frac{2}{3}$$

Thus,

$$\frac{1}{3} + 5.75 = \frac{1}{3} + \frac{23}{4} = \frac{1}{3} + \frac{23}{2 \cdot 2}$$   LCD = 2 · 2 · 3 = 12

$$= \frac{2 \cdot 2 \cdot 1}{2 \cdot 2 \cdot 3} + \frac{23 \cdot 3}{2 \cdot 2 \cdot 3}$$   Supply missing numbers

$$= \frac{2 \cdot 2 + 23 \cdot 3}{2 \cdot 2 \cdot 3}$$   Add numerators over LCD

$$= \frac{4 + 69}{12}$$

$$= \frac{73}{12}$$

$$= 6\frac{1}{12}.$$

If we had used the decimal 0.33 to approximate $\frac{1}{3}$, we would have found the sum as follows.

$$\frac{1}{3} + 5.75 \approx 0.33 + 5.75$$

$$= 6.08$$

Check:   Does $6\frac{1}{12} = 6.08$?

Since $\frac{1}{12} = 0.08\overline{3}$ (dividing 1 by 12),

$$6\frac{1}{12} = 6.08\overline{3}.$$

Thus, using 0.33 to approximate $\frac{1}{3}$ makes $\frac{1}{3} + 5.75 \approx 6.08$ correct to the nearest hundredth. In some problems, this approximation may be accurate enough, and we can avoid adding fractions.

Answer: $1\frac{7}{12}$ [It is best to change 2.25 to a fraction.]

---

**EXAMPLE 3   APPLICATION TO SEWING**

A certain fabric costs \$4.29 per yard. How much will $3\frac{1}{8}$ yards cost?
   Multiply $3\frac{1}{8} \times 4.29$. Since $\frac{1}{8} = 0.125$ has a terminating decimal form, find the product using decimals.

$$\begin{array}{r} 3.125 \\ \times\ 4.29 \\ \hline 28125 \\ 6250\phantom{0} \\ 12500\phantom{00} \\ \hline 13.40625 \end{array}$$   $3\frac{1}{8} = 3.125$

Rounding to the nearest cent, the fabric will cost \$13.41.

**PRACTICE EXERCISE 3**

Brad works $8\frac{5}{6}$ hours repairing a television set. If he receives \$13.75 per hour, how much will he be paid?

Answer: \$121.46

## ② ORDER OF OPERATIONS FOR DECIMALS

The order of operations was first considered in Section 2.8 relative to whole numbers and again in Section 4.5 when fractions were involved. The same order applies to decimals and to decimals mixed with fractions. For easy reference and review, the basic rules are repeated here.

### Order of Operations

Operations should be performed in the following order.

1. Operate within parentheses.
2. Evaluate all powers and roots in any order.
3. Multiply and divide, in order, from left to right.
4. Add and subtract, in order, from left to right.

---

**EXAMPLE 4**   **USING THE ORDER OF OPERATIONS ON DECIMALS**

Perform the indicated operations.

(a) $2.5 + 3.7 - 1.9 = 6.2 - 1.9$     Add first

$\qquad\qquad\qquad = 4.3$          Then subtract

(b) $(1.8)(3.4) - 2.24 = 6.12 - 2.24$     Multiply first

$\qquad\qquad\qquad\quad = 3.88$          Then subtract

(c) $6.75 \div (1.5 + 3.5) = 6.75 \div 5$     Add inside parentheses first

$\qquad\qquad\qquad\quad = 1.35$          Then divide

(d) $(1.2)^2 + 2.8 \div 1.4 = 1.44 + 2.8 \div 1.4$     Square first

$\qquad\qquad\qquad\qquad = 1.44 + 2$          Divide next

$\qquad\qquad\qquad\qquad = 3.44$          Then add

---

**PRACTICE EXERCISE 4**

Perform the indicated operations.

(a) $4.32 \div 1.6 + 6.7$

(b) $15.45 - (2.3)(4.1)$

(c) $4.3 \cdot (6.7 - 5.1)$

(d) $8.76 - (2.4)^2 \div 0.8$

Answers: (a) **9.4**   (b) **6.02**
(c) **6.88**   (d) **1.56**

---

**EXAMPLE 5**   **ORDER OF OPERATIONS ON FRACTIONS AND DECIMALS**

Perform the indicated operations.

(a) $\dfrac{1}{2} + (3.5)(2.4) = \dfrac{1}{2} + 8.4$     Multiply first

$\qquad\qquad\qquad = 0.5 + 8.4$     Change $\frac{1}{2}$ to a decimal

$\qquad\qquad\qquad = 8.9$          Then add

(b) $\left(\dfrac{1}{4}\right)^2 + (1.5)^2 - 2.3 = \dfrac{1}{16} + 2.25 - 2.3$     Square first

$\qquad\qquad\qquad\qquad = 0.0625 + 2.25 - 2.3$     Change $\frac{1}{16}$ to a decimal

$\qquad\qquad\qquad\qquad = 2.3125 - 2.3$     Add first

$\qquad\qquad\qquad\qquad = 0.0125$          Then subtract

(c) $3.2 \div \dfrac{1}{4} - 6\dfrac{4}{5} = 3.2 \times 4 - 6.8$     Multiply by the reciprocal of $\frac{1}{4}$, 4

$\qquad\qquad\qquad\quad = 12.8 - 6.8$     Multiply first

$\qquad\qquad\qquad\quad = 6$          Then subtract

---

**PRACTICE EXERCISE 5**

Perform the indicated operations.

(a) $\left(\dfrac{2}{3}\right)\left(\dfrac{3}{5}\right) \div 0.8$

(b) $4.8 \div \left(\dfrac{1}{4}\right)^2 + 6.9$

(c) $\left(\dfrac{7}{8}\right)(0.12) - 0.005$

Answers: (a) **0.5**   (b) **83.7**
(c) **0.1**

## ③ ARRANGING DECIMALS IN ORDER OF SIZE

The next topic involves comparing decimals. For example, we might want to arrange

$$0.3, \quad 0.31, \quad 0.301$$

in order of size from left to right, the smallest first. To make such comparisons, first write each number with the same number of decimal places attaching zeros as needed.

$$0.300, \quad 0.310, \quad 0.301$$

Since the largest number of decimal places in the given numbers is three (in 0.301), we have written each number with three decimal places. We ignore the decimal points and compare sizes as if comparing whole numbers. Left to right, smallest first, the arrangement is

$$300, \quad 301, \quad 310.$$

Thus, the original decimals arranged by size are

$$0.3, \quad 0.301, \quad 0.31.$$

---

| EXAMPLE 6 ARRANGING DECIMALS IN ORDER OF SIZE | PRACTICE EXERCISE 6 |
|---|---|

Arrange the following decimals in order of size, with the smallest on the left.

$$0.05, \quad 3.5, \quad 3.05, \quad 3.4, \quad 0.5$$

First write all numbers with two decimal places.

$$0.05, \quad 3.50, \quad 3.05, \quad 3.40, \quad 0.50$$

Ignoring decimals, arrange

$$5, \quad 350, \quad 305, \quad 340, \quad 50.$$

The two-place decimals, arranged by size, are

$$0.05, \quad 0.50, \quad 3.05, \quad 3.40, \quad 3.50,$$

so the original decimals are arranged by size as follows.

$$0.05, \quad 0.5, \quad 3.05, \quad 3.4, \quad 3.5$$

**PRACTICE EXERCISE 6**

Arrange the following decimals in order of size with the smallest on the left.

$$1.06, \quad 1.6, \quad 1.16, \quad 1.106, \quad 0.16$$

Answer: **0.16, 1.06, 1.106, 1.16, 1.6**

---

Decimals can be compared with fractions by changing the fractions to decimals.

---

| EXAMPLE 7 ARRANGING FRACTIONS AND DECIMALS | PRACTICE EXERCISE 7 |
|---|---|

Arrange the following numbers in order of size with the smallest on the left.

$$0.26, \quad 2.6, \quad \frac{1}{4}, \quad \frac{2}{5}, \quad 0.39$$

**PRACTICE EXERCISE 7**

Arrange the following numbers in order of size with the smallest on the left.

$$0.76, \quad 7\frac{1}{2}, \quad \frac{3}{4}, \quad 0.7, \quad \frac{4}{5}$$

First change the fractions to decimals. Since $\frac{1}{4} = 0.25$ and $\frac{2}{5} = 0.4$, the list becomes:

$$0.26, \quad 2.6, \quad 0.25, \quad 0.4, \quad 0.39$$

Next write all the numbers with two decimal places.

$$0.26, \quad 2.60, \quad 0.25, \quad 0.40, \quad 0.39$$

Ignoring the decimals, we must arrange

$$26, \quad 260, \quad 25, \quad 40, \quad 39.$$

The two-place decimals arranged by size are

$$0.25, \quad 0.26, \quad 0.39, \quad 0.40, \quad 2.60,$$

so the original numbers are arranged by size as follows:

$$\frac{1}{4}, \quad 0.26, \quad 0.39, \quad \frac{2}{5}, \quad 2.6.$$

Answer: $0.7, \frac{3}{4}, 0.76, \frac{4}{5}, 7\frac{1}{2}$

## ④ COMPARING UNIT PRICES

Comparing sizes of decimals is important in unit-pricing problems. The **unit price** of an item is

$$\text{unit price} = \frac{\text{total price}}{\text{number of units}}.$$

For example, if a 16-ounce (abbreviated **oz**) can of juice sells for 89¢,

$$\text{unit price} = \frac{\text{total price}}{\text{number of units}} = \frac{89¢}{16 \text{ oz}} \approx 5.56¢ \text{ per oz.}$$

Also, if a box of 300 facial tissues sells for 58¢, the unit price is

$$\frac{\text{total price}}{\text{number of units}} = \frac{58¢}{300} \approx 0.19¢ \text{ per tissue.}$$

The wise shopper compares unit prices of various brands of an item.

| EXAMPLE 8 COMPARING UNIT PRICES |
| --- |

The following table contains information about four brands of detergent. Comparing unit prices, we see that Brand Z is the best buy (assuming that the quality of all products is the same).

| Brand | Ounces | Total Price | Unit Price |
| --- | --- | --- | --- |
| W | 18 | $0.87 | $0.87 \div 18 \approx \$0.048$ |
| X | 20 | $0.93 | $0.93 \div 20 \approx \$0.047$ |
| Y | 16 | $0.78 | $0.78 \div 16 \approx \$0.049$ |
| Z | 25 | $1.16 | $1.16 \div 25 \approx \$0.046$ |

| PRACTICE EXERCISE 8 |
| --- |

A 360-sheet roll of paper towels sells for 69¢ while a 440-sheet roll sells for 79¢. Assuming equal quality, which is the better buy?

Answer: The 440-sheet roll at approximately 0.18¢ per sheet is a better buy than the 360-sheet roll at 0.19¢ per sheet.

## 5.7 EXERCISES A

*Perform the indicated operations by changing all decimals to fractions.*

**1.** $\dfrac{1}{4} + 2.6$

**2** $3.25 - \dfrac{3}{8}$

**3.** $5\dfrac{1}{7} + 7.4$

**4.** $\dfrac{2}{3} \times 1.75$

**5.** $4.1 \div \dfrac{5}{6}$

**6.** $2\dfrac{2}{3} \times 3.25$

*Perform the indicated operations by first changing all fractions to decimals, correct to two decimal places. Give answers to the nearest hundredth.*

**7.** $\dfrac{1}{4} + 2.6$

**8** $3.25 - \dfrac{3}{8}$

**9.** $5\dfrac{1}{7} + 7.4$

**10.** $\dfrac{2}{3} \times 1.75$

**11.** $4.1 \div \dfrac{5}{6}$

**12.** $2\dfrac{2}{3} \times 3.25$

*Solve.*

**13.** Benny walked 3.2 hours at a rate of $2\dfrac{1}{2}$ miles per hour. How far did he walk?

**14.** Kerry earns $5.85 per hour working as a dance instructor. She substituted for Janine one week and received $2\dfrac{1}{2}$ times her usual hourly rate. What was her pay per hour that week?

**15.** A tailor has 44 yards of material. If $3\dfrac{2}{3}$ yards are required to make a coat, how many coats can be made?

**16.** Two hikers gave the number of miles that they hiked on Saturday as 7.25 miles and $6\dfrac{7}{8}$ miles. How much farther did the one hike than the other?

*Perform the indicated operations. Answers may vary somewhat due to rounding differences.*

**17.** $3.7 - 2.1 + 8.8$

**18.** $(4.3)(6.5) + 7.29$

**19.** $9.45 \div (1.4)(2.5)$

**20** $(6.2)^2 - 6.6 \div 1.1$

**21.** $\dfrac{1}{4} + (3.5)(6.4)$

**22.** $(2.5)^2 + \left(\dfrac{1}{3}\right)^2 - 1.9$

**23.** $7.2 \div \dfrac{1}{2} - 2\dfrac{1}{3}$

**24.** $(3.7)^2 \div \left(\dfrac{1}{4}\right)^2 + \dfrac{1}{8}$

*Arrange the given numbers in order of size with the smallest on the left.*

**25.** 0.085, 0.85, 8.5, 8.05

**26.** 0.047, 0.048, 0.08, 0.04

**27.** $0.51, \frac{1}{2}, \frac{7}{16}, 0.15, 5.1$

*5 .4315*

**28** $\frac{2}{3}, 0.65, 6.05, \frac{7}{10}, 6.5$

*.67      .7*

*Solve.*

**29.** A $5\frac{1}{2}$-pound ham sells for $17.38. What is the price per pound (the unit price)?

**30.** If $16\frac{1}{4}$ ounces of drink sell for 87¢, what is the price per ounce (the unit price) to the nearest tenth of a cent?

*In Exercises 31–32, complete the unit price column in each table and decide which item is the best buy.*

**31.**

| Brand | Ounces | Total Price | Unit Price |
|-------|--------|-------------|------------|
| A | 12 | $1.27 | |
| B | 8 | $0.83 | |
| C | 16 | $1.61 | |

**32.**

| Brand | Number of Tissues | Total Price | Unit Price |
|-------|-------------------|-------------|------------|
| Western | 200 | 47¢ | |
| Eastern | 250 | 56¢ | |
| Southern | 400 | 93¢ | |

## FOR REVIEW

**33.** The table below shows Wilma's grades during one quarter. Complete the table and calculate her GPA.

| Course | Credits | Grade | Value | Course Value |
|--------|---------|-------|-------|--------------|
| Geology | 4 | A | 4 | |
| Calculus | 4 | B | 3 | |
| Western Art | 3 | D | 1 | |
| Sociology | 3 | A | 4 | |
| Psychology | 3 | C | 2 | |

**34.** At the start of a trip, Gordy noticed that his odometer reading was 32,043.6. At the end of the trip it was 35,581.0. If he used 165.3 gallons of gas, how many miles per gallon did he get, to the nearest tenth?

**35.** It costs $17.85 a day plus 15¢ per mile for every mile driven over 100 to rent a car. How much will it cost to rent the car for three days if it is driven a total of 437 miles?

---

**ANSWERS:** 1. $2\frac{17}{20}$  2. $2\frac{7}{8}$  3. $12\frac{19}{35}$  4. $1\frac{1}{6}$  5. $4\frac{23}{25}$  6. $8\frac{2}{3}$  7. 2.85  8. 2.87  9. 12.54  10. 1.17  11. 4.94  12. 8.68  13. 8 miles  14. $14.63  15. 12  16. 0.375 or $\frac{3}{8}$ mile  17. 10.4  18. 35.24  19. 2.7  20. 32.44  21. 22.65  22. $4.46\overline{1}$  23. $12.0\overline{6}$  24. 219.165  25. 0.085, 0.85, 8.05, 8.5  26. 0.04, 0.047, 0.048, 0.08  27. 0.15, $\frac{7}{16}$, $\frac{1}{2}$, 0.51, 5.1  28. 0.65, $\frac{2}{3}$, $\frac{7}{10}$, 6.05, 6.5  29. $3.16 per pound  30. 5.4¢ per ounce  31. Brand A is $0.106 per ounce; Brand B is $0.104 per ounce; Brand C is $0.101 per ounce. Thus, Brand C is the best buy.  32. Western is 0.235¢ per tissue; Eastern is 0.224¢ per tissue; Southern is 0.233¢ per tissue. Thus, Eastern is the best buy.  33. GPA = 2.88  34. 21.4 mpg  35. $104.10

## 5.7 EXERCISES B

*Perform the indicated operations by changing all decimals to fractions.*

**1.** $\frac{3}{4} + 4.2$

**2.** $2.25 - \frac{7}{8}$

**3.** $6.2 + 3\frac{4}{5}$

**4.** $\dfrac{1}{3} \times 6.25$

**5.** $8.4 \div \dfrac{1}{6}$

**6.** $2.75 \times 4\dfrac{1}{3}$

*Perform the indicated operations by first changing all fractions to decimals, correct to two decimal places. Give answers to the nearest hundredth.*

**7.** $\dfrac{3}{4} + 4.2$

**8.** $2.25 - \dfrac{7}{8}$

**9.** $6.2 + 3\dfrac{4}{5}$

**10.** $\dfrac{1}{3} \times 6.25$

**11.** $8.4 \div \dfrac{1}{6}$

**12.** $2.75 \times 4\dfrac{1}{3}$

*Solve.*

**13.** A certain fabric costs \$3.79 per yard. How much will $5\dfrac{3}{8}$ yards cost?

**14.** Kathy Shipp has \$295.50 in a passbook savings account and $3\dfrac{1}{4}$ times that much in a money market fund. How much does she have in the money market fund?

**15.** Claude paid \$27,423.52 for $51\dfrac{3}{4}$ acres of land. Rounded to the nearest cent, what was the cost per acre?

**16.** One tank holds 21.6 gallons of fuel and another holds $12\dfrac{1}{8}$ gallons. What is the combined fuel capacity of the two tanks?

*Perform the indicated operations. Answers may vary somewhat due to rounding differences.*

**17.** $5.9 + 6.2 - 8.7$

**18.** $12.8 + (3.9)(4.2)$

**19.** $3.85 \div (1.4)(2.5)$

**20.** $(4.9)^2 - 8.4 \div 1.2$

**21.** $\dfrac{1}{8} + (2.1)^2$

**22.** $(4.1)^2 + \left(\dfrac{1}{4}\right)^2 - 2.7$

**23.** $6.8 \div \dfrac{1}{4} + 3\dfrac{1}{5}1$

**24.** $(1.6)^2 \div \left(\dfrac{1}{2}\right)^2 - \dfrac{3}{8}$

*Arrange the given numbers in order of size with the smallest on the left.*

**25.** 0.074, 7.04, 0.74, 70.4

**26.** 0.95, 0.095, 0.59, 5.9

**27.** $\dfrac{1}{4}$, 0.23, $\dfrac{1}{5}$, 2.25, 0.02

**28.** 0.32, 3.2, $\dfrac{3}{10}$, $\dfrac{1}{3}$, 3.02

*Solve.*

**29.** A $12\dfrac{1}{2}$-ounce bottle of shampoo costs \$1.00. What is the price per ounce (unit price)?

**30.** A roll of aluminum foil costing 39¢ contains 25.5 square feet of foil. To the nearest tenth of a cent, what is the price per square foot?

*In Exercises 31–32 complete the unit price column in each table and decide which item is the best buy.*

**31.**

| Brand | Grams | Total Price | Unit Price |
|-------|-------|-------------|------------|
| X | 40 | 84¢ | |
| Y | 35 | 77¢ | |
| Z | 30 | 57¢ | |

**32.**

| Brand of Corn | Ounces | Total Price | Unit Price |
|---------------|--------|-------------|------------|
| High Quality | 19 | \$0.85 | |
| Top Crop | $14\dfrac{1}{2}$ | \$0.76 | |
| Generic | 12.9 | \$0.69 | |

## FOR REVIEW

**33.** The table below shows Norman's grades during the spring quarter. Complete the table and calculate his GPA.

| Course | Credits | Grade | Value | Course Value |
|--------|---------|-------|-------|--------------|
| GLG 101 | 4 | B | 3 | |
| MAT 105 | 3 | A | 4 | |
| SOC 240 | 3 | B | 3 | |
| PSY 150 | 3 | C | 2 | |
| PE 100 | 1 | A | 4 | |

**34.** Karen bought 2 sweaters at $12.95 each and a pair of shoes for $31.90. The sales tax on her purchases amounted to $2.76. She put the items on layaway by paying $\frac{1}{3}$ down. How much must she pay when she picks up the items?

**35.** It costs $18.50 a day to rent a paint sprayer. If paint costs $12.95 a gallon and four gallons will be required, how much will it cost Juan to paint his fence if he uses the sprayer for $1\frac{1}{2}$ days?

## 5.7 EXERCISES C

**1.** On a recent trip George used $28\frac{1}{2}$ gallons of gas which was purchased at 122.9¢ per gallon. How much did George spend for gas on the trip?
[Answer: $35.03]

**2.** Complete the unit price column and decide which brand of soap is the best buy.

| Brand | Ounces | Total Price | Unit Price |
|-------|--------|-------------|------------|
| Klean | 5 | 36¢ | |
| Skrub | 6 | 39¢ | |
| Shine | 4.5 | 35¢ | |
| Zust | $3\frac{7}{8}$ | 29¢ | |

## 5.8 THE CALCULATOR

### STUDENT GUIDEPOSTS

**1** Basic Operations on a Calculator      **2** Special Functions on a Calculator

### **1** BASIC OPERATIONS ON A CALCULATOR

You may have wondered why calculators have not been discussed until now. The reason is simple. The basic skills of arithmetic must be well understood. While the calculator can help perform complex arithmetic calculations, it does *not* tell what calculations to perform.

Although simple computations can easily be done by hand, lengthy computations, especially those involving decimals, are greatly simplified by using a calculator.

Since there are many calculators on the market, and since they all have different features, it is difficult to give a thorough discussion of how to use any particular one of them. The best advice is to study the instruction manual which comes with your model.

Although differences exist, most calculators consist of three major parts:

**1.** A **display** on which numbers are shown.

**2.** A **keyboard** with buttons called **keys,** which include **number keys** (the ten digits and a decimal point), and **operation keys** (+, −, ×, ÷, and =).

**3.** A **register** in which numbers are stored (not displayed) and used in computations.

Number keys place numbers in the display, and operation keys transfer numbers to the register or calculate results using numbers already in the register. Also, there is a **clear key** which "clears" the calculator of number or operation entries. The clear key removes all numbers used in previous calculations and places the number 0 in the display so that a new calculation can begin.

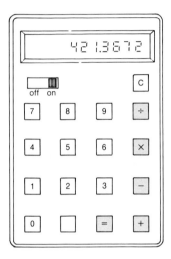

**Figure 5.9**

A typical calculator is shown in Figure 5.9. The operation keys appear as ☐, the decimal point key is ⎡·⎦, the number keys appear as ☐, and ⎡C⎦ is the clear key. The number

$$421.3672$$

is shown in the display and is entered into the calculator by pressing the following keys in order from left to right.

⎡4⎦ ⎡2⎦ ⎡1⎦ ⎡·⎦ ⎡3⎦ ⎡6⎦ ⎡7⎦ ⎡2⎦

Making calculations is merely a matter of pressing the right keys in the right order. For example, to find the sum

$$3 + 8$$

the keys to press, and the order in which they are pressed, are shown below.

| ⎡3⎦ | ⎡+⎦ | ⎡8⎦ | ⎡=⎦ |
|---|---|---|---|
| *First* | *Second* | *Third* | *Fourth* |
| This places the number 3 in the display. | This transfers 3 to the register where it is ready to be added to the next number. | This places 8 in the display. | This adds the number in the display (8) to the number in the register (3) and places the result (11) in the display. |

To find the difference

$$15 - 6,$$

press the following keys in order.

$$\boxed{1}\ \boxed{5}\ \boxed{-}\ \boxed{6}\ \boxed{=}$$

The display will show the difference, which is 9.

To find the product

$$12.3 \times 1.05,$$

press the following keys in the order shown.

$$\boxed{1}\ \boxed{2}\ \boxed{\cdot}\ \boxed{3}\ \boxed{\times}\ \boxed{1}\ \boxed{\cdot}\ \boxed{0}\ \boxed{5}\ \boxed{=}$$

The display will show the product, which is 12.915.

Lengthy division problems are greatly simplified by using a calculator. For example, to find

$$4.23738 \div 4.13,$$

press, in order,

$$\boxed{4}\ \boxed{\cdot}\ \boxed{2}\ \boxed{3}\ \boxed{7}\ \boxed{3}\ \boxed{8}\ \boxed{\div}\ \boxed{4}\ \boxed{\cdot}\ \boxed{1}\ \boxed{3}\ \boxed{=}.$$

The quotient, 1.026, is shown in the display.

A calculator can find sums of more than two addends. For example, to find

$$1.3 + 2.06 + 4.32 + 11.5$$

press, in order,

$$\boxed{1}\ \boxed{\cdot}\ \boxed{3}\ \boxed{+}\ \boxed{2}\ \boxed{\cdot}\ \boxed{0}\ \boxed{6}\ \boxed{+}\ \boxed{4}\ \boxed{\cdot}\ \boxed{3}\ \boxed{2}\ \boxed{+}\ \boxed{1}\ \boxed{1}\ \boxed{\cdot}\ \boxed{5}\ \boxed{=}.$$

The sum, 19.18, will appear in the display.

Most calculators are limited by the number of digits that can be displayed, usually eight.

---

### EXAMPLE 1  BASIC OPERATIONS ON A CALCULATOR

Perform the following calculations using your calculator.

**(a)** $5.3 + 7.6$

The sequence of keys to press is shown below.

$$\boxed{5}\ \boxed{\cdot}\ \boxed{3}\ \boxed{+}\ \boxed{7}\ \boxed{\cdot}\ \boxed{6}\ \boxed{=}$$

The sum, 12.9, will appear in the display.

**(b)** $8.5 - 1.9$

$$\boxed{8}\ \boxed{\cdot}\ \boxed{5}\ \boxed{-}\ \boxed{1}\ \boxed{\cdot}\ \boxed{9}\ \boxed{=}$$

The difference, 6.6, will appear in the display when the above keys are pressed.

**(c)** $12.9 \times 18.1$

$$\boxed{1}\ \boxed{2}\ \boxed{\cdot}\ \boxed{9}\ \boxed{\times}\ \boxed{1}\ \boxed{8}\ \boxed{\cdot}\ \boxed{1}\ \boxed{=}$$

The product is 233.49.

**(d)** $3.948 \div 3.76$

$$\boxed{3}\ \boxed{\cdot}\ \boxed{9}\ \boxed{4}\ \boxed{8}\ \boxed{\div}\ \boxed{3}\ \boxed{\cdot}\ \boxed{7}\ \boxed{6}\ \boxed{=}$$

The quotient is 1.05.

### PRACTICE EXERCISE 1

Perform the following calculations using your calculator.

**(a)** $2.1 + 3.5$

**(b)** $6.7 - 5.9$

**(c)** $14.5 \times 8.6$

**(d)** $69.1932 \div 5.29$

**(e)** $10.7 - 2.4 + 8.5$

$$\boxed{1}\,\boxed{0}\,\boxed{.}\,\boxed{7}\,\boxed{-}\,\boxed{2}\,\boxed{.}\,\boxed{4}\,\boxed{+}\,\boxed{8}\,\boxed{.}\,\boxed{5}\,\boxed{=}$$

The answer is 16.8.

**(f)** $\dfrac{3}{4} + 2.6$

$$\boxed{3}\,\boxed{\div}\,\boxed{4}\,\boxed{+}\,\boxed{2}\,\boxed{.}\,\boxed{6}\,\boxed{=}$$

When the first four keys have been pressed, the display will show 0.75, which is $\frac{3}{4}$ as a decimal. Continuing gives the sum 3.35.

**(e)** $2.5 + 3.6 - 1.8$

**(f)** $\dfrac{1}{4} + 3.7$

Answers: (a) **5.6**  (b) **0.8**
(c) **124.7**  (d) **13.08**  (e) **4.3**
(f) **3.95**

## ② SPECIAL FUNCTIONS ON A CALCULATOR

Some calculators are more sophisticated and have other special keys. We will discuss four of these keys now. The first is $\boxed{x^2}$, the key used for squaring a number. Simply enter the number to be squared and press $\boxed{x^2}$.

---

| EXAMPLE 2    SQUARING A NUMBER ON A CALCULATOR |
| --- |

Square each number.

**(a)** 12

The steps used to find $(12)^2$ are:

$$\boxed{1}\,\boxed{2}\,\boxed{x^2}$$

The display will show 144. Of course you could find $(12)^2$ as follows,

$$\boxed{1}\,\boxed{2}\,\boxed{\times}\,\boxed{1}\,\boxed{2}\,\boxed{=}$$

but the squaring key cuts down the number of entries considerably.

**(b)** 31.75

Use these steps.

$$\boxed{3}\,\boxed{1}\,\boxed{.}\,\boxed{7}\,\boxed{5}\,\boxed{x^2}$$

The answer in the display is 1008.0625.

| PRACTICE EXERCISE 2 |
| --- |

Square each number.

**(a)** 21

**(b)** 18.09

Answers: (a) **441**  (b) **327.2481**

---

| **HINT** |
| --- |

Since a calculator display is (usually), limited to eight digits, when a number with many decimal places is squared, the calculator will give only a rounded approximation of the result. For example, if you square 10.62358, the result should have 10 decimal places, but your calculator will round this off and give 112.86045, correct to five decimal places.

---

The next special key we consider is the square root key, $\boxed{\sqrt{\phantom{x}}}$, used for finding the square root of a number. For many numbers, the calculator will give only an approximation of the square root.

| **EXAMPLE 3** FINDING SQUARE ROOTS ON A CALCULATOR | **PRACTICE EXERCISE 3** |
|---|---|

Find the square root of each number.

**(a)** 16

We know already that $\sqrt{16} = 4$, but use your calculator to verify this by following these steps.

$$\boxed{1}\ \boxed{6}\ \boxed{\sqrt{\phantom{x}}}$$

The display shows 4.

**(b)** 13.25

Use these steps.

$$\boxed{1}\ \boxed{3}\ \boxed{.}\ \boxed{2}\ \boxed{5}\ \boxed{\sqrt{\phantom{x}}}$$

The display will show 3.6400549, which is $\sqrt{13.25}$ correct to seven decimal places. In a practical situation, we might use 3.6 or 3.64 to approximate $\sqrt{13.25}$.

Find the square root of each number.

**(a)** 81

**(b)** 28.76

Answers: (a) 9   (b) 5.3628351

---

///////////// **CAUTION** ///////////////

In Example 3(b) we found $\sqrt{13.25}$ to be 3.6400549. This means that $(3.6400549)^2$ should be 13.25. Suppose we use 3.64 as an approximation of $\sqrt{13.25}$. If we square 3.64 by pressing

$$\boxed{3}\ \boxed{.}\ \boxed{6}\ \boxed{4}\ \boxed{x^2},$$

the display shows 13.2496, which is a close approximation of 13.25. Remember that when using a calculator, approximate values are often obtained so don't be concerned if your answers differ slightly from those obtained by someone else.

///////////

The third special key $\boxed{y^x}$ is the exponential key and is used to find powers of numbers other than squares (for which you will use $\boxed{x^2}$). When using this key, the base $y$ is always entered first followed by the exponent $x$. The following example will make this clear.

| **EXAMPLE 4** FINDING POWERS ON A CALCULATOR | **PRACTICE EXERCISE 4** |
|---|---|

Find each power.

**(a)** $2^4$

We know that $2^4 = 2 \cdot 2 \cdot 2 \cdot 2 = 16$ but use your calculator to verify this by following these steps.

$$\boxed{2}\ \underset{\substack{\text{This puts 2}\\ \text{in for } y}}{\boxed{y^x}}\ \underset{\substack{\text{This puts 4}\\ \text{in for } x}}{\boxed{4}}\ \boxed{=}$$

The display shows 16, which is $2^4$.

Find each power.

**(a)** $5^3$

**(b)** $(1.09)^{10}$

Follow these steps.

$$\boxed{1}\ \boxed{\cdot}\ \boxed{0}\ \boxed{9}\ \boxed{y^x}\ \boxed{1}\ \boxed{0}\ \boxed{=}$$

The display shows 2.3673637, which is $(1.09)^{10}$.

**(b)** $(4.36)^7$

Finally, some calculators have parentheses keys $\boxed{(}$ and $\boxed{)}$. These keys are used to group operations and maintain the order of operations we have learned. For example, consider

$$10 \div 2 + 3 \quad \text{and} \quad 10 \div (2 + 3).$$

We know that we divide before adding so

$$10 \div 2 + 3 = 5 + 3 = 8.$$

However, with the parentheses in the second problem, we must add first and then divide 10 by the result.

$$10 \div (2 + 3) = 10 \div 5 = 2$$

Since your calculator uses the same order of operations, if you follow these steps

$$\boxed{1}\ \boxed{0}\ \boxed{\div}\ \boxed{2}\ \boxed{+}\ \boxed{3}\ \boxed{=},$$

the display gives 8. To find $10 \div (2 + 3)$, use the parentheses keys

$$\boxed{1}\ \boxed{0}\ \boxed{\div}\ \boxed{(}\ \boxed{2}\ \boxed{+}\ \boxed{3}\ \boxed{)}\ \boxed{=}$$

and the display shows 2, which we know is the correct result based on our order of operations.

---

**EXAMPLE 5    USING THE PARENTHESES KEYS**

Calculate each number.

**(a)** $6 \times (8 + 3)$

Use these steps.

$$\boxed{6}\ \boxed{\times}\ \boxed{(}\ \boxed{8}\ \boxed{+}\ \boxed{3}\ \boxed{)}\ \boxed{=}$$

The display shows 66. Verify this without using a calculator.

**(b)** $12.8 \div (6.1 - 2.3) + 5.5$

Follow these steps.

$$\boxed{1}\ \boxed{2}\ \boxed{\cdot}\ \boxed{8}\ \boxed{\div}\ \boxed{(}\ \boxed{6}\ \boxed{\cdot}\ \boxed{1}\ \boxed{-}\ \boxed{2}\ \boxed{\cdot}\ \boxed{3}\ \boxed{)}\ \boxed{+}\ \boxed{5}\ \boxed{\cdot}\ \boxed{5}\ \boxed{=}$$

The answer in the display is 8.8684211, which is approximately 8.9, rounded to the nearest tenth.

**PRACTICE EXERCISE 5**

Calculate each number.

**(a)** $20 \div (7 + 3)$

**(b)** $4.3 \times (7.1 + 2.8) - 3.6$

## HINT

This brief introduction to the calculator, together with your operator's manual, should help you discover the power and usefulness of calculating devices. However, do not become so dependent on your calculator that you reach for it to make simple calculations such as $2 + 5$, $11 - 4$, $12 \div 3$, $\sqrt{9}$, $4^2$, and so forth. Such calculations can be performed mentally in far less time than it would take to make the appropriate entries. However, more complicated calculations, especially those involving decimals such as $14.362 \div 1.089$, $(6.4508)^2$, $\sqrt{4.78}$, and so forth, are greatly simplified with a calculator. In other words, you should be "calculator wise" and learn when to use and when *not* to use your calculator. Approximating calculations can also help you discover possible errors made by pressing a wrong key.

## 5.8 EXERCISES A

*What calculation is made by pressing the keys as shown from left to right, and what is the result?*

1. $\boxed{1}\ \boxed{0}\ \boxed{+}\ \boxed{3}\ \boxed{5}\ \boxed{=}$

2. $\boxed{6}\ \boxed{2}\ \boxed{-}\ \boxed{3}\ \boxed{1}\ \boxed{=}$

3. $\boxed{3}\ \boxed{8}\ \boxed{\times}\ \boxed{2}\ \boxed{4}\ \boxed{=}$

4. $\boxed{5}\ \boxed{5}\ \boxed{\div}\ \boxed{1}\ \boxed{1}\ \boxed{=}$

5. $\boxed{2}\ \boxed{\cdot}\ \boxed{3}\ \boxed{+}\ \boxed{1}\ \boxed{\cdot}\ \boxed{4}\ \boxed{=}$

6. $\boxed{4}\ \boxed{\cdot}\ \boxed{7}\ \boxed{+}\ \boxed{\cdot}\ \boxed{9}\ \boxed{=}$

7. $\boxed{6}\ \boxed{\cdot}\ \boxed{1}\ \boxed{2}\ \boxed{\times}\ \boxed{5}\ \boxed{\cdot}\ \boxed{3}\ \boxed{=}$

8. $\boxed{1}\ \boxed{\cdot}\ \boxed{3}\ \boxed{\div}\ \boxed{4}\ \boxed{\cdot}\ \boxed{5}\ \boxed{=}$

9. $\boxed{6}\ \boxed{2}\ \boxed{\cdot}\ \boxed{3}\ \boxed{4}\ \boxed{\div}\ \boxed{4}\ \boxed{2}\ \boxed{\cdot}\ \boxed{3}\ \boxed{5}\ \boxed{7}\ \boxed{=}$

10. $\boxed{1}\ \boxed{\cdot}\ \boxed{5}\ \boxed{+}\ \boxed{3}\ \boxed{\cdot}\ \boxed{6}\ \boxed{2}\ \boxed{+}\ \boxed{5}\ \boxed{4}\ \boxed{\cdot}\ \boxed{7}\ \boxed{+}\ \boxed{8}\ \boxed{0}\ \boxed{=}$

11. $\boxed{1}\ \boxed{\cdot}\ \boxed{6}\ \boxed{x^2}$

12. $\boxed{2}\ \boxed{5}\ \boxed{\cdot}\ \boxed{3}\ \boxed{7}\ \boxed{\sqrt{\phantom{x}}}$

13. $\boxed{1}\ \boxed{1}\ \boxed{y^x}\ \boxed{4}\ \boxed{=}$

14. $\boxed{6}\ \boxed{\div}\ \boxed{(}\ \boxed{2}\ \boxed{+}\ \boxed{1}\ \boxed{)}\ \boxed{=}$

15. $\boxed{6}\ \boxed{\div}\ \boxed{2}\ \boxed{+}\ \boxed{1}\ \boxed{=}$

16. $\boxed{2}\ \boxed{5}\ \boxed{\sqrt{\phantom{x}}}\ \boxed{+}\ \boxed{3}\ \boxed{x^2}\ \boxed{=}$

*Perform the following calculations using your calculator.*

17. $\begin{array}{r} 15.2 \\ +\ 1.03 \\ \hline \end{array}$

18. $\begin{array}{r} 402.11 \\ +\ 6.85 \\ \hline \end{array}$

19. $\begin{array}{r} 542.309 \\ +\ 62.155 \\ \hline \end{array}$

**20.**    18.6
         − 9.2

**21.**    703.12
         − 41.35

**22.**    625.157
         − 142.668

**23.**    21.7
         × 4.2

**24.**    62.37
         × 4.71

**25.**    603.97
         × 78.56

**26.** 3.2)‾8.2

**27.** 6.21)‾0.467613

**28.** 40.3)‾209.765

**29.** $\dfrac{1}{4} - 0.05$

**30.** $7 + \dfrac{5}{16}$

**31.** $\dfrac{2}{5} \div 1.2$

**32.** $0.2 + 5.8 - 3.1$

**33.** $2.6 \times 5.3 \div 1.2$

**34** $5 \div 1.2 + 6.5$

**35.** $(6.98)^2$

**36.** $\sqrt{4.29}$

**37.** $(10.3)^5$

**38.** $4.8 \div (3.2 - 1.7)$

**39.** $(2.6 + 4.3) \times (1.3 - 0.8)$

**40.** $(7.04)^2 - \sqrt{8.15}$

**41.** $(3.8)^8 \div (6.3 + 10.4)$

*Solve.*

**42.** The sales tax on an item is found by multiplying the cost by 0.0775. How much tax would be charged on a new car which costs $6745.83?

**43.** To find the property tax on an apartment building, the assessed value of the building is multiplied by the tax rate, 0.163. How much tax would have to be paid on an apartment building which has an assessed value of $823,475.80?

**44.** Find the area of a rectangle which is 4.237 m long and 3.075 m wide.

**45.** To find the monthly interest which must be paid on his home mortgage, Carlos must multiply the mortgage balance by 0.007285. If his balance in June is $38,276.55, how much interest must he pay?

## FOR REVIEW

**46.** Arrange the given numbers in order of size with the smallest on the left.

$$3.1, \quad \frac{3}{10}, \quad 0.31, \quad 0.013, \quad \frac{1}{3}$$

**47.** A 12-ounce bottle of soda sells for 45¢. What is the price per ounce (unit price)?

ANSWERS:  1. $10 + 35 = 45$  2. $62 - 31 = 31$  3. $38 \times 24 = 912$  4. $55 \div 11 = 5$  5. $2.3 + 1.4 = 3.7$  6. $4.7 + 0.9 = 5.6$  7. $6.12 \times 5.3 = 32.436$  8. $1.3 \div 4.5 = 0.2\overline{8}$  9. $62.34 \div 42.357 = 1.4717756$  10. $1.5 + 3.62 + 54.7 + 80 = 139.82$  11. $(1.6)^2 = 2.56$  12. $\sqrt{25.37} = 5.0368641$  13. $(11)^4 = 14,641$  14. $6 \div (2 + 1) = 2$  15. $6 \div 2 + 1 = 4$  16. $\sqrt{25} + 3^2 = 14$  17. $16.23$  18. $408.96$  19. $604.464$  20. $9.4$  21. $661.77$  22. $482.489$  23. $91.14$  24. $293.7627$  25. $47,447.883$  26. $2.5625$  27. $0.0753$  28. $5.2050868$  29. $0.2$  30. $7.3125$  31. $0.\overline{3}$  32. $2.9$  33. $11.48\overline{3}$  34. $10.\overline{6}$  35. $48.7204$  36. $2.0712315$  37. $115,927.41$  38. $3.2$  39. $3.45$  40. $46.70678$  41. $2603.4683$  42. $\$522.80$  43. $\$134,226.56$  44. $13.028775$ sq m  45. $\$278.84$  46. $0.013, \frac{3}{10}, 0.31, \frac{1}{3}, 3.1$  47. $3.75¢$ per ounce

## 5.8 EXERCISES B

*What calculation is being made by pressing the keys as shown left to right, and what is the result?*

**1.** $\boxed{2}\,\boxed{0}\,\boxed{+}\,\boxed{4}\,\boxed{6}\,\boxed{=}$  **2.** $\boxed{8}\,\boxed{2}\,\boxed{-}\,\boxed{7}\,\boxed{3}\,\boxed{=}$

**3.** $\boxed{4}\,\boxed{1}\,\boxed{\times}\,\boxed{3}\,\boxed{5}\,\boxed{=}$  **4.** $\boxed{5}\,\boxed{6}\,\boxed{\div}\,\boxed{8}\,\boxed{=}$

**5.** $\boxed{6}\,\boxed{\cdot}\,\boxed{5}\,\boxed{+}\,\boxed{2}\,\boxed{\cdot}\,\boxed{9}\,\boxed{=}$  **6.** $\boxed{9}\,\boxed{\cdot}\,\boxed{8}\,\boxed{+}\,\boxed{\cdot}\,\boxed{3}\,\boxed{=}$

**7.** $\boxed{4}\,\boxed{\cdot}\,\boxed{3}\,\boxed{5}\,\boxed{\times}\,\boxed{1}\,\boxed{\cdot}\,\boxed{2}\,\boxed{=}$  **8.** $\boxed{5}\,\boxed{\cdot}\,\boxed{2}\,\boxed{\div}\,\boxed{1}\,\boxed{\cdot}\,\boxed{3}\,\boxed{=}$

**9.** $\boxed{9}\,\boxed{0}\,\boxed{\cdot}\,\boxed{0}\,\boxed{4}\,\boxed{3}\,\boxed{8}\,\boxed{5}\,\boxed{\div}\,\boxed{2}\,\boxed{8}\,\boxed{\cdot}\,\boxed{4}\,\boxed{0}\,\boxed{5}\,\boxed{=}$

**10.** $\boxed{5}\,\boxed{\cdot}\,\boxed{6}\,\boxed{-}\,\boxed{3}\,\boxed{\cdot}\,\boxed{0}\,\boxed{5}\,\boxed{+}\,\boxed{2}\,\boxed{\cdot}\,\boxed{9}\,\boxed{9}\,\boxed{=}$

**11.** $\boxed{4}\,\boxed{\cdot}\,\boxed{2}\,\boxed{8}\,\boxed{x^2}$  **12.** $\boxed{3}\,\boxed{\cdot}\,\boxed{4}\,\boxed{0}\,\boxed{9}\,\boxed{\sqrt{\phantom{x}}}$

**13.** $\boxed{8}\,\boxed{\cdot}\,\boxed{5}\,\boxed{y^x}\,\boxed{4}\,\boxed{=}$  **14.** $\boxed{1}\,\boxed{8}\,\boxed{\div}\,\boxed{(}\,\boxed{3}\,\boxed{+}\,\boxed{6}\,\boxed{)}\,\boxed{=}$

**15.** $\boxed{1}\,\boxed{8}\,\boxed{\div}\,\boxed{3}\,\boxed{+}\,\boxed{6}\,\boxed{=}$  **16.** $\boxed{7}\,\boxed{x^2}\,\boxed{-}\,\boxed{4}\,\boxed{\cdot}\,\boxed{2}\,\boxed{\sqrt{\phantom{x}}}\,\boxed{=}$

*Perform the following calculations using your calculator.*

**17.** $\begin{array}{r} 19.3 \\ + 1.08 \\ \hline \end{array}$

**18.** $\begin{array}{r} 709.44 \\ + 5.87 \\ \hline \end{array}$

**19.** $\begin{array}{r} 652.993 \\ + 43.875 \\ \hline \end{array}$

**20.** $\begin{array}{r} 19.3 \\ - 5.7 \\ \hline \end{array}$

**21.** $\begin{array}{r} 204.53 \\ - 54.88 \\ \hline \end{array}$

**22.** $\begin{array}{r} 476.259 \\ - 341.563 \\ \hline \end{array}$

**23.** $\begin{array}{r} 42.8 \\ \times 6.5 \\ \hline \end{array}$

**24.** $\begin{array}{r} 59.47 \\ \times 2.38 \\ \hline \end{array}$

**25.** $\begin{array}{r} 705.84 \\ \times 23.75 \\ \hline \end{array}$

**26.** $6.5\overline{)30.55}$

**27.** $4.58\overline{)0.273884}$

**28.** $20.7\overline{)109.36224}$

**29.** $\dfrac{1}{5} - 0.19$

**30.** $12 + \dfrac{15}{16}$

**31.** $\dfrac{3}{8} \div 2.5$

**32.** $0.9 + 4.07 - 2.58$

**33.** $4.7 \times 2.9 \div 1.02$

**34.** $6 \div 2.4 + 7.3$

**35.** $(4.09)^2$

**36.** $\sqrt{11.37}$

**37.** $(21.4)^6$

**38.** $6.3 \times (1.7 + 2.5)$

**39.** $(4.3 - 1.9) \times (8.5 + 2.7)$

**40.** $(21.08)^2 - \sqrt{14.3}$

**41.** $(2.18)^6 \div (9.2 - 4.7)$

*Solve.*

**42.** The sales tax on a purchase can be found by multiplying the selling price by 0.0425. How much tax would be charged on a new boat which sells for $24,325.65?

**43.** The property tax on a large apartment building can be found by multiplying the tax rate, 0.197, by the assessed value of the building. How much tax would have to be paid if the assessed value is $179,840.50?

**44.** Find the area of a rectangular garden which measures 7.235 yards long and 5.108 yards wide.

**45.** To find the amount of interest paid in a given month, multiply the monthly balance by 0.008275. How much interest will Daphne pay in August if her loan balance is $42,179.85?

## FOR REVIEW

**46.** Arrange the given numbers in order of size with the smallest on the left.

$$0.88, \quad \frac{6}{7}, \quad \frac{13}{14}, \quad 0.92, \quad \frac{9}{10}$$

**47.** Janet earns $12.30 an hour. How much will she be paid for working $8\frac{3}{4}$ hours?

## 5.8 EXERCISES C

*Solve.*

**1.** The area of a rectangle is 106.005 square yards, and its width is 9.25 yards. Divide the area by the width to find the length of the rectangle.
[Answer: 11.46 yd]

**2.** During five consecutive weeks, Mr. Whorton discovered that the overhead expenses on his business were $497.36, $643.21, $825.13, $503.07, and $601.58. What were the total expenses and the average expense per week?

**3.** In a certain triangle, two sides are 4.28 cm and 7.15 cm. The third side is given by

$$\sqrt{(4.28)^2 + (7.15)^2}.$$

Find the third side, correct to the nearest hundredth of a centimeter.
[Answer: 8.33 cm]

**4.** The amount of money in a certain savings account at the end of 10 years when $1500 is invested initially is given by $(1.02)^{40}(1500)$. Find the amount.

## CHAPTER 5 REVIEW

### KEY WORDS

**5.1** A **decimal fraction** is a fraction with denominator a power of 10. Decimal fractions are usually called simply **decimals.** The period used in a decimal is called a **decimal point.**

**5.2** If the division process ends with a remainder of zero, the decimal is a **terminating decimal.**

If the division process does not end, there is a repeating block of digits, and the decimal is a **repeating decimal.**

**5.3** The number of digits to the right of the decimal point is called the number of **decimal places** in the number.

**5.8** On a calculator, the **display** shows numbers, the **register** stores numbers used in computations, the **number keys** are the buttons for the ten digits and the decimal point, and the **operation keys** are used to perform operations, such as adding, subtracting, multiplying, and dividing.

### KEY CONCEPTS

**5.1** **1.** Each decimal has a fractional form and a decimal form.

**2.** A decimal point separates the ones digit from the tenths digit.

**5.2** Every fraction can be written as a decimal that either terminates or repeats.

**5.3** When rounding a decimal, if the first digit to the right of the desired position is less than 5, drop it and all digits to the right of it. If the first digit to the right is greater than or equal to 5, increase the digit in the desired position by one and drop all digits to the right of it. *Never* round off one digit at a time.

**5.4** **1.** To add decimals, line up the decimal points and add like whole numbers.

**2.** To subtract decimals, line up the decimal points and subtract like whole numbers.

**5.5** **1.** The product of two decimals has the same number of decimal places as the sum of the decimal places in the two factors.

**2.** To multiply by a power of 10, move the decimal point to the *right* the same number of decimal places as the number of zeros in the power of 10.

**3.** To multiply by 0.1, 0.01, or 0.001, move the decimal point to the *left* the same number of decimal places as decimal places in 0.1, 0.01, or 0.001.

**5.6** **1.** When dividing a decimal by a decimal, move the decimal point in the divisor and dividend the same number of places to obtain a whole number divisor. The decimal point in the quotient is directly above the new decimal point in the dividend.

**2.** To divide by a power of 10, move the decimal point to the *left* the same number of decimal places as the number of zeros in the power of 10.

**3.** To divide by 0.1, 0.01, or 0.001, move the decimal point to the *right* the same number of decimal places as decimal places in 0.1, 0.01, 0.001.

**5.7** The unit price of an item is

$$\text{unit price} = \frac{\text{total price}}{\text{number of units}}.$$

**5.8** Only use a calculator for complex problems; do not become so dependent on it that you use it for simple calculations that can be found mentally.

### REVIEW EXERCISES

*Part I*

**5.1** **1.** What de we call fractions that have a power of 10 in the denominator?

*Give the* formal *word name for each decimal.*

**2.** 23.4

**3.** 7.0005

*Write each number in decimal notation.*

**4.** four and seven hundredths

**5.** two hundred thirty-three and three hundred two ten thousandths

**6.** Write $27.16 using a word name as it would appear on a check.

**5.2** *Change each decimal to a fraction (or mixed number).*

**7.** 0.63          **8.** 2.7          **9.** 5.125

*Change each fraction to a decimal.*

**10.** $\dfrac{1}{9}$          **11.** $\dfrac{5}{3}$          **12.** $\dfrac{3}{16}$

**5.3** *Give the number of decimal places in each number.*

**13.** 2.103          **14.** 4.003          **15.** 0.01010

*Round each decimal to the given number of decimal places or position.*

**16.** 3.255; two      **17.** 6.4908; thousandth      **18.** 26.05; tenth      **19.** $37.51; dollar

**20.** $0.\overline{84}$; tenth      **21.** $0.\overline{84}$; hundredth      **22.** $0.\overline{84}$; three      **23.** $0.426; cent

*Solve.*

**24.** What is the approximate cost of 19 pounds of grapes at 99¢ per pound?

**25.** Alvin has a new car loan in the amount of $7895.60. If he plans to pay off the loan in 40 equal monthly payments, what is the approximate amount he must pay each month?

**5.4**    **26.** Find the sum.

$$6.1 + 6.01 + 0.601 + 0.0601 + 60.1$$

**27.** The daily receipts for one week at Tony's Tastee Tacos were $421.36, $347.20, $289.17, $655.27, and $504.08. What was the total for the week?

**28.** Find the total deposit on the following portion of a deposit slip.

| DEPOSIT | | |
|---|---|---|
| | Dollars | Cents |
| CASH | 371 | 05 |
| CHECKS | 480 | 77 |
| | 19 | 40 |
| | 2037 | 06 |
| TOTAL | | |

**29.** Find the perimeter of the given figure.

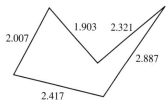

*Find each difference.*

**30.** $42.6 - 3.08$      **31.** $100 - 74.36$      **32.** $2 - 1.003$

**33.** Ev received a check for $235.96. He spent $45.30 for a pair of shoes and $38.07 for a power saw. How much did he have left?

**34.** Becky bought a sweater for $18.35 and a pair of earrings for $8.29. The sales tax on her purchases amounts to $1.07. How much change will she receive if she pays with two $20 bills?

**5.5** *Find each product.*

**35.** 13.06
  × 2.7

**36.** 3.1108
  × 2.06

**37.** 0.4001
  × 1.3

**38.** The EPA-estimated miles per gallon for a car is 23.7 mpg. Its gas tank capacity is 11.6 gallons. To the nearest mile, what is the driving range of the car?

**39.** Mr. Swokowski is paid $11.37 per hour. His timecard for one week is shown in the figure. How much was he paid that week?

| Day | Hours worked |
|-----|--------------|
| M   | 8.1          |
| T   | 5.3          |
| W   | 4.7          |
| T   | 10.3         |
| F   | 7.8          |

*Without making the actual computation, use an estimate to decide which of the given answers is correct.*

**40.** 2.9 × 24.007    **(a)** 0.696    **(b)** 6.96    **(c)** 69.6    **(d)** 696

*Solve.*

**41.** A car gets 28.9 mpg and has a gas tank with capacity 12.2 gallons. What is the approximate driving range of the car? The exact driving range? Isn't the exact driving range also an *approximate* driving range?

**42.** What is the approximate cost of 9 pounds of candy selling for $1.87 per pound? The exact cost?

*Find the products.*

**43.** 10.785 × 1000

**44.** 10.785 × 0.1

**45.** 10.785 × 0.01

**5.6** *Find each quotient.*

**46.** 2.9)‾6.177‾

**47.** 42)‾222.6‾

**48.** 0.032)‾1.3312‾

**49.** 4.13)‾15.6‾ (to the nearest tenth)

**50.** 0.089)‾4.62‾ (to the nearest hundredth)

**51.** Terry drove 746.2 miles and used 34.9 gallons of gas. To the nearest tenth, how many miles per gallon did he get?

**52.** The table shows Betty's grades during one quarter. Complete the table and calculate her GPA correct to the nearest hundredth.

| Course | Credits | Grade | Value | Course Value |
|--------|---------|-------|-------|--------------|
| Math    | 4 | B | 3 | |
| French  | 1 | F | 0 | |
| Biology | 4 | C | 2 | |
| English | 3 | A | 4 | |
| Russian | 3 | D | 1 | |

**53.** What is the approximate cost per pound of 9 pounds of candy selling for $18.45? The exact cost?

**54.** The Morins plan to carpet their master bedroom. If the room is rectangular in shape and measures 4.8 yards long by 3.9 yards wide, and the total cost of the job is $383.76, what is the approximate cost of one square yard of carpet? The exact cost?

*Without making actual computations, use an estimate to decide which of the given answers is correct.*

**55.** 401.87 ÷ 39.6    **(a)** 1.015    **(b)** 10.15    **(c)** 101.5    **(d)** 1015

*Find the quotients.*

**56.** $43.89 \div 100$        **57.** $1.497 \div 1000$        **58.** $286.1 \div 10$

**5.7** *Find the sums (to the nearest hundredth).*

**59.** $\dfrac{2}{3} + 7.25$        **60.** $6\dfrac{1}{4} + 8.39$        **61.** $\dfrac{1}{6} + 3.78$

**62.** Clyde earns \$12.43 an hour. How much will he be paid for working $7\frac{1}{4}$ hours?

**63.** If $18\frac{1}{2}$ ounces of cereal sell for \$1.09, what is the price per ounce (the unit price) to the nearest tenth of a cent?

**64.** The Meyers paid \$113,750 for their house. If the house has 2000 sq ft, how much did they pay per sq ft?

**65.** Arrange 0.74, 7.04, 7.4, 0.074, and 70.4 in order of size with the smallest on the left.

**66.** Complete the unit price column in the table and determine which item is the best buy.

| Brand | Pounds | Total price | Unit price |
|-------|--------|-------------|------------|
| X | 7 | \$2.99 | |
| Y | 11 | \$4.71 | |
| Z | 8 | \$3.32 | |

*Perform the indicated operations.*

**67.** $6.3 \div \dfrac{1}{3} - 1\dfrac{1}{2}$        **68.** $(2.6)^2 - \dfrac{7}{8} \times \dfrac{3}{14}$

**5.8** *What calculation is made by pressing the keys as shown from left to right, and what is the result?*

**69.** $\boxed{8}\ \boxed{\cdot}\ \boxed{7}\ \boxed{\div}\ \boxed{1}\ \boxed{2}\ \boxed{=}$        **70.** $\boxed{2}\ \boxed{1}\ \boxed{\cdot}\ \boxed{6}\ \boxed{-}\ \boxed{4}\ \boxed{\cdot}\ \boxed{3}\ \boxed{7}\ \boxed{=}$

**71.** $\boxed{1}\ \boxed{\cdot}\ \boxed{0}\ \boxed{0}\ \boxed{9}\ \boxed{\times}\ \boxed{3}\ \boxed{\cdot}\ \boxed{5}\ \boxed{=}$        **72.** $\boxed{6}\ \boxed{+}\ \boxed{2}\ \boxed{\cdot}\ \boxed{5}\ \boxed{-}\ \boxed{4}\ \boxed{=}$

**73.** $\boxed{2}\ \boxed{\cdot}\ \boxed{3}\ \boxed{\times}\ \boxed{6}\ \boxed{+}\ \boxed{8}\ \boxed{\cdot}\ \boxed{9}\ \boxed{=}$        **74.** $\boxed{9}\ \boxed{\times}\ \boxed{9}\ \boxed{\times}\ \boxed{9}\ \boxed{=}$

**75.** $\boxed{4}\ \boxed{\cdot}\ \boxed{2}\ \boxed{x^2}$        **76.** $\boxed{3}\ \boxed{\cdot}\ \boxed{2}\ \boxed{7}\ \boxed{\sqrt{\phantom{x}}}$

**77.** $\boxed{1}\ \boxed{\cdot}\ \boxed{5}\ \boxed{y^x}\ \boxed{6}\ \boxed{=}$        **78.** $\boxed{6}\ \boxed{\div}\ \boxed{(}\ \boxed{2}\ \boxed{+}\ \boxed{3}\ \boxed{)}\ \boxed{-}\ \boxed{1}\ \boxed{=}$

*Solve.*

**79.** The sales tax on an item is found by multiplying the cost by 0.0425. How much tax would be charged on a new motor home which costs \$14,137.58?

**80.** Find the area of a rectangle 5.621 m long by 3.749 m wide.

**Part II**

*Perform the indicated operations.*

**81.**    $\begin{array}{r} 4.1035 \\ \times\ \ 2.3 \\ \hline \end{array}$        **82.**    $\begin{array}{r} 6.15 \\ 12.034 \\ +\ \ 9.2 \\ \hline \end{array}$        **83.**    $\begin{array}{r} 142.75 \\ -\ 38.016 \\ \hline \end{array}$

**84.** $1.1596 \times 100$        **85.** $1.1596 \div 100$        **86.** $1.1596 \times 0.1$

**87.** $1.1596 \div 0.001$        **88.** $53\overline{)113.42}$        **89.** $0.35\overline{)4.68}$ (to the nearest tenth)

**90.** $3\frac{3}{4} + 2.48$

**91.** $4.67 - 1\frac{1}{5}$

**92.** $(2.6)^2 \times 4\frac{1}{2}$

**93.** Round 6.485 to the nearest **(a)** hundredth, **(b)** tenth, and **(c)** unit.

**94.** Change $\frac{11}{3}$ to a decimal.

**95.** Change 0.85 to a fraction.

**96.** Write $102.38 using a word name as it would appear on a check.

**97.** Estimate the product $6098.5 \times 41.89$.

*Solve.*

**98.** For every $1000 of assessed value on a store, the owner must pay $4.89 in real estate taxes. How much tax must be paid on a store with an assessed value of $125,600?

**99.** Beth received checks of $20.50, $16.75, and $42.30. Out of this total she bought a sweater for $29.95. How much money did she have left?

**100.** Jackie bought 3 books for $8.95 each and paid with two $20 bills. How much change did she get back?

**101.** Which is a better buy, a 24-ounce bottle of soap for $1.20 or a 38-ounce bottle for $2.28?

**102.** Jeff traveled 55 mph for $5\frac{1}{2}$ hours. If he used 12.8 gallons of gas during this time, what was his mileage per gallon, to the nearest tenth?

**103.** What is the approximate cost of 40 pounds of steak at $4.95 per pound? What is the exact cost?

*Without making the actual computation, use an estimate to decide which of the given answers is correct.*

**104.** $398.097 \div 49.7$    **(a)** 0.801    **(b)** 8.01    **(c)** 80.1    **(d)** 801

*Use a calculator to perform the following operations.*

**105.** $62.308 \times 401.375$

**106.** $0.0435\overline{)2.6513}$

**107.** $(3.4)^2 \div 1.7$

**108.** $(1.05)^5 \times 1050$

**109.** $6.35 \times (4.1 - 3.8)$

**110.** $\sqrt{6.9 \times 11.4}$

---

ANSWERS: 1. decimal fractions or decimals 2. twenty-three and four tenths 3. seven and five ten thousandths 4. 4.07 5. 233.0302 6. Twenty-seven and $\frac{16}{100}$ dollars 7. $\frac{63}{100}$ 8. $2\frac{7}{10}$ 9. $5\frac{1}{8}$ 10. $0.\overline{1}$ 11. $1.\overline{6}$ 12. 0.1875 13. 3 14. 3 15. 5 16. 3.26 17. 6.491 18. 26.1 19. $38 20. 0.8 21. 0.85 22. 0.848 23. $0.43 24. $20.00 [using 20 × 100¢] 25. $200 [using 8000 ÷ 40] 26. 72.8711 27. $2217.08 28. $2908.28 29. 11.535 30. 39.52 31. 25.64 32. 0.997 33. $152.59 34. $12.29 35. 35.262 36. 6.408248 37. 0.52013 38. 275 mi 39. $411.59 40. (c) 41. approximate range: 360 mi [using 30 × 12]; exact range: 352.6 mi; yes 42. $18.00 [using 9 × 2]; $16.83 43. 10,785 44. 1.0785 45. 0.10785 46. 2.13 47. 5.3 48. 41.6 49. 3.8 50. 51.91 51. 21.4 mpg 52. GPA = 2.33 53. $2 [20 ÷ 10]; $2.05 54. $20; $20.50 55. (b) 10.15 56. 0.4389 57. 0.001497 58. 28.61 59. 7.92 60. 14.64 61. 3.95 62. $90.12 63. 5.9¢ 64. $56.88 65. 0.074, 0.74, 7.04, 7.4, 70.4 66. Brand X is $0.427; Brand Y is $0.428; Brand Z is $0.415. Thus, Brand Z is the best buy. 67. 17.4 68. 6.5725 69. $8.7 \div 12 = 0.725$ 70. $21.6 - 4.37 = 17.23$ 71. $1.009 \times 3.5 = 3.5315$ 72. $6 + 2.5 - 4 = 4.5$ 73. $2.3 \times 6 + 8.9 = 22.7$ 74. $9 \times 9 \times 9 = 729$ 75. $(4.2)^2 = 17.64$ 76. $\sqrt{3.27} = 1.8083141$ 77. $(1.5)^6 = 11.390625$ 78. $6 \div (2 + 3) - 1 = 0.2$ 79. $600.85 80. 21.073129 sq m 81. 9.43805 82. 27.384 83. 104.734 84. 115.96 85. 0.011596 86. 0.11596 87. 1159.6 88. 2.14 89. 13.4 90. 6.23 91. 3.47 92. 30.42 93. (a) 6.49 (b) 6.5 (c) 6 94. $3.\overline{6}$ 95. $\frac{17}{20}$ 96. One hundred two and $\frac{38}{100}$ dollars 97. 240,000 [using 6000 × 40] 98. $614.18 99. $49.60 100. $13.15 101. The 24-ounce bottle at 5¢ an ounce is better than the 38-ounce bottle at 6¢ an ounce. 102. 23.6 mpg 103. approximate cost: $200; actual cost: $198 104. (b) 105. 25,008.874 106. 60.949425 107. 6.8 108. 1340.0957 109. 1.905 110. 8.8690473

**1.** Give the formal word name for 73.005.

1. _____

**2.** Write "ninety-two and six hundredths" in decimal notation.

2. _____

**3.** Change 0.16 to a fraction reduced to lowest terms.

3. _____

**4.** Change $\frac{7}{16}$ to a decimal.

4. _____

**5.** Round 6.355 to two decimal places.

5. _____

**6.** Round 45.05 to the nearest tenth.

6. _____

**7.** Round $0.\overline{74}$ to the nearest hundredth.

7. _____

*Perform the indicated operations.*

**8.** $9.1 + 0.91 + 9.01 + 91.0$

8. _____

**9.** $11 - 5.003$

9. _____

*Solve.*

**10.** Mr. and Mrs. Forbes stayed at the Sands Hotel in Reno. Their room cost $126.87 and they charged $48.53 in food and beverages to their room account. If Mr. Forbes paid the bill with two $100 bills, how much change did he receive?

10. _____

**11.** What is the *approximate* cost per pound of dog food if a 32-pound sack sells for $15.36? What is the *exact* cost per pound?

11. _____

*Perform the indicated operations.*

**12.**  71.05
     $\times$ 1.8

**13.**  4.7$\overline{)47.376}$

12. _____

13. _____

**14.**  6.357 $\times$ 100

**15.**  6.357 $\times$ 0.1

14. _____

15. _____

**16.**  6.357 $\div$ 1000

**17.**  6.357 $\div$ 0.1

16. _____

17. _____

*Solve.*

**18.** Hector buys 14.5 pounds of apples at $0.43 per pound, and 11.2 pounds of sausage at $1.59 per pound. To the nearest cent, how much do these purchases total?

18. _____

**19.** Dinah drove 956.3 miles and used 38.6 gallons of gas. To the nearest tenth, how many miles per gallon did she get?

19. _____

**20.** A bricklayer uses 415 bricks at a cost of 9.3¢ per brick. What is the approximate cost of the bricks? The exact cost (to the nearest cent)?

20. _____

**21.** Perform the indicated operations.

$$(3.5)(4.1) - \left(\frac{1}{2}\right)^2$$

21. _____

**22.** What calculation is being made by pressing the keys shown from left to right? Determine the answer.

$\boxed{8}\,\boxed{.}\,\boxed{0}\,\boxed{3}\,\boxed{\div}\,\boxed{2}\,\boxed{.}\,\boxed{5}\,\boxed{=}$

22. _____

# Applications of Ratio, Proportion, and Percent

## 6.1 RATIO AND PROPORTION

STUDENT GUIDEPOSTS

1 Ratio
2 Proportion
3 Solving Proportions

### 1 RATIO

The word *ratio* is commonly used in many practical situations. For example, an excellent recipe for honey-butter is two cups of butter and three cups of honey. In this recipe, the *ratio* of butter to honey is 2 to 3. A **ratio** is a way to compare two numbers, like 2 parts of butter for every 3 parts of honey. A ratio can also be expressed as a quotient of two numbers. For example, the ratio of the number 5 to the number 8 is the quotient or fraction $\frac{5}{8}$. We call $\frac{5}{8}$ the **fraction form** for the ratio of 5 to 8. Another way of writing this ratio is

$$5:8, \text{ read ''the ratio of 5 to 8.''}$$

---

#### EXAMPLE 1 WRITING RATIOS IN FRACTION FORM

Write each ratio in fraction form.

(a) The ratio of 7 to 10    $\dfrac{7}{10}$

(b) The ratio 10:5    $\dfrac{10}{5}$ which reduces to $\frac{2}{1}$ (see Chapter 3.)

#### PRACTICE EXERCISE 1

Write each ratio in fraction form.

(a) The ratio of 5 to 11

(b) The ratio of 4 to 12

Answers: (a) $\frac{5}{11}$   (b) $\frac{4}{12}$ or $\frac{1}{3}$ (reduced)

---

Ratios occur in many applications such as tax rates, rate of speed, gas mileage, and unit cost. These are shown in the following table.

| Application | Ratio Involved |
|---|---|
| $20 tax on a $400 purchase | $\dfrac{\$20}{\$400} = \dfrac{\$1}{\$20}$ means $1 tax on each $20 spent |
| 300 miles in 6 hours | $\dfrac{300 \text{ mi}}{6 \text{ hr}} = \dfrac{50 \text{ mi}}{1 \text{ hr}} = 50\dfrac{\text{mi}}{\text{hr}} = 50 \text{ mph}$ |

100 miles on 5 gallons of gas
$$\frac{100 \text{ mi}}{5 \text{ gal}} = \frac{20 \text{ mi}}{1 \text{ gal}} = 20\frac{\text{mi}}{\text{gal}} = 20 \text{ mpg}$$

$8 for 4 pounds of meat
$$\frac{\$8}{4 \text{ lb}} = \$2 \text{ per lb}$$

40 children and 5 adults
$$\frac{40 \text{ children}}{5 \text{ adults}} = 8 \text{ children per adult}$$

## HINT

The word *per* is often used to describe ratios; it means "for each." For example, $2 per lb means $2 for each pound. Also, *mph* is the abbreviation for "miles per hour," and *mpg* stands for "miles per gallon."

| EXAMPLE 2   WRITING RATIOS | PRACTICE EXERCISE 2 |
|---|---|

A company employs 18 women and 12 men.

**(a)** What is the ratio of men to women in the company?

$$\frac{12 \text{ men}}{18 \text{ women}} = \frac{12}{18} = \frac{2}{3}$$

The ratio of men to women is $\frac{2}{3}$ or 2 to 3 or 2:3.

**(b)** What is the ratio of women employed to the total number of employees?

There are $18 + 12 = 30$ total employees.

$$\frac{18 \text{ women}}{30 \text{ employees}} = \frac{18}{30} = \frac{3}{5}$$

The ratio of women to total employees is $\frac{3}{5}$ or 3 to 5 or 3:5.

There are 200 floor-level seats and 120 balcony seats in an auditorium.

**(a)** What is the ratio of balcony seats to floor seats?

**(b)** What is the ratio of floor seats to the total number of seats?

Answers:  (a) $\frac{3}{5}$ or 3 to 5 or 3:5
(b) $\frac{5}{8}$ or 5 to 8 or 5:8

| EXAMPLE 3   RATIOS IN A GEOMETRY PROBLEM | PRACTICE EXERCISE 3 |
|---|---|

Consider the triangle in Figure 6.1. Find the ratio of the length of the shortest side to the length of the longest side.

Find the ratio of the length of the longest side to the length of the next longest side in the triangle in Figure 6.1.

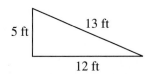

**Figure 6.1**

Since the shortest side is 5 ft and the longest side is 13 ft, the desired ratio is $\frac{5}{13}$ or 5 to 13 or 5:13.

Answer: $\frac{13}{12}$ or 13 to 12 or 13:12

## ❷ PROPORTION

Any statement that says two quantities are equal is called an **equation.** For example, $2 + 1 = 3$ and $\frac{1}{2} = 0.5$ are equations. An equation which states that two ratios are equal is called a **proportion.** For example,

$$\frac{1}{2} = \frac{4}{8}$$

is a proportion. When a pair of numbers such as 1 and 2 has the same ratio as a pair such as 4 and 8, the numbers are called **proportional.** Recall from Section 3.2 that two fractions (ratios) are equal whenever their cross products are equal. The cross products in the above proportion are:

$$2 \cdot 4 \longleftarrow \frac{1}{2} \bowtie \frac{4}{8} \dashrightarrow 1 \cdot 8.$$

Since $2 \cdot 4 = 1 \cdot 8$, we know that the fractions are equal.

Proportions can be used to solve many problems. For example, if 3 hours are required to travel 165 miles, we might ask: "How many hours will be required to travel 275 miles?" The proportion

$$\frac{\text{first time} \xrightarrow{\text{in hours}}}{\text{first distance} \xrightarrow{\text{in miles}}} \frac{3}{165} = \frac{\square}{275} \xleftarrow{\text{in hours}} \text{second time} \atop \xleftarrow{\text{in miles}} \text{second distance}$$

completely describes this problem. That is, the ratio of the first time to the first distance must equal the ratio of the second time to the second distance. We need to find the missing number (represented by the $\square$). In section 3.2 we were able to find such numbers in similar problems by inspection. There is a more direct method. Since

$$\frac{3}{165} = \frac{\square}{275},$$

we know that the cross products are equal. Thus,

$$3 \cdot 275 = 165 \cdot \square$$
$$825 = 165 \cdot \square.$$

To determine $\square$, we use the related division sentence first introduced in Section 3.4. Recall that every multiplication sentence such as $2 \cdot 3 = 6$ has two related division sentences

$$3 = 6 \div 2 = \frac{6}{2} \quad \text{and} \quad 2 = 6 \div 3 = \frac{6}{3}.$$

In the work above, the multiplication sentence or equation

$$825 = 165 \cdot \square$$

can be replaced with the related division sentence or equation

$$\square = \frac{825}{165},$$

which reduces to $\square = 5$. Thus,

$$\frac{\square}{275} \text{ becomes } \frac{5}{275}, \qquad \text{Replace } \square \text{ with 5}$$

and we see that 5 is indeed the desired number since

$$\frac{3}{165} = \frac{5}{275}. \qquad \text{Both cross products are 825}$$

Often a letter (instead of $\square$) is used to represent the missing number in a proportion. Suppose we use $a$. Then

$$\frac{3}{165} = \frac{\square}{275}$$

can be written

$$\frac{3}{165} = \frac{a}{275},$$

and we need to find the value of $a$. The method is exactly the same as above, using $a$ instead of $\square$.

$$3 \cdot 275 = 165 \cdot a \qquad \text{Cross products are equal}$$
$$875 = 165 \cdot a \qquad \text{Multiply 3 and 275}$$
$$a = \frac{825}{165} \qquad \text{Related division sentence}$$
$$a = 5 \qquad \text{Divide 825 by 165}$$

## ❸ SOLVING PROPORTIONS

The process of finding a missing number in a given proportion is called **solving the proportion.** The preceding discussion suggests a method for doing this.

---

**To Solve a Proportion Involving a Missing Number $a$**

1. Set the cross products equal to each other (that is, **equate** the cross products).
2. Then $a$ is equal to the numerical cross product divided by the multiplier of $a$. That is, the related division sentence gives the desired value of $a$.

---

| **EXAMPLE 4** SOLVING PROPORTIONS | **PRACTICE EXERCISE 4** |
|---|---|

Solve the proportions.

**(a)** $\dfrac{a}{4} = \dfrac{12}{16}$

$$16 \cdot a = 4 \cdot 12 \qquad \text{Cross products are equal}$$
$$16 \cdot a = 48 \qquad \text{Simplify}$$
$$a = \frac{48}{16} \qquad \text{Divide by 16, the multiplier of } a\text{; this is the related}$$
$$\qquad\qquad\qquad \text{division sentence}$$
$$a = 3$$

Thus, the missing value is 3, and

$$\frac{3}{4} = \frac{12}{16}. \qquad \text{To check, find the cross products:}$$
$$\qquad\qquad 3 \cdot 16 = 48 \text{ and } 4 \cdot 12 = 48$$

**(b)** $\dfrac{2}{a} = \dfrac{4}{10}$

$$2 \cdot 10 = 4 \cdot a \qquad \text{Equate cross products}$$
$$20 = 4 \cdot a \qquad \text{Simplify}$$
$$\frac{20}{4} = a \qquad \text{Divide 20 by 4}$$
$$5 = a$$

Solve the proportions.

**(a)** $\dfrac{a}{2} = \dfrac{15}{10}$

**(b)** $\dfrac{6}{21} = \dfrac{2}{a}$

Thus, the missing value is 5, and

$$\frac{2}{5} = \frac{4}{10}.$$

To check, find the cross products:
2 · 10 = 20 and 5 · 4 = 20

Answers: (a) 3 (b) 7

## HINT

Here is another way to remember how to find the missing number $a$ in a proportion. Consider the proportion $\frac{a}{4} = \frac{12}{16}$ in Example 4(a). We found the cross-product equation $16 \cdot a = 4 \cdot 12$, or $16 \cdot a = 48$. Since $16 \cdot a$ and 48 are equal numbers, if they are both divided by the same number 16, the multiplier of $a$, we could assume that the quotients are also equal. That is,

$$\frac{16 \cdot a}{16} = \frac{48}{16}.$$

Since $\frac{16 \cdot a}{16} = \frac{16}{16} \cdot \frac{a}{1} = 1 \cdot a = a$, by dividing both sides of the cross-product equation by 16 we obtain the related division sentence $a = \frac{48}{16} = 3$. We will study this process in more detail in Chapter 9.

### EXAMPLE 5   A PROPORTION IN A PROBLEM IN POLITICS

In an election the winning candidate won by a ratio of 5 to 4. If she received 600 votes, how many votes did the losing candidate receive?

If $a$ represents the number of votes received by the losing candidate, the proportion

winning ratio → $\dfrac{5}{4} = \dfrac{600}{a}$ ← votes for winner
← votes for loser

describes the problem.

$$5 \cdot a = 4 \cdot 600 \qquad \text{Equate cross products}$$
$$5 \cdot a = 2400 \qquad \text{Simplify}$$
$$a = \frac{2400}{5} \qquad \text{Divide by 5}$$
$$a = 480 \qquad \text{Simplify}$$

Thus, the loser received 480 votes.

### PRACTICE EXERCISE 5

In Professor Walter's algebra class, the ratio of girls to boys is 4 to 3. If there are 18 boys in the class, how many girls are there?

Let $a$ represent the number of girls in the class.

$$\frac{4}{3} = \frac{a}{18}$$

Answer: 24 girls

## 6.1   EXERCISES A

*Write the fraction form for the given ratios.*

**1.** 1 to 8

**2.** 2 to 7

**3.** 6 to 4

**4.** 12 to 3

**5.** 6 to 1

**6.** 4 to 4

*Write the ratio that is indicated by the given fraction.*

**7.** $\dfrac{1}{4}$

**8.** $\dfrac{3}{5}$

**9.** $\dfrac{5}{2}$

**10.** $\dfrac{13}{6}$

**11.** $\dfrac{6}{13}$

**12.** $\dfrac{9}{9}$

*Write as a ratio and reduce.*

**13.** 200 miles in 4 hours

**14.** $10 for 5 pounds of candy

**15.** 24 children in 8 families

**16.** 125 miles on 5 gallons of gas

**17.** 75¢ for 3 candy bars

**18.** 88 ounces in 11 glasses

**19.** 2700 words on 12 pages

**20.** $1350 earned in 3 weeks

**21.** 750 television sets in 500 households (reduce to a one-household comparison)

**22.** 10 cups of flour in 4 cakes (reduce to a one-cake comparison)

*Solve the proportions.*

**23.** $\dfrac{a}{3} = \dfrac{14}{21}$

**24.** $\dfrac{9}{a} = \dfrac{3}{1}$

**25.** $\dfrac{4}{7} = \dfrac{a}{28}$

**26.** $\dfrac{40}{35} = \dfrac{2}{a}$

**27.** $\dfrac{a}{13} = \dfrac{3}{39}$

**28.** $\dfrac{12}{a} = \dfrac{4}{1}$

**29.** $\dfrac{60}{72} = \dfrac{a}{6}$

**30.** $\dfrac{2}{1} = \dfrac{1}{a}$

*Solve.*

**31.** If it takes 4 hours to travel 208 miles, how many hours would it take to travel 364 miles?

**32.** Representative Wettaw won an election by a ratio of 8 to 5. If he received 10,400 votes, how many votes did his opponent receive?

**33** If a wire 70 ft long weighs 84 pounds, how much will 110 ft of the same wire weigh?

**34.** If $\frac{1}{2}$ inch on a map represents 10 miles, how many miles are represented by $6\frac{1}{2}$ inches?

**35.** If 3 coats cost a total of $156, how many coats can be bought for $364?

**36** A ranger wishes to estimate the number of antelope in a preserve. He catches 58 antelope, tags their ears, and returns them to the preserve. Some time later, he catches 29 antelope and discovers that 7 of them are tagged. Estimate the number of antelope in the preserve.

**37.** A baseball pitcher gave up 60 earned runs in 180 innings. Estimate the number of earned runs he will give up every 9 innings. This number is called his *earned run average*.

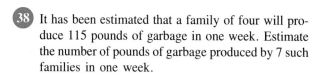 It has been estimated that a family of four will produce 115 pounds of garbage in one week. Estimate the number of pounds of garbage produced by 7 such families in one week.

## FOR REVIEW

*The following exercises will help you prepare for the next section. Change each fraction or mixed number to a decimal.*

**39.** $\dfrac{13}{4}$     **40.** $2\dfrac{5}{11}$     **41.** $\dfrac{4}{5}$

*Change each decimal to a fraction or mixed number.*

**42.** 0.3     **43.** 0.14     **44.** 5.6

**45.** Change $\dfrac{4}{7}$ to a decimal, rounded to the nearest hundredth.

ANSWERS:   1. $\frac{1}{8}$   2. $\frac{2}{7}$   3. $\frac{6}{4}$ or $\frac{3}{2}$   4. $\frac{12}{3}$ or $\frac{4}{1}$   5. $\frac{6}{1}$   6. $\frac{4}{4}$ or $\frac{1}{1}$   7. 1 to 4   8. 3 to 5   9. 5 to 2   10. 13 to 6   11. 6 to 13   12. 9 to 9 or 1 to 1   13. 50 mph   14. $2 per pound   15. 3 children per family   16. 25 mpg   17. 25¢ per candy bar   18. 8 ounces per glass   19. 225 words per page   20. $450 per week   21. 1.5 televisions per household   22. 2.5 cups of flour per cake   23. 2   24. 3   25. 16   26. $\frac{7}{4}$   27. 1   28. 3   29. 5   30. $\frac{1}{2}$   31. 7 hours   32. 6500 votes   33. 132 pounds   34. 130 miles   35. 7 coats   36. about 240 antelope   37. 3   38. 805 pounds   39. 3.25   40. $2.\overline{45}$   41. 0.8   42. $\frac{3}{10}$   43. $\frac{7}{50}$   44. $5\frac{3}{5}$   45. 0.57

## 6.1 EXERCISES B

*Write the fraction form for the given ratios.*

**1.** 1 to 12     **2.** 3 to 7     **3.** 9 to 3

**4.** 15 to 6     **5.** 13 to 1     **6.** 6 to 6

*Write the ratio that is indicated by the given fractions.*

**7.** $\dfrac{1}{5}$     **8.** $\dfrac{4}{7}$     **9.** $\dfrac{7}{2}$

**10.** $\dfrac{15}{4}$     **11.** $\dfrac{4}{15}$     **12.** $\dfrac{8}{8}$

*Write as a ratio and reduce.*

**13.** 120 feet in 4 seconds     **14.** $12 for 4 pounds of nuts

**15.** 320 fish in 16 aquariums     **16.** 720 kilometers on 60 liters of gasoline

**17.** $7 for 4 tubes of toothpaste     **18.** 360 gallons in 3 tanks

**19.** 250 chairs to 150 people

**20.** $2700 in 4 savings accounts

**21.** 450 cars in 200 families (reduce to a one-family comparison)

**22.** 18 hours to split 4 cords of wood (reduce to a one-cord comparison)

*Solve the proportions.*

**23.** $\dfrac{a}{2} = \dfrac{11}{22}$

**24.** $\dfrac{7}{a} = \dfrac{14}{10}$

**25.** $\dfrac{3}{11} = \dfrac{a}{22}$

**26.** $\dfrac{50}{45} = \dfrac{10}{a}$

**27.** $\dfrac{a}{14} = \dfrac{28}{56}$

**28.** $\dfrac{15}{a} = \dfrac{3}{5}$

**29.** $\dfrac{56}{64} = \dfrac{a}{8}$

**30.** $\dfrac{3}{1} = \dfrac{1}{a}$

*Solve.*

**31.** If it takes 6 hours to travel 210 miles, how many hours will it take to travel 385 miles?

**32.** Peter Horn won a city council election by a ratio of 7 to 4. If he received 4522 votes, how many votes did his opponent receive?

**33.** If a tree 15 ft tall casts a shadow 4 ft long, how long will the shadow of a 40 ft building be at the same instant?

**34.** If $\frac{1}{4}$ inch on a map represents 20 miles, how many inches are required to represent 150 miles?

**35.** If 4 cassette tapes cost a total of $31.80, how many tapes could be bought for $71.55?

**36.** A game and fish officer wishes to estimate the number of trout in a lake. He catches 94 trout, tags their fins, and returns them to the lake. Some time later, he catches 50 trout and discovers that 12 of them are tagged. Estimate the number of trout in the lake.

**37.** A softball pitcher gave up 24 earned runs in 210 innings. The number of earned runs given up every 9 innings is called a pitcher's earned run average. Calculate her earned run average (to the nearest hundredth).

**38.** It has been estimated that a family of three will produce 85 pounds of garbage in one week. Estimate the number of pounds of garbage produced by 10 such families in one week.

## FOR REVIEW

*The following exercises will help you prepare for the next section. Change each fraction or mixed number to a decimal.*

**39.** $\dfrac{13}{8}$

**40.** $1\dfrac{8}{9}$

**41.** $\dfrac{3}{20}$

*Change each decimal to a fraction or mixed number.*

**42.** 0.7

**43.** 0.32

**44.** 2.05

**45.** Change $\frac{3}{7}$ to a decimal, rounded to the nearest hundredth.

## 6.1 EXERCISES C

**1.** In a recent survey of 6274 adults, it was found that 3726 were female and 2548 were male. Find each ratio and use a calculator to express the fraction as a decimal rounded to the nearest hundredth.

**(a)** What is the ratio of males to total adults in the survey?

**(b)** What is the ratio of females to total adults in the survey?

**(c)** What is the male-to-female-ratio?

**(d)** What is the female-to-male-ratio?

**2.** A car travels 468.9 miles on 21.7 gallons of gas. Traveling at the same rate, how many gallons of gas would the car need to go 750 miles (to the nearest tenth)?

[Answer: 34.7 gallons]

## 6.2 PERCENT AND PERCENT CONVERSIONS

### STUDENT GUIDEPOSTS

1 The Meaning of Percent

2 Changing a Percent to a Fraction

3 Changing a Percent to a Decimal

4 Changing a Decimal to a Percent

5 Changing a Fraction to a Percent

### 1 THE MEANING OF PERCENT

The idea of percent and the percent symbol (%) are used widely in our society. We may pay a five-percent (5%) sales tax, read that Mr. Gomez won the election with 52% of the vote, or hear that a basketball player hit 80% of his free throws.

The word **percent** means "per hundred." That is, it refers to the number of parts in each one hundred parts. For example, when the tax rate is 5%, we pay 5¢ tax on each 100¢ purchase.

$$\text{Percent: } 5\% \qquad \text{Fraction or ratio: } \frac{5}{100}$$

In Figure 6.2, the five shaded coins (5¢) correspond to 5% tax on the one hundred coins (100¢).

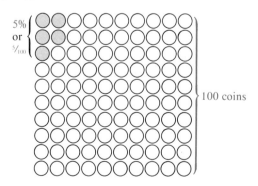

**Figure 6.2**

Thus, $\qquad$ 5% is $\dfrac{5}{100}$ $\quad$ or $\quad$ $5 \times \dfrac{1}{100}$ $\quad$ or $\quad$ $5 \times 0.01.$

## ② CHANGING A PERCENT TO A FRACTION

If Mr. Gomez received 52 out of every 100 votes, then he got 52% of the vote.

$$\text{Percent: } 52\% \qquad \text{Fraction or ratio: } \frac{52}{100}$$

### HINT

The words "out of" can be replaced with a division bar. Notice above how "52 out of every 100 votes" became the fraction $\frac{52}{100}$.

A basketball player could have hit 8 out of 10 (or 80 out of 100) free throws which we shall see is 80%.

$$\text{Percent: } 80\% \qquad \text{Fraction or ratio: } \frac{80}{100} = \frac{8}{10}$$

We also hear on the news that the price of gasoline is 250% of its price several years ago.

$$\text{Percent: } 250\% \qquad \text{Fraction or ratio: } \frac{250}{100} = 2.5$$

Thus, 250% means that the price of gasoline is 2.5 times what it was several years ago.

There are also percents less than 1%. For example, the cost of living might increase 0.8% during the month of December.

$$\text{Percent: } 0.8\% \qquad \text{Fraction or ratio: } \frac{0.8}{100} = \frac{(0.8)(10)}{(100)(10)} = \frac{8}{1000}$$

In each of these examples, the percent was changed to a fraction by removing the percent symbol (%) and dividing by 100. Since dividing by 100 is the same as multiplying by $\frac{1}{100}$, we can form the following rule.

### To Change a Percent to a Fraction

1. Replace the % symbol with either $\left( \times \frac{1}{100} \right)$ or $( \div 100 )$.
2. Evaluate and then reduce the fraction.

---

### EXAMPLE 1  CHANGING PERCENTS TO FRACTIONS

Change each percent to a fraction.

**(a)** $38\% = 38 \times \frac{1}{100} = \frac{38}{100} = \frac{19 \cdot 2}{50 \cdot 2} = \frac{19}{50}$   Replace % with $\left( \times \frac{1}{100} \right)$ and reduce the fraction

**(b)** $500\% = 500 \times \frac{1}{100} = \frac{500}{100} = \frac{5 \cdot 100}{1 \cdot 100} = \frac{5}{1} = 5$

**(c)** $0.3\% = 0.3 \div 100 = \frac{0.3}{100} = \frac{(0.3)(10)}{(100)(10)}$   Multiply by $\frac{10}{10}$ to get a whole number in the numerator; this is the same as moving the decimal point one place to the right in both the numerator and denominator

$$= \frac{3}{1000}$$

### PRACTICE EXERCISE 1

Change each percent to a fraction.

**(a)** 25%

**(b)** 300%

**(c)** 0.5%

**(d)** $2\frac{3}{4}\% = \left(2\frac{3}{4}\right)\left(\frac{1}{100}\right)$

$\qquad = \frac{11}{4} \cdot \frac{1}{100} = \frac{11}{400}$    Note: $2\frac{3}{4} = \frac{4 \cdot 2 + 3}{4} = \frac{11}{4}$

**(e)** $66\frac{2}{3}\% = \left(66\frac{2}{3}\right)\left(\frac{1}{100}\right)$

$\qquad = \left(\frac{200}{3}\right)\left(\frac{1}{100}\right)$    Note: $66\frac{2}{3} = \frac{3 \cdot 66 + 2}{3} = \frac{200}{3}$

$\qquad = \frac{2 \cdot \cancel{100}}{3 \cdot \cancel{100}} = \frac{2}{3}$

**(f)** $100\% = 100 \div 100 = \frac{100}{100} = 1$

**(d)** $3\frac{1}{5}\%$

**(e)** $33\frac{1}{3}\%$

**(f)** $1000\%$

Answers: (a) $\frac{1}{4}$  (b) $3$  (c) $\frac{1}{200}$
(d) $\frac{4}{125}$  (e) $\frac{1}{3}$  (f) $10$

---

## ❸ CHANGING A PERCENT TO A DECIMAL

Percents are frequently given as decimals. Changing to a decimal is exactly the same as changing to a fraction; simply remove the % symbol and multiply by $\frac{1}{100}$, which is 0.01 in decimal form.

### To Change a Percent to a Decimal

Replace the % symbol with ($\times$ 0.01) and multiply.

**HINT**

Multiplying a number by 0.01 is the same as moving the decimal point two places to the left. Thus, to change 5% to a decimal write 0.05.

| EXAMPLE 2  CHANGING PERCENTS TO DECIMALS | PRACTICE EXERCISE 2 |

Change each percent to a decimal.

**(a)** $29\% = 29 \times 0.01$   Remember that 29. is the same as 29, then move
$\qquad = 0.29. = 0.29$   the decimal point two places to the left

**(b)** $39.5\% = 39.5 \times 0.01$   Move the decimal point two places to the left
$\qquad = 0.39.5 = 0.395$

**(c)** $0.7\% = (0.7)(0.01)$

$\qquad = 0.00.7 = 0.007$   Put two zeros before 7

**(d)** $823\% = (823)(0.01) = 8.23. = 8.23$

Change each percent to a decimal.

**(a)** 65%

**(b)** 26.5%

**(c)** 0.3%

**(d)** 550%

**(e)** $2\frac{3}{4}\% = 2.75\%$ $\qquad \frac{3}{4} = 0.75$

$\qquad\qquad = (2.75)(0.01)$

$\qquad\qquad = 0.\underset{\smile}{02}.75 = 0.0275$

**(f)** $0.01\% = (0.01)(0.01) = 0.\underset{\smile}{00}.01 = 0.0001$

**(e)** $7\frac{3}{5}\%$

**(f)** $1\%$

Answers: (a) **0.65** (b) **0.265**
(c) **0.003** (d) **5.5** (e) **0.076**
(f) **0.01**

---

## HINT

Remember that the % symbol means "divide by 100 or multiply by $\frac{1}{100} = 0.01$." Thus, after the % symbol is removed and the change to decimal notation is made, you will have a smaller number. It may help if you keep in mind a simple example, such as

$$50\% = 0.5 \qquad \text{or} \qquad 50\% \text{ means } \frac{1}{2}.$$

---

## ❹ CHANGING A DECIMAL TO A PERCENT

Some practical problems require changing from percent to fractions and decimals. Others require converting from a decimal or a fraction to a percent. To change a decimal to a percent we multiply by 100, the reverse of dividing by 100, which was used to change a percent to a decimal.

> ### To Change a Decimal to a Percent
> Multiply the decimal by 100 and attach the % symbol.

---

## HINT

Multiplying a number by 100 is the same as moving the decimal point two places to the right. Thus, to change a decimal to a percent, move the decimal point two places to the right and attach the % symbol.

---

| **EXAMPLE 3** CHANGING DECIMALS TO PERCENTS | **PRACTICE EXERCISE 3** |

Change each decimal to a percent.

**(a)** $0.31 = (0.31)(100)\% = \underset{\smile}{31}.\% = 31\%$ $\qquad$ Move the decimal point two places to the right

**(b)** $3.25 = (3.25)(100)\% = \underset{\smile}{325}.\% = 325\%$

**(c)** $1 = (1)(100)\% = \underset{\smile}{100}.\% = 100\%$ $\qquad$ Attach two zeros when moving the decimal point

**(d)** $0.01 = (0.01)(100)\% = \underset{\smile}{01}.\% = 1\%$

**(e)** $0.007 = (0.007)(100)\% = \underset{\smile}{00}.7\% = 0.7\%$

**(f)** $56.2 = (56.2)(100)\% = \underset{\smile}{5620}.\% = 5620\%$

Change each decimal to a percent.

**(a)** 0.62

**(b)** 2.75

**(c)** 2

**(d)** 0.02

**(e)** 0.002

**(f)** 35.3

Answers: (a) **62%** (b) **275%**
(c) **200%** (d) **2%** (e) **0.2%**
(f) **3530%**

## HINT

To keep from moving the decimal point the wrong way, keep in mind a simple example, such as $0.5 = 50\%$.

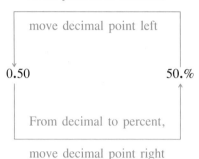

From percent to decimal,

move decimal point left

0.50                    50.%

From decimal to percent,

move decimal point right

Consider the first letter of the words **D**ecimal and **P**ercent. Since **D** comes before **P** in the alphabet, the letters **DP** can remind us that to change from **D**ecimal to **P**ercent, the decimal point is moved from left to right while to change from **P**ercent to **D**ecimal, it is moved from right to left.

## 5 CHANGING A FRACTION TO A PERCENT

We change fractions to percents by multiplying the fraction by 100 and attaching the % symbol, just as was done with decimals.

| To Change a Fraction to a Percent |
|---|
| Multiply the fraction by 100 and attach the % symbol. |

*or change to a decimal → change decimal to a %*

| **EXAMPLE 4**   CHANGING FRACTIONS TO PERCENTS | **PRACTICE EXERCISE 4** |
|---|---|

Change each fraction to a percent.

**(a)** $\dfrac{4}{5} = \dfrac{4}{5} \times 100\%$

$= \dfrac{4 \times \overset{20}{\cancel{100}}}{\cancel{5}}\%$

$= (4 \times 20)\% = 80\%$

**(b)** $\dfrac{1}{8} = \dfrac{1}{8} \times 100\% = \dfrac{100}{8}\%$

$= 12.5\%$     Divide 100 by 8

This could also be written as $12\frac{1}{2}\%$.

**(c)** $\dfrac{2}{3} = \dfrac{2}{3} \times 100\% = \dfrac{200}{3}\%$

$= 66.\overline{6}\%$     Divide 200 by 3

Since $0.\overline{6} = \frac{2}{3}$, we could also write this as $66\frac{2}{3}\%$.

Change each fraction to a percent.

**(a)** $\dfrac{3}{10}$

**(b)** $\dfrac{5}{8}$

**(c)** $\dfrac{1}{3}$

Answers: (a) 30%   (b) 62.5% or $62\frac{1}{2}\%$   (c) $33.\overline{3}\%$ or $33\frac{1}{3}\%$

In Example 4(**c**) we could have rounded the percent to 66.7% (correct to the nearest tenth of a percent) or to 66.67% (correct to the nearest hundredth of a percent) in a practical situation.

| **EXAMPLE 5** FINDING APPROXIMATE PERCENTS | **PRACTICE EXERCISE 5** |

Change $\frac{2}{7}$ to an approximate percent by rounding to the nearest percent, nearest tenth of a percent, and nearest hundredth of a percent.

Change $\frac{5}{7}$ to an approximate percent by rounding to the nearest percent, nearest tenth of a percent, and nearest hundredth of a percent.

$$
\begin{array}{r}
.28571 \\
7\overline{)2.00000} \quad \text{Divide 2 by 7}\\
\underline{1\;4}\\
60\\
\underline{56}\\
40\\
\underline{35}\\
50\\
\underline{49}\\
10\\
\underline{7}\\
3
\end{array}
$$

$\dfrac{2}{7} \approx 0.29 = 29\%$    To the nearest percent

$\dfrac{2}{7} \approx 0.286 = 28.6\%$    To the nearest tenth of a percent

$\dfrac{2}{7} \approx 0.2857 = 28.57\%$    To the nearest hundredth of a percent

Answer: 71%, 71.4%; 71.43%

Some fractions, decimals, and percents which occur often are listed in the table. The numbers on each line are equal. You should memorize this table for use in the exercises.

| Fraction | Decimal | Percent | Fraction | Decimal | Percent |
|---|---|---|---|---|---|
| $\frac{1}{20}$ | 0.05 | 5% | $\frac{1}{2}$ | 0.5 | 50% |
| $\frac{1}{10}$ | 0.1 | 10% | $\frac{3}{5}$ | 0.6 | 60% |
| $\frac{1}{8}$ | 0.125 | 12.5% | $\frac{5}{8}$ | 0.625 | 62.5% |
| $\frac{1}{6}$ | $0.1\overline{6}$ | $16.\overline{6}\% = 16\frac{2}{3}\%$ | $\frac{2}{3}$ | $0.\overline{6}$ | $66.\overline{6}\% = 66\frac{2}{3}\%$ |
| $\frac{1}{5}$ | 0.2 | 20% | $\frac{7}{10}$ | 0.7 | 70% |
| $\frac{1}{4}$ | 0.25 | 25% | $\frac{3}{4}$ | 0.75 | 75% |
| $\frac{3}{10}$ | 0.3 | 30% | $\frac{4}{5}$ | 0.8 | 80% |
| $\frac{1}{3}$ | $0.\overline{3}$ | $33.\overline{3}\% = 33\frac{1}{3}\%$ | $\frac{5}{6}$ | $0.8\overline{3}$ | $83.\overline{3}\% = 83\frac{1}{3}\%$ |
| $\frac{3}{8}$ | 0.375 | 37.5% | $\frac{7}{8}$ | 0.875 | 87.5% |
| $\frac{2}{5}$ | 0.4 | 40% | $\frac{9}{10}$ | 0.9 | 90% |
| | | | 1 | 1.0 | 100% |

**Table of Equal Fractions, Decimals, and Percents**

## 6.2  EXERCISES A

**1.** The word *percent* means "per _____."

**2.** To change a percent to a fraction drop the % symbol and divide by 100 or multiply by _____.

**3.** To change a percent to a decimal replace the % symbol with ($\times$ 0.01) and multiply. This is the same as moving the decimal point two places to the _____.

*Change each percent to a fraction.*

**4.** 87%  **5.** 16%  **6.** 50%  **7.** 1%

**8.** 117%  **9.** 125%  **10.** 1000%  **11.** $\frac{1}{4}$%

**12.** 0.7%  **13.** 12.5%  **14.** $16\frac{2}{3}$%  **15.** 0.05%

*Change each percent to a decimal.*

**16.** 92%  **17.** 60%  **18.** 1%  **19.** 145%

**20.** 1000%  **21.** 0.1%  **22.** 37.35%  **23.** 200%

**24.** $\frac{1}{4}$%  **25.** 800%  **26.** 4392.5%  **27.** $6\frac{3}{4}$%

**28.** In the final two games of the season, a quarterback completed 58 passes in 100 attempts.
   **(a)** What fractional part of his attempts did he complete?
   **(b)** Write the fraction in **(a)** as a percent.
   **(c)** Write the percent in **(b)** as a decimal.

**29.** On a recent placement exam, 45 students out of 100 who took the exam placed in algebra.
   **(a)** What fractional part of the students placed in algebra?
   **(b)** Write the fraction in **(a)** as a percent.
   **(c)** Write the percent in **(b)** as a decimal.

**30.** Of all college basketball players, fewer than 0.8% will play in the NBA. Write 0.8% as a decimal.

*Change each decimal to a percent.*

**31.** 0.07  **32.** 0.375  **33.** 3.75

**34.** 37.5  **35.** 375  **36.** 0.009

**37.** 0.0009  **38.** 0.32  **39.** 0.032

*Change each fraction to a percent.*

**40.** $\dfrac{13}{10}$        **41.** $\dfrac{7}{4}$        **42.** $\dfrac{5}{3}$

**43.** $\dfrac{7}{50}$        **44.** $\dfrac{3}{25}$        **45.** $\dfrac{1}{1000}$

**46** $\dfrac{100}{3}$        **47.** $\dfrac{7}{2}$        **48.** $\dfrac{106}{200}$

*Find the approximate percent to the nearest tenth of a percent.*

**49** $\dfrac{4}{7}$        **50.** $\dfrac{2}{9}$        **51.** $\dfrac{15}{13}$

*Complete the table given one value in each row.*

| | Fraction | Decimal | Percent | | Fraction | Decimal | Percent |
|---|---|---|---|---|---|---|---|
| **52.** | $\dfrac{3}{10}$ | —— | —— | **53.** | —— | 0.6 | —— |
| **54.** | —— | —— | 62.5% | **55.** | 1 | —— | —— |
| **56.** | —— | 0.875 | —— | **57.** | —— | —— | $66\dfrac{2}{3}\%$ |
| **58.** | $\dfrac{1}{20}$ | —— | —— | **59.** | —— | $0.\overline{3}$ | —— |
| **60.** | —— | —— | $83\dfrac{1}{3}\%$ | **61.** | $\dfrac{1}{2}$ | —— | —— |
| **62.** | —— | $0.1\overline{6}$ | —— | **63.** | —— | —— | 12.5% |

*Give the answers to the following exercises as a fraction, as a decimal, and as a percent.*

**64.** Of 30 students in an algebra class, 18 are girls. What portion of the students is girls?

**65.** What portion of the students in Exercise 64 is boys?

**66.** A chemist mixed 12 parts of acid with 13 parts of distilled water to obtain a particular solution. What part of the solution is acid?

**67.** What part of the solution in Exercise 66 is distilled water?

## FOR REVIEW

**68.** Write the fraction form for the ratio 7 to 8.

**69.** Write the ratio that is indicated by $\frac{12}{11}$.

*Solve the proportions.*

**70.** $\dfrac{12}{a} = \dfrac{6}{2}$

**71.** $\dfrac{7}{1} = \dfrac{1}{a}$

*Solve.*

**72.** If it takes 8 pounds of fertilizer to cover 3000 square feet of lawn, how many pounds will be necessary to cover 7500 square feet?

**73.** If it takes 8 pounds of fertilizer to cover 300 square feet of lawn, how many square feet can be covered by 35 pounds of fertilizer?

*The following exercises review topics covered in Section 3.4 to help you prepare for the next section. Solve.*

**74.** If one-half of a number is 225, find the number.

**75.** After driving 720 miles, the Blows had completed $\frac{3}{5}$ of their trip. How long was their trip?

**ANSWERS:** 1. hundred  2. $\frac{1}{100}$  3. left  4. $\frac{87}{100}$  5. $\frac{4}{25}$  6. $\frac{1}{2}$  7. $\frac{1}{100}$  8. $\frac{117}{100}$  9. $\frac{5}{4}$  10. 10  11. $\frac{1}{400}$  12. $\frac{7}{1000}$  13. $\frac{1}{8}$
14. $\frac{1}{6}$  15. $\frac{1}{2000}$  16. 0.92  17. 0.6  18. 0.01  19. 1.45  20. 10  21. 0.001  22. 0.3735  23. 2  24. 0.0025  25. 8
26. 43.925  27. 0.0675  28. (a) $\frac{58}{100}$ (or $\frac{29}{50}$, reduced to lowest terms) (b) 58% (c) 0.58  29. (a) $\frac{45}{100}$ (or $\frac{9}{20}$, reduced to lowest terms) (b) 45% (c) 0.45  30. 0.008  31. 7%  32. 37.5%  33. 375%  34. 3750%  35. 37,500%  36. 0.9%
37. 0.09%  38. 32%  39. 3.2%  40. 130%  41. 175%  42. $166\frac{2}{3}\%$  43. 14%  44. 12%  45. 0.1%  46. $3333\frac{1}{3}\%$
47. 350%  48. 53%  49. 57.1%  50. 22.2%  51. 115.4%  52. 0.3; 30%  53. $\frac{3}{5}$; 60%  54. $\frac{5}{8}$; 0.625  55. 1.0;
100%  56. $\frac{7}{8}$; 87.5%  57. $\frac{2}{3}$; 0.$\overline{6}$  58. 0.05; 5%  59. $\frac{1}{3}$; $33\frac{1}{3}\%$  60. $\frac{5}{6}$; 0.8$\overline{3}$  61. 0.5; 50%  62. $\frac{1}{6}$; $16\frac{2}{3}\%$  63. $\frac{1}{8}$;
0.125  64. $\frac{18}{30}$ or $\frac{3}{5}$ (reduced); 0.6; 60%  65. $\frac{12}{30}$ or $\frac{2}{5}$ (reduced); 0.4; 40%  66. $\frac{12}{25}$; 0.48; 48%  67. $\frac{13}{25}$; 0.52; 52%
68. $\frac{7}{8}$  69. 12 to 11  70. 4  71. $\frac{1}{7}$  72. 20 pounds  73. 1312.5 square feet  74. 450  75. 1200 miles

## 6.2 EXERCISES B

**1.** The word which means "per hundred" is _____.

**2.** To change a percent to a decimal, remove the % symbol and move the decimal point to the left _____ places.

**3.** To change a percent to a fraction multiply by 0.01 or divide by _____.

*Change each percent to a fraction.*

**4.** 98%

**5.** 22%

**6.** 20%

**7.** 6%

**8.** 110%

**9.** 375%

**10.** 2000%

**11.** $\dfrac{3}{4}\%$

**12.** 0.2%

**13.** 23.6%

**14.** $83\dfrac{1}{3}\%$

**15.** 0.15%

*Change each percent to a decimal.*

**16.** 73%          **17.** 80%          **18.** 8%          **19.** 235%

**20.** 5000%          **21.** 0.8%          **22.** 62.45%          **23.** 700%

**24.** $\frac{3}{4}$%          **25.** 100%          **26.** 7325.4%          **27.** $10\frac{2}{5}$%

**28.** In a sample of 100 radios, 2 were found to be defective.
   **(a)** What fractional part of the radios was defective?
   **(b)** Write the fraction in **(a)** as a percent.
   **(c)** Write the percent in **(b)** as a decimal.

**29.** Of 100 students in Weitzel Elementary School, 25 were found to have at least one cavity in a tooth.
   **(a)** What fractional part of the students had a cavity?
   **(b)** Write the fraction in **(a)** as a percent.
   **(c)** Write the percent in **(b)** as a decimal.

**30.** During the month of June, of the cars sold at Tyrrell Chevrolet-Buick, 42.5% were Buicks. Write 42.5% as a decimal.

*Change each decimal to a percent.*

**31.** 0.09          **32.** 0.435          **33.** 4.35

**34.** 43.5          **35.** 435          **36.** 0.008

**37.** 0.0008          **38.** 0.62          **39.** 0.062

*Change each fraction to a percent.*

**40.** $\frac{25}{10}$          **41.** $\frac{11}{4}$          **42.** $\frac{4}{3}$

**43.** $\frac{11}{50}$          **44.** $\frac{4}{25}$          **45.** $\frac{1}{2000}$

**46.** $\frac{400}{3}$          **47.** $\frac{13}{2}$          **48.** $\frac{48}{200}$

*Find the approximate percent to the nearest tenth of a percent.*

**49.** $\frac{6}{7}$          **50.** $\frac{5}{9}$          **51.** $\frac{17}{13}$

*Complete the table given one value in each row.*

| | Fraction | Decimal | Percent | | Fraction | Decimal | Percent |
|---|---|---|---|---|---|---|---|
| **52.** | $\frac{3}{4}$ | _____ | _____ | **53.** | _____ | 0.2 | _____ |
| **54.** | _____ | _____ | 90% | **55.** | $\frac{1}{10}$ | _____ | _____ |

| | Fraction | Decimal | Percent | | Fraction | Decimal | Percent |
|---|---|---|---|---|---|---|---|
| **56.** | ___ | 0.8 | ___ | **57.** | ___ | ___ | 25% |
| **58.** | $\frac{1}{8}$ | ___ | ___ | **59.** | ___ | 0.7 | ___ |
| **60.** | ___ | ___ | 37.5% | **61.** | $\frac{5}{6}$ | ___ | ___ |
| **62.** | ___ | 0.4 | ___ | **63.** | ___ | ___ | $33\frac{1}{3}\%$ |

*Give the answers to the following exercises as a fraction, as a decimal, and as a percent.*

**64.** On a 200-mile trip, Velma drove 140 miles and her friend Ingrid drove 60 miles. What portion of the trip did Velma drive?

**65.** What portion of the trip in Exercise 64 did Ingrid drive?

**66.** Of the 80 classes taught by the mathematics department at a community college, 55 were taught during the day and the rest were taught in the evening. What part of the classes was taught during the day?

**67.** What part of the classes in Exercise 66 was taught in the evening?

## FOR REVIEW

**68.** Write the fraction form for the ratio 4 to 1.

**69.** Write the ratio which is indicated by $\frac{10}{10}$.

*Solve the proportions.*

**70.** $\dfrac{35}{a} = \dfrac{7}{1}$

**71.** $\dfrac{1}{12} = \dfrac{a}{1}$

*Solve.*

**72.** If it takes 9 ribbons to type 3500 pages of a manuscript, how many ribbons are needed to type 700 pages?

**73.** If it takes 9 ribbons to type 3500 pages of manuscript, how many pages can be typed using 3 ribbons?

*The following exercises review topics covered in Section 3.4 to help you prepare for the next section. Solve.*

**74.** If one-fourth of a number is 110, find the number.

**75.** After working for 7 hours, Sarah had completed $\frac{2}{5}$ of a job. How long will it take to do the whole job?

## 6.2 EXERCISES C

**1.** The price of a coat is reduced 35%.

   **(a)** What fractional part of the original price amounts to the reduction?

   **(b)** What fractional part of the original price is the new sale price?

   **(c)** What percent of the original price must a buyer pay?

   $\left[\text{Answer: } \textbf{(a) } \frac{7}{20} \textbf{ (b) } \frac{13}{20} \textbf{ (c) } 65\%\right]$

**2.** The sales-tax rate in Washburn, Indiana, is 3.5%. Write 3.5% as a decimal and as a fraction.

**3.** Each employee in the State Retirement System must contribute 6.35% of his salary to the retirement fund. Write 6.35% as a decimal and as a fraction.

*Use a calculator to write each fraction as a percent rounded to the nearest hundredth of a percent.*

**4.** $\dfrac{23}{41}$          **5.** $\dfrac{7}{101}$          **6.** $\dfrac{123}{477}$          **7.** $\dfrac{853}{329}$

[Answer: 25.79%]

## 6.3  PERCENT PROBLEMS

### STUDENT GUIDEPOSTS

**1** The Basic Percent Problem
**2** Solving Percent Problems (Method 1)
**3** Using Proportions to Solve Percent Problems (Method 2)
**4** Applications of Percent

### 1 THE BASIC PERCENT PROBLEM

In Section 3.4 we solved problems of the type:

> On a two-day trip of 440 miles, Rob drove $\frac{1}{2}$ the distance the first day. How far did he drive the first day?

Letting ☐ represent the distance driven the first day, and recalling that the word "of" translates to multiplication, and the word "is" translates to equals,

$$\boxed{\text{distance 1st day}} \quad \text{is} \quad \frac{1}{2} \quad \text{of} \quad 440$$
$$\downarrow \qquad\qquad \downarrow \quad \downarrow \quad \downarrow \quad \downarrow$$
$$\square \qquad = \left(\frac{1}{2}\right) \cdot (440).$$

Thus,

$$\square = \frac{1}{2} \cdot 440 = 220 \text{ miles.}$$

Using the notion of percent, the same problem might be stated as follows:

> On a two-day trip of 440 miles, Rob drove 50% of the distance the first day. How far did he drive the first day?

By changing 50% to the fraction $\frac{1}{2}$, this problem can be solved as above. Similarly, 50% could be changed to the decimal 0.5.

$$\boxed{\text{distance 1st day}} \quad \text{is} \quad 50\% \quad \text{of} \quad 440$$
$$\downarrow \qquad\qquad \downarrow \quad \downarrow \quad \downarrow$$
$$\square \qquad = (0.5) \cdot (440)$$

Thus,

$$\square = (0.5)(440) = 220 \text{ miles.}$$

This problem is an example of the basic percent problem that takes the form

$$A \text{ is } P\% \text{ of } B \qquad \text{or} \qquad P\% \text{ of } B \text{ is } A.$$

When we solve a percent problem we must find one of $A$, $P$, or $B$, given the other two. In a percent problem, we can identify $A$, $P$, and $B$ using the following:

> $A$ is the *amount* and it is next to the word *is*,
> $P$ is the *percent* and is followed by the % symbol,
> $B$ is the *base* and it follows the word *of*.

For example, consider the following statement.

$$\begin{array}{cccccc} 48 & \text{is} & 25\% & \text{of} & 192 \\ \downarrow & \downarrow & \downarrow\;\downarrow & \downarrow & \downarrow \\ A & = & P\;\% & \cdot & B \end{array}$$

This equation is called the **basic percent equation** and will be used in solving percent problems.

## ② SOLVING PERCENT PROBLEMS (METHOD 1)

We now illustrate the first method for solving a basic percent problem, "$A$ is $P\%$ of $B$," by finding the amount $A$ when the percent $P$ and base $B$ are given.

| EXAMPLE 1   SOLVING FOR THE AMOUNT $A$ |
| --- |

Solve by using the basic percent equation $A = P\% \cdot B$.

**(a)** What is 5% of $22.40?

In this case, $A$ is the unknown, $P = 5$, and $B = 22.40$. Thus we have

$$\begin{array}{ccccc} \boxed{\text{What number}} & \text{is} & 5\% & \text{of} & \boxed{22.40}\;? \\ \downarrow & & \downarrow & \downarrow & \downarrow \\ A & = & P\% & \cdot & B \\ \downarrow & & \downarrow & & \downarrow \\ A & = & (0.05) & \cdot & 22.40 \end{array}$$

Notice that *is* translates to $=$, *of* translates to $\cdot$ (times or multiplication), and 5% has been changed to decimal form, 0.05. To find $A$ we simply multiply.

$$A = (0.05) \cdot (22.40) = 1.12$$

Thus, $A$ is $1.12, that is, $1.12 is 5% of $22.40.

**(b)** 120% of 450 is what number?

This translates as follows:

$$\begin{array}{ccccc} \boxed{120\,\%} & \text{of} & \boxed{450} & \text{is} & \boxed{\text{what number}}\;? \\ \downarrow & \downarrow & \downarrow & \downarrow & \downarrow \\ P\% & \cdot & B & = & A \\ \downarrow & & \downarrow & & \downarrow \\ 1.2 & \cdot & 450 & = & A \\ & & 540 & = & A \qquad \text{\small Multiply 1.2 by 450} \end{array}$$

Thus, $A$ is 540, that is, 120% of 450 is 540.

| PRACTICE EXERCISE 1 |
| --- |

Solve.

**(a)** What is 10% of $120.20?

**(b)** 210% of 70 is what number?

Answers: (a) $12.02   (b) 147

---

╱╱╱╱╱╱╱╱╱╱╱╱╱ **CAUTION** ╱╱╱╱╱╱╱╱╱╱╱╱╱

In any percent problem, $P\%$ must always be changed to a decimal or fraction form in order to carry out the necessary calculations. When you remove the % symbol, be sure to multiply by 0.01. For instance, in Example 1(a) it would have been wrong to write

$$A = (5) \cdot (22.40).$$

╱╱╱╱╱╱╱╱╱

## HINT

Part (a) of Example 1 shows that the percent of a number is smaller than the number when the percent is less than 100%, while part (b) shows that the percent of a number is larger than the number when the percent is greater than 100%. These facts can help us determine whether an answer to a problem is reasonable or not.

Solving for $P$ or $B$ in a percent problem is a bit more challenging than solving for $A$ since in both cases the related division sentence must be used. Suppose now that $B$ is the unknown. For example, consider

8 is 40% of what number?

This sentence can be translated to the basic percent equation $A = P\% \cdot B$ just like we did before.

$$
\begin{array}{ccccc}
\text{8} & \text{is} & \text{40\%} & \text{of} & \boxed{\text{what number}}\,? \\
\downarrow & \downarrow & \downarrow & \downarrow & \downarrow \\
A & = & P\% & \cdot & B \\
\downarrow & & \downarrow & & \downarrow \\
8 & = & 0.40 & \cdot & B
\end{array}
$$

$A = 8$ and $P\% = 40\%$, which becomes 0.4

We can find $B$ by writing the related division sentence

$$\frac{8}{0.4} = B. \quad \text{Divide 8 by the multiplier of } B, 0.4$$

Dividing 8 by 0.4 gives

$$B = 20. \quad \frac{8}{0.4} = 20$$

---

### EXAMPLE 2   SOLVING FOR THE BASE $B$

Solve by using the basic percent equation $A = P\% \cdot B$.

**(a)** 60 is 20% of what number?

In this case, $B$ is the unknown, $P = 20$, and $A = 60$. Thus, we have

$$
\begin{array}{ccccc}
\text{60} & \text{is} & \text{20\%} & \text{of} & \boxed{\text{what number}}\,? \\
\downarrow & \downarrow & \downarrow & \downarrow & \downarrow \\
A & = & P\% & \cdot & B \\
\downarrow & & \downarrow & & \downarrow \\
60 & = & (0.2) & \cdot & B
\end{array}
$$

Translate to the related division sentence and divide.

$$B = \frac{60}{0.2} = 300$$

Thus, 60 is 20% of 300. Does this seem reasonable?

**(b)** 2.5% of what number is 12.5?

Although the order of wording has been reversed, this is the same type of problem.

$$
\begin{array}{ccccc}
\text{2.5\%} & \text{of} & \boxed{\text{what number}} & \text{is} & \text{12.5}\,? \\
\downarrow & \downarrow & \downarrow & \downarrow & \downarrow \\
P\% & \cdot & B & = & A \\
\downarrow & & \downarrow & & \downarrow \\
(0.025) & \cdot & B & = & 12.5
\end{array}
$$

---

### PRACTICE EXERCISE 2

Solve.

**(a)** 392 is 70% of what number?

**(b)** 10.5% of what number is 89.25?

The related division sentence gives

$$B = \frac{12.5}{0.025} = 500.$$

Thus, 2.5% of 500 is 12.5. Is this reasonable?

Finally, we consider finding the percent $P\%$ when the amount $A$ and base $B$ are given. For example, suppose we are asked:

What percent of 30 is 6?

First we translate to the basic percent equation $A = P\% \cdot B$, or in this case $P\% \cdot B = A$.

$$\boxed{\text{What percent}} \ \text{of} \ 30 \ \text{is} \ 6 \ ?$$

$$P\% \qquad \cdot \quad B = A$$
$$P\% \qquad \cdot \quad 30 = 6$$

Since $P\%$ is $P(0.01)$ changed to a decimal, substituting we have:

$P(0.01) \cdot 30 = 6$     $P\%$ is $P(0.01)$

$P(0.3) = 6$     $(0.01) \cdot 30 = 0.3$

$P = \dfrac{6}{0.3}$     Translate to the related division sentence

$P = 20$     $\dfrac{6}{0.3} = 20$

Thus, $P\%$ is 20%, and 20% of 30 is 6.

---

| **EXAMPLE 3**   SOLVING FOR THE PERCENT $P\%$ | **PRACTICE EXERCISE 3** |

Solve by using the basic percent equation $P\% \cdot B = A$.

Solve.

**(a)** What percent of 300 is 4.5?
Find the following:

**(a)** What percent of 150 is 7.5?

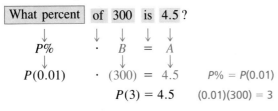

$$\boxed{\text{What percent}} \ \text{of} \ 300 \ \text{is} \ 4.5 \ ?$$
$$P\% \qquad \cdot \quad B = A$$
$$P(0.01) \quad \cdot \ (300) = 4.5 \qquad P\% = P(0.01)$$
$$P(3) = 4.5 \qquad (0.01)(300) = 3$$

Translate to the related division sentence.

$$P = \frac{4.5}{3}$$     Related division sentence

$$P = 1.5$$     $\dfrac{4.5}{3} = 1.5$

Thus, $P\%$ is 1.5%, and 1.5% of 300 is 4.5. Does this seem reasonable?

**(b)** 320 is what percent of 240?

Translate the following:

$$\begin{array}{ccccccc} \boxed{320} & \text{is} & \boxed{\text{what percent}} & \text{of} & \boxed{240}\,? \\ \downarrow & \downarrow & \downarrow & & \downarrow & \downarrow \\ A & = & P\% & \cdot & B \\ \downarrow & & \downarrow & & \downarrow \\ 320 & = & P(0.01) & \cdot & 240 \end{array}$$

$$320 = P(2.4) \qquad (0.01)(240) = 2.4$$

$$\frac{320}{2.4} = P \qquad \text{Related division sentence}$$

$$133.\overline{3} = P \qquad \text{Divide 320 by 2.4}$$

Thus, $P\%$ is $133.\overline{3}\%$ or $133\frac{1}{3}\%$. Since 320 was larger than 240, it is reasonable to assume that the desired percent should be greater than 100%.

**(b)** 280 is what percent of 112?

**Answers:** (a) 5%   (b) 250%

---

## ❸ USING PROPORTIONS TO SOLVE PERCENT PROBLEMS (METHOD 2)

We now consider a second method that can be used to solve percent problems. Since percent is the number of parts per hundred, a percent such as 60% can be written as a ratio in the following way:

$$\frac{60}{100} \quad \begin{array}{l} \leftarrow \text{Sixty parts} \\ \leftarrow \text{Per hundred} \end{array}$$

Also, the fraction $\frac{3}{5}$, changed to percent, is 60% (dividing 3 by 5 and converting to percent). We may express this fact by writing the proportion

$$\frac{3}{5} = \frac{60}{100}.$$

Suppose we are asked to solve the following percent problem:

What is 60% of 5?

Since, as shown above,

$$\frac{3}{5} = \frac{60}{100},$$

we know that 3 is 60% of 5. If we did not know the answer, we could solve the proportion

$$\frac{A}{5} = \frac{60}{100}$$

for $A$, the amount, to obtain 3.

Now we ask another question:

3 is 60% of what number?

This can be written as the proportion

$$\frac{3}{B} = \frac{60}{100},$$

where $B$, the base, is now the number to be found. (Of course, in this case we know it is 5.)

Finally, consider the question:

What percent of 5 is 3?

In this case we have the proportion

$$\frac{3}{5} = \frac{P}{100}$$

where $P$, the percent, is what we are looking for (the 60 in 60%).

In general, combining what we have learned from these three percent questions, we have

$$\begin{array}{c} \text{Amount} \to A \\ \text{Base} \quad\;\; \to B \end{array} = \frac{P}{100}. \leftarrow \text{Percent}$$

We call this the **basic percent proportion.** Every percent problem can be stated using this proportion where either $A$, $B$, or $P$ is missing.

---

### HINT

When reading a percent problem, notice that the amount, $A$, is next to the word *is* and the base, $B$, is next to the word *of*. Thus, the percent proportion

$$\frac{A}{B} = \frac{P}{100}$$

can be remembered by writing

$$\frac{\text{is}}{\text{of}} = \frac{\%}{100}.$$

---

We now solve the percent problems in Examples 1, 2, and 3 using percent proportions.

---

**EXAMPLE 4** SOLVING FOR THE AMOUNT $A$

Solve by using the basic percent proportion.

**(a)** What is 5% of $22.40?

Start with the basic percent proportion

$$\frac{A}{B} = \frac{P}{100}$$

with $A$ the unknown, $B = 22.40$, and $P = 5$.

$$\frac{A}{22.40} = \frac{5}{100}$$

$100A = (5)(22.40)$     Equate cross products

$A = \dfrac{(5)(22.40)}{100}$     Divide by the multiplier of $A$, 100

$A = 1.12$     Simplify

Thus, $A$ is $1.12.

---

**PRACTICE EXERCISE 4**

Solve.

**(a)** What is 10% of $120.20?

**(b)** 120% of 450 is what number?

Substitute 120 for $P$ and 450 for $B$ in the percent proportion

$$\frac{A}{B} = \frac{P}{100}.$$

$$\frac{A}{450} = \frac{120}{100}$$

$100A = (120)(450)$     Equate cross products

$A = \dfrac{(120)(450)}{100}$     Divide by 100

$A = 540$     Simplify

Thus, 120% of 450 is 540.

**(b)** 210% of 70 is what number?

$x = 210\% \cdot 70$

$x = 147.00$

$147$

Answers: (a) $12.02    (b) 147

---

//////////// **CAUTION** ////////////

When using the percent proportion *do not* write the percent as a decimal or a fraction. This is a difference between the percent proportion and the percent equation.

//////////

---

**EXAMPLE 5** SOLVING FOR THE BASE $B$

Solve by using the basic percent proportion.

**(a)** 60 is 20% of what number?

Substitute 60 for $A$ and 20 for $P$ in the basic percent proportion

$$\frac{A}{B} = \frac{P}{100}.$$

$$\frac{60}{B} = \frac{20}{100}$$

$(60)(100) = (20)B$     Equate cross products

$\dfrac{(60)(100)}{20} = B$     Divide by 20

$300 = B$

Thus, 60 is 20% of 300.

**(b)** 2.5% of what number is 12.5?

Substitute 12.5 for $A$ and 2.5 for $P$ in the basic percent proportion.

$$\frac{12.5}{B} = \frac{2.5}{100}$$

$(12.5)(100) = (2.5)B$     Equate cross products

$\dfrac{(12.5)(100)}{2.5} = B$     Divide by 2.5

$500 = B$

Thus, 2.5% of 500 is 12.5.

**PRACTICE EXERCISE 5**

Solve.

**(a)** 392 is 70% of what number?

**(b)** 10.5% of what number is 89.25?

Answers: (a) 560    (b) 850

| **EXAMPLE 6   SOLVING FOR THE PERCENT P%** | **PRACTICE EXERCISE 6** |

Solve by using the basic percent proportion.

**(a)** What percent of 300 is 4.5?

Substitute 4.5 for $A$ and 300 for $B$ in the percent proportion

$$\frac{A}{B} = \frac{P}{100}.$$

$$\frac{4.5}{300} = \frac{P}{100}$$

$$(4.5)(100) = (300)P \qquad \text{Equate cross products}$$

$$\frac{(4.5)(100)}{300} = P \qquad \text{Divide by 300}$$

$$1.5 = P$$

Thus, 1.5% of 300 is 4.5.

**(b)** 320 is what percent of 240?

Substitute 320 for $A$ and 240 for $B$ in the percent proportion.

$$\frac{320}{240} = \frac{P}{100}$$

$$(320)(100) = (240)P \qquad \text{Equate cross products}$$

$$\frac{(320)(100)}{240} = P \qquad \text{Divide by 240}$$

$$133.\overline{3} = P$$

Thus, $P\%$ is $133.\overline{3}\%$ or $133\frac{1}{3}\%$.

**Solve.**

**(a)** What percent of 150 is 7.5?

**(b)** 280 is what percent of 112?

Answers:  (a)  5%    (b)  250%

---

### $\boxed{\textbf{H I N T}}$

Consider both the equation method (Method 1) and the proportion method (Method 2) for solving percent problems and use the one that you prefer or that your instructor recommends. In the examples that follow, we present both methods in a side-by-side approach to help you decide which to use.

---

### ❹ APPLICATIONS OF PERCENT

We conclude this section with three examples of typical percent applications.

| **EXAMPLE 7   A PERCENT APPLICATION IN CHEMISTRY** | **PRACTICE EXERCISE 7** |

If 14 g (grams) of pure sulfuric acid is mixed with water and the resulting solution is 28% acid, what is the total weight of the solution?

We know that 14 g is 28% of the solution. Thus, we must answer the question:

$$14 \text{ is } 28\% \text{ of what number?}$$

Minnie spends $120 a week to feed her family. If this is 24% of the total weekly family income, what is the weekly income of her family?

Method 1 (Equation Method)      Method 2 (Proportion Method)

$$A = P\% \cdot B$$
$$14 = (0.28)B$$
$$\frac{14}{0.28} = B$$
$$50 = B$$

$$\frac{A}{B} = \frac{P}{100}$$
$$\frac{14}{B} = \frac{28}{100}$$
$$(14)(100) = (28)B$$
$$\frac{(14)(100)}{28} = B$$
$$50 = B$$

Thus, 14 g is 28% of 50 g of the solution.

Answer: $500

---

| **EXAMPLE 8   A PERCENT APPLICATION IN FAMILY BUDGETING** | **PRACTICE EXERCISE 8** |

The income of the Ross family is $1250 per month. If they spend 22% of their income for food, how much do they pay for food each month?

We must find 22% of the income, $1250. Thus the question is:

What is 22% of $1250?

Method 1                Method 2

$$A = P\% \cdot B$$
$$A = (0.22)(1250)$$
$$A = 275$$

$$\frac{A}{B} = \frac{P}{100}$$
$$\frac{A}{1250} = \frac{22}{100}$$
$$100A = (22)(1250)$$
$$A = \frac{(22)(1250)}{100} = 275$$

Thus, $275 is 22% of $1250, and the family spends $275 for food each month. If we come up with $2750 for an answer, since $2750 is more than $1250 but 22% is less than 100%, we would have known we had made an error.

At a recent concert attended by 840 people, 45% of those in attendance were children. How many children were at the concert?

Answer: 378 children

---

| **EXAMPLE 9   AN APPLICATION OF PERCENT IN SPORTS** | **PRACTICE EXERCISE 9** |

A basketball player hit 16 out of 20 free throws in a game. What was her shooting percent?

The question to answer is

16 is what percent of 20?

Method 1                Method 2

$$A = P\% \cdot B$$
$$16 = (P\%)(20)$$
$$16 = P(0.01)(20)$$
$$16 = P(0.2)$$
$$\frac{16}{0.2} = P$$
$$80 = P$$

$$\frac{A}{B} = \frac{P}{100}$$
$$\frac{16}{20} = \frac{P}{100}$$
$$(16)(100) = (20)P$$
$$\frac{(16)(100)}{20} = P = 80$$

Thus, her shooting percent was 80%. That is, 16 is 80% of 20.

During the World Series, Reggie Jackson had 10 hits in 30 times at bat. What was his batting percentage in the series?

Answer: $33.\overline{3}\%$ (Batting percentages or averages are often expressed as three-place decimals so we might give 0.333 for the answer.)

## 6.3 EXERCISES A

*Solve the percent problems. Before checking, ask yourself if your answer seems reasonable.*

**1.** What is 20% of 150?

**2.** What is 140% of 20?

**3.** 70% of 600 is what number?

**4.** 10 is 20% of what number?

**5.** 200 is 40% of what number?

**6.** 4.5 is 150% of what number?

**7.** What percent of 50 is 10?

**8.** What percent of 10 is 50?

**9.** 75 is what percent of 225?

**10.** What is 35% of 70?

**11.** 6.2 is what percent of 24.8?

**12.** 0.15 is 60% of what number?

**13.** There were 1600 votes cast in an election. Mr. Lawrence received 62% of the votes. How many votes did he receive? [*Hint:* What is 62% of 1600?]

**14.** Flagstaff, Arizona, received 120 inches of snow one year. This was 150% of the normal snowfall. What is the normal annual snowfall in Flagstaff? [*Hint:* 120 is 150% of what number?]

$A = .62 \cdot 50$

**15** A baseball team won 62% of its games and played a total of 50 games. How many games did the team win?

**16.** In an acid solution weighing 120 g there are 35 g of acid. To the nearest percent, what is the percent of acid in the solution?

$2860 = .58 \cdot Bt$

**17.** Ms. McShane received 2860 of the votes in an election. This was about 58% of the votes cast. Approximately how many votes were cast?

**18** A student answered 22 questions correctly on a test having 30 questions. To the nearest percent, what percent of her answers were correct?

$A = .08 \cdot 3600$

**19.** What is the commission on a sale of $3600 if the commission rate is 8%?

**20.** If an account worth $1500 increased to $1800 in one year, what was the interest rate for the year?

$2640 = .12 \cdot Bt$

**21.** Wanda received an increase in salary of $2640. If this was a 12% raise, what was her original salary?

**22** Henry's weight decreased from 220 pounds to 187 pounds on a diet. What was the percent decrease in his weight?

## FOR REVIEW

$\dfrac{2860}{B} = \dfrac{58}{100}$

**23.** Change 35% to a fraction.

$.58B = 286000$

**24.** Change 13.5% to a decimal.

**25.** Tonya has deposited $2000 in a savings account which earns 8.75% interest. Write 8.75% as a decimal.

*Change to percents.*

**26.** 0.17

**27.** 0.005

**28.** 0.05

**29.** 0.5

**30.** $\dfrac{3}{8}$

**31.** $\dfrac{7}{9}$

**32.** In a two-person school board election, Dr. Yard received 2130 votes out of a total of 3360 votes cast.

(a) What fractional part of the votes did Dr. Yard receive?

(b) Express the fraction in (a) as a decimal rounded to the nearest thousandth.

(c) Express the decimal in (b) as a percent rounded to the nearest tenth of a percent.

(d) What percent of the votes (to the nearest tenth of a percent) did Dr. Yard's opponent receive?

ANSWERS: 1. 30  2. 28  3. 420  4. 50  5. 500  6. 3  7. 20%  8. 500%  9. $33\frac{1}{3}$%  10. 24.5  11. 25%
12. 0.25  13. 992 votes  14. 80 in  15. 31 games  16. 29%  17. 4931 votes  18. 73%  19. $288  20. 20%
21. $22,000  22. 15%  23. $\frac{7}{20}$  24. 0.135  25. 0.0875  26. 17%  27. 0.5%  28. 5%  29. 50%  30. 37.5%
31. 77.$\overline{7}$%  32. (a) $\frac{71}{112}$ (b) 0.634 (c) 63.4% (d) 36.6%

## 6.3 EXERCISES B

*Solve the percent problems. Before checking, ask yourself if your answer seems reasonable.*

**1.** What is 16% of 400?

**2.** What is 420% of 5?

**3.** 35% of 900 is what number?

**4.** 30 is 5% of what number?

**5.** 120 is 0.2% of what number?

**6.** 12.5 is 625% of what number?

**7.** What percent of 80 is 24?

**8.** What percent of 520 is 130?

**9.** 15 is what percent of 8?

**10.** 55% of 4000 is what number?

**11.** 0.6 is 12% of what number?

**12.** 9.8 is 280% of what number?

**13.** If there is a 0.05% impurity rate in a water sample of 820 grams, how many grams of impurities are in the sample?

**14.** During one year Phoenix had a total of 168 days when the temperature exceeded 100°. This was 120% of normal. Normally how many days per year does the temperature exceed 100° in Phoenix?

**15.** There were 10,850 votes cast for the two people in an election, and the winner received 62% of the total. How many votes did she receive?

**16.** If 6 liters of acid are mixed with 9 liters of water, what is the percent of acid in the solution?

**17.** Burford answered 66 questions correctly on a test and received a score of 44%. How many questions were on the test?

**18.** At a recent campus sold-out showing of the movie *Rambo First Blood Part II*, 485 members of the audience were under 30 years of age. If the theater seats 615, to the nearest tenth of a percent, what percent of those present was under 30 years old?

**19.** What is the commission rate if Rollie earns a commission of $102.24 on sales of $568?

**20.** From one year to the next, the price of ground beef went from 89¢ per pound to $1.09 per pound. What was the percent increase (to the nearest tenth of a percent)?

**21.** Patrick received a raise of 14%. What is his new salary if his former salary was $36,000?

**22.** A retailer bought a lamp for $35.00 and marked up the price 60%. What was the markup and what was the selling price?

## FOR REVIEW

**23.** Change 82% to a fraction.

**24.** Change $6\frac{1}{2}$% to a decimal.

**25.** Rod purchased a bike on sale at 30% off the regular price. Change 30% to a decimal.

*Change to percents.*

**26.** 0.31

**27.** 0.004

**28.** 0.04

**29.** 0.4                    **30.** $\dfrac{7}{8}$                    **31.** $\dfrac{1}{6}$

**32.** Of the 1092 hours of prime-time television programming in a recent year, it was estimated that 648 hours were viewed by children under the age of 16.

(a) What fractional part of the total prime-time hours was viewed by children?

(b) Express the fraction in (a) as a decimal rounded to the nearest thousandth.

(c) Express the decimal in (b) as a percent rounded to the nearest tenth.

(d) What percent of the total hours (to the nearest tenth of a percent) was not viewed by children?

## 6.3 EXERCISES C

*Use a calculator and give answers to the nearest hundredth or nearest hundredth of a percent.*

**1.** What is 12.7% of 257?
[Answer: 32.64]

**2.** 27 is 6.3% of what number?

**3.** What percent of 83 is 122?

**4.** During one year the Browns spent 10.3% of their income on entertainment and vacations. If they spent $3485.20 on these two items, to the nearest cent, what was their income?
[Answer: $33,836.89]

*Write each of the following as a percent proportion and solve. Give answers rounded to the nearest tenth or nearest tenth of a percent.*

**5.** 6.9 is what percent of 72.7?
[Answer: 9.5%]

**6.** 27.3 is 14.2% of what number?

**7.** What is 102.6% of 38.5?

## 6.4 TAX PROBLEMS

━━━━━━ **STUDENT GUIDEPOSTS** ━━━━━━

1 Sales Tax Problems                 3 Income Tax Problems
2 Social Security Tax Problems

In the next five sections we will study a variety of applications of percent all of which can be translated into the basic percent equation

$$A = P\% \cdot B$$

or the basic percent proportion

$$\frac{A}{B} = \frac{P}{100}.$$

We will give a dual presentation in many of the examples. Use the method you prefer.

### 1 SALES TAX PROBLEMS

In most states a tax is charged on purchases made in retail stores. The rate of this tax, called a **sales tax,** varies from location to location. In a state with a **sales-tax rate** of 5%, the tax on a purchase of $10.00 is given by

$$\text{sales tax} = \boxed{5\%} \text{ of } \boxed{\$10.00} \, . \qquad \text{Use } A = P\% \cdot B$$
$$= (0.05) \cdot (\$10.00)$$
$$= \$0.50$$

When you buy an item for $10.00, you must pay a

$$\text{total price} = \boxed{\text{price before tax}} + \boxed{\text{sales tax}} \, .$$
$$= \$10.00 + \$0.50$$
$$= \$10.50$$

We now present a summary of the necessary concepts related to a sales-tax problem.

---

### To Solve a Sales-Tax Problem

1. Use the basic percent equation

$$A = P\% \cdot B$$
$$\text{sales tax} = \text{tax rate } (0.01) \cdot \text{price before tax}$$

*or* use the basic percent proportion

$$\frac{A}{B} = \frac{P}{100}$$
$$\frac{\text{sales tax}}{\text{price before tax}} = \frac{\text{tax rate}}{100} \, .$$

2. The total price is given by

$$\text{total price} = \text{price before tax} + \text{sales tax.}$$

---

### HINT

In many sales-tax problems, a calculator can be useful since you are performing decimal calculations.

---

### EXAMPLE 1  FINDING SALES TAX AND TOTAL PRICE

If the sales-tax rate is 5% in Tucson, Arizona, what is the sales tax on a TV selling for $625.40? What is the total price?

Find the sales tax.

Method 1 (Equation Method)

$$A = P \% \cdot B$$
$$\text{sales tax} = \text{tax rate}(0.01) \cdot \text{price before tax}$$
$$= (0.05) \cdot (\$625.40)$$
$$= \$31.27$$

### PRACTICE EXERCISE 1

If the sales-tax rate is 4% in Blue Key, Florida, what is the sales tax on a pair of water skis selling for $315.50? What is the total price including tax?

Method 2 (Proportion Method)

$$\frac{A}{B} = \frac{P}{100}$$

$$\frac{\text{sales tax}}{\text{price before tax}} = \frac{\text{tax rate}}{100}$$

$$\frac{\text{sales tax}}{\$625.40} = \frac{5}{100}$$

$$(\text{sales tax})(100) = (5)(\$625.40)$$

$$\text{sales tax} = \frac{(5)(\$625.40)}{100}$$

$$= \$31.27$$

Now add the price before tax and the sales tax to obtain the total price.

$$\text{total price} = \text{price before tax} + \text{sales tax}$$
$$= \quad \$625.40 \quad + \quad \$31.27$$
$$= \$656.67$$

Thus, one must pay a total of $656.67 for the TV set.

///////////////// **CAUTION** ///////////////

Always check to see if your answer seems reasonable. A 5% tax is $5 for each $100. If we had found the tax to be $312.70, we would have realized that this was much too big. (The decimal point is too far to the right.) Similarly, we know that an answer of $3.13 is wrong because the decimal point is too far to the left.

/////////////

The sales-tax rate in a particular city can be found if we know the price before tax and the amount of sales tax paid.

| **EXAMPLE 2** FINDING THE SALES-TAX RATE | **PRACTICE EXERCISE 2** |
|---|---|

What is the sales-tax rate if the tax is $2.50 on the purchase of a $62.50 sport coat?

The question asked is: $2.50 is what percent of $62.50?

What is the sales-tax rate in Denver if the tax is $24.77 on the purchase of a desk with a selling price of $495.40?

Method 1

$$A = P\% \cdot B$$
$$\$2.50 = P\% \cdot \$62.50$$
$$2.50 = P(0.01)(62.50)$$
$$2.50 = P(0.625)$$
$$\frac{2.50}{0.625} = P$$
$$4 = P$$

Method 2

$$\frac{A}{B} = \frac{P}{100}$$
$$\frac{\$2.50}{\$62.50} = \frac{P}{100}$$
$$(2.50)(100) = (62.50)P$$
$$\frac{(2.50)(100)}{62.50} = P$$
$$4 = P$$

Thus, the sales-tax rate is 4%. Since

$$(0.04)(62.50) = 2.50, \; 4\% \text{ does check.}$$

Finally, when the sales-tax rate and the sales tax are given, we can find the price of an item before tax.

| EXAMPLE 3  FINDING THE PRICE OF AN ITEM | PRACTICE EXERCISE 3 |

A tax of $1.26 was charged on a purchase of cosmetics in a city where the tax rate is 3.5%.

**(a)** What was the price of the cosmetics before tax?

The question asked is: $1.26 is 3.5% of what number?

Method 1

$$A = P\% \cdot B$$
$$\$1.26 = (3.5)(0.01)B$$
$$1.26 = (0.035)B$$
$$\frac{1.26}{0.035} = B$$
$$36 = B$$

Method 2

$$\frac{A}{B} = \frac{P}{100}$$
$$\frac{\$1.26}{B} = \frac{3.5}{100}$$
$$(1.26)(100) = (3.5)B$$
$$\frac{(1.26)(100)}{3.5} = B = 36$$

The cosmetics sold for $36.00 before tax.

**(b)** What was the total price paid?

$$\text{total price} = \boxed{\text{price before tax}} + \boxed{\text{sales tax}}$$
$$= \quad \$36.00 \quad + \quad \$1.26 = \$37.26$$

A tax of $5.76 was charged on a purchase of a pair of boots in Dallas where the tax rate is 4.5%.

**(a)** What was the selling price (price before tax) of the boots?

**(b)** What was the total price paid?

Answers: (a) $128.00
(b) $133.76

## 2 SOCIAL SECURITY TAX PROBLEMS

The Social Security tax is used by the federal government to provide income for retired people. In 1989 the tax rate was 7.51% on the first $48,000 of earnings. A wage earner who made $48,000 or more in 1989 paid 7.51% of $48,000, which was

$$(0.0751)(\$48,000) = \$3604.80,$$

in Social Security taxes. A person who made less than $48,000 paid 7.51% of whatever was earned. We summarize this in the following rule.

---
### To Find the Amount of Social Security Tax (1989)

1. If total earnings are less than $48,000:
   Use the percent equation

$$A \quad = \quad P \quad \% \quad \cdot \quad B$$
$$\text{Social Security tax} = (7.51)(0.01) \cdot \text{total earnings}$$

   *or* use the percent proportion

$$\frac{A}{B} = \frac{P}{100}.$$
$$\frac{\text{Social Security tax}}{\text{total earnings}} = \frac{7.51}{100}$$

2. If total earnings are $48,000 or more, the Social Security tax is fixed at $3604.80.
---

| EXAMPLE 4   FINDING SOCIAL SECURITY TAX | PRACTICE EXERCISE 4 |

Clara earned $39,600 in 1989. How much Social Security tax did she pay?

Since Clara made less than $48,000, use Method 1 or Method 2.

Benito earned $22,480 in 1989. How much Social Security tax did he pay?

Method 1

$$A = P\% \cdot B$$

S. S. tax $= (7.51)(0.01) \cdot$ total earnings

$\quad\quad\quad = (7.51)(0.01) \cdot (\$39,600)$

$\quad\quad\quad = (0.0751)(\$39,600)$

$\quad\quad\quad = \$2973.96$

*Because*
*7% = .07*
*so*
*7.51% = .0751*

Method 2

$$\frac{A}{B} = \frac{P}{100}$$

$$\frac{\text{S. S. tax}}{\text{total earnings}} = \frac{7.51}{100}$$

$$\frac{\text{S. S. tax}}{\$39,600} = \frac{7.51}{100}$$

$$(\text{S. S. tax})(100) = (7.51)(\$39,600)$$

$$\text{S. S. tax} = \frac{(7.51)(\$39,600)}{100}$$

$$= \$2973.96$$

Thus, Clara had to pay $2973.96 in Social Security tax in 1989. Clearly a calculator is helpful in solving problems like this.

**Answer: $1688.25 (to the nearest cent)**

## ③ INCOME TAX PROBLEMS

Income tax is usually the biggest tax that a person pays each year. Many taxpayers find the tax by looking up their taxable income (total earnings less deductions) in a table and reading the tax. However, for some, determining federal or state income taxes requires use of percent.

The tax rate schedule for single taxpayers from the 1989 federal income tax forms is shown on the next page. (You can obtain other tables at your post office.) Suppose your taxable income was $19,580. Look down the two columns on the left until you find the two numbers between which your income falls. In this case, $19,580 is between $18,550 and $44,900. The tax is calculated by adding 28% of the difference between $19,580 and $18,550 to $2,782.50. First subtract $18,550 from $19,580.

$$\$19,580 - \$18,550 = \$1030$$

Take 28% of $1030,

$$(0.28)(\$1030) = \$288.40,$$

and add this amount to $2,782.50 to obtain the total income tax to be paid.

$$\$2,782.50 + \$288.40 = \$3070.90$$

Thus, $3070.90 is your income tax in 1989.

## 1989 Tax Rate Schedule

**Schedule X —Single**

| If line 5 is: Over— | But not over— | The tax is: | of the amount over— |
|---|---|---|---|
| $0 | $18,550 | ------- 15% | $0 |
| 18,550 | 44,900 | **$2,782.50 + 28%** | 18,550 |
| 44,900 | 93,130 | **10,160.50 + 33%** | 44,900 |
| 93,130 | ------- | Use Worksheet below to figure your tax. | |

*Handwritten:*
2,782.50 is 15% of $18,550
2,782.50 + 28% · (44,900 − 18,550) is $10,160.50

### EXAMPLE 5   FINDING INCOME TAX

Paul is a single taxpayer who uses the tax rate schedule. In 1989 his taxable income was $9800. How much income tax did he pay?

Paul's taxable income, $9800, was between $0 and $18,550. He paid 15% of his taxable income which was

$$(0.15)(\$9800) = \$1470.$$

### PRACTICE EXERCISE 5

Rose is a single taxpayer who used the tax rate schedule in 1989. If her taxable income was $46,000, how much income tax did she pay?

Answer: **$10,523.50**

## 6.4   EXERCISES A

*Solve the following problems.*

1. What is the sales tax on a price before tax of $200 if the tax rate is 5%? What is the total cost?

2. The price before tax of a dress is $62.80 and the tax rate is 5%. What is the tax and the total price?

3. Find the tax and total price of a shirt if the price before tax is $15.20, and the tax rate is 2.5%.

4. If $0.43 tax is charged on a price of $21.50, what is the tax rate?

5. The tax rate is 2.5% and the tax is $3.05. What is the price before tax?

6. If the tax on a coffee pot selling for $34.50 before tax is $0.52, what is the tax rate (to the nearest tenth of a percent)?

7. The tax rate in Hudson, New York, is 5% and the tax charged on an item is $12.50. What is its price before tax?

8. A used car lists for $6827.50. If the sales-tax rate is 4.5%, find the tax and the total price.

**9.** Susan had an income from wages of $15,800 in 1989. How much Social Security tax did she pay?

**10** Toni made $52,350 in 1989. What Social Security tax did she pay?

**11.** Mark was a single taxpayer with a taxable income of $24,200 in 1989. What income tax did he pay?

**12.** On total wages of $20,800 in 1989, Jim had a taxable income of $16,600. How much income tax did he pay on his taxable income? How much Social Security tax did he pay on his total wages?

## FOR REVIEW

*Solve the percent problems.*

**13.** What is 18% of 245?

**14.** 5.2 is 8% of what number?

**15.** 676 is what percent of 520?

**16.** In a recent poll, 52% of those contacted were female. If 182 females were in the poll, how many were contacted?

**17.** Maria answered 76 of 88 questions correctly on a test. What was her percent score to the nearest percent?

ANSWERS:   1. $10; $210   2. $3.14; $65.94   3. $0.38; $15.58   4. 2%   5. $122   6. 1.5%   7. $250   8. $307.24; $7134.74   9. $1186.58   10. $3604.80 (she only paid tax on the first $48,000)   11. $4364.50   12. $2490; $1562.08   13. 44.1   14. 65   15. 130%   16. 350   17. 86%

## 6.4  EXERCISES B

*Solve the following problems.*

**1.** What is the sales tax on a price before tax of $300 if the tax rate is 4%? What is the total cost?

**2.** If a TV sells for $620 (before tax) and the sales-tax rate is 5%, what is the tax? What is the total cost?

**3.** Find the tax and total price of a toaster if the price before tax is $22.40, and the tax rate is 3.5%?

**4.** If a tax of $18.90 is charged on a purchase of $525.00, what is the tax rate?

**5.** The sales-tax rate in Mountain, Colorado, is 4%. If the tax on a pair of ski boots was $3.58, what was the price before tax?

**6.** The tax on a sofa costing $685.50 before tax is $41.13. What is the tax rate?

**7.** The tax rate in Pacifica, California, is 6%, and the tax charged on an item is $33.63. What is the price before tax of the item?

**8.** A new boat lists for $17,435.75. If the sales-tax rate is 5.5%, find the tax and the total cost of the boat.

**9.** The Social Security tax rate was 7.51% in 1989. How much Social Security tax did Claude pay on a salary of $18,200?

**10.** Margie had an income of $54,650 in 1989. How much Social Security tax did she pay?

**11.** Ruben was a single taxpayer with a taxable income of $33,200 in 1989. How much income tax did he pay?

**12.** On total wages of $31,200 in 1989, Kent had a taxable income of $23,400. How much income tax did he pay on his taxable income? How much Social Security tax did he pay?

**FOR REVIEW**

*Solve the percent problems.*

**13.** What is 120% of 65?

**14.** 40.8 is 12% of what number?

**15.** 48 is what percent of 32?

**16.** It has been estimated that 18% of the residents of a county cannot read or write. If there are 26,580 people living in the county, about how many of them cannot read or write?

**17.** Because of illness, Troy could only work 27 hours of his normal 40-hour work week. What percent of the normal week did he work?

## 6.4 EXERCISES C

*Solve.*

**1.** A washer normally selling for $389.95 is put on sale at 20% off the regular price. If the sales-tax rate is 4.5%, to the nearest cent, how much will the washer cost including the sales tax?

**2.** Paul buys a typewriter for $459.95 and a desk for $189.98. If the sales tax on the purchase of these items is $26.65, to the nearest tenth of a percent, what is the sales-tax rate?
[Answer: 4.1%]

## 6.5 COMMISSION AND DISCOUNT PROBLEMS

### STUDENT GUIDEPOSTS

**1** Commission Problems    **2** Discount Problems

We next look at two quantities which, like sales tax, are calculated by taking a percent of the selling price.

### 1 COMMISSION PROBLEMS

Some salespersons receive all or part of their income as a percent of their sales. This is called a **commission.** The **commission rate** is the percent of the sales that the person receives.

| To Solve a Commission Problem |
|---|
| Use the percent equation |

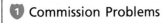 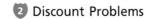

$$A \quad = \quad P \quad \% \cdot \quad B$$

commission = commission rate(0.01) · total sales

*or* use the percent proportion

$$\frac{A}{B} = \frac{P}{100}$$

$$\frac{\text{commission}}{\text{total sales}} = \frac{\text{commission rate}}{100}.$$

---

| **EXAMPLE 1**  FINDING A COMMISSION |
| --- |

Elaine sells furniture and receives a 20% commission on the selling price. In April, her sales totaled $8360. What was her commission?

Method 1

$$A \quad = \quad P \quad \% \quad \cdot \quad B$$

$$\text{commission} = \text{commission rate}(0.01) \cdot \text{total sales}$$

$$= \quad 20(0.01) \quad \cdot \quad \$8360$$

$$= (0.20)(\$8360)$$

$$= \$1672$$

Method 2

$$\frac{A}{B} = \frac{P}{100}$$

$$\frac{\text{commission}}{\text{total sales}} = \frac{\text{commission rate}}{100}$$

$$\frac{\text{commission}}{\$8360} = \frac{20}{100}$$

$$(100)\text{commission} = (20)(\$8360)$$

$$\text{commission} = \frac{(20)(\$8360)}{100}$$

$$= \$1672$$

Thus, Elaine received a commission of $1672.00 for April.

| **PRACTICE EXERCISE 1** |
| --- |

Rich receives a 3% commission on all life insurance policy sales. In September his sales totaled $195,400. What was his commission?

Answer: $5862

---

As with sales-tax problems, we can find the commission rate if we know the total sales and the commission.

---

| **EXAMPLE 2**  FINDING A COMMISSION RATE |
| --- |

Shane received a $123 commission on sales of $820 in men's clothing. What was his commission rate?

The question to answer is: $123 is what percent of $820?

Method 1

$$A \quad = \quad P \quad \% \quad \cdot \quad B$$

$$\text{commission} = \text{commission rate}(0.01) \cdot \text{total sales}$$

$$123 \quad = \text{commission rate}(0.01) \cdot \quad (820)$$

$$123 = \text{commission rate }(8.2)$$

$$\frac{123}{8.2} = \text{commission rate}$$

$$15 = \text{commission rate}$$

| **PRACTICE EXERCISE 2** |
| --- |

Adrienne received a commission of $98.75 on sales of $790 in an appliance store. What was her commission rate?

Method 2

$$\frac{A}{B} = \frac{P}{100}$$

$$\frac{\text{commission}}{\text{total sales}} = \frac{\text{commission rate}}{100}$$

$$\frac{123}{820} = \frac{\text{commission rate}}{100}$$

$$(123)(100) = (820)\text{commission rate}$$

$$\frac{(123)(100)}{820} = \text{commission rate}$$

$$15 = \text{commission rate}$$

Shane's commission rate is 15%.

Answer: 12.5%

---

### EXAMPLE 3 FINDING TOTAL SALES

What are the total sales on which Mario received a commission of $336 if his commission rate is 12%?

Since the commission is known but not the amount of the sales, the question is: $336 is 12% of what number?

Method 1

$$A = P\% \cdot B$$

$$\text{commission} = \text{commission rate}(0.01) \cdot \text{total sales}$$

$$\$336 = 12(0.01) \cdot \text{total sales}$$

$$\$336 = (0.12)\text{total sales}$$

$$\frac{\$336}{0.12} = \text{total sales}$$

$$\$2800 = \text{total sales}$$

Method 2

$$\frac{A}{B} = \frac{P}{100}$$

$$\frac{\text{commission}}{\text{total sales}} = \frac{\text{commission rate}}{100}$$

$$\frac{\$336}{\text{total sales}} = \frac{12}{100}$$

$$(\$336)(100) = (12)\text{total sales}$$

$$\frac{(\$336)(100)}{12} = \text{total sales}$$

$$\$2800 = \text{total sales}$$

The sales total is $2800.

### PRACTICE EXERCISE 3

If Louis received a commission of $46.04 working in a hardware store where he is paid at a commission rate of 8%, what were his total sales?

Answer: $575.50

| **EXAMPLE 4  MONTHLY INCOME INVOLVING A COMMISSION** | **PRACTICE EXERCISE 4** |

An automobile salesperson receives a monthly salary of $500 plus 2% of the first $20,000 sales and 4% of sales above $20,000. What is the monthly income on sales of $38,000?

First, calculate 2% of $20,000.

$$(0.02)(\$20,000) = \$400$$

Next, calculate 4% of the difference $38,000 − $20,000 = $18,000.

$$(0.04)(\$18,000) = \$720$$

The total income is $500 plus the two commissions.

$$\$500 + \$400 + \$720 = \$1620$$

The total income for the month is $1620.

A real estate salesman receives a monthly salary of $850 plus 0.75% of the first $100,000 in sales and 1.5% of sales above $100,000. How much did he earn in a month when he sold $225,000 worth of property?

Answer: $3925

## ② DISCOUNT PROBLEMS

A **discount** is a reduction in the regular price of an item when it is put on sale. The **discount rate** is the percent of the regular price that the item is reduced. The **sale price** is the price after the discount. For example, if an appliance store reduces a $600 refrigerator to a sale price of $450, the discount is

$$\$600 - \$450 = \$150.$$

The discount rate relative to the original price is

$$\frac{150}{600} = \frac{1 \cdot \cancel{150}}{4 \cdot \cancel{150}} = \frac{1}{4} = 0.25$$

which is 25%. We are usually given the regular price and the discount rate and are asked to find this discount. In this problem, given the $600 and the 25%, we could have calculated the discount.

$$(0.25)(\$600) = \$150$$

---

### To Solve a Discount Problem

1. Use the percent equation

$$\begin{array}{ccccccc} A & = & P & \% & \cdot & B \end{array}$$
discount = discount rate(0.01) · regular price

*or* use the percent proportion

$$\frac{A}{B} = \frac{P}{100}$$

$$\frac{\text{discount}}{\text{regular price}} = \frac{\text{discount rate}}{100}.$$

2. The sale price is given by

sale price = regular price − discount.

## EXAMPLE 5 FINDING A DISCOUNT

What is the discount and the sale price of a coat which is regularly priced $69.50 if the discount rate is 20%?

First find the discount.

Method 1

$$A \quad = \quad P \quad \% \quad \cdot \quad B$$

discount = discount rate(0.01) · regular price

$$= \quad 20(0.01) \quad \cdot \quad \$69.50$$

$$= (0.20)(\$69.50)$$

$$= \$13.90$$

Method 2

$$\frac{A}{B} = \frac{P}{100}$$

$$\frac{discount}{regular\ price} = \frac{discount\ rate}{100}$$

$$\frac{discount}{\$69.50} = \frac{20}{100}$$

$$(100)discount = (20)(\$69.50)$$

$$discount = \frac{(20)(\$69.50)}{100}$$

$$= \$13.90$$

Then

$$sale\ price = regular\ price - discount$$

$$= \quad 69.50 \quad - \quad 13.90$$

$$= 55.60.$$

During the sale, the coat can be bought for $55.60.

## EXAMPLE 6 FINDING A SALE PRICE

A retailer first discounted a $450 washing machine 10%. When the machine was not purchased, he discounted it 30% of the sale price. What was the second sale price and the overall rate of discount?

The first discount is

$$(0.10)(\$450) = \$45,$$

and the first sale price is

$$\$450 - \$45 = \$405.$$

Since the second discount is taken on the sale price of $405, find 30% of $405.

$$(0.30)(\$405) = \$121.50$$

The second discount is $121.50, and the second sale price is

$$\$405.00 - \$121.50 = \$283.50.$$

Since the second discount was taken on $405 and not on $450, the discount rates cannot be added. Add the discounts.

$$\$45.00 + \$121.50 = \$166.50$$

The total discount is $166.50 on a regular price of $450. The discount rate is given by:

$$\$166.50 = \text{discount rate}(0.01) \cdot \$450$$

$$\frac{\$166.50}{\$4.50} = \text{discount rate}$$

$$37 = \text{discount rate}$$

The overall discount rate is 37%, which is less than the 40% obtained by adding 10% and 30%. The reason for the difference is that the 30% discount was taken on $405 instead of on $450.

Answer: $578; 32% [Note that the overall discount is not the sum of the two discounts which is 35%.]

## 6.5  EXERCISES A

*Solve the following problems.*

1. An automobile salesperson receives a 2% commission on the price of each car sold. What is the commission on an $8500 sale?

2. Mary receives a 15% commission on all clothing sales. If her sales were $3872.16 during the month of April, what was her commission?

3. Alphonse was paid a commission of $142.65 on sales of $475.50. What was his commission rate?

4. Angela received a commission of $639.12 on her furniture sales for the month of June. If the commission rate is 12%, what were her total sales?

$639.12 = 12\% \cdot ?$

5. The Sun Realty Company receives a 6% commission on all real estate sales. On July 16 the company sold a house for $82,560 and a lot for $21,320. What was the total commission on these sales?

103,880

6. Lynne receives a 1.5% commission each month on all farm equipment sales of $20,000 or less. For all sales over $20,000, she is paid a 3% commission on the amount over $20,000. What is her commission on sales of $45,280 during the month of May?

7. Considering the total commission that Lynne received in Exercise 6, what is her overall commission rate?

8. A retailer discounted all the coats in her shop by 25%. What is the discount and the sale price on a coat regularly priced $128.60?

9. A refrigerator normally sells for $720, but an appliance dealer put it on sale at an 18% discount. What is the discount and the sale price?

10. If a $40 pair of shoes is on sale for $34, what is the discount rate?

11. The price of a shirt is decreased by $3.50. If this is a 20% discount, what is the normal price of the shirt?

12. In a closeout sale, a carpet dealer has reduced all carpet prices by 40%. If the normal price of a pattern is $16.90 per square yard, how much would it cost to buy the carpet for a room which is 4 yards wide and 5 yards long?

13. Find the discount rate if a car normally priced at $6250 is on sale for $5687.50.

14. In the spring a department store reduced the price of coats at a discount rate of 25%. When some coats were left, the store reduced their price 20% of the sale price. What was the second sale price of a coat that was originally priced $142.80?

15. What was the overall rate of discount on the coat in Exercise 14?

16. A toaster normally selling for $42.50 is on sale for 12% off the normal price. What total price must Sean pay if, after the discount is made, a 5% sales tax is added?

## FOR REVIEW

17. The sales-tax rate in Twin Falls, Idaho, is 4%. How much tax will Paula pay on the purchase of three posters each selling for $2.98 before tax? What is the total price of the posters?

18. Jeff was charged $3.29 in sales tax on the purchase of a pair of basketball shoes. What is the sales-tax rate if the before tax selling price of the shoes was $65.80?

19. Maria had an income from wages of $19,600 in 1989. How much Social Security tax did she pay if the rate was 7.51% that year?

20. In 1989 Juanita had a taxable income of $15,743. If she used the tax table given in Section 6.4, how much income tax did she pay?

---

ANSWERS: (Answers may vary slightly due to round-off differences.)  1. $170   2. $580.82   3. 30%   4. $5326   5. $6232.80   6. $1058.40   7. 2.3% (to the nearest $\frac{1}{10}$%)   8. $32.15; $96.45   9. $129.60; $590.40   10. 15%   11. $17.50   12. $202.80   13. 9%   14. $85.68   15. 40%   16. $39.27   17. $0.36; $9.30   18. 5%   19. $1471.96   20. $2361.45

## 6.5   EXERCISES B

*Solve the following problems.*

1. The Land Realty Company charges a commission of 6% on all sales. How much commission will the company receive for selling a ranch for $260,000?

2. Beth receives a 12% commission on all sales. If her sales totaled $2475.50 one week, what was her commission?

3. Jill received a commission of $402.60 on sales of $7320.00. What was her commission rate?

4. Art received a commission of $9817.50 on sales of computer hardware for the month of December. If his commission rate is 15%, what were his total sales?

5. The Vacation Center pays its employees an 8% commission on all sales. During June Shannon sold a motorhome for $36,580 and a travel trailer for $14,230. What was her total commission on these sales?

6. Wayne receives a 2% commission on sales of $24,000 or less. For sales over $24,000, he receives 4% of the amount over $24,000. What is his total commission on sales of $36,200?

7. Considering the total commission that Wayne receives in Exercise 6, what is his overall commission rate (to the nearest tenth of a percent)?

8. Dana put all swim suits in her store on sale at 40% off the regular price. What is the discount and sale price of a suit normally priced at $35.00?

9. A video recorder normally sells for $650, but it is put on sale at a 20% discount. What is the discount and the sale price?

10. What is the discount rate if an appliance normally selling for $42.50 is on sale for $33.15?

11. The price of a steam iron is decreased by $7.35. If this is a 30% discount, what is the normal selling price of the iron?

12. Carpet normally selling for $19.80 per square yard is being reduced by 30%. How much would it cost to carpet a room which is 3 yards wide and 4 yards long?

13. Find the rate of discount if a truck, originally priced at $13,450, is sold for $11,567.

14. A set of dishes was first discounted 20%. When the dishes did not sell, they were reduced by 30% of the sale price. What was the second sale price if the dishes originally sold for $160.50?

15. What is the overall rate of discount on the dishes in Exercise 14?

16. A jacket normally selling for $35 is on sale for 15% off the normal price. What total price must Amber pay if, after the discount is made, a 4% sales tax is added?

## FOR REVIEW

**17.** The sales-tax rate in Orchard, Washington, is 3%. How much tax will Bunni be charged on the purchase of four books each selling for $3.98 before tax? What is the total price of the books?

**18.** Stephanie was charged $1.42 in sales tax on the purchase of a gift for her father. What is the sales-tax rate if the gift was priced at $35.50 before tax?

**19.** The Schulzes had an income from wages of $29,500 in 1989. How much Social Security tax did they pay if the rate was 7.51% that year?

**20.** In 1989 Lupe had a taxable income of $38,200. If she used the tax table given in Section 6.4, how much tax did she pay?

## 6.5 EXERCISES C

**1.** Jenny sold a sweater for $39.95, a pair of pants for $34.50, and a jacket for $79.98. If she received a commission of $19.30 on these sales, to the nearest tenth of a percent, what is the commission rate?

**2.** A sofa was first discounted 25%. When it did not sell, it was reduced again by 30% of the sale price. When again it did not sell, it was reduced a third time by 40% of the second sale price. What was the third sale price if the sofa originally was priced at $785.50? To the nearest tenth of a percent, what was the overall rate of discount?
[Answer: $247.43, 68.5%]

## 6.6 INTEREST PROBLEMS

=== STUDENT GUIDEPOSTS ===

1 Simple Interest

2 Equal Payments on a Loan

3 Compound Interest

### 1 SIMPLE INTEREST

The money paid for the use of money is called **interest.** When we borrow money, we pay interest to the lender. When we invest money in a bank account we receive interest from the bank. The money borrowed or invested is called the **principal** and the money paid for the use is a percent of the principal. This percent is called the **rate of interest.** For example, suppose you borrow $500 for one year and the interest rate is 12% per year. To find the interest charged, find 12% of $500.

$$(0.12)(\$500) = \$60$$

At the end of the year you must return the $500 plus the interest, $60, for a total of

$$\$500 + \$60 = \$560.$$

If you had needed the $500 for two years, the interest rate would have been

$$2(0.12) = 0.24 = 24\%.$$

The total interest for two years is

$$(0.24)(\$500) = \$120.$$

At the end of two years you would pay

$$\$500 + \$120 = \$620.$$

Interest calculated this way is called **simple interest.** Interest is charged only on the principal, and not on the interest earned. In general, the following formula is used to calculate simple interest. Remember that the interest rate must always be changed from a percent to a decimal.

---

### Simple Interest

If $P$ represents a principal, $R$ is a yearly rate of interest converted to a decimal, $T$ is the time in years, and $I$ is simple interest, then

$$I = P \cdot R \cdot T.$$

In addition, if $A$ is the amount to be repaid or amount in the account, then

$$A = P + I.$$

---

### EXAMPLE 1   THE AMOUNT TO REPAY ON A LOAN

Find the interest on $1500 for 2 years if the interest rate is 9% per year. How much must be repaid at the end of the 2-year period?

First find the simple interest. We are given that $P = \$1500$, $R = 9\% = 0.09$, and $T = 2$. Substitute these values.

$$
\begin{aligned}
I = \quad & P \quad \cdot \quad R \quad \cdot \quad T \\
= \ & (\$1500) \cdot (0.09) \cdot (2) \\
= \ & \$270
\end{aligned}
$$

The interest is $270. Next we find $A$, the amount to be repaid.

$$
\begin{aligned}
A = \quad & P \ + \ I \\
= \ & \$1500 + \$270 \\
= \ & \$1770
\end{aligned}
$$

At the end of 2 years, $1770 must be repaid.

### PRACTICE EXERCISE 1

Find the interest on $3500 for 3 years if the interest rate is 12% per year. How much must be repaid at the end of the 3-year period?

Answer: $1260; $4760

---

### EXAMPLE 2   INTEREST ON A TIME LESS THAN ONE YEAR

What interest must be paid on a 6-month loan of $2800 if the interest rate is 16% per year? What amount must be repaid in 6 months?

Since 6 months is half of a year, we use 0.5 for $T$, along with $2800 for $P$, and $16\% = 0.16$ for $R$.

$$
\begin{aligned}
I = \quad & P \quad \cdot \quad R \quad \cdot \quad T \\
= \ & (\$2800) \cdot (0.16) \cdot (0.5) \\
= \ & \$224
\end{aligned}
$$

The interest for 6 months is $224. Next we find $A$, the amount to be repaid.

$$
\begin{aligned}
A = \quad & P \ + \ I \\
= \ & \$2800 + \$224 \\
= \ & \$3024
\end{aligned}
$$

Thus, in 6 months, $3024 must be repaid.

### PRACTICE EXERCISE 2

What interest must be paid on an 8-month loan of $1400 if the interest rate is 9% per year? What amount must be repaid in 8 months?

Since 8 months is $\frac{8}{12}$ or $\frac{2}{3}$ of a year, use $\frac{2}{3}$ for the number of years when finding the simple interest.

Answer: $84; $1484

| **EXAMPLE 3   MONTHLY INTEREST ON A SAVINGS ACCOUNT** | **PRACTICE EXERCISE 3** |

Heather put $3622.16 in a savings account for one month. If the interest paid is 9% per year, how much interest will Heather be paid at the end of the month?

Since 1 month is $\frac{1}{12}$ of a year, use $\frac{1}{12}$ for $T$, along with $3622.16 for $P$, and 9% = 0.09 for $R$.

$$I = \quad P \quad \cdot \quad R \quad \cdot \quad T$$

$$= (\$3622.16) \cdot (0.09) \cdot \left(\frac{1}{12}\right)$$

$$= \$27.17 \quad \text{Rounded to the nearest cent}$$

Heather will be paid $27.17 in interest at the end of the month.

Wally put $4150.15 in a savings account for two months. If the interest paid is 18% per year, how much interest will Wally earn at the end of two months (to the nearest cent)?

Answer:  $124.50

## ❷ EQUAL PAYMENTS ON A LOAN

On some purchases the customer pays the amount of the purchase, together with simple interest charged, in monthly payments. For example, suppose you buy a refrigerator costing $750, pay $50 down, and wish to pay off the rest, with interest, over 18 months. The amount after the down payment is

$$\$750 - \$50 = \$700.$$

If the interest rate $R$ is 12% = 0.12 per year and you have 18 months to pay, then $T = 1.5$ (18 months is 12 months + 6 months = 1 year + 0.5 year = 1.5 years), and the principal is $P = \$700$.

$$I = \quad P \quad \cdot \quad R \quad \cdot \quad T$$
$$= (\$700) \cdot (0.12) \cdot (1.5)$$
$$= \$126$$

Add the principal and interest to obtain the amount $A$ to be repaid.

$$A = \quad P \quad + \quad I$$
$$= \$700 + \$126$$
$$= \$826$$

Thus, the amount to be paid in 18 equal payments is $826. To find the amount of each payment, divide $826 by 18.

$$\frac{\$826}{18} = \$45.89 \quad \text{Monthly payment, rounded to nearest cent}$$

| **To Find the Amount of Each Payment on a Loan** |
1. Find the total interest over the period.
2. Add the amount borrowed to the interest to find the total due.
3. Divide the total due by the number of payments.

### EXAMPLE 4 Finding a Monthly Payment

Sam buys a television set for \$587.50. He pays 10% down and the rest in 24 equal monthly payments. If the simple interest is 16%, what are his monthly payments?

First, find his down payment by taking 10% of \$587.50.

$$(0.10)(\$587.50) = \$58.75$$

The balance due is found by subtracting the down payment from the price of the set.

$$\$587.50 - \$58.75 = \$528.75 \quad \text{Balance due}$$

Use the balance due, \$528.75 as the principal $P$, $16\% = 0.16$ as $R$, and 2 for $T$ ($T$ is always in years and 24 months is 2 years).

$$\begin{aligned} I &= P \cdot R \cdot T \\ &= (\$528.75) \cdot (0.16) \cdot (2) \\ &= \$169.20 \end{aligned}$$

Then the total due to be paid in the monthly payments is

$$\$528.75 + \$169.20 = \$697.95.$$

Divide the total due by 24 to find the amount of each monthly payment.

$$\text{monthly payment} = \frac{\$697.95}{24} = \$29.08 \quad \text{Rounded to nearest cent}$$

Sam must pay \$29.08 each month for 24 months to pay off the television set.

### Practice Exercise 4

The Wilsons purchase two rooms of furniture for \$2850.80. They pay 20% down and the remainder in 36 equal monthly payments. If the simple interest charged is 12%, what are their monthly payments?

Answer: \$86.16

## 3 COMPOUND INTEREST

For many loans or savings accounts, interest is paid on interest as well as the principal. This type of interest is called **compound interest.** For example, if you borrow \$1000 for a year at 10% per year simple interest, the interest paid is

$$(0.10)(\$1000) = \$100.$$

If the 10% per year is compounded every 6 months (semiannually), the interest is calculated every 6 months and added to the principal. Since there are 2 compounding periods every year, the yearly interest rate 10% is divided by 2 to obtain an interest rate of

$$\frac{10\%}{2} = 5\%$$

for every 6-month period. The interest on \$1000 for the first 6 months is

$$(0.05)(\$1000) = \$50.$$

With compound interest, the principal on which interest is paid for the second 6 months is the original principal plus the first 6-month interest,

$$\$1000 + \$50 = \$1050.$$

The interest on \$1050 for the second 6 months is

$$(0.05)(\$1050) = \$52.50.$$

$\frac{r}{t} = n$

P • n = new Principle
X time
\$\$

$A = P(1 + \frac{r}{n})^{nt}$
where
A = Amount
P = Principle
R = Rate
n = # of times compounded per year
t = time

The total interest for the year is the sum of the interests on the first and second 6-month periods.

$$\$50.00 + \$52.50 = \$102.50.$$

Notice that the compound interest amounted to $2.50 more than the simple interest.

This process could be continued to find the interest compounded semiannually over more than one year.

### To Solve a Compound Interest Problem

1. Determine the interest rate for a compounding period by dividing the yearly rate by the number of periods per year.
2. Find the interest for the first period.
3. Use the sum of the interest found in Step 2 and the principal as the principal for the next period.
4. Continue until all periods have been accounted for.

Interest can be compounded over different periods. The most common types are shown in the table below where $R$ is the annual or yearly interest rate.

| Type of compounding | Number of compounding periods each year | Interest rate per period |
|---|---|---|
| compounded annually | 1 | $R$ |
| compounded semiannually | 2 | $\dfrac{R}{2}$ |
| compounded quarterly | 4 | $\dfrac{R}{4}$ |
| compounded monthly | 12 | $\dfrac{R}{12}$ |

### HINT

Remember that when interest is compounded, the interest used for each compounding period is

$$\frac{\text{Annual Interest Rate}}{\text{Number of Periods Per Year}}.$$

For example, if the annual interest rate is 12%, and compounding is semiannually, quarterly, or monthly, use 6%, 3%, or 1%, respectively, as the interest rate each period.

---

### EXAMPLE 5  INTEREST COMPOUNDED ANNUALLY

Richard invests $5000 in an account which pays 8% compounded annually. How much will he have in his account at the end of 2 years?

Calculate the interest for the first year. Since there is only one period per year, the interest rate is 8% = 0.08.

$$(0.08)(\$5000) = \$400$$

### PRACTICE EXERCISE 5

Lori invests $3000 in an account which pays 11% interest compounded annually. How much will she have in the account at the end of 3 years?

The principal on which the second year interest is calculated is thus

$$\$5000 + \$400 = \$5400.$$

The interest for the second year is

$$(0.08)(\$5400) = \$432.$$

The total in the account at the end of 2 years is

$$\$5400 + \$432 = \$5832.$$

Answer: **$4102.89**

---

| **EXAMPLE 6  INTEREST COMPOUNDED SEMIANNUALLY** | **PRACTICE EXERCISE 6** |

Joe borrowed $600 for $1\frac{1}{2}$ years. Interest is 12% per year compounded semiannually. What amount must he pay back in $1\frac{1}{2}$ years?

    Since semiannually means twice a year, the interest rate for a compounding period is

$$\frac{12\%}{2} = 6\%,$$

and the number of periods is 3. Interest for the first 6 months (the first period) is

$$(0.06)(\$600) = \$36.$$

The principal to be used for the second period is

$$\$600 + \$36 = \$636.$$

Interest for the second period is

$$(0.06)(\$636) = \$38.16.$$

The principal to be used for the third period is

$$\$636.00 + \$38.16 = \$674.16.$$

Interest for the third period is

$$(0.06)(\$674.16) = \$40.45. \quad \text{To the nearest cent}$$

The final amount to be paid is

$$\$674.16 + \$40.45 = \$714.61.$$

Verna borrowed $1500 for 1 year. Interest on the loan is 12% per year, compounded quarterly (four times a year). What amount must she pay back at the end of the year?

    Remember that 12% per year compounded quarterly is 3% (one fourth of 12%) per period.

Answer: **$1688.26**

---

## 6.6 EXERCISES A

*Solve the following problems.*

**1.** Hans borrowed $600 for one year at a bank where the interest rate is 12%. How much interest did he have to pay at the end of the year?

**2.** Hannah put $750 in a savings account which pays 6% interest per year. What did she have in her account at the end of the year?

**3.** Peter bought a snow blower for $450 and paid $50 down. The remaining amount he paid off at the end of 6 months. If interest is 14% per year (7% for 6 months), how much did he have to pay when he paid off the loan?

**4** Lisa borrowed $800 for expenses at the beginning of the school year. If simple interest was 12% per year, how much did she have to pay if she paid off the loan at the end of 9 months?

**5.** Harold bought a $2526.50 stereo system and paid 10% down. The remaining money plus interest was to be paid at the end of 2 years. If interest is 14% per year simple interest, how much interest (to the nearest cent) did Harold have to pay?

**6.** Lynn borrowed $5482.50 for 2 years at 12% simple interest. What did she have to pay at the end of the period to pay off the loan?

**7.** If Lynn in Exercise 6 had paid off the loan in 24 equal monthly payments, what would her monthly payment have been (to the nearest cent)?

**8** Robert will pay off a $1213.40 debt in 18 monthly payments. Simple interest is 16% per year. What are his payments (to the nearest cent)?

**9.** Susan borrowed $500 from a loan shark who charges 15% per month simple interest. What interest did she have to pay for one year? What was her monthly payment if she paid off the loan in 12 months?

**10.** Sylvester invests $2500 in an account which pays 7% per year compounded annually. What does he have in the account at the end of 2 years?

**11** Maria borrows $625.50 at 12% compounded semi-annually. What must she pay (to the nearest cent) to pay off the loan at the end of 18 months?

**12.** Patricia has $7000 to invest in an account which pays 10% compound interest. How much is in her account at the end of one year if the interest is compounded semiannually? How much is in her account at the end of one year if the compounding is quarterly (4 times per year)?

## FOR REVIEW

**13.** Joseph receives a 20% commission on all appliances he sells. His total sales for June were $7218.50. What was his commission?

**14.** Michael's, a shop for men, put suits on sale at a 33% discount. What is the sale price of a $285 suit?

ANSWERS:   1. $72   2. $795   3. $428   4. $872   5. $636.68   6. $6798.30   7. $283.26   8. $83.59   9. $900; $116.67   10. $2862.25   11. $744.98   12. $7717.50; $7726.69   13. $1443.70   14. $190.95

## 6.6   EXERCISES B

*Solve the following problems.*

**1.** Brad borrowed $1800 for one year at his credit union where the interest rate is 10%. How much interest did he pay at the end of the year?

**2.** Boyd put $900 in a savings account which pays 9% interest per year. What was the value of his account at the end of one year?

3. Katherine bought a microwave oven for $480 and paid $80 down. She paid off the remaining amount plus interest at the end of 6 months. If the interest rate she was charged was 8%, how much did she pay?

4. Dave Markee borrowed $700 to fix his car. If he was charged 12% interest per year, how much did he have to pay back at the end of 8 months?

5. Colleen bought a coat for $252.20 and paid 20% down. The remaining amount is to be paid off at the end of 6 months. If the simple interest rate is 10% per year, what will she have to pay to pay off the loan?

6. Calvin borrowed $628 for 2 years. If the simple interest rate is 12% per year, what amount will he have to pay at the end of the 2-year period?

7. If Calvin in Exercise 6 pays off the loan in 24 equal monthly payments, how much will he pay each month?

8. Bruce will pay off a $782.50 debt in 18 monthly payments. If simple interest is 14% per year, what is his monthly payment?

9. Vinnie, the loan shark, charges 20% *per month* simple interest. How much interest will he collect on a loan of $1000 which is paid off in one year? If Vinnie collects the loan with interest in 12 equal monthly payments, how much does he receive each month?

10. Maria invests $5800 in an account that pays 8% per year compounded semiannually. How much is in her account at the end of a year?

11. Virgil borrows $840.50 at 10% interest compounded semiannually. What must he pay (to the nearest cent) to pay off the loan at the end of 30 months?

12. William borrows $540.00 at 12% compound interest per year. What total amount must he pay at the end of the year if the compounding is semiannually? How much if the compounding is quarterly?

## FOR REVIEW

13. Bridgette earned $123.65 in commissions on sales in a dress shop. If she works on a 10% commission rate, what were her total sales?

14. Mr. Miller, a senior citizen, receives a discount of 15% on all bills for food at the Village Waffle Shoppe. If the bill comes to $123.40 when he takes his family out to eat, how much must he pay?

## 6.6 EXERCISES C

1. Find the simple interest earned on $13,455 at 11.25% per year for 20 months.

2. Carl invests $8575 in an account which pays 11.5% interest compounded semiannually. How much is in his account at the end of 18 months?
[Answer: $10,140.87]

*It can be shown that the amount A in an account if a principal P is deposited at 12% interest compounded monthly for 5 years is given by*

$$A = P(1.01)^{60}.$$

*Use a calculator with a* $\boxed{y^x}$ *key and this formula to find the amount in an account with the value of P given in Exercises 3–5.*

3. $P = \$1000$

4. $P = \$5000$

5. $P = \$10,000$

# 6.7 PERCENT INCREASE AND DECREASE PROBLEMS

The percent by which a quantity increases is called the **percent increase.** Similarly, the percent by which a quantity decreases is called the **percent decrease.** For example, if a salary increased 7%, we find the increase by taking 7% of the *original* salary. Thus, if you earn $20,000 a year and receive a 7% raise, the increase in salary or your raise is

$$(0.07)(\$20,000) = \$1400$$

and your new salary will be

$$\text{new salary} = \underline{\text{former salary}} + \underline{\text{raise}}$$
$$= \quad \$20,000 \quad + \$1400$$
$$= \$21,400.$$

If school enrollment decreases by 6%, we find the decrease by taking 6% of the *original* number of students. For example, if BSU had 10,500 students last year and enrollment went down 6% this year, the decrease in enrollment was

$$(0.06)(10,500) = 630 \text{ students.}$$

The enrollment this year is

$$\text{this year's enrollment} = \underline{\text{last year's enrollment}} - \underline{\text{decrease}}$$
$$= \quad 10,500 \quad - \quad 630$$
$$= 9870 \text{ students.}$$

These are examples of percent increase and percent decrease problems.

---

### To Solve a Percent Increase (Decrease) Problem

1. Use the percent equation

$$\begin{array}{ccccc} A & = & P\% & \cdot & B \end{array}$$
increase (decrease) = percent increase (decrease) · original amount

*or* use the percent proportion

$$\frac{A}{B} = \frac{P}{100}$$

$$\frac{\text{increase (decrease)}}{\text{original amount}} = \frac{\text{percent increase (decrease)}}{100}.$$

2. The new amount is given by

new amount = original amount + increase

*or*

new amount = original amount − decrease.

---

| EXAMPLE 1   PERCENT DECREASE APPLICATION | PRACTICE EXERCISE 1 |
|---|---|

A labor union votes to take a 5% decrease in wages in order to keep the company from going bankrupt. If the average salary before the decrease was $17,500 per year, what is the new salary?

First find the decrease in salary.

Due to declining enrollments, teachers in the Kachina District are faced with a 3% decrease in wages. If the average salary before the decrease is $14,700, what will be the new average salary?

Method 1

$$A = P\% \cdot B$$

decrease = percent decrease · original salary

$$= (0.05) \cdot (\$17,500)$$

$$= \$875$$

Method 2

$$\frac{A}{B} = \frac{P}{100}$$

$$\frac{\text{decrease}}{\text{original salary}} = \frac{\text{percent decrease}}{100}$$

$$\frac{\text{decrease}}{\$17,500} = \frac{5}{100}$$

$$\text{decrease}(100) = (5)(\$17,500)$$

$$\text{decrease} = \frac{(5)(\$17,500)}{100}$$

$$= \$875$$

Thus, the decrease in salary is $875. Next find the new salary.

new salary = original salary − decrease in salary

$$= \$17,500 - \$875$$

$$= \$16,625$$

The new average salary is $16,625.

Answer: $14,259

---

**EXAMPLE 2** PERCENT INCREASE APPLICATION

State University expects an increase in enrollment of 2.5% next year. If the present enrollment is 15,276, what will be the approximate number of students next year?

First find the increase in enrollment.

Method 1

$$A = P\% \cdot B$$

increase = percent increase · original enrollment

$$= (0.025) \cdot (15,276)$$

$$= 381.9$$

Method 2

$$\frac{A}{B} = \frac{P}{100}$$

$$\frac{\text{increase}}{\text{original enrollment}} = \frac{\text{percent increase}}{100}$$

$$\frac{\text{increase}}{15,276} = \frac{2.5}{100}$$

$$\text{increase}(100) = (2.5)(15,276)$$

$$\text{increase} = \frac{(2.5)(15,276)}{100}$$

$$= 381.9$$

**PRACTICE EXERCISE 2**

The placement director expects an increase of 3.5% in the number of students taking the placement exam in mathematics during the fall semester. If 2480 students took the exam last fall, what will be the approximate number of students tested this fall?

We must round to the nearest whole number so the increase will be approximately 382 students. The new enrollment is found by adding this year's enrollment to the increase.

$$\begin{array}{r} 15276 \\ +\quad 382 \\ \hline 15658 \end{array}$$  Original enrollment
Enrollment increase
New enrollment

There will be approximately 15,658 students next year.

Answer: 2567 students

## HINT

Remember that percent increase or decrease is always taken on the original quantity. To find the new quantity *add* the increase to the original amount or *subtract* the decrease from the original amount.

### EXAMPLE 3  FINDING A PERCENT DECREASE

The price of a dozen eggs dropped from 75¢ to 67¢. What was the percent decrease?

First find the amount of the decrease.

$$75¢ - 67¢ = 8¢$$

Since 8¢ is the amount of the decrease, and 75¢ is the original price, we must find the percent decrease, $P$.

Method 1

$$A = P\% \cdot B$$

decrease = percent decrease · original price

$$8 = P(0.01) \cdot (75)$$

$$8 = P(0.75)$$

$$\frac{8}{0.75} = P$$

$$10.\overline{6} = P$$

Method 2

$$\frac{A}{B} = \frac{P}{100}$$

$$\frac{\text{decrease}}{\text{original price}} = \frac{\text{percent decrease}}{100}$$

$$\frac{8}{75} = \frac{P}{100}$$

$$(8)(100) = (75)P$$

$$\frac{(8)(100)}{75} = P$$

$$10.\overline{6} = P$$

Thus, the percent decrease was $10.\overline{6}\%$ or $10\frac{2}{3}\%$.

### PRACTICE EXERCISE 3

The price of a six-pack of soda dropped from $2.25 to $1.85. What was the percent decrease, to the nearest tenth of a percent?

Answer: 17.8%

| **EXAMPLE 4** FINDING AN ORIGINAL PRICE | **PRACTICE EXERCISE 4** |

The price of gasoline increased 10%. The amount of this increase was 12¢ per gallon. What was the original price of a gallon of gasoline?

    We know the percent increase and the amount of the increase, so we must find the original price.

Method 1

$A$    =    $P\%$    ·    $B$

increase = percent increase · original price

$12$ = $(0.10)$ · original price

$\dfrac{12}{0.10}$ = original price

$120$ = original price

Method 2

$\dfrac{A}{B} = \dfrac{P}{100}$

$\dfrac{\text{increase}}{\text{original price}} = \dfrac{\text{percent increase}}{100}$

$\dfrac{12}{\text{original price}} = \dfrac{10}{100}$

$(12)(100) = (10)\text{original price}$

$\dfrac{(12)(100)}{10} = \text{original price}$

$120 = \text{original price}$

Thus, the original price of a gallon of gasoline was 120¢ or $1.20.

During the winter the price of home heating oil increased 20%. If the amount of increase was 18¢ per gallon, what was the original price per gallon?

| **EXAMPLE 5** FINDING A PERCENT INCREASE | **PRACTICE EXERCISE 5** |

The Smith family bought a house two years ago for $85,000. It is now valued at $102,680. To the nearest percent, what is the percent increase in value?

    To find the amount of the increase, subtract the original value from the new value.

| | |
|---|---|
| $102680 | New value |
| −   85000 | Original value |
| $ 17680 | Value increase |

We must find the percent increase, $P$.

Method 1

$A$    =    $P\%$    ·    $B$

increase = percent increase · original value

$17,680$ = $P(0.01)$ · $85,000$

$17,680 = P(850)$

$\dfrac{17,680}{850} = P$

$20.8 = P$

Paul Pendergast has a classic 1957 Chevrolet which he purchased for $2400 several years ago. Today it is valued at $3700. To the nearest tenth of a percent, what is the percent increase in value?

Method 2

$$\frac{A}{B} = \frac{P}{100}$$

$$\frac{\text{increase}}{\text{original value}} = \frac{\text{percent increase}}{100}$$

$$\frac{17{,}680}{85{,}000} = \frac{P}{100}$$

$$(17{,}680)(100) = (85{,}000)P$$

$$\frac{(17{,}680)(100)}{85{,}000} = P$$

$$20.8 = P$$

Thus, to the nearest percent, the percent increase is about 21%.

Answer: 54.2%

---

////////////// **C A U T I O N** ///////////////

In problems such as the one in Example 5, be sure to compare the increase in value to the *original* value, not the new value. In such cases always ask:

**increase is what percent of *original value*?**

//////////

---

An interesting application of percent increase and decrease has to do with changes in automobile fuel economy. Such things as temperature, humidity, wind conditions, and road conditions have an effect on the amount of fuel used.

Suppose we assume the percent increases and decreases in the following tables for the given conditions. An increase in fuel economy means more miles per gallon (mpg) and a decrease means fewer mpg.

| Condition | % Increase | | Condition | % Decrease |
|-----------|-----------|---|-----------|-----------|
| tailwind | 15% | | headwind | 17% |
| smooth road | 12% | | rough road | 25% |
| high humidity | 6% | | low temperature | 10% |

---

**EXAMPLE 6**  APPLICATIONS OF FUEL ECONOMY

**(a)** If Michael's car normally gets 30 mpg, what should it get with a tailwind?

First use the table to find that a tailwind increases economy by 15%. Find 15% of 30 mpg.

$$(0.15)(30) = 4.5$$

A tailwind gives Michael an additional 4.5 mpg for a total of

$$30 + 4.5 = 34.5 \text{ mpg.}$$

Thus, Michael's car gets about 34.5 mpg with a tailwind.

**(b)** Suppose Mary drives on a rough road under high humidity conditions. What is the percent change in economy? If her normal mileage is 20 mpg, what is her mileage under the given conditions?

**PRACTICE EXERCISE 6**

**(a)** If Michael's car normally gets 30 mpg, what should it get when he drives on a smooth road?

Since the rough road means a 25% decrease and high humidity means a 6% increase, find the difference:

$$25\% - 6\% = 19\%.$$

Under these conditions there is a 19% decrease in economy. Taking 19% of 20,

$$(0.19)(20) = 3.8 \text{ mpg},$$

the amount of decrease is 3.8 mpg. The mileage under these conditions is

$$20 - 3.8 = 16.2 \text{ mpg}.$$

Thus, Mary's mileage is approximately 16.2 mpg.

**(b)** If Mary drives against a headwind but on a smooth road, what is the percent change in economy? What is her mileage under these conditions if normally she gets 20 mpg?

Answers: (a) 33.6 mpg  (b) 5% decrease; 19 mpg

## 6.7   EXERCISES A

*Solve the following problems.*

**1.** Laura was given a 9% increase in salary. What is her new salary if her old salary was $28,250?

**2.** The average salary of the workers for Farms, Incorporated, was $12,320. After the union negotiated a 12% increase, what is the average new salary?

**3.** The executives of the Paper Corporation agree to take an 8% salary cut in order to help the financial situation of the company. If the president was making $125,300 per year, what is his new salary?

**4.** City College expects a decrease in enrollment of 3.5% next year. If the current enrollment is 23,254, what will be the approximate enrollment next year?

**5** The population of East, Texas, increased from 329 to 415. To the nearest percent, what was the percent increase?

**6.** Over a period of years, the price of a gallon of gasoline increased from $0.32 to $1.27. To the nearest percent, what was the percent increase?

**7.** Because of oversupply, the price of lettuce dropped from 92¢ per head to 39¢ per head. To the nearest tenth of a percent, what was the percent decrease?

**8** Forrest received a 15% increase in salary. The amount of the increase was $1824. What was his original salary?

9. By attending a weight-control class, Mary lost 24 pounds. If she now weighs 115 pounds, what was her original weight and her percent decrease to the nearest tenth of one percent?

10. The value of a new car decreases 27% the first year. If the decrease amounted to $2300, what was the original value?

*Refer to the mileage tables in this section for Exercises 11–14.*

11. The normal gasoline mileage of a car is 30 mpg. What is its mileage on a smooth road?

12. Henry normally gets 8 mpg with his truck. If he drives on a rough road, what mileage will he get?

13. Jane is driving with a tailwind but the temperature is low. What will be her mileage if she normally gets 42 mpg?

14. Sue is driving into a headwind under conditions of high humidity. She has a decrease of 4.4 mpg in her mileage. What is her normal mileage?

15. Driving at high speeds reduces your mileage. Michael normally gets about 33 mpg, but by driving fast, he only got 27 mpg. To the nearest tenth of a percent, what was his percent decrease?

16. By keeping his car in good mechanical condition, Arnie can get an increase in mileage of 13%. Good driving habits will give a 9% increase in mileage. By keeping his car in good condition and driving correctly, what mileage can Arnie get if his normal rate is 30 mpg?

## FOR REVIEW

17. Sara buys a stereo for $728.50 and puts 10% down, paying the rest in 18 monthly payments. If the simple interest is 16%, what are her payments (to the nearest cent)?

18. Henry puts $1500 in a savings account which pays 8% compounded semiannually. What will be the value of his account at the end of 2 years?

ANSWERS:   1. $30,792.50   2. $13,798.40   3. $115,276   4. 22,440   5. 26%   6. 297%   7. 57.6%   8. $12,160   9. 139; 17.3%   10. $8518.52   11. 33.6 mpg   12. 6 mpg   13. 44.1 mpg   14. 40 mpg   15. 18.2%   16. 36.6 mpg   17. $45.17   18. $1754.79

## 6.7   EXERCISES B

*Solve the following problems.*

**1.** Maxine received a salary of $18,560 one year. The next year she received a raise of 8%. What was her new salary?

**2.** The employees of the Soft Shoe Company negotiated an 11% salary increase. If their average salary was $14,200, what is it now?

**3.** Several years ago, the price of gasoline was $0.40 per gallon. What was the price after a 190% increase?

**4.** A major corporation is forced to decrease its number of employees by 4.5% next year. If it now employs 3540 workers, how many will be employed next year?

**5.** Northeast University's enrollment increased from 24,300 to 27,800. To the nearest tenth, what was the percent increase?

**6.** Due to overstocking, a retailer is forced to decrease the price of an item from $1.85 to $1.25. To the nearest tenth of a percent, what is the percent decrease?

**7.** A car dealer lowered the price of a new car from $14,800 to $13,320. To the nearest percent, what was the percent decrease?

**8.** A new car decreased in value a total of $1920 the first year. If this was a 24% decrease, what was the original value?

**9.** After six weeks of basic training, a new recruit in the army lost 14 pounds. If he now weighs 176 pounds, what was his original weight and the percent decrease in weight (to the nearest tenth of a percent)?

**10.** The value of a Navajo Indian rug increased 15% during the first year. If the increase amounted to $114.87, what was the original price?

*Refer to the mileage tables in this section for Exercises 11–14.*

**11.** The normal gasoline mileage for a motorhome is 12 mpg. What is the mileage with a tailwind?

**12.** Danielle normally gets 26 mpg with her car. If she drives during a period of low temperature, what mileage will she get?

**13.** Rufus is driving with a headwind on a smooth road. What will his mileage be if he usually gets 18 mpg?

**14.** Randy is driving into a headwind, on a rough road, and at a low temperature. If he has a decrease of 13 mpg in his mileage, what is his normal mileage?

**15.** By driving at the speed limit, you can improve your mileage from what would be obtained by driving over the limit. If Marla got 14 mpg while exceeding the limit and 19 mpg while driving at the speed limit, what was the percent increase (to the nearest tenth of a percent)?

**16.** When a car is tuned properly, its gas mileage can be improved by 5%. If Ronnie was getting 24 mpg before a tune-up, what mileage can he expect to get after the car has been serviced?

## FOR REVIEW

**17.** Den bought a set of tools for $565.50 by putting 20% down and paying the rest in 12 monthly payments. If he is charged 14% simple interest, what are his payments?

**18.** Montana plans to put $3000 into a savings account for two years. She can choose one account which pays 11% simple interest or another which pays 10% interest compounded semiannually. Which account will return the most interest?

## 6.7 EXERCISES C

*Solve.*

1. A new car decreases in value by approximately 27.5% each year. If a car is purchased for $12,685.35, what is its approximate value two years later?

2. By decreasing the thermostat from 72° to 68°, a company can reduce its heating bill by 35.5%. If the bill last month, prior to turning down the thermostat, was $478.63, what can the company expect to pay for heat this month?
   [Answer: $308.72]

## 6.8 MARKUP AND MARKDOWN PROBLEMS

A special type of percent increase problem involves marking up prices in a business. Any wholesale or retail business must sell an item for more than its cost. This extra amount goes to pay operating expenses and give the owners a return on their investment. In many businesses, the difference between the cost and selling price is a fixed percent of the cost.

The amount by which the selling price is increased over the cost is called the **markup** on the item. The markup is divided by the cost to get the **percent markup.** For example, if the cost of an item is $4 and the selling price is $5, then the markup on the item is

$$\$5 - \$4 = \$1.$$

The percent markup is found by dividing $1 (markup) by $4 (cost).

$$\frac{\text{markup}}{\text{cost}} = \frac{1}{4} = 0.25 = 25\%$$

Thus, the percent markup is 25%.

In most businesses, the percent markup is given and the amount of the markup must be found. Suppose a 30% markup is needed to make a business prosper. When a pair of shoes costing $20 is to be sold, the sales staff can calculate the markup as follows:

$$(0.30)(\$20) = \$6.$$

The markup is $6 and the selling price is

$$\$20 + \$6 = \$26.$$

Thus, the store will sell the pair of shoes for $26.

On the other hand, suppose a wholesaler bought too many of a particular item. He might have to reduce his selling price to sell them. This is a percent decrease or a discount problem, and the **percent markdown** is based on the selling price of the item. For example, a 30% reduction would be found by taking 30% of the selling price. The amount of the price reduction is called the **markdown.**

### To Solve a Markup (Markdown) Problem

1. Use the percent equation

$$A \qquad = \qquad P\% \qquad \cdot \qquad B$$

markup (markdown) = percent markup (markdown) · cost (price)

*or* use the percent proportion

$$\frac{A}{B} = \frac{P}{100}$$

$$\frac{\text{markup (markdown)}}{\text{cost (price)}} = \frac{\text{percent markup (markdown)}}{100}.$$

**2.** To find the new selling price use

$$\text{selling price} = \text{cost} + \text{markup}$$

*or*

$$\text{new selling price} = \text{old selling price} - \text{markdown}.$$

## HINT

In a markup problem, the markup is always a percent of the original cost, *not* a percent of the new selling price.

---

### EXAMPLE 1  USING MARKUP TO FIND THE SELLING PRICE

An appliance wholesaler has found that his percent markup must be 25%. If he buys a washing machine for $285, what is the selling price?

**Method 1**

$$A \quad = \quad P\% \quad \cdot \quad B$$

$$\text{markup} = \text{percent markup} \cdot \text{cost}$$

$$\downarrow \qquad\qquad \downarrow \qquad\qquad \downarrow$$

$$\text{markup} = \quad (0.25) \quad \cdot \$285$$

$$= \$71.25$$

**Method 2**

$$\frac{A}{B} = \frac{P}{100}$$

$$\frac{\text{markup}}{\text{cost}} = \frac{\text{percent markup}}{100}$$

$$\frac{\text{markup}}{\$285} = \frac{25}{100}$$

$$\text{markup}(100) = (25)(\$285)$$

$$\text{markup} = \frac{(25)(\$285)}{100}$$

$$= \$71.25$$

The markup is $71.25, and the selling price is the sum of the markup and the cost.

$$\begin{array}{ll} \$285.00 & \text{Cost} \\ + \quad 71.25 & \text{Markup} \\ \hline \$356.25 & \text{Selling price} \end{array}$$

The dealer must sell the washing machine for $356.25.

### PRACTICE EXERCISE 1

In order to stay in business, the owner of a clothing store must have a percent markup of 85%. If he buys a suit for $125.00, what is the selling price of the suit?

Answer: $231.25

---

### EXAMPLE 2  FINDING A MARKUP AND THE PERCENT MARKUP

The Shoe Store buys one type of shoe for $26.50 per pair and sells them for $34.45. What is the markup and the percent markup?

Find the markup first by finding the difference between the selling price and the cost.

$$\begin{array}{l} \$34.45 \\ - \quad 26.50 \\ \hline \$ \ 7.95 \end{array} \quad \text{Selling price} - \text{cost} = \text{markup}$$

The markup is $7.95. The percent markup is found by answering the question: $7.95 is what percent of $26.50?

### PRACTICE EXERCISE 2

A retailer buys a radio for $34.50 and sells it for $48.30. What is the markup and the percent markup?

Let $P$ be the percent markup.

| Method 1 | Method 2 |
|---|---|

**Method 1**

$$A = P\% \cdot B$$

markup = percent markup · cost

$$7.95 = P(0.01) \cdot 26.50$$

$$7.95 = P(0.265)$$

$$\frac{7.95}{0.265} = P$$

$$30 = P$$

**Method 2**

$$\frac{A}{B} = \frac{P}{100}$$

$$\frac{\text{markup}}{\text{cost}} = \frac{\text{percent markup}}{100}$$

$$\frac{7.95}{26.50} = \frac{P}{100}$$

$$(7.95)(100) = (26.50)P$$

$$\frac{(7.95)(100)}{26.50} = P$$

$$30 = P$$

The percent markup is 30%.

Answer: **$13.80; 40%**

---

| **EXAMPLE 3** FINDING THE COST AND SELLING PRICE | **PRACTICE EXERCISE 3** |
|---|---|

A clothing wholesaler has a 40% markup. If the markup on a suit is $86.50, what is the cost? What is the selling price?

A store owner operates on a 45% markup. If the markup on a stereo is $108, what is the cost? What is the selling price?

**Method 1**

$$A = P\% \cdot B$$

markup = percent markup · cost

$$86.50 = (0.40) \cdot \text{cost}$$

$$\frac{86.50}{0.40} = \text{cost}$$

$$216.25 = \text{cost}$$

**Method 2**

$$\frac{A}{B} = \frac{P}{100}$$

$$\frac{\text{markup}}{\text{cost}} = \frac{\text{percent markup}}{100}$$

$$\frac{86.50}{\text{cost}} = \frac{40}{100}$$

$$(86.50)(100) = \text{cost}(40)$$

$$\frac{(86.50)(100)}{40} = \text{cost}$$

$$216.25 = \text{cost}$$

The cost to the wholesaler is $216.25.

$$\text{selling price} = \text{cost} + \text{markup}$$

$$= \$216.25 + \$86.50$$

$$= \$302.75$$

Thus, the selling price is $302.75.

Answer: **$240; $348**

---

| **EXAMPLE 4** USING MARKDOWN TO FIND A NEW SELLING PRICE | **PRACTICE EXERCISE 4** |
|---|---|

The wholesale dealer in Example 3 must reduce the selling price of his suits by 30% in order to make room for new suits. What is the markdown and the new selling price of the suit that sold for $302.75?

When the stereo did not sell, the owner in Practice Exercise 3 was forced to reduce the selling price by 30%. What was the markdown and the new selling price of the stereo which sold for $348?

**Method 1**

$$A = P\% \cdot B$$

markdown = percent markdown · price

$$\text{markdown} = (0.30) \cdot \$302.75$$

$$= \$90.83 \quad \text{Nearest cent}$$

Method 2

$$\frac{A}{B} = \frac{P}{100}$$

$$\frac{markdown}{price} = \frac{percent\ markdown}{100}$$

$$\frac{markdown}{\$302.75} = \frac{30}{100}$$

$$markdown(100) = (\$302.75)(30)$$

$$markdown = \frac{(\$302.75)(30)}{100}$$

$$= \$90.83$$

Then we have

new selling price = old selling price − markdown.

$$= \$302.75 - \$90.83$$

$$= \$211.92$$

The suit now sells for $211.92.

Answer: **$104.40; $243.60**

Notice that the new selling price in Example 4 is actually less than the cost given in Example 3.

$$\begin{array}{r} \$216.25 \\ -\ \ \ 211.92 \\ \hline \$\ \ \ 4.33 \end{array}$$

The suit was sold for $4.33 less than the cost. (The dealer lost on the sale.) This means that the 30% markdown ($90.83) is more than the 40% markup ($86.50). That is,

30% of $302.75  ($90.83)

is more than

40% of $216.25  ($86.50).

This shows that we need to keep in mind the amount on which the percent is taken as well as the percent rate when working with markup and markdown problems.

---

| **EXAMPLE 5**  MARKUP AND MARKDOWN TOGETHER |
| --- |

The Green Grocery Company bought lettuce for 30¢ per head. The markup on the cost was 80%. After a few days, the price of the remaining lettuce was reduced by 60% of the selling price.

**(a)** What was the markup and the first selling price?

To find the markup, take 80% of 30¢.

$$(0.80)(30¢) = 24¢$$

The first selling price was

$$30¢ + 24¢ = 54¢.$$

| **PRACTICE EXERCISE 5** |
| --- |

Booker's Books bought a book for $3.00. The markup on cost was 75%. When the book did not sell, it was reduced 40% of the selling price.

**(a)** What was the markup and the first selling price?

**(b)** What was the markdown and the second selling price?

The markdown is 60% of 54¢ (selling price).

$$(0.60)(54¢) = 32¢ \text{ (to the nearest cent)}$$

To find the second selling price, subtract 32¢ from 54¢.

$$54¢ - 32¢ = 22¢$$

**(c)** What was the loss on the lettuce sold after the markdown?

In order to sell the lettuce before it spoiled, the grocer sold it for 22¢. This was a loss of

$$30¢ - 22¢ = 8¢$$

per head. The grocer sold at an 8¢ loss to prevent a loss of 30¢ on each head spoiled.

**(d)** What percent of the cost was the loss?

The question we must answer is:

$$
\begin{array}{ccccc}
8¢ & \text{is} & \boxed{\text{what percent}} & \text{of} & 30¢ ? \\
\downarrow & \downarrow & \downarrow & & \downarrow \quad \downarrow \\
A & = & P\% & \cdot & B \\
\downarrow & & \downarrow & & \downarrow \\
8 & = & P(0.01) & \cdot & 30 \\
8 & = & P(0.3) & & \\
\dfrac{8}{0.3} & = & P & & \text{Related division sentence} \\
26.\overline{6} & = & P & &
\end{array}
$$

Thus, the percent loss based on the cost of the lettuce was $26.\overline{6}\%$.

**(b)** What was the markdown and the second selling price?

**(c)** What was the profit on the book sold after the markdown?

**(d)** What percent of the cost was the profit?

Answers: (a) $2.25; $5.25
(b) $2.10; $3.15   (c) $0.15
(d) 5%

---

## HINT

You have probably noticed that in many of the percent applications we have considered, a basic relationship is used which can be summarized by using $A = P\% \cdot B$ or $\dfrac{A}{B} = \dfrac{P}{100}$ in the following forms.

$$\textbf{Change = percent} \cdot \textbf{original amount}$$

*or*

$$\frac{\textbf{change}}{\textbf{original amount}} = \frac{\textbf{percent}}{\textbf{100}}$$

Now that you have experience with the basic terminology used in the various applications, you may want to remember this one formula and use it to cover a multitude of situations. For example, in a markup problem, *change* is *markup*, *percent* is *percent markup*, and *original amount* is *cost*.

## 6.8 EXERCISES A

*Solve the following problems.*

1. If the markup is 20%, what is the selling price of a set of glasses that costs $26?

2. Ace Appliances bought a refrigerator for $216.00 and sold it for $280.80. What were the markup and the percent markup?

3. Roberta's Interior Design made $3.50 per yard on carpet sold to a customer. If the markup was 40%, what was the cost and what was the selling price?

4. To raise money for a project, the Student Club bought flowers for $2.00 per dozen and sold them on campus for $5.00 per dozen. What was the percent markup?

5. Because of a high spoilage rate, the markup on fruit must be 120%. What is the selling price (to the nearest cent) of peaches that cost 22¢ per pound?

6. Fred's TV Shop bought a color television set for $582 and sold it for $727.50. What was the percent markup?

7. The selling price of a toaster is $36.20. It is put on sale with a markdown (discount) rate of 30%. What is the reduced selling price?

8. A coat which sells for $125 is put on sale for $95. What is the markdown rate?

9. The markdown on a loaf of bread is 36¢ and the markdown rate is 40%. What was the old selling price? What is the new selling price?

10. The Tough Tool company bought a set of tools for $120. The markup was 20%. When the set did not sell, the price was reduced at a markdown rate of 15%. What was the first selling price and what was the second selling price?

11. If the tool set in Exercise 10 sold at the second price, what was the profit? What percent of the cost was the profit?

12. Mario's Men's Shop bought a suit for $180. The markup was 30%. Because of changing styles, the suit was sold at a discount of 25%. What was the loss? What percent of the cost was the loss?

## FOR REVIEW

13. Under normal conditions, Martha's car gets 28.5 mpg. While driving against a headwind, the mileage was 25.3 mpg. To the nearest tenth, what was the percent decrease?

14. Jenny got a 12% raise. If her old salary was $16,250, what is her new salary?

ANSWERS: 1. $31.20   2. $64.80; 30%   3. $8.75; $12.25   4. 150%   5. 48¢   6. 25%   7. $25.34   8. 24%   9. 90¢; 54¢   10. $144; $122.40   11. $2.40; 2%   12. $4.50; 2.5%   13. 11.2%   14. $18,200

## 6.8   EXERCISES B

*Solve the following problems.*

1. Hard Hardware bought a shipment of nails for $820.40. If the markup is 15%, what is the selling price?

2. Oranges cost a grocer 30¢ per pound and the markup is 70%. What is the selling price?

3. The Fresh Grocery Store buys tomatoes for 32¢ per pound and sells them for 68¢ per pound. What is the percent markup?

4. The Video Store made $12.90 on the sale of a video tape. If the markup was 30%, what were the cost and the selling price of the tape?

5. Fast Eddie bought a watch for $8.70 and sold it at a markup rate of 250%. What was the selling price of the watch?

6. The Fan Shop buys a ceiling fan for $125 and sells it for $195. What is the percent markup?

7. The selling price of a washer is $345.80. It is put on sale at a markdown rate of 35%. What is the reduced selling price?

8. A camera which sells for $289.75 is put on sale for $195.50. To the nearest tenth of a percent, what is the markdown rate?

9. The markup on books is 25%. There was a $2.05 markup on a novel. What was the original cost?

10. Lettuce was bought for 30¢ per head and sold at an 80% markup. The lettuce that did not sell the first day was reduced at a markdown rate of 50%. What was the first selling price and what was the second selling price?

11. If the lettuce in Exercise 10 sold at the second price, what was the loss? What percent of the cost was the loss?

12. A ball costing $10 has a markup rate of 40% and is then put on sale at a markdown of 20%. What is the profit? What is the percent profit based on the cost?

## FOR REVIEW

13. By driving at high speeds, Mark reduced his mileage by 18%. However, a tailwind gave him a 5% increase in mileage. Under these conditions, what was his mileage if normally it is 30 mpg?

14. Last year Bill earned $42,560.00, and this year his salary will be $47,241.60. What was the percent raise that Bill received?

## 6.8   EXERCISES C

*Solve.*

1. Some store owners operate on a 50%–10% markup rate. The cost of an item is first marked up 50%, and the resulting price is marked up an additional 10% to obtain the selling price. Under these conditions, what would be the selling price of a men's suit which costs the owner $105?

2. A lamp which was purchased by a dealer for $28.70 was to be sold at a 70.5% markup. When it did not sell, the dealer reduced the price at a markdown rate of 50%. Did the dealer make a profit or take a loss when the lamp was then sold? What was the percent profit or loss (to the nearest tenth of a percent)?
[Answer: loss of $4.23, 14.7%]

# CHAPTER 6 REVIEW

## KEY WORDS

**6.1** A **ratio** is a comparison of two numbers using division.

An equation which states that two ratios are equal is a **proportion.**

When two pairs of numbers have the same ratio, the numbers are **proportional.**

**Solving a proportion** is the process of finding a missing number in a proportion.

**6.2** The word **percent** means "per hundred" and refers to the number of parts in one hundred parts.

**6.3** In the basic percent problem "*A* is *P*% of *B*," *A* is the **amount,** *P* is the **percent,** and *B* is the **base.**

**6.4** A tax charged on purchases is a **sales tax** and the percent of tax is the **sales-tax rate.**

**6.5** Income received as a percent of total sales is called a **commission** and the percent used is the **commission rate.**

A **discount** is a reduction in the regular price of an item based on a **discount rate** which is a percent of the *regular* price. The **sale price** is the price after the discount.

**6.6** Money paid for the use of money is called **interest.** The money borrowed or invested is the **principal,** and the **rate of interest** is the percent applied to the principal.

Interest found using only the principal and not using previous interest along with the principal is called **simple interest.**

Interest earned or paid on previous interest along with the principal is called **compound interest.**

**6.7** The percent by which a quantity increases (decreases) is called the **percent increase (decrease).**

**6.8** The amount by which the selling price of an item is increased over the cost is called **markup,** and the percent used to calculate this increase is the **percent markup.**

The **markdown** of an item is the price reduction based on the **markdown rate** applied to the selling price of the item.

## KEY CONCEPTS

**6.1** **1.** Ratios can be expressed in three ways, $2:3$ or 2 to 3 or $\frac{2}{3}$. The latter form is the fractional form of the ratio.

**2.** To solve a proportion involving a missing number $a$, equate the cross products and divide the numerical cross product by the multiplier of $a$.

**6.2** **1.** To change a percent to a fraction, replace the % symbol with either $\left( \times \frac{1}{100} \right)$ or $(\div 100)$. Then evaluate and reduce the fraction.

**2.** To change a percent to a decimal, replace the % symbol with $(\times 0.01)$ and multiply. This is the same as moving the decimal point two places to the left.

**3.** To change a decimal to a percent, multiply the decimal by 100 and attach the % symbol. This is the same as moving the decimal point two places to the right.

**4.** To change a fraction to a percent, multiply the fraction by 100 and attach the % symbol.

**6.3** **1.** The basic percent equation is

$$A = P\% \cdot B.$$

To solve for $P$ or $B$, use the related division sentences

$$P\% = \frac{A}{B} \quad \text{or} \quad B = \frac{A}{P\%}.$$

**2.** The basic percent proportion is

$$\frac{A}{B} = \frac{P}{100}.$$

Solve for any of $A$, $B$, or $P$ by solving the proportion.

**6.4** **1.** sales tax = tax rate (0.01) · price before tax

*or*

$$\frac{\text{sales tax}}{\text{price before tax}} = \frac{\text{tax rate}}{100}$$

*and*

total price = price before tax + sales tax

**2.** In 1989, the Social Security-tax rate was 7.51% on the first $48,000 of earnings.

Social Security tax = (7.51)(0.01) · total earnings

*or*

$$\frac{\text{Social Security tax}}{\text{total earnings}} = \frac{7.51}{100}$$

**3.** Use the tax-rate schedule given in Section 6.4 to calculate income taxes paid in 1989 by a single taxpayer.

**6.6** **1.** If $P$ is a principal, $R$ is a yearly rate of interest converted to a decimal, $T$ is the time in years, and $I$ is simple interest, then $I = P \cdot R \cdot T$. If $A$ is the amount to be repaid or the amount in the account, then $A = P + I$.

**2.** To find the amount of each payment on a loan, divide the sum of the principal and interest by the number of payments.

**3.** In a compound interest problem, the interest rate used for each period is

$$\frac{\text{Annual Interest Rate}}{\text{Number of Periods Per Year}}.$$

For example, if the annual interest rate is 10%, if we compound annually (1 time a year), semiannually (2 times a year), or quarterly (4 times a year), the interest rate per period is 10%, 5%, and 2.5%, respectively.

**4.** In a compound interest problem, the principal for the second period is the original principal plus the interest charged for the first period. This process continues over each period.

**6.7** **1.** increase (decrease) = percent increase (decrease) · original amount

*or*

$$\frac{\text{increase (decrease)}}{\text{original amount}} = \frac{\text{percent increase (decrease)}}{100}$$

*and*

new amount = original amount + increase

*or*

new amount = original amount − decrease

**2.** Remember that in a percent increase (decrease) problem, the percent is taken on the original amount *not* the new amount.

**6.8** **1.** markup (markdown) = percent markup (markdown) · cost (price)

*or*

$$\frac{\text{markup (markdown)}}{\text{cost (price)}} = \frac{\text{percent markup (markdown)}}{100}$$

*and*

selling price = cost + markup

*or*

new selling price = price − markdown

**2.** Percent markup is based on cost.

**3.** Percent markdown is based on selling price.

## REVIEW EXERCISES

*Part I*

**6.1** **1.** Write the fraction form for the ratio 8 to 5.

**2.** Write the ratio which is indicated by the fraction $\frac{1}{8}$.

*Solve the proportions.*

**3.** $\dfrac{a}{8} = \dfrac{1}{2}$

**4.** $\dfrac{7}{21} = \dfrac{6}{a}$

**5.** A boat can travel 140 miles on 35 gallons of gas. How far can it travel on 56 gallons of gas?

**6.** If $\frac{1}{2}$ inch on a map represents 20 miles, how many miles will be represented by $5\frac{1}{2}$ inches?

**6.2**    **7.** *Percent* means "per _____."

*Change from percents to fractions.*

**8.** 23%          **9.** 324%          **10.** 0.6%          **11.** $6\frac{1}{4}\%$

*Change from percents to decimals.*

**12.** 84%          **13.** 267%          **14.** $\frac{3}{4}\%$          **15.** 0.3%

*Change to percents.*

**16.** 0.63          **17.** 0.032          **18.** 3.2          **19.** $\frac{3}{5}$

*Solve the following percent problems.*

**6.3**    **20.** What is 30% of 620?

**21.** 6 is 0.5% of what number?

**22.** What percent of 72 is 9?

**23.** On a test with 300 questions, Susan got 86% correct. How many did she answer correctly?

**24.** A basketball player hit 19 of 26 shots. To the nearest percent, what percent of his shots did he make?

**25.** There are 16 grams of alcohol in a solution which is 25% alcohol. How much does the solution weigh?

**6.4**    **26.** What is the sales tax and total cost of a $17.50 shirt if the tax rate is 6%?

**27.** If a sales tax of $13.97 is charged on the purchase of a $254 dryer, what is the sales-tax rate?

**28.** Raymond had an income from wages of $17,350 in 1989. The Social Security tax rate was 7.51% on all wages up to $48,000 that year. What Social Security tax did Raymond pay?

**29.** Diane had a taxable income of $47,300 in 1989. Use the tax table in Section 6.4 to determine how much income tax she paid.

**6.5**    **30.** Peter got a commission of $1530 for selling an antique car. If the selling price was $10,200, what was his commission rate?

**31.** Wanda receives 2% commission on all car sales of $25,000 or less. She receives 3% commission on all sales above $25,000. What is her commission on sales of $42,500?

**32.** A lawn mower normally sells for $216. At the end of the summer, it was put on sale at a 30% discount. What was the sale price?

**33.** Find the discount rate if a $622.50 ring is on sale for $510.45.

**6.6**    **34.** Phillip bought a watch for $285. He paid 10% down and paid the remaining amount, plus interest, at the end of 6 months. If simple interest is 14% per year, how much did he pay at the end of 6 months?

**35.** William invests $3500 in an account which pays 8% compounded semiannually. How much will he have in the account at the end of 2 years?

**6.7**    **36.** State University expects a 4.5% decrease in students next year. If the current enrollment is 26,820, what will be the enrollment next year?

**37.** The price of tomatoes increased from 42¢ per pound to 67¢ per pound. To the nearest tenth of a percent, what was the percent increase?

**6.8**    **38.** A dealer bought a microwave oven for $682.50 and sold it for $819.00. What was the percent markup?

**39.** An electric blanket which normally sells for $42.00 is put on sale for $29.40. What is the markdown rate?

*Part II*

*Solve.*

**40.** Sam's car usually gets 28 mpg. He is driving on a rough road which will decrease his mileage by 14%, but a tailwind is increasing his mileage by 9%. What mileage should he get?

**41.** If the markup on a motorcycle is 15%, what is the selling price if the cost is $1020?

**42.** The average salary of the workers at Steel, Incorporated, was $19,280. After the union negotiated an 11% increase, what was the average salary?

**43.** Shelley borrowed $722.50 to buy a stereo. If simple interest is 15% per year, what are her monthly payments if she pays for 18 months?

**44.** Faye receives a commission of 20% on clothing sales. She sold $2868.50 worth of clothing during the month of May. What was her commission?

**45.** The sales-tax rate in Newton, Vermont, is 4%. If a tax of $2.58 is charged on a dress, what is the price of the dress?

**46.** The Shirt Shop bought one type of shirt for $14.50. The markup was 28%. Because of oversupply, the shirts were put on sale at a markdown of 25%. What was the percent of profit or loss based on the original cost?

**47.** Bernadine makes $428 a week as a secretary. Her boss told her that starting next week she will receive a raise of $8\frac{1}{2}$% per week. What will be her new weekly salary?

**48.** If $1000 increased to $1120 in a savings account one year, what was the simple interest rate for the year?

**49.** If the sales-tax rate in Stephentown, New York, is 6%, what tax would be charged on a purchase of $12.50?

*Complete the table given one value in each row.*

| | Fraction | Decimal | Percent | | Fraction | Decimal | Percent |
|---|---|---|---|---|---|---|---|
| **50.** | $\frac{1}{4}$ | ___ | ___ | **51.** | ___ | 0.42 | ___ |
| **52.** | ___ | ___ | 35% | **53.** | ___ | 1.3 | ___ |
| **54.** | $\frac{4}{3}$ | ___ | ___ | **55.** | ___ | ___ | 0.5% |

**56.** Solve the proportion.

$$\frac{a}{3} = \frac{5}{15}$$

**57.** A car can go 450 miles on 18 gallons of gas. How far can the car go on 32 gallons of gas?

**58.** 6 is what percent of 20?

**59.** What is 12% of 37?

**60.** 80.75 is 85% of what number?

---

ANSWERS: 1. $\frac{8}{5}$  2. 1:8 or 1 to 8  3. 4  4. 18  5. 224 miles  6. 220 miles  7. hundred  8. $\frac{23}{100}$  9. $\frac{81}{25}$  10. $\frac{3}{500}$  11. $\frac{1}{16}$  12. 0.84  13. 2.67  14. 0.0075  15. 0.003  16. 63%  17. 3.2%  18. 320%  19. 60%  20. 186  21. 1200  22. 12.5%  23. 258  24. 73%  25. 64 grams  26. $1.05; $18.55  27. 5.5%  28. $1302.99  29. $10,952.50  30. 15%  31. $1025  32. $151.20  33. 18%  34. $274.46  35. $4094.50  36. 25,613  37. 59.5%  38. 20%  39. 30%  40. 26.6 mpg  41. $1173  42. $21,400.80  43. $49.17  44. $573.70  45. $64.50  46. 4% (loss)  47. $464.38  48. 12%  49. $0.75  50. 0.25; 25%  51. $\frac{21}{50}$; 42%  52. $\frac{7}{20}$; 0.35  53. $\frac{13}{10}$; 130%  54. 1.$\overline{3}$; 133.$\overline{3}$% or $133\frac{1}{3}$%  55. $\frac{1}{200}$; 0.005  56. 1  57. 800 miles  58. 30%  59. 4.44  60. 95

**1.** Write the fraction form for the ratio 13 to 7.

1. _____

**2.** Solve the proportion. $\dfrac{12}{a} = \dfrac{72}{18}$

2. _____

**3.** A steel cable 35 feet long weighs 105 pounds. How much will 93 feet of the same cable weigh?

3. _____

**4.** Change 95% to a fraction in lowest terms.

4. _____

**5.** Change 0.8% to a decimal.

5. _____

**6.** Change 5.06 to a percent.

6. _____

**7.** Change $\frac{9}{5}$ to a percent.

7. _____

*Solve.*

**8.** 12 is what percent of 60?

8. _____

**9.** What is 18% of 120?

9. _____

**10.** 160 is 400% of what number?

10. _____

**11.** The Social Security tax rate was 7.51% in 1989. How much Social Security tax did Winston pay on a salary of $15,400?

11. _____

**CONTINUED**

12. A toaster with a normal price of $29.50 is to be discounted 30%. What is the sale price?

12. _____

13. Herbert bought a TV for $720 and paid 25% down. If simple interest is 13%, how much will he have to pay at the end of a year to pay off the debt?

13. _____

14. While driving with a headwind, Tina's car got 22.8 mpg instead of the normal 27.9 mpg. To the nearest tenth, what was the percent decrease?

14. _____

15. A refrigerator was bought for $820 and put on sale at a markup rate of 20%. It was then reduced at a markdown rate of 15%. What was the percent profit or loss based on cost?

15. _____

16. Wylie Smith, sports information director, makes a profit for NAU of 40¢ on the sale of each basketball program. If the markup rate is 25%, what is the cost of each program, and what is the selling price?

16. _____

17. It costs $29.90 for a tune-up at Andy's Garage. If Andy offers a Saturday special at 30% off the regular price, how much will a tune-up cost on Saturday?

17. _____

18. If an account worth $12,500 increased to $14,000 in one year, what was the simple interest rate for the year?

18. _____

# Introduction to Statistics

## 7.1 CIRCLE GRAPHS AND BAR GRAPHS

STUDENT GUIDEPOSTS

1 Statistics and Numerical Data     3 Bar Graphs
2 Circle Graphs

### 1 STATISTICS AND NUMERICAL DATA

**Statistics** is the branch of mathematics that deals with collecting, organizing, and summarizing numerical data in order to describe or interpret it. The world of sports relies on statistics such as batting averages, pass completions, free-throw percentages, and points scored. The number of people under age 30 in the United States, the average study time of junior college students, and the Michigan election results are all statistics.

Tables provide one way to present data in a manner that is easy to interpret. Suppose Tanya, a college student, has decided to keep a record of her monthly expenses. She might summarize the data in the table.

| Monthly expenses | |
| --- | --- |
| *Item* | *Amount spent* |
| Rent | $200 |
| Food | $150 |
| Entertainment | $100 |
| Clothes | $ 35 |
| Other | $ 15 |
| TOTAL | $500 |

### 2 CIRCLE GRAPHS

Sometimes it is hard to interpret a large set of data. One way to simplify the process is to ''picture'' the data in some way. A **circle graph** is often used to represent parts of a whole unit. For example, the data collected by Tanya and presented in the table above could be organized and displayed in a circle graph. The circle graph in Figure 7.1 represents her entire budget, and each pie-shaped part, or sector, represents the number of dollars she spent on each item.

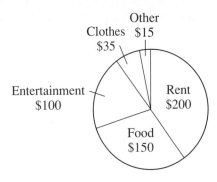

**Figure 7.1** Circle Graph

By reading information from a circle graph, various comparisons can be made in the form of ratios. Consider the circle graph in Figure 7.1 again. Since the total amount Tanya spent during the month was $500, the ratio of rent to total expenses is

$$\frac{\$200}{\$500} = \frac{2}{5},$$

and the ratio of clothing expenses to total expenses is

$$\frac{\$35}{\$500} = \frac{7}{100}.$$

Converting these ratios to percents, Tanya sees that she spent 40% of her budget on rent and 7% on clothes.

Tanya can also find the ratio of one expense to another. For example, the ratio of food expenses to entertainment expenses is

$$\frac{\$150}{\$100} = \frac{3}{2},$$

and the ratio of rent to clothing expenses is

$$\frac{\$200}{\$35} = \frac{40}{7}.$$

Thus, for every $3 spent on food, Tanya spends $2 on entertainment, and for every $40 spent on rent, she spends $7 on clothes.

| **EXAMPLE 1** READING A CIRCLE GRAPH | **PRACTICE EXERCISE 1** |

The circle graph in Figure 7.2 shows the distribution of Mr. Green's monthly income.

Use the circle graph in Figure 7.2 to answer the following.

**(a)** If Mr. Green's income is $2800 each month, what amount does he pay in federal income tax?

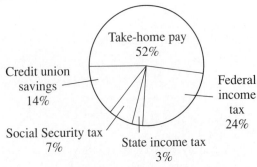

**Figure 7.2** Circle Graph

**(a)** If Mr. Green's income is $2800 each month, determine his take-home pay.

    Since the percent of take-home pay is 52%, find

$$52\% \text{ of } \$2800.$$

$$\underset{0.52}{\downarrow} \quad \underset{\cdot}{\downarrow} \quad \underset{2800}{\downarrow} = 1456$$

Thus, Mr. Green takes home $1456 each month.

**(b)** What is the ratio of federal income tax to state income tax?
The ratio is

$$\frac{24\%}{3\%} = \frac{8}{1}.$$

Thus, for every $8 paid in federal income tax Mr. Green pays $1 in state income tax.

**(b)** What is the ratio of credit union savings to Social Security tax?

Answers: (a) $672    (b) $\frac{2}{1}$

## ❸ BAR GRAPHS

Another way to present data is with a **bar graph.** Suppose 100 people have been polled regarding their preference among four brands of toothpaste, Brand A, Brand B, Brand C, and Brand D. The results of the poll are given in the bar graph in Figure 7.3.

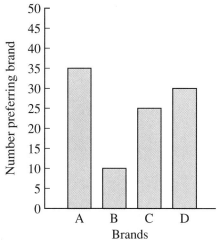

**7.3**  Bar Graph

    The graph shows that 35 people chose Brand A, 10 chose Brand B, 25 chose Brand C, and 30 chose Brand D.

    Bar graphs are quite useful when comparing data. It is easy to see at a glance which brand was most preferred (Brand A) and which was least preferred (Brand B). Also, Brand A was preferred to Brand B by a ratio of

$$\frac{35}{10} = \frac{7}{2}.$$

That is, for every 7 people who chose Brand A, 2 people chose Brand B.

| EXAMPLE 2 READING A BAR GRAPH | PRACTICE EXERCISE 2 |

Use the bar graph in Figure 7.3 to determine the ratio of people preferring Brand C to those preferring Brand B.

Since 25 people preferred Brand C and 10 people preferred Brand B, the ratio is $\frac{25}{10} = \frac{5}{2}$.

Use the bar graph in Figure 7.3 to determine the ratio of those preferring Brand A over Brand C.

Answer: $\frac{7}{5}$

A double-bar graph is often used to compare two sets of data. The double-bar graph in Figure 7.4 shows the total enrollment in beginning algebra classes compared to intermediate algebra classes at a community college during a four-year period.

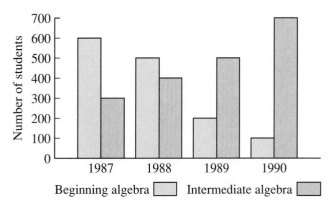

**Figure 7.4**   Double-Bar Graph

| EXAMPLE 3 READING A DOUBLE-BAR GRAPH | PRACTICE EXERCISE 3 |

Use the double-bar graph in Figure 7.4 to answer the following.

**(a)** What was the number of beginning algebra students in 1988?

Since the bar on the left in 1988 rises to the 500 level, there were 500 students in beginning algebra in 1988.

**(b)** Which year had the greatest number of intermediate algebra students?

The right bar rises highest in 1990. Thus, the greatest number of students in intermediate algebra occurred in 1990. In fact, that number was 700 students.

**(c)** What could you say about the number of beginning algebra students over the four-year period?

Since the number went from 600 to 500 to 200 to 100, we can see that the number of beginning algebra students declined over the four-year period.

Use the double-bar graph in Figure 7.4 to answer the following.

**(a)** What was the number of intermediate algebra students in 1989?

**(b)** Which year had more beginning algebra students, 1988 or 1989?

**(c)** What could you say about the number of intermediate algebra students over the four-year period?

Answers: (a) 500    (b) 1988
(c) The number increased.

## 7.1  EXERCISES A

*Exercises 1–8 refer to the circle graph below which shows a breakdown of the amounts spent by the Warrens on an addition to their house.*

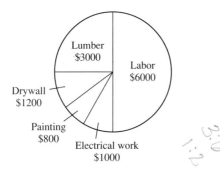

1. What was the total cost of the addition?

2. What was the largest expense item on the project?

3. What was the ratio of the cost of labor to the cost of lumber?

4. What was the ratio of the cost of painting to the cost of drywall?

5. What was the ratio of the cost of lumber to the cost of electrical work?

6. Lumber accounted for what percent of the total cost?

7. Labor accounted for what percent of the total cost?

8. Electrical work accounted for what percent of the total cost?

*Exercises 9–20 refer to the circle graph below which shows the distribution of the 42,500 students enrolled in various schools in Coconino County.*

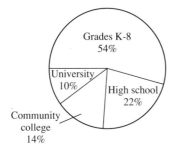

9. Which sector represents the largest number of students?

10. Which sector represents the smallest number of students?

11. How many students are in grades K–8?

12. How many students are in high school?

13. How many students are enrolled in the community college?

14. How many students are enrolled in the university?

**15** How many students are in grades K–8 and in high school?

**16.** How many students are in the community college and in the university?

**17.** What is the ratio of high school students to university students?

**18.** What is the ratio of students in grades K–8 to students in high school?

**19.** What is the ratio of high school students to students in the university and the community college?

**20.** How many students are in school beyond the high school level?

*Exercises 21–32 refer to the bar graph below which shows the number of students by major course of study in the College of Arts and Science.*

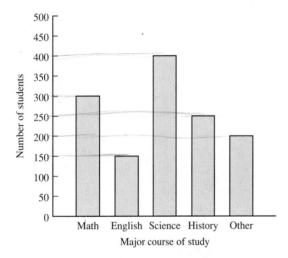

**21.** How many students are in the College of Arts and Science?

**22** What is the ratio of math majors to the total number of students in the college?

**23.** What percent of students in the college are math majors (to the nearest tenth of a percent)?

**24.** What is the ratio of science majors to the total number of students in the college?

**25** What percent of students in the college are science majors (to the nearest tenth of a percent)?

**26.** What percent of students are math and science majors (to the nearest tenth of a percent)?

**27.** What is the ratio of science majors to history majors?

**28.** What is the ratio of English majors to math majors?

**29.** What percent of students have a major other than math, science, English, or history (to the nearest tenth of a percent)?

**30.** What is the total of math and English majors?

**31.** What percent of the students are math and English majors (to the nearest tenth of a percent)?

**32.** What is the ratio of math and science majors to English and history majors?

*Exercises 33–40 refer to the double-bar graph below which shows the number of male and female athletes at a university over a four-year period.*

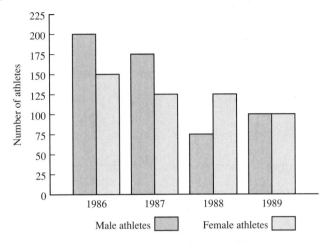

**33.** Which year had the greatest number of male athletes?

**34.** Which year had the smallest number of male athletes?

**35.** Which year had the greatest number of female athletes?

**36.** Which year had the smallest number of female athletes?

**37.** What was the ratio of male athletes to female athletes in 1986?

**38.** What was the ratio of female athletes to male athletes in 1988?

**39.** Which year had the same number of male and female athletes?

**40.** Would it be fair to conclude from the graph that some athletics programs were forced to be dropped? Explain.

---

ANSWERS: 1. $12,000   2. labor   3. $\frac{2}{1}$ or 2 to 1   4. $\frac{2}{3}$ or 2 to 3   5. $\frac{3}{1}$ or 3 to 1   6. 25%   7. 50%   8. $8.\overline{3}$% or $8\frac{1}{3}$%   9. grades K–8   10. university students   11. 22,950   12. 9350   13. 5950   14. 4250   15. 32,300   16. 10,200   17. $\frac{11}{5}$ or 11 to 5   18. $\frac{27}{11}$ or 27 to 11   19. $\frac{11}{12}$ or 11 to 12   20. 10,200   21. 1300   22. $\frac{3}{13}$ or 3 to 13   23. 23.1%   24. $\frac{4}{13}$ or 4 to 13   25. 30.8%   26. 53.8%   27. $\frac{8}{5}$ or 8 to 5   28. $\frac{1}{2}$ or 1 to 2   29. 15.4%   30. 450   31. 34.6%   32. $\frac{7}{4}$ or 7 to 4   33. 1986   34. 1988   35. 1986   36. 1989   37. $\frac{4}{3}$ or 4 to 3   38. $\frac{5}{3}$ or 5 to 3   39. 1989   40. Yes; the number of males and females seemed to decline over the period.

## 7.1  EXERCISES B

*Exercises 1–8 refer to the circle graph below which gives the approximate land area of the seven continents in millions of square miles.*

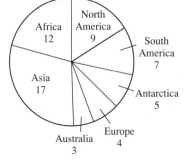

1. What is the approximate total land area of the seven continents?

2. What is the largest continent by area?

3. What is the ratio of the area of Africa to the area of North America?

4. What is the ratio of the area of Australia to the area of South America?

5. What is the ratio of the area of Europe to the area of Asia?

6. To the nearest tenth of a percent, the area of Asia is what percent of the total land area?

7. To the nearest tenth of a percent, the area of North America is what percent of the total land area?

8. North and South America together account for what percent of the total land area (to the nearest tenth of a percent)?

*Exercises 9–20 refer to the circle graph below which shows how a university spends its funds totalling $25,000,000 in a given year.*

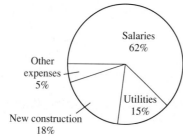

9. Which sector represents the greatest expenditure?

10. Which sector represents the smallest expenditure?

11. How much was spent on salaries?

12. How much was spent on utilities?

13. How much was spent on new construction?

14. How much was spent on other expenses?

15. How much was spent on salaries and utilities?

16. How much was spent on utilities and new construction?

17. What is the ratio of utility expenses to new-construction expenses?

18. What is the ratio of salaries to utilities?

19. What is the ratio of salaries to the expenses for new construction and utilities?

20. How much was spent in all categories besides salaries?

*Exercises 21–32 refer to the bar graph at the right which shows the number of new cars sold by a dealer during a five-month period.*

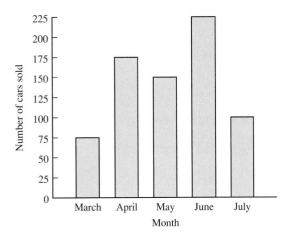

**21.** How many cars were sold in this five-month period?

**22.** What is the ratio of cars sold in March to the total number sold in the five-month period?

**23.** What percent of the cars sold in the five-month period were sold in March (to the nearest tenth of a percent)?

**24.** What is the ratio of cars sold in June to the total number sold in the five-month period?

**25.** What percent of the cars sold in the five-month period were sold in June (to the nearest tenth of a percent)?

**26.** What percent of the cars sold in the five-month period were sold in June and July (to the nearest tenth of a percent)?

**27.** What is the ratio of cars sold in June to cars sold in May?

**28.** What is the ratio of cars sold in March to cars sold in July?

**29.** What percent of the cars sold in the five-month period were sold in a month other than July (to the nearest tenth of a percent)?

**30.** What is the total of the cars sold in March and April?

**31.** What percent of the cars sold in the five-month period were sold in March and April (to the nearest tenth of a percent)?

**32.** What is the ratio of the cars sold in March, April, and May to the cars sold in June and July?

*Exercises 33–40 refer to the double-bar graph at the right which shows the quarterly profits for a large corporation in millions of dollars for the years 1989 and 1990.*

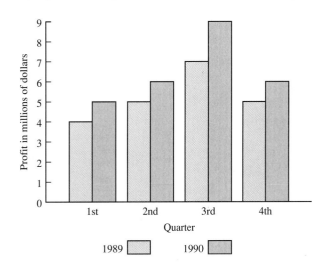

**33.** Which quarter showed the greatest profit in 1989?

**34.** Which quarter showed the smallest profit in 1989?

**35.** Which quarter showed the greatest profit in 1990?

**36.** Which quarter showed the smallest profit in 1990?

**37.** What was the profit in the 2nd quarter of 1989?

**38.** What was the profit in the 3rd quarter of 1990?

**39.** What was the total profit in all four quarters in 1990?

**40.** Would it be fair to say that during both years there was a steady increase in profits through the first three quarters and a decline during the fourth quarter? Explain.

## 7.1 EXERCISES C

*A die was rolled 600 times and the number of times that each number (1, 2, 3, 4, 5, or 6) came up was recorded. The results of this experiment are shown in the bar graph at the right. The* statistical probability *of obtaining a particular number is defined as the ratio of the number of times that number came up on the die to the total number of times the die is rolled. Use this information in Exercises 1–6 to give the statistical probability of rolling the particular number.*

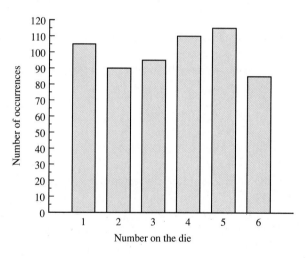

**1.** 1          **2.** 2          **3.** 3          **4.** 4          **5.** 5          **6.** 6

$\left[\text{Answer: } \frac{7}{40}\right]$

## 7.2 BROKEN-LINE GRAPHS AND HISTOGRAMS

═══ STUDENT GUIDEPOSTS ═══

**1** Broken-Line Graphs          **3** Frequency Polygons
**2** Histograms

### **1** BROKEN-LINE GRAPHS

A **broken-line graph** is often used to show a trend in data collected over a period of time. For example, the broken-line graph in Figure 7.5 shows the number of traffic citations issued by a police department over a six-month period. Each dot shows the number of tickets given during that month.

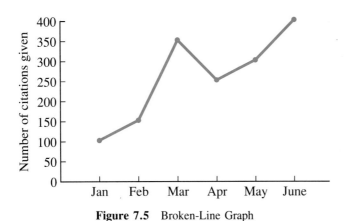

**Figure 7.5** Broken-Line Graph

| EXAMPLE 1 READING A BROKEN-LINE GRAPH |
|---|

Refer to the broken-line graph in Figure 7.5.

**(a)** In which month were 350 tickets given out?

Since the dot above March is at the 350 level, 350 tickets were issued in March.

**(b)** How would you briefly describe the trend in giving out tickets over the six-month period?

Basically the number of tickets given increased from month-to-month with the exception of March to April.

| PRACTICE EXERCISE 1 |
|---|

Refer to the broken-line graph in Figure 7.5.

**(a)** How many citations were issued in April?

**(b)** How many tickets were given out during the six-month period?

Answers: (a) **250**   (b) **1550**

Two broken-line graphs in the same figure are often used to show a comparison between two sets of data. For example, the broken-line graphs in Figure 7.6 show the quarterly profits realized by a company during the years 1988 and 1989. By showing both graphs together, it is easy to see at a glance how one year compared with the other.

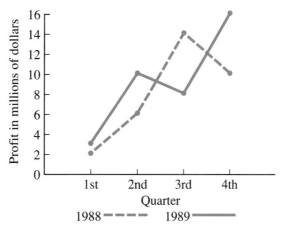

**Figure 7.6** Double-Line Graph

| EXAMPLE 2 READING A DOUBLE-LINE GRAPH |
|---|

Refer to the graph in Figure 7.6.

**(a)** What was the profit made during the first quarter of 1989?

Since the dot on the solid line is halfway between the 2 and 4 million marks, the profit this quarter was 3 million dollars.

| PRACTICE EXERCISE 2 |
|---|

Refer to the graph in Figure 7.6.

**(a)** What was the profit made during the 3rd quarter of 1988?

**(b)** Between which two consecutive quarters in 1988 was there a decrease in profit?

   The dashed line graph goes down from 14 million dollars to 10 million dollars between the 3rd and 4th quarters of 1988.

**(c)** In which year did the company show the greater profit?

   Adding up the quarterly profits in 1988 we obtain

$$2 + 6 + 14 + 10 = 32 \text{ million dollars.}$$

   Adding up the quarterly profits in 1989 we obtain

$$3 + 10 + 8 + 16 = 37 \text{ million dollars.}$$

Thus, the greater profit was made in 1989.

**(b)** Between which two consecutive quarters in 1989 did the profit increase the most?

**(c)** Which year had the better profit change between the 3rd and 4th quarters?

Answers: (a) **14 million dollars**
(b) **between the 3rd and 4th quarters** (c) **1989**

## ② HISTOGRAMS

Suppose sixty men in an organization list their heights in inches. The resulting data might be somewhat difficult to understand due to the large number of heights. To make the data easier to interpret, the heights could be divided into groups called **class intervals.** We might use 66 inches–68 inches as one interval, 68 inches–70 inches as another, and so forth. The number of heights in each class interval is called the class frequency. The table below summarizes the data collected.

| Height in inches (class intervals) | Number of men (class frequency) |
| --- | --- |
| 66–68 | 4 |
| 68–70 | 12 |
| 70–72 | 24 |
| 72–74 | 18 |
| 74–76 | 2 |

Suppose we assume that a person whose height is exactly 68 inches is in the class interval 68–70 and not in 66–68, and similarly for persons whose height is 70 inches, 72 inches, and 74 inches. Then each of the sixty men has his height in exactly one of the class intervals. A special type of bar graph, a **histogram** shown in Figure 7.7, is used to picture the information in the table. The width of each bar represents the range of values in that class interval, and the length or height of the bar corresponds to the **class frequency.**

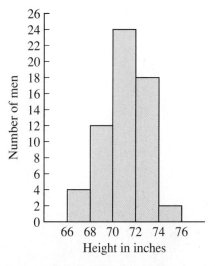

**Figure 7.7** Histogram

| EXAMPLE 3   READING A HISTOGRAM | PRACTICE EXERCISE 3 |
|---|---|

Use the histogram in Figure 7.7 to answer the following.

**(a)** How many men are 70 inches or taller?

We must add the class frequencies for the class intervals 70–72, 72–74, and 74–76.

$$24 + 18 + 2 = 44$$

There are 44 men 70 inches or taller.

**(b)** Find the ratio of men 68 to 70 inches tall to those 72 to 74 inches tall.

Since there are 12 men with heights between 68 and 70 inches and 18 men with heights between 72 and 74 inches, the ratio is

$$\frac{12}{18} = \frac{2}{3} \text{ or 2 to 3.}$$

**(c)** What percent of the men in the organization are between 68 and 70 inches in height?

Since there are 12 men 68 to 70 inches tall out of a total of 60 men in the organization, the question is: 12 is what percent of 60?

Method 1 (Equation Method)

$$\begin{array}{ccc} A = & P\% & \cdot B \\ \downarrow & \downarrow & \downarrow \\ 12 = & P(0.01) & \cdot 60 \end{array}$$

$$12 = P(0.60)$$

The related division sentence is:

$$P = \frac{12}{0.60}$$

$$P = 20$$

Method 2 (Proportion Method)

$$\frac{A}{B} = \frac{P}{100}$$

$$\frac{12}{60} = \frac{P}{100}$$

$$12(100) = 60(P)$$

The related division sentence is:

$$P = \frac{12(100)}{60}$$

$$P = 20$$

Thus, 20% of the men are between 68 and 70 inches tall.

Use the histogram in Figure 7.7 to answer the following.

**(a)** How many men are shorter than 72 inches?

**(b)** Find the ratio of men with heights between 66 and 68 inches to those between 74 and 76 inches.

**(c)** What percent of the men in the organization are between 68 and 74 inches in height?

Answers: (a) 40   (b) $\frac{2}{1}$ or 2 to 1
(c) 90%

## ❸ FREQUENCY POLYGONS

Sometimes data that is divided into class intervals is displayed using a special kind of broken-line graph. For instance, suppose we find the midpoint of each class interval, called the **class midpoint,** in the example above. The class midpoint of the class interval 66–68 inches is 67 inches, and the remaining class midpoints are 69 inches, 71 inches, 73 inches, and 75 inches. Locate the class midpoints in the histogram in Figure 7.7, and place a dot at the corresponding point on the top of each bar. When these dots are connected, as shown in Figure 7.8, the resulting broken-line graph is called a **frequency polygon.**

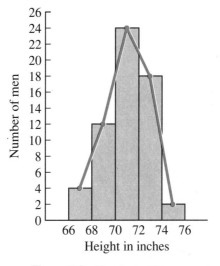

**Figure 7.8**   Frequency Polygon

---

**HINT**

Histograms and frequency polygons are used when data collected consists of a wide range of values such as numbers obtained by measurements.

---

## 7.2   EXERCISES A

*The number of personal computers sold by a dealer in a six-week period is shown in the broken-line graph at the right. Use this graph in Exercises 1–10.*

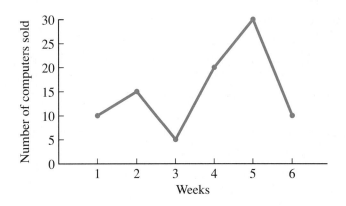

1. How many computers were sold in week 2?

2. During which week(s) were 10 computers sold?

3. How many computers were sold during the six-week period?

4. What was the ratio of computers sold in week 5 to week 3?

5. During which week did the dealer sell the most computers?

6. During which week did the dealer sell the least computers?

7. Between which two consecutive weeks did the sales increase the most?

8. Between which two consecutive weeks did the sales decrease the most?

9. What percent of the total sales were made during week 5?

10. What percent of the total sales were made during week 1?

*The double-line graph at the right shows the comparison between new subscribers signed up for cable television during a four-month period in 1988 and 1989. Use this graph in Exercises 11–18.*

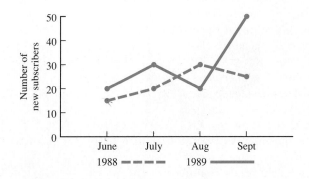

**11.** How many new subscribers were there in June of 1988?

**12.** How many new subscribers were there in August of 1989?

**13.** How many new subscribers were there in the four months of 1988?

**14.** How many new subscribers were there in the four months of 1989?

**15.** To the nearest percent, what was the percent increase or decrease in new subscribers from 1988 to 1989 during the four-month period?

**16.** What was the ratio of new subscribers in June 1988 to June 1989?

**17.** What was the ratio of new subscribers in July 1989 to August 1989?

**18.** Between which two consecutive months of 1989 did the company have the greatest increase in new subscribers?

*The hourly wages of all the employees of Mutter Manufacturing Company are presented in the histogram at the right. Use this information in Exercises 19–28. Assume that $5 is in the class interval $5–$7 not in $3–$5, similarly for $7, $9, and $11.*

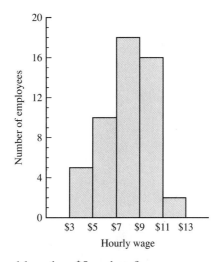

**19.** How many employees earn at least $3 and less than $5 an hour?

**20.** How many employees earn at least $7 and less than $9 an hour?

**21.** How many employees earn $7 an hour or more?

**22** How many employees earn less than $9 an hour?

**23.** How many employees are there in Mutter Manufacturing Company?

**24.** What is the ratio of employees earning $9 or more an hour to the total number of employees?

**25** What percent of the employees earn $9 or more an hour (to the nearest tenth of a percent)?

**26.** What is the ratio of employees earning less than $7 an hour to those who earn $9 or more an hour?

**27.** What values would be used as the class midpoints of the class intervals to construct the frequency polygon for this data?

**28.** Draw the frequency polygon in the figure above.

## FOR REVIEW

*Exercises 29–32 refer to the circle graph below which shows the distribution of Chuck's gross monthly income of $2500.*

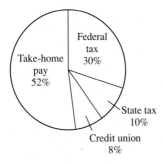

**29.** What is Chuck's monthly take-home pay?

**30.** How much federal tax does Chuck pay each month?

**31.** How much state tax does Chuck pay each month?

**32.** What is the ratio of federal tax to take-home pay?

ANSWERS: 1. 15  2. weeks 1 and 6  3. 90  4. $\frac{6}{1}$ or 6 to 1  5. week 5  6. week 3  7. Between weeks 3 and 4
8. Between weeks 5 and 6  9. 33.$\overline{3}$% or 33$\frac{1}{3}$%  10. 11.$\overline{1}$% or 11$\frac{1}{9}$%  11. 15  12. 20  13. 90  14. 120  15. 33%
increase  16. $\frac{3}{4}$ or 3 to 4  17. $\frac{3}{2}$ or 3 to 2  18. August and September  19. 5  20. 18  21. 36  22. 33  23. 51
24. $\frac{6}{17}$ or 6 to 17  25. 35.3%  26. $\frac{5}{6}$ or 5 to 6  27. $4, $6, $8, $10, $12  28.
29. $1300  30. $750  31. $250  32. $\frac{15}{26}$ or 15 to 26

## 7.2 EXERCISES B

*The number of yards gained passing by quarterback Greg Wyatt in five home games of the 1989 season is shown in the broken-line graph below. Use this graph in Exercises 1–10.*

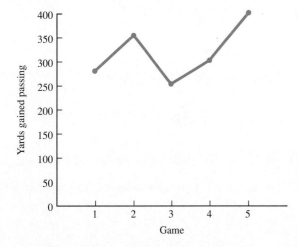

1. How many yards passing did Greg have in game 1?

2. During which game did Greg have 400 yards passing?

3. How many total yards passing did Greg have in the five home games?

4. What was the ratio of yards passing in game 2 to game 3?

5. In which game did Greg have the most yards passing?

6. In which game did Greg have the fewest yards passing?

7. Between which two consecutive home games did his yards passing increase the most?

8. Between which two consecutive home games did his yards passing decrease?

9. To the nearest tenth of a percent, what percent of Greg's total passing yards in the five home games was gained in week 5?

10. To the nearest tenth of a percent, what percent of Greg's total passing yards in the five home games was gained in week 1?

*The double-line graph below shows the number of entrants (in thousands) to Sunset Crater National Monument during four months in the summers of 1988 and 1989. Use this graph in Exercises 11–18.*

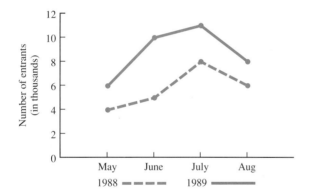

11. How many monument entrants were there in June of 1989?

12. How many monument entrants were there in June of 1988?

13. How many monument entrants were there in the four-month period in 1988?

14. How many monument entrants were there in the four-month period in 1989?

15. To the nearest percent, during the four-month period what was the percent increase in monument entrants from 1988 to 1989?

16. What was the ratio of monument entrants in June 1989 to June 1988?

17. What was the ratio of monument entrants in July 1988 to June 1988?

18. Between which two consecutive months in 1989 did the number of monument entrants increase the most?

*During a past period of years, the weather service has recorded the annual precipitation received by Morgan, Colorado, and the results are given in the histogram at the right. Use this information in Exercises 19–28. Assume that 5 inches is in the class interval 5 inches–10 inches, not in 0 inches–5 inches, similarly for 10 inches, 15 inches, and 20 inches.*

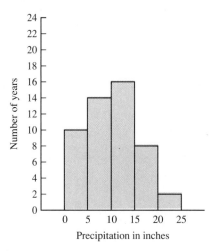

**19.** In how many years was the precipitation total less than 5 inches?

**20.** In how many years was the precipitation total at least 10 inches and less than 15 inches?

**21.** In how many years did Morgan receive 10 inches or more of precipitation?

**22.** In how many years did Morgan receive less than 15 inches of precipitation?

**23.** What was the total number of years in the recording period?

**24.** What is the ratio of the number of years of 15 to 20 inches of precipitation to the number of years of 20 to 25 inches?

**25.** In what percent of the total number of years did Morgan receive 20 inches or more of precipitation?

**26.** What is the ratio of the number of years of less than 5 inches of precipitation to the number of years of 10 inches or more of precipitation?

**27.** What values would be used as the class midpoints of the class intervals to construct the frequency polygon for this data?

**28.** Draw the frequency polygon in the figure above.

## FOR REVIEW

*Exercises 29–32 refer to the bar graph at the right which shows the number of students who preferred various cable television networks in a recent survey.*

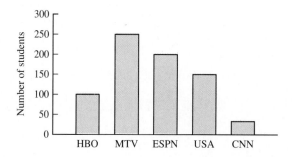

**29.** How many students preferred HBO?

**30.** How many students were in the survey?

**31.** To the nearest tenth of a percent, what percent of the students preferred ESPN?

**32.** What was the ratio of students preferring MTV to CNN?

## 7.2 EXERCISES C

*The letters in our alphabet occur with different frequencies in the words that we use. The letter "e" occurs most frequently and the letter "z" occurs least frequently. In fact, it has been shown statistically that in a large sample of words, the words on a page in a text for example, approximately 13% of the letters will be "e", 7% will be "a", 6% will be "s", and 8% will be "n". Select a page in one of your textbooks (a page with many words), and count the number of times e, a, s, and n occur. Also, count the total number of letters on the page.*

**1.** Make a table showing the number of times each of the four letters occurs and transfer the information to a bar graph.

**2.** Which of these letters occurred most frequently?

**3.** Which of these letters occurred least frequently?

**4.** How do the percents of occurrence for the letters e, a, s, and n in your data compare to the percents given above?

## 7.3 MEAN, MEDIAN, AND MODE

### STUDENT GUIDEPOSTS

**1** Mean (Average)       **3** Mode
**2** Median

### 1 MEAN (AVERAGE)

In Sections 7.1 and 7.2 numerical data were presented using circle graphs, bar graphs, broken-line graphs, and histograms. Often it is necessary to look for one number or statistic to characterize a whole collection of numbers. For example, suppose Jenny has scores of 69, 79, and 98 on three chemistry tests. To find out how she is doing in the course, Jenny might calculate the *mean* or *average* of her scores.

#### Mean (Average)

The **mean** or **average** of several values is the ratio of the sum of the values to the number of values.

$$\text{mean (average)} = \frac{\text{sum of values}}{\text{number of values}}$$

To find the mean of 69, 79, and 98, add these three scores and divide by 3 (the number of scores).

$$\text{mean} = \frac{69 + 79 + 98}{3} \quad \begin{matrix} \longleftarrow \text{ Sum of the scores} \\ \longleftarrow \text{ Number of scores} \end{matrix}$$

$$= \frac{246}{3} = 82$$

Making 69, 79, and 98 on three tests is the same as making

$$82, \quad 82, \quad \text{and} \quad 82$$

on three tests. The mean or average gives Jenny a better idea of how well she is doing in chemistry. Her professor may use the average to compute her course grade.

| EXAMPLE 1 FINDING A MEAN IN A BUSINESS | PRACTICE EXERCISE 1 |

The daily sales at George's Pie Place were $285.85, $192.50, $222.82, $315.16, $427.45, and $396.80 for a six-day week. To get a better idea of how well his business is doing, George would like to know the average daily sales for the week. What is the average?

Add the sales for the 6 days and divide by 6.

$285.85
  192.50
  222.82
  315.16
  427.45
  396.80
$1840.58      $\dfrac{\$1840.58}{6} \approx \$306.76$      To the nearest cent

George's shop is averaging about $306.76 per day in sales.

In Example 1, if George had known that the total sales for the 6-day week were $1840.58, he would only have had to divide by 6. The next example illustrates this type of problem.

| EXAMPLE 2 FINDING A MEAN IN INDUSTRY | PRACTICE EXERCISE 2 |

Suppose tests by General Motors show that one of its cars will go 420 miles on a tank of gasoline. What is the average miles per gallon (mpg) of the car if the tank holds 12.5 gal?

In this problem the total miles traveled on the 12.5 gal is given. Thus, all that is needed to find the average is to divide 420 mi by 12.5 gal.

$$\frac{420 \text{ mi}}{12.5 \text{ gal}} = 33.6 \, \frac{\text{mi}}{\text{gal}} = 33.6 \text{ mpg}$$

## ② MEDIAN

The mean of a collection of numbers can often be influenced by one very large or very small number (compared to the others). As a result, a mean is often a poor measure of the characteristics of a set of data. For example, the mean of the numbers

$$1, \quad 2, \quad 3, \quad 4, \quad 90$$

is

$$\frac{1 + 2 + 3 + 4 + 90}{5} = \frac{100}{5} = 20.$$

Since four of the five numbers are very small compared to the fifth number, to use 20 to describe this set of data might be somewhat misleading. Actually, a number closer to 1, 2, 3, or 4 than 20 would provide a better description. Another way to describe a set of data is to look at the value in the middle after the numbers are arranged by size. For example, consider the following:

$$1, \quad 2, \quad \underset{\underset{\text{middle}}{\uparrow}}{3}, \quad 4, \quad 90.$$

The value in the middle is 3, and certainly 3 gives a better description of the data set than the mean 20. The middle value of a collection of numbers that is arranged in order of size is the *median* of the data.

Suppose Russell made scores of 72, 65, and 88 on his first three tests. To find the middle score, or median score, arrange them in order of increasing value.

$$65, \quad \underset{\underset{\text{middle}}{\uparrow}}{72}, \quad 88$$

The median score is 72.

But what about the middle score if Russell had four test scores, 72, 65, 88, and 95? Once again arrange them in order.

$$65, \quad 72, \quad \underset{\underset{\text{middle}}{\uparrow}}{88}, \quad 95$$

In this case, there is no middle score. To approximate a middle score, average the *two* scores in the middle.

$$\text{median} = \frac{72 + 88}{2} = 80$$

---

### Median

The middle value of several values is called the **median.** To find the median:

1. Arrange the values in increasing order from left to right.
2. If there is an odd number of values, the median is the middle one.
3. If there is an even number of values, the median is the average of the *two* values in the middle.

---

### EXAMPLE 3  FINDING A MEDIAN

Find the median of the given numbers.

**(a)** 5, 12, 2, 14, 8, 7, 9

First arrange the numbers in order.

$$2, \quad 5, \quad 7, \quad 8, \quad 9, \quad 12, \quad 14$$

There are seven numbers, and the middle one is the fourth from the left. The median is 8.

**(b)** 125, 105, 216, 205, 110, 175

Arrange the numbers in order.

$$105, \quad 110, \quad 125, \quad 175, \quad 205, \quad 216$$

The two numbers in the middle are 125 and 175. Find their average.

$$\frac{125 + 175}{2} = 150$$

Thus, 150 is the median of the numbers.

### PRACTICE EXERCISE 3

Find the median of the given numbers.

**(a)** 9, 20, 2, 18, 27

**(b)** 206, 140, 180, 375

Answers: (a) 18   (b) 193

---

| **EXAMPLE 4** AN APPLICATION OF MEAN AND MEDIAN |
|---|

In Greenborough, the number of traffic accidents per day is recorded. The following are the numbers for one week.

$$12, \quad 18, \quad 22, \quad 10, \quad 8, \quad 15, \quad 27$$

Find the average number of accidents per day and the median number per day.

First find the average.

$$\frac{12 + 18 + 22 + 10 + 8 + 15 + 27}{7} = \frac{112}{7} = 16$$

To find the median, arrange the numbers in order.

$$8, \quad 10, \quad 12, \quad 15, \quad 18, \quad 22, \quad 27$$

The median is 15 which is close to the mean of 16.

---

| PRACTICE EXERCISE 4 |
|---|

Jeanine made scores of 22, 24, 15, 27, 19, and 30 on six quizzes in geology lab. Find the average score and the median score on these quizzes.

Answer: average score: 22.8, median score: 23

---

| **H I N T** |
|---|

Remember to arrange the numbers by size before selecting the middle one as the median. For example, the median of

$$5, \quad 1, \quad 2, \quad 3, \quad 8$$

is 3, not 2. This is easy to see after rewriting the numbers as 1, 2, 3, 5, 8.

---

| **EXAMPLE 5** FINDING THE MEAN AND MEDIAN USING A TABLE |
|---|

The Lopez Engineering Company has 30 employees. Six of them make $18,500 per year, 5 make $19,200 per year, 2 make $22,800 per year, 10 make $24,500 per year, 4 make $32,800 per year, 2 make $38,000 per year, and 1 makes $52,400 per year. What is the mean or average salary of the 30 employees? What is the median salary?

A good way to find the sum of the salaries is to set up a table.

| Salary | Number of Employees | Salary Times Number of Employees |
|---|---|---|
| 18,500 | 6 | 111,000 |
| 19,200 | 5 | 96,000 |
| 22,800 | 2 | 45,600 |
| 24,500 | 10 | 245,000 |
| 32,800 | 4 | 131,200 |
| 38,000 | 2 | 76,000 |
| 52,400 | 1 | 52,400 |
| Totals | 30 | $757,200 |

*Frequency Table*

When each salary is multiplied by the number of employees making that salary and the results are added, the total is $757,200. This is divided by 30, the total number of employees, to get the mean or average salary.

$$\frac{\$757,200}{30} = \$25,240$$

---

| PRACTICE EXERCISE 5 |
|---|

A new car dealer has 40 cars in stock. Five are valued at $18,000, 12 at $15,000, 10 at $12,000, 7 at $9000, and 6 at $8000. What is the mean or average value of a car on his lot? What is the median value?

Since there are 30 employees, the median salary is the average of the middle two salaries once the salaries are arranged by size. This will be the average of the 15th and 16th salaries in the arrangement. Without actually listing all the salaries, we can find the 15th and 16th ones by looking at the table. Use the column for the number of employees and count down 15. Since the 15th and 16th positions are included in the 10 employees in the fourth row from the top, the 15th and 16th salary values are both $24,500, making the average of the two also $24,500. Thus, the median salary is $24,500.

Answer:  mean = $12,525; median = $12,000

### ❸ MODE

The final measure used to describe a set of data is called the *mode*. Consider the numbers

$$1, \quad 5, \quad 101, \quad 101.$$

The mean is

$$\frac{1 + 5 + 101 + 101}{4} = \frac{208}{4} = 52,$$

and the median is

$$\frac{5 + 101}{2} = \frac{106}{2} = 53.$$

But does 52 or 53 give an accurate reflection of the data set? Since two of the four numbers are 101, much greater than 52 or 53, it might be better to describe this data set using 101, the number that occurs most frequently in the set.

*If the set had been 1,5,101,102 it would have had no mode*

Number

Numbers

| Mode |

The value that occurs most frequently in a set of values is the **mode** of the data. If there are two values that occur more frequently than all the others, the data has two modes and is called **bimodal.** If no value occurs more than once in the set, the data has **no mode.**

| EXAMPLE 6   FINDING A MODE |

Find the mode of the given numbers.

**(a)**  5, 8, 1, 8, 7, 9, 8, 5

Since 8 occurs three times, 5 occurs twice, and all the rest of the numbers occur only once, the mode is 8.

**(b)**  2, 7, 3, 7, 8, 8, 1, 5

Since 7 and 8 both occur twice and all other numbers occur only once, the set of numbers has two modes, 7 and 8. It is a bimodal set.

**(c)**  3, 9, 7, 5, 2, 1, 11

Since every number occurs exactly one time, there is no mode.

| PRACTICE EXERCISE 6 |

Find the mode of the given numbers.

**(a)**  1, 1, 1, 5, 2, 2, 9, 8

**(b)**  6, 6, 2, 1, 5, 2, 9, 10

**(c)**  2, 8, 5, 12, 9, 7

Answers:  (a)  1     (b)  6 and 2 (bimodal)     (c)  no mode

## 7.3 EXERCISES A

*Find the mean of the given numbers. Give answers to the nearest tenth when appropriate.*

**1.** 7, 19, 13

**2.** 7, 9, 8, 12

**3.** 12, 22, 18, 25, 9

**4.** 102, 95, 116, 118, 125, 106

**5** 7.2, 6.1, 8.6, 9.2, 5.5, 7.2, 8.8

**6.** 42, 43, 58, 55, 61, 41, 48, 48

*Find the median.*

**7.** 7, 19, 13

**8.** 7, 9, 8, 12

**9.** 12, 22, 25, 18, 9

**10.** 102, 95, 116, 118, 125, 106

**11** 7.2, 6.1, 8.6, 9.2, 5.5, 7.2, 8.8

**12.** 42, 43, 58, 55, 61, 41, 48, 48

*Find the mode.*

**13.** 5, 6, 1, 6, 3, 2

**14.** 2, 3, 9, 2, 9, 7, 9, 11

**15.** 8, 3, 8, 5, 7, 3, 2

**16.** 2, 11, 17, 9, 6, 41

**17.** 4.3, 2.7, 4.3, 12.8, 15.1

**18.** 149, 150, 211, 150, 435, 147

*Solve.*

**19.** Suppose the Ford Motor Company's tests show that one of their cars will go 480 miles on a tank of gasoline. If the tank holds 13.6 gal, what is the average miles per gallon (mpg)?

**20.** A basketball player scored 252 points in 18 games. What is his average per game?

**21.** During a six-day week, the sales at Joseph's Shop for Men were $482.16, $316.58, $299.81, $686.47, $819.72, and $992.16. What were the average daily sales?

**22.** The heights of the starting five players on the Upstate University basketball team are 78 inches, 82 inches, 72 inches, 84 inches, and 73 inches. What is the average height? What is the median height?

**23** The table at the right gives information on the yearly salaries of the employees at one plant of Mutter Manufacturing. Find the average salary. Find the median salary. What is the mode of the salaries?

| Salary | Number of Employees | Salary Times Number of Employees |
|--------|--------------------|---------------------------------|
| 16,200 | 8 | |
| 20,800 | 10 | |
| 28,600 | 7 | |
| 32,000 | 2 | |
| 46,000 | 1 | |
| Totals | | |

## FOR REVIEW

*The profit (in millions of dollars) made by the ABC Corporation over a six-year interval is shown in the broken-line graph below. Use this graph in Exercises 24–30.*

**24.** What profit was made in 1985?

**25.** What profit was made in 1984?

**26.** What was the total profit made in 1986 and 1987?

**27.** What was the total profit made during the six-year interval?

**28.** What was the ratio of the profit in 1986 to that in 1987?

**29.** What percent (to the nearest tenth) of the total profit in the six-year interval was made in 1985?

**30.** What percent (to the nearest tenth) of the total profit in the six-year interval was made in the years 1984 and 1985?

---

ANSWERS:   1. 13   2. 9   3. 17.2   4. 110.3   5. 7.5   6. 49.5   7. 13   8. 8.5   9. 18   10. 111   11. 7.2   12. 48   13. 6   14. 9   15. 8 and 3 (bimodal)   16. no mode   17. 4.3   18. 150   19. 35.3 mpg   20. 14   21. $599.48   22. 77.8; 78   23. $23,135.71; $20,800; $20,800   24. 5 million dollars   25. 5.5 million dollars   26. 9 million dollars   27. 24.5 million dollars   28. $\frac{2}{1}$ or 2 to 1   29. 20.4%   30. 42.9%

## 7.3  EXERCISES B

*Find the mean of the given numbers. Give answers to the nearest tenth when appropriate.*

**1.** 11, 21, 19

**2.** 8, 12, 14, 22

**3.** 14, 35, 16, 19, 43

**4.** 130, 87, 102, 96, 145, 110

**5.** 6.3, 4.7, 8.2, 9.6, 7.1, 8.3, 2.5

**6.** 51, 57, 62, 65, 71, 83, 90, 97

*Find the median.*

**7.** 5, 12, 37

**8.** 6, 9, 10, 14

**9.** 5, 21, 27, 17, 19

**10.** 107, 93, 110, 84, 87, 130

**11.** 6.3, 4.7, 8.2, 9.6, 7.1, 8.3, 2.5

**12.** 51, 57, 62, 65, 90, 83, 71, 97

*Find the mode.*

**13.** 11, 17, 5, 17, 6, 21, 17

**14.** 6, 3, 6, 8, 11, 3, 21, 3

**15.** 2, 12, 5, 2, 8, 5, 7, 1

**16.** 3, 9, 21, 8, 7, 26, 5

**17.** 4.8, 7.6, 8.1, 9.3, 7.6, 2.4

**18.** 159, 276, 423, 159, 342, 177

*Solve.*

**19.** Tests show that a motorhome will travel 535.5 miles on a tank of gasoline. If the tank holds 42.5 gallons, what is the average miles per gallon (mpg)?

**20.** Quarterback Craig Austin completed 438 passes in 11 games. What is the average number of completions per game (to the nearest tenth)?

**21.** The number of TV sets sold per day by a dealer over a 12-day period is 2, 4, 6, 2, 10, 3, 6, 8, 12, 8, 22, and 12. What is the median number of sets sold per day? What is the average number of sets sold per day (to the nearest tenth)?

**22.** Maria scored 73, 85, 65, 92, 78, and 87 on six tests in math. What is her average score? What is her median score?

**23.** The Mutter Manufacturing record of sick-leave days available to employees is recorded below. What is the average number of days available? What is the median number? What is the mode?

| Number of Days Sick Leave Available | Number of Employees | Number of Days Times Number of Employees |
|---|---|---|
| 14 | 4 | |
| 21 | 12 | |
| 28 | 8 | |
| 35 | 4 | |

Totals

## FOR REVIEW

*The number of years of teaching experience by the faculty at CSU is shown in the histogram at the right. Use this graph in Exercises 24–30.*

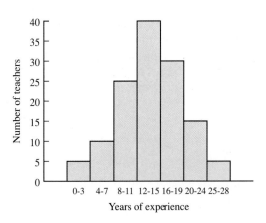

**24.** How many teachers have 0–3 years of experience?

**25.** How many teachers have 8–15 years of experience?

**26.** How many teachers have 12 or more years of experience?

**27.** How many teachers have 7 or less years of experience?

**28.** How many total teachers are represented by this data?

**29.** What percent (to the nearest tenth) of the total number of teachers have 16 or more years of experience?

**30.** What is the ratio of the number of teachers with 16–19 years of experience to those with 7 or less years of experience?

## 7.3    EXERCISES C

*Solve.*

**1.** Suppose Marni made scores of 65, 72, and 88 on the first three tests in math. What score did she make on the fourth test if the average score on the four was 80? [Answer: 95]

**2.** Suppose Burford made scores of 32, 38, 51, and 82 on four of five exams. If the median score of the five exams is 51, could the score on Burford's fifth test have been 35?

*Which of the mean, median, or mode do you think is the best measure to describe each set of numbers in Exercises 3–6?*

**3.** 1, 1, 1, 1, 20, 100

**4.** 1, 1, 99, 99 [Answer: mode (bimodal)]

**5.** 1, 2, 3, 1000

**6.** 1, 2, 3, 4, 5

## CHAPTER 7 REVIEW

### KEY WORDS

**7.1** **Statistics** is a branch of mathematics that deals with collecting, organizing, and summarizing numerical data in order to describe or interpret it.

A **circle graph** uses pie-shaped sectors of a circle to represent different parts of a quantity.

A **bar graph** is used to compare various parts of a data set.

**7.2** A **broken-line graph** can be used to show trends in data collected over a period of time.

A **histogram** is a special type of bar graph used to display data that is divided into groups called **class intervals.** The **class frequency** of

an interval is represented by the length or height of each bar.

**7.3** The **mean** or **average** of several values is the sum of the values divided by the number of values.

The **median** of a data set is the middle value or the average of the two middle values when the values are arranged in order of size.

The **mode** of a set of values is the value that occurs most frequently in the set.

### KEY CONCEPTS

**7.1** Numerical data can be displayed or pictured using a circle graph or a bar graph.

**7.2** A histogram is often used when a large number of data items can best be described by dividing the data into groups.

**7.3** **1.** Before finding the median of a collection of numbers, be sure to arrange the numbers in order of size.

**2.** If there is an even number of values, the median is the mean (average) of the two middle values.

**3.** A bimodal set of data has two values with the greatest frequency of occurrence. If all numbers in a collection occur exactly once, the data set has no mode.

## REVIEW EXERCISES

*Part I*

**7.1** *Exercises 1–4 refer to the circle graph at the right which shows preference in vacation locations of 300 adults surveyed in a random poll. Give answers rounded to the nearest tenth.*

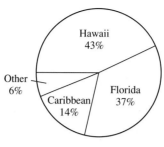

**1.** How many of those surveyed preferred Hawaii?

**2.** How many of those surveyed preferred the Caribbean?

**3.** What is the ratio of those preferring Hawaii to those preferring Florida?

**4.** How many of those surveyed preferred a vacation location other than Hawaii?

**7.2** *Exercises 5–10 refer to the broken-line graph at the right which shows the number of new homes sold in a subdivision over a six-month period.*

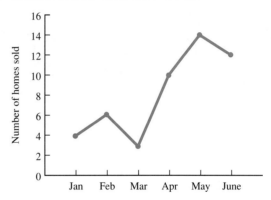

**5.** How many homes were sold in April?

**6.** How many homes were sold in May and June together?

**7.** Between which two consecutive months did sales of new homes show the greatest increase?

**8.** How many homes were sold in the six-month period?

**9.** What was the ratio of homes sold in April to those sold in March?

**10.** What percent (to the nearest tenth) of the homes sold in the six-month period were sold in January?

*Exercises 11–16 refer to the histogram at the right which relates the number of cars to miles per gallon in an EPA survey. Assume that a car making 5 mpg, for example, is included in the range 5–10 mpg and not 0–5 mpg, similarly for 10 mpg, 15 mpg, and so forth.*

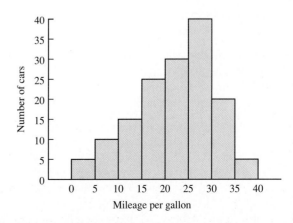

**11.** How many cars got at least 5 mpg and less than 10 mpg?

**12.** How many cars got 30 mpg or better?

**13.** How many cars got less than 20 mpg?

**14.** How many cars were represented in the survey?

**15.** What percent of the total number of cars got 35 mpg or better?

**16.** What is the ratio of cars that got 30 mpg or better to those that got less than 5 mpg?

**7.3**  *Find the mean and median. Give answers to the nearest tenth.*

**17.** 16, 12, 19, 22, 8

**18.** 375, 206, 311, 222, 320, 312

**19.** Maxine's car will go 589 miles on 15.5 gallons of gasoline. What is Maxine's average miles per gallon?

*Find the mode.*

**20.** 6, 5, 1, 3, 5, 7, 12, 5

**21.** 2, 9, 3, 2, 7, 3, 4, 12

**22.** The Thai Villa Restaurant had sales of $1020, $926, $820, $772, $1428, and $1362 over a six-day period. What were the average daily sales to the nearest dollar? What was the median?

**23.** The following table gives information of the monthly salaries of the employees of Lloyd's Shoppe. Find the average salary. Find the median salary. What is the mode of the salaries?

| Salary | Number of Employees | Salary Times Number of Employees |
|--------|---------------------|----------------------------------|
| $ 820  | 4                   |                                  |
| $ 860  | 8                   |                                  |
| $ 920  | 5                   |                                  |
| $ 980  | 2                   |                                  |
| $1620  | 1                   |                                  |
| Totals |                     |                                  |

**Part II**

*Exercises 24–27 refer to the bar graph at the right, which gives the number of adults choosing a particular sport as their favorite to view on television.*

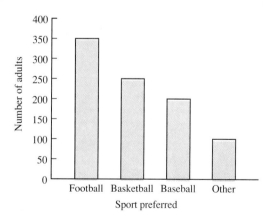

**24.** How many adults were in the survey?

**25.** What is the ratio of those preferring basketball to the total number surveyed?

**26.** What percent of those surveyed preferred basketball (to the nearest tenth of a percent)?

**27.** What is the ratio of those who prefer football to those who prefer baseball?

*Find the mean, median, and mode.*

**28.** 12, 3, 9, 5, 5

**29.** 6.1, 2.5, 3.8, 9.2, 100.7

**30.** 18, 12, 17, 29, 40, 12

**31.** 4, 7, 4, 9, 7, 6, 8, 10

*Exercises 32–40 refer to the double-bar graph below which compares the number of students enrolled in four math courses at State University in 1988 and 1989.*

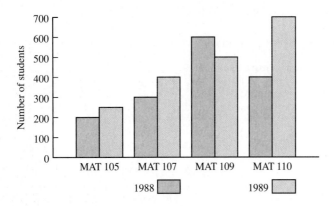

**32.** How many students were in MAT 107 in 1989?

**33.** How many students were in MAT 109 in 1988?

**34.** How many students were in these four math courses in 1988?

**35.** How many students were in these four math courses in 1989?

**36.** What was the ratio of students in MAT 110 in 1989 to those in MAT 110 in 1988?

**37.** What percent of students in all four classes in 1988 were in MAT 109?

**38.** During which year were there more students in MAT 109?

**39.** What was the percent increase in students taking MAT 110 from 1988 to 1989?

**40.** How many total students took MAT 105 in these two years?

---

ANSWERS:  1. 129  2. 42  3. $\frac{43}{37}$ or 43 to 37  4. 171  5. 10  6. 26  7. Between March and April  8. 49  9. $\frac{10}{3}$ or 10 to 3  10. 8.2%  11. 10  12. 25  13. 55  14. 150  15. 3.$\overline{3}$% or $3\frac{1}{3}$%  16. $\frac{5}{1}$ or 5 to 1  17. 15.4; 16  18. 291; 311.5  19. 38 mpg  20. 5  21. 2 and 3 (bimodal)  22. $1055; $973  23. $917; $860; $860  24. 900  25. $\frac{5}{18}$ or 5 to 18  26. 27.8%  27. $\frac{7}{4}$ or 7 to 4  28. 6.8; 5; 5  29. 24.46; 6.1; no mode  30. 21.$\overline{3}$; 17.5; 12  31. 6.875; 7; 4 and 7  32. 400  33. 600  34. 1500  35. 1850  36. $\frac{7}{4}$ or 7 to 4  37. 40%  38. 1988  39. 75%  40. 450

*Problems 1–3 refer to the circle graph below which shows the distribution of responses from 1000 adults about television viewing preference.*

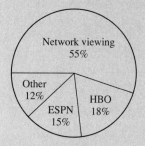

**1.** How many of those polled preferred HBO?

1. _____

**2.** How many of those polled preferred ESPN?

2. _____

**3.** What is the ratio of those who preferred network viewing to ESPN?

3. _____

*Problems 4–8 refer to the bar graph below which shows the preferences of 150 adults in type of restaurant food.*

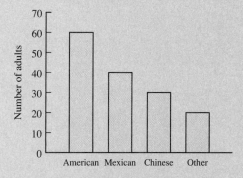

**4.** How many of those questioned preferred Mexican food?

4. _____

**5.** How many of those questioned preferred food other than American food?

5. _____

**6.** What is the ratio of those preferring American food to those preferring Chinese food?

6. _____

**7.** What percent of those interviewed preferred Mexican food?

7. _____

**8.** What percent of those interviewed preferred some kind of food besides American?

8. _____

*The weights of all the members of a football team are presented in the histogram below. Use this information in problems 9–12. Assume that 175 pounds is in the class 175–200 pounds, not in 150–175 pounds, similarly for 200 pounds, 225 pounds, and 250 pounds.*

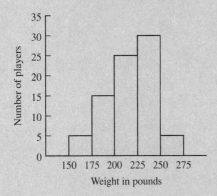

9. How many players weigh at least 200 and less than 225 pounds?

9. _____

10. How many players are on the team?

10. _____

11. What is the ratio of players weighing 225 pounds or more to the total number of players on the team?

11. _____

12. What percent of the players weigh less than 200 pounds?

12. _____

13. Find the mean of the given numbers.

    11,  22,  32,  17,  18

13. _____

14. Find the median of the given numbers.

    11,  22,  32,  17,  18

14. _____

15. Find the mode of the given numbers.

    14,  8,  9,  17,  8,  17,  23,  8,  5

15. _____

16. The weights of the starting four defensive linemen for PSU are 246 pounds, 253 pounds, 257 pounds, and 262 pounds.
    **(a)** What is the average weight?  **(b)** What is the median weight?
    **(c)** What is the mode of the weights?

16. (a) _____

    (b) _____

    (c) _____

# Measurement and Geometry

## 8.1 THE ENGLISH SYSTEM

### 1 ENGLISH UNITS OF MEASURE

A standard system of units of measure is a useful tool for people living and working together. This chapter discusses two systems of measurement, the English system and the metric system. While many countries throughout the world use the metric system, the United States uses the English system.

| Units of Measure: English System | |
|---|---|
| Length | 1 foot (ft) = 12 inches (in) |
| | 1 yard (yd) = 3 feet (ft) |
| | 1 mile (mi) = 5280 feet (ft) |
| Volume | 1 pint (pt) = 16 fluid ounces (fl oz) |
| | 1 quart (qt) = 2 pints (pt) |
| | 1 gallon (gal) = 4 quarts (qt) |
| Weight | 1 pound (lb) = 16 ounces (oz) |
| | 1 ton = 2000 pounds (lb) |
| Time | 1 minute (min) = 60 seconds (sec) |
| | 1 hour (hr) = 60 minutes (min) |
| | 1 day (da) = 24 hours (hr) |
| | 1 week (wk) = 7 days (da) |
| | 1 year (yr) = 365 days (da) |
| | = 52 weeks (wk) |
| | = 12 months (mo) |

As the table shows, each type of measurement (length, volume, and so on) has several units in the English system. The weight of this book might be given in ounces, while the weight of a student would usually be given in pounds.

## ② UNIT FRACTION

To work with units of measure, we need to know how to convert from one unit to another. A good way is to use **unit fractions,** fractions which have value 1. Some examples of unit fractions are

$$\frac{12 \text{ in}}{1 \text{ ft}}, \quad \frac{1 \text{ ft}}{12 \text{ in}}, \quad \frac{4 \text{ qt}}{1 \text{ gal}}, \quad \frac{1 \text{ gal}}{4 \text{ qt}}, \quad \frac{16 \text{ oz}}{1 \text{ lb}}, \quad \frac{1 \text{ lb}}{16 \text{ oz}}.$$

Each of the fractions has value 1 because the numerator and denominator are equal. It is important to show the unit of measure as well as the number. Obviously, $\frac{12}{1}$ is *not* 1, but $\frac{12 \text{ in}}{1 \text{ ft}}$ is 1 since

$$1 \text{ ft} = 12 \text{ in}.$$

Thus, to change 4 yd to feet, write

$$
\begin{aligned}
4 \text{ yd} &= 4 \text{ yd} \times 1 & &\text{Multiply by 1} \\
&= \frac{4 \text{ yd}}{1} \times \frac{3 \text{ ft}}{1 \text{ yd}} & &\text{Divide out yards} \\
&= 4 \times 3 \text{ ft} \\
&= 12 \text{ ft}.
\end{aligned}
$$

With this method, the yards are divided out as if they were numbers.

---

## HINT

The fact that the yards do divide out tells us that we were right to multiply by 3 ft instead of dividing by 3 ft. If we had tried to multiply

$$\frac{4 \text{ yd}}{1} \times \frac{1 \text{ yd}}{3 \text{ ft}},$$

the yards would not have divided out and the change of units could not be made.

---

### EXAMPLE 1  CONVERTING FEET TO INCHES

Convert 16 feet to inches (16 ft = _____ in).
    Since

$$1 \text{ ft} = 12 \text{ in},$$

we need to use one of the unit fractions

$$\frac{12 \text{ in}}{1 \text{ ft}} \quad \text{or} \quad \frac{1 \text{ ft}}{12 \text{ in}}.$$

To divide out the ft in 16 ft, use $\frac{12 \text{ in}}{1 \text{ ft}}$.

$$
\begin{aligned}
16 \text{ ft} &= 16 \text{ ft} \times 1 & &\text{Multiply by 1} \\
&= \frac{16 \text{ ft}}{1} \times \frac{12 \text{ in}}{1 \text{ ft}} & &\text{Divide out ft} \\
&= 16 \times 12 \text{ in} \\
&= 192 \text{ in}
\end{aligned}
$$

### PRACTICE EXERCISE 1

Convert 23 feet to inches.

Answer: 276 in

| EXAMPLE 2   CONVERTING INCHES TO FEET | PRACTICE EXERCISE 2 |
|---|---|

Convert 72 inches to feet (72 in = _____ ft).

    For this problem use the unit fraction $\frac{1 \text{ ft}}{12 \text{ in}}$.

Convert 108 inches to feet.

$$72 \text{ in} = 72 \text{ in} \boxed{\times 1} \qquad \text{Multiply by 1}$$

$$= \frac{72 \cancel{\text{ in}}}{1} \times \boxed{\frac{1 \text{ ft}}{12 \cancel{\text{ in}}}} \qquad \text{Divide out in}$$

$$= \frac{72}{12} \text{ ft}$$

$$= 6 \text{ ft}$$

| EXAMPLE 3   CONVERSION IN WEIGHT AND VOLUME | PRACTICE EXERCISE 3 |
|---|---|

Complete the conversion in units.

Complete the conversion in units.

**(a)** 22 lb = _____ oz.

    The unit fraction is $\frac{16 \text{ oz}}{1 \text{ lb}}$.

**(a)** 15 lb = _____ oz.

$$22 \text{ lb} = 22 \text{ lb} \boxed{\times 1} \qquad \text{Multiply by 1}$$

$$= \frac{22 \cancel{\text{ lb}}}{1} \times \boxed{\frac{16 \text{ oz}}{1 \cancel{\text{ lb}}}} \qquad \text{Divide out lb}$$

$$= 22 \times 16 \text{ oz}$$

$$= 352 \text{ oz}$$

**(b)** 96 fl oz = _____ pt.

    With the unit fraction $\frac{1 \text{ pt}}{16 \text{ fl oz}}$, the fl oz divide out.

**(b)** 144 fl oz = _____ pt.

$$96 \text{ fl oz} = 96 \text{ fl oz} \boxed{\times 1}$$

$$= \frac{96 \cancel{\text{ fl oz}}}{1} \times \boxed{\frac{1 \text{ pt}}{16 \cancel{\text{ fl oz}}}}$$

$$= \frac{96}{16} \text{ pt}$$

$$= 6 \text{ pt}$$

**(c)** 6 pt = _____ qt.

**(c)** 9 pt = _____ qt.

$$6 \text{ pt} = 6 \text{ pt} \boxed{\times 1}$$

$$= \frac{6 \cancel{\text{ pt}}}{1} \times \boxed{\frac{1 \text{ qt}}{2 \cancel{\text{ pt}}}}$$

$$= \frac{6}{2} \text{ qt}$$

$$= 3 \text{ qt}$$

    Parts (b) and (c) of Example 3 together change 96 fl oz to 3 qt. Some unit conversions may need two or more unit fractions.

| EXAMPLE 4 Using Several Unit Fractions | PRACTICE EXERCISE 4 |
|---|---|

Complete the change in units.
5 wk = _____ min.

$$5 \text{ wk} = 5 \text{ wk} \times 1$$

$$= \frac{5 \text{ wk}}{1} \times \frac{7 \text{ da}}{1 \text{ wk}} \qquad \text{Convert 5 wk to 35 da}$$

$$= \frac{35 \text{ da}}{1} \times \frac{24 \text{ hr}}{1 \text{ da}} \qquad \text{Convert 35 da to 840 hr}$$

$$= \frac{840 \text{ hr}}{1} \times \frac{60 \text{ min}}{1 \text{ hr}} \qquad \text{Convert 840 hr to 50,400 min}$$

$$= 840 \times 60 \text{ min}$$

$$= 50,400 \text{ min}$$

The steps can be combined.

$$5 \text{ wk} = \frac{5 \text{ wk}}{1} \times \frac{7 \text{ da}}{1 \text{ wk}} \times \frac{24 \text{ hr}}{1 \text{ da}} \times \frac{60 \text{ min}}{1 \text{ hr}}$$

$$= 5 \times 7 \times 24 \times 60 \text{ min}$$

$$= 50,400 \text{ min}.$$

**Practice Exercise 4**

Complete the change in units.
8 wk = _____ min

Answer: 80,640

## ❸ HOUSEHOLD UNITS

Some units of measure used in the kitchen are given in this table.

| Household Units |
|---|
| 1 tablespoon (tbsp) = 3 teaspoons (tsp) |
| 1 cup = 16 tablespoons (tbsp) |
| = 8 fluid ounces (fl oz) |
| 1 pint (pt) = 2 cups |

| EXAMPLE 5 Converting Household Units | PRACTICE EXERCISE 5 |
|---|---|

Complete the change in units.

**(a)** 7 tbsp = _____ tsp.

$$7 \text{ tbsp} = 7 \text{ tbsp} \times 1$$

$$= \frac{7 \text{ tbsp}}{1} \times \frac{3 \text{ tsp}}{1 \text{ tbsp}} \qquad \frac{3 \text{ tsp}}{1 \text{ tbsp}} \text{ is a unit fraction}$$

$$= 7 \times 3 \text{ tsp} = 21 \text{ tsp}$$

**(b)** 40 tbsp = _____ pt.

$$40 \text{ tbsp} = \frac{40 \text{ tbsp}}{1} \times \frac{1 \text{ cup}}{16 \text{ tbsp}} \times \frac{1 \text{ pt}}{2 \text{ cups}}$$

$$= 40 \times \frac{1}{16} \times \frac{1}{2} \text{ pt} = \frac{40}{(16)(2)} \text{ pt}$$

$$= \frac{2 \cdot 2 \cdot 2 \cdot 5}{2 \cdot 2 \cdot 2 \cdot 2 \cdot 2} \text{ pt} = \frac{5}{4} \text{ pt} = 1\frac{1}{4} \text{ pt}$$

**Practice Exercise 5**

Complete the change in units.

**(a)** 15 tbsp = _____ tsp.

**(b)** 120 tbsp = _____ pt.

Answers: (a) 45 (b) 3.75

## ④ COMPOUND UNITS

Remember from Chapter 6 that *percent* means "per hundred." We can also say that

$$\frac{12 \text{ in}}{1 \text{ ft}} \quad \text{means} \quad 12 \text{ inches per foot.}$$

Thus, 50 miles per hour can be written as

$$\frac{50 \text{ mi}}{1 \text{ hr}}.$$

Most of the time the 1 in the denominator is not written. 50 miles per hour may be written in any of the following ways.

$$\frac{50 \text{ mi}}{\text{hr}} \qquad 50\frac{\text{mi}}{\text{hr}} \qquad 50 \text{ mi/hr} \qquad 50 \text{ mph}$$

Other units can be written these ways.

$$12 \text{ feet per second} \quad \text{means} \quad \frac{12 \text{ ft}}{\text{sec}} = 12\frac{\text{ft}}{\text{sec}} = 12 \text{ ft/sec}$$

$$8 \text{ gallons per minute} \quad \text{means} \quad \frac{8 \text{ gal}}{\text{min}} = 8\frac{\text{gal}}{\text{min}} = 8 \text{ gal/min}$$

| EXAMPLE 6 CONVERTING UNITS OF RATE | PRACTICE EXERCISE 6 |
|---|---|

Complete the change in units.

$$30\frac{\text{mi}}{\text{hr}} = \underline{\qquad}\frac{\text{ft}}{\text{sec}}.$$

Combine the steps.

$$30\frac{\text{mi}}{\text{hr}} = \frac{30 \text{ mi}}{\text{hr}} = \frac{30 \text{ mi}}{\text{hr}} \times \frac{5280 \text{ ft}}{1 \text{ mi}} \times \frac{1 \text{ hr}}{60 \text{ min}} \times \frac{1 \text{ min}}{60 \text{ sec}} \qquad \begin{array}{l}\text{Divide out}\\ \text{mi, hr,}\\ \text{and min}\end{array}$$

$$= \frac{30 \times 5280 \text{ ft}}{60 \times 60 \text{ sec}}$$

$$= \frac{44 \text{ ft}}{\text{sec}} = 44\frac{\text{ft}}{\text{sec}}$$

Complete the change in units.

$$90\frac{\text{mi}}{\text{hr}} = \underline{\qquad}\frac{\text{ft}}{\text{sec}}$$

Answer: 132

| EXAMPLE 7 APPLICATION OF CONVERSION | PRACTICE EXERCISE 7 |
|---|---|

Claudia wants to buy 15 inches of fringe which costs $1.08 per yard. How much must she pay?

First convert dollars per yard to dollars per foot, then convert dollars per foot to dollars per inch.

$$\frac{1.08 \text{ dollars}}{\text{yd}} = \frac{1.08 \text{ dollars}}{1 \text{ yd}} \times \frac{1 \text{ yd}}{3 \text{ ft}} \times \frac{1 \text{ ft}}{12 \text{ in}} \qquad \text{Unit fractions}$$

$$= \frac{1.08 \text{ dollars}}{3 \times 12 \text{ in}}$$

$$= \frac{0.03 \text{ dollars}}{\text{in}}$$

Metal tubing costs $3.60 per yard. How much must Larry pay for 28 inches?

Since Claudia wants 15 in of the fringe, multiply 15 in by $\frac{0.03 \text{ dollars}}{\text{in}}$.

$$\frac{15 \text{ in}}{1} \times \frac{0.03 \text{ dollars}}{\text{in}} = \$0.45$$

She must pay \$0.45 for the fringe.                    Answer: \$2.80

| **EXAMPLE 8** RATE APPLICATION | **PRACTICE EXERCISE 8** |
|---|---|

Herman is filling a tank with water from a larger tank. If 2 pints per minute flow through the filling tube, how many gallons per minute flow through?

Bernice is told that a machine can fill 120 pints in one minute. How many gallons per minute is this?

Change pints per minute to gallons per minute.

$$2\frac{\text{pt}}{\text{min}} = \frac{2 \text{ pt}}{\text{min}} = \frac{2 \text{ pt}}{1 \text{ min}} \times \frac{1 \text{ qt}}{2 \text{ pt}} \times \frac{1 \text{ gal}}{4 \text{ qt}}$$

$$= \frac{2 \text{ gal}}{2 \times 4 \text{ min}} = \frac{1 \text{ gal}}{4 \text{ min}} = \frac{1}{4}\frac{\text{gal}}{\text{min}}$$

Thus, $\frac{1}{4}$ of a gallon flows through the tube per minute.

Answer: 15 gal per min

## 8.1 EXERCISES A

*Study the lists of equal units before completing the following.*

**1.** 5 ft = _____ in

**2.** 72 in = _____ ft

**3.** 72 in = _____ yd

**4.** 21 da = _____ wk

**5.** 64 oz = _____ lb

**6.** 36 hr = _____ da

**7.** 72 fl oz = _____ pt

**8.** 4500 lb = _____ ton

**9.** 6 wk = _____ da

**10.** 3 mi = _____ ft

**11.** 18 qt = _____ gal

**12.** 16 mo = _____ yr

**13.** 1 hr = _____ sec

**14** 5400 sec = _____ hr

**15.** 96 in = _____ yd

**16.** 7 yd = _____ in

**17.** 3 ton = _____ oz

**18.** 96 fl oz = _____ qt

**19.** 2 mi = _____ in

**20.** 3 wk = _____ min

**21.** 7200 sec = _____ da

**22.** $15\frac{\text{mi}}{\text{hr}}$ = _____ $\frac{\text{ft}}{\text{sec}}$

**23** $88\frac{\text{ft}}{\text{sec}}$ = _____ $\frac{\text{mi}}{\text{hr}}$

**24.** $5\frac{\text{cents}}{\text{in}}$ = _____ $\frac{\text{dollars}}{\text{yd}}$

**25.** $2\frac{\text{gal}}{\text{min}}$ = _____ $\frac{\text{qt}}{\text{sec}}$

**26.** $200\frac{\text{lb}}{\text{da}}$ = _____ $\frac{\text{ton}}{\text{wk}}$

**27.** $32\frac{\text{oz}}{\text{hr}}$ = _____ $\frac{\text{lb}}{\text{min}}$

*Solve.*

**28.** George buys 30 in of material which costs $2.16 per yard. How much does he have to pay?

**29.** The speed limit is 55 miles per hour and Maria is driving at 5280 feet per minute. Is she driving over the speed limit? If so, by how much?

**30.** Richard is filling a tank at the rate of 4 pints per minute. How many gallons per hour is this?

**31** A farmer needs to buy 3 miles of wire. What is the total cost if the wire costs 20¢ per foot?

## FOR REVIEW

*The following exercises review material from Chapter 4 to help prepare you for the next section.*

**32.**
$$6\frac{1}{2}$$
$$+\,9\frac{1}{3}$$

**33.**
$$42\frac{2}{5}$$
$$-\,16\frac{3}{4}$$

**34.**
$$162\frac{5}{12}$$
$$-\,117\frac{17}{18}$$

ANSWERS: 1. 60  2. 6  3. 2  4. 3  5. 4  6. 1.5  7. 4.5  8. 2.25  9. 42  10. 15,840  11. 4.5  12. $1\frac{1}{3}$  13. 3600
14. 1.5  15. $2\frac{2}{3}$  16. 252  17. 96,000  18. 3  19. 126,720  20. 30,240  21. $\frac{1}{12}$  22. 22  23. 60  24. 1.80  25. $\frac{2}{15}$
26. 0.7  27. $\frac{1}{30}$  28. $1.80  29. yes; $5\frac{\text{mi}}{\text{hr}}$  30. $30\frac{\text{gal}}{\text{hr}}$  31. $3168  32. $15\frac{5}{6}$  33. $25\frac{13}{20}$  34. $44\frac{17}{36}$

## 8.1 EXERCISES B

*Complete the change in units.*

**1.** 9 ft = _____ in

**2.** 132 in = _____ ft

**3.** 132 in = _____ yd

**4.** 63 da = _____ wk

**5.** 88 oz = _____ lb

**6.** 96 hr = _____ da

**7.** 108 fl oz = _____ pt

**8.** 15,000 lb = _____ ton

**9.** 13 wk = _____ da

**10.** 10 mi = _____ ft

**11.** 27 qt = _____ gal

**12.** 37 mo = _____ yr

**13.** $\frac{1}{2}$ hr = _____ sec

**14.** 400 sec = _____ hr

**15.** 144 in = _____ yd

**16.** 12 yd = _____ in

**17.** 0.1 ton = _____ oz

**18.** 160 fl oz = _____ qt

**19.** 0.4 mi = _____ in

**20.** $\frac{1}{2}$ wk = _____ min

**21.** 86,400 sec = _____ da

**22.** $75\dfrac{mi}{hr} =$ _____ $\dfrac{ft}{sec}$

**23.** $11\dfrac{ft}{sec} =$ _____ $\dfrac{mi}{hr}$

**24.** $20\dfrac{cents}{in} =$ _____ $\dfrac{dollars}{yd}$

**25.** $30\dfrac{gal}{min} =$ _____ $\dfrac{qt}{sec}$

**26.** $700\dfrac{lb}{da} =$ _____ $\dfrac{ton}{wk}$

**27.** $640\dfrac{oz}{hr} =$ _____ $\dfrac{lb}{min}$

*Solve.*

**28.** Sirloin steak is priced at $2.79 per pound. How much would the Moores have to pay for 35 lb?

**29.** Burt is driving at the rate of 55 feet per second. If he is driving where the speed limit is 40 miles per hour, how does his speed compare to the limit?

**30.** A water pump will deliver 210 gallons per minute. How many quarts per second is this?

**31.** Electrical wiring costs 5¢ per inch. How much would 25 yards of this wire cost?

## FOR REVIEW

*The following exercises review material from Chapter 4 to help prepare you for the next section.*

**32.**
$$14\frac{3}{7}$$
$$+\ \ 8\frac{5}{14}$$

**33.**
$$22\frac{1}{3}$$
$$-\ 11\frac{4}{5}$$

**34.**
$$216\frac{29}{60}$$
$$-\ 129\frac{17}{20}$$

## 8.1 EXERCISES C

*Use a calculator to find the following change in units. Give answers correct to the nearest tenth.*

**1.** 627 in = _____ yd

**2.** $55\dfrac{mi}{hr} =$ _____ $\dfrac{ft}{sec}$
[Answer: 80.7]

**3.** 143 da = _____ yr

**4.** Mount McKinley in Alaska is the highest mountain in the United States at 20,320 feet above sea level. Give the height in miles.
[Answer: 3.8]

## 8.2 ARITHMETIC OF MEASUREMENT NUMBERS

**STUDENT GUIDEPOSTS**

1. Simplifying Measurement Units
2. Adding Measurement Units
3. Subtracting Measurement Units
4. Multiplying Measurement Units
5. Dividing Measurement Units

### 1 SIMPLIFYING MEASUREMENT UNITS

Suppose the answer to a time-conversion problem is 64 minutes. When a time answer is more than 1 hour (60 min), the result is usually written in hours and minutes.

$$64 \text{ min} = \textbf{60 min} + 4 \text{ min}$$
$$= \textbf{1 hr } 4 \text{ min} \qquad \text{Plus sign left out}$$

This is read ''one hour four minutes.'' This form of the answer gives a better idea of the size of some measurements, especially when the numbers are large. For example, we have a better idea of the time if we are told a job will take 8 hr rather than 480 min.

Other measurement units can also be simplified. For example, to change 32 inches to feet and inches, write 32 as the sum of a multiple of 12 and a number less than 12.

$$32 \text{ in} = \textbf{24 in} + 8 \text{ in}$$
$$= \textbf{2 ft } 8 \text{ in} \qquad 24 \text{ in} = 2 \text{ ft}$$

This is read ''two feet eight inches.''

---

| **EXAMPLE 1**   SIMPLIFYING UNITS | **PRACTICE EXERCISE 1** |
|---|---|

Simplify the following measurements.

**(a)** 3 hr 92 min

To change 92 minutes to hours and minutes, write 92 as the sum of a multiple of 60 and a number less than 60.

$$3 \text{ hr } \textbf{92 min} = 3 \text{ hr} + \textbf{60 min} + \textbf{32 min} \qquad 92 = 60 + 32$$
$$= 3 \text{ hr} + \textbf{1 hr} + 32 \text{ min} \qquad 60 \text{ min} = 1 \text{ hr}$$
$$= \textbf{4 hr } 32 \text{ min} \qquad \text{Add 3 hr and 1 hr}$$

**(b)** 3 yd 8 ft

$$3 \text{ yd } \textbf{8 ft} = 3 \text{ yd} + \textbf{6 ft} + \textbf{2 ft} \qquad 8 = 6 + 2$$
$$= 3 \text{ yd} + \textbf{2 yd} + 2 \text{ ft} \qquad 6 \text{ ft} = 2 \text{ yd}$$
$$= \textbf{5 yd } 2 \text{ ft}$$

*Practice Exercise 1:*

Simplify the following measurements.

**(a)** 7 hr 116 min

**(b)** 5 yd 14 ft

Answers: (a) **8 hr 56 min**
(b) **9 yd 2 ft**

---

| **EXAMPLE 2**   EXPANDING UNITS | **PRACTICE EXERCISE 2** |
|---|---|

Write 125 inches as yards, feet, and inches.

First, change 125 inches to feet and inches by writing it as the sum of a multiple of 12 and a number less than 12.

$$125 \text{ in} = \textbf{120 in} + 5 \text{ in} \qquad 120 \text{ is a multiple of 12}$$
$$= \textbf{10(12) in} + 5 \text{ in}$$
$$= \textbf{10 ft } 5 \text{ in} \qquad 12 \text{ in} = 1 \text{ ft}$$

Now change 10 ft to yards and feet.

$$\textbf{10 ft } 5 \text{ in} = \textbf{9 ft} + \textbf{1 ft} + 5 \text{ in} \qquad 10 = 9 + 1$$
$$= \textbf{3 yd } 1 \text{ ft } 5 \text{ in} \qquad 9 \text{ ft} = 3 \text{ yd}$$

*Practice Exercise 2:*

Write 400 inches as yards, feet, and inches.

Answer: **11 yd 0 ft 4 in** =
**11 yd 4 in**

---

| **EXAMPLE 3**   SIMPLIFYING VOLUME UNITS | **PRACTICE EXERCISE 3** |
|---|---|

Simplify 9 gal 18 qt 9 pt.

First change 9 pints to quarts and pints.

$$9 \text{ pt} = \textbf{8 pt} + 1 \text{ pt}$$
$$= \textbf{4 qt } 1 \text{ pt} \qquad 2 \text{ pt} = 1 \text{ qt}$$

*Practice Exercise 3:*

Simplify 4 gal 34 qt 23 pt.

Adding 4 qt to 18 qt gives 22 qt. To change 22 qt to gallons and quarts, write 22 as the sum of a multiple of 4 and a number less than 4.

$$22 \text{ qt } 1 \text{ pt} = 20 \text{ qt} + 2 \text{ qt} + 1 \text{ pt}$$
$$= 5(4 \text{ qt}) + 2 \text{ qt} + 1 \text{ pt}$$
$$= 5 \text{ gal } 2 \text{ qt } 1 \text{ pt} \qquad 4 \text{ qt} = 1 \text{ gal}$$

Since

$$9 \text{ gal} + 5 \text{ gal} = 14 \text{ gal},$$

the final answer is 14 gal 2 qt 1 pt.

**Answer: 15 gal 1 qt 1 pt**

---

## ❷ ADDING MEASUREMENT UNITS

In these simplifications, we have been adding measurements. In Example 3 we wrote

$$9 \text{ gal} + 5 \text{ gal} = 14 \text{ gal}.$$

Only measurements in the same unit can be added. Thus, 9 gal can be added to 5 gal, but 14 gal cannot be added to 2 qt without first converting one of the units to the other.

In some cases, we need to add measurements such as 8 ft 5 in and 6 ft 4 in. This problem is like adding mixed numbers. We can change both to inches and add or we can add them as mixed measurement units.

---

### EXAMPLE 4  ADDING UNITS

Add.

**(a)** 8 ft 5 in and 6 ft 4 in
Add each unit separately.

$$
\begin{array}{r}
8 \text{ ft} \quad 5 \text{ in} \\
+ \quad 6 \text{ ft} \quad \underline{4 \text{ in}} \\
\text{sum of feet} \longrightarrow \quad 14 \text{ ft} \quad 9 \text{ in} \longleftarrow \text{sum of inches}
\end{array}
$$

The sum is 14 ft 9 in.

**(b)** 3 hr 28 min and 7 hr 54 min

$$
\begin{array}{r}
3 \text{ hr} \quad 28 \text{ min} \\
+ \quad 7 \text{ hr} \quad \underline{54 \text{ min}} \\
\text{sum of hours} \longrightarrow \quad 10 \text{ hr} \quad 82 \text{ min} \longleftarrow \text{sum of minutes}
\end{array}
$$

Since 82 min is more than 1 hr, simplify the sum.

$$10 \text{ hr } \mathbf{82 \text{ min}} = 10 \text{ hr} + \mathbf{60 \text{ min}} + \mathbf{22 \text{ min}}$$
$$= 10 \text{ hr} + 1 \text{ hr} + 22 \text{ min} \qquad 60 \text{ min} = 1 \text{ hr}$$
$$= 11 \text{ hr } 22 \text{ min}$$

The sum is 11 hr 22 min.

### PRACTICE EXERCISE 4

Add.

**(a)** 17 ft 7 in and 5 ft 2 in

**(b)** 14 hr 42 min and 5 hr 37 min

Answers: **(a)** 22 ft 9 in
**(b)** 20 hr 19 min

## ❸ SUBTRACTING MEASUREMENT UNITS

To subtract measurement units, subtract only like units.

| EXAMPLE 5   SUBTRACTING UNITS | PRACTICE EXERCISE 5 |
|---|---|

Subtract 6 ft 4 in from 8 ft 5 in.

$$
\begin{array}{r}
8 \text{ ft} \quad 5 \text{ in} \\
- \ 6 \text{ ft} \quad 4 \text{ in} \\
\hline
2 \text{ ft} \quad 1 \text{ in}
\end{array}
$$

difference of feet ⟶ 2 ft   1 in ⟵ difference of inches

The difference is 2 ft 1 in.

Subtract 10 ft 4 in from 14 ft 11 in.

Answer:  4 ft 7 in

As with subtraction of mixed numbers, sometimes we need to borrow. With measurement units, this means changing one unit of the larger measure to units of the smaller.

| EXAMPLE 6   SUBTRACTING WITH BORROWING | PRACTICE EXERCISE 6 |
|---|---|

Subtract.     17 ft 2 in
          − 9 ft 8 in

Since 2 inches is less than the 8 inches to be subtracted, borrow 1 ft (12 in) from 17 ft.

$$
\begin{aligned}
17 \text{ ft } 2 \text{ in} &= 16 \text{ ft} + 1 \text{ ft} + 2 \text{ in} \\
&= 16 \text{ ft} + 12 \text{ in} + 2 \text{ in} \\
&= 16 \text{ ft } 14 \text{ in}
\end{aligned}
$$

Now subtract.

$$
\begin{array}{r}
17 \text{ ft } 2 \text{ in} = \quad 16 \text{ ft} \quad 14 \text{ in} \\
- \ 9 \text{ ft } 8 \text{ in} = - \quad 9 \text{ ft} \quad 8 \text{ in} \\
\hline
7 \text{ ft} \quad 6 \text{ in}
\end{array}
$$

The difference is 7 ft 6 in.

Subtract.     25 ft 1 in
          − 11 ft 9 in

Answer:  13 ft 4 in

## ❹ MULTIPLYING MEASUREMENT UNITS

Some multiplication problems involve measurement units. For example, suppose it requires 2 yd 1 ft 7 in of material to make one suit. To learn how much material is required for 4 suits, multiply 2 yd 1 ft 7 in by 4. As with addition and subtraction, all units could be converted to inches and multiplied by 4, but the same result can be reached by multiplying 2 yd by 4, 1 ft by 4, and 7 in by 4, and then simplifying the result.

$$
\begin{aligned}
4 \times (2 \text{ yd } 1 \text{ ft } 7 \text{ in}) &= 4(2 \text{ yd} + 1 \text{ ft} + 7 \text{ in}) \\
&= 8 \text{ yd} + 4 \text{ ft} + 28 \text{ in} \qquad \text{Distributive property}
\end{aligned}
$$

Now simplify. First change 28 in to feet: 28 in = 24 in + 4 in = 2 ft + 4 in. Then, rewrite the product.

8 yd + 4 ft + 28 in = 8 yd + 4 ft + 2 ft + 4 in     Replace 28 in by 2 ft + 4 in

$$= 8 \text{ yd} + 6 \text{ ft} + 4 \text{ in} \qquad \text{Add 4 ft and 2 ft}$$
$$= 8 \text{ yd} + 2 \text{ yd} + 4 \text{ in} \qquad \text{3 ft} = 1 \text{ yd}$$
$$= 10 \text{ yd } 4 \text{ in}$$

Thus, 10 yd 4 in is needed for the 4 suits. This could be written as 10 yd 0 ft 4 in.

| EXAMPLE 7 MULTIPLYING BY A NON-MEASUREMENT NUMBER | PRACTICE EXERCISE 7 |
|---|---|

If a tank holds 7 gal 3 qt 1 pt of liquid, how much would be used to fill 6 tanks of the same size?

Multiply 7 gal 3 qt 1 pt by 6.

$$6 \times (7 \text{ gal } 3 \text{ qt } 1 \text{ pt}) = 6(7 \text{ gal} + 3 \text{ qt} + 1 \text{ pt})$$

| | |
|---|---|
| $= 42 \text{ gal} + 18 \text{ qt} + 6 \text{ pt}$ | |
| $= 42 \text{ gal} + 18 \text{ qt} + 3 \text{ qt}$ | 2 pt = 1 qt |
| $= 42 \text{ gal} + 21 \text{ qt}$ | 18 qt + 3 qt = 21 qt |
| $= 42 \text{ gal} + 20 \text{ qt} + 1 \text{ qt}$ | 20 qt = 5(4 qt) |
| $= 42 \text{ gal} + 5 \text{ gal} + 1 \text{ qt}$ | 4 qt = 1 gal |
| $= 47 \text{ gal } 1 \text{ qt}$ | 42 gal + 5 gal<br>  = 47 gal |

For a survival test each person was given 1 gal 2 qt 1 pt of water. How much water was needed to supply the 15 people taking the test?

Answer: 24 gal 1 qt 1 pt

It is possible to multiply measurement units by measurement units.

| EXAMPLE 8 MULTIPLYING MEASUREMENT UNITS | PRACTICE EXERCISE 8 |
|---|---|

Find the area of a rectangle which is 4 ft wide and 6 ft long.

Recall from Chapter 2 that

$$\text{area} = \text{length} \times \text{width}$$
$$= 6 \text{ ft} \times 4 \text{ ft}$$

To work the problem find not only $6 \times 4$ but also ft $\times$ ft.

$$\text{area} = 6 \text{ ft} \times 4 \text{ ft}$$
$$= (6 \times 4)(\text{ft} \times \text{ft})$$
$$= 24 \text{ ft}^2$$
$$= 24 \text{ sq ft}$$

The area is 24 sq ft. (The unit sq ft or $\text{ft}^2$ is used for area. We will say more about it later in this chapter.)

Find the area of a rectangular garden which is 10 yd wide and 17 yd long.

Answer: 170 yd$^2$

## 5 DIVIDING MEASUREMENT UNITS

To divide a measurement by an ordinary number, first divide the largest unit by the number. If there is a remainder, add it to the next unit and divide again.

| **EXAMPLE 9    DIVIDING BY A NON-MEASUREMENT NUMBER** | **PRACTICE EXERCISE 9** |

Divide 17 lb 6 oz by 3.

First divide 17 lb by 3.

$$\begin{array}{r} 5 \text{ lb} \\ 3\overline{)17 \text{ lb}} \\ \underline{15} \\ 2 \text{ lb} \end{array}$$

Thus, 17 lb ÷ 3 = 5 lb with a remainder of 2 lb. Change 2 lb to ounces.

$$2 \text{ lb} = 2(16) \text{ oz} = \underline{32 \text{ oz}}$$

Add the 6 oz from the original measurement to 32 oz.

$$\underline{32 \text{ oz}} + 6 \text{ oz} = 38 \text{ oz}$$

Now divide 38 oz by 3.

$$\begin{array}{r} 12 \text{ oz} \\ 3\overline{)38 \text{ oz}} \\ \underline{3} \\ 8 \\ \underline{6} \\ 2 \text{ oz} \end{array}$$

The answer to this division is 12 oz with remainder 2 or $12\frac{2}{3}$ oz. The final answer to the problem (17 lb 6 oz) ÷ 3 is

$$5 \text{ lb } 12\frac{2}{3} \text{ oz.}$$

Divide 42 lb 13 oz by 5.

Answer:  8 lb 9 oz

---

Some problems involve division of a measurement by a measurement. Suppose you want to know how many shirts costing $16 each can be bought for $48.

$$\frac{48 \text{ dollars}}{16 \text{ dollars}} = \frac{3 \times \cancel{16 \text{ dollars}}}{\cancel{16 \text{ dollars}}} = 3$$

The dollars cancel and the answer is the whole number (with no measurement unit) 3. You can buy 3 shirts for $48.

| **EXAMPLE 10    DIVIDING MEASUREMENT UNITS** | **PRACTICE EXERCISE 10** |

How many 4-ft shelves can be cut from a board 16 ft long?

Divide 16 ft by 4 ft.

$$\frac{16 \text{ ft}}{4 \text{ ft}} = \frac{4 \times \cancel{4} \cancel{\text{ft}}}{\cancel{4} \cancel{\text{ft}}} = 4$$

Thus, 4 shelves can be cut from the board.

One uniform can be made from 3 yd of material. How many uniforms can be made from 51 yd of the material?

Answer:  17 uniforms

## 8.2  EXERCISES A

*Simplify.*

**1.** 83 min

**2.** 3 hr 92 min

**3.** 5 yd 8 ft

**4.** 8 gal 17 qt

**5.** 14 lb 73 oz

**6.** 4 yd 7 ft 23 in

*Perform the indicated operations and simplify.*

**7.**   8 lb   3 oz
     + 4 lb 10 oz

**8.**   5 hr 52 min
     + 13 hr 41 min

**9.**   3 gal 3 qt
     + 4 gal 2 qt

**10.**   6 yd 2 ft 11 in
      + 7 yd 1 ft  9 in

**11.**   7 yd 2 ft
      − 3 yd 1 ft

**12.**   8 hr 30 min
      − 2 hr 26 min

**13.**   8 gal 1 qt
      − 4 gal 3 qt

**14.**   5 da 3 hr  8 min
      − 4 da 6 hr 40 min

**15.**   9 yd 2 ft  3 in
      − 2 yd 2 ft 10 in

**16.** 3 × (5 hr 16 min)

**17** 5 × (4 gal 3 qt)

**18.** 4 × (2 yd 2 ft 11 in)

**19.** 6 × (2 hr 18 min 14 sec)

**20.** 3 in × 12 in

**21.** 7 yd × 12 yd

**22.** (5 lb 12 oz) ÷ 2

**23** (7 hr 33 min) ÷ 3

**24.** 18 ft ÷ 3 ft

*Solve.*

**25.** How many $6 ski caps can be bought for $24?

**26.** A chemist has 5 gal 3 qt of one liquid to be mixed with 4 gal 2 qt of another. How much will he have after he mixes them?

**27.** Henry took 6 gal 2 qt of water on a camping trip. After using 2 gal 4 qt, how much water did he have left?

**28** If it takes 1 lb 12 oz of flour to make one loaf of bread, how much will it take to make 7 loaves?

**29.** Find the area of a rectangle which is 16 inches long and 5 inches wide.

**30.** If it takes 1 lb 12 oz of flour to make one loaf of bread, how much will it take to make a loaf one half the size?

## FOR REVIEW

*Complete the unit conversion.*

**31.** 16 ton = _____ lb

**32.** $44 \dfrac{\text{ft}}{\text{sec}} =$ _____ $\dfrac{\text{mi}}{\text{hr}}$

**33.** $50 \dfrac{\text{cents}}{\text{in}} =$ _____ $\dfrac{\text{dollars}}{\text{ft}}$

ANSWERS:  1. 1 hr 23 min  2. 4 hr 32 min  3. 7 yd 2 ft  4. 12 gal 1 qt  5. 18 lb 9 oz  6. 6 yd 2 ft 11 in
7. 12 lb 13 oz  8. 19 hr 33 min  9. 8 gal 1 qt  10. 14 yd 1 ft 8 in  11. 4 yd 1 ft  12. 6 hr 4 min  13. 3 gal 2 qt
14. 20 hr 28 min  15. 6 yd 2 ft 5 in  16. 15 hr 48 min  17. 23 gal 3 qt  18. 11 yd 2 ft 8 in  19. 13 hr 49 min 24 sec
20. 36 sq in  21. 84 sq yd  22. 2 lb 14 oz  23. 2 hr 31 min  24. 6  25. 4  26. 10 gal 1 qt  27. 3 gal 2 qt
28. 12 lb 4 oz  29. 80 sq in  30. 14 oz  31. 32,000  32. 30  33. 6

## 8.2  EXERCISES B

*Simplify.*

**1.** 137 min

**2.** 7 hr 72 min

**3.** 4 yd 19 ft

**4.** 2 gal 25 qt

**5.** 6 lb 105 oz

**6.** 5 yd 14 ft 42 in

*Perform the indicated operations and simplify.*

**7.**  9 lb 11 oz
  + 5 lb  1 oz

**8.**  12 hr 39 min
  + 10 hr 32 min

**9.**  9 gal 3 qt
  + 7 gal 3 qt

**10.**  12 yd 1 ft  5 in
  + 18 yd 2 ft 10 in

**11.**  4 yd 2 ft
  − 1 yd 1 ft

**12.**  19 hr 49 min
  − 15 hr 32 min

**13.**  17 gal 1 qt
  − 13 gal 2 qt

**14.**  22 da  2 hr 36 min
  − 21 da 22 hr 48 min

**15.**  6 yd 1 ft 1 in
  − 3 yd 2 ft 2 in

**16.** 4 × (3 hr 10 min)

**17.** 10 × (3 gal 2 qt)

**18.** 6 × (7 yd 1 ft 8 in)

**19.** 3 × (5 hr 40 min 25 sec)

**20.** 14 in × 22 in

**21.** 5 yd × 9 yd

**22.** (7 lb 8 oz) ÷ 3

**23.** (10 hr 20 min) ÷ 4

**24.** 33 ft ÷ 11 ft

*Solve.*

**25.** Steak costs $3 per pound. How many pounds can Ralph buy with $18?

**26.** For a company picnic, Ward brought 7 gal 2 qt of iced tea. If Sarah brought 10 gal 3 qt, how much tea did they have?

**27.** If the people at the picnic in Exercise 26 drank 15 gal 3 qt of the tea, how much was left?

**28.** Billy Joe has a chili recipe that requires 2 lb 6 oz of ground beef. How much meat should he buy to make 5 times his recipe?

**29.** Vicki needs to paint a wall which is 8 feet high and 12 feet wide. How many square feet of wall is this to paint?

**30.** A cookie recipe requires 1 lb 2 oz of flour. How much flour would be used if only one third of the recipe is to be made?

## FOR REVIEW

*Complete the unit conversion.*

**31.** 0.5 ton = _____ lb

**32.** 33 $\dfrac{\text{ft}}{\text{sec}}$ = _____ $\dfrac{\text{mi}}{\text{hr}}$

**33.** 200 $\dfrac{\text{cents}}{\text{in}}$ = _____ $\dfrac{\text{dollars}}{\text{ft}}$

## 8.2 EXERCISES C

*Solve.*

**1.** If material costs $2.70 per yard, how much does 8 feet of the material cost?
[Answer: $7.20]

**2.** Wire costs $13.85 per yard. How much does the wire cost per foot (to the nearest cent)?

## 8.3 THE METRIC SYSTEM

### STUDENT GUIDEPOSTS

**1** Length in Meters
**2** Volume in Liters
**3** Weight in Grams
**4** Units of Measure

Most of the world uses the metric system of measurement. It is a decimal system, and changes in units can be made by moving the decimal point. Thus, it is much easier to convert from one unit to another in the metric system than it is in the English system (just as easy as changing $1.27 to 127¢).

### 1 LENGTH IN METERS

The standard of length in the metric system is the **meter.** The opening in a normal doorway is about 2 meters high. Thus, a meter is about half the height of an opening in a doorway. A twin bed is about a meter wide and most newborn babies are about $\frac{1}{2}$ meter tall.

The actual length of the scale in Figure 8.1 is 0.1 meter. The distance between each mark is 0.01 meter.

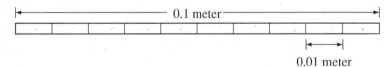

**Figure 8.1**

### 2 VOLUME IN LITERS

If we make a box as shown in Figure 8.2 (not actual size), the box holds one **liter.** This is the standard unit for capacity in the metric system. If you are thirsty on a summer day, you can probably drink a liter of liquid. It is a little less than three cans of soft drink.

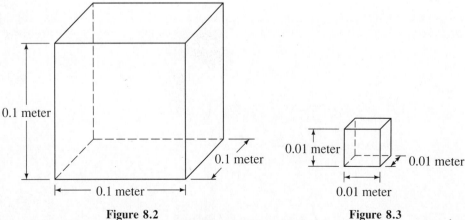

**Figure 8.2**                    **Figure 8.3**

## ❸ WEIGHT IN GRAMS

The standard for mass or weight at sea level is the **gram.** Figure 8.3 shows a cube of side 0.01 meter. If this cube were filled with water, the weight of the water would be 1 gram. One large paper clip (or two small ones) weighs about one gram and a large steak might weigh 1000 grams.

## ❹ UNITS OF MEASURE

As with the English system, the various size units such as 1000 grams have names, and the naming is standard. For example, the names for 1000 meters, 1000 liters, and 1000 grams all start with the same prefix, *kilo*. Thus,

$$1000 \text{ meters} = 1 \text{ } kilo\text{meter},$$
$$1000 \text{ liters} = 1 \text{ } kilo\text{liter},$$
$$1000 \text{ grams} = 1 \text{ } kilo\text{gram}.$$

The following table lists the prefixes and symbols used in the metric system.

| Number of units | 1000 | 100 | 10 | 1 | $\frac{1}{10}$ or 0.1 | $\frac{1}{100}$ or 0.01 | $\frac{1}{1000}$ or 0.001 |
|---|---|---|---|---|---|---|---|
| Prefix | kilo | hecto | deka | unit | deci | centi | milli |
| Length | km | hm | dam | meter(m) | dm | cm | mm |
| Volume | kl | hl | dal | liter(L) | dl | cl | ml |
| Weight | kg | hg | dag | gram(g) | dg | cg | mg |

The different size units are used in different applications. For example, the size of a room is given in meters, the distance between Miami and Washington, D.C., in kilometers, and bolt sizes in millimeters. In Figure 8.4, the paper clip and bolt are shown actual size.

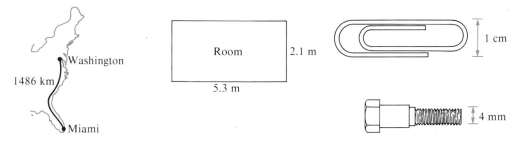

**Figure 8.4**

In Figure 8.5 (drawn to actual size), Figure 8.1 has been repeated using the new names we have learned and a few more details. This should give you a good idea of the sizes of some units of length and their relationship. A meter is 10 times the total length of the scale in Figure 8.5. A kilometer is about the length of five city blocks.

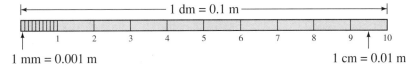

**Figure 8.5**

The most commonly used measures of length are the kilometer, meter, centimeter, and millimeter. These same prefixes—kilo, centi, and milli—are used for volume and weight also.

In volume units, the liter is used to measure gasoline and beverages. The kiloliter measures the capacity of large tanks, and the milliliter measures liquids in a laboratory.

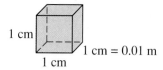

1 cm
1 cm
1 cm = 0.01 m

**Figure 8.6**

The cube in Figure 8.6, which is 1 cm on a side, has volume 1 ml and this much water weighs 1 g. Thus, the gram is a small unit of measurement. It is used in the laboratory where an even smaller unit, the milligram, is also needed. The unit that would suit many common weight applications is the kilogram. As has been mentioned, a large steak weighs about 1 kg (1000 g).

Now that we have some idea of the size of metric units, we turn to conversion problems. Instead of using decimal conversion units such as 1 mm = 0.001 m, it will be easier to write all unit fractions in terms of 10, 100, or 1000. The following table gives the equal units.

| Length | Volume | Weight |
|---|---|---|
| 1 km  = 1000 m | 1 kl  = 1000 L | 1 kg  = 1000 g |
| 1 hm  =  100 m | 1 hl  =  100 L | 1 hg  =  100 g |
| 1 dam =   10 m | 1 dal =   10 L | 1 dag =   10 g |
| 1 m   =   10 dm | 1 L   =   10 dl | 1 g   =   10 dg |
| 1 m   =  100 cm | 1 L   =  100 cl | 1 g   =  100 cg |
| 1 m   = 1000 mm | 1 L   = 1000 ml | 1 g   = 1000 mg |

## H I N T

Remember that multiplying and dividing by powers of ten can be done by moving the decimal point. Consider these examples.

$$16.2 \times 100 = 1620 \qquad \text{Move two places to the right}$$

$$16.2 \div 100 = 0.162 \qquad \text{Move two places to the left}$$

---

**EXAMPLE 1** CONVERTING METERS TO CENTIMETERS

5.2 m = _____ cm.

Since 1 m = 100 cm, the unit fraction is $\frac{100 \text{ cm}}{1 \text{ m}}$.

$$5.2 \text{ m} = 5.2 \text{ m} \times 1$$
$$= \frac{5.2 \text{ m}}{1} \times \frac{100 \text{ cm}}{1 \text{ m}}$$
$$= 5.2 \times 100 \text{ cm}$$
$$= 520 \text{ cm}$$

**PRACTICE EXERCISE 1**

13.25 m = _____ cm.

5.2 is multiplied by 100 which moves the decimal point 2 places to the right. That is,

$$5.2 \text{ m} = 520. \text{ cm.}$$
$$\underset{2}{\overset{\curvearrowright}{}}$$

---

| **EXAMPLE 2** CONVERTING MILLILITERS TO DECILITERS | **PRACTICE EXERCISE 2** |

362 ml = _____ dl.
　　The unit fractions are $\frac{1 \text{ L}}{1000 \text{ ml}}$ and $\frac{10 \text{ dl}}{1 \text{ L}}$.

$$362 \text{ ml} = \frac{362 \text{ ml}}{1} \times \frac{1 \text{ L}}{1000 \text{ ml}} \times \frac{10 \text{ dl}}{1 \text{ L}}$$

$$= 362 \times \frac{1}{1000} \times \frac{10}{1} \text{ dl}$$

$$= 362 \times \frac{10}{1000} \text{ dl}$$

$$= 362 \times \frac{1}{100} \text{ dl}$$

$$= 3.62 \text{ dl}$$

362 is divided by 100 which moves the decimal point 2 places to the left.

64,800 ml = _____ dl.

---

　　In Example 2, multiplying by one unit fraction moved the decimal point 3 places to the left and multiplying by the other moved it 1 place to the right. The same change of units could have been made by moving the decimal point 2 places to the left. This method of conversion is shown in the next example.

---

| **EXAMPLE 3** CONVERTING BY MOVING DECIMAL PLACES | **PRACTICE EXERCISE 3** |

362 ml = _____ dl.
　　First show all the units in a diagram.

| 1000 L | 100 L | 10 L | **1 L** | 0.1 L | 0.01 L | 0.001 L |
|---|---|---|---|---|---|---|
| **(kl)** | **(hl)** | **(dal)** | | **(dl)** | **(cl)** | **(ml)** |

Start at **ml** and make 2 moves to the *left* to get to **dl.** Thus, to change ml to dl move the decimal point 2 places to the *left*.

$$362 \text{ ml} = 3.62 \text{ dl} = 3.62 \text{ dl}$$
$$\underset{2}{\overset{\curvearrowleft}{}}$$

0.057 ml = _____ dl.

---

| **EXAMPLE 4** CONVERTING HECTOGRAMS TO CENTIGRAMS | **PRACTICE EXERCISE 4** |

0.0256 hg = _____ cg.

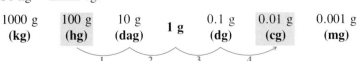

| 1000 g | 100 g | 10 g | **1 g** | 0.1 g | 0.01 g | 0.001 g |
|---|---|---|---|---|---|---|
| **(kg)** | **(hg)** | **(dag)** | | **(dg)** | **(cg)** | **(mg)** |

0.000882 hg = _____ cg.

Start at hg and move 4 times to the *right* to get to cg. Thus, to change hg to cg, move the decimal point 4 places to the *right*.

$$0.0246 \text{ hg} = 00256. \text{ cg} = 256 \text{ cg}$$
$$\underset{4}{\phantom{0.0256}}$$

Answer: 8.82

---

| **EXAMPLE 5**   CONVERTING MILLIMETERS TO HECTOMETERS | **PRACTICE EXERCISE 5** |

3280 mm = _____ hm.

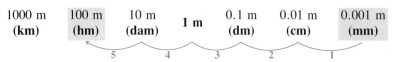

Start at mm and make 5 moves to the *left* to get to hm. Thus, to change mm to hm, move the decimal point 5 places to the *left*.

$$3280 \text{ mm} = 0.03280 \text{ hm}$$
$$\underset{5}{\phantom{0.03280}}$$

426,000 mm = _____ hm.

Answer: 4.26

---

The time units for the metric system are the same as for the English system. However, the temperature scale is different. Both metric and English temperature will be discussed in the next section when we convert from one system to the other.

## 8.3  EXERCISES A

*Answer true or false. If the answer is false, tell why.*

**1.** The system of measurement used in most of the world is the metric system.

**2.** The opening in a normal doorway is about 10 meters high.

**3.** A liter is the space in a box which is 0.1 meter or one decimeter on a side.

**4.** The weight of water that can be put in a box which is 0.01 meter (1 centimeter) on a side is one gram.

**5.** A large steak might weigh about one gram.

**6.** To change units of length in the metric system, just move the decimal point.

*Complete the following by moving the decimal point as shown in the examples.*

**7.** 1 km = _____ m

**8.** 1 hl = _____ L

**9.** 1 dag = _____ g

**10.** 1 m = _____ dm

**11.** 1 mg = _____ g

**12.** 1 g = _____ mg

**13.** 62 g = _____ kg

**14.** 3.2 cm = _____ m

**15.** 16 dal = _____ L

**16.** 10.3 cm = _____ hm

**17** 0.00721 kg = _____ dag

**18.** 63,000 ml = _____ kl

**19.** 1700 dm = _____ hm

**20.** 16.25 cg = _____ mg

**21.** 0.000212 hl = _____ ml

**22.** 92.6 mm = _____ cm

**23** 4728 dg = _____ hg

**24.** 5,000,000 cl = _____ dal

**25.** 0.0052 hm = _____ mm

**26.** 0.07 dag = _____ mg

**27.** 3200 dl = _____ kl

**28.** 0.0953 km = _____ dm

**29.** 42 kg = _____ hg

**30.** 7.2 hl = _____ cl

**31.** $30 \frac{cm}{sec} = $ _____ $\frac{m}{hr}$

**32** $20 \frac{cg}{da} = $ _____ $\frac{g}{wk}$

**33.** $360 \frac{kl}{hr} = $ _____ $\frac{ml}{min}$

*Perform the indicated operations.*

**34.**  2.05 m
  + 7.97 m

**35.**  382.6 ml
  + 23.8 ml

**36.**  983 kg
  − 724 kg

**37.**  75.36 cm
  − 28.71 cm

**38.** 16.3 m
  × 5

**39.** 16.3 m
  × 5 m

*Solve.*

**40.** How many 22-ml samples can be obtained from 330 ml of a chemical solution?

**41** If steak costs $6 per kilogram, how much would 600 grams of steak cost?

**42.** Gabi is traveling at $60 \frac{km}{hr}$. How many meters per minute is this?

**43.** A rancher needs 10 km of wire. How much will it cost him if the wire is 40¢ per meter?

## FOR REVIEW

*Perform the indicated operations and simplify.*

**44.**  8 ft 11 in
  + 1 ft  3 in

**45.**  8 lb 12 oz
  − 3 lb 15 oz

**46.**  12 gal 2 qt 1 pt
  − 6 gal 3 qt 1 pt

**47.** 5 × (7 yd 2 ft 10 in)

**48.** (11 hr 40 min) ÷ 5

**49.** 42 ft ÷ 7 ft

ANSWERS:  1. true  2. false (about 2 m)  3. true  4. true  5. false (about one kilogram)  6. true  7. 1000
8. 100  9. 10  10. 10  11. 0.001  12. 1000  13. 0.062  14. 0.032  15. 160  16. 0.00103  17. 0.721  18. 0.063
19. 1.7  20. 162.5  21. 21.2  22. 9.26  23. 4.728  24. 5000  25. 520  26. 700  27. 0.32  28. 953  29. 420
30. 72,000  31. 1080  32. 1.4  33. 6,000,000  34. 10.02 m  35. 406.4 ml  36. 259 kg  37. 46.65 cm  38. 81.5 m
39. 81.5 m² (m × m = m²)  40. 15  41. $3.60  42. 1000 $\frac{m}{min}$  43. $4000  44. 10 ft 2 in  45. 4 lb 13 oz
46. 5 gal 3 qt  47. 39 yd 2 ft 2 in  48. 2 hr 20 min  49. 6

## 8.3 EXERCISES B

*Answer true or false. If the answer is false, tell why.*

1. The United States is one of the few countries in the world that uses the English system.

2. The opening in a normal doorway is about 2 meters high.

3. The liter is used to measure gasoline in the metric system.

4. One gram of meat is more than one pound of meat.

5. A large steak might weigh about one kilogram.

6. Units of weight in the metric system can be changed by moving the decimal point.

*Complete the following by moving the decimal point as shown in the examples.*

7. 40 km = _____ m

8. 6 hl = _____ L

9. 0.05 dag = _____ g

10. 50 m = _____ dm

11. 0.061 mg = _____ g

12. 0.061 g = _____ mg

13. 420 g = _____ kg

14. 2300 cm = _____ m

15. 2.5 dal = _____ L

16. 6320 cm = _____ hm

17. 8.7 kg = _____ dag

18. 493,000 ml = _____ kl

19. 0.75 dm = _____ hm

20. 0.88 cg = _____ mg

21. 5.5 hl = _____ ml

22. 250 mm = _____ cm

23. 98,750 dg = _____ hg

24. 0.44 cl = _____ dal

25. 927 hm = _____ mm

26. 0.000082 dag = _____ mg

27. 200 dl = _____ kl

28. 0.000024 km = _____ dm

29. 0.081 kg = _____ hg

30. 4400 hl = _____ cl

31. $2 \dfrac{cm}{sec} = $ _____ $\dfrac{m}{hr}$

32. $500 \dfrac{cg}{da} = $ _____ $\dfrac{g}{wk}$

33. $0.06 \dfrac{kl}{hr} = $ _____ $\dfrac{ml}{min}$

*Perform the indicated operations.*

34.  8.09 km
   + 3.66 km

35.  24.89 dl
   + 13.77 dl

36.  402 g
   − 389 g

37.  4.735 m
   − 1.908 m

38. 213.2 cm
   × 8

39. 213.2 cm
   × 8 cm

*Solve.*

40. Dr. Warren has 42 g of salt. How many 7-dg samples can she get from her supply?

41. Peaches are marked at $1.20 per kilogram. How much will 4000 g of peaches cost?

42. A tank is being filled at the rate of 25 liters per minute. What is this in kiloliters per hour?

43. A perfume is worth $2.60 per milliliter. How much would 5 liters cost?

## FOR REVIEW

*Perform the indicated operations and simplify.*

**44.**  29 ft 7 in
        + 18 ft 6 in

**45.**  2 lb 3 oz
        − 1 lb 7 oz

**46.**  8 gal 1 qt
        − 2 gal 3 qt 1 pt

**47.** 8 × (4 yd 1 ft 6 in)

**48.** (15 hr 36 min) ÷ 6

**49.** 132 ft ÷ 11 ft

## 8.3 EXERCISES C

*Complete the following using an appropriate diagram.*

**1.** $50 \dfrac{dl}{min} = \underline{\hspace{1cm}} \dfrac{hl}{hr}$

[Answer: 3]

**2.** $140 \dfrac{kg}{wk} = \underline{\hspace{1cm}} \dfrac{dag}{hr}$

**3.** $120 \dfrac{cm}{hr} = \underline{\hspace{1cm}} \dfrac{km}{sec}$

**4.** If 1 mi = 1.61 km and a river is flowing at a rate of $4.2 \frac{mi}{hr}$, how fast is it flowing in kilometers per hour?

$\left[\text{Answer: } 6.762 \frac{km}{hr}\right]$

## 8.4 CONVERSIONS BETWEEN MEASUREMENT SYSTEMS

### STUDENT GUIDEPOSTS

**1** Converting Units of Length, Volume, and Weight

**2** Speed in Both Systems

**3** Temperature in Both Systems

### 1 CONVERTING UNITS OF LENGTH, VOLUME, AND WEIGHT

If the United States ever changes to the metric system, it will not be necessary to change units from one system to another. But while both systems are in use it is important to know the relationship between the two.

In the table below are some useful conversion units of measure. Notice that these units are all approximations.

| Length, Volume, and Weight | | |
|---|---|---|
| | *Metric to English* | *English to Metric* |
| Length units | 1 km ≈ 0.621 mi | 1 mi ≈ 1.61 km |
| | 1 m ≈ 39.37 in | 1 in ≈ 2.54 cm |
| Volume units | 1 L ≈ 1.06 qt | 1 qt ≈ 0.946 L |
| Weight units | 1 kg ≈ 2.20 lb | 1 lb ≈ 454 g |

| HINT |
| --- |

Since the conversion units are approximate, we can get slightly different answers by converting different ways. For example,

$$111 \text{ km} \approx \frac{111 \text{ km}}{1} \times \frac{0.621 \text{ mi}}{1 \text{ km}}$$

$$\approx 68.93 \text{ mi.}$$

However,

$$111 \text{ km} \approx 111 \text{ km} \times \frac{1 \text{ mi}}{1.61 \text{ km}}$$

$$\approx 68.94 \text{ mi.}$$

Do not be concerned with small differences like this in answers.

---

Since we are familiar with the English units, these conversions will help us get a better idea of the size of the metric units. For example, on a quarter-mile track, a kilometer is about 2.5 times around. If two towns are 10 miles apart, they are about 16 km apart. A meter (39.37 in) is a little more than a yard (36 in). Figure 8.7 shows the relationship between centimeters and inches.

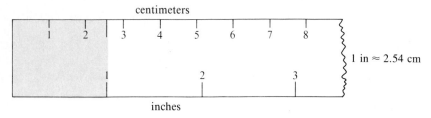

**Figure 8.7**

There is a little more milk in a liter container than there is in a quart, as shown in Figure 8.8.

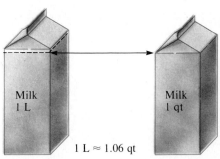

**Figure 8.8**

We now know that the 1-kg steak mentioned in Section 8.3 weighs 2.2 lb. Figure 8.9 shows the relative sizes of 1 lb and 1 kg of the same material.

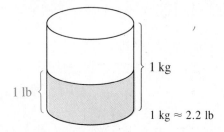

**Figure 8.9**

The following examples show some conversions between the two systems. In most of these problems a calculator is helpful.

| EXAMPLE 1   CONVERTING MEASURES OF LENGTH |
|---|

**(a)** 20 km = _____ mi.

Since 0.621 mi ≈ 1 km, the unit fraction is $\frac{0.621 \text{ mi}}{1 \text{ km}}$.

$$20 \text{ km} \approx \frac{20 \text{ km}}{1} \times \boxed{\frac{0.621 \text{ mi}}{1 \text{ km}}} \qquad \text{Multiply by the unit fraction}$$

$$= 20 \times (0.621) \text{ mi}$$

$$\approx 12.4 \text{ mi} \qquad \text{Rounded to the nearest tenth}$$

**(b)** 35 mi = _____ km.

Multiply by the unit fraction $\frac{1.61 \text{ km}}{1 \text{ mi}}$.

$$35 \text{ mi} \approx \frac{35 \text{ mi}}{1} \times \boxed{\frac{1.61 \text{ km}}{1 \text{ mi}}} \qquad \text{Multiply by the unit fraction}$$

$$= 35 \times 1.61 \text{ km}$$

$$\approx 56.4 \text{ km} \qquad \text{Rounded to the nearest tenth}$$

**(c)** $18\frac{3}{4}$ in = _____ cm.

The unit fraction is $\frac{2.54 \text{ cm}}{1 \text{ in}}$.

$$18\frac{3}{4} = 18.75 \text{ in}$$

$$\approx \frac{18.75 \text{ in}}{1} \times \frac{2.54 \text{ cm}}{1 \text{ in}}$$

$$= (18.75)(2.54) \text{ cm}$$

$$\approx 47.63 \text{ cm} \qquad \text{Rounded to the nearest hundredth}$$

| PRACTICE EXERCISE 1 |
|---|

**(a)** 44 km = _____ mi.

**(b)** 120 mi = _____ km.

**(c)** $216\frac{1}{2}$ in = _____ cm.

Answers: (a) 27.3   (b) 193.2
(c) 549.9

| EXAMPLE 2   CONVERSION OF SPEED |
|---|

Joe is driving at a rate of 95 $\frac{\text{km}}{\text{hr}}$. What is his speed in miles per hour?

$$95 \frac{\text{km}}{\text{hr}} \approx \frac{95 \text{ km}}{\text{hr}} \times \frac{0.621 \text{ mi}}{1 \text{ km}}$$

$$\approx 59 \frac{\text{mi}}{\text{hr}} \qquad \text{Rounded to the nearest mile per hour}$$

| PRACTICE EXERCISE 2 |
|---|

Vera is driving 50 $\frac{\text{km}}{\text{hr}}$ in a 30 $\frac{\text{mi}}{\text{hr}}$ zone. Approximately how many miles per hour is she traveling over the speed limit?

Answer: She is about $1\frac{\text{mi}}{\text{hr}}$ over the speed limit.

## ❷ SPEED IN BOTH SYSTEMS

Figure 8.10 shows a comparison of some common speed limits. Approximations are to the nearest kilometer per hour. A speed of $55\frac{\text{mi}}{\text{hr}}$ is about $89\frac{\text{km}}{\text{hr}}$.

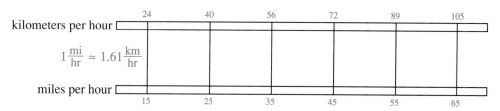

**Figure 8.10**

| EXAMPLE 3 CONVERSION OF MILLILITERS TO QUARTS | PRACTICE EXERCISE 3 |

520 ml = _____ qt.

Since the unit fraction that we know is $\frac{1.06\ qt}{1\ L}$, first change 520 ml to liters.

| **1 L** | 0.1 L (dl) | 0.01 L (cl) | 0.001 L (ml) |

$$520\ ml = 0.520\ L$$

$$0.520\ L \approx \frac{0.52\ L}{1} \times \frac{1.06\ qt}{1\ L}$$

$$= (0.52)(1.06)\ qt$$

$$\approx 0.55\ qt \qquad \text{Rounded to the nearest hundredth}$$

1020 ml = _____ qt.

Answer: **1.08**

| EXAMPLE 4 APPLICATION OF VOLUME CONVERSION | PRACTICE EXERCISE 4 |

The gasoline tank on Roberto's car holds 12.6 gal. How many liters will it hold?

$$12.6\ gal = \frac{12.6\ gal}{1} \times \frac{4\ qt}{1\ gal} \qquad \text{Change 12.6 gal to quarts}$$

$$= (12.6)(4)\ qt$$

$$= 50.4\ qt$$

$$50.4\ qt \approx \frac{50.4\ qt}{1} \times \frac{0.946\ L}{1\ qt} \qquad \text{Change quarts to liters}$$

$$= (50.4)(0.946)\ L$$

$$\approx 47.7\ L \qquad \text{To the nearest tenth}$$

Deb has a foreign car with a 40-L gas tank. How many gallons does the tank hold?

Answer: **10.6 gal**

| EXAMPLE 5 WEIGHT CONVERSION | PRACTICE EXERCISE 5 |

68.2 hg = _____ lb.

The unit fraction is $\frac{2.20\ lb}{1\ kg}$. First, change 62.8 hg to kilograms.

| 1000 g (1 kg) | 100 g (1 hg) | 68.2 hg = 6.82 kg |

$$6.82\ kg \approx \frac{6.82\ kg}{1} \times \frac{2.20\ lb}{1\ kg} \qquad \text{Change kilograms to pounds}$$

$$= (6.82)(2.20)\ lb$$

$$= 15.0\ lb \qquad \text{To the nearest tenth}$$

2240 dg = _____ lb.

Answer: **0.49**

| EXAMPLE 6 APPLICATION OF WEIGHT CONVERSION | PRACTICE EXERCISE 6 |

A football team's starting fullback weighs 200 lb. How many kilograms does he weigh?

$$200\ lb \approx \frac{200\ lb}{1} \times \frac{454\ g}{1\ lb}$$

$$= (200)(454)\ g$$

$$= 90,800\ g$$

A foreign basketball player trying to break into the NBA reported his weight as 97 kg. How many pounds does he weigh?

Change 90,800 g to kilograms.

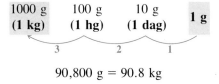

90,800 g = 90.8 kg

Answer: 213 lb

The following table gives comparisons of common weights of people in pounds and kilograms (to the nearest kilogram).

| Pounds | Kilograms |
|--------|-----------|
| 100 | 45 |
| 125 | 57 |
| 150 | 68 |
| 175 | 79 |
| 200 | 91 |
| 225 | 102 |
| 250 | 114 |

### ③ TEMPERATURE IN BOTH SYSTEMS

The temperature scale used in the metric system is the **Celsius** (°C) scale. Water freezes at 0°C (32°F) and boils at 100°C (212°F). Figure 8.11 shows a comparison between degrees **Fahrenheit** (°F) and degrees Celsius. (In the past Celsius was called *centigrade*.)

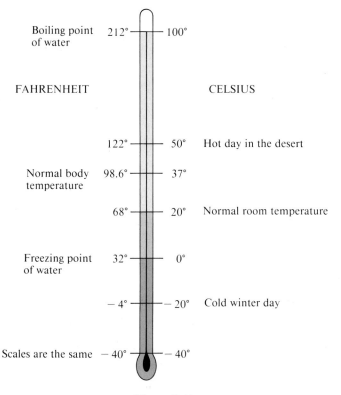

**Figure 8.11**

The equal temperatures on the scale in Figure 8.11 can be found once we know the relationship between the two scales. For example, using the following rules we can find that 20°C = 68°F.

| To Convert from °C to °F |
| --- |

Multiply $C$, the temperature in °C, by $\frac{9}{5} = 1.8$ and add 32.

$$F = \frac{9}{5}C + 32 \quad \text{or} \quad F = 1.8C + 32$$

---

### EXAMPLE 7  CONVERTING CELSIUS TO FAHRENHEIT

20°C = _____ °F.

$$F = \frac{9}{5}C + 32$$

$$F = \frac{9}{5}20 + 32 \qquad C = 20$$

$$= \frac{9 \cdot 20}{5} + 32 \qquad \text{Multiply before adding}$$

$$= \frac{9 \cdot 4 \cdot 5}{5} + 32$$

$$= 36 + 32 = 68$$

Thus, 20°C is 68°F.

**PRACTICE EXERCISE 7**

35°C = _____ °F.

Answer: 95

---

| To Convert from °F to °C |
| --- |

First subtract 32 from $F$, the temperature in °F, then multiply by $\frac{5}{9}$.

$$C = \frac{5}{9}(F - 32)$$

---

### EXAMPLE 8  CONVERTING FAHRENHEIT TO CELSIUS

122°F = _____ °C.

$$C = \frac{5}{9}(F - 32)$$

$$= \frac{5}{9}(122 - 32) \qquad F = 122$$

$$= \frac{5}{9}(90) \qquad \text{Subtract inside parentheses first}$$

$$= \frac{5 \cdot \cancel{9} \cdot 10}{\cancel{9}} = 50 \qquad \text{Then multiply}$$

Thus, 122°F is 50°C.

**PRACTICE EXERCISE 8**

86°F = _____ °C.

Answer: 30

| **EXAMPLE 9**    **APPLICATION OF TEMPERATURE CONVERSION** | **PRACTICE EXERCISE 9** |
|---|---|

You are going to a ballgame and the weatherman says the temperature will be 30°C. What kind of clothes should you wear?

If you have a better understanding of °F, change °C to °F using

$$F = \frac{9}{5} \cdot C + 32.$$

$$F = \frac{9}{5} \cdot 30 + 32 \quad C = 30$$

$$= \frac{9 \cdot 30}{5} + 32$$

$$= \frac{9 \cdot \cancel{5} \cdot 6}{\cancel{5}} + 32$$

$$= 54 + 32 = 86$$

Thus, 30°C = 86°F, and you should wear light clothing.

Would you go swimming in water which has a temperature of 25°C?

Answer: 77°F water would be fine for swimming.

## 8.4 EXERCISES A

**1.** Which is longer, 1 mi or 1 km?

**2.** Which is longer, 3 cm or 1 in?

**3.** Which is more, 1 qt or 1 L?

**4.** Which is more, 1 kg or 3 lb?

*Complete.*

**5.** 16 km = _____ mi

**6.** 128 mi = _____ km

**7.** 2.6 m = _____ in

**8.** 72 in = _____ cm

**9.** 3 km = _____ ft

**10.** 3.2 ft = _____ cm

**11.** $80 \frac{\text{km}}{\text{hr}}$ = _____ $\frac{\text{mi}}{\text{hr}}$

**12.** $20 \frac{\text{mi}}{\text{hr}}$ = _____ $\frac{\text{km}}{\text{hr}}$

**13.** 20 qt = _____ L

**14** 40 L = _____ gal

**15.** 50 kg = _____ lb

**16.** 5 lb = _____ g

**17.** 0.25 kg = _____ oz

**18.** 625 lb = _____ kg

**19** 40°C = _____ °F

**20.** 50°F = _____ °C

**21.** 200°C = _____ °F

**22** 200°F = _____ °C

*Solve.*

**23.** Patricia bought a French car with the capacity of the gas tank given as 48 L. How many gallons of gasoline will the tank hold?

**24.** Winston was driving in Mexico where the speed limit was $100 \frac{\text{km}}{\text{hr}}$. What is this in miles per hour?

**25.** The average weight of the starting line of a football team is 258 lb. What is this in kilograms?

**26.** Would you put your hand in water at 95°C?

## FOR REVIEW

*Complete the following using an appropriate diagram.*

**27.** 720 mm = _____ cm

**28.** 0.45 g = _____ mg

**29.** 0.0021 kl = _____ ml

---

ANSWERS:  (Some answers are rounded.)  1. 1 mile  2. 3 cm  3. 1 L  4. 3 lb  5. 9.9  6. 206  7. 102.4  8. 183  9. 9837  10. 97.5  11. 50  12. 32  13. 18.9  14. 10.6  15. 110  16. 2270  17. 8.8  18. 284  19. 104  20. 10  21. 392  22. $93\frac{1}{3}$  23. 12.7 gal  24. $62\frac{\text{mi}}{\text{hr}}$  25. 117 kg  26. No (95°C = 203°F)  27. 72  28. 450  29. 2100

## 8.4  EXERCISES B

**1.** Which is longer, 1 dm or 1 in?

**2.** Which is longer, 10 mi or 17 km?

**3.** Which is more, 1 gal or 4 L?

**4.** Which is more, 500 g or 1 lb?

*Complete.*

**5.** 125 km = _____ mi

**6.** 48 mi = _____ km

**7.** 0.056 m = _____ in

**8.** 9.5 in = _____ cm

**9.** 0.05 km = _____ ft

**10.** 0.15 ft = _____ cm

**11.** $25\dfrac{\text{km}}{\text{hr}}$ = _____ $\dfrac{\text{mi}}{\text{hr}}$

**12.** $100\dfrac{\text{mi}}{\text{hr}}$ = _____ $\dfrac{\text{km}}{\text{hr}}$

**13.** 0.5 qt = _____ L

**14.** 8.5 L = _____ gal

**15.** 0.2 kg = _____ lb

**16.** $\dfrac{1}{2}$ lb = _____ g

**17.** 0.0046 kg = _____ oz

**18.** 85 lb = _____ kg

**19.** 15°C = _____ °F

**20.** 77°F = _____ °C

**21.** 1000°C = _____ °F

**22.** 1000°F = _____ °C

*Solve.*

**23.** How many liters will a 19-gal gas tank hold?

**24.** Patricia's speedometer reads in kilometers per hour. How fast can she legally drive where the speed limit is $45\frac{\text{mi}}{\text{hr}}$?

**25.** Willy must have his weight down to 82 kg to compete in Europe. What is this in pounds?

**26.** Instructions call for a temperature control to be at 65°C. What is this in °F?

## FOR REVIEW

*Complete the following using an appropriate diagram.*

**27.** 0.5 mm = _____ cm

**28.** 122 g = _____ mg

**29.** 0.111 kl = _____ ml

## 8.4 EXERCISES C

*Use a calculator to complete the following and give answers rounded to the nearest hundredth.*

**1.** 17.65 pt = _____ L

[Answer: 8.35]

**2.** $3\frac{1}{9}$ m = _____ in

**3.** 4.65 kg = _____ oz

**4.** $75\dfrac{km}{hr}$ = _____ $\dfrac{ft}{sec}$

**5.** 77.2°C = _____ °F

[Answer: 170.96]

**6.** 15.5°F = _____ °C

## 8.5 RECTANGLES

---

### STUDENT GUIDEPOSTS

1 Perimeter of a Rectangle

2 Perimeter of a Square

3 Area in Square Units

4 Area of a Rectangle

5 Area of a Square

6 Conversion of Units of Area

7 Special Units for Area (Optional)

---

### 1 PERIMETER OF A RECTANGLE

Units of measure are used in describing geometric figures. Figure 8.12 shows a **rectangle** which is a four-sided figure with opposite sides parallel and adjacent sides perpendicular. We use $l$ for length and $w$ for width.

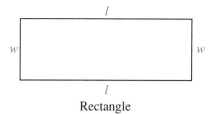

Rectangle

**Figure 8.12**

The **perimeter** of a rectangle, or any other geometric figure, is the distance around the figure. Thus, the perimeter of a rectangle is

$$l + w + l + w = l + l + w + w$$
$$= 2l + 2w.$$

---

#### Perimeter of a Rectangle

The perimeter ($P$) of a rectangle is twice the length plus twice the width.

$$P = 2l + 2w$$

---

| **EXAMPLE 1** CALCULATING A PERIMETER | **PRACTICE EXERCISE 1** |

Find the perimeter of the rectangle shown in Figure 8.13 with length 5 cm and width 3 cm.

$$P = 2l + 2w$$
$$= 2(5 \text{ cm}) + 2(3 \text{ cm})$$
$$= 10 \text{ cm} + 6 \text{ cm}$$
$$= 16 \text{ cm}$$

Thus, the perimeter is 16 cm.

Find the perimeter of a rectangle with length 9 mm and width 8 mm.

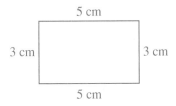

**Figure 8.13**

Answer: 34 mm

| **EXAMPLE 2** APPLICATION OF PERIMETER | **PRACTICE EXERCISE 2** |

Find the length of a fence needed to enclose a rectangular pasture which is 72 m wide and 90 m long.

    The perimeter of the pasture is the length of the fence needed.

$$P = 2l + 2w$$
$$= 2(90 \text{ m}) + 2(72 \text{ m})$$
$$= 180 \text{ m} + 144 \text{ m}$$
$$= 324 \text{ m}$$

Thus, 324 m of fence are needed.

A farmer needs to fence a rectangular pasture which is 520 ft long and 470 ft wide. How long must the fence be?

Answer: 1980 ft

## ❷ PERIMETER OF A SQUARE

A special type of rectangle, called a **square,** is shown in Figure 8.14. All its sides have the same length. If $s$ is the length of a side, then the perimeter of a square is

$$P = 2l + 2w$$
$$= 2s + 2s$$
$$= 4s.$$

```
        s
    ┌─────────┐
  s │         │ w = s   Square
    └─────────┘
       l = s
```

**Figure 8.14**

### Perimeter of a Square

The perimeter of a square is four times the length of a side.

$$P = 4s$$

---

**EXAMPLE 3   CALCULATING THE PERIMETER OF A SQUARE**

Find the perimeter of a square whose sides are 15.3 m.

$$P = 4s$$
$$= 4(\textbf{15.3 m})$$
$$= 61.2 \text{ m}$$

15.3 m

15.3 m    15.3 m

15.3 m

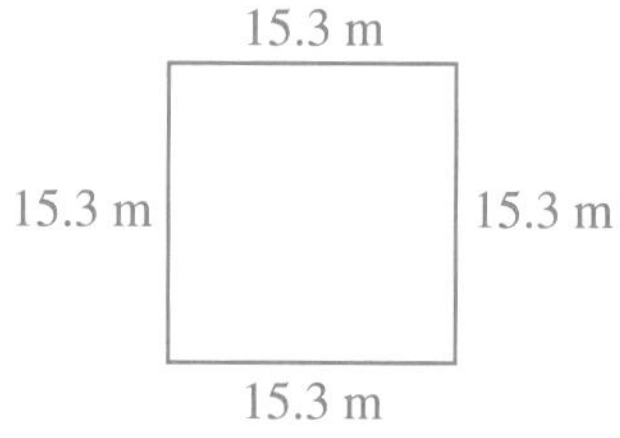

**Figure 8.15**

**PRACTICE EXERCISE 3**

A square has side $3\frac{3}{4}$ ft. Find the perimeter.

Answer: 15 ft

---

### $\boxed{\textbf{H I N T}}$

Remember the definition of perimeter as the distance around the figure. This will allow you to calculate the perimeter even if you don't remember a formula. Consider Example 3.

$$P = 15.3 \text{ m} + 15.3 \text{ m} + 15.3 \text{ m} + 15.3 \text{ m} = 61.2 \text{ m}.$$

---

### ③ AREA IN SQUARE UNITS

Two special squares are shown in Figure 8.16. The area of a square which has side 1 cm is one **square centimeter** and is written 1 cm$^2$. Likewise, if the side is 1 in, the area is one **square inch,** 1 in$^2$. Similarly, if some other unit of length is used, the area is 1 square unit, or 1 unit$^2$.

1 cm

1 cm

1 in

1 in

Area = $A$ = 1 square centimeter (sq cm)  
      = 1 cm$^2$

Area = $A$ = 1 square inch (sq in)  
      = 1 in$^2$

**Figure 8.16**

## ④ AREA OF A RECTANGLE

Square units of measure are used for the areas of rectangles. The rectangle in Figure 8.17 has area 10 cm² since there are 10 of the 1-cm² squares contained in it. In general, this is what is meant by *area*. Notice that 10 cm² can be found by multiplying 5 cm by 2 cm.

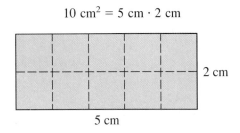

$$10 \text{ cm}^2 = 5 \text{ cm} \cdot 2 \text{ cm}$$

2 cm

5 cm

**Figure 8.17**

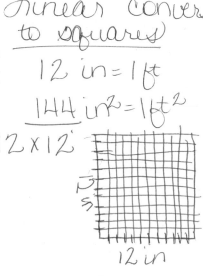

*Handwritten note:*
Linear conversion to squares)
12 in = 1 ft
144 in² = 1 ft²
12 × 12

12 in

| Area of a Rectangle |
| --- |

The area ($A$) of a rectangle is the length times the width.

$$A = lw$$

Notice that the multiplication of letters can be written $lw$ as well as $l \cdot w$ or $l \times w$.

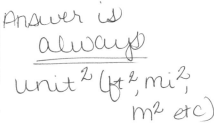

*Handwritten note:*
Answer is always
unit² (ft², mi², m² etc)

---

| EXAMPLE 4 CALCULATING THE AREA OF A RECTANGLE |
| --- |

Find the area of the rectangle with width 6 m and length 12 m, shown in Figure 8.18. (Sometimes we say "the rectangle that is 6 m by 12 m.")

$$
\begin{aligned}
A &= lw \\
&= (12 \text{ m})(6 \text{ m}) \\
&= (12)(6) \text{ m}^2 \\
&= 72 \text{ m}^2
\end{aligned}
$$

The area is 72 square meters.

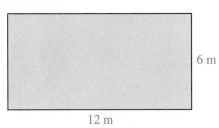

6 m

12 m

**Figure 8.18**

| PRACTICE EXERCISE 4 |
| --- |

Find the area of a rectangle which is 17 cm by 20 cm.

Answer: 340 cm²

---

| EXAMPLE 5 APPLICATION OF AREA |
| --- |

How many square yards of carpet are needed to cover the floor of a room which is 4.2 yd by 5.5 yd? How much will it cost if carpet is $22 per square yard?

The amount of carpet needed is the area of the floor of the room (Figure 8.19).

| PRACTICE EXERCISE 5 |
| --- |

A rectangular room is 3.8 m by 4.5 m. How much will it cost to carpet the room if installed carpet costs $26 per square meter?

$$A = lw$$
$$= (5.5 \text{ yd})(4.2 \text{ yd})$$
$$= (5.5)(4.2) \text{ yd}^2$$
$$= 23.1 \text{ yd}^2$$

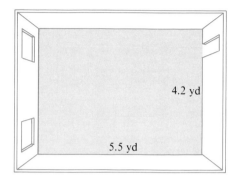

**Figure 8.19**

The cost is $22 per square yard times 23.1 square yards.

$$\text{Cost} = 22 \frac{\text{dollars}}{\cancel{\text{yd}^2}} \times \frac{23.1 \ \cancel{\text{yd}^2}}{1}$$
$$= \$508.20$$

Answer: **$444.60**

## ⑤ AREA OF A SQUARE

Since all the sides of a square have the same measure $s$, the area of a square is

$$l \cdot w = s \cdot s = s^2.$$

| Area of a Square |
| --- |
| The area of a square is the square of a side.<br><br>$$A = s^2$$ |

| EXAMPLE 6   CALCULATING THE AREA OF A SQUARE | PRACTICE EXERCISE 6 |
| --- | --- |

Find the area of the square with side 15 mi, shown in Figure 8.20.

$$A = s^2$$
$$= (15 \text{ mi})^2$$
$$= 225 \text{ mi}^2$$

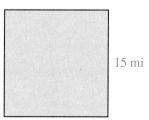

15 mi

**Figure 8.20**

The area is 225 square miles.

The side of a square measures 42 mm. Find the area of the square.

Answer: **1764 mm²**

| **EXAMPLE 7   APPLICATION OF AREA** | **PRACTICE EXERCISE 7** |

Find the area of the region in Figure 8.21.

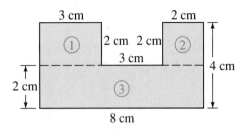

**Figure 8.21**

The region can be divided into three rectangles in several different ways. We will choose one way.

①  is 2 cm by 3 cm     Area ① = 3 cm × 2 cm = 6 cm²
②  is 2 cm by 2 cm     Area ② = 2 cm × 2 cm = 4 cm²
③  is 2 cm by 8 cm     Area ③ = 8 cm × 2 cm = 16 cm²

The total area is the sum of the areas of the three rectangles.

$$A = 6 \text{ cm}^2 + 4 \text{ cm}^2 + 16 \text{ cm}^2 = 26 \text{ cm}^2.$$

Find the area of the region shown.

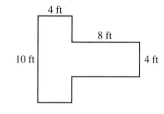

Answer: 72 ft²

## ❻ CONVERSION OF UNITS OF AREA

The conversion factor for converting any square unit is the square of the length factor. Thus,
$$1 \text{ ft}^2 = (12 \text{ in})^2 = 144 \text{ in}^2.$$

### ┃HINT┃

You do not have to remember that
$$1 \text{ ft}^2 = 144 \text{ in}^2$$

or that
$$1 \text{ yd}^2 = 9 \text{ ft}^2.$$

Just remember, for example, that
$$1 \text{ yd} = 3 \text{ ft}$$

and thus
$$(1 \text{ yd})^2 = (3 \text{ ft})^2$$
$$1 \text{ yd}^2 = 9 \text{ ft}^2.$$

| **EXAMPLE 8   CONVERTING SQUARE UNITS** | **PRACTICE EXERCISE 8** |

**(a)**  2.5 yd² = _____ ft²

$$2.5 \text{ yd}^2 = \frac{2.5 \text{ yd}^2}{1} \times \left( \frac{3 \text{ ft}}{1 \text{ yd}} \right)^2 = \frac{2.5 \text{ yd}^2}{1} \times \frac{9 \text{ ft}^2}{1 \text{ yd}^2}$$
$$= (2.5)(9) \text{ ft}^2$$
$$= 22.5 \text{ ft}^2$$

**(a)**  $7\frac{2}{3}$ yd² = _____ ft².

**(b)** 720 in² = _____ ft²

$$720 \text{ in}^2 = \frac{720 \text{ in}^2}{1} \times \left(\frac{1 \text{ ft}}{12 \text{ in}}\right)^2 = \frac{720 \text{ in}^2}{1} \times \frac{1 \text{ ft}^2}{144 \text{ in}^2}$$

$$= \frac{720}{144} \text{ ft}^2$$

$$= 5 \text{ ft}^2$$

**(b)** 288 in² = _____ ft².

Answers: (a) 69   (b) 2

The conversion factors for all square units in the metric system are the squares of the length conversion units.

$$1 \text{ m}^2 = (100 \text{ cm})^2$$
$$= 10,000 \text{ cm}^2$$
$$1 \text{ cm}^2 = (0.01 \text{ m})^2$$
$$= 0.0001 \text{ m}^2$$

## HINT

These conversions can also be made by moving the decimal point *twice* the number of places for the length conversion.

---

**EXAMPLE 9   CONVERSION OF METRIC SYSTEM AREA UNITS**

**PRACTICE EXERCISE 9**

**(a)** 65,000 cm² = _____ hm².

**(a)** 3,280,000 cm² = _____ hm².

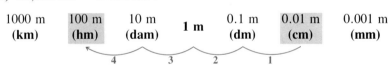

As shown, to change from cm to hm, the decimal point is moved 4 places to the left. Thus, to change from cm² to hm² the decimal point is moved

$$2(4) = 8 \text{ places}$$

to the left.

65,000 cm² = 0.00065000 hm²   Put in three more zeros

8 places

= 0.00065 hm²

**(b)** 0.032 m² = _____ mm².

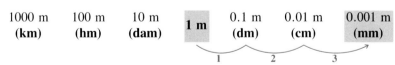

**(b)** 0.55 m² = _____ mm².

To change m² to mm², the decimal point is moved 2(**3**) = 6 places to the right.

0.032 m² = 0032000. mm²

6 places

= 32,000 mm²

Answers: (a) 0.0328   (b) 550,000

## ➐ SPECIAL UNITS FOR AREA (OPTIONAL)

There are special units for area in both English and metric systems. Land is usually measured in acres in the English system and in hectares in the metric system. These measures are defined in terms of units we already know.

| Special Units of Area in the English System |
| :---: |
| 1 acre = 43,560 ft$^2$ |
| 640 acres = 1 square mile (1 mi$^2$) |

The acre is shown in terms of both square feet and square miles. To determine how many square yards there are in an acre, convert 43,560 ft$^2$ to square yards. Since one square yard is 3 ft by 3 ft, as shown in Figure 8.22,

$$1 \text{ yd}^2 = (3 \text{ ft})^2 = 9 \text{ ft}^2.$$

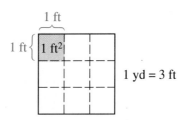

**Figure 8.22**

Thus, to convert 1 acre = 43,560 ft$^2$ to yd$^2$ use the unit fraction $\left(\dfrac{1 \text{ yd}}{3 \text{ ft}}\right)^2 =$

$\dfrac{1 \text{ yd}^2}{9 \text{ ft}^2}$.

$\dfrac{23 \text{ mi}^2}{1} = \left(\dfrac{640 \text{ acres}}{1 \text{ mi}}\right)^2$

$$1 \text{ acre} = 43,560 \text{ ft}^2$$
$$= \frac{43,560 \text{ ft}^2}{1} \times \frac{1 \text{ yd}^2}{9 \text{ ft}^2}$$
$$= \frac{43,560}{9} \text{ yd}^2$$
$$= 4840 \text{ yd}^2$$

### EXAMPLE 10  APPLICATION OF ACRES

Brand X Cattle Company owns 23 square miles (mi$^2$) of ranch land. How many acres is this?

Since 640 acres = 1 mi$^2$, use the unit fraction $\frac{640 \text{ acres}}{1 \text{ mi}^2}$. why not $\left(\dfrac{640 \text{ acres}}{1 \text{ mi}}\right)^?$ because acres is already squared

$$23 \text{ mi}^2 = \frac{23 \text{ mi}^2}{1} \times \frac{640 \text{ acres}}{1 \text{ mi}^2}$$
$$= (23)(640) \text{ acres}$$
$$= 14,720 \text{ acres}$$

### PRACTICE EXERCISE 10

Stephanie bought a lot that has an area of 65,340 ft$^2$. How many acres is this?

Answer: 1.5 acres

Special units of area in the metric system are given in the following table.

## Special Units of Area in the Metric System

1 are (a) = 100 m$^2$
1 hectare (ha) = 100 are = 10,000 m$^2$
1 centare (ca) = 0.01 are = 1 m$^2$

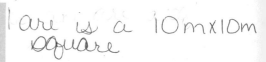

*1 are is a 10m×10m square*

The unit *are* is pronounced "air." The hectare is used for measurement of land and is about the same as 2.5 acres in the English system. The centare is used for areas like the floor area of a building while cm$^2$ and mm$^2$ are used for smaller objects.

---

**EXAMPLE 11**   **CONVERSION TO HECTARES**

62,500 m$^2$ = _____ ha.

$$62,500 \text{ m}^2 = \frac{62,500 \text{ m}^2}{1} \times \frac{1 \text{ ha}}{10,000 \text{ m}^2}$$

$$= \frac{62,500}{10,000} \text{ ha}$$

$$= 6.25 \text{ ha}$$

**PRACTICE EXERCISE 11**

2340 m$^2$ = _____ ha.

Answer: 0.234

---

## 8.5  EXERCISES A

*Find the perimeter of the rectangle with the given width and length.*

**1.** 6 ft by 11 ft

**2.** 2.5 m by 3.6 m

**3.** 102 mm by 30.5 mm

**4.** 6.8 mi by $9\frac{1}{2}$ mi

**5.** 7.2 cm by 7.2 cm

**6.** $22\frac{1}{2}$ mi by $22\frac{1}{2}$ mi

*Find the area of the rectangle with the given width and length.*

**7.** 6 ft by 11 ft

**8.** 2.5 m by 3.6 m

**9.** 102 mm by 30.5 mm

**10.** 6.8 mi by $9\frac{1}{2}$ mi

**11.** 7.2 cm by 7.2 cm

**12.** $22\frac{1}{2}$ mi by $22\frac{1}{2}$ mi

*Complete.*

**13.** 7 yd$^2$ = _____ ft$^2$

**14** 1020 cm$^2$ = _____ m$^2$

**15.** 0.01 mi$^2$ = _____ ft$^2$

**16.** 6,195,200 yd$^2$ = _____ mi$^2$

**17.** 7.5 ft$^2$ = _____ in$^2$

**18.** 108 in$^2$ = _____ ft$^2$

**19.** 0.001 km$^2$ = _____ m$^2$

**20.** 44,000 m$^2$ = _____ hm$^2$

**21.** 16 cm$^2$ = _____ mm$^2$

**22.** 103,500 m$^2$ = _____ km$^2$

**23** 4,352,000 mm$^2$ = _____ hm$^2$

**24.** 60 hm$^2$ = _____ m$^2$

*Solve.*

**25.** Find the perimeter and area of the given figure.

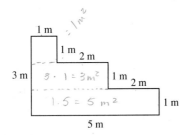

**26.** Fencing costs $3.20 per yard. If a pasture 62 yd by 205 yd is to be fenced, how much will it cost?

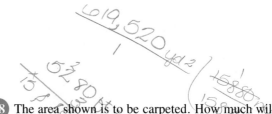

**27.** A rose garden is 7.2 m by 10.8 m. If a square with side 2.5 m is used in the middle of the garden for a fountain, what area remains for roses?

**28** The area shown is to be carpeted. How much will it cost if carpet costs $16.50 per square yard?

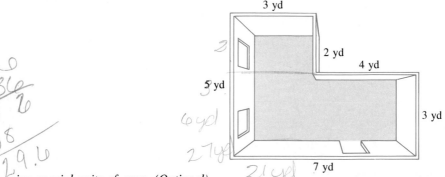

*Make the conversions using special units of area. (Optional)*

**29.** 2 acres = _____ ft² 

**30.** 2048 acres = _____ mi²

**31.** 2420 yd² = _____ acres

**32.** 6 ha = _____ a

**33.** 44,000 m² = _____ ha

**34.** 60 ha = _____ m²

## FOR REVIEW

*Complete.*

**35.** 36.2 m = _____ ft

**36.** $6\frac{1}{2}$ gal = _____ L

**37.** 368 g = _____ lb

---

ANSWERS: 1. 34 ft 2. 12.2 m 3. 265 mm 4. 32.6 mi 5. 28.8 cm 6. 90 mi 7. 66 ft² 8. 9.0 m² 9. 3111 mm² 10. 64.6 mi² 11. 51.84 cm² 12. 506.25 mi² 13. 63 14. 0.102 15. 278,784 16. 2 17. 1080 18. 0.75 19. 1000 20. 4.4 21. 1600 22. 0.1035 23. 0.0004352 24. 600,000 25. 16 m; 9 m² 26. $1708.80 27. 71.51 m² 28. $445.50 29. 87,120 30. 3.2 31. 0.5 32. 600 33. 4.4 34. 600,000 35. 118.8 (to the nearest tenth) 36. 24.6 (to the nearest tenth using 1 qt = 0.946 L) 37. 0.81 (to the nearest hundredth)

## 8.5 EXERCISES B

*Find the perimeter of the rectangle with the given width and length.*

**1.** 10 ft by 17 ft

**2.** 8.2 m by 9.5 m

**3.** 2500 mm by 45.5 mm

**4.** 2.2 mi by $6\frac{1}{2}$ mi

**5.** 40.2 cm by 40.2 cm

**6.** $10\frac{1}{2}$ mi by $10\frac{1}{2}$ mi

*Find the area of the rectangle with the given width and length.*

**7.** 10 ft by 17 ft

**8.** 8.2 m by 9.5 m

**9.** 2500 mm by 45.5 mm

**10.** 2.2 mi by $6\frac{1}{2}$ mi

**11.** 40.2 cm by 40.2 cm

**12.** $10\frac{1}{2}$ mi by $10\frac{1}{2}$ mi

*Complete.*

**13.** 25 yd$^2$ = _____ ft$^2$

**14.** 200 cm$^2$ = _____ m$^2$

**15.** 0.005 mi$^2$ = _____ ft$^2$

**16.** 619,520 yd$^2$ = _____ mi$^2$

**17.** $\frac{1}{6}$ ft$^2$ = _____ in$^2$

**18.** 475.2 in$^2$ = _____ ft$^2$

**19.** 0.0002 km$^2$ = _____ m$^2$

**20.** 7,500,000 m$^2$ = _____ hm$^2$

**21.** 0.12 cm$^2$ = _____ mm$^2$

**22.** 4500 m$^2$ = _____ km$^2$

**23.** 800,000,000 mm$^2$ = _____ hm$^2$

**24.** 0.05 hm$^2$ = _____ m$^2$

*Solve.*

**25.** The play area shown needs to have a new fence around it. How long will the fence need to be?

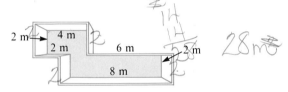

**26.** A rancher needs to fence an area which is 2500 yd by 4200 yd. If fencing costs $2.60 per yard, how much will it cost?

**27.** A flower garden is in the shape of a square which is 22.5 m on a side. If a walk 1.5 m wide is put across the garden parallel to one of the sides, how much area is left for flowers?

**28.** The area shown is carpeted with carpeting costing $24.50 per square yard. How much did it cost?

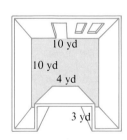

*Make the conversions using special units of area. (Optional)*

**29.** 0.6 acres = _____ ft$^2$

**30.** 1152 acres = _____ mi$^2$

**31.** 3630 yd$^2$ = _____ acres

**32.** 0.37 ha = _____ a

**33.** 7,500,000 m$^2$ = _____ ha

**34.** 0.05 ha = _____ m$^2$

**FOR REVIEW**

*Complete.*

**35.** 0.86 m = _____ ft

**36.** $20\frac{1}{4}$ gal = _____ L

**37.** 1650 g = _____ lb

## 8.5 EXERCISES C

*Use a calculator in the following exercises and give all answers rounded to the nearest hundredth.*

1. Find the perimeter and area of a rectangle with length 6.28 feet and width 42.3 inches. [*Hint:* Notice that the units are different.]

2. Brand Y Cattle Company owns a ranch which measures 6 km by 9.2 km. How many hectares does the company own?
[Answer: 5520 ha]

## 8.6 PARALLELOGRAMS, TRIANGLES, AND TRAPEZOIDS

### STUDENT GUIDEPOSTS

1 Parallelogram

2 Area of a Parallelogram

3 Area of a Triangle

4 Area of a Trapezoid

### 1 PARALLELOGRAM

A four-sided figure whose opposite sides are parallel and equal in length is called a **parallelogram.** A rectangle is a special case of this type of figure. Several parallelograms are shown in Figure 8.23.

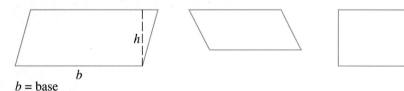

$b$ = base
$h$ = height

Parallelograms

**Figure 8.23**

As with other figures the perimeter of a parallelogram is the distance around it. Since opposite sides of a parallelogram are equal, we can find the perimeter as in Example 1.

---

| EXAMPLE 1   PERIMETER OF A PARALLELOGRAM |
| --- |

Find the perimeter of the parallelogram shown in Figure 8.24.

$$P = 2(10.5 \text{ m}) + 2(6.5 \text{ m})$$
$$= 34 \text{ m}$$

6.5 m

10.5 m

**Figure 8.24**

---

| PRACTICE EXERCISE 1 |
| --- |

Find the perimeter of a parallelogram with base $7\frac{1}{2}$ ft and side $4\frac{1}{2}$ ft.

Answer:  24 ft

## ② AREA OF A PARALLELOGRAM

To discover how to find the area of a parallelogram, cut off one end and put it on the other end. (You might try this.) This forms a rectangle (Figure 8.25) with length $l = b$ and width $w = h$. Thus, the area of the parallelogram (which is the area of the rectangle) is the base times the height.

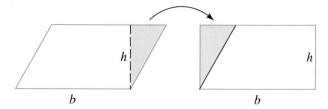

**Figure 8.25**

### Area of a Parallelogram

The area of a parallelogram with base $b$ and height $h$ is

$$A = bh.$$

| **EXAMPLE 2** AREA OF A PARALLELOGRAM | **PRACTICE EXERCISE 2** |
|---|---|

Find the area of the parallelogram, shown in Figure 8.26, with height 15.2 cm and base 20.5 cm.

$$A = bh$$
$$= (20.5 \text{ cm})(15.2 \text{ cm})$$
$$= (20.5)(15.2) \text{ cm}^2$$
$$= 311.6 \text{ cm}^2$$

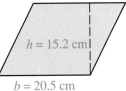

**Figure 8.26**

Find the area of a parallelogram with base 10.8 mm and height 12.5 mm.

Answer: $135 \text{ mm}^2$

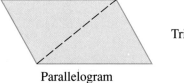

Parallelogram

Triangle

Triangle

**Figure 8.27**

## ③ AREA OF TRIANGLE

If the parallelogram in Figure 8.27 is cut along the dotted line, the area of each of the two triangles formed is one half the area of the parallelogram. Thus, a triangle can be made into a parallelogram by adding to it a triangle of the same size. This is shown in Figure 8.28, where triangle $B$ is the same size and shape as the original triangle $A$. Since the heights of the triangle and the parallelogram are the same, the area of the triangle must be one half the area of the parallelogram.

$b = \text{base}$
$h = \text{height}$

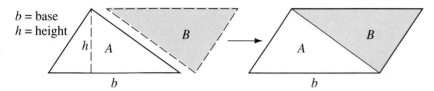

**Figure 8.28**

## Area of a Triangle

The area of a triangle with base $b$ and height $h$ is

$$A = \frac{1}{2} bh.$$

---

| **EXAMPLE 3 CALCULATING AREA OF A TRIANGLE** | **PRACTICE EXERCISE 3** |

Find the area of the triangle shown in Figure 8.29.

$$A = \frac{1}{2} bh$$

$$= \frac{1}{2} (5.5 \text{ cm})(6.8 \text{ cm})$$

$$= \frac{1}{2} (5.5)(6.8) \text{ cm}^2$$

$$= 18.7 \text{ cm}^2$$

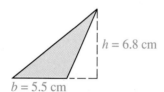

**Figure 8.29**

Find the area of a triangle with base 22.2 m and height 16.5 m.

Answer: 183.2 m$^2$

---

| **EXAMPLE 4 APPLICATION OF AREA** | **PRACTICE EXERCISE 4** |

A wall to be painted is in the shape shown in Figure 8.30.

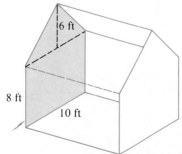

**Figure 8.30**

$8 \cdot 10 = 80 \text{ ft}^2$
$10 \cdot 6 = 60 \div 2 = 30 \text{ ft}^2$
$110 \text{ ft}^2$

The wall shown in the figure is to be painted.

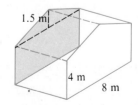

**(a)** What is the area to be painted?

**(a)** How many square feet must be painted?

First find the area of the rectangle.

$$A = lw = 10 \text{ ft} \cdot 8 \text{ ft} = 80 \text{ ft}^2$$

The base of the triangle is 10 ft and the height is 6 feet.

$$A = \frac{1}{2} bh = \frac{1}{2} (10 \text{ ft})(6 \text{ ft}) = 30 \text{ ft}^2$$

The total area to be covered is

$$80 \text{ ft}^2 + 30 \text{ ft}^2 = 110 \text{ ft}^2.$$

**(b)** How much paint will it take if one gallon covers 220 ft$^2$?

Multiply 110 ft$^2$ by $\frac{1 \text{ gal}}{220 \text{ ft}^2}$.

$$\frac{110 \not{\text{ft}^2}}{1} \cdot \frac{1 \text{ gal}}{220 \not{\text{ft}^2}} = \frac{\not{110} \text{ gal}}{2(\not{110})} = \frac{1}{2} \text{ gal}$$

Thus, it will take $\frac{1}{2}$ gallon to cover the area.

**(b)** If one liter of paint covers 19 m$^2$, how many liters of paint are required?

Answers: **(a)** 38 m$^2$   **(b)** 2 L

## 4 AREA OF A TRAPEZOID

A **trapezoid** is a four-sided figure with two parallel sides. See Figure 8.31. The parallel lines are called the *bases* and *h* is the *height*.

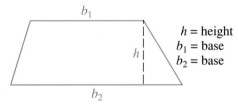

$h$ = height
$b_1$ = base
$b_2$ = base

Trapezoid

**Figure 8.31**

If we take a trapezoid the same shape and size as the given one, turn it over and put the two trapezoids together, as in Figure 8.32, it forms a parallelogram. The area of the parallelogram is the base times the height. The base is $b_1 + b_2$ and the height is $h$. The area of the trapezoid is one half their product.

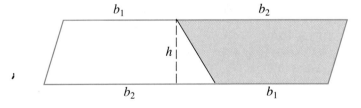

**Figure 8.32**

### Area of a Trapezoid

The area $A$ of a trapezoid with height $h$ and bases $b_1$ and $b_2$ is

$$A = \frac{1}{2} (b_1 + b_2)h.$$

| EXAMPLE 5 CALCULATING THE AREA OF A TRAPEZOID | PRACTICE EXERCISE 5 |

Find the area of the trapezoid shown in Figure 8.33.

$$A = \frac{1}{2}(b_1 + b_2)h$$

$$= \frac{1}{2}(24 \text{ mm} + 16 \text{ mm})(12 \text{ mm})$$

$$= \frac{1}{2}(40 \text{ mm})(12 \text{ mm})$$

$$= \frac{1}{2}(40)(12) \text{ mm}^2 = 240 \text{ mm}^2$$

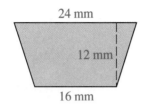

**Figure 8.33**

Find the area of the trapezoid with bases 6 inches and 14 inches and height 9 inches.

Answer:  90 in²

## 8.6   EXERCISES A

*Find the area of each figure.* Square answer

**1.**

3 m
6 m

**2.**
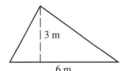
3 m
6 m

**3.**
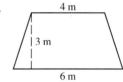
4 m
3 m
6 m

**4.**
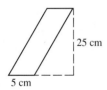
25 cm
5 cm

**5.**
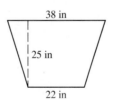
38 in
25 in
22 in

**6.**
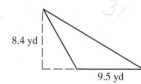
8.4 yd
9.5 yd

**7.**
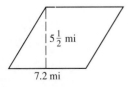
$5\frac{1}{2}$ mi
7.2 mi

**8.**
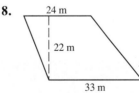
24 m
22 m
33 m

**9.**
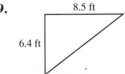
8.5 ft
6.4 ft

**10.**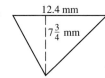

**11**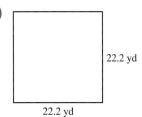

**12.**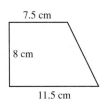

*Find the perimeter of each figure.*

**13.**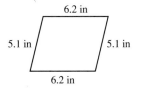

**14.**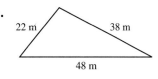

**15.**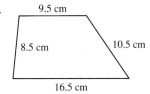

**16** Find the area of the figure.

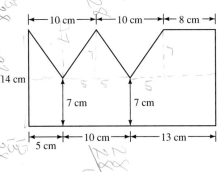

**17.** The recreation area sketched in the figure is to be carpeted. How much will it cost if carpet is $14.50 per square yard?

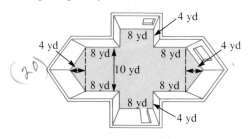

**18.** For a play, 10 walls shaped like the one shown in the figure must be painted on both sides. How much will it cost if a gallon of paint covers 160 ft² and costs $12.50 per gallon?

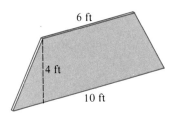

**19.** If fencing costs $1.25 per foot, find the cost to fence the area shown.

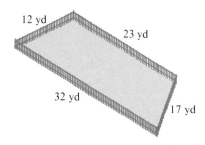

## FOR REVIEW

*Complete.*

**20.** 288 in² = _____ ft²

**21.** 82,460 m² = _____ hm²

**22.** 0.3 mi² = _____ yd²

ANSWERS:  1. 18 m²  2. 9 m²  3. 15 m²  4. 125 cm²  5. 750 in²  6. 39.9 yd²  7. 39.6 mi²  8. 627 m²  9. 27.2 ft²  10. 48.05 mm²  11. 492.84 yd²  12. 76 cm²  13. 22.6 in  14. 108 m  15. 45 cm  16. 322 cm²  17. $4988  18. $50  19. $315  20. 2  21. 8.246  22. 929,280

## 8.6 EXERCISES B

*Find the area of each figure.*

**1.**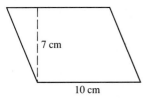

7 cm
10 cm

**2.**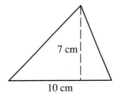

7 cm
10 cm

**3.**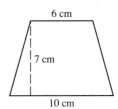

6 cm
7 cm
10 cm

**4.**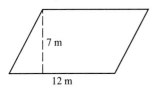

7 m
12 m

**5.**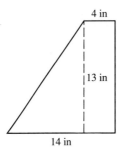

4 in
13 in
14 in

**6.**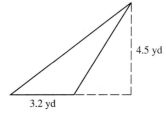

4.5 yd
3.2 yd

**7.**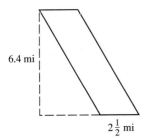

6.4 mi
$2\frac{1}{2}$ mi

**8.**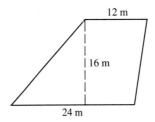

12 m
16 m
24 m

**9.**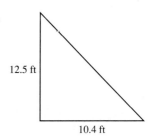

12.5 ft
10.4 ft

**10.**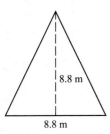

8.8 m
8.8 m

**11.**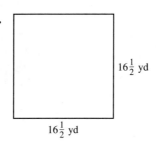

$16\frac{1}{2}$ yd
$16\frac{1}{2}$ yd

**12.**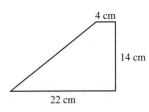

4 cm
14 cm
22 cm

*Find the perimeter of each figure.*

**13.**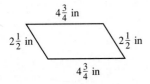

$4\frac{3}{4}$ in
$2\frac{1}{2}$ in    $2\frac{1}{2}$ in
$4\frac{3}{4}$ in

**14.**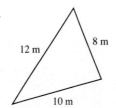

12 m    8 m
10 m

**15.**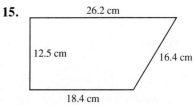

26.2 cm
12.5 cm    16.4 cm
18.4 cm

**16.** Find the area of the figure.

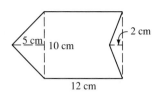

**17.** The patio area shown is to be covered with outdoor carpeting. What will be the total cost if the carpeting can be installed for $22.50 per square meter?

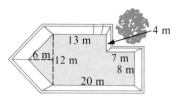

**18.** The sheet of metal shown is to be insulated on both sides. If 6 m² of insulation cost $10.20, how much will it cost to do the job?

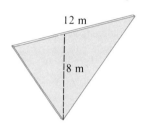

**19.** A large ranch is to be fenced. If fencing costs $3.75 per meter, find the cost to fence the area shown.

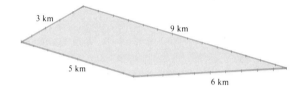

## FOR REVIEW

*Complete.*

**20.** 1440 in² = _____ ft²

**21.** 100,000 m² = _____ hm²

**22.** 0.04 mi² = _____ yd²

## 8.6 EXERCISES C

**1.** A patio is in the shape of a trapezoid with bases 8.1 yd and 6.7 yd and height 5.8 yd. Assuming no waste, to the nearest cent how much will it cost to cover the patio with outdoor carpeting which costs $2.15 a square foot?
[Answer: $830.50]

**2.** A sheet of metal is in the shape of a triangle with base 1500 cm and height 800 cm, and has a piece removed from the center of it in the shape of a parallelogram with base 125 cm and height 85 cm. How much will it cost to cover the metal on both sides with rustproofing paint which costs $0.45 per square meter?

## 8.7 CIRCLES

### STUDENT GUIDEPOSTS

**1** Basic Terms for the Circle

**2** Formula for Circumference

**3** Formula for Area

### 1 BASIC TERMS FOR THE CIRCLE

On a **circle,** as shown in Figure 8.34, all points are the same distance from the **center** (*O*). The distance from the center to each point is the **radius** (*r*). Twice the radius is the **diameter** (*d*). The distance around the circle is called the **circumference** (*c*).

*there are an infinite # of radii in a circle*

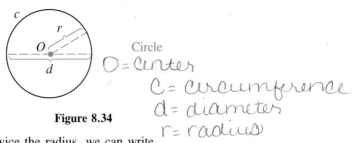

**Figure 8.34**

Since the diameter is twice the radius, we can write

$$d = 2r \quad \text{and} \quad r = \frac{d}{2}. \quad r = \frac{1}{2}d$$

---

| **EXAMPLE 1   DIAMETER AND RADIUS** | **PRACTICE EXERCISE 1** |
|---|---|

**(a)** Find the diameter of the circle with radius 2.3 cm, shown in Figure 8.35.

$$d = 2r = 2(2.3 \text{ cm})$$
$$= 4.6 \text{ cm}$$

**Figure 8.35**

**(b)** Find the radius of a circle with diameter 10.8 in (see Figure 8.36).

$$r = \frac{d}{2} = \frac{10.8 \text{ in}}{2}$$
$$= 5.4 \text{ in}$$

**Figure 8.36**

**(a)** Find the diameter of a circle with radius $5\frac{1}{2}$ inches.

**(b)** Find the radius of a circle with diameter 22.4 meters.

Answers:  (a)  11 in    (b)  11.2 m

---

## ❷ FORMULA FOR CIRCUMFERENCE

If the circumference, $c$, of any circle is divided by its diameter, $d$, the result is a constant called $\pi$ (the Greek letter *pi*). Suppose, for example, that we measure the circle formed by the top of a can. If the circumference $c$ is 6.2 inches, and the diameter $d$ is 2.0 inches, then

$$\frac{c}{d} = \frac{6.2 \text{ in}}{2.0 \text{ in}}$$
$$= 3.1.$$

For any circle measured, $\frac{c}{d}$ will always be close to 3.1, because $\frac{c}{d}$ is always equal to $\pi$. This number has been calculated to many decimal places, but it is usually approximated as 3.14 or $\frac{22}{7}$.

Since $\frac{c}{d} = \pi$, the related multiplication sentence is $c = \pi d$.

### Circumference of a Circle

The circumference of a circle with diameter $d$ and radius $r$ is

$$c = \pi d = 2\pi r. \quad \textit{Because } 2r = 1d$$

---

| **EXAMPLE 2** **FINDING THE CIRCUMFERENCE OF A CIRCLE** | **PRACTICE EXERCISE 2** |
|---|---|

**(a)** Find the circumference of a circle with diameter 32 mm. Use 3.14 for $\pi$.

$$c = \pi d$$
$$\approx 3.14(32 \text{ mm})$$
$$\approx 100.5 \text{ mm} \quad \text{Rounded to the nearest tenth}$$

**(b)** Find the circumference of a circle with radius 21 in. Use $\frac{22}{7}$ for $\pi$.

$$c = 2\pi r$$
$$\approx 2 \cdot \frac{22}{7} \cdot 21 \text{ in} = 2(22)\frac{21}{7} \text{ in}$$
$$= 2(22)(3) \text{ in} \quad 21 \div 7 = 3$$
$$= 132 \text{ in}$$

**(a)** Find the circumference of a circle with diameter 24 m. Use 3.14 for $\pi$.

**(b)** Find the circumference of a circle with radius 63 inches. Use $\frac{22}{7}$ for $\pi$.

Answers: (a) 75.36 m    (b) 396 in

---

### ❸ FORMULA FOR AREA

Next we find the area of a circle.

### Area of a Circle

The area $A$ of a circle is given by

$$A = \pi r^2.$$

We can get some idea of why this is true by looking at Figure 8.37. Suppose we cut the circle in Figure 8.37 (a) along all the radii (plural of radius). When we put the top half of the circle above the bottom half as in Figure 8.37 (b) and then slide them together as in Figure 8.37 (c), the figure formed is approximately a parallelogram with height $r$. Its base is $\pi r$ since $2\pi r$ is the circumference and half the circle is used on top and half on the bottom. Thus,

$$A = bh$$
$$= (\pi r)(r) = \pi r^2.$$

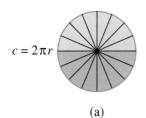

(a)

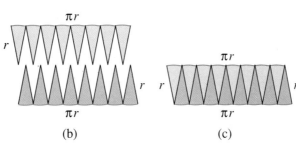

(b)          (c)

**Figure 8.37**

| EXAMPLE 3 FINDING THE AREA OF A CIRCLE | PRACTICE EXERCISE 3 |
|---|---|

**(a)** Find the area of a circle with radius 9 km. Use 3.14 for $\pi$.

$$A = \pi r^2$$
$$\approx 3.14(9 \text{ km})^2$$
$$= (3.14)(9^2) \text{ km}^2$$
$$\approx 254 \text{ km}^2 \qquad \text{To the nearest unit}$$

**(b)** Find the area of a circle with diameter 70 ft. Use $\frac{22}{7}$ for $\pi$.

$$r = \frac{d}{2} = \frac{70 \text{ ft}}{2} = 35 \text{ ft}$$

$$A = \pi r^2 \approx \frac{22}{7} (35 \text{ ft})^2$$

$$= \frac{(22)(35)(\overset{5}{\cancel{35}})}{\underset{1}{\cancel{7}}} \text{ ft}^2 \qquad 35 \div 7 = 5$$

$$= (22)(35)(5) \text{ ft}^2 = 3850 \text{ ft}^2$$

**(a)** Find the area of a circle with radius 15 cm. Use 3.14 for $\pi$.

**(b)** Find the area of a circle with diameter 210 inches. Use $\frac{22}{7}$ for $\pi$.

Answers: (a) 706.5 cm² 
(b) 34,650 in²

| EXAMPLE 4 APPLICATION OF AREA | PRACTICE EXERCISE 4 |
|---|---|

In the machine part shown in Figure 8.38, each circular hole has radius 3 cm. Find the area of metal which is left. Use 3.14 for $\pi$.

First calculate the area of the trapezoid.

$$A = \frac{1}{2} (b_1 + b_2)h$$

$$= \frac{1}{2} (16 \text{ cm} + 24 \text{ cm}) \, 14 \text{ cm}$$

$$= \frac{1}{2} (40 \text{ cm})(14 \text{ cm})$$

$$= \frac{1}{2} (40)(14) \text{ cm}^2$$

$$= (20)(14) \text{ cm}^2 = 280 \text{ cm}^2$$

A triangle with base 10 inches and height 5 inches has three holes of radius 1 inch drilled through it. What is the area left in the triangle after the holes are drilled? Use 3.14 for $\pi$.

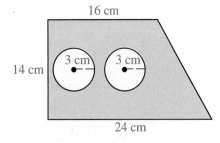

**Figure 8.38**

Now find the area of each circle.

$$A = \pi r^2 \approx 3.14(3 \text{ cm})^2$$
$$= (3.14)(9) \text{ cm}^2 = 28.26 \text{ cm}^2$$

Thus, the two circles have area

$$2(28.26 \text{ cm}^2) = 56.52 \text{ cm}^2.$$

The area of metal is the area of the trapezoid minus the area of the circles.

$$\begin{array}{r} 280.00 \\ - \phantom{0}56.52 \\ \hline 223.48 \end{array}$$

The area left is 223.48 cm$^2$.

Answer: 15.58 in$^2$

---

| **EXAMPLE 5** UNIT PRICE APPLICATION | **PRACTICE EXERCISE 5** |

A 10-inch diameter pizza costs $3.75, and a 14-inch diameter pizza of the same type costs $5.50. Which pizza costs less per square inch? (Use 3.14 for $\pi$.)

A 8-inch diameter peach pie costs $4.20 and a 12-inch diameter pie costs $8.50. Which pie costs less per square inch? Use 3.14 for $\pi$.

The radius of the 10-in pizza is 5 in and the radius of the 14-in pizza is 7 in.

Area of 5-inch radius pizza: $A = \pi r^2$
$$\approx (3.14)(5 \text{ in})^2$$
$$= (3.14)(25) \text{ in}^2$$
$$= 78.5 \text{ in}^2$$

Area of 7-in radius pizza: $A = \pi r^2$
$$\approx (3.14)(7 \text{ in})^2$$
$$= (3.14)(49) \text{ in}^2$$
$$\approx 153.9 \text{ in}^2 \qquad \text{To nearest tenth}$$

Cost of 5-in radius pizza per square inch (unit price):

$$\frac{\$3.75}{78.5 \text{ in}^2} = \$0.048 \text{ per square inch.} \qquad \text{Nearest tenth of a cent}$$

Cost of 7-in radius pizza per square inch (unit price):

$$\frac{\$5.50}{153.9 \text{ in}^2} = \$0.036 \text{ per square inch.} \qquad \text{Nearest tenth of a cent}$$

Thus, the 14-in pizza costs less per square inch.

Answer: The 8-inch pie costs approximately $0.084 per square inch, and the 12-inch pie costs approximately $0.075 per square inch. Thus the 12-inch pie is the best buy.

---

# 8.7 EXERCISES A

*Find the diameter of the circle with the given radius.*

**1.** $r = 11$ in

**2.** $r = 5.85$ cm

**3.** $r = \dfrac{3}{4}$ ft

*Find the radius of the circle with the given diameter.*

**4.** $d = 24$ m

**5.** $d = \dfrac{5}{2}$ in

**6.** $d = 7.24$ mm

*Find the circumference and area of the circle. (Use $\frac{22}{7}$ for $\pi$.)*

**7.** $r = 7$ yd

**8.** $d = 28$ cm

**9.** $r = \dfrac{7}{2}$ ft

**10.** $d = 98$ mm

**11** $r = 7.7$ mi

**12.** $d = \dfrac{42}{5}$ cm

*Find the circumference and area of the circle. (Use 3.14 for $\pi$.)*

**13.** $r = 6.8$ km

**14.** $d = 92$ in

**15.** $r = 2.7$ cm

**16.** $d = 35.8$ yd

**17** $r = 0.05$ cm

**18.** $d = 1.02$ in

*Solve.*

**19** How much more area is there for water to pass through in a $\frac{3}{4}$-in diameter water hose than there is in a $\frac{1}{2}$-in diameter water hose? (Use 3.14 for $\pi$.)

**20.** A 12-inch diameter pizza costs $4.50. A 16-inch diameter pizza costs $7.50. Which pizza costs less per square inch (unit price)?

*Find the area of metal left on each machine part shown which has circular holes drilled in it. (Use 3.14 for $\pi$.)*

**21.**

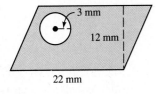

**22**

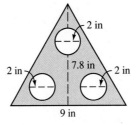

**23.**

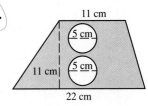

## FOR REVIEW

**24.** The area shown is to be fenced. If fencing costs $4.20 per meter, how much will the project cost?

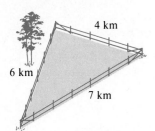

**25.** It will cost $8.50 per square yard to insulate the area shown. How much will the job cost if both sides of the area must be insulated?

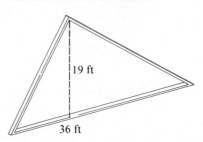

ANSWERS: (Some answers are rounded.)  1. 22 in   2. 11.7 cm   3. $\frac{3}{2}$ ft   4. 12 m   5. $\frac{5}{4}$ in   6. 3.62 mm
7. 44 yd; 154 yd²  8. 88 cm; 616 cm²   9. 22 ft; $\frac{77}{2}$ ft²   10. 308 mm; 7546 mm²   11. 48.4 mi; 186.34 mi²
12. 26.4 cm; 55.44 cm²   13. 42.7 km; 145.2 km²   14. 288.9 in; 6644.2 in²   15. 17.0 cm; 22.9 cm²   16. 112.4 yd;
1006.1 yd²   17. 0.314 cm; 0.00785 cm²   18. 3.20 in; 0.817 in²   19. 0.246 in²   20. 12-in: $0.0398 per in²; 16-in:
$0.0373 per in²; 16-in pizza costs less   21. 235.74 mm²   22. 25.68 in²   23. 142.25 cm²   24. $71,400   25. $646

## 8.7   EXERCISES B

*Find the diameter of the circle with the given radius.*

**1.** $r = 28$ cm

**2.** $r = 2.75$ ft

**3.** $r = \dfrac{7}{6}$ in

*Find the radius of the circle with the given diameter.*

**4.** $d = 120$ mm

**5.** $d = \dfrac{2}{3}$ in

**6.** $d = 28.6$ m

*Find the circumference and the area of the circle. (Use $\frac{22}{7}$ for $\pi$.)*

**7.** $r = 14$ m

**8.** $d = 98$ ft

**9.** $r = \dfrac{14}{11}$ cm

**10.** $d = 350$ in

**11.** $r = 5.67$ yd

**12.** $d = 84$ mm

*Find the circumference and area of the circle. (Use 3.14 for $\pi$.)*

**13.** $r = 25$ dm

**14.** $d = 32.8$ ft

**15.** $r = 420$ mm

**16.** $d = 1.64$ in

**17.** $r = 0.08$ m

**18.** $d = 9.66$ yd

*Solve.*

**19.** A circular garden has radius 9 m. If a 1-m-wide circular walk is put around it, what is the area of the walk? Use 3.14 for $\pi$. [*Hint:* Find the area of a circle with 10-m radius and subtract the area of the garden.]

**20.** A 14-inch diameter pizza costs $6.00 and a 16-inch diameter pizza costs $8.00. Which pizza costs less per square inch? Use 3.14 for $\pi$.

*Find the area of metal left on each machine part shown which has circular holes drilled in it. Use 3.14 for $\pi$.*

**21.**

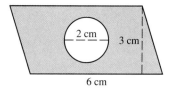

**22.**

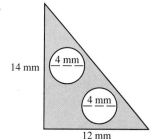

**23.**

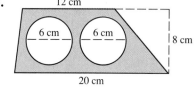

## FOR REVIEW

*Solve.*

**24.** A farmer's field is in the shape of the trapezoid shown. How much will it cost to fence the area if fencing costs $1.55 per foot?

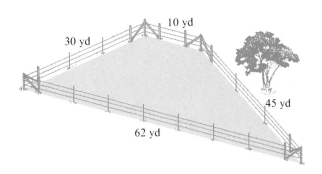

**25.** The area shown is to be painted with two coats of paint. If it costs $3.62 per square meter for the paint job, what will be the total cost for both coats?

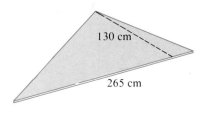

## 8.7 EXERCISES C

*Solve.*

**1.** Find the area (to the nearest hundredth) of the metal left on the machine part shown which has three circular holes drilled in it. Use 3.14 for $\pi$.
[Answer: 1257.77 cm²]

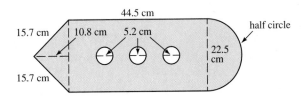

**2.** Find the exterior perimeter (to the nearest tenth) of the machine part shown in Exercise 1.

## 8.8 RECTANGULAR SOLIDS AND VOLUME

### STUDENT GUIDEPOSTS

1. Volume in Cubic Units
2. Volume of a Rectangular Solid
3. Volume of a Cube
4. Converting Cubic Units
5. Surface Area of a Rectangular Solid
6. Surface Area of a Cube

### 1 VOLUME IN CUBIC UNITS

To measure length we use units like 1 cm and 1 inch. When we talk about area we use units like 1 cm² and 1 in². To find volume we will use cubic units like 1 cm³ and 1 in³.

The **volume** of a cube which is 1 cm by 1 cm by 1 cm is one **cubic centimeter** and is written 1 cm³. Similarly, a cube which is 1 in by 1 in by 1 in is one

**cubic inch,** 1 in³. See Figure 8.39. If some other unit were used for measuring, the volume of the cube would be 1 cubic unit, or 1 unit³.

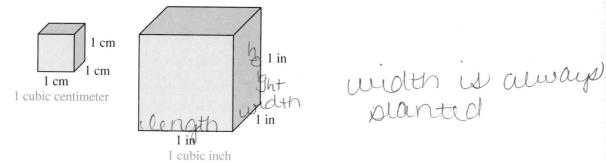

1 cm
1 cm
1 cm
1 cubic centimeter

height
width
length
1 in
1 in
1 in
1 cubic inch

*width is always slanted*

**Figure 8.39**

### ② VOLUME OF A RECTANGULAR SOLID

The cubic unit of measure is used for the volume of solids. The solid in Figure 8.40(a) has volume 20 cm³ since there are 10 cm³ on the bottom layer and 10 cm³ on the top layer. In Figure 8.40(b) there is one more layer (10 cm³) than in Figure 8.40(a), so the volume is 30 cm³. These volumes can be found without a diagram by multiplying the area of the base by the height.

$$20 \text{ cm}^3 = \underbrace{5 \text{ cm} \cdot 2 \text{ cm}}_{\text{area of base}} \cdot \underbrace{2 \text{ cm}}_{\text{height}}$$

$$30 \text{ cm}^3 = \underbrace{5 \text{ cm} \cdot 2 \text{ cm}}_{\text{area of base}} \cdot \underbrace{3 \text{ cm}}_{\text{height}}$$

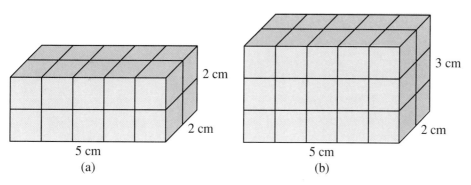

2 cm
2 cm
5 cm
(a)

3 cm
2 cm
5 cm
(b)

**Figure 8.40**

> **Volume of a Rectangular Solid**
>
> The volume $V$ of a rectangular solid is the area of the base $lw$ times the height $h$.
> $$V = lwh$$

---

| **EXAMPLE 1  CALCULATING VOLUME** | **PRACTICE EXERCISE 1** |
|---|---|

Find the volume of the rectangular solid in Figure 8.41 on the following page which is 12 m by 7 m by 20 m.

$$\begin{aligned} V &= lwh \\ &= (12 \text{ m})(7 \text{ m})(20 \text{ m}) \\ &= (12)(7)(20) \text{ m}^3 \\ &= 1680 \text{ m}^3 \end{aligned}$$

Find the volume of a rectangular solid which is 5 cm by 12 cm by 30 cm.

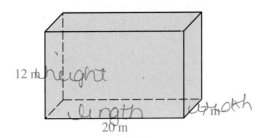

**Figure 8.41**

Answer: 1800 cm$^3$

---

### EXAMPLE 2   APPLICATION TO CONSTRUCTION

A highway construction truck has a bed which is 5.5 yd long, 2.5 yd wide, and 2 yd deep. How many cubic yards of gravel will it hold?
    Find the volume of the truck bed.

$$V = lwh$$
$$= (5.5 \text{ yd})(2.5 \text{ yd})(2 \text{ yd})$$
$$= (5.5)(2.5)(2) \text{ yd}^3$$
$$= 27.5 \text{ yd}^3$$

The truck will hold 27.5 yd$^3$ of gravel.

### PRACTICE EXERCISE 2

A trailer for hauling cotton is 18 ft by 8 ft by 10 ft. What volume of cotton will it hold?

Answer: 1440 ft$^3$

---

## ③ VOLUME OF A CUBE

For a **cube,** $l = w = h$. We talk about *edges* of the solid, and call the length of each edge $e$.

### Volume of a Cube

The volume $V$ of a cube is the cube of the edge.

$$V = e^3$$

---

### EXAMPLE 3   FINDING THE VOLUME OF A CUBE

Find the volume of the cube in Figure 8.42 which has an edge of 1.5 ft.

$$V = (1.5 \text{ ft})^3$$
$$= (1.5 \text{ ft})(1.5 \text{ ft})(1.5 \text{ ft})$$
$$= 3.375 \text{ ft}^3$$

### PRACTICE EXERCISE 3

Find the volume of a cube which has an edge of $3\frac{1}{2}$ inches.

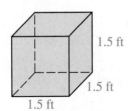

1.5 ft
1.5 ft
1.5 ft

**Figure 8.42**

Answer: $42\frac{7}{8}$ in$^3$

**④ CONVERTING CUBIC UNITS**

To convert from one cubic unit to another, we must *cube* the length conversion factor. For example, in Figure 8.43 there are 9 cubes on each of the 3 layers. Thus, there are $9 \cdot 3 = 27$ cubes. This is the same as

$$1 \text{ yd}^3 = (3 \text{ ft})^3 = 3^3 \text{ ft}^3 = 27 \text{ ft}^3.$$

Similarly,

$$1 \text{ ft}^3 = (12 \text{ in})^3 = 1728 \text{ in}^3.$$

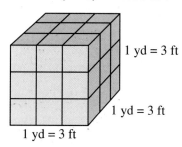

1 yd = 3 ft

1 yd = 3 ft

1 yd = 3 ft

**Figure 8.43**

| **HINT** |
| --- |

In the metric system, cubic units can be converted by moving the decimal point *three* times the number of decimal places as for length conversion.

---

| **EXAMPLE 4**   CUBIC UNITS IN THE METRIC SYSTEM | **PRACTICE EXERCISE 4** |
| --- | --- |

**(a)** $72,500 \text{ cm}^3 = $ _____ $\text{m}^3$.

| 1000 m | 100 m | 10 m | **1 m** | 0.1 m | 0.01 m | 0.001 m |
| --- | --- | --- | --- | --- | --- | --- |
| **(km)** | **(hm)** | **(dam)** | | **(dm)** | **(cm)** | **(mm)** |

2 places left

As shown above, to change from cm to m the decimal point is moved 2 places to the left. Thus, to change from $\text{cm}^3$ to $\text{m}^3$, the decimal point is moved

$$3(2) = 6 \text{ places}$$

to the left. This is the same as $1 \text{ m}^3 = (100 \text{ cm})^3 = 1,000,000 \text{ cm}^3$.

$$72,500 \text{ cm}^3 = 0.072500 \text{ m}^3$$

6 places

$$= 0.0725 \text{ m}^3$$

**(b)** $0.532 \text{ cm}^3 = $ _____ $\text{mm}^3$.

| 1000 m | 100 m | 10 m | **1 m** | 0.1 m | 0.01 m | 0.001 m |
| --- | --- | --- | --- | --- | --- | --- |
| **(km)** | **(hm)** | **(dam)** | | **(dm)** | **(cm)** | **(mm)** |

1 place right

To change $\text{cm}^3$ to $\text{mm}^3$, move the decimal point $3(1) = 3$ places to the right.

$$0.532 \text{ cm}^3 = 0532. \text{ mm}^3$$

3 places

$$= 532 \text{ mm}^3$$

**(a)** $3,280,000 \text{ cm}^3 = $ _____ $\text{m}^3$.

**(b)** $25.6 \text{ cm}^3 = $ _____ $\text{mm}^3$.

Answers:  (a)  3.28    (b)  25,600

Since $1 L = (1 dm)^3 = (10 cm)^3 = 1000 cm^3$, the following relationships between $cm^3$ and ml are true.

$$1000 cm^3 = 1000 ml = 1 L$$
$$1 cm^3 = 1 ml = 0.001 L.$$

The abbreviation cc is sometimes used for *cubic centimeters* ($cm^3$). Thus,

$$1 cc = 1 cm^3 = 1 ml.$$

Remember that $1 cm^3$ of water weighs 1 g and thus

1 ml of water weighs 1 g;
1 L of water weighs 1 kg.

---

| EXAMPLE 5 APPLICATION OF VOLUME UNITS | PRACTICE EXERCISE 5 |
|---|---|

A small tank 328 mm by 154 mm by 80 mm is filled with water. How much does the water weigh?

The volume of the tank is

$$V = lwh$$
$$= (328 mm)(154 mm)(80 mm)$$
$$= (328)(154)(80) mm^3$$
$$= 4,040,960 mm^3.$$

Convert this to cubic centimeters ($cm^3$ or cc).

| 1000 m **(km)** | 100 m **(hm)** | 10 m **(dam)** | **1 m** | 0.1 m **(dm)** | 0.01 m **(cm)** | 0.001 m **(mm)** |
|---|---|---|---|---|---|---|

Move the decimal point $3(1) = 3$ places to the left.

$$4,040,960 mm^3 = 4040.960 cm^3$$
$$= 4040.96 cc$$

Since 1 cc of water weighs 1 g, there are 4040.96 g of water in the tank. This is a little more than 4 kg (4.04096 kg).

A tank which is 2.5 dm by 3.8 dm by 4.2 dm is filled with water. How much does the water weigh?

Answer: 39,900 g = 39.9 kg

---

## ⑤ SURFACE AREA OF A RECTANGULAR SOLID

Another type of problem involving a rectangular solid is to find the area of the surface. If we can do this, we can find the amount of material used to make a tank like the one in Example 5 or the amount of paint needed to paint a building.

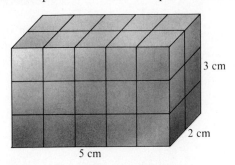

**Figure 8.44**

The **surface area** of an object is the area of all its surfaces. Look at the solid in Figure 8.44 which was used to count cubic centimeters earlier. Now, we count square centimeters on its surface. There are

$$5 \text{ cm} \cdot 3 \text{ cm} = 15 \text{ cm}^2$$

on the front face. But the back is just like the front, so there are

$$2 \cdot 15 \text{ cm}^2 = 30 \text{ cm}^2 \qquad 2lh$$

on the front and back together. On the two ends there are

$$2(2 \text{ cm})(3 \text{ cm}) = 12 \text{ cm}^2. \qquad 2wh$$

On the bottom there are

$$5 \text{ cm} \cdot 2 \text{ cm} = 10 \text{ cm}^2,$$

and the same number on the top. The total of the top and bottom is

$$2(10 \text{ cm}^2) = 20 \text{ cm}^2. \qquad 2lw$$

The total surface area is the sum of all surfaces.

| | | |
|---|---|---|
| Area of front and back: | 30 cm² | 2lh |
| Area of two ends: | 12 cm² | 2wh |
| Area of bottom and top: | 20 cm² | 2lw |
| Total surface area: | 62 cm² | |

### Surface Area of a Rectangular Solid

The total surface area ($A$) of a rectangular solid is

$$A = 2lh + 2wh + 2lw.$$

*[handwritten notes in right margin: lh = front & back panels, wh = left and right panels, lw = top and bottom panels]*

*[handwritten note: a = 2(front and back) + 2(left and right) + 2(top and bottom)]*

---

### EXAMPLE 6   FINDING THE SURFACE AREA

Find the total surface area of a rectangular solid which is 6 m by 3 m by 7 m.

$$A = 2lh + 2wh + 2lw$$
$$= 2(6 \text{ m})(7 \text{ m}) + 2(3 \text{ m})(7 \text{ m}) + 2(6 \text{ m})(3 \text{ m})$$
$$= 2(6)(7)\text{m}^2 + 2(3)(7)\text{m}^2 + 2(6)(3)\text{m}^2$$
$$= 84 \text{ m}^2 + 42 \text{ m}^2 + 36 \text{ m}^2$$
$$= 162 \text{ m}^2$$

### PRACTICE EXERCISE 6

Find the surface area of a rectangular solid which is 12 ft by 16 ft by 20 ft.

Answer:  1504 ft²

---

### EXAMPLE 7   APPLICATION OF SURFACE AREA

How many square inches of glass are needed to make a fish tank which is 15 in by 12 in by 10 in if the top is left open?

We must find the surface area of the four sides and the bottom. The tank is shown in Figure 8.45 on the following page. The area of the front and back together is

$$2 \cdot 15 \text{ in} \cdot 10 \text{ in} = 300 \text{ in}^2.$$

The area of the two ends is

$$2 \cdot 12 \text{ in} \cdot 10 \text{ in} = 240 \text{ in}^2.$$

### PRACTICE EXERCISE 7

A metal box with no top is to be made with a base 10 cm by 25 cm and a height of 8 cm. How many square centimeters of metal are required?

The area of the bottom is

$$15 \text{ in} \cdot 12 \text{ in} = 180 \text{ in}^2.$$

The area of glass needed is the sum of all these areas.

| | |
|---|---|
| Area of front and back: | 300 in² |
| Area of the two ends: | 240 in² |
| Area of bottom: | 180 in² |
| Total area: | 720 in² |

720 in² of glass are used to make the tank.

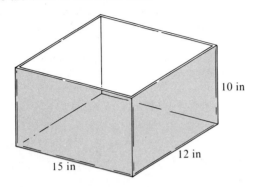

**Figure 8.45**

---

### EXAMPLE 8 CONSTRUCTION APPLICATION

The four vertical sides of a storage building are to be painted. One gallon of paint will cover 150 ft² of area. How much paint will be used if the building is 16 ft by 12 ft by 8 ft?

The area of the front and back is twice the length times the height.

$$2 \cdot 16 \text{ ft} \cdot 8 \text{ ft} = 256 \text{ ft}^2$$

The area of the two ends is twice the width times the height.

$$2 \cdot 12 \text{ ft} \cdot 8 \text{ ft} = 192 \text{ ft}^2$$

The total area is the sum of these two areas.

| | |
|---|---|
| Area of front and back: | 256 ft² |
| Area of two ends: | 192 ft² |
| Total area: | 448 ft² |

Since the paint covers 150 ft² per gal, the unit fraction is $\frac{1 \text{ gal}}{150 \text{ ft}^2}$.

$$\frac{448 \text{ ft}^2}{1} \times \frac{1 \text{ gal}}{150 \text{ ft}^2} = \frac{448}{150} \text{ gal}$$

$$\approx 2.99 \text{ gal} \qquad \text{To the nearest hundredth}$$

Thus, it will take about 3 gal to paint the building.

### PRACTICE EXERCISE 8

The outside walls of a workshop are to be covered with siding which costs $1.50 per square foot. How much will it cost if the building is 16 ft wide, 18 ft long, and 8 ft high?

Answer: $816

## ⑥ SURFACE AREA OF A CUBE

If a rectangular solid is a cube, then $l = w = h = e$.

### Surface Area of a Cube

The surface area of a cube with edge $e$ is  or  $6e^2$

$$A = 6e^2.$$

| **EXAMPLE 9**   FINDING SURFACE AREA OF A CUBE | **PRACTICE EXERCISE 9** |

Find the surface area of a cube with edge 14 cm.

$$A = 6e^2$$
$$= 6(14 \text{ cm})^2$$
$$= 6(196) \text{ cm}^2$$
$$= 1176 \text{ cm}^2$$

Find the surface area of a cube with edge 200 mm.

Answer: 240,000 mm²

### HINT

If you do not remember a formula for surface area, just add up the areas of all the surfaces.

## 8.8 EXERCISES A

1. The volume of a cube 1 m by 1 m by 1 m is _____.

2. The volume of a rectangular solid is given by the formula $V = $ _____.

3. The volume of a cube is given by the formula $V = $ _____.

4. Complete:   $1 \text{ cm}^3 = $ _____ ml.

5. Complete:   $1000 \text{ cc} = $ _____ L.

6. 1 ml of water weighs _____ g.

*Find the volume of the rectangular solid.*

**7.** 10 m by 5 m by 6 m

**8.** 2.3 cm by 1.2 cm by 4.5 cm

**⑨** $9\frac{1}{2}$ in by 3.4 in by $8\frac{3}{4}$ in

*Find the volume of the cube with the given edge length.*

**10.** 8 m

**11.** 10.2 cm

**12.** $1\frac{3}{5}$ in

*Complete.*

**13.** $1 \text{ yd}^3 = $ _____ $\text{ft}^3$

**14.** $1 \text{ ft}^3 = $ _____ $\text{in}^3$

**15.** $8 \text{ ft}^3 = $ _____ $\text{in}^3$

**16** 324 ft³ = _____ yd³          **17** 0.0085 hm³ = _____ m³          **18.** 6.2 mm³ = _____ cm³

**19.** 7 cc = _____ ml          **20** 525 L = _____ m³          **21.** 5 ml = _____ mm³

*Find the total surface area of each rectangular solid.*

**22.** 6 ft by 2 ft by 5 ft          **23.** $1\frac{1}{2}$ in by $\frac{1}{2}$ in by 4 in

**24.** 10.2 yd by $8\frac{1}{2}$ yd by 4.5 yd          **25** 15.5 cm by 6.4 cm by 20.2 cm

*Find the surface area of the cube with the given edge.*

**26.** 5 ft          **27.** 21 mm          **28.** $1\frac{1}{2}$ yd

*Solve.*

**29.** Frank's truck is to be used to carry topsoil. If the truck bed is 2 yd by 1.5 yd by 0.8 yd, how many cubic yards of topsoil will it hold?

**30** A tank is in the shape of a cube. How many grams of water will it hold if the edge of the cube is 22 cm?

**31** How many square centimeters of glass did it take to make the bottom and sides of the tank in Exercise 30?

**32.** Elizabeth owns a rectangular building which is 30 ft by 12 ft by 8 ft. How much will it cost to put siding on the building if the siding costs $0.75 per square foot?

## FOR REVIEW

*Find the circumference and area of each circle. Use* 3.14 *for* $\pi$.

**33.** $r = 11$ cm          **34.** $d = 8.2$ in

ANSWERS: 1. 1 m³   2. *lwh*   3. $e^3$   4. 1   5. 1   6. 1   7. 300 m³   8. 12.42 cm³   9. 282.625 in³   10. 512 m³  11. 1061.208 cm³   12. 4.096 in³   13. 27   14. 1728   15. 13,824   16. 12   17. 8500   18. 0.0062   19. 7   20. 0.525  21. 5000   22. 104 ft²   23. $17\frac{1}{2}$ in²   24. 341.7 yd²   25. 1083.16 cm²   26. 150 ft²   27. 2646 mm²   28. $13\frac{1}{2}$ yd²  29. 2.4 yd³   30. 10,648 g   31. 2420 cm²   32. $504   33. 69.08 cm; 379.94 cm²   34. 25.75 in; 52.78 in² (nearest hundredth)

## 8.8 EXERCISES B

**1.** The volume of a cube 1 ft by 1 ft by 1 ft is _____.

**2.** The surface area of a rectangular solid is given by the formula $A =$ _____.

**3.** The surface area of a cube is given by the formula $A =$ _____.

**4.** Complete: 1 L = _____ cm$^3$.

**5.** Complete: 1 dm$^3$ = _____ L.

**6.** The volume of 1000 grams of water is _____ L.

*Find the volume of the rectangular solid.*

**7.** 6 m by 12 m by 15 m

**8.** 9.2 cm by 1.5 cm by 7.5 cm

**9.** $6\frac{1}{2}$ in by 5.5 in by $3\frac{1}{5}$ in

*Find the volume of the cube with the given edge length.*

**10.** 12 m

**11.** 0.1 cm

**12.** $2\frac{1}{6}$ in

*Complete.*

**13.** 7 yd$^3$ = _____ ft$^3$

**14.** 14 ft$^3$ = _____ in$^3$

**15.** 567 ft$^3$ = _____ yd$^3$

**16.** 0.25 ft$^3$ = _____ in$^3$

**17.** 820 mm$^3$ = _____ cm$^3$

**18.** 0.0000725 hm$^3$ = _____ dm$^3$

**19.** 250 cc = _____ ml

**20.** 0.07 ml = _____ mm$^3$

**21.** 1200 L = _____ m$^3$

*Find the total surface area of each rectangular solid.*

**22.** 3 ft by 5 ft by 12 ft

**23.** $4\frac{1}{2}$ in by $\frac{1}{4}$ in by 6 in

**24.** 8.6 yd by $4\frac{1}{5}$ yd by 7.5 yd

**25.** 12.2 cm by 100 cm by 6.5 cm

*Find the surface area of the cube with the given edge.*

**26.** 17 ft

**27.** 0.01 mm

**28.** $4\frac{1}{6}$ yd

*Solve.*

**29.** A large earth mover has a bed which is 5 yd by 4 yd by 10 yd. How many cubic yards of dirt will it hold?

**30.** A water tank is 22.5 cm by 10 cm by 5.2 cm. How many grams of water will it hold?

**31.** A box which is a cube with edge 8.5 dm is to be covered with foil. How much foil is required?

**32.** A rectangular tank 4 m by 6 m by 5 m (5 m is the height) needs to be insulated on the sides and top. If insulation costs $2.15 per square meter, how much will it cost?

## FOR REVIEW

*Find the circumference and area of each circle. Use 3.14 for π.*

**33.** $r = 27$ cm

**34.** $d = 32.4$ in

## 8.8 EXERCISES C

**1.** A Buick Park Avenue has a 3.8 L (engine displacement) V-6 engine. How many cubic inches is this? [*Hint:* 1 cm³ ≈ 0.061 in³]

**2.** Lake Louise has a surface area of 4.2 square miles with an average depth of 22.5 feet. What is the approximate number of cubic feet of water in Lake Louise?
[Answer: 2,634,500,000 ft³]

## 8.9 CYLINDERS AND SPHERES

═══════ STUDENT GUIDEPOSTS ═══════

**①** Volume of a Cylinder     **③** Volume of a Sphere
**②** Surface Area of a Cylinder     **④** Surface Area of a Sphere

### ① VOLUME OF A CYLINDER

A **cylinder** (see Figure 8.46) is a solid with circular ends of the same radius. For example, a can of beans is in the shape of a cylinder.

Remember that the volume of a rectangular solid is the area of the base times the height. The same is true for a cylinder. Since the base is a circle with area $\pi r^2$, we have the following rule.

---

**Volume of a Cylinder**

The volume $V$ of a cylinder is the area of the base, $\pi r^2$, times the height, $h$.

$$V = \pi r^2 h$$

---

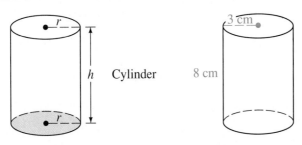

**Figure 8.46**       **Figure 8.47**

---

| **EXAMPLE 1**   FINDING THE VOLUME OF A CYLINDER | **PRACTICE EXERCISE 1** |
|---|---|

Find the volume of the cylinder in Figure 8.47 which has a base of radius 3 cm and is 8 cm high. Use 3.14 for $\pi$.

Since $r = 3$ cm and $h = 8$ cm,

$$\begin{aligned} V &= \pi r^2 h \\ &\approx 3.14(\textbf{3 cm})^2(\textbf{8 cm}) \\ &= (3.14)(9)(8) \text{ cm}^3 \qquad \text{cm}^2 \times \text{cm} = \text{cm}^3 \\ &= 226.08 \text{ cm}^3. \end{aligned}$$

Find the volume of a cylinder with radius of the base 6 cm and height 20 cm. Use 3.14 for $\pi$.

**Answer: 2260.8 cm³**

| EXAMPLE 2 APPLICATION OF VOLUME | PRACTICE EXERCISE 2 |

A can of blueberry pie filling has diameter 3 in and height 4.5 in. How many cans of filling are needed to fill a 9-in diameter pie pan 1 in deep?

First find the volume of pie filling in one can.

$$V = \pi r^2 h$$

$$\approx 3.14(1.5 \text{ in})^2(4.5 \text{ in})$$

$$= (3.14)(2.25)(4.5) \text{ in}^3$$

$$\approx 31.8 \text{ in}^3 \qquad \text{Volume in one can to nearest tenth}$$

Now find the volume of pie filling in a 9-in diameter (4.5-in radius) pie pan.

$$V = \pi r^2 h$$

$$\approx 3.14(4.5 \text{ in})^2(\mathbf{1 \text{ in}}) \qquad \text{Filling is 1 in deep}$$

$$= (3.14)(20.25)(1) \text{ in}^3$$

$$\approx 63.6 \text{ in}^3 \qquad \text{Volume in pie pan to nearest tenth}$$

Since one can has 31.8 in$^3$ of filling, multiply by the unit fraction $\frac{1 \text{ can}}{31.8 \text{ in}^3}$.

$$\frac{63.6 \text{ in}^3}{1} \times \frac{1 \text{ can}}{31.8 \text{ in}^3} = \frac{63.6}{31.8} \text{ can}$$

$$= 2 \text{ cans}$$

It will take 2 cans of filling for the pie.

## ② SURFACE AREA OF A CYLINDER

To see how to find the surface area of a cylinder, consider a can without top or bottom. See Figure 8.48(a). If it is cut along the seam and pressed flat, as in Figure 8.48(b), a rectangle is formed. The length of the rectangle is the circumference of the circle and the width is the height. Thus, the surface area of the side of the can is $2\pi rh$.

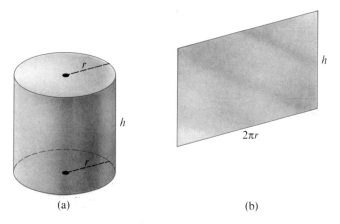

(a)  (b)

**Figure 8.48**

Since the area of the top of the can is $\pi r^2$, and the area of the bottom is also $\pi r^2$, we have the following rule.

| Surface Area of a Cylinder |
|---|

The surface area $A$ of a cylinder is

$$A = 2\pi rh + 2\pi r^2.$$

Side + top & bottom

---

| EXAMPLE 3 FINDING THE SURFACE AREA | PRACTICE EXERCISE 3 |
|---|---|

Find the surface areas of a cylinder with radius 7 ft and height 9 ft.

$$A = 2\pi rh + 2\pi r^2$$
$$\approx 2(3.14)(7 \text{ ft})(9 \text{ ft}) + 2(3.14)(7 \text{ ft})^2$$
$$= 2(3.14)(7)(9) \text{ ft}^2 + 2(3.14)(49) \text{ ft}^2$$
$$= 703.36 \text{ ft}^2$$

Find the surface area of a cylinder with radius 16 ft and height 4 ft.

Answer: 2009.6 ft$^2$

---

| EXAMPLE 4 APPLICATION TO CONSTRUCTION | PRACTICE EXERCISE 4 |
|---|---|

The side and top of a cylindrical water tank are to be painted. The tank has radius 3 m and height 8 m. How many liters of paint will be used if a liter covers 8 m$^2$?

To find the surface area of the side, find $2\pi rh$.

$$2\pi rh \approx 2(3.14)(3 \text{ m})(8 \text{ m})$$
$$= 2(3.14)(3)(8) \text{ m}^2 = 150.72 \text{ m}^2$$

To find the area of the top, find $\pi r^2$.

$$\pi r^2 \approx 3.14(3 \text{ m})^2$$
$$= (3.14)(9) \text{ m}^2 = 28.26 \text{ m}^2$$

To find the total, add.

| | |
|---|---|
| Area of side: | 150.72 m$^2$ |
| Area of top: | 28.26 m$^2$ |
| Total area: | 178.98 m$^2$ |

The area to be covered is about 179 m$^2$. Multiply this by $\frac{1 \text{ L}}{8 \text{ m}^2}$.

$$\frac{179 \text{ m}^2}{1} \times \frac{1 \text{ L}}{8 \text{ m}^2} = \frac{179}{8} \text{ L}$$
$$\approx 22.4 \text{ L} \quad \text{To the nearest tenth}$$

It will take about 22.4 L to paint the tank.

A rustproofing must be applied to the top and sides of a cylindrical storage tank. The radius of the base is 10 m and the height is 12 m. How many liters of the rustproofing will be needed if one liter covers 20 m$^2$?

Answer: 53.38 L

---

## ③ VOLUME OF A SPHERE

A **sphere** (Figure 8.49) is a solid in the shape of a ball. The radius of a sphere is the distance from its center to the surface. The diameter is twice the radius. If a sphere is cut through the center, the cut surface is a circle.

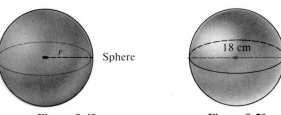

**Figure 8.49**          **Figure 8.50**

### Volume of a Sphere

The volume $V$ of a sphere is

$$V = \frac{4}{3}\pi r^3.$$

---

**EXAMPLE 5**  FINDING THE VOLUME OF A SPHERE

Find the volume of the sphere in Figure 8.50 with diameter 18 cm.

If the diameter is 18 cm, then the radius, $r$, is $\frac{18}{2}$ cm or 9 cm.

$$V = \frac{4}{3}\pi r^3 \approx \frac{4}{3}(3.14)(9 \text{ cm})^3$$

$$= \frac{4}{3}(3.14)729 \text{ cm}^3 = 3052.08 \text{ cm}^3$$

**PRACTICE EXERCISE 5**

Find the volume of a sphere with diameter 10 ft.

Answer:  523.3 ft$^3$

---

**EXAMPLE 6**  APPLICATION OF VOLUME OF A SPHERE

A spherical tank with radius 2 m is filled with a liquid chemical. How many liters are in the tank?

By definition, a liter is a cube 0.1 m on edge. 0.1 m is 1 dm, so if 2 m is converted to 20 dm, the volume of the sphere will be in liters.

$$V = \frac{4}{3}\pi r^3$$

$$\approx \frac{4}{3}(3.14)(20 \text{ dm})^3 \qquad \text{2 m = 20 dm}$$

$$= \frac{4}{3}(3.14)8000 \text{ dm}^3$$

$$\approx 33{,}493 \text{ dm}^3 \qquad \text{To the nearest unit}$$

$$= 33{,}493 \text{ L} \qquad \text{1 L is 1 dm}^3$$

**PRACTICE EXERCISE 6**

How many liters of chemical can be stored in a sphere with radius 80 cm?

Answer:  2143.6 L

---

### ④ SURFACE AREA OF A SPHERE

We can also determine the surface area of a sphere.

### Surface Area of a Sphere

The surface area $A$ of a sphere is

$$A = 4\pi r^2.$$

| EXAMPLE 7 APPLICATION TO CONSTRUCTION | PRACTICE EXERCISE 7 |

A spherical tank with radius 3.2 m is to be covered with insulation. If the insulation costs $2.50 per square meter, how much will the material cost?

$$A = 4\pi r^2$$
$$\approx 4(3.14)(3.2 \text{ m})^2$$
$$= 4(3.14)(10.24) \text{ m}^2$$
$$\approx 128.61 \text{ m}^2 \qquad \text{To the nearest hundredth}$$

Since the insulation costs $2.50 per square meter, the cost is

$$\frac{128.61 \text{ m}^2}{1} \times \frac{\$2.50}{1 \text{ m}^2} \approx \$321.53. \qquad \text{To the nearest cent}$$

A spherical tank with diameter 6.4 m is to be prepared for painting. If the labor costs are $4.50 per square meter, how much will the job cost?

Answer: $578.76

---

## H I N T

When approximate decimal numbers are used, there will frequently be minor differences in answers. You will also see some answers that are rounded as in Examples 6 and 7.

---

## 8.9  EXERCISES A

*Find the volume of each cylinder. Use 3.14 for $\pi$.*

**1.** $r = 6$ cm, $h = 5$ cm

**2** $d = 24.4$ in, $h = 30$ in

**3.** $d = 11.4$ m, $h = 4.4$ m

*Find the surface area of each cylinder. Use 3.14 for $\pi$.*

**4.** $r = 6$ cm, $h = 5$ cm

**5** $d = 24.4$ in, $h = 30$ in

**6.** $d = 11.4$ m, $h = 4.4$ m

*Find the volume of each sphere. Use 3.14 for $\pi$.*

**7.** $r = 12$ mm

**8** $r = \dfrac{3}{4}$ in

**9.** $d = 5.6$ yd

*Find the surface area of each sphere. Use 3.14 for $\pi$.*

**10.** $r = 12$ mm

**11.** $r = \dfrac{3}{4}$ in

**12** $d = 5.6$ yd

*Solve. Use 3.14 for $\pi$.*

**13** A can of cherry pie filling has diameter 8 cm and height 12 cm. How many cans (to the nearest tenth) are needed to fill a 20-cm diameter pie pan 3 cm deep?

**14.** A cylindrical storage tank has radius 3.8 ft and height 9.8 ft. How many gallons (to the nearest tenth) of paint are needed to paint the tank (including top and bottom) if one gallon covers 150 ft$^2$?

**15.** A spherical tank has radius 8.6 m. If a rust-preventing material costs $1.50 per square meter, what is the cost to rustproof the tank?

**16.** A spherical tank with radius 32 cm is filled with gasoline. How many liters are in the tank?

## FOR REVIEW

*Find the volume and surface area of each rectangular solid.*

**17.** 10 m by 5 m by 12 m

**18.** 2.2 ft by 1.2 ft by 6.5 ft

---

ANSWERS: 1. 565.2 cm$^3$  2. 14,020.7 in$^3$ (nearest tenth)  3. 448.9 m$^3$ (nearest tenth)  4. 414.48 cm$^2$  5. 3233.2 in$^2$ (nearest tenth)  6. 361.5 m$^2$ (nearest tenth)  7. 7234.56 mm$^3$  8. 1.766 in$^3$ (nearest thousandth)  9. 91.91 yd$^3$ (nearest hundredth)  10. 1808.64 mm$^2$  11. 7.065 in$^2$  12. 98.47 yd$^2$ (nearest hundredth)  13. 1.6 cans  14. 2.2 gal  15. $1393.41  16. 137.19 L (nearest hundredth)  17. 600 m$^3$; 460 m$^2$  18. 17.16 ft$^3$; 49.48 ft$^2$

## 8.9 EXERCISES B

*Find the volume of each cylinder. Use 3.14 for π.*

**1.** $r = 9$ cm, $h = 15$ cm

**2.** $d = 6.8$ m, $h = 10.5$ m

**3.** $d = 20.4$ in, $h = 8.5$ in

*Find the surface area of each cylinder. Use 3.14 for π.*

**4.** $r = 9$ cm, $h = 15$ cm

**5.** $d = 6.8$ m, $h = 10.5$ m

**6.** $d = 20.4$ in, $h = 8.5$ in

*Find the volume of the sphere. Use 3.14 for π.*

**7.** $r = 30$ mm

**8.** $r = \dfrac{2}{5}$ in

**9.** $d = 9.8$ yd

*Find the surface area of each sphere. Use 3.14 for π.*

**10.** $r = 30$ mm

**11.** $r = \dfrac{2}{5}$ in

**12.** $d = 9.8$ yd

*Solve. Use 3.14 for π.*

**13.** Mercury is stored in a cylinder of radius 16 cm and height 20 cm. How many cylindrical tubes with radius 0.5 cm and height 100 cm can be filled from a full supply?

**14.** A cylindrical tank is to be made from sheet metal. If the tank is to have radius 2.2 ft and height 12 ft, what will the metal cost if it is $16.50 per square foot?

**15.** A spherical tank with diameter 36 ft is to be insulated. If insulation costs $0.50 per square foot, how much will the job cost?

**16.** How many liters of chemical can be stored in a sphere with radius 650 mm?

## FOR REVIEW

*Find the volume and surface area of each rectangular solid.*

**17.** 20 m by 6 m by 14 m

**18.** 1.6 ft by 4.5 ft by 8.8 ft

## 8.9   EXERCISES C

**1.** Find the volume of the wood remaining when the rectangular block shown at right has a hole bored through it. Use 3.14 for $\pi$ and give the answer rounded to the nearest hundredth.
[Answer: 0.46 ft$^3$]

**2.** Find the total surface area of the wood remaining in the block in Exercise 1.

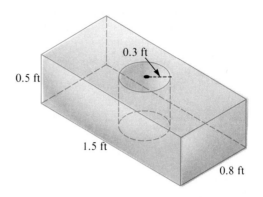

## CHAPTER 8 REVIEW

### KEY WORDS

**8.1**   The **English system** is the measurement system used in the United States.

A **unit fraction** is a fraction which has value 1 and is used to make measurement conversions.

**8.3**   The **metric system** is the system of measurement used in most of the countries of the world. The **meter** is the basic unit for length, the **liter** is the basic unit for volume, and the **gram** is the basic unit for weight.

**8.4**   The **Celsius** scale is the temperature scale used in the metric system.

The **Fahrenheit** scale is the temperature scale used in the English system.

**8.5**   A **rectangle** is a four-sided figure with opposite sides parallel and adjacent sides perpendicular.

The **perimeter** of a figure is the distance around the figure.

A **square** is a rectangle with all sides equal.

A **square unit** is the area of a square which measures one unit on each side.

**8.6**   A **parallelogram** is a four-sided figure whose opposite sides are parallel and equal.

A **trapezoid** is a four-sided figure with two parallel sides.

**8.7**   A **circle** is a figure with all points the same distance from a point called the **center** of the circle.

The **radius** of a circle is the distance from the center to any point on the circle.

The **diameter** of a circle is twice the radius.

The **circumference** of a circle is the distance around the circle.

**8.8**   A **cube** is a rectangular solid with length, width, and height the same.

A **cubic unit** is the volume of a cube which has measure one unit on each edge.

The **surface area** of an object is the area of all of its surfaces.

**8.9**   A **cylinder** is a solid with circular ends of the same radius.

A **sphere** is a solid in the shape of a ball.

### KEY CONCEPTS

**8.3**   **1.** In the metric system, conversions in units can be made by moving the decimal point.

**2.** A meter is about half the height of a doorway opening. A liter is the volume of a cube 0.1 m (1 dm) on a side. A gram is the weight of water that can be put in a cube 0.01 m (1 cm) on a side.

**8.4**   **1.** To change °C to °F, use $F = \frac{9}{5}C + 32$,

**2.** To change °F to °C, use $C = \frac{5}{9}(F - 32)$.

**8.5**   **1.** Rectangle:   $P = 2l + 2w$,     $A = lw$
Square:   $P = 4s$     $A = s^2$

**2.** Area is given in square units. In the metric system, a change in units is made by moving the decimal point twice the number of places used for changing length units.

**8.6**   **1.** Parallelogram:   $A = bh$
     **2.** Triangle:   $A = \frac{1}{2}bh$
     **3.** Trapezoid:   $A = \frac{1}{2}(b_1 + b_2)h$

**8.7**   Circle:   $d = 2r$,   $c = \pi d = 2\pi r$,
     $A = \pi r^2$

**8.8**   **1.** Rectangular solid:   $V = lwh$,

$A = 2lh + 2wh + 2lw$
Cube:   $V = e^3$,   $A = 6e^2$

**2.** Volume is given in cubic units. In the metric system, a change in units is made by moving the decimal point three times the number of places used for changing length units.

**8.9**   **1.** Cylinder:   $V = \pi r^2 h$,   $A = 2\pi rh + 2\pi r^2$
     **2.** Sphere:   $V = \frac{4}{3}\pi r^3$,   $A = 4\pi r^2$

## REVIEW EXERCISES

*Part I*

**8.1**   *Complete.*

   **1.** 7 ft = _____ in
   **2.** 30 in = _____ ft
   **3.** 48 fl oz = _____ pt

   **4.** 90 min = _____ hr
   **5.** 3 ton = _____ lb
   **6.** 18 pt = _____ gal

**8.2**   *Perform the indicated operation and simplify.*

   **7.**   10 ft 3 in
     +  7 ft 2 in

   **8.**   10 ft 3 in
     −  7 ft 2 in

   **9.**   18 hr 22 min
     +  5 hr 56 min

   **10.**   18 hr 22 min
     −  5 hr 56 min

   **11.**   12 yd 2 ft 2 in
     +  6 yd 2 ft 8 in

   **12.**   12 yd 2 ft 2 in
     −  6 yd 2 ft 8 in

   **13.** $3 \times (5 \text{ gal } 3 \text{ qt } 1 \text{ pt})$

   **14.** 8 ft × 3 ft

   **15.** (8 lb 13 oz) ÷ 3

   **16.** 18 dollars ÷ 3 dollars

**8.3**   *Complete.*

   **17.** 860 m = _____ km
   **18.** 0.582 m = _____ cm
   **19.** 483 ml = _____ dl

   **20.** 48,300 mg = _____ kg
   **21.** 0.00035 km = _____ dm
   **22.** 5.6 L = _____ cl

*Perform the indicated operation.*

   **23.**   15.2 m
     + 75.3 m

   **24.**   831.6 L
     − 416.9 L

   **25.** 42.8 g
     × 7

**8.4**   *Complete.*

   **26.** 12 km = _____ mi
   **27.** 420 in = _____ m
   **28.** $50 \dfrac{\text{mi}}{\text{hr}} = $ _____ $\dfrac{\text{km}}{\text{hr}}$

   **29.** 22 qt = _____ L
   **30.** 18 L = _____ gal
   **31.** 16 lb = _____ g

   **32.** Would water at 65°C feel cool to your hand?

   **33.** If your Italian car has a 54-liter gas tank, how many gallons will it hold?

**8.5**   *Find the perimeter and area of the rectangle or square.*

   **34.** 9 cm by 20 cm
   **35.** $6\frac{1}{2}$ ft by 4.2 ft
   **36.** 3.2 mi by 3.2 mi

*Complete.*

**37.** 25 yd$^2$ = _____ ft$^2$     **38.** 3060 cm$^2$ = _____ m$^2$     **39.** 72 in$^2$ = _____ ft$^2$

**8.6** *Find the area of each figure.*

**40.** 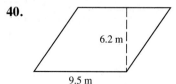     **41.** 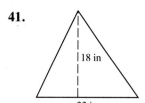     **42.**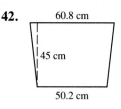

*Find the perimeter of each figure.*

**43.** 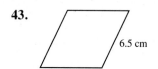     **44.** 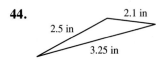     **45.**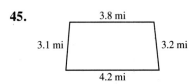

**8.7** *Find the circumference and area of each circle. Use 3.14 for $\pi$ and round answers to the nearest tenth.*

**46.** $r = 6.8$ km     **47.** $d = 22$ in     **48.** $r = 8\frac{1}{2}$ ft

**8.8** *Find the volume of each rectangular solid.*

**49.** 2.2 ft by 1.5 ft by 6.4 ft     **50.** 52 cm by 22 cm by 8 cm

*Find the volume of each cube with the given edge.*

**51.** 16 m     **52.** 8.2 in

*Complete.*

**53.** 3.2 yd$^3$ = _____ ft$^3$     **54.** 4800 cm$^3$ = _____ m$^3$

*Find the surface area of each rectangular solid.*

**55.** 2.2 ft by 1.5 ft by 6.4 ft     **56.** 52 cm by 22 cm by 8 cm

*Find the surface area of each cube with the given edge.*

**57.** 16 m     **58.** 8.2 in

**59.** How many grams of water will a tank 10 cm by 8 cm by 2.5 cm hold?

**60.** How many square centimeters of glass did it take to make the sides and bottom of the tank in Exercise 59?

**8.9** *Find the volume of each cylinder. Use 3.14 for $\pi$.*

**61.** $r = 8$ cm, $h = 12$ cm     **62.** $d = 2.2$ in, $h = 1.5$ in

*Find the surface area of each cylinder.*

**63.** $r = 8$ cm, $h = 12$ cm     **64.** $d = 2.2$ in, $h = 1.5$ in

*Find the volume of each sphere.*

**65.** $r = 10$ ft     **66.** $d = 8.8$ m

*Find the surface area of each sphere.*

**67.** $r = 10$ ft

**68.** $d = 8.8$ m

**69.** A cylindrical tank with $r = 120$ cm and $h = 450$ cm is filled with gasoline. How many liters are in the tank?

**70.** A spherical tank with radius 12 ft is to be insulated. If insulation costs \$0.50 per square foot, how much will it cost to insulate the tank?

**Part II**

*Complete.*

**71.** $0.032$ kg = _____ oz

**72.** $7.8$ ft = _____ cm

**73.** $114°F$ = _____°C

**74.** $6.35$ dg = _____ dag

**75.** $9900$ ml = _____ hl

**76.** $0.0008$ g = _____ cg

**77.** $1$ hr = _____ sec

**78.** $3200$ oz = _____ ton

**79.** $22 \dfrac{\text{ft}}{\text{sec}} = $ _____ $\dfrac{\text{mi}}{\text{hr}}$

**80.** $6$ yd$^2$ = _____ ft$^2$

**81.** $6$ yd$^3$ = _____ ft$^3$

**82.** $2000$ cm$^3$ = _____ m$^3$

**83.** A 12-in diameter pizza costs \$5.00 and a 16-in diameter pizza costs \$8.00. Which pizza costs less per square inch (unit price)?

**84.** An electrician needs 60 yd of wire. What will be the total cost if the wire is 50¢ per foot?

**85.** What will it cost to carpet the area shown if carpet costs \$22.50 per square yard?

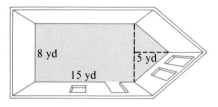

**86.** The area shown is to be carpeted. How much will it cost if carpet is \$19.50 per square yard?

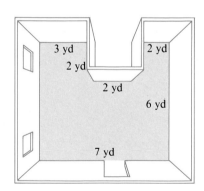

*In Exercises 87–88, find the area shaded. Use 3.14 for $\pi$.*

**87.**

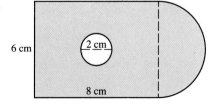

**88.**

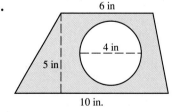

**89.** Find the surface area of the sphere with diameter 10 meters. Use 3.14 for $\pi$.

**90.** Find the volume of a cylinder with radius 3.2 feet and height 7.5 feet. Use 3.14 for $\pi$.

**91.** If it takes 2 lb 4 oz of flour to make a loaf of bread, how much will it take to make 5 loaves?

**92.** The Giant Cattle Ranch is a rectangle 22 mi by 42 mi. What is the area in acres?

*Complete.*

**93.** 456,000 m² = _____ ha

**94.** 35 a = _____ ha

**95.** 3 mi² = _____ acres

---

ANSWERS: (Some answers are rounded.)  1. 84  2. 2.5  3. 3  4. 1.5  5. 6000  6. 2.25  7. 17 ft 5 in
8. 3 ft 1 in  9. 24 hr 18 min  10. 12 hr 26 min  11. 19 yd 1 ft 10 in  12. 5 yd 2 ft 6 in  13. 17 gal 2 qt 1 pt
14. 24 ft²  15. 2 lb 15 oz  16. 6  17. 0.86  18. 58.2  19. 4.83  20. 0.0483  21. 3.5  22. 560  23. 90.5 m
24. 414.7 L  25. 299.6 g  26. 7.452  27. 10.668  28. 80.5  29. 20.8  30. 4.77  31. 7264  32. no (it would feel hot:
65°C = 149°F)  33. 14.31 gal  34. 58 cm; 180 cm²  35. 21.4 ft; 27.3 ft²  36. 12.8 mi; 10.24 mi²  37. 225  38. 0.306
39. 0.5  40. 58.9 m²  41. 198 in²  42. 2497.5 cm²  43. 26 m  44. 7.85 in  45. 14.3 mi  46. 42.7 km; 145.2 km²
47. 69.1 in; 379.9 in²  48. 53.4 ft; 226.9 ft²  49. 21.12 ft³  50. 9152 cm³  51. 4096 m³  52. 551.368 in³  53. 86.4
54. 0.0048  55. 53.96 ft²  56. 3472 cm²  57. 1536 m²  58. 403.44 in²  59. 200 g  60. 170 cm²  61. 2411.52 cm³
62. 5.7 in³  63. 1004.8 cm²  64. 17.96 in²  65. 4186.7 ft³  66. 356.6 mi³  67. 1256 ft²  68. 243.2 m²
69. 20,347.2 L  70. $904.32  71. 1.13  72. 237.7  73. 46  74. 0.0635  75. 0.099  76. 0.08  77. 3600  78. 0.1
79. 15  80. 54  81. 162  82. 0.002  83. 16-in costs less (12-in: $0.044 per square inch; 16-in: $0.040 per square inch)
84. $90  85. $3150  86. $741  87. 58.99 cm²  88. 27.44 in²  89. 314 m²  90. 241.2 ft³ (to the nearest tenth)
91. 11 lb 4 oz  92. 591,360 acres  93. 45.6  94. 0.35  95. 1920

*Complete.*

**1.** 40 oz = _____ lb

1. _____

**2.** 2800 ml = _____ hl

2. _____

**3.** 6.5 kg = _____ lb

3. _____

**4.** 65°C = _____ °F

4. _____

**5.** 3 ft$^2$ = _____ in$^2$

5. _____

**6.** 8 m$^3$ = _____ cm$^3$

6. _____

**7.** Perform the indicated operation and simplify.

7. _____

$$\begin{array}{r} 12 \text{ yd } 2 \text{ ft } 3 \text{ in} \\ - \ \ 3 \text{ yd } 2 \text{ ft } 8 \text{ in} \\ \hline \end{array}$$

*Solve.*

**8.** A liquid weighs 2 g per cubic centimeter. How much would 5 L weigh?

8.

**9.** Find the area of the given figure.

6 ft

4 ft

9.

**10.** Find the perimeter of the given figure.

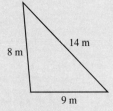

8 m    14 m

9 m

10. _____

**11.** Find the volume of a cylinder with radius 8 cm and height 10 cm.

11. _____

**12.** Find the surface area of a sphere with radius 20 inches.

12. _____

*Solve.*

**13.** Gravel costs $1.50 per cubic meter. How much would a truckload cost if the bed of the truck is 4.2 m by 5.0 m by 2.5 m?

13. _____

**14.** A machine part is in the shape shown in the figure. What is the area of metal left after the two circular holes are drilled out?

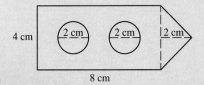

4 cm    2 cm    2 cm    2 cm

8 cm

14. _____

# Introduction to Algebra

## 9.1 INTEGERS AND THE NUMBER LINE

### STUDENT GUIDEPOSTS

1 Integers

2 Equal, Less Than, and Greater Than

3 Absolute Value

In Chapter 1 we used a number line to help picture the whole numbers. The number 0 was located on a line, a unit length was selected, and numbers were marked off to the right of 0 as shown in Figure 9.1.

**Figure 9.1**

### 1 INTEGERS

There are no numbers to the left of 0 in Figure 9.1. What meaning could be given to a number to the left of 0? One example of such a number is a temperature of 5° below zero on a cold day, which is sometimes represented as −5° (read ''negative 5 degrees''). Others are shown in the following table.

| Measurement | Number |
| --- | --- |
| 5° above zero | 5°  (or +5°) |
| 5° below zero | −5° |
| 100 ft above sea level | 100 ft  (or +100 ft) |
| 100 ft below sea level | −100 ft |
| $16 deposit into an account | $16  (or +$16) |
| $16 check written on an account | −$16 |

Putting a minus sign in front of each of the counting numbers gives the **negative integers.** The collection of negative integers together with the counting numbers (sometimes called **positive integers**) and zero is the set of **integers.** The whole numbers are often called **nonnegative integers.** The number line in Figure 9.2 shows negative integers, positive integers, and zero. Note that 1 can be written as +1, 2 as +2, and so on. Also, −1 is read as ''negative one,'' for example.

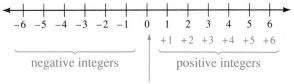

negative integers

positive integers

neither positive nor negative

**Figure 9.2**

## ② EQUAL, LESS THAN, AND GREATER THAN

Two numbers on a number line are **equal** (for example, $3 + 2 = 5$) if they correspond to the same point. One number is **less than** (**<**) a second number if the first is to the left of the second on a number line. The number to the right is said to be **greater than** (**>**) the other.

---

### ┌──────┐ H I N T └──────┘

The point of $<$ or $>$ is always toward the *smaller* number.

---

### EXAMPLE 1 RELATIONSHIPS BETWEEN NUMBERS

Some relationships between numbers on the number line in Figure 9.3 are given below.

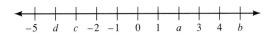

**Figure 9.3**

**(a)** 1 is less than 3      $1 < 3$

**(b)** $a$ is greater than 0      $a > 0$

**(c)** $-2$ is less than 0      $-2 < 0$

**(d)** $-1$ is greater than $d$      $-1 > d$

**(e)** $-5$ is less than 3      $-5 < 3$

**(f)** $b$ is greater than $c$      $b > c$

**(g)** $c$ is equal to $-3$      $c = -3$

### PRACTICE EXERCISE 1

The following relationships are determined by the number line in Figure 9.3. Write each expression using $<$, $>$, or $=$.

**(a)** 3 is greater than 1

**(b)** 0 is less than $a$

**(c)** 0 is greater than $-2$

**(d)** $d$ is less than $-1$

**(e)** 3 is greater than $-5$

**(f)** $c$ is less than $b$

**(g)** $a$ is equal to 2

Answers: (a) $3 > 1$   (b) $0 < a$
(c) $0 > -2$   (d) $d < -1$
(e) $3 > -5$   (f) $c < b$
(g) $a = 2$

## ③ ABSOLUTE VALUE

In order to add, subtract, multiply, and divide integers, we will need to use *absolute value*.

### Absolute Value

The **absolute value** of a number $a$ is the number of units from zero to $a$ on a number line. The absolute value of $a$ is written $|a|$.

| EXAMPLE 2   DETERMINING ABSOLUTE VALUE | PRACTICE EXERCISE 2 |
|---|---|

Use the number line in Figure 9.4 to determine the absolute value of each number.

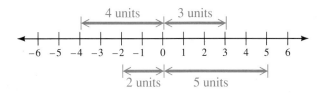

**Figure 9.4**

**(a)** 3 is 3 units from 0. Thus, $|3| = 3$.

**(b)** 5 is 5 units from 0. Thus, $|5| = 5$.

**(c)** 0 is 0 units from 0. Thus, $|0| = 0$.

**(d)** $-2$ is 2 units from 0. Thus, $|-2| = 2$.

**(e)** $-4$ is 4 units from 0. Thus, $|-4| = 4$.

Use the number line in Figure 9.4 to determine the absolute value of each number.

**(a)** $|4|$

**(b)** $|6|$

**(c)** $|-6|$

**(d)** $|-3|$

**(e)** $|-1|$

Answers: **(a)** 4  **(b)** 6  **(c)** 6
**(d)** 3  **(e)** 1

## HINT

Intuitively, the absolute value of a positive number or zero is the number itself. The absolute value of a negative number is the positive number formed by removing the minus sign. Thus, *the absolute value of zero is zero and of any other number is positive, never negative.*

## 9.1 EXERCISES A

*Answer true or false. If the answer is false, tell why.*

**1.** The numbers $-1, -2, -3, -4$, and so on are called negative integers.

**2.** The numbers 1, 2, 3, 4, and so on are called positive integers.

**3.** The numbers, $0, 1, -1, 2, -2, 3, -3$, and so on are called integers.

**4.** The word name for $-8$ is positive eight.

**5.** A $32 check written on your checking account could be represented by $-\$32$.

**6.** $23°$ below zero can be written as $23°$.

**7.** A number is less than another if it is to the left of the other on a number line.

*Place the correct symbol (=, <, or >) between the integers in each pair. (Refer to the given number line if necessary.)*

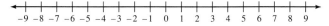

**8.** 3  7          **9.** 7  3          **10.** 0  3          **11.** 0  $-3$          **12.** $-1$  $-3$

**13.** $-3$  6        **14.** 9  $-9$        **15.** $-7$  $-6$        **16.** $-100$  3        **17.** $-5$  $-5$

*Is the given statement true or false? If the answer is false, tell why.*

**18.** $6 > 0$        **19.** $-3 > 0$        **20.** $-8 < -2$        **21.** $-5 = 5$

**22.** $100 > -4$        **23.** $-9 < -1$        **24.** $-8 > 1$        **25.** $-100 > -2$

*Find the absolute values.*

**26.** $|6|$        **27.** $|0|$        **28.** $|-9|$        **29.** $|83|$        **30.** $|-83|$

## FOR REVIEW

*The following exercises review material from Chapter 1 to help prepare you for the next section. Add or subtract the whole numbers.*

**31.** $9 + 7$        **32.** $9 - 7$        **33.** $21 + 13$

**34.** $21 - 13$        **35.** $37 + 0$        **36.** $37 - 37$

ANSWERS:  1. true   2. true   3. true   4. false (negative)   5. true   6. false ($-23°$)   7. true   8. $3 < 7$   9. $7 > 3$
10. $0 < 3$   11. $0 > -3$   12. $-1 > -3$   13. $-3 < 6$   14. $9 > -9$   15. $-7 < -6$   16. $-100 < 3$   17. $-5 = -5$
18. true   19. false ($-3 < 0$)   20. true   21. false ($-5 < 5$)   22. true   23. true   24. false ($-8 < 1$)   25. false
($-100 < -2$)   26. 6   27. 0   28. 9   29. 83   30. 83   31. 16   32. 2   33. 34   34. 8   35. 37   36. 0

## 9.1   EXERCISES B

*Answer true or false. If the answer is false, tell why.*

**1.** The positive integers are the same as the counting numbers.

**2.** The negative integers can be obtained by taking the negatives of the positive integers.

**3.** The integers include the positive integers, the negative integers, and zero.

**4.** The word name for $-3$ is negative three.

**5.** The temperature $18°$ below zero can be written $+18°$.

**6.** A \$10 deposit in a checking account can be written $-\$10$.

**7.** A \$20 withdrawal from a savings account can be written $-\$20$.

*Place the correct symbol ($=$, $<$, or $>$) between the integers in each pair. (Refer to the given number line if necessary.)*

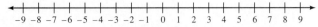

**8.** $4 \quad 8$        **9.** $9 \quad 2$        **10.** $0 \quad 5$        **11.** $0 \quad -2$        **12.** $-2 \quad 4$

**13.** $4 \quad -4$        **14.** $-87 \quad 2$        **15.** $-8 \quad -8$        **16.** $4 \quad 0$        **17.** $-4 \quad 0$

*Is the given statement true or false? If the answer is false, tell why.*

**18.** $6 < -3$        **19.** $5 > 0$        **20.** $-5 > 0$        **21.** $-10 < 10$

**22.** $-7 > 1$        **23.** $0 < 10$        **24.** $-9 = 9$        **25.** $-9 = -9$

*Find the absolute values.*

**26.** $|8|$          **27.** $|-0|$          **28.** $|-7|$          **29.** $|-75|$          **30.** $|75|$

## FOR REVIEW

*The following exercises review material from Chapter 1 to help prepare you for the next section. Add or subtract the whole numbers.*

**31.** $8 + 3$          **32.** $8 - 3$          **33.** $32 + 15$

**34.** $32 - 15$          **35.** $29 + 0$          **36.** $29 - 29$

## 9.1 EXERCISES C

**1.** If $a$ is a positive integer, is $-a$ positive or negative?

**2.** If $a$ is a negative integer, is $-a$ positive or negative?

**3.** For what integer $a$ is $a = -a$?

**4.** For what integers $a$ is $|a| = a$?
[Answer: $a$ is nonnegative]

## 9.2 ADDING INTEGERS

> ### STUDENT GUIDEPOSTS
> **1** Adding on a Number Line     **3** Adding Integers with Unlike Signs
> **2** Adding Integers with Like Signs     **4** Adding More Than Two Integers

### **1** ADDING ON A NUMBER LINE

We saw how to add nonnegative integers (whole numbers) in Chapter 1. We shall review this process as we learn how to add any two integers. Consider the addition problem $3 + 2$ on the number line in Figure 9.5.

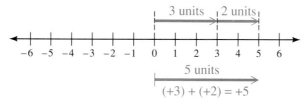

**Figure 9.5**

To find $3 + 2$, start at 0, move 3 units to the *right,* and then move 2 more units to the *right*. We are then 5 units to the *right* of zero. Thus, $3 + 2 = +5$ (or 5).

    We can also find $3 + (-2)$ on a number line. Start at 0, move 3 units to the *right,* and then 2 units to the *left*. This puts us 1 unit to the *right* of 0, as shown in Figure 9.6. Thus, $3 + (-2) = +1$ (or 1).

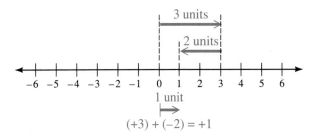

**Figure 9.6**

When adding integers using a number line, move to the left when a number is negative and to the right when it is positive.

| EXAMPLE 1   ADDING INTEGERS WITH LIKE SIGNS | PRACTICE EXERCISE 1 |

Find $(-3) + (-2)$.

Draw a number line. Start at 0, move 3 units to the left for $-3$, and then 2 more units to the left for $-2$. As seen in Figure 9.7, we end up 5 units to the left of 0, at $-5$. Thus, $(-3) + (-2) = -5$.

Find $(-6) + (-4)$.

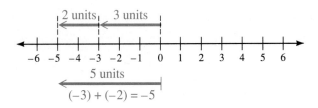

**Figure 9.7**

**Answer:** $-10$

| EXAMPLE 2   ADDING INTEGERS WITH UNLIKE SIGNS | PRACTICE EXERCISE 2 |

Find $(-3) + 2$.

Draw a number line. Start at 0, move 3 units to the left, and then 2 units to the right. As Figure 9.8 shows, we are then at $-1$. Thus, $(-3) + 2 = -1$.

Find $(-6) + 4$.

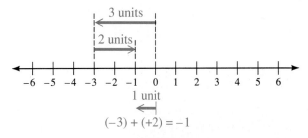

**Figure 9.8**

**Answer:** $-2$

| EXAMPLE 3   ADDING INTEGERS WITH UNLIKE SIGNS | PRACTICE EXERCISE 3 |

Find $2 + (-3)$.

Draw a number line. Start at 0, move 2 units to the right, and then 3 units to the left. (See Figure 9.9.) This puts us at $-1$. Thus, $2 + (-3) = -1$.

Find $4 + (-6)$.

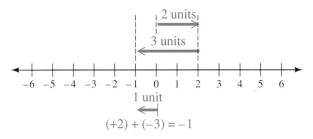

**Figure 9.9**                    Answer:  $-2$

## ② ADDING INTEGERS WITH LIKE SIGNS

The number-line method of adding numbers shows us what addition means, but it takes too much time. For example, we found $(-3) + (-2)$ to be $-5$ using a number line (Figure 9.7). However, we could have added $3 + 2$ and attached a minus sign:

$$(-3) + (-2) = -(3 + 2) = -5.$$

This is an example of the following rule.

| Adding Integers with Like Signs |
|---|

To add two integers with like signs:

1. Add their absolute values, ignoring the signs.
2. The sum has the same sign as the numbers being added.

## ③ ADDING INTEGERS WITH UNLIKE SIGNS

When the signs of the numbers being added are not the same, we need another rule. For $3 + (-2) = 1$ (Figure 9.6), we could have found the difference between 3 and 2 and attached a plus sign:

$$3 + (-2) = +(3 - 2) = +1 = 1.$$

Also, for $(-3) + 2 = 2 + (-3) = -1$ (Figures 9.8 and 9.9), we could have subtracted 2 from 3 and attached a minus sign:

$$(-3) + 2 = 2 + (-3) = -(3 - 2) = -1.$$

These are examples of the following rule.

| Adding Integers with Unlike Signs |
|---|

To add two integers with unlike signs:

1. Ignore the signs and subtract the smaller absolute value from the larger absolute value.
2. The result has the same sign as the number with the larger absolute value.
3. If the absolute values are the same, the sum is zero.

## HINT

Remember that if a number (other than zero) has no sign this is the same as having a plus sign attached. For example, 5 and $+5$ are the same.

---

### EXAMPLE 4 ADDITION USING THE RULES

Add.

**(a)** $5 + 8 = 13$

**(b)** $(-5) + (-8) = -(5 + 8) = -13$

**(c)** $5 + (-8) = -(8 - 5) = -3$      $|-8| = 8 > 5 = |5|$

**(d)** $(-5) + 8 = +(8 - 5) = +3 = 3$      $|-5| = 5 < 8 = |8|$

**(e)** $7 + (-7) = 7 - 7 = 0$      $|7| = 7 = |-7|$

**(f)** $(-6) + 0 = -(6 - 0) = -6$

**(g)** $0 + 4 = 4$

**(h)** $(-3) + (-2) + (-17) = -(3 + 2 + 17) = -22$

### PRACTICE EXERCISE 4

Add.

**(a)** $9 + 7$

**(b)** $(-9) + (-7)$

**(c)** $9 + (-7)$

**(d)** $(-9) + 7$

**(e)** $6 + (-6)$

**(f)** $(-8) + 0$

**(g)** $0 + 8$

**(h)** $(-4) + (-5) + (-11)$

Answers: (a) 16   (b) $-16$
(c) 2   (d) $-2$   (e) 0   (f) $-8$
(g) 8   (h) $-20$

---

## ❹ ADDING MORE THAN TWO INTEGERS

Taken together, the commutative and associative laws of addition allow us to add several signed numbers by reordering and regrouping the numbers to obtain the sum in the most convenient manner.

### Adding More Than Two Integers with Mixed Signs

1. Add all the positive integers.
2. Add all the negative integers.
3. Add the resulting pair of integers (one positive and one negative) using the rule above for adding two numbers with unlike signs.

---

### EXAMPLE 5 ADDING MORE THAN TWO INTEGERS

Add.

$$(-3) + 6 + 8 + (-4) + (-7) + (-3) + 9$$
$$= [6 + 8 + 9] + [(-3) + (-4) + (-7) + (-3)]$$
$$= [6 + 8 + 9] + [-(3 + 4 + 7 + 3)]$$
$$= [23] + [-17]$$
$$= 23 - 17 = +6 = 6$$

### PRACTICE EXERCISE 5

Add.

$$(-8) + 7 + 5 + (-4) + 12 + (-6) + (-10)$$

Answer: $-4$

### EXAMPLE 6 ADDITION USING COLUMNS

Add the following column of numbers.

| | Positive Integers | Negative Integers | Sum |
|---|---|---|---|
| −321 | 201 | −321 | −517 |
| 201 | 64 | − 78 | 404 |
| 64 | 123 | −118 | −113 |
| − 78 | 16 | −517 | |
| 123 | 404 | | |
| −118 | | | |
| 16 | | | |
| −113 | | | |

Thus, from the last addition, −113 is the answer.

### PRACTICE EXERCISE 6

Add the following column of numbers.

$$
\begin{array}{r}
218 \\
-112 \\
-108 \\
72 \\
18 \\
-332 \\
-989 \\
804 \\
\hline
\end{array}
$$

Answer: −429

### EXAMPLE 7 APPLICATION OF ADDITION

John had $932 in his bank account on May 1. He deposited $326 and $791 during the month and wrote checks for $816, $315, and $940 during May. What was his balance at the end of the month?

The deposits can be considered as positive numbers and the checks written as negative numbers. Thus, add 932, 326, 791, −816, −315, and −940.

| 932 | − 816 | |
|---|---|---|
| 326 | − 315 | −2071 |
| 791 | − 940 | 2049 |
| 2049 Sum of deposits | −2071 Sum of checks | − 22 Final sum |

John is overdrawn (in the red) $22 on his account.

### PRACTICE EXERCISE 7

Rhonda was in a contest. She started the day with 950 points and lost 120 points, 215 points, and 445 points. Then she won 65 points, 515 points, and 140 points. How many points did she have at the end of the day?

Answer: 890

## 9.2 EXERCISES A

*Perform the indicated operations.*

**1.** 4 + 3

**2.** (−4) + (−3)

**3.** 4 + (−3)

**4.** (−4) + 3

**5.** 4 + 0

**6.** 0 + 4

**7.** (−4) + 0

**8.** 0 + (−4)

**9.** 7 + (−7)

**10.** (−7) + 7

**11.** (−7) + (−7)

**12.** (−8) + (−5)

**13.** 9 + (−1)

**14.** (−1) + 9

**15.** 14 + 21

**16.** 19 + (−25)

**17.** (−16) + 14

**18.** (−4) + (+3)

**19.** $(-3) + 2 + (-7) + (-8) + 9$

**20.** $5 + 8 + 9 + (-3) + (-7)$

**21.** $(-8) + (-6) + 9 + (-3) + 5 + (-6)$

**22.** $(-4) + (-2) + 6 + 7 + (-1)$

*Add.*

**23.**
```
  23
 -16
 -36
   5
  81
```

**24.**
```
 -16
 -81
 -14
 -70
 -93
```

**25.**
```
 -64
  25
  38
 -17
 -83
```

**26.** On March 1, Allen had $64 in his bank account. During the month, he wrote checks for $81, $108, and $192 and made deposits of $123 and $140. What was the value of his account at the end of the month?

**27** At 5:30 A.M., it was $-42°$F. By 2:00 P.M. the same day, the temperature had risen $57°$ to the recorded high temperature. Shortly thereafter, a cold front passed through dropping the temperature $41°$ by 5:00 P.M. What was the temperature at 5:00 P.M.?

## FOR REVIEW

*Place the correct symbol (=, <, or >) between the integers in each pair.*

**28.** $-5 \quad -4$

**29.** $0 \quad -6$

**30.** $9 \quad -5$

*Find the absolute value.*

**31.** $|6|$

**32.** $|-8|$

**33.** $|-24|$

---

ANSWERS: 1. 7  2. $-7$  3. 1  4. $-1$  5. 4  6. 4  7. $-4$  8. $-4$  9. 0  10. 0  11. $-14$  12. $-13$  13. 8  14. 8  15. 35  16. $-6$  17. $-2$  18. $-1$  19. $-7$  20. 12  21. $-9$  22. 6  23. 57  24. $-274$  25. $-101$  26. $-54$ dollars  27. $-26°$  28. $-5 < -4$  29. $0 > -6$  30. $9 > -5$  31. 6  32. 8  33. 24

## 9.2 EXERCISES B

*Perform the indicated operations.*

**1.** $5 + 2$

**2.** $(-5) + (-2)$

**3.** $5 + (-2)$

**4.** $(-5) + 2$

**5.** $5 + 0$

**6.** $0 + 5$

**7.** $(-5) + 0$

**8.** $0 + (-5)$

**9.** $5 + (-5)$

**10.** $(-5) + 5$

**11.** $(-5) + (-5)$

**12.** $(-9) + (-3)$

**13.** $11 + (-2)$

**14.** $(-2) + 11$

**15.** $15 + 25$

**16.** $18 + (-23)$

**17.** $(-15) + 19$

**18.** $(-8) + (+7)$

**19.** $(-5) + 4 + (-9) + (-10) + 6$

**20.** $5 + 9 + (-3) + 1 + (-10)$

**21.** $(-5) + (-7) + 10 + (-3) + 4 + (-7)$

**22.** $(-6) + (-3) + 6 + 10 + (-2)$

*Add.*

**23.**    26
 −14
 −51
  3
  92

**24.** − 17
 − 20
 − 43
 −  5
 −110

**25.** −78
  32
  51
 −60
 −92

**26.** On July 4, Uncle Sam had $97 in his checking account. During the next week he wrote checks for $15, $102, and $312, and made deposits of $145 and $170. What was the value of his account at the end of the week?

**27.** At 8:00 A.M., Leah Johnson had a temperature of 99.8°. It rose another 3.1° before falling 4.3° by noon. What was her temperature at noon?

## FOR REVIEW

*Place the correct symbol (=, <, or >) between the integers in each pair.*

**28.** −6 −7

**29.** 0 −10

**30.** 11 −2

*Find the absolute value.*

**31.** $|-19|$

**32.** $|42|$

**33.** $|-98|$

## 9.2 EXERCISES C

*To display a negative number on a calculator use the change of sign key (usually* $\boxed{+/-}$ *). For example, to display −5, first enter* $\boxed{5}$ *then press* $\boxed{+/-}$ *to change the sign to −5. Perform the following calculations on a calculator.*

**1.** −19 + 7
[Answer: −12]

**2.** −123 + 341

**3.** −451 + 127

**4.** −19 + (−7)
[Answer: −26]

**5.** −123 + (−341)

**6.** −451 + (−127)

## 9.3 SUBTRACTING INTEGERS

### STUDENT GUIDEPOSTS

**1** Defining Subtraction Using Addition

**2** Using Addition and Subtraction Together

## **1** DEFINING SUBTRACTION USING ADDITION

Consider the subtraction problem

$$6 - 2 = \text{what number?}$$

We solved this problem in Chapter 1 by changing to the related addition sentence

$$2 + (\text{what number}) = 6?$$

Since $2 + 4 = 6$, the number is 4 so that $6 - 2 = 4$. Now consider the problem

$$2 - 6 = \text{what number?}$$

In a similar manner, we are really asking

$$6 + (\text{what number}) = 2?$$

The desired number must be negative, and in fact,

$$6 + (-4) = 2.$$

Thus, $2 - 6 = -4$. Notice that $2 + (-6)$ is also $-4$, so

$$2 - 6 = 2 + (-6).$$

As a result, we can define subtraction of integers by using addition. This is illustrated further in the following examples where we change the sign of the number being subtracted, the *subtrahend,* and then add using the techniques of Section 9.2.

$$5 - (+3) = 5 + (-3) \qquad \text{Change +3 to −3 and subtraction to addition}$$
$$= 5 - 3 = 2 \qquad \text{Add}$$
$$3 - (+5) = 3 + (-5) \qquad \text{Change +5 to −5 and subtraction to addition}$$
$$= -(5 - 3) = -2 \qquad \text{Add}$$
$$5 - (-3) = 5 + (+3) \qquad \text{Change −3 to +3 and subtraction to addition}$$
$$= 5 + 3 = 8 \qquad \text{Add}$$
$$-5 - (+3) = -5 + (-3) \qquad \text{Change +3 to −3 and subtraction to addition}$$
$$= -(5 + 3) = -8 \qquad \text{Add}$$
$$-5 - (-3) = -5 + (+3) \qquad \text{Change −3 to +3 and subtraction to addition}$$
$$= -(5 - 3) = -2 \qquad \text{Add}$$

This leads to the following rule.

---

### Subtracting Integers

1. Change the sign of the *subtrahend* and change subtraction to addition.
2. Use the rules for addition in Section 9.2 to add.

---

**EXAMPLE 1  SUBTRACTING INTEGERS**

Subtract.

**(a)** $7 - (+4) = 7 + (-4)$     Change +4 to −4 and add
$$= 7 - 4 = 3$$

**(b)** $7 - (-4) = 7 + 4 = 11$     Change −4 to 4 and add

**(c)** $(-7) - (+4) = (-7) + (-4)$     Change sign and add
$$= -(7 + 4) = -11$$

**(d)** $(-7) - (-4) = (-7) + 4$     Change sign and add
$$= -(7 - 4) = -3$$

**(e)** $9 - 9 = 9 + (-9) = 0$

**(f)** $9 - (-9) = 9 + 9 = 18$

**PRACTICE EXERCISE 1**

Subtract.

**(a)** $11 - (+2)$

**(b)** $11 - (-2)$

**(c)** $(-11) - (+2)$

**(d)** $(-11) - (-2)$

**(e)** $15 - 15$

**(f)** $15 - (-15)$

**(g)** $-15 - 15$

**(h)** $-15 - (-15)$

**(g)** $(-9) - 9 = (-9) + (-9) = -(9 + 9) = -18$
**(h)** $(-9) - (-9) = (-9) + 9 = 0$

## ② USING ADDITION AND SUBTRACTION TOGETHER

Many problems involve both addition and subtraction. Our procedure will be to change each subtraction to addition and carry out the addition as we did in Section 9.2. For example,

$$2 - (+3) + 6 = 2 + (-3) + 6$$
$$= (2 + 6) + (-3)$$
$$= 8 + (-3) = 5.$$

Notice that once the subtraction was changed to addition, the problem was one like we had done in the previous section.

---

| EXAMPLE 2  PROBLEMS WITH BOTH OPERATIONS |
| --- |

Perform the indicated operations.

**(a)** $7 + (-3) - (+2)$

First change the problem completely to an addition problem and then add.

$$7 + (-3) - (+2) = 7 + (-3) + (-2)$$
$$= 7 + [(-3) + (-2)]$$
$$= 7 + (-5)$$
$$= 7 - 5 = 2$$

**(b)** $(-3) - (+2) - (5) - (-6)$

Here we must change three subtractions to addition. Remember that $5 = +5$.

$$(-3) - (+2) - (5) - (-6) = -3 + (-2) + (-5) + (+6)$$
$$= 6 + [(-3) + (-2) + (-5)]$$
$$= 6 + [-10]$$
$$= -(10 - 6) = -4$$

| PRACTICE EXERCISE 2 |
| --- |

Perform the indicated operations.

**(a)** $5 + (-6) - (+7)$

**(b)** $(-2) - (-5) + (+2) - (7)$

---

## 9.3 EXERCISES A

*Perform the indicated operations.*

**1.** $4 - (+3)$

**2.** $4 - 3$

**3.** $-4 - (-3)$

**4.** $4 - (-3)$

**5.** $(-4) - 3$

**6.** $4 - 0$

**7.** $0 - 4$

**8.** $(-4) - 0$

**9.** $0 - (-4)$

**10.** $7 - (-7)$

**11.** $(-7) - 7$

**12.** $(-7) - (-7)$

**13.** $(-8) - (-5)$          **14.** $9 - (-1)$          **15.** $(-1) - 9$

**16.** $14 - 21$          **17.** $19 - (-25)$          **18.** $(-16) - 14$

**19.** $-8 + (-3)$          **20.** $8 + (-3)$          **21.** $-8 + 3$

**22.** $-9 - (-2) + (-3)$          **23.** $10 - (-2) - (-8)$          **24.** $-6 + (+2) - (+17)$

**25.** $16 + (-3) - (-2) - 7$          **26.** $-21 + 8 + (-6) - (-10)$          **27.** $-40 + (-20) - (30) - (-10)$

**28.** $0 - 5 + 5 - (-5)$          **29.** $62 - (-88) + (-90) - (-24)$          **30.** $-57 - 63 - (-22) - 18$

**31** When a cold front came through Cut Bank, Montana, the temperature dropped from 35° above zero to 13° below zero $(-13°)$. What was the total change in temperature?

**32.** A helicopter is 600 ft above sea level and a submarine directly below it is 250 ft below sea level $(-250$ ft$)$. How far apart are they?

## FOR REVIEW

*The following exercises review material from Chapter 2 to help prepare you for the next section. Perform the indicated operations.*

**33.** $16 \times 11$          **34.** $(23)(100)$          **35.** $(42)(68)$

**36.** $75 \div 5$          **37.** $360 \div 10$          **38.** $88 \div 22$

ANSWERS: 1. 1  2. 1  3. −1  4. 7  5. −7  6. 4  7. −4  8. −4  9. 4  10. 14  11. −14  12. 0  13. −3  14. 10  15. −10  16. −7  17. 44  18. −30  19. −11  20. 5  21. −5  22. −10  23. 20  24. −21  25. 8  26. −9  27. −80  28. 5  29. 84  30. −116  31. −48°  32. 850 ft  33. 176  34. 2300  35. 2856  36. 15  37. 36  38. 4

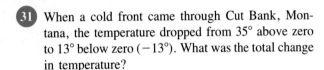

## 9.3 EXERCISES B

*Perform the indicated operations.*

**1.** $8 - (+7)$          **2.** $8 - 7$          **3.** $-8 - (-7)$

**4.** $8 - (-7)$          **5.** $(-8) - 7$          **6.** $10 - 0$

**7.** $0 - 10$          **8.** $(-10) - 0$          **9.** $0 - (-10)$

**10.** $6 - (-6)$          **11.** $(-6) - 6$          **12.** $(-6) - (-6)$

**13.** $(-11) - (-4)$          **14.** $(-13) - (-1)$          **15.** $(-1) - 7$

**16.** $(-12) - 24$          **17.** $18 - (-27)$          **18.** $(-17) - 11$

**19.** $-10 + (-21)$          **20.** $16 + (-10)$          **21.** $-17 + 10$

**22.** $-11 - (-5) + (-8)$          **23.** $14 - (-6) - (-12)$          **24.** $-9 + (+4) - (+19)$

**25.** $4 + (-10) - (-5) - 8$          **26.** $-16 - 7 + (-4) + (-1)$          **27.** $-10 - (-10) - 50 - (-60)$

**28.** $14 - 0 + 14 - (-14)$    **29.** $-58 - (-28) + (-92) - (-88)$    **30.** $44 - 68 - 48 - (-89)$

**31.** In a five-hour period of time, the temperature in Empire, N.Y. dropped from 41° to 12° below zero. What was the total temperature change?

**32.** A balloon is 1200 ft above sea level, and a diving bell directly below it is 340 ft beneath the surface of the water. How far apart are the two?

## FOR REVIEW

*The following exercises review material from Chapter 2 to help prepare you for the next section. Perform the indicated operations.*

**33.** $25 \times 22$    **34.** $(48)(10)$    **35.** $(16)(85)$

**36.** $44 \div 4$    **37.** $2200 \div 100$    **38.** $51 \div 17$

## 9.3 EXERCISES C

*Perform the indicated operations.*

**1.** $(-962) - (-508)$    **2.** $-4862 - 40,002$    **3.** $429 - (-198) + (-375)$
[Answer: 252]

## 9.4 MULTIPLYING AND DIVIDING INTEGERS

### STUDENT GUIDEPOSTS

**1** Multiplying Two Integers
**2** Dividing Two Integers
**3** Multiplying More Than Two Integers

### 1 MULTIPLYING TWO INTEGERS

We learned how to multiply whole numbers in Chapter 2. To learn to multiply integers, we need to decide what to do about the signs. Look at the following products and try to see a pattern.

Decreases by 1 each time ⟶    Decreases by 2 each time

$$4 \cdot 2 = 8$$
$$3 \cdot 2 = 6 \qquad \text{6 is 2 less than 8}$$
$$2 \cdot 2 = 4$$
$$1 \cdot 2 = 2$$
$$0 \cdot 2 = 0$$
$$(-1) \cdot 2 = -2 \qquad \text{$-2$ is 2 less than 0}$$
$$(-2) \cdot 2 = -4 \qquad \text{$-4$ is 2 less than $-2$}$$
$$(-3) \cdot 2 = -6$$
$$(-4) \cdot 2 = -8$$

We know the products $4 \cdot 2$, $3 \cdot 2$, $2 \cdot 2$, $1 \cdot 2$, and $0 \cdot 2$, and we can see that these products decrease by 2 each time. For this pattern to continue, a negative number times a positive number must be negative. For example,

$$(-1) \cdot 2 = -2, \quad (-2) \cdot 2 = -4,$$
$$(-3) \cdot 2 = -6, \quad (-4) \cdot 2 = -8.$$

Similarly, a positive number times a negative number must be negative. We use this to find a pattern in the following products.

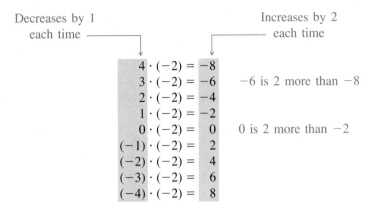

Decreases by 1 each time ⟶

Increases by 2 each time ⟶

$$4 \cdot (-2) = -8$$
$$3 \cdot (-2) = -6 \qquad \text{-6 is 2 more than -8}$$
$$2 \cdot (-2) = -4$$
$$1 \cdot (-2) = -2$$
$$0 \cdot (-2) = 0 \qquad \text{0 is 2 more than -2}$$
$$(-1) \cdot (-2) = 2$$
$$(-2) \cdot (-2) = 4$$
$$(-3) \cdot (-2) = 6$$
$$(-4) \cdot (-2) = 8$$

For this pattern to continue, a negative number times a negative number must be a positive number. For example,

$$(-1) \cdot (-2) = 2, \qquad (-2) \cdot (-2) = 4,$$
$$(-3) \cdot (-2) = 6, \qquad (-4) \cdot (-2) = 8.$$

These results are summarized in the next rule.

### To Multiply Two Integers

1. Multiply the absolute values of the integers.
2. If the integers have like signs, the product is positive.
3. If the integers have unlike signs, the product is negative.
4. If one number (or both numbers) is zero, the product is zero.

---

### EXAMPLE 1   MULTIPLYING INTEGERS

Multiply.

**(a)** $(+5) \cdot (+2) = 5 \cdot 2 = +10 = 10$

Like signs

**(b)** $(-5) \cdot (-2) = +(5 \cdot 2) = +10 = 10$

Like signs

**(c)** $(+5) \cdot (-2) = 5 \cdot (-2) = -(5 \cdot 2) = -10$   Unlike signs

**(d)** $(-5) \cdot (+2) = (-5) \cdot 2 = -(5 \cdot 2) = -10$   Unlike signs

**(e)** $0 \cdot (-8) = 0$

**(f)** $0 \cdot 0 = 0$

**(g)** $(-8) \cdot (-10) = 80$

**(h)** $(-8) \cdot 10 = -80$

### PRACTICE EXERCISE 1

Multiply.

**(a)** $(9) \cdot (7)$

**(b)** $(-9) \cdot (-7)$

**(c)** $(9) \cdot (-7)$

**(d)** $(-9) \cdot (7)$

**(e)** $0 \cdot (-10)$

**(f)** $(-12) \cdot 0$

**(g)** $(-12) \cdot (-20)$

**(h)** $(-12) \cdot 20$

Answers: **(a)** 63   **(b)** 63
**(c)** −63   **(d)** −63   **(e)** 0   **(f)** 0
**(g)** 240   **(h)** −240

---

### HINT

Remember, when multiplying two integers, that *like signs have a positive product and unlike signs have a negative product.*

| | |
|---|---|
| **EXAMPLE 2**  APPLICATION OF MULTIPLICATION | **PRACTICE EXERCISE 2** |

Willie wrote 3 checks for $75 each and 4 checks for $105 each. By how much did this change his bank balance?

The 3 checks for $75 could be calculated as follows:

$$(3)(-75) = -225.$$

There are 4 checks for $105, so

$$(4)(-105) = -420.$$

Thus, his account is changed by

$$-225 + (-420) = -225 - 420 = -645 \text{ dollars.}$$

His balance is $645 less.

Monica withdrew two $50 bills and five $100 bills from her savings account. By how much did this change her account?

Answer: −$600

## ② DIVIDING TWO INTEGERS

We have seen that the only difference between multiplying whole numbers and multiplying integers is finding the sign of the answer. Earlier, division was defined in terms of multiplication; for example,

$$\frac{15}{3} = 5 \quad \text{because} \quad 3 \cdot 5 = 15.$$

Therefore, we can use the same rules for signs when dividing integers that we used for multiplication. For example,

$$\frac{15}{-3} = -5 \quad \text{because} \quad (-3) \cdot (-5) = 15.$$

### To Divide Two Integers

1. Divide the absolute values of the integers.
2. If the integers have like signs, the quotient is positive.
3. If the integers have unlike signs, the quotient is negative.
4. Zero divided by any number (except zero) is zero, and division of any number by zero is undefined.

| | |
|---|---|
| **EXAMPLE 3**  DIVIDING INTEGERS | **PRACTICE EXERCISE 3** |

Divide.

**(a)** $(+6) \div (+3) = +(6 \div 3) = 2$  Like signs

**(b)** $(-6) \div (-3) = +(6 \div 3) = 2$  Like signs

**(c)** $(+6) \div (-3) = 6 \div (-3) = -(6 \div 3) = -2$  Unlike signs

**(d)** $(-6) \div (+3) = (-6) \div 3 = -(6 \div 3) = -2$  Unlike signs

**(e)** $0 \div (-3) = 0$

**(f)** $(-3) \div 0$ is undefined

Divide.

**(a)** $15 \div 5$

**(b)** $(-15) \div (-5)$

**(c)** $15 \div (-5)$

**(d)** $(-15) \div 5$

**(e)** $0 \div (-7)$

**(f)** $(-7) \div 0$

**(g)** $(-88) \div (-8) = 88 \div 8 = 11$

**(h)** $(88) \div (-8) = -(88 \div 8) = -11$

**(g)** $(-110) \div (-11)$

**(h)** $(110) \div (-11)$

Answers: (a) 3    (b) 3    (c) −3
(d) −3    (e) 0    (f) undefined
(g) 10    (h) −10

---

## HINT

Remember that the rule of signs for division is the same as for multiplication: *like signs have a positive quotient and unlike signs have a negative quotient.*

---

## EXAMPLE 4    APPLICATION OF DIVISION

Mary, Sue, and Maria share an apartment with total rent of $720. By how much is each woman's account changed if they each write a check for an equal share of the rent?

Since the rent is a decrease in their accounts, use −720. To find the shares, divide −720 by 3.

$$-720 \div 3 = -240$$

Thus, each woman's account is decreased by $240.

### PRACTICE EXERCISE 4

Four men agree to share equally the expenses for a fishing trip. If the total expenses are $1120, by how much will each man's checking account be changed?

Answer:  −$280

---

## ❸ MULTIPLYING MORE THAN TWO INTEGERS

When multiplying more than two integers, multiply in pairs and keep track of the signs.

## EXAMPLE 5    MULTIPLYING SEVERAL INTEGERS

Multiply.

**(a)** $(3)(-2)(-5) = (-6)(-5) = 30$        $(3)(-2) = -6$

**(b)** $(-1)(-1)(-2)(-3)(-4) = (1)(-2)(-3)(-4)$        $(-1)(-1) = 1$

$$= (-2)(-3)(-4)$$        $(1)(-2) = -2$

$$= (6)(-4)$$        $(-2)(-3) = 6$

$$= -24$$

**(c)** $(-1)(-1)(-1) = (1)(-1) = -1$

**(d)** $(-1)(-1)(-1)(-1) = (1)(-1)(-1) = (-1)(-1) = 1$

### PRACTICE EXERCISE 5

Multiply.

**(a)** $(4)(2)(-3)(-1)$

**(b)** $(-2)(3)(-5)(-6)(2)$

**(c)** $(-2)(-2)(-2)(-2)$

**(d)** $(-2)(-2)(-2)(-2)(-2)$

Answers: (a) 24    (b) −360
(c) 16    (d) −32

---

## HINT

Notice that products involving an odd number of minus signs are negative while those with an even number are positive.

## 9.4 EXERCISES A

*Perform the indicated operations.*

**1.** $(2)(6)$

**2.** $(-2)(-6)$

**3.** $(2)(-6)$

**4.** $(-2)(6)$

**5.** $(0)(-6)$

**6.** $(-6)(0)$

**7.** $(-5) \cdot 7$

**8.** $(-3)(-9)$

**9.** $(10)(-4)$

**10.** $(-10)(8)$

**11.** $(-10)(-12)$

**12.** $(-9)(-12)$

**13.** $(-9)(12)$

**14.** $(9)(-12)$

**15.** $(12)(-12)$

**16.** $(-12)(-12)$

**17.** $(-15) \cdot 20$

**18.** $(-6) \div (-2)$

**19.** $6 \div (-2)$

**20.** $(-6) \div 2$

**21.** $0 \div (-6)$

**22.** $(-6) \div 0$

**23.** $(-8) \div 2$

**24.** $(-10) \div 5$

**25.** $(-12) \div (-4)$

**26.** $\dfrac{99}{-11}$

**27.** $\dfrac{-36}{-4}$

**28.** $\dfrac{-12}{-12}$

**29.** $\dfrac{-12}{12}$

**30.** $\dfrac{0}{-38}$

**31.** $\dfrac{-100}{5}$

**32.** $\dfrac{-48}{-16}$

**33.** $\dfrac{52}{-13}$

**34.** $(-1)(2)(-3)$

**35.** $(-1)(-2)(-3)(-4)$

**36.** $(-1)(-2)(-3)(4)(-5)(6)(-1)$

**37** Tony had to write four $12 checks during the month of June. By how much did this change his bank balance?

**38.** Martha had 423 points in a contest, but then she received 20 penalty points 3 times. What was her total after the 3 penalties?

## FOR REVIEW

*Perform the indicated operations.*

**39.** $(-12) + 9$

**40.** $(-12) + (-9)$

**41.** $15 + (-8)$

**42.** $(-12) - 9$

**43.** $(-12) - (-9)$

**44.** $15 - (-8)$

---

ANSWERS: 1. 12  2. 12  3. −12  4. −12  5. 0  6. 0  7. −35  8. 27  9. −40  10. −80  11. 120  12. 108
13. −108  14. −108  15. −144  16. 144  17. −300  18. 3  19. −3  20. −3  21. 0  22. undefined  23. −4
24. −2  25. 3  26. −9  27. 9  28. 1  29. −1  30. 0  31. −20  32. 3  33. −4  34. 6  35. 24  36. −720
37. −48 dollars  38. 363  39. −3  40. −21  41. 7  42. −21  43. −3  44. 23

## 9.4 EXERCISES B

*Perform the indicated operations.*

**1.** (5)(9)

**2.** (−5)(−9)

**3.** (5)(−9)

**4.** (−5)(9)

**5.** (0)(−7)

**6.** (−7)(0)

**7.** (−3)(8)

**8.** (−2)(−7)

**9.** (12)(−2)

**10.** (−12)(2)

**11.** (−12)(−11)

**12.** (−6)(−13)

**13.** (−6)(13)

**14.** (6)(−13)

**15.** (11)(−11)

**16.** (−11)(−11)

**17.** (−15)(30)

**18.** (−9) ÷ (−3)

**19.** 9 ÷ (−3)

**20.** (−9) ÷ 3

**21.** 0 ÷ (−5)

**22.** (−5) ÷ 0

**23.** (−12) ÷ 4

**24.** (−15) ÷ 3

**25.** (−16) ÷ (−8)

**26.** $\dfrac{88}{-11}$

**27.** $\dfrac{-24}{-8}$

**28.** $\dfrac{-15}{-15}$

**29.** $\dfrac{-15}{15}$

**30.** $\dfrac{0}{-25}$

**31.** $\dfrac{-100}{10}$

**32.** $\dfrac{-56}{-8}$

**33.** $\dfrac{52}{-26}$

**34.** (−1)(−2)(−3)

**35.** (−1)(2)(−3)(4)

**36.** (−1)(−2)(3)(−1)(5)(−4)

**37.** During each of 5 consecutive hours, the temperature dropped 7°. By how much did this change the temperature?

**38.** Hank Anderson received 330 points for a performance, but then was penalized 25 points each for four rule infractions. What was his point total after the penalties were imposed?

## FOR REVIEW

*Perform the indicated operations.*

**39.** (−17) + 6

**40.** (−17) + (−6)

**41.** 17 + (−6)

**42.** (−17) − 6

**43.** (−17) − (−6)

**44.** 17 − (−6)

## 9.4 EXERCISES C

*Use a calculator in Exercises 1–3.*

**1.** (−421)(235)

**2.** (−338)(−127)

**3.** (−1771) ÷ (253)

**4.** What number must be multiplied by −5 to obtain −35?

[Answer: 7]

**5.** What number must be divided by −5 to obtain −35?

# 9.5 RATIONAL NUMBERS

## **1** RATIONAL NUMBERS

When dividing integers in Section 9.4, we did not find quotients such as

$$\frac{3}{2}, \quad \frac{-12}{17}, \quad \frac{-3}{-7}, \quad \frac{113}{-13}.$$

We have already studied fractions like $\frac{3}{2}$, but the other fractions in the list are new because of the signs.

The set of all fractions formed by dividing pairs of integers (division by zero excluded) is the set of **rational numbers.** These numbers can be matched with points on a number line. For example, the numbers

$$\frac{1}{2}, \quad -\frac{1}{2}, \quad \frac{-5}{-2}, \quad \frac{3}{4}, \quad \frac{-8}{3}, \quad \frac{-200}{-50}, \quad \frac{0}{-8} \quad \frac{23}{4}, \quad \text{and} \quad -\frac{23}{4}$$

are shown on the number line in Figure 9.10.

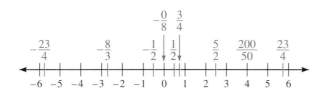

**Figure 9.10**

## **2** IRRATIONAL AND REAL NUMBERS

You might assume that the rational numbers would fill up a number line. However, there are many points left over which correspond to numbers called **irrational numbers.** Two well-known irrational numbers are $\sqrt{2}$ and $\pi$. The rational numbers together with the irrational numbers are called the **real numbers.**

We work with rational numbers in both fractional and decimal forms. Note that every integer is also a rational number since it can be written as the quotient of itself and 1. For example, $5 = \frac{5}{1}$ is rational. The properties of order and absolute value that were defined for integers also hold for all rational numbers. Thus, if $a$ and $b$ are rational numbers,

| | |
|---|---|
| $a = b$ | if $a$ and $b$ represent the same point on a number line, |
| $a < b$ | if $a$ is to the left of $b$ on a number line, |
| $a > b$ | if $a$ is to the right of $b$ on a number line, |
| $\lvert a \rvert$ | is the number of units from 0 to $a$ on a number line. |

| EXAMPLE 1 ORDER OF RATIONAL NUMBERS | PRACTICE EXERCISE 1 |

Consider the number line in Figure 9.11.

**Figure 9.11**

(a) $2.5 < \frac{11}{3}$ since 2.5 is to the left of $\frac{11}{3}$.

(b) $-2\frac{1}{3} < -0.5$ since $-2\frac{1}{3}$ is to the left of $-0.5$.

(c) $-\frac{23}{4} < 5.25$ since $-\frac{23}{4}$ is to the left of 5.25.

(d) $\frac{3}{4} > a$ since $\frac{3}{4}$ is to the right of $a$.

(e) $b > -3.5$ since $b$ is to the right of $-3.5$.

(f) $a < b$ since $a$ is to the left of $b$.

(g) $|2.5| = 2.5$ since 2.5 is 2.5 units from 0.

(h) $\left|-\frac{23}{4}\right| = \frac{23}{4}$ since $-\frac{23}{4}$ is $\frac{23}{4}$ units from 0.

Refer to the number line in Figure 9.11 to answer true or false.

(a) $-0.5 < \frac{11}{3}$

(b) $-2\frac{1}{3} > -0.5$

(c) $2.5 < 5.25$

(d) $-\frac{23}{4} < -2\frac{1}{3}$

(e) $a > -3.5$

(f) $b > -0.5$

(g) $\left|\frac{11}{3}\right| = \frac{11}{3}$

(g) $|-3.5| = -3.5$

Answers: (a) true   (b) false
(c) true   (d) true   (e) false
(f) true   (g) true   (h) false

## ❸ ADDING RATIONAL NUMBERS

To add rational numbers, use the rules for adding fractions and the rules of signs for adding integers.

| EXAMPLE 2 ADDING RATIONAL NUMBERS | PRACTICE EXERCISE 2 |

Add.

(a) $\dfrac{2}{3} + \dfrac{1}{6} = \dfrac{4}{6} + \dfrac{1}{6} = \dfrac{5}{6}$

(b) $\left(-\dfrac{3}{4}\right) + \dfrac{1}{4} = \dfrac{-3+1}{4} = \dfrac{-2}{4} = -\dfrac{1}{2}$

(c) $\dfrac{3}{5} + \left(-\dfrac{4}{7}\right) = \dfrac{3 \cdot 7}{5 \cdot 7} + \dfrac{(-4)(5)}{5 \cdot 7}$   LCD $= 5 \cdot 7 = 35$

$= \dfrac{21}{35} + \dfrac{-20}{35}$

$= \dfrac{21 - 20}{35} = \dfrac{1}{35}$

(d) $-7.3 + (-5.4) = -(7.3 + 5.4) = -12.7$

Add.

(a) $\dfrac{2}{5} + \dfrac{3}{10}$

(b) $\dfrac{3}{8} + \left(-\dfrac{7}{8}\right)$

(c) $\left(-\dfrac{5}{6}\right) + \dfrac{5}{8}$

(d) $-8.7 + (-10.5)$

Answers: (a) $\frac{7}{10}$   (b) $-\frac{1}{2}$   (c) $-\frac{5}{24}$
(d) $-19.2$

## ❹ SUBTRACTING RATIONAL NUMBERS

Subtraction is also the same as for integers. That is, to subtract, change the sign on the subtrahend and add.

| EXAMPLE 3 SUBTRACTING RATIONAL NUMBERS | PRACTICE EXERCISE 3 |

Subtract.

**(a)** $\dfrac{1}{6} - \dfrac{5}{12} = \dfrac{2}{12} + \left(-\dfrac{5}{12}\right) = \dfrac{2 + (-5)}{12} = \dfrac{-3}{12} = -\dfrac{1}{4}$

**(b)** $-\dfrac{7}{6} - \left(-\dfrac{5}{18}\right) = -\dfrac{7 \cdot 3}{6 \cdot 3} + \dfrac{5}{18} + \dfrac{-21}{18} + \dfrac{5}{18} + \dfrac{-16}{18} = -\dfrac{8}{9}$

**(c)** $-\dfrac{9}{10} - \dfrac{8}{15} = -\dfrac{9 \cdot 3}{10 \cdot 3} + \dfrac{-8 \cdot 2}{15 \cdot 2} = \dfrac{-27 + (-16)}{30} = -\dfrac{43}{30}$

**(d)** $-4.32 - (-7.14) = -4.32 + 7.14 = 2.82$

Subtract.

**(a)** $\dfrac{2}{7} - \dfrac{9}{14}$

**(b)** $-\dfrac{3}{4} - \left(-\dfrac{7}{12}\right)$

**(c)** $-\dfrac{4}{21} - \dfrac{5}{14}$

**(d)** $-8.12 - (-10.05)$

Answers: (a) $-\frac{5}{14}$ (b) $-\frac{1}{6}$ (c) $-\frac{23}{42}$ (d) $1.93$

| EXAMPLE 4 APPLICATION OF ADDITION | PRACTICE EXERCISE 4 |

Marge had $685.35 in her account on March 1. During March, she deposited one check for $438.16 and another for $723.92. The checks that she wrote on her account were for $23.47, $492.88, and $219.36. What was the balance in her account at the end of the month?

Think of the beginning balance and the deposits as positive rational numbers and the checks as negative rational numbers. Add the positives, add the negatives, and then add the results.

| 685.35 | | $-\;\;23.47$ | | | |
| 438.16 | | $-492.88$ | | 1847.43 | |
| 723.92 | Sum of | $-219.36$ | Sum of | $-\;\;735.71$ | Sum of |
| 1847.43 | positives | $-735.71$ | negatives | 1111.72 | results |

At the end of March, Marge had $1111.72 in her account.

Walter started the month with $215.18 in his checking account. He deposited $172.40 and $309.08 during the month. He wrote checks of $35.60, $185.09, $372.85, and $192.47. What was his balance at the end of the month?

Answer: $-\$89.35$ (overdrawn by $89.35)

## ⑤ MULTIPLYING AND DIVIDING RATIONAL NUMBERS

When multiplying and dividing rational numbers, use the same rules for signs as with integers. Thus, if two numbers with the same sign are multiplied or divided, the result is positive. When the signs are different, the product or quotient is negative.

| EXAMPLE 5 MULTIPLYING RATIONAL NUMBERS | PRACTICE EXERCISE 5 |

Multiply.

**(a)** $\dfrac{2}{3} \cdot \dfrac{9}{4} = \dfrac{\overset{1}{\cancel{2}}}{\underset{1}{\cancel{3}}} \cdot \dfrac{\overset{3}{\cancel{9}}}{\underset{2}{\cancel{4}}} = \dfrac{1 \cdot 3}{1 \cdot 2} = \dfrac{3}{2}$     Both signs positive

**(b)** $\left(-\dfrac{8}{15}\right) \cdot \left(-\dfrac{35}{4}\right) = \dfrac{8}{15} \cdot \dfrac{35}{4}$     Both signs negative

$\qquad\qquad = \dfrac{\overset{2}{\cancel{8}}}{\underset{3}{\cancel{15}}} \cdot \dfrac{\overset{7}{\cancel{35}}}{\underset{1}{\cancel{4}}} = \dfrac{2 \cdot 7}{3 \cdot 1} = \dfrac{14}{3}$

Multiply.

**(a)** $\dfrac{5}{12} \cdot \dfrac{4}{15}$

**(b)** $\left(-\dfrac{7}{18}\right) \cdot \left(-\dfrac{3}{14}\right)$

**(c)** $(-3.2)(6) = -(3.2)(6) = -19.2$ One positive and one negative

**(d)** $(7.5)(-1.2) = -(7.5)(1.2) = -9.0$

**(c)** $(8)(-2.5)$

**(d)** $(-3.2)(-8.5)$

Answers: (a) $\frac{1}{9}$ (b) $\frac{1}{12}$ (c) $-20$ (d) $27.2$

## EXAMPLE 6 DIVIDING RATIONAL NUMBERS

Divide.

**(a)** $\frac{6}{5} \div \frac{3}{25} = \frac{6}{5} \cdot \frac{25}{3} = \frac{\overset{2}{\cancel{6}}}{\underset{1}{\cancel{5}}} \cdot \frac{\overset{5}{\cancel{25}}}{\underset{1}{\cancel{3}}} = \frac{2 \cdot 5}{1 \cdot 1} = \frac{10}{1} = 10$

**(b)** $\left(\frac{18}{7}\right) \div \left(-\frac{10}{21}\right) = -\left(\frac{18}{7} \cdot \frac{21}{10}\right) = -\left(\frac{\overset{9}{\cancel{18}}}{\underset{1}{\cancel{7}}} \cdot \frac{\overset{3}{\cancel{21}}}{\underset{5}{\cancel{10}}}\right) = -\left(\frac{27}{5}\right) = -\frac{27}{5}$

**(c)** $\frac{-7.5}{-3} = \frac{7.5}{3} = 2.5$

**(d)** $(-21.3) \div (10) = -2.13$

## PRACTICE EXERCISE 6

Divide.

**(a)** $\frac{7}{12} \div \frac{14}{9}$

**(b)** $\left(-2\frac{1}{2}\right) \div \left(\frac{15}{8}\right)$

**(c)** $\frac{-23.4}{-5.2}$

**(d)** $87.5 \div (-100)$

Answers: (a) $\frac{3}{8}$ (b) $-\frac{4}{3}$ (c) $4.5$ (d) $-0.875$

## EXAMPLE 7 APPLICATION TO FINANCE

If the \$816.25 cost of renting a houseboat is to be divided equally among 5 people and one has \$94.62 in his account, what would be his balance if he pays a full share?

$$\begin{array}{r} 163.25 \\ 5\overline{)816.25} \\ \underline{5}\phantom{16.25} \\ 31\phantom{6.25} \\ \underline{30}\phantom{6.25} \\ 16\phantom{.25} \\ \underline{15}\phantom{.25} \\ 12\phantom{5} \\ \underline{10}\phantom{5} \\ 25 \end{array}$$

Divide 816.25 by 5

$$\begin{array}{r} -163.25 \\ \underline{94.62} \\ -68.63 \end{array}$$

Add 94.62 and −163.25

He would be in the red or overdrawn by \$68.63.

## PRACTICE EXERCISE 7

Shana wanted \$35.62 to buy a pair of shoes. She and three of her friends won a prize of \$132.60. If each of the four gets an equal share, did Shana have enough for the shoes?

Answer: −\$2.47 (She needs \$2.47 more.)

## 6 MULTIPLYING MORE THAN TWO RATIONAL NUMBERS

We may multiply more than two rational numbers. Remember that an even number of negative signs gives a positive product and an odd number gives a negative product.

## EXAMPLE 8 MULTIPLYING SEVERAL RATIONAL NUMBERS

Multiply.

**(a)** $\left(-\frac{3}{2}\right)\left(\frac{2}{3}\right)\left(-\frac{7}{3}\right)\left(-\frac{3}{7}\right) = -\frac{\cancel{3} \cdot \cancel{2} \cdot \cancel{7} \cdot \cancel{3}}{\cancel{2} \cdot \cancel{3} \cdot \cancel{3} \cdot \cancel{7}} = -1$ Odd number of minus signs (3)

## PRACTICE EXERCISE 8

Multiply.

**(a)** $\left(-\frac{2}{5}\right)\left(-\frac{5}{4}\right)\left(\frac{4}{3}\right)\left(-\frac{3}{8}\right)$

**(b)** $\left(-\dfrac{1}{2}\right)\left(-\dfrac{1}{2}\right)\left(-\dfrac{1}{2}\right)\left(-\dfrac{1}{2}\right) = \dfrac{1}{16}$ 
Even number of minus signs (4)

**(b)** $\left(-\dfrac{1}{3}\right)\left(-\dfrac{1}{3}\right)\left(-\dfrac{1}{3}\right)\left(-\dfrac{1}{3}\right)$

Answers: (a) $-\dfrac{1}{4}$ (b) $\dfrac{1}{81}$

## 9.5 EXERCISES A

**1.** Find $\dfrac{7}{2}$, $-\dfrac{7}{2}$, $1.5$, $-1.5$, $3\dfrac{1}{2}$, $-3\dfrac{1}{2}$, $\dfrac{17}{3}$, and $-\dfrac{17}{3}$ on the given number line.

*Place the correct symbol ( =, <, or > ) between the two numbers.*

**2.** $0 \quad 1$

**3.** $-1.5 \quad 1.5$

**4.** $\dfrac{7}{2} \quad 1.5$

**5.** $\dfrac{17}{3} \quad \dfrac{7}{2}$

**6.** $-\dfrac{17}{3} \quad -\dfrac{7}{2}$

**7.** $0 \quad -1.5$

*Answer true or false. If the answer is false, tell why.*

**8.** $-3\dfrac{1}{2} = -\dfrac{7}{2}$

**9.** $-1.5 < -\dfrac{7}{2}$

**10.** $2.6 > -1.5$

**11.** $-2.6 > -1.5$

**12.** $-1.2 < -0.5$

**13.** $-5\dfrac{1}{4} < -\dfrac{19}{4}$

*Find the absolute values.*

**14.** $\left|\dfrac{7}{2}\right|$

**15.** $|-1.5|$

**16.** $|1.5|$

**17.** $\left|-\dfrac{17}{3}\right|$

**18.** $\left|-\dfrac{99}{2}\right|$

**19.** $\left|\dfrac{-6}{-5}\right|$

**20.** $\left|-\dfrac{100}{3}\right|$

**21.** $|-9.8|$

*Perform the indicated operations.*

**22.** $\dfrac{1}{2} + \left(-\dfrac{1}{4}\right)$

**23.** $2.1 + (-7.3)$

**24.** $\left(-\dfrac{3}{8}\right) + \left(-\dfrac{3}{4}\right)$

**25.** $-\dfrac{2}{3} + \dfrac{1}{4}$

**26.** $\dfrac{6}{5} + \left(-\dfrac{2}{3}\right)$

**27.** $(-3.8) + (-9.3)$

**28.** $6.3 + (-8.8)$

**29.** $\left(-\dfrac{2}{15}\right) + \left(-\dfrac{7}{3}\right)$

**30.** $\dfrac{1}{2} - \left(-\dfrac{1}{4}\right)$

**31.** $2.1 - (-7.3)$

**32.** $\left(-\dfrac{3}{8}\right) - \left(-\dfrac{3}{4}\right)$

**33.** $-\dfrac{2}{3} - \dfrac{1}{4}$

**34.** $\dfrac{6}{5} - \left(-\dfrac{2}{3}\right)$

**35.** $(-4.6) - (-3.9)$

**36.** $(-16) - \dfrac{3}{2}$

**37.** $\left(\dfrac{1}{2}\right)\left(-\dfrac{1}{4}\right)$

**38.** $\left(-\dfrac{12}{5}\right)\left(\dfrac{15}{4}\right)$

**39.** $(-3)(-1.2)$

**40.** $(7.3)(-10)$

**41.** $\left(\dfrac{24}{7}\right)\left(-\dfrac{21}{16}\right)$

**42.** $(-22)\left(\dfrac{1}{4}\right)$

**43.** $\left(-\dfrac{17}{3}\right)\left(\dfrac{6}{5}\right)$

**44.** $(-3.2)(6.5)$

**45.** $\dfrac{1}{2} \div \left(-\dfrac{1}{4}\right)$

**46.** $(-12.6) \div (-4.2)$

**47.** $(-4.2) \div 8.4$

**48.** $\left(-\dfrac{25}{3}\right) \div \left(-\dfrac{5}{21}\right)$

**49.** $\left(-\dfrac{14}{27}\right) \div \dfrac{7}{18}$

**50.** $\dfrac{-27.5}{-10}$

**51.** $\dfrac{21.3}{-3}$

**52.** $\left(-\dfrac{3}{2}\right)\left(-\dfrac{7}{9}\right)\left(\dfrac{1}{2}\right)\left(-\dfrac{16}{35}\right)$

**53.** $\left(-\dfrac{6}{5}\right)\left(-\dfrac{5}{6}\right)\left(-\dfrac{12}{7}\right)\left(-\dfrac{7}{12}\right)$

**54.** $\left(-\dfrac{5}{2}\right)\left(-\dfrac{5}{2}\right)\left(-\dfrac{5}{2}\right)$

*Solve.*

**55** Henry's bank balance was $217.93 on December 1. He made one deposit of $493.16 and wrote checks for $16.25, $92.87, and $388.14 during December. What was his balance at the end of the month?

**56.** If the 12 members of a club share equally in the expenses of hosting a party and the total cost is $195.60, what is the amount to be paid by each member? What will be Sue's balance if she writes a check and has only $10.20 in her account?

**57.** Linda Warren had an investment worth $3216.10 at the first of the year. If the value increased by $116.25 each month for 5 months and then decreased by $135.42 each month for 7 months, what was the value of the account at the end of the year?

**58.** A grocery store sold 126 lb of peaches at a total loss of $17.64. In terms of signed numbers, what was the loss per pound?

## FOR REVIEW

*Perform the indicated operations.*

**59.** $(-8)(-9)$

**60.** $(-10)(10)$

**61.** $(-2)(-12)$

**62.** $(-18) \div 3$

**63.** $(-65) \div (-13)$

**64.** $100 \div (-5)$

ANSWERS:    **1.**

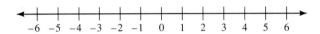

**2.** $0 < 1$   **3.** $-1.5 < 1.5$   **4.** $\frac{7}{2} > 1.5$   **5.** $\frac{17}{3} > \frac{7}{2}$   **6.** $-\frac{17}{3} < -\frac{7}{2}$   **7.** $0 > -1.5$   **8.** true   **9.** false (>)   **10.** true
**11.** false (<)   **12.** true   **13.** true   **14.** $\frac{7}{2}$   **15.** 1.5   **16.** 1.5   **17.** $\frac{17}{3}$   **18.** $\frac{99}{2}$   **19.** $\frac{6}{5}$   **20.** $\frac{100}{3}$   **21.** 9.8   **22.** $\frac{1}{4}$
**23.** $-5.2$   **24.** $-\frac{9}{8}$   **25.** $-\frac{5}{12}$   **26.** $\frac{8}{15}$   **27.** $-13.1$   **28.** $-2.5$   **29.** $-\frac{37}{15}$   **30.** $\frac{3}{4}$   **31.** 9.4   **32.** $\frac{3}{8}$   **33.** $-\frac{11}{12}$   **34.** $\frac{28}{15}$
**35.** $-0.7$   **36.** $-\frac{35}{2}$   **37.** $-\frac{1}{8}$   **38.** $-9$   **39.** 3.6   **40.** $-73$   **41.** $-\frac{9}{2}$   **42.** $-\frac{11}{2}$   **43.** $-\frac{34}{5}$   **44.** $-20.8$   **45.** $-2$   **46.** 3
**47.** $-\frac{1}{2}$   **48.** 35   **49.** $-\frac{4}{3}$   **50.** 2.75   **51.** $-7.1$   **52.** $-\frac{4}{15}$   **53.** 1   **54.** $-\frac{125}{8}$   **55.** $213.83   **56.** $16.30, -$6.10
**57.** $2849.41   **58.** $-$0.14   **59.** 72   **60.** $-100$   **61.** 24   **62.** $-6$   **63.** 5   **64.** $-20$

## 9.5  EXERCISES B

**1.** Find $\frac{3}{2}$, $-\frac{3}{2}$, 2.5, $-2.5$, $3\frac{1}{4}$, $-3\frac{1}{4}$, $\frac{13}{4}$, and $-\frac{13}{4}$ on the given number line.

*Place the correct symbol ( =, <, or >) between the two numbers.*

**2.** 1.6   $-1.6$

**3.** $-5.1$   2.2

**4.** $-\dfrac{15}{2}$   $-7\dfrac{1}{2}$

**5.** $-\dfrac{8}{3}$   $-0.5$

**6.** $0$   $-\dfrac{1}{9}$

**7.** 2.5   $\dfrac{9}{4}$

*Answer true or false. If the answer is false, tell why.*

**8.** $0.5 < -7$

**9.** $\dfrac{3}{2} > 1\dfrac{1}{2}$

**10.** $-3\dfrac{3}{4} = -\dfrac{15}{4}$

**11.** $\dfrac{12}{5} < \dfrac{13}{5}$

**12.** $-2.2 > -2\dfrac{2}{5}$

**13.** $6.25 < 6\dfrac{1}{4}$

*Find the absolute values.*

**14.** $\left|\dfrac{9}{4}\right|$

**15.** $|-3.2|$

**16.** $|3.2|$

**17.** $\left|-\dfrac{12}{5}\right|$

**18.** $\left|-\dfrac{8}{111}\right|$

**19.** $\left|\dfrac{-8}{-9}\right|$

**20.** $\left|-\dfrac{3}{100}\right|$

**21.** $|-2.61|$

*Perform the indicated operations.*

**22.** $\dfrac{1}{2} + \left(-\dfrac{3}{4}\right)$

**23.** $1.3 + (-2.5)$

**24.** $\left(-\dfrac{1}{2}\right) + \left(-\dfrac{2}{5}\right)$

**25.** $-\dfrac{3}{8} + \dfrac{3}{4}$

**26.** $\dfrac{3}{2} + \left(-\dfrac{2}{5}\right)$

**27.** $(-7.2) + (-2.5)$

**28.** $4.8 + (-7.9)$

**29.** $\left(-\dfrac{4}{21}\right) + \left(-\dfrac{4}{7}\right)$

**30.** $\dfrac{2}{9} - \left(-\dfrac{1}{3}\right)$

**31.** $3.1 - (-1.7)$

**32.** $\left(-\dfrac{2}{5}\right) - \left(-\dfrac{7}{15}\right)$

**33.** $-\dfrac{1}{6} - \dfrac{3}{8}$

**34.** $\dfrac{9}{2} - \left(-\dfrac{5}{4}\right)$

**35.** $(-7.2) - (-8.3)$

**36.** $(-7) - \dfrac{1}{5}$

**37.** $\left(\dfrac{1}{3}\right)\left(-\dfrac{1}{6}\right)$

**38.** $\left(-\dfrac{11}{4}\right)\left(\dfrac{3}{22}\right)$

**39.** $(-5)(-2.5)$

**40.** $(9.1)(-100)$

**41.** $\left(\dfrac{48}{33}\right)\left(-\dfrac{22}{24}\right)$

**42.** $(-24)\left(\dfrac{1}{15}\right)$

**43.** $\left(-\dfrac{12}{5}\right)\left(\dfrac{15}{14}\right)$

**44.** $(-8.2)(7.5)$

**45.** $\dfrac{1}{3} \div \left(-\dfrac{1}{6}\right)$

**46.** $(-18.2) \div (-9.1)$

**47.** $(-6.8) \div 13.6$

**48.** $\left(-\dfrac{15}{14}\right) \div \left(-\dfrac{10}{7}\right)$

**49.** $\left(-\dfrac{9}{24}\right) \div \dfrac{21}{4}$

**50.** $\dfrac{-33.5}{-10}$

**51.** $\dfrac{24.4}{-4}$

**52.** $\left(-\dfrac{1}{2}\right)\left(-\dfrac{2}{3}\right)\left(\dfrac{3}{7}\right)\left(-\dfrac{14}{3}\right)$

**53.** $\left(-\dfrac{4}{9}\right)\left(-\dfrac{9}{4}\right)\left(-\dfrac{11}{6}\right)\left(-\dfrac{6}{11}\right)$

**54.** $\left(-\dfrac{2}{3}\right)\left(-\dfrac{2}{3}\right)\left(-\dfrac{2}{3}\right)$

*Solve.*

**55.** Barb's bank balance was $418.83 on May 1. She made one deposit of $623.09 and three withdrawals of $27.52, $119.62, and $250.00. What was her balance at the end of the month?

**56.** A weekend camping trip cost a total of $175.50. If five people are sharing the cost equally, what is each person's share? What will be Mack's bank balance if he has $26.09 in his account and writes a check for his full share?

**57.** Walter McDonald invested $1216.25 at the first of the year. The account increased in value for 4 months by $92.16 per month. If the account then decreased in value by $113.07 per month for 8 months, what was the value of the account at the end of the year?

**58.** A department store sold 216 shirts at a loss of $1.19 per shirt. What was the total loss in terms of signed numbers?

## FOR REVIEW

*Perform the indicated operations.*

**59.** $(-12)(-7)$

**60.** $(-40)(4)$

**61.** $(-5)(-25)$

**62.** $(-35) \div 7$

**63.** $(-121) \div (-11)$

**64.** $240 \div (-20)$

## 9.5 EXERCISES C

*Use a calculator in Exercises 1–4.*

**1.** $(-21.305) - (-18.413)$    **2.** $(103.65) + (-98.73)$    **3.** $(-1.008)(3.405)$    **4.** $(-7.6153) \div (3.01)$
[Answer: $-2.53$]

*Perform the indicated operations.*

**5.** $\left(7\frac{1}{8}\right)\left(-2\frac{3}{4}\right)$

**6.** $\left(-7\frac{1}{8}\right) \div \left(-2\frac{3}{4}\right)$
$\left[\text{Answer: } 2\frac{13}{22}\right]$

## 9.6 PROPERTIES OF RATIONAL NUMBERS

=== STUDENT GUIDEPOSTS ===

1 Properties in Common with Whole Numbers
2 Negatives and Reciprocals
3 Variables
4 Algebraic Expressions
5 Double-Negative Property
6 Terms, Factors, and Coefficients
7 Using the Distributive Laws
8 Removing or Clearing Parentheses

### 1 PROPERTIES IN COMMON WITH WHOLE NUMBERS

The rational numbers have all the properties that we considered for whole numbers. If $a$, $b$, and $c$ represent rational numbers, they have the following properties.

$a + b = b + a$ — Commutative law of addition
$a \cdot b = b \cdot a$ — Commutative law of multiplication
$a + (b + c) = (a + b) + c$ — Associative law of addition
$a \cdot (b \cdot c) = (a \cdot b) \cdot c$ — Associative law of multiplication
$a + 0 = 0 + a = a$ — Identity for addition
$a \cdot 1 = 1 \cdot a = a$ — Identity for multiplication
$a(b + c) = a \cdot b + a \cdot c$ — Distributive law

Notice that 0 and 1 are called **identities** because we get back what we started with in each case. For example,

$$5 + 0 = 5 \quad \text{and} \quad 1 \cdot 5 = 5.$$

### 2 NEGATIVES AND RECIPROCALS

In addition to these properties, rational numbers have the following properties.

$a + (-a) = 0$ — Negative for any number
$a \cdot \dfrac{1}{a} = 1$ — Reciprocal for any number except zero

These properties were not discussed with whole numbers since whole numbers do not have negatives and reciprocals. As examples,

$$-5 \text{ is the negative of } 5,$$
$$5 \text{ is the negative of } -5,$$
$$\frac{2}{3} \text{ is the reciprocal of } \frac{3}{2}, \text{ and}$$
$$\frac{3}{2} \text{ is the reciprocal of } \frac{2}{3}.$$

## ③ VARIABLES

Notice that in the discussion of properties we have used letters to represent numbers. Such letters are called **variables.** In algebra variables are used extensively. Variables allow us to change word statements into brief algebraic statements. For example, using $x$ as a variable,

three times a number becomes $3 \cdot x$,    and
five more than twice a number becomes $2 \cdot x + 5$.

Expressions such as $3 \cdot x$ can be written as $3x$ and $2 \cdot x + 5$ as $2x + 5$. Thus $3x$ means three *times x*.

## ④ ALGEBRAIC EXPRESSIONS

Both $3x$ and $2x + 5$ are examples of *algebraic expressions*. An **algebraic expression** contains variables and numbers combined by the basic operations of addition, subtraction, multiplication, and division. It is important to be able to **evaluate** (find the value of) an algebraic expression when specific values of the variables are given.

---

### To Evaluate an Algebraic Expression

1. Replace each variable (letter) with its specified value.

2. Evaluate the numerical expressions using the same order of operations as used for whole numbers as follows:
   a. Operate within parentheses first.
   b. Find all powers and square roots next.
   c. Multiply and divide, in order, from left to right.
   d. Add and subtract, in order, from left to right.

---

### EXAMPLE 1    EVALUATING EXPRESSIONS

**(a)** Evaluate $5a - (b + c)$ for $a = 2$, $b = 4$, and $c = 3$.

$$\begin{aligned}
5a - (b + c) &= 5(2) - (4 + 3) & \text{Replace each letter with its given value} \\
&= 10 - 7 & \text{Add inside parentheses and multiply} \\
&= 3 & \text{Then subtract}
\end{aligned}$$

**(b)** Evaluate $3xy + 2z$ for $x = 5$, $y = 2$, and $z = -3$.

$$\begin{aligned}
3xy + 2z &= 3(5)(2) + 2(-3) & \text{Replace variables with numbers} \\
&= 30 + (-6) & \text{Multiply first} \\
&= 24 & \text{Then add}
\end{aligned}$$

### PRACTICE EXERCISE 1

**(a)** Evaluate $7ab - (b - c)$ for $a = 5$, $b = 2$, and $c = 7$.

$7(5 \cdot 2) - (2 + 7)$
$7 \cdot 10 - (-5)$
$70 + (+5)$
$75$

**(b)** Evaluate $9x + 3xyz - 6z$ for $x = -2$, $y = -1$, and $z = 4$.

$9(-2) + 3(-2)(-1)(4) - 6(4)$
$-18 + -6(-4) - 24$
$-18 + -24 + 24$
$-18$

**(c)** Evaluate $3[2(a + 2) - 3(b + c)]$ for $a = -2$, $b = 4$, and $c = -8$.

$$3[2(a + 2) - 3(b + c)] = 3[2(-2 + 2) - 3(4 + (-8))] \quad \text{Replace variables with numbers}$$

$$= 3[2(0) - 3(-4)] \quad \text{Add inside parentheses first}$$

$$= 3[0 + 12]$$
$$= 3[12]$$
$$= 36$$

**(c)** Evaluate
$-2[5(a - b) - 7(c + 2)]$ for $a = -5$, $b = -2$, and $c = -3$.

$-2[5(-5+2)-7(-3+2)]$
$-2[5(-3)-7(-1)]$
$-2[-15+7]$
$-2[-8]$

Answers: (a) 75 (b) −18
(c) 16   16

---

### 5 DOUBLE-NEGATIVE PROPERTY

The following example uses the **double negative property, $-(-a) = a$.**

**EXAMPLE 2  USING THE DOUBLE-NEGATIVE PROPERTY**

Evaluate for $a = -3$ *and* $b = -2$.

**(a)** $a - b = (-3) - (-2) = (-3) + 2 = -1$   Use parentheses when substituting

**(b)** $b - [-a] = (-2) - [-(-3)] = (-2) - [3]$   Use $-(-a) = a$
$\qquad = -2 - 3 = -5$

**(c)** $2a - 3b = 2(-3) - 3(-2) = -6 - (-6) = -6 + 6 = 0$

**(d)** $|a - 2b| = |(-3) - 2(-2)| = |(-3) - (-4)|$
$\qquad = |(-3) + 4| = |1| = 1$

**PRACTICE EXERCISE 2**

Evaluate for $x = -4$ and $y = -6$.

**(a)** $y - x$

**(b)** $-x - (-y)$

**(c)** $4y - 2x$

**(d)** $|-x + 3y|$

Answers: (a) −2 (b) −2
(c) −16 (d) 14

---

### 6 TERMS, FACTORS, AND COEFFICIENTS

A part of an algebraic expression which is a product of numbers and variables and is separated from the rest of the expression by plus (or minus) signs is called a **term.** The numbers and letters which are multiplied in a term are called **factors** of the term. The numerical factor is called the **(numerical) coefficient** of the term. For example,

$$3x, \quad 4a + 7, \quad 3x + 7y - z, \quad 4x + 5a + 3 - 8x$$

are algebraic expressions with one, two, three, and four terms, respectively. In $4a + 7$, the term $4a$ has factors 4 and $a$, and 4 is the coefficient of the term.

Two terms are **similar** or **like terms** if they contain the same variables. In $4x + 5a + 3 - 8x$, the terms $4x$ and $-8x$ are like terms. Note that the minus sign goes with the term and thus the coefficient of $-8x$ is $-8$.

### 7 USING THE DISTRIBUTIVE LAWS

The distributive law of multiplication over addition is very important in algebra. Earlier we saw, for example, that

$$2(3 + 5) = 2 \cdot 3 + 2 \cdot 5 = 6 + 10 = 16$$

which gives the same result as when we add first inside the parentheses and then multiply

$$2(3 + 5) = 2(8) = 16.$$

We can also distribute when variables are involved. For example,

$$2(x + y) = 2x + 2y.$$

Multiplication can also be distributed over subtraction since subtraction is really addition of a negative. For example,

$$2(4 - 3) = 2(4 + (-3)) = 2(4) + 2(-3) = 8 - 6 = 2.$$

Notice that $2(4 - 3) = 2(1) = 2$, and also that

$$2(4 - 3) = 2(4) - 2(3) = 8 - 6 = 2.$$

Similarly, when variables are used,

$$2(x - y) = 2x - 2y.$$

Thus, we really have two distributive laws to work with,

$$a(b + c) = ab + ac \quad \text{and} \quad a(b - c) = ab - ac.$$

When the terms of an expression have a common factor, the distributive laws can be used in the reverse order to **remove the common factor** by a process called **factoring.** For example,

$$2x + 2y = 2(x + y) \quad \text{and} \quad 2x - 2y = 2(x - y).$$

The right side of each equation is the factored form of the expression on the left obtained by removing the common factor 2 using the distributive laws.

---

**EXAMPLE 3   FACTORING USING THE DISTRIBUTIVE LAWS**

Use the distributive laws to factor.

(a) $3x + 3y = 3(x + y)$ — The distributive law in reverse order

(b) $4a - 4b = 4(a - b)$ — 4 and $(a - b)$ are factors of $4a - 4b$

(c) $5u + 5v - 5w = 5(u + v - w)$

(d) $3x + 6 = 3 \cdot x + 3 \cdot 2$ — 6 is $3 \cdot 2$
$= 3(x + 2)$ — Factor out 3

(e) $8a - 8 = 8 \cdot a - 8 \cdot 1$ — Express 8 as $8 \cdot 1$
$= 8(a - 1)$ — Factor out 8

(f) $-3x - 3y = (-3)x + (-3)y$ — Factor out $-3$
$= (-3)(x + y)$

---

**PRACTICE EXERCISE 3**

Use the distributive laws to factor.

(a) $7a + 7b$

(b) $8x - 8y$

(c) $4u - 4v + 4w$

(d) $4a + 12$

(e) $3x - 3$

(f) $-5a - 5b$

Answers: (a) $7(a + b)$
(b) $8(x - y)$   (c) $4(u - v + w)$
(d) $4(a + 3)$   (e) $3(x - 1)$
(f) $-5(a + b)$

---

**HINT**

Any factoring problem can be checked by multiplying. For example, since

$$3(x + y) = 3x + 3y \quad \text{and} \quad 4(a - b) = 4a - 4b,$$

the factoring in the first two parts of Example 3 is correct.

---

When an expression contains like terms, it can be simplified by **collecting or combining like terms.** This process is illustrated in the next example.

| **EXAMPLE 4** **COLLECTING LIKE TERMS** | **PRACTICE EXERCISE 4** |

Use the distributive laws to collect like terms.

**(a)** $3x + 7x = (3 + 7)x$ — Factor out $x$
$= 10x$ — Add 3 and 7

**(b)** $6y + 2y - y + 5 = 6 \cdot y + 2 \cdot y - 1 \cdot y + 5$ — $-y = -1 \cdot y$
$= (6 + 2 - 1)y + 5$ — Factor $y$ out of first
$= 7y + 5$ — three terms

The terms $7y$ and $5$ *cannot* be collected since they are not like terms. Thus, $7y + 5$ *is not* $12y$. You can see this more easily if you replace $y$ by some number. For example, when $y = 2$,

$7y + 5 = 7(2) + 5 = 14 + 5 = 19$    but    $12y = 12(2) = 24$.

**(c)** $4a + 7b - a + 6b = 4a - a + 7b + 6b$ — Commutative law
$= 4 \cdot a - 1 \cdot a + 7 \cdot b + 6 \cdot b$ — $-a = -1 \cdot a$
$= (4 - 1)a + (7 + 6)b$ — Distributive law
$= 3a + 13b$

With practice, some steps can be left out.

**(d)** $0.07x + x = (0.07) \cdot x + 1 \cdot x$ — $x = 1 \cdot x$
$= (0.07 + 1)x$ — Distributive law
$= 1.07x$

**(e)** $3x + 2y - 5z$ has no like terms to collect.

Use the distributive laws to collect like terms.

**(a)** $4x + 3x$

**(b)** $4a - 2a + 5a + 9$

**(c)** $2x + 3y + 4x - 8y$

**(d)** $y + 0.12y$

**(e)** $4a + 2b - c$

Answers:  (a) $7x$   (b) $7a + 9$
(c) $6x - 5y$   (d) $1.12y$   (e) no
like terms to collect

---

## ⑧ REMOVING OR CLEARING PARENTHESES

The distributive laws and the fact that $-x = (-1) \cdot x$ give the following equalities.

$$-(a + b) = (-1)(a + b) = (-1)(a) + (-1)(b) = -a - b,$$
$$-(a - b) = (-1)(a - b) = (-1)(a) - (-1)(b) = -a + b,$$
$$-(-a - b) = (-1)(-a - b) = (-1)(-a) - (-1)(b) = a + b.$$

These are special cases of the following rule.

### To Simplify An Expression by Removing Parentheses

1. When a minus sign precedes a set of parentheses, remove the parentheses and use the minus sign to change the sign of every term.

2. When a plus sign precedes a set of parentheses, remove the parentheses and do *not* change any of the signs of the terms.

---

| **EXAMPLE 5** **REMOVING PARENTHESES** | **PRACTICE EXERCISE 5** |

Simplify by removing parentheses.

**(a)** $-(x + 1) = -x - 1$    Change all signs

**(b)** $-(x - 1) = -x + 1$    Change all signs

Simplify by removing parentheses.

**(a)** $-(a + b)$

**(b)** $-(b - a)$

(c) $-(-x + y + 5) = +x - y - 5 = x - y - 5$    Change all signs

(d) $+(-x + y + 5) = -x + y + 5$    Change *no* signs

(e) $x - (y - 3) = x - y + 3$    Change all signs in parentheses

(c) $-(-2x + 2y - 5)$

(d) $+(-2x + 2y - 5)$

(e) $a - (-2b + 5)$

Answers: (a) $-a - b$   (b) $a - b$
(c) $2x - 2y + 5$   (d) $-2x + 2y - 5$   (e) $a + 2b - 5$

---

The process of removing parentheses is sometimes called **clearing parentheses.**

---

**EXAMPLE 6**   COLLECTING LIKE TERMS

Clear parentheses and collect like terms.

(a) $3x + (2x - 7) = 3x + 2x - 7$    Do not change signs
$= (3 + 2)x - 7$    Collect or combine like terms
$= 5x - 7$

(b) $y - (4y - 4) = y - 4y + 4$    Change all signs within parentheses
$= (1 - 4)y + 4$    Collect or combine like terms
$= -3y + 4$

(c) $3 - (5a + 2) + 7a = 3 - 5a - 2 + 7a$    Change all signs within parentheses
$= 7a - 5a + 3 - 2$    Commutative law
$= 2a + 1$    Collect or combine like terms

(d) $3x - (-2x - 7) + 5 = 3x + 2x + 7 + 5$    Change all signs within parentheses
$= (3 + 2)x + 7 + 5$
$= 5x + 12$

**PRACTICE EXERCISE 6**

Clear parentheses and collect like terms.

(a) $4x + (5x - 2y)$

(b) $2x - (7x - 2y)$

(c) $4a - (3b - 3a)$

(d) $5y - (-4y - 1) + 7$

Answers: (a) $9x - 2y$
(b) $-5x + 2y$   (c) $7a - 3b$
(d) $9y + 8$

---

**CAUTION**

Change *all* signs within the parentheses.

$-(-2x - 7) = 2x + 7$   *not*   $2x - 7.$

---

## 9.6 EXERCISES A

*Find the negative of each of the given numbers.*

**1.** 25     **2.** $-25$     **3.** $\dfrac{1}{3}$     **4.** $-\dfrac{2}{5}$     **5.** $-\dfrac{1}{3}$     **6.** 0

*Find the reciprocal of each of the given numbers.*

**7.** 5     **8.** $\dfrac{1}{5}$     **9.** $\dfrac{3}{5}$     **10.** $-\dfrac{2}{3}$     **11.** 1     **12.** 0

*Evaluate for a = −2, b = 3, and c = −5.*

**13.** $a + b$

**14.** $a - b$

**15.** $a - (-b)$

**16.** $3a - b$

**17.** $2a + 3b$

**18.** $a + b - c$

**19** $a - (b - c)$

**20.** $abc - 3$

**21.** $2[a - (b + c)]$

**22** $\dfrac{a}{b} - \dfrac{b}{c}$

**23** $\dfrac{a}{b} + \left(-\dfrac{a}{b}\right)$

**24.** $\dfrac{a + b + c}{-a - b - c}$

*How many terms does each expression have?*

**25.** $4x + 3y$

**26.** $-2x - 3y + 4 - z$

**27.** $3w$

*Use the distributive laws to factor.*

**28.** $4x + 4y$

**29.** $5x - 5y$

**30.** $3a + 3b + 3c$

**31.** $3a + 9$

**32.** $10x + 20y$

**33.** $4a + 2 + 12b$

*Multiply.*

**34.** $4(x + y)$

**35.** $5(x - y)$

**36.** $3(a + b + c)$

**37.** $3(a + 3)$

**38.** $10(x + 2y)$

**39.** $2(2a + 1 + 6b)$

*Use the distributive laws to collect or combine like terms.*

**40.** $7x + 2x$

**41.** $3y - y$

**42.** $5a - 2a + 3$

**43.** $3x + 5x - 10x$

**44.** $-7y + 2y - 3 + 2$

**45.** $9a - 2b + 3a + 7b$

**46** $\dfrac{1}{2}x + 2y + 2x + \dfrac{1}{4}y$

**47.** $2y + 5 + 2y + 5$

**48.** $\dfrac{3}{4}a - \dfrac{2}{3}b - \dfrac{1}{8}a - b + 1$

*Remove parentheses and collect or combine like terms.*

**49.** $-(x + 4)$

**50.** $-(x - 4)$

**51.** $+(3x + 1)$

**52.** $+(5 - 3x)$

**53.** $+(-5 - a)$

**54.** $+(-2x + 5)$

**55.** $-(-x + y - 1)$

**56** $x + (-x + y + 1)$

**57.** $2x + (3x + 1)$

**58.** $+(a + 1) + (1 + a)$

**59** $2[x - (4x + 3)]$

**60.** $-3[2y - (4y + 5)]$

## FOR REVIEW

*Answer true or false. If the answer is false, tell why.*

**61.** $\dfrac{3}{8} > -\dfrac{8}{3}$

**62.** $-0.5 < -1.5$

**63.** $-\dfrac{7}{2} < -2.5$

*Find the absolute value.*

**64.** $|-12|$

**65.** $\left|\dfrac{-8}{-3}\right|$

**66.** $|-2.17|$

*Perform the indicated operations.*

**67.** $\dfrac{2}{3} - \left(-\dfrac{1}{9}\right)$

**68.** $(-8.2) - (-7.4)$

**69.** $\left(-\dfrac{11}{6}\right)\left(\dfrac{3}{22}\right)$

**70.** $(-12.8) \div (-8)$

*The following exercises review material from Section 2.8 to help prepare you for the next section. Evaluate.*

**71.** $\dfrac{3^5}{3^2}$

**72.** $4^2 \cdot 5^2$

**73.** $(4.5)^2$

**74.** $\dfrac{4^3}{2^3}$

---

ANSWERS: 1. $-25$  2. $25$  3. $-\frac{1}{3}$  4. $\frac{2}{5}$  5. $\frac{1}{5}$  6. $0$  7. $\frac{1}{3}$  8. $5$  9. $\frac{5}{3}$  10. $-\frac{3}{2}$  11. $1$  12. undefined  13. $1$
14. $-5$  15. $1$  16. $-9$  17. $5$  18. $6$  19. $-10$  20. $27$  21. $0$  22. $-\frac{1}{15}$  23. $0$  24. $-1$  25. $2$  26. $4$  27. $1$
28. $4(x + y)$  29. $5(x - y)$  30. $3(a + b + c)$  31. $3(a + 3)$  32. $10(x + 2y)$  33. $2(2a + 1 + 6b)$  34. $4x + 4y$
35. $5x - 5y$  36. $3a + 3b + 3c$  37. $3a + 9$  38. $10x + 20y$  39. $4a + 2 + 12b$  40. $9x$  41. $2y$  42. $3a + 3$  43. $-2x$
44. $-5y - 1$  45. $12a + 5b$  46. $-\frac{3}{2}x - \frac{7}{4}y$  47. $0$  48. $\frac{5}{8}a - \frac{5}{3}b + 1$  49. $-x - 4$  50. $-x + 4$  51. $3x + 1$
52. $5 - 3x$  53. $5 + a$  54. $2x - 5$  55. $x - y + 1$  56. $2x - y + 1$  57. $-x - 1$  58. $0$  59. $-6x + 6$  60. $6y + 15$
61. true  62. false ($>$)  63. true  64. $12$  65. $\frac{8}{3}$  66. $2.17$  67. $\frac{7}{9}$  68. $-0.8$  69. $-\frac{1}{4}$  70. $1.6$  71. $27$  72. $400$
73. $400$  74. $8$

## 9.6 EXERCISES B

*Find the negative of each of the given numbers.*

**1.** $12$

**2.** $-12$

**3.** $-\dfrac{1}{3}$

**4.** $\dfrac{2}{3}$

**5.** $6.2$

**6.** $-0$

*Find the reciprocal of each of the given numbers.*

**7.** $9$

**8.** $\dfrac{1}{9}$

**9.** $\dfrac{4}{7}$

**10.** $-\dfrac{9}{5}$

**11.** $0.1$

**12.** $10$

*Evaluate for $a = -3$, $b = 2$, and $c = -4$.*

**13.** $a + c$

**14.** $a - c$

**15.** $a - (-c)$

**16.** $2a + b$

**17.** $4a + 2b$

**18.** $a - b - c$

**19.** $a - (b - c)$

**20.** $ab - bc$

**21.** $3[b - (a - c)]$

**22.** $\dfrac{c}{b} - \dfrac{b}{a}$

**23.** $\dfrac{c}{b} + \left(-\dfrac{c}{b}\right)$

**24.** $\dfrac{-a - b - c}{a + b + c}$

*How many terms does each expression have?*

**25.** $2x + 5y$       **26.** $-3x - y + 5 - 7z$       **27.** $2a$

*Use the distributive laws to factor.*

**28.** $2a + 2b$       **29.** $4x - 4y$       **30.** $5a + 5b + 5c$

**31.** $2a + 4$       **32.** $10a + 40b$       **33.** $6x + 2 + 8y$

*Multiply.*

**34.** $2(a + b)$       **35.** $4(x - y)$       **36.** $5(a + b + c)$

**37.** $2(a + 2)$       **38.** $10(a + 4b)$       **39.** $2(3x + 1 + 4y)$

*Use the distributive laws to collect or combine like terms.*

**40.** $5x + 3x$       **41.** $-3x + 8x$       **42.** $2a - 4a + 7$

**43.** $6x - x + 4x$       **44.** $3y + 5 - 3y - 5$       **45.** $3a - 2b + 7a - 5b$

**46.** $\frac{1}{2}x - 3y + \frac{3}{2}x + 2y$       **47.** $2a - b - 4a + b$       **48.** $-\frac{1}{3}x + \frac{1}{2} - \frac{2}{3}x + \frac{1}{3}y$

*Remove parentheses and collect or combine like terms.*

**49.** $-(x + 3)$       **50.** $-(x - 3)$       **51.** $+(2x + 1)$

**52.** $+(2x - 1)$       **53.** $-(-3 - x)$       **54.** $-(-3 + x)$

**55.** $-(-x + y - 2)$       **56.** $x - (-x - y + 1)$       **57.** $4x - (2x - 1)$

**58.** $-(x + y) - (x - y)$       **59.** $4[2x - (3x + 4)]$       **60.** $-5[4y - (-2y - 3)]$

## FOR REVIEW

*Answer true or false. If the answer is false, tell why.*

**61.** $-\frac{11}{4} > \frac{9}{4}$       **62.** $-3.2 < -2.6$       **63.** $-3.5 > -\frac{15}{2}$

*Find the absolute value.*

**64.** $|-20|$       **65.** $\left|\frac{-6}{5}\right|$       **66.** $|-101|$

*Perform the indicated operations.*

**67.** $\frac{3}{4} - \left(-\frac{5}{12}\right)$       **68.** $(-4.1) - 2.6$       **69.** $\left(\frac{9}{15}\right)\left(-\frac{5}{6}\right)$       **70.** $(-4.8) \div 4$

*The following exercises review material from Section 2.8 to help prepare you for the next section. Evaluate.*

**71.** $\frac{4^4}{4^2}$       **72.** $3^2 \cdot 7^2$       **73.** $(3.7)^2$       **74.** $\frac{6^3}{3^3}$

## 9.6 EXERCISES C

*Use a calculator and the distributive laws to collect or combine like terms.*

**1.** $1.05x + 3.62x$

**2.** $4.57x - 2.31x$

**3.** $6.15x - 2.33x + 4.17x$
[Answer: $7.99x$]

*Evaluate for $a = -3.25$.*

**4.** $2a + 1.78$

**5.** $-a + 3.25$
[Answer: $6.5$]

**6.** $-a - 3.25$

## 9.7 INTEGER EXPONENTS AND SCIENTIFIC NOTATION

=== STUDENT GUIDEPOSTS ===

**1** Exponential Notation
**2** Product Rule
**3** Quotient Rule
**4** Power Rule
**5** Product and Quotient to a Power
**6** Zero Power of a Number
**7** Negative Exponents
**8** Summary of Rules of Exponents
**9** Scientific Notation
**10** Calculations in Scientific Notation

### **1** EXPONENTIAL NOTATION

In Section 2.8 **exponential notation** was introduced. For example, $3 \cdot 3 \cdot 3 \cdot 3$ was written as $3^4$.

$$\underbrace{3 \cdot 3 \cdot 3 \cdot 3}_{4 \text{ factors}} = \underset{\text{base}}{3}^{4} \longleftarrow \text{exponent}$$

3 is the **base**, 4 the **exponent**, and $3^4$ the **exponential expression** (read "3 to the fourth power"). Similarly,

$$a \cdot a \cdot a = a^3$$

is called the **third power** or **cube** of $a$;

$$a \cdot a = a^2$$

is called the **second power** or **square** of $a$. The **first power** of $a$ is $a^1$ which is usually written simply as $a$.

| EXAMPLE 1 WRITING WITH EXPONENTS | PRACTICE EXERCISE 1 |
|---|---|

Write in exponential notation.

**(a)** $\underbrace{5 \cdot 5 \cdot 5}_{3 \text{ factors}} = 5^3$

**(b)** $\underbrace{x \cdot x \cdot x \cdot x \cdot x \cdot x}_{6 \text{ factors}} = x^6$

**(c)** $\underbrace{2 \cdot 2}_{2} \cdot \underbrace{y \cdot y \cdot y \cdot y}_{4} = 2^2 \cdot y^4$

Write in exponential notation.

**(a)** $7 \cdot 7 \cdot 7 \cdot 7$

**(b)** $y \cdot y \cdot y \cdot y \cdot y \cdot y$

**(c)** $3 \cdot 3 \cdot 3 \cdot a \cdot a \cdot a \cdot a \cdot a$

Answers: (a) $7^4$ (b) $y^7$ (c) $3^3 \cdot a^5$

| | |
|---|---|
| **EXAMPLE 2    WRITING WITHOUT EXPONENTS** | **PRACTICE EXERCISE 2** |

Write without using exponents.

**(a)** $4^5 = \underbrace{4 \cdot 4 \cdot 4 \cdot 4 \cdot 4}_{5 \text{ factors}}$     The product is 1024

**(b)** $3y^2 = 3 \cdot \underbrace{y \cdot y}_{2 \text{ } y\text{'s as factors}}$     3 is *not* squared

**(c)** $(3y)^2 = \underbrace{(3y)(3y)}_{2 \text{ } (3y)\text{'s}}$     3 *is* squared

Write without exponents.

**(a)** $2^5$

**(b)** $5y^4$

**(c)** $(5y)^4$

Answers: (a) $2 \cdot 2 \cdot 2 \cdot 2 \cdot 2 = 32$
(b) $5 \cdot y \cdot y \cdot y \cdot y$
(c) $(5y) \cdot (5y) \cdot (5y) \cdot (5y)$

## ❷ PRODUCT RULE

When terms containing exponential expressions are combined by multiplication, division, or taking powers, the work can be simplified by using the basic properties of exponents. For example,

$$a^3 \cdot a^2 = \underbrace{(a \cdot a \cdot a)}_{3 \text{ factors}}\underbrace{(a \cdot a)}_{2 \text{ factors}} = \underbrace{a \cdot a \cdot a \cdot a \cdot a}_{5 \text{ factors}} = a^5.$$

When two exponential expressions *with the same base* are multiplied, the product is that base raised to the sum of the exponents on the original expressions.

> **Product Rule for Exponents**
>
> If *a* is any number, and *m* and *n* are positive integers,
>
> $$a^m \cdot a^n = a^{m+n}.$$
>
> (To multiply powers with the same base, add exponents.)

| | |
|---|---|
| **EXAMPLE 3    USING THE PRODUCT RULE** | **PRACTICE EXERCISE 3** |

Find the product.

**(a)** $a^3 \cdot a^4 = a^{3+4} = a^7$

**(b)** $5^3 \cdot 5^7 = 5^{3+7} = 5^{10}$     *Not* $25^{10}$

**(c)** $2^3 \cdot 2^2 \cdot 2^6 = 2^{3+2+6} = 2^{11}$  The rule applies to more than two factors

**(d)** $3x^3 \cdot x^2 = 3x^{3+2} = 3x^5$     The 3 is not raised to the powers

Find the product.

**(a)** $x^5 \cdot x^6$

**(b)** $8^2 \cdot 8^5$

**(c)** $3^2 \cdot 3^3 \cdot 3^4$

**(d)** $4x^4x^6$

Answers: (a) $x^{11}$  (b) $8^7$  (c) $3^9$
(d) $4x^{10}$

## ❸ QUOTIENT RULE

When two powers with the same base are divided, for example,

$$\frac{a^5}{a^2} = \frac{\overbrace{a \cdot a \cdot a \cdot a \cdot a}^{5 \text{ factors}}}{\underbrace{a \cdot a}_{2 \text{ factors}}} = \frac{a \cdot a \cdot a \cdot \cancel{a} \cdot \cancel{a}}{\cancel{a} \cdot \cancel{a}} = \underbrace{a \cdot a \cdot a}_{3 \text{ factors}} = a^3,$$

the quotient can be found by raising the base to the difference of the exponents $(5 - 2 = 3)$.

### Quotient Rule for Exponents

If $a$ is any number except zero and $m$, $n$, and $m - n$ are positive integers, then

$$\frac{a^m}{a^n} = a^{m-n}.$$

(To divide powers with the same base, subtract exponents.)

---

### EXAMPLE 4   USING THE QUOTIENT RULE

Find the quotient.

**(a)** $\dfrac{a^7}{a^3} = a^{7-3} = a^4$

**(b)** $\dfrac{5^8}{5^5} = 5^{8-5} = 5^3$

**(c)** $\dfrac{2^3}{3^4}$   Cannot be simplified using the rule of exponents since the bases are different

**(d)** $\dfrac{3x^3}{x^2} = 3x^{3-2} = 3x^1 = 3x$     $x^1 = x$

### PRACTICE EXERCISE 4

Find the quotient.

**(a)** $\dfrac{y^8}{y^2}$

**(b)** $\dfrac{3^9}{3^7}$

**(c)** $\dfrac{5^2}{4^9}$

**(d)** $\dfrac{a^5}{2a^2}$

Answers: (a) $y^6$   (b) $3^2 = 9$
(c) cannot be simplified using the rules of exponents    (d) $\frac{a^3}{2}$

---

### ④ POWER RULE

When raising a power to a power, for example,

$$(a^2)^3 = \underbrace{(a^2) \cdot (a^2) \cdot (a^2)}_{3 \text{ factors}} = (a \cdot a) \cdot (a \cdot a) \cdot (a \cdot a) = \underbrace{a \cdot a \cdot a \cdot a \cdot a \cdot a}_{6 \text{ factors}} = a^6.$$

the resulting exponential expression can be found by raising the base to the product of the exponents ($2 \cdot 3 = 6$).

### Power Rule for Exponents

If $a$ is any number, and $m$ and $n$ are positive integers,

$$(a^m)^n = a^{m \cdot n}.$$

(To raise a power to a power, multiply the exponents.)

---

### ⚠ CAUTION

Do not confuse the power rule with the product rule. For example,

$$(a^5)^2 = a^{5 \cdot 2} = a^{10}$$

but

$$a^5 a^2 = a^{5+2} = a^7.$$

| EXAMPLE 5   USING THE POWER RULE | PRACTICE EXERCISE 5 |

Find the powers.

**(a)** $(a^3)^8 = a^{3 \cdot 8} = a^{24}$

**(b)** $2(x^3)^2 = 2x^{3 \cdot 2} = 2x^6$

Find the powers.

**(a)** $(x^2)^5$

**(b)** $5(y^4)^3$

Answers: (a) $x^{10}$   (b) $5y^{12}$

## 5 PRODUCT AND QUOTIENT TO A POWER

A product or quotient of expressions is often raised to a power. For example,

$$(3x)^4 = \underbrace{(3x) \cdot (3x) \cdot (3x) \cdot (3x)}_{4 \text{ factors}} = \underbrace{3 \cdot 3 \cdot 3 \cdot 3}_{4 \text{ factors}} \cdot \underbrace{x \cdot x \cdot x \cdot x}_{4 \text{ factors}} = 3^4 x^4,$$

and

$$\left(\frac{2}{y}\right)^3 = \underbrace{\frac{2}{y} \cdot \frac{2}{y} \cdot \frac{2}{y}}_{3 \text{ factors}} = \frac{\overbrace{2 \cdot 2 \cdot 2}^{3 \text{ factors}}}{\underbrace{y \cdot y \cdot y}_{3 \text{ factors}}} = \frac{2^3}{y^3}.$$

These illustrate the next rule.

### Product and Quotient to a Power

If $a$ and $b$ are any numbers, and $n$ is a positive integer, then

$$(a \cdot b)^n = a^n \cdot b^n \qquad \text{and} \qquad \left(\frac{a}{b}\right)^n = \frac{a^n}{b^n} \quad (b \text{ not zero}).$$

| EXAMPLE 6   POWER OF A PRODUCT OR QUOTIENT | PRACTICE EXERCISE 6 |

Simplify.

**(a)** $(2y)^5 = 2^5 \cdot y^5 = 2^5 y^5 = 32y^5$

**(b)** $(3ab)^4 = 3^4 \cdot a^4 \cdot b^4$     Raise each factor to the fourth power
$$= 81a^4 b^4$$

**(c)** $\left(\dfrac{y}{4}\right)^3 = \dfrac{y^3}{4^3} = \dfrac{y^3}{64}$

**(d)** $\left(\dfrac{2x}{3}\right)^5 = \dfrac{(2x)^5}{3^5}$    $\left(\dfrac{a}{b}\right)^n = \dfrac{a^n}{b^n}$

$$= \dfrac{2^5 x^5}{3^5} \qquad (ab)^n = a^n b^n$$

$$= \dfrac{32x^5}{243}$$

Simplify.

**(a)** $(3x)^4$

**(b)** $(2xy)^3$

**(c)** $\left(\dfrac{5}{x}\right)^4$

**(d)** $\left(\dfrac{3x}{2y}\right)^2$

Answers: (a) $81x^4$   (b) $8x^3 y^3$
(c) $\frac{625}{x^4}$   (d) $\frac{9x^2}{4y^2}$

## 🟢 ZERO POWER OF A NUMBER

We know that if $a$ is not zero,

$$\frac{a^m}{a^n} = a^{m-n}.$$

Suppose we let $m = n$. Then

$$\frac{a^m}{a^m} = a^{m-m} = a^0 \quad \text{and also} \quad \frac{a^m}{a^m} = 1.$$

(Any number divided by itself is 1.) This suggests the following definition.

---

### Zero Exponent

If $a$ is any number except zero,

$$a^0 = 1.$$

---

### EXAMPLE 7 USING ZERO EXPONENTS

Simplify.

**(a)** $5^0 = 1$

**(b)** $(2a^2b^3)^0 = 1$ (assuming $a \neq 0$ and $b \neq 0$)

### PRACTICE EXERCISE 7

Simplify.

**(a)** $9^0$

**(b)** $(4xy^2)^0$

Answers: (a) **1** (b) **1** (assuming $x \neq 0$ and $y \neq 0$)

---

## 🟢 NEGATIVE EXPONENTS

Again, consider

$$\frac{a^m}{a^n} = a^{m-n} \quad (a \neq 0).$$

What happens when $n > m$? For example, if we let $n = 5$ and $m = 2$, we would have

$$\frac{a^m}{a^n} = \frac{a^2}{a^5}$$
$$= a^{2-5} = a^{-3}.$$

If we look at the problem another way, we have

$$\frac{a^2}{a^5} = \frac{\cancel{a} \cdot \cancel{a}}{\cancel{a} \cdot \cancel{a} \cdot a \cdot a \cdot a}$$
$$= \frac{1}{a \cdot a \cdot a}$$
$$= \frac{1}{a^3}.$$

Thus, we conclude that $a^{-3} = \frac{1}{a^3}$. This suggests a way to define exponential expressions with negative integer exponents.

## Negative Exponents

If $a \neq 0$ and $n$ is a positive integer ($-n$ is a negative integer), then

$$a^{-n} = \frac{1}{a^n}.$$

### EXAMPLE 8    USING NEGATIVE EXPONENTS

Simplify and write without negative exponents.

**(a)** $5^{-3} = \dfrac{1}{5^3} = \dfrac{1}{125}$

**(b)** $2^{-3} \cdot 2^2 = 2^{-3+2} = 2^{-1} = \dfrac{1}{2}$

**(c)** $\dfrac{3^{-1}}{3^2} = 3^{-1-2} = 3^{-3} = \dfrac{1}{3^3} = \dfrac{1}{27}$

### PRACTICE EXERCISE 8

Simplify and write without negative exponents.

**(a)** $7^{-2}$

**(b)** $4^{-5} \cdot 4^4$

**(c)** $\dfrac{5^{-2}}{5}$

Answers:  (a) $\frac{1}{49}$   (b) $\frac{1}{4}$   (c) $\frac{1}{125}$

## ⑧ SUMMARY OF RULES OF EXPONENTS

All the rules of exponents stated for positive integer exponents apply to all integer exponents: positive, negative, and zero. These are summarized here.

### Rules for Exponents

Let $a$ and $b$ be any two numbers, $m$ and $n$ any two integers.

1. $a^m \cdot a^n = a^{m+n}$

2. $\dfrac{a^m}{a^n} = a^{m-n}$   $(a \neq 0)$

3. $(a^m)^n = a^{mn}$

4. $(a \cdot b)^n = a^n b^n$

5. $\left(\dfrac{a}{b}\right)^n = \dfrac{a^n}{b^n}$   $(b \neq 0)$

6. $a^0 = 1$   $(a \neq 0)$

7. $a^{-n} = \dfrac{1}{a^n}$   $(a \neq 0)$

### EXAMPLE 9    USING SEVERAL RULES

Simplify and express without using negative exponents.

**(a)** $(2a)^{-1} = \dfrac{1}{(2a)^1} = \dfrac{1}{2a}$      $(2a)^{-1}$ is *not* $-2a$

**(b)** $2a^{-1} = 2\dfrac{1}{a} = \dfrac{2}{a}$      Compare with part (a)

**(c)** $y^3 y^{-5} = y^{3+(-5)} = y^{-2} = \dfrac{1}{y^2}$

**(d)** $\dfrac{a^{-2}}{a^3} = a^{-2-3} = a^{-5} = \dfrac{1}{a^5}$

### PRACTICE EXERCISE 9

Simplify and express without using negative exponents.

**(a)** $(3x)^{-2}$

**(b)** $3x^{-2}$

**(c)** $x^{-4} x^2$

**(d)** $\dfrac{b^{-3}}{b^5}$

Answers:  (a) $\frac{1}{9x^2}$   (b) $\frac{3}{x^2}$   (c) $\frac{1}{x^2}$
(d) $\frac{1}{b^8}$

## ⑨ SCIENTIFIC NOTATION

One important use of integer exponents is in *scientific notation*. For example, a scientist might use the number

$$235{,}000{,}000{,}000{,}000{,}000{,}000$$

but instead of writing out all the zeros, he or she would write

$$2.35 \times 10^{20}.$$

This short form is easier to use in computations. Likewise, the number

$$0.000000000057$$

could be written as

$$5.7 \times 10^{-11}.$$

A number is written in **scientific notation** if it is the product of a number between 1 and 10 and a power of 10.

---

### To Write a Number in Scientific Notation

1. Move the decimal point to the position immediately to the right of the first nonzero digit.

2. Multiply by a power of ten which is equal in absolute value to the number of decimal places moved. The exponent on 10 is positive if the original number is greater than 10 and negative if the number is less than 1.

---

### EXAMPLE 10   USING SCIENTIFIC NOTATION

Write in scientific notation.

(a) $\underset{\text{6 places}}{2{,}500{,}000} = 2.5 \times 10^{6}$     Count 6 decimal places

first nonzero digit

(b) $\underset{\text{6 places}}{0.0000025} = 2.5 \times 10^{-6}$     Count 6 decimal places

first nonzero digit

(c) $\underset{\text{9 places}}{4{,}321{,}000{,}000} = 4.321 \times 10^{9}$

(d) $\underset{\text{11 places}}{0.00000000001} = 1 \times 10^{-11}$

### PRACTICE EXERCISE 10

Write in scientific notation.

(a) 480,000

(b) 0.000048

(c) 795,000,000

(d) 0.00000000052

Answers: (a) $4.8 \times 10^{5}$
(b) $4.8 \times 10^{-5}$   (c) $7.95 \times 10^{8}$
(d) $5.2 \times 10^{-10}$

---

### HINT

Notice that

$$32 = 3.2 \times 10^{1}$$

and

$$3.2 = 3.2 \times 10^{0} \qquad \text{Since } 10^{0} = 1$$

## 🔟 CALCULATIONS IN SCIENTIFIC NOTATION

Scientific notation not only shortens the notation for many numbers, but also helps in calculations which involve very large or very small numbers.

---

### EXAMPLE 11   CALCULATING WITH SCIENTIFIC NOTATION

Perform the indicated operations using scientific notation.

(a) $(30{,}000)(2{,}000{,}000) = (3 \times 10^4)(2 \times 10^6)$
$\qquad\qquad\qquad\qquad = (3 \cdot 2) \times (10^4 \times 10^6)$   Change order
$\qquad\qquad\qquad\qquad = 6 \times 10^{10}$   Add exponents

(b) $(2.4 \times 10^{-12})(4.0 \times 10^{11}) = (2.4)(4.0) \times (10^{-12} \times 10^{11})$
$\qquad\qquad\qquad\qquad\qquad = 9.6 \times 10^{-1}$

(c) $\dfrac{3.2 \times 10^{-1}}{1.6 \times 10^5} = \dfrac{3.2}{1.6} \times \dfrac{10^{-1}}{10^5} = 2 \times 10^{-6}$

---

### PRACTICE EXERCISE 11

Perform the indicated operations using scientific notation.

(a) $(0.0005)(300{,}000)$

(b) $(3.6 \times 10^{10})(2.5 \times 10^{-18})$

(c) $\dfrac{4.8 \times 10^{14}}{1.2 \times 10^{-6}}$

Answers: (a) $1.5 \times 10^2$
(b) $9 \times 10^{-8}$   (c) $4 \times 10^{20}$

---

## 9.7   EXERCISES A

*Write in exponential notation.*

**1.** $x \cdot x \cdot x$ 　　　　**2.** $2 \cdot 2 \cdot y \cdot y \cdot y$ 　　　　**3.** $4 \cdot a \cdot a \cdot a$ 　　　　**4.** $(3b)(3b)$

*Write without using exponents.*

**5.** $x^4$ 　　　　**6.** $a^7$ 　　　　**7.** $2y^4$ 　　　　**8.** $(2y)^4$

*Simplify and write without negative exponents.*

**9.** $x^2 \cdot x^5$ 　　　　**10.** $a^3 \cdot a^2 \cdot a^4$ 　　　　**11.** $2y^2 \cdot y^8$

**12.** $\dfrac{a^4}{a^3}$ 　　　　**13.** $\dfrac{2y^5}{y^3}$ 　　　　**14.** $(a^3)^4$

**15.** $(2x^3)^4$ 　　　　**16.** $2(x^3)^4$ 　　　　**17.** $\left(\dfrac{2}{x^3}\right)^4$

**18.** $\dfrac{a^3}{b^5}$ 　　　　**19.** $5^0$ 　　　　**20.** $0^0$

**㉑** $(2x)^{-1}$ 　　　　**㉒** $2x^{-1}$ 　　　　**23.** $\dfrac{2x^7}{x^9}$

**24.** $3y^4 y^{-7}$ 　　　　**25.** $(3y)^{-2}$ 　　　　**26.** $3y^{-2}$

**27.** $a^{-4}$

**28.** $\dfrac{2x^{-1}}{x}$

**29.** $\dfrac{a^{-2}}{a^4}$

*Evaluate when $a = -2$ and $b = 3$.*

**30** $3a^2$

**31** $(3a)^2$

**32.** $-3a^2$

**33.** $(-3a)^2$

**34.** $a^2 + b^2$

**35.** $-b^2$

**36.** $(-b)^2$

**37** $a^2 - b^2$

**38** $(a - b)^2$

*Write in scientific notation.*

**39.** 300,000

**40.** 0.00003

**41.** 82,700,000

**42.** 0.0000000000756

*Perform the indicated operations using scientific notation.*

**43.** $(4 \times 10^5)(1 \times 10^6)$

**44** $(40{,}000{,}000)(20{,}000)$

**45.** $\dfrac{3.3 \times 10^{12}}{1.1 \times 10^{-2}}$

**46.** $\dfrac{0.0000006}{0.03}$

## FOR REVIEW

*Remove parentheses and collect or combine like terms.*

**47.** $-(x - 3)$

**48.** $2x - (4x - 1)$

**49.** $-(1 - x) + (x + 1)$

**50.** $2[x - (3x - 2)]$

*The following exercises review material from Section 1.5 to help prepare you for the next chapter. Write the related addition or subtraction sentence.*

**51.** $x + 5 = 3$

**52.** $a - 5 = 3$

**53.** $x + \dfrac{1}{5} = \dfrac{2}{5}$

**54.** $a - \dfrac{1}{3} = \dfrac{2}{3}$

**55.** $x + 0.2 = -1.5$

**56.** $a - 6.8 = -10.2$

---

**ANSWERS:** 1. $x^3$ 2. $2^2 y^3$ 3. $4a^3$ 4. $(3b)^2$ 5. $x \cdot x \cdot x \cdot x$ 6. $a \cdot a \cdot a \cdot a \cdot a \cdot a \cdot a$ 7. $2 \cdot y \cdot y \cdot y \cdot y$
8. $(2y)(2y)(2y)(2y)$ 9. $x^7$ 10. $a^9$ 11. $2y^{10}$ 12. $a^1 = a$ 13. $2y^2$ 14. $a^{12}$ 15. $16x^{12}$ 16. $2x^{12}$ 17. $\frac{16}{x^{12}}$ 18. cannot
be simplified further 19. 1 20. undefined 21. $\frac{1}{2x}$ 22. $\frac{2}{x}$ 23. $\frac{2}{x^2}$ 24. $\frac{3}{y^3}$ 25. $\frac{1}{9y^2}$ 26. $\frac{3}{y^2}$ 27. $\frac{1}{a^4}$ 28. $\frac{2}{x^2}$ 29. $\frac{1}{a^6}$
30. 12 31. 36 32. $-12$ 33. 36 34. 13 35. $-9$ 36. 9 37. $-5$ 38. 25 39. $3 \times 10^5$ 40. $3 \times 10^{-5}$
41. $8.27 \times 10^7$ 42. $7.56 \times 10^{-11}$ 43. $4 \times 10^{11}$ 44. $8 \times 10^{11}$ 45. $3 \times 10^{14}$ 46. $2 \times 10^{-5}$ 47. $-x + 3$ 48. $-2x + 1$
49. $2x$ 50. $-4x + 4$ 51. $x = 3 - 5$ 52. $a = 3 + 5$ 53. $x = \frac{2}{5} - \frac{1}{5}$ 54. $a = \frac{2}{3} + \frac{1}{3}$ 55. $x = -1.5 - 0.2$
56. $a = -10.2 + 6.8$

## 9.7  EXERCISES B

*Write in exponential notation.*

**1.** $y \cdot y \cdot y \cdot y$      **2.** $4 \cdot 4 \cdot z \cdot z \cdot z$      **3.** $3www$      **4.** $(2u)(2u)$

*Write without using exponents.*

**5.** $y^6$      **6.** $a^9$      **7.** $4y^3$      **8.** $(4y)^3$

*Simplify and write without negative exponents.*

**9.** $y^2 \cdot y^7$      **10.** $x^3 \cdot x^2 \cdot x^6$      **11.** $2z^2 \cdot z^5$

**12.** $\dfrac{b^5}{b^2}$      **13.** $\dfrac{2y^7}{y^4}$      **14.** $(w^3)^5$

**15.** $(2c^2)^4$      **16.** $2(y^3)^5$      **17.** $\left(\dfrac{2}{a^2}\right)^3$

**18.** $\dfrac{x^3}{y^4}$      **19.** $7^0$      **20.** $0^0$

**21.** $(5y)^{-1}$      **22.** $5y^{-1}$      **23.** $\dfrac{3a^3}{a^7}$ ——

**24.** $4z^4z^{-9}$      **25.** $(5w)^{-2}$      **26.** $5w^{-2}$

**27.** $a^{-3}$      **28.** $\dfrac{3x^{-2}}{x^2}$      **29.** $\dfrac{y^{-4}}{y^3}$

*Evaluate when $x = -3$ and $y = 2$.*

**30.** $4x^2$      **31.** $(4x)^2$      **32.** $-4x^2$

**33.** $(-4x)^2$      **34.** $x^2 + y^2$      **35.** $-x^2$

**36.** $(-x)^2$      **37.** $y^2 - x^2$      **38.** $(y - x)^2$

*Write in scientific notation.*

**39.** 33,000      **40.** 0.000004      **41.** 93,500,000      **42.** 0.000000000254

*Perform the indicated operations using scientific notation.*

**43.** $(5 \times 10^3)(1 \times 10^4)$      **44.** $(60,000,000)(50,000)$      **45.** $\dfrac{4.4 \times 10^{15}}{1.1 \times 10^{-3}}$      **46.** $\dfrac{0.00000008}{0.04}$

## FOR REVIEW

*Remove parentheses and collect or combine like terms.*

**47.** $-(a - 6)$      **48.** $4y - (6y - 3)$      **49.** $-(2 - y) + (2 + y)$      **50.** $3[w - (2w - 4)]$

*The following exercises review material from Section 1.5 to help prepare you for the next chapter. Write the related addition or subtraction sentence.*

**51.** $x + 9 = 6$

**52.** $a - 8 = 1$

**53.** $x + \frac{1}{4} = \frac{1}{2}$

**54.** $a - \frac{1}{6} = \frac{1}{3}$

**55.** $x + 1.8 = -2.2$

**56.** $a - 5.1 = -8.3$

## 9.7  EXERCISES C

*Simplify and write without negative exponents.*

**1.** $2x^{-2}x^5x^3$

**2.** $\dfrac{x^{-2}y^4}{x^3y^{-5}}$

**3.** $(2a^{-2})^3$

$\left[\text{Answer: } \frac{8}{a^6}\right]$

**4.** $(3x^2)^{-3}$

**5.** $\left(\dfrac{2y}{x^3}\right)^2$

**6.** $\left(\dfrac{2y}{x^3}\right)^{-2}$

$\left[\text{Answer: } \frac{x^6}{4y^2}\right]$

## CHAPTER 9 REVIEW

### KEY WORDS

**9.1**  The set of **integers** is the set of whole numbers and their negatives.

The **absolute value** of a number $a$ is the number of units from zero to $a$ on a number line.

**9.5**  The **rational numbers** are all fractions formed by dividing pairs of integers (excluding dividing by zero).

The **irrational numbers** are all numbers on a number line that are not rational.

The **real numbers** are the rational numbers together with the irrational numbers.

**9.6**  **Variables** are letters used to represent numbers.

An **algebraic expression** contains variables and numbers combined by the basic operations of addition, subtraction, multiplication, and division.

A **term** is a part of an algebraic expression that is a product of numbers and variables and is separated from the rest of the expression by plus (or minus) signs.

**Factors** are numbers or letters that are multiplied in a term.

The **coefficient** of a term is the numerical factor. **Similar** or **like terms** contain the same variable.

**9.7**  An **exponential expression** is an expression containing numbers or variables raised to a power.

An **exponent** is a number that tells how many times a number is multiplied by itself.

A number is written in **scientific notation** if it is the product of a number between 1 and 10 and a power of 10.

## KEY CONCEPTS

**9.2**  **1.** To add two integers with like signs, add their absolute values. The sum has the same sign as the numbers being added.

**2.** To add two integers with unlike signs, take the absolute value of each and subtract the smaller from the larger. Give the result the sign of the number with the larger absolute value. If the absolute values are the same, the sum is zero.

**9.3**  To subtract two integers, change the sign of the subtrahend and add.

**9.4**  When two integers have like signs, their product or quotient is positive. When they have unlike signs, their product or quotient is negative.

**9.6**  **1.** Commutative Laws:
$a + b = b + a$    and
$a \cdot b = b \cdot a$

**2.** Associative Laws:
$(a + b) + c = a + (b + c)$    and
$(a \cdot b) \cdot c = a \cdot (b \cdot c)$

**3.** Identities:
$a + 0 = 0 + a = a$    and
$a \cdot 1 = 1 \cdot a = a$

**4.** Negatives:
$a + (-a) = (-a) + a = 0$

**5.** Reciprocals:
$a \cdot \dfrac{1}{a} = \dfrac{1}{a} \cdot a = 1$

**6.** To evaluate an expression involving grouping symbols, start with the innermost symbols of grouping first.

**7.** Distributive Laws:
$a(b + c) = a \cdot b + a \cdot c$    and
$a(b - c) = a \cdot b - a \cdot c$

**8.** When removing parentheses preceded by a minus sign, change *all* signs within the parentheses.

**9.7**  **1.** Rules of Exponents:
$a^m \cdot a^n = a^{m+n}$

$\dfrac{a^m}{a^n} = a^{m-n} \quad (a \neq 0)$

$(a^m)^n = a^{mn}$

$(a \cdot b)^n = a^n \cdot b^n$

$\left(\dfrac{a}{b}\right)^n = \dfrac{a^n}{b^n} \quad (b \neq 0)$

$a^0 = 1 \quad (a \neq 0)$

$a^{-n} = \dfrac{1}{a^n} \quad (a \neq 0)$

**2.** To put a number in scientific notation, write it as the product of a number between 1 and 10 and a power of 10.

## REVIEW EXERCISES

*Part I*

**9.1**  *Is the given statement true or false? If the statement is false, tell why.*

**1.** $-3 > -1$    **2.** $-7 < 1$    **3.** $-6 < -2$

**4.** $|-8| = 8$    **5.** $|0| = 0$    **6.** $|1| = 1$

**9.2**  *Perform the indicated operations.*

**7.** $(-5) + 2$    **8.** $(-5) + (-2)$    **9.** $5 + (-2)$

**10.** $(-4) + 0$    **11.** $6 + (-6)$    **12.** $(-12) + (-12)$

**13.** $16 + (-4)$    **14.** $8 + (-12)$    **15.** $(-5) + (-2)$

**9.3**  *Perform the indicated operations.*

**16.** $(-5) - (-2)$    **17.** $5 - (-2)$    **18.** $(-4) - 0$

**19.** $6 - (-6)$    **20.** $20 - (-35)$    **21.** $(-9) - (-16)$

**22.** $(-3) + 8 - (+3) + (-9) + (-1) - 0$    **23.** $64 + (-81) - (+92) + (-3)$

**9.4**   *Perform the indicated operations.*

**24.** $(3)(-6)$

**25.** $(-8)(-3)$

**26.** $0 \cdot (-8)$

**27.** $(-12)(10)$

**28.** $(-12)(-10)$

**29.** $(-5)(4)$

**30.** $(-6) \div (-2)$

**31.** $(-6) \div 2$

**32.** $6 \div (-2)$

**33.** $(-25) \div 5$

**34.** $0 \div 18$

**35.** $(-18) \div 0$

**36.** $\dfrac{48}{-6}$

**37.** $\dfrac{-24}{3}$

**38.** $\dfrac{-600}{-60}$

**9.5**   *Is the given statement true or false? If the statement is false, tell why.*

**39.** $\dfrac{2}{3} < \dfrac{4}{3}$

**40.** $-1.5 > -1$

**41.** $|-1.2| = -1.2$

*Perform the indicated operations.*

**42.** $\dfrac{1}{3} + \dfrac{4}{3}$

**43.** $\dfrac{2}{3} + \left(-\dfrac{4}{3}\right)$

**44.** $(-1.3) + (-8.2)$

**45.** $\left(-\dfrac{3}{8}\right) - \left(-\dfrac{1}{4}\right)$

**46.** $9.1 - (-7.8)$

**47.** $4 - \left(-\dfrac{4}{5}\right)$

**48.** $\left(-\dfrac{3}{4}\right)\left(-\dfrac{2}{9}\right)$

**49.** $\left(\dfrac{16}{5}\right)\left(-\dfrac{35}{8}\right)$

**50.** $(-4)(-2.4)$

**51.** $\left(\dfrac{5}{3}\right) \div \left(-\dfrac{2}{9}\right)$

**52.** $\left(-\dfrac{6}{5}\right) \div \left(-\dfrac{24}{15}\right)$

**53.** $\dfrac{24.8}{-4}$

**54.** Maria's account had $182.32 in it on June 1. She deposited $842.12 and wrote checks for $32.56, $423.80, and $67.20 during the month. What was her balance at the end of June? Could she pay cash for a $600 stereo?

**9.6**   *Find the negative of each of the given numbers.*

**55.** $9$

**56.** $-\dfrac{1}{2}$

**57.** $0$

*Find the reciprocal of each of the given numbers.*

**58.** $-4$

**59.** $\dfrac{7}{4}$

**60.** $-\dfrac{7}{4}$

*Evaluate for $a = 4$, $b = -2$, and $c = -3$.*

**61.** $a + b + c$

**62.** $2a - 3b - c$

**63.** $b - (c - a)$

**64.** $\dfrac{a}{b} - 3c$

**65.** $2[b - (a - c)]$

**66.** $|b - a|$

*How many terms does each expression have?*

**67.** $2x + 3y + 5$

**68.** $4z$

**69.** $2a + b + c - 5d$

*Use the distributive laws to factor.*

**70.** $6x - 6y$

**71.** $5a + 5b + 5c$

**72.** $ab + ac$

**73.** $5x + 10y$

**74.** $6x + 9$

**75.** $-8x - 8y$

*Multiply.*

**76.** $8(x + y)$

**77.** $3(x - y)$

**78.** $6(a - b + 2c)$

**79.** $-7(x + y - z)$

**80.** $3(2a - 5b - 3c)$

**81.** $-3(2a - 5b - 3c)$

*Remove the parentheses and collect or combine like terms.*

**82.** $x - (2x + 1)$

**83.** $x - (2x - 1)$

**84.** $3x + (2x - 5)$

**85.** $-(a - 1) + (1 - a)$

**86.** $3[2x - (3x - 2)]$

**87.** $-3[(2 - y) - 3y]$

**9.7** *Simplify and write without negative exponents.*

**88.** $(2x^2)^{-2}$

**89.** $\dfrac{a^{-2}}{a^3}$

**90.** $\dfrac{3^2 x^{-1}}{x^4}$

**91.** $2^0 y^3 y^{-3}$

*Evaluate when $a = -3$ and $b = 1$.*

**92.** $4a^2$

**93.** $(4a)^2$

**94.** $-4a^2$

**95.** $|a^2 + b^2|$

**96.** $-(-a)$

**97.** $(-2a)^{-1}$

*Write in scientific notation.*

**98.** $360,000$

**99.** $0.000000125$

**100.** $2.6$

**Part II**

*Evaluate for $x = -3$ and $y = -5$.*

**101.** $x^2 + y$

**102.** $3[x - (2 - y)]$

**103.** $(y - x)^2$

*Use the distributive laws to factor.*

**104.** $2x + 4y$

**105.** $35 - 5b$

**106.** $6x + 9y - 12z$

*Perform the indicated operations.*

**107.** $(-8)(-10)$

**108.** $-4 - (-2)$

**109.** $(-12) \div 4$

**110.** $6 - (-3) + (-4)$

**111.** $\left(-\dfrac{2}{5}\right) \div \dfrac{4}{35}$

**112.** $-\dfrac{3}{4} - \left(-\dfrac{1}{2}\right)$

**113.** $\left(-\dfrac{4}{5}\right)\left(\dfrac{55}{12}\right)$

**114.** $(-16) + (-8) - 24$

**115.** $3.2 + (-6.1)$

**116.** $0 \div \dfrac{3}{2}$

**117.** $(2.4)(-1.5)$

**118.** $24 \div (-2.4)$

*Use scientific notation to calculate.*

**119.** $\dfrac{0.00016}{0.08}$

**120.** $(2,400,000)(0.00015)$

*Remove parentheses and collect or combine like terms.*

**121.** $-2[3y - (2 - y) + 6]$

**122.** $-(4x + 1) - (-4x + 1)$

*Simplify and write without negative exponents.*

**123.** $\dfrac{2^0 x^{-1}}{x^2}$

**124.** $\dfrac{3^{-2} y^{-3}}{y}$

**125.** $4^{-1} x^{-2} y^5$

*Is the given statement true or false? If the statement is false, tell why.*

**126.** $-16 > -2$

**127.** $\dfrac{2}{7} < \dfrac{3}{11}$

**128.** $|-3.7| = 3.7$

**129.** If Henry has \$137.75 in the bank, can he buy 6 shirts which cost \$22.75 each including tax?

**130.** Diana had \$2316.20 in an account. If during the month the account earned \$25.65 interest and she withdrew \$420.16 twice, what was her balance at the end of the month?

---

**ANSWERS:**  1. false ($<$)  2. true  3. true  4. true  5. true  6. true  7. $-3$  8. $-7$  9. 3  10. $-4$  11. 0  12. $-24$  13. 12  14. $-4$  15. $-7$  16. $-3$  17. 7  18. $-4$  19. 12  20. 55  21. 7  22. $-8$  23. $-112$  24. $-18$  25. 24  26. 0  27. $-120$  28. 120  29. $-20$  30. 3  31. $-3$  32. $-3$  33. $-5$  34. 0  35. undefined  36. $-8$

37. $-8$  38. 10  39. true  40. false ($<$)  41. false $|-1.2| = 1.2$  42. $\frac{5}{3}$  43. $-\frac{2}{3}$  44. $-9.5$  45. $-\frac{1}{8}$  46. 16.9

47. $\frac{24}{5}$  48. $\frac{1}{6}$  49. $-14$  50. 9.6  51. $-\frac{15}{2}$  52. $\frac{3}{4}$  53. $-6.2$  54. \$500.88; no, she would have $-$\$99.12 balance

55. $-9$  56. $\frac{1}{2}$  57. 0  58. $-\frac{1}{4}$  59. $\frac{4}{7}$  60. $-\frac{4}{7}$  61. $-1$  62. 17  63. 5  64. 7  65. $-18$  66. 6  67. 3  68. 1

69. 4  70. $6(x - y)$  71. $5(a + b + c)$  72. $a(b + c)$  73. $5(x + 2y)$  74. $3(2x + 3)$  75. $-8(x + y)$  76. $8x + 8y$
77. $3x - 3y$  78. $6a - 6b + 12c$  79. $-7x - 7y + 7z$  80. $6a - 15b - 9c$  81. $-6a + 15b + 9c$  82. $-x - 1$
83. $-x + 1$  84. $5x - 5$  85. $2 - 2a$  86. $-3x + 6$  87. $-6 + 12y$  88. $\frac{1}{4x^4}$  89. $\frac{1}{a^5}$  90. $\frac{9}{x^5}$  91. 1  92. 36  93. 144
94. $-36$  95. 10  96. $-3$  97. $\frac{1}{6}$  98. $3.6 \times 10^5$  99. $1.25 \times 10^{-7}$  100. $2.6 \times 10^0$  101. 4  102. $-30$  103. 4
104. $2(x + 2y)$  105. $5(7 - b)$  106. $3(2x + 3y - 4z)$  107. 80  108. $-2$  109. $-3$  110. 5  111. $-\frac{7}{2}$  112. $-\frac{1}{4}$
113. $-\frac{11}{3}$  114. $-48$  115. $-2.9$  116. 0  117. $-3.6$  118. $-10$  119. $2 \times 10^{-3}$  120. $3.6 \times 10^2$  121. $-8y - 8$
122. $-2$  123. $\frac{1}{x^3}$  124. $\frac{1}{9y^4}$  125. $\frac{y^5}{4x^2}$  126. false ($<$)  127. false ($>$)  128. true  129. yes (shirts cost \$136.50)
130. \$1501.53

*Answer true or false. If the answer is false, tell why.*

**1.** $-1.3 < -\dfrac{3}{2}$

1. _____

**2.** $|-11| = 11$

2. _____

*Perform the indicated operations.*

**3.** $(-7) + 8$

3. _____

**4.** $(-5) - (-4) + (-2)$

4. _____

**5.** $(-6)(4)$

5. _____

**6.** $(-24) \div 3$

6. _____

**7.** $0 \div (-2)$

7. _____

**8.** $\left(-\dfrac{3}{4}\right) - \left(-\dfrac{2}{3}\right)$

8. _____

**9.** $\left(-\dfrac{7}{5}\right) \div \left(-\dfrac{14}{25}\right)$

9. _____

**10.** $(-3,2)(1.5)$

10. _____

**11.** Find the negative of $-2$.

11. _____

**12.** Find the reciprocal of $\frac{7}{12}$.

12. _____

**13.** Factor:  $12x - 8y$

13. _____

**14.** Multiply:  $7(a - 3b)$

14. _____

**15.** Remove parentheses and collect like terms.
$7x - 3(x - 1)$

15. _____

*Simplify and write without negative exponents.*

**16.** $5^0 a^3 a^{-5}$

16. _____

**17.** $\dfrac{3^{-1} x^{-1}}{x^2}$

17. _____

**18.** $(a^{-1} b^2)^2$

18. _____

*Evaluate for $x = -2$ and $y = 3$.*

**19.** $x + 2y$

19. _____

**20.** $|x - y|$

20. _____

**21.** $x - (2x + y)$

21. _____

**22.** Write 1,420,000 in scientific notation.

22. _____

**23.** Write 0.0000793 in scientific notation.

23. _____

**24.** Sam has \$216.18 in his account. How much does he have left after writing three checks for \$47.31 each?

24. _____

# Solving Equations

## 10.1 LINEAR EQUATIONS AND THE ADDITION-SUBTRACTION RULE

**STUDENT GUIDEPOSTS**

1. Equations
2. Solving Equations
3. Types of Equations
4. Addition-Subtraction Rule

### 1 EQUATIONS

An **equation** is a statement that two quantities or expressions are equal. The two quantities are written with an equal sign (=) between them. The expression to the left of the equal sign is called the **left side** of the equation and the expression to the right of the equal sign is the **right side** of the equation.

$$x + 5 = 3 + 4$$
left side    right side

Some equations are true, some are false, and for some, the truth value cannot be determined. For example,

$3 + 4 = 7$ is true,

$3 + 4 = 7 - 5$ is false, and

$x + 3 = 5$ is neither true nor false since the value of $x$ is not known.

### 2 SOLVING EQUATIONS

The equation $x + 3 = 5$ contains a variable. A variable was defined in Section 9.6 as a letter which represents a number. When the variable in an equation can be replaced by a number which makes the resulting equation true, that number is called a **solution** of the equation. The process of finding all solutions is called **solving the equation.**

### 3 TYPES OF EQUATIONS

In this chapter, we are concerned with **linear equations,** that is, equations in which the variable occurs to the first power only. The equation

$$x + 3 = 7$$

has 4 as a solution because when $x$ is replaced by 4,

$$+ 3 = 7$$

559

is a true equation. 6 is not a solution since $6 + 3 = 7$ is false. An equation such as this, which is true for some replacements of the variable and false for others, is called a **conditional equation.**

The equation

$$x + 2 = x + 2$$

has many solutions. In fact, every number is a solution. An equation like this is called an **identity.** The equation

$$x + 2 = x$$

has no solutions. An equation like this is called a **contradiction.**

Some simple equations may be solved by inspection. Others require more sophisticated techniques. Consider the equations

$$x - 3 = 5 \quad \text{and} \quad x = 8.$$

Both have 8 as a solution. However, it is easier to see that 8 is a solution to the second. When two equations have the same solution the equations are **equivalent.**

## ➍ ADDITION-SUBTRACTION RULE

Whenever an equation has the variable by itself on one side of the equal sign and a number by itself on the other, the solution to the equation is obvious. The process of solving an equation consists of changing it into an equivalent equation of this type. The first rule for solving an equation is the addition-subtraction rule.

> ### · The Addition-Subtraction Rule
>
> If the same number or expression is added to or subtracted from both sides of an equation, an equivalent equation is obtained.

### HINT

In Section 1.5 we introduced the related addition and subtraction sentences or equations. For example,

$$x - 3 = 5 \quad \text{has related addition sentence} \quad x = 5 + 3,$$

and

$$y + 2 = 9 \quad \text{has related subtraction sentence} \quad y = 9 - 2.$$

Thus, the addition-subtraction rule is simply a restatement of these concepts within the context of solving equations.

---

### EXAMPLE 1    USING THE ADDITION-SUBTRACTION RULE

Solve.

$$x - 3 = 5$$
$$x - 3 + 3 = 5 + 3 \qquad \text{Add 3 to both sides}$$
$$x + 0 = 8$$
$$x = 8 \qquad \text{0 is the additive identity}$$

The solution is 8.

### PRACTICE EXERCISE 1

Solve. $x - 8 = 4$

Answer: 12

| EXAMPLE 2   USING THE ADDITION-SUBTRACTION RULE | PRACTICE EXERCISE 2 |
|---|---|

Solve.

$$y + 5 = 9$$
$$y + 5 - 5 = 9 - 5 \qquad \text{Subtract 5 from both sides}$$
$$y + 0 = 4$$
$$y = 4 \qquad \text{0 is the additive identity}$$

The solution is 4.

Solve. $y + 10 = 2$

Answer: $-8$

Get in the habit of checking all indicated solutions to an equation. To check, replace the variable in the original equation with the solution you found. The solution is correct if a true equation results.

The solution in Example 1 is correct because

$$8 - 3 = 5. \qquad \text{Replace } x \text{ with 8 in } x - 3 = 5$$

The solution in Example 2 is correct because

$$4 + 5 = 9. \qquad \text{Replace } y \text{ with 4 in } y + 5 = 9$$

| EXAMPLE 3   USING THE ADDITION-SUBTRACTION RULE | PRACTICE EXERCISE 3 |
|---|---|

Solve and check.

$$5 = z + 8$$
$$5 - 8 = z + 8 - 8 \qquad \text{Subtract 8 from both sides}$$
$$-3 = z + 0$$
$$-3 = z \qquad \text{Check the solution } -3$$
$$\text{Check:} \quad 5 \overset{?}{=} -3 + 8 \qquad \text{Replace } z \text{ with } -3 \text{ in the original equation}$$
$$5 = 5 \qquad \text{The solution } -3 \text{ does check}$$

Solve. $9 = z - 7$

Answer: 16

**HINT**

The addition-subtraction rule can be remembered with the phrase "add or subtract the same thing on both sides."

## 10.1 EXERCISES A

*Give an example of each of the following.*

**1.** A true equation

**2.** A contradiction

**3.** A linear equation

*Solve the following equations by direct observation or inspection.*

**4.** $x + 3 = 5$

**5.** $2y = 8$

**6.** $2 + z = 2 - z$

**7.** $2 + x = 2 + x$       **8.** $y + 5 = y + 1$       **9.** $2z = 5z$

*Use the equation $y - 1 = 5$ to answer Exercises 10–15.*

**10.** What is the variable?

**11.** What is the left side of the equation?

**12.** What is the right side of the equation?

**13.** What is the solution?

**14.** Is it an identity?

**15.** Is it a contradiction?

*Use the addition-subtraction rule to solve. Check all solutions.*

**16.** $x + 3 = 7$       **17.** $y - 4 = 9$       **18.** $z + 12 = -4$

**19.** $x - 9 = -6$       **20.** $x - 5 = -4$       **21.** $y - 3 = 10$

**22.** $4 = x + 1$       **23.** $-7 = y - 3$       **24.** $-3 = z + 2$

**25.** $\dfrac{1}{3} + x = 5$       **26** $-\dfrac{1}{4} + z = \dfrac{3}{4}$       **27.** $\dfrac{1}{2} + y = -3$

**28.** $x - 2.1 = 3.4$       **29** $-2.6 = z - 3.9$       **30.** $-4.2 = y + 1.7$

*Exercises 31–33 review the related division sentence introduced in Section 3.4 to help you prepare for the next section. Write the related division sentence.*

**31.** $2 \cdot 3 = 6$       **32.** $\dfrac{1}{2} \cdot 10 = 5$       **33.** $9 \cdot \dfrac{2}{3} = 6$

ANSWERS: 1–3. Answers will vary. 4. 2   5. 4   6. 0   7. Any number is a solution.   8. no solution   9. 0   10. $y$ 11. $y - 1$   12. 5   13. 6   14. no   15. no   16. 4   17. 13   18. $-16$   19. 3   20. 1   21. 13   22. 3   23. $-4$   24. $-5$ 25. $\frac{14}{3}$   26. 1   27. $-\frac{7}{2}$   28. 5.5   29. 1.3   30. $-5.9$   31. $3 = \frac{6}{2}$ or $2 = \frac{6}{3}$   32. $10 = 5 \div \frac{1}{2}$ or $\frac{1}{2} = \frac{5}{10}$   33. $\frac{2}{3} = \frac{6}{9}$ or $9 = 6 \div \frac{2}{3}$

## 10.1 EXERCISES B

*Give an example of each of the following.*

**1.** A false equation

**2.** An identity

**3.** An equation which is neither true nor false

*Solve the following equations by direct observation or inspection.*

**4.** $x + 5 = 12$       **5.** $3y = 15$       **6.** $4 + z = 4 - z$

**7.** $1 + x = 1 + x$

**8.** $y + 2 = y + 8$

**9.** $8z = 3z$

*Use the equation $x + 3 = 19$ to answer Exercises 10–15.*

**10.** What is the variable?

**11.** What is the left side of the equation?

**12.** What is the right side of the equation?

**13.** What is the solution?

**14.** Is it an identity?

**15.** Is it a contradiction?

*Solve and check.*

**16.** $x + 2 = 12$

**17.** $y - 3 = 10$

**18.** $z + 13 = -2$

**19.** $x - 11 = -9$

**20.** $x - 7 = -1$

**21.** $y - 14 = -12$

**22.** $5 = x + 7$

**23.** $-3 = y - 3$

**24.** $-4 = z + 3$

**25.** $\dfrac{1}{3} + x = 8$

**26.** $\dfrac{1}{3} + y = \dfrac{7}{3}$

**27.** $-\dfrac{1}{4} + z = 1$

**28.** $x - 2.3 = 5.7$

**29.** $-5.7 = y - 2.3$

**30.** $-3.8 = z - 1.2$

*Exercises 31–33 review the related division sentence introduced in Section 3.4 to help you prepare for the next section. Write the related division sentence.*

**31.** $4 \cdot 5 = 20$

**32.** $\dfrac{1}{3} \cdot 12 = 4$

**33.** $8 \cdot \dfrac{3}{4} = 6$

## 10.1 EXERCISES C

*Solve.*

**1.** $x + 2\dfrac{1}{2} = 5$

$\left[\text{Answer: } 2\tfrac{1}{2}\right]$

**2.** $y - 3\dfrac{2}{3} = 4$

**3.** $1\dfrac{3}{8} + z = \dfrac{1}{4}$

**4.** $x + 0.526 = 2.378$
[Answer: 1.852]

**5.** $y - 1.775 = 0.239$

**6.** $4.258 + z = 1.007$

## 10.2 THE MULTIPLICATION-DIVISION RULE

In Section 10.1 we changed equations into equivalent equations by adding or subtracting the same number on both sides. The same technique works for multiplication and division and follows from the concept of the related division sentence discussed in Section 3.4. For example,

$$2 \cdot 5 = 10 \qquad \text{has related division sentence} \qquad 5 = \frac{10}{2} \text{ or } 2 = \frac{10}{5}.$$

The same is true for equations involving variables. For example,

$$2x = 10 \quad \text{has related division sentence} \quad x = \frac{10}{2},$$

which gives $x = 5$ as the solution to the equation. These remarks lead to the next equation-solving rule.

---

### Multiplication-Division Rule

If both sides of an equation are multiplied or divided by the same non-zero number or expression, an equivalent equation is obtained.

---

### EXAMPLE 1 USING THE MULTIPLICATION-DIVISION RULE

Solve.

$$3x = 12$$

$$\frac{1}{3} \cdot 3x = \frac{1}{3} \cdot 12 \qquad \text{Multiply each side by } \tfrac{1}{3}, \text{ the reciprocal of 3}$$

$$1 \cdot x = 4 \qquad \tfrac{1}{3} \cdot 3 = 1 \text{ and } \tfrac{1}{3} \cdot 12 = \tfrac{12}{3} = 4$$

$$x = 4$$

The solution is 4.

Check: $3 \cdot 4 \overset{?}{=} 12 \qquad$ Replace $x$ with 4 in $3x = 12$

$\qquad\quad 12 = 12$

### PRACTICE EXERCISE 1

Solve. $9x = 36$

Answer: 4

---

Another way to solve the equation in Example 1 is to divide both sides of the equation by 3.

$$3x = 12$$

$$\frac{3x}{3} = \frac{12}{3} \qquad \text{Divide by 3}$$

$$\frac{3}{3} \cdot x = 4$$

$$1 \cdot x = 4$$

$$x = 4 \qquad \text{The same solution}$$

---

### EXAMPLE 2 USING THE MULTIPLICATION-DIVISION RULE

Solve.

$$\frac{1}{2}x = 5$$

$$2 \cdot \frac{1}{2}x = 2 \cdot 5 \qquad \text{Multiply each side by 2, the reciprocal of } \tfrac{1}{2}$$

$$1 \cdot x = 10$$

$$x = 10$$

The solution is 10.

### PRACTICE EXERCISE 2

Solve. $\frac{1}{5}x = 7$

Check: $\frac{1}{2} \cdot 10 \stackrel{?}{=} 5$     Replace $x$ with 10 in the original equation

$5 = 5$     Answer: 35

---

| EXAMPLE 3 AN EQUATION INVOLVING DECIMALS | PRACTICE EXERCISE 3 |
|---|---|

Solve.

$$-4.5z = 36$$
$$\frac{-4.5z}{-4.5} = \frac{36}{-4.5}$$     Divide each side by $-4.5$
$$1 \cdot z = -8$$
$$z = -8$$

The solution is $-8$. Check.

Solve. $-2.8y = -56$

Answer: 20

---

### **HINT**

From the above examples we see that it is most helpful to multiply or divide both sides by a number that will make the coefficient of the variable 1. The multiplication-division rule can be remembered as "multiply or divide both sides of the equation by the same thing (zero excluded)."

---

## 10.2 EXERCISES A

*Solve and check.*

**1.** $4x = 16$

**2.** $2y = 12$

**3.** $20 = 4z$

**4.** $-11x = 66$

**5.** $15y = -75$

**6.** $-48 = 6z$

**7.** $\frac{1}{3}x = 2$

**8.** $\frac{1}{5}y = -2$

**9.** $-\frac{1}{8}z = 5$

**10.** $\frac{2}{3}x = 8$

**11** $\frac{3}{4}y = -6$

**12.** $-\frac{2}{7}z = 10$

**13.** $2x = \frac{1}{3}$

**14.** $3y = -\frac{3}{2}$

**15.** $-4z = \frac{2}{3}$

**16.** $2x = 14.6$

**17** $-2.5y = 7.5$

**18.** $-0.7z = -5.6$

## FOR REVIEW

**19.** $x + 12 = -10$

**20.** $y - \dfrac{3}{2} = 4$

**21.** $z + 2.7 = 7.3$

**22.** $x + \dfrac{3}{4} = 2$

**23.** $y - \dfrac{2}{5} = \dfrac{1}{10}$

**24.** $z + \dfrac{1}{3} = \dfrac{3}{5}$

*Exercises 25–28 review material from Section 9.6 to help you prepare for the next section. Remove parentheses and collect or combine like terms.*

**25.** $2 + 4(x + 1)$

**26.** $3x - 2(x + 4)$

**27.** $4x - (x - 1) + x$

**28.** $2(4x + 1) - 3x$

---

ANSWERS:  1. 4  2. 6  3. 5  4. −6  5. −5  6. −8  7. 6  8. −10  9. −40  10. 12  11. −8  12. −35  13. $\frac{1}{6}$  14. $-\frac{1}{2}$  15. $-\frac{1}{6}$  16. 7.3  17. −3  18. 8  19. −22  20. $\frac{11}{2}$  21. 4.6  22. $\frac{5}{4}$  23. $\frac{1}{2}$  24. $\frac{4}{15}$  25. $4x + 6$  26. $x - 8$  27. $4x + 1$  28. $5x + 2$

## 10.2 EXERCISES B

*Solve and check.*

**1.** $7x = 49$

**2.** $4y = 20$

**3.** $72 = 6z$

**4.** $-11x = 44$

**5.** $12y = -96$

**6.** $-63 = -z$

**7.** $\dfrac{2}{3}x = 14$

**8.** $\dfrac{y}{5} = -3$

**9.** $-\dfrac{2}{7}z = 8$

**10.** $\dfrac{3x}{5} = 6$

**11.** $\dfrac{4y}{3} = -2$

**12.** $\dfrac{8z}{-5} = -16$

**13.** $3x = \dfrac{1}{4}$

**14.** $4y = -\dfrac{4}{5}$

**15.** $-2z = \dfrac{6}{7}$

**16.** $3x = -25.5$

**17.** $1.2y = 8.4$

**18.** $-1.2z = -4.8$

## FOR REVIEW

*Solve.*

**19.** $-21 + x = -25$

**20.** $-\dfrac{5}{6} + y = \dfrac{7}{12}$

**21.** $1.9 + z = -6.7$

**22.** $x + \dfrac{3}{5} = 1$

**23.** $y - \dfrac{1}{4} = \dfrac{3}{8}$

**24.** $z + \dfrac{1}{2} = \dfrac{5}{3}$

*Exercises 25–28 review material from Section 9.6 to help you prepare for the next section. Remove parentheses and collect or combine like terms.*

**25.** $3 + 2(x + 5)$

**26.** $4x - 3(x + 1)$

**27.** $5x - (x - 2) + x$

**28.** $3(7x - 1) + 3$

## 10.2 EXERCISES C

*Solve.*

**1.** $2\dfrac{1}{3}x = -14$

[Answer: $-6$]

**2.** $-0.035y = 0.0875$

**3.** $-2.418z = -12.09$

## 10.3 SOLVING EQUATIONS BY COMBINING RULES

=== STUDENT GUIDEPOSTS ===

**1** Using Both Rules to Solve an Equation

**2** Equations with Like Terms

**3** Equations with Parentheses

**4** Summary of the Method for Solving Linear Equations

### **1** USING BOTH RULES TO SOLVE AN EQUATION

To solve an equation, we isolate the variable on one side so the solution can be found by inspection. Sometimes both the addition-subtraction rule and the multiplication-division rule are needed to do this.

### **HINT**

When both rules are needed to solve an equation, generally, the addition-subtraction rule is used before the multiplication-division rule, as in the following example.

---

**EXAMPLE 1  USING BOTH RULES**

Solve.

$$3x - 5 = 13$$
$$3x - 5 + 5 = 13 + 5 \qquad \text{Add 5 to both sides}$$
$$3x + 0 = 18$$
$$3x = 18$$
$$\frac{1}{3} \cdot 3x = \frac{1}{3} \cdot 18 \qquad \text{Multiply both sides by } \tfrac{1}{3}$$
$$1 \cdot x = 6$$
$$x = 6$$

The solution is 6.

**PRACTICE EXERCISE 1**

Solve. $4x + 7 = 13$

Check:  $3 \cdot 6 - 5 \stackrel{?}{=} 13$

$18 - 5 \stackrel{?}{=} 13$

$13 = 13$

Answer: $\frac{3}{2}$

## ❷ EQUATIONS WITH LIKE TERMS

Some equations have like terms on one or both sides. In order to solve these equations, collect or combine all like terms before applying the addition-subtraction and multiplication-division rules.

| EXAMPLE 2    COMBINING LIKE TERMS FIRST | PRACTICE EXERCISE 2 |

Solve.

$$7 = 3y + 2y$$

$$7 = 5y \qquad \text{Combine like terms}$$

$$\frac{1}{5} \cdot 7 = \frac{1}{5} \cdot 5y \qquad \text{Multiply both sides by } \frac{1}{5}$$

$$\frac{7}{5} = 1 \cdot y$$

$$\frac{7}{5} = y$$

The solution is $\frac{7}{5}$. Check this by substituting in the original equation.

Solve. $5z - 8z = 9$

Answer: $-3$

Some equations have terms with the variable on both sides. The addition-subtraction rule can be used to get them together on the same side so that they can be combined.

| EXAMPLE 3    LIKE TERMS WITH ADDITION AND SUBTRACTION | PRACTICE EXERCISE 3 |

Solve.

$$5x - 3 = 2x + 6$$

$$5x - 3 + 3 = 2x + 6 + 3 \qquad \text{Add 3 to both sides}$$

$$5x + 0 = 2x + 9$$

$$5x = 2x + 9$$

$$5x - 2x = 2x - 2x + 9 \qquad \text{Subtract } 2x \text{ from both sides}$$

$$3x = 0 + 9$$

$$3x = 9$$

$$\frac{1}{3} \cdot 3x = \frac{1}{3} \cdot 9 \qquad \text{Multiply both sides by } \frac{1}{3}$$

$$1 \cdot x = 3$$

$$x = 3$$

The solution is 3.

Solve. $3y + 11 = 7y - 5$

Check:   $5 \cdot 3 - 3 \overset{?}{=} 2 \cdot 3 + 6$

$\quad\quad\quad 15 - 3 \overset{?}{=} 6 + 6$

$\quad\quad\quad\quad 12 = 12$

Answer:  4

---

We see from Example 3 that it is important to get all terms involving the variable on one side of the equation, and then use the multiplication-division rule to make the coefficient of the variable 1. Remember to collect or combine like terms whenever it is appropriate.

| **EXAMPLE 4**   SOLVING A MORE COMPLEX EQUATION | **PRACTICE EXERCISE 4** |

Solve.

Solve. $8 - 3x + 7 = -6x + 29 - 4x$

$$-3 - x + 10 = 4x + 14 + 2x$$

$\quad\quad -x + 7 = 6x + 14 \quad\quad$ Collect or combine like terms

$\quad -x + 7 - 7 = 6x + 14 - 7 \quad$ Subtract 7 from both sides

$\quad\quad\quad -x = 6x + 7$

$\quad -x - 6x = 6x - 6x + 7 \quad$ Subtract 6x from both sides

$\quad\quad\quad -7x = 7$

$\left(-\dfrac{1}{7}\right)(-7)x = \left(-\dfrac{1}{7}\right) \cdot 7 \quad$ Multiply both sides by $-\frac{1}{7}$

$\quad\quad\quad 1 \cdot x = -1$

$\quad\quad\quad\quad x = -1$

Check the solution, $-1$.

Answer:  2

## ③ EQUATIONS WITH PARENTHESES

To solve an equation containing parentheses, first remove the parentheses using the distributive property. The resulting equation can then be solved using previous methods.

| **EXAMPLE 5**   SOLVING AN EQUATION WITH PARENTHESES | **PRACTICE EXERCISE 5** |

Solve.

Solve. $5(x + 3) - 10 = -5$

$\quad\quad 3(x - 1) + 5 = -4 \quad\quad$ Remove parentheses by multiplying

$\quad\quad 3x - 3 + 5 = -4 \quad\quad\quad$ both $x$ and 1 by 3

$\quad\quad\quad 3x + 2 = -4$

$\quad 3x + 2 - 2 = -4 - 2 \quad$ Subtract 2

$\quad\quad\quad\quad 3x = -6$

$\quad\quad \dfrac{1}{3} \cdot 3x = \dfrac{1}{3} \cdot (-6) \quad$ Multiply by $\frac{1}{3}$

$\quad\quad\quad 1 \cdot x = -2$

$\quad\quad\quad\quad x = -2$

The solution is $-2$. Check in the original equation.

Answer:  $-2$

## HINT

Remember to remove all parentheses first using the distributive law, then collect or combine like terms and solve.

---

| EXAMPLE 6   AN EQUATION WITH PARENTHESES | PRACTICE EXERCISE 6 |

Solve.

$$5y - (2y - 3) = 8 \qquad \text{Remove parentheses and change signs}$$
$$5y - 2y + 3 = 8$$
$$3y + 3 = 8$$
$$3y + 3 - 3 = 8 - 3$$
$$3y = 5$$
$$\frac{1}{3} \cdot 3y = \frac{1}{3} \cdot 5$$
$$y = \frac{5}{3}$$

The solution is $\frac{5}{3}$. Check in the original equation.

Solve. $9x - (3x + 2) = 7$

$$6x + 2 - 2 = 7 - 2$$
$$\frac{6x}{6} = \frac{5}{6}$$
$$x = 5/6$$

Answer: $\frac{5}{6}$

---

| EXAMPLE 7   AN EQUATION WITH TWO SETS OF PARENTHESES | PRACTICE EXERCISE 7 |

Solve.

$$x - 5(x - 2) = 14 - 2(x + 3)$$
$$x - 5x + 10 = 14 - 2x - 6 \qquad \text{Remove parentheses and change signs}$$
$$-4x + 10 = 8 - 2x \qquad \text{Collect or combine like terms}$$
$$-4x + 2x + 10 = 8 - 2x + 2x \qquad \text{Add } 2x$$
$$-2x + 10 = 8$$
$$-2x + 10 - 10 = 8 - 10 \qquad \text{Subtract 10}$$
$$-2x = -2$$
$$\left(-\frac{1}{2}\right) \cdot (-2)x = \left(-\frac{1}{2}\right) \cdot (-2)$$
$$x = 1$$

The solution is 1.

Check:  $1 - 5(1 - 2) \overset{?}{=} 14 - 2(1 + 3)$
$$1 - 5(-1) \overset{?}{=} 14 - 2(4)$$
$$1 + 5 \overset{?}{=} 14 - 8$$
$$6 = 6$$

Solve. $2x - 3(x + 4) = 5 - 4(x - 1)$

$$2x - 3x - 12 = 5 - 4x + 4$$
$$-1x - 12 = 9 - 4x \;^{+4x}$$
$$3x - 12 \;^{+12} = 9 \;^{+12}$$
$$3x = 21$$
$$x = 7$$

Answer: 7

## ④ SUMMARY OF THE METHOD FOR SOLVING LINEAR EQUATIONS

We now summarize the equation-solving techniques we have learned.

### To Solve a Linear Equation

1. Simplify both sides by removing parentheses and combining like terms.
2. Use the addition-subtraction rule to isolate all variable terms on one side and all constant terms on the other side. Combine like terms when possible.
3. Use the multiplication-division rule to obtain a variable with coefficient 1.
4. Check the solution by substituting in the original equation.
5. If an identity results, the original equation has every real number as a solution. If a contradiction results, there is no solution.

Finally, we conclude this section with examples of two special types of equations.

### EXAMPLE 8    SPECIAL TYPES OF EQUATIONS

Solve.

**(a)**     $2(x + 1) = 2x + 2$

$\qquad 2x + 2 = 2x + 2$    Remove parentheses

$2x - 2x + 2 = 2x - 2x + 2$    Subtract $2x$ from both sides

Since we obtain an identity, $2 = 2$, equivalent to the original equation, the original equation is also an identity and has every number as a solution.

**(b)**  $1 - (x - 2) = 2 - x$

$\qquad 1 - x + 2 = 2 - x$    Remove parentheses and watch the sign

$\qquad 3 - x = 2 - x$    Combine like terms

$\qquad 3 - x + x = 2 - x + x$    Add $x$ to both sides

$\qquad\qquad 3 = 2$

Since we obtain a contradiction, $3 = 2$, the original equation is also a contradiction and has no solution.

### PRACTICE EXERCISE 8

Solve.

**(a)**  $3(x - 3) = 1 + 3x - 10$

**(b)**  $2x - (1 + x) = x$

Answers: **(a) every real number (identity)    (b) no solution (contradiction)**

## 10.3 EXERCISES A

*Solve.*

**1.** $2x + 1 = 5$

**2.** $3y - 1 = 8$

**3.** $-4z + 3 = 5$

**4.** $3x + 6 = 36$

**5.** $-2y + 1 = 13$

**6.** $5 - 7z = 33$

**7.** $2x + 3x = 10$

**8.** $3y - y = 7$

**9.** $3z = 5 - 7z$

**10.** $4x + 7 = x + 10$

**11.** $6 + 8y = 8y + 6$

**12.** $3z - 6 = 2z + 9$

**13.** $32 - 8x = 3 - 9x$

**14.** $-7 + 21y + 2 = 16$

**15.** $z + 3z = 8 - 2z + 10$

**16.** $1 + 3x + 8 = x + 9 + 2x$

**17.** $2 - x + 5 = x + 3 - 2x$

**18** $x + 9 + 6x = 2 + x + 1$

**19.** $6y - 4y + 1 = 12 - 2y - 11$

**20.** $2.1x - 3.2 = -8.4x - 45.2$

**21** $3y + \dfrac{5}{2}y + \dfrac{3}{2} = \dfrac{1}{2}y + \dfrac{5}{2}y$

*Remove parentheses and solve.*

**22.** $3(2x + 1) = 21$

**23.** $10 = 5(y - 20)$

**24.** $2(5z + 1) = 42$

**25.** $2(3 + 4x) - 5 = 49$

**26.** $5y - (2y + 8) = 16$

**27** $6z - (3z + 8) = 10$

**28.** $3x - 4(x + 2) = 5$

**29.** $2y - (13 - 2y) = 59$

**30.** $3z - (z + 4) = 0$

**31.** $5(x + 4) - 4(x + 3) = 0$

**32.** $4(y - 3) - 6(y + 1) = 0$

**33** $9z - (3z - 18) = 36$

**34** $7(x - 5) = 10 - (x + 1)$

**35.** $(y - 9) - (y + 6) = 4y$

**36.** $8(2z + 1) = 4(7z + 7)$

**37.** $5 + 3(x + 2) = 3 - (x + 2)$

**38.** $\dfrac{1}{3}(6x - 9) = \dfrac{1}{2}(8x - 4)$

**39.** $5(x + 1) - 4x = x - 5$

**40.** $4 - (2x + 3) = x + 1 - 3x$

**41.** $2(x - 1) = 3(x + 1) - x - 1$

**42** $3(2 - 4x) = 4(2x - 1) - 2(1 + x)$

**43.** $2 + y = 3 - 2[1 - 2(y + 1)]$

## FOR REVIEW

*Solve.*

**44.** $-\dfrac{3}{4}x = 18$

**45.** $6y = -\dfrac{3}{2}$

**46.** $\dfrac{x}{-5} = -2$

---

ANSWERS:   1. 2   2. 3   3. $-\frac{1}{2}$   4. 10   5. $-6$   6. $-4$   7. 2   8. $\frac{7}{2}$   9. $\frac{1}{2}$   10. 1   11. every number (identity)
12. 15   13. $-29$   14. 1   15. 3   16. every number (identity)   17. no solution (contradiction)   18. $-1$   19. 0
20. $-4$   21. $-\frac{3}{5}$   22. 3   23. 22   24. 4   25. 6   26. 8   27. 6   28. $-13$   29. 18   30. 2   31. $-8$   32. $-9$   33. 3
34. $\frac{11}{2}$   35. $-\frac{15}{4}$   36. $-\frac{5}{3}$   37. $-\frac{5}{2}$   38. $-\frac{1}{2}$   39. no solution (contradiction)   40. every number (identity)   41. no
solution (contradiction)   42. $\frac{2}{3}$   43. $-1$   44. $-24$   45. $-\frac{1}{4}$   46. 10

## 10.3   EXERCISES B

*Solve.*

**1.** $3x + 1 = 7$

**2.** $4y - 2 = 10$

**3.** $-2z + 5 = 9$

**4.** $2x + 3 = 11$

**5.** $2y - 5 = -3$

**6.** $3 - 4z = -5$

**7.** $3x + 7x = 40$

**8.** $4y - y = 27$

**9.** $5z = 2 - 9z$

**10.** $8x + 2 = x + 16$

**11.** $2y + 3 = 3 + 2y$

**12.** $3z + 6 = 2z - 9$

**13.** $25 - 7x = 4 - 10x$

**14.** $-2 + 8y + 27 = 1$

**15.** $z + 2z = 3 - 2z + 17$

**16.** $1 + 5x + 4 = 2x + 3 + 3x$

**17.** $7 - x + 1 = x + 7 - 2x$

**18.** $x + 3 + 4x = 2 + x + 1$

**19.** $3y - y + 10 = 14 - 5y + 3$

**20.** $6.4 - 4.2x = 16.8x + 90.4$

**21.** $2x - \dfrac{1}{3} + \dfrac{2}{3}x = \dfrac{4}{9} + \dfrac{1}{3}x$

*Clear parentheses and solve.*

**22.** $2(3x + 1) = 14$

**23.** $9 = 3(4y - 1)$

**24.** $7(2z - 1) = -35$

**25.** $5(2 + 3x) - 1 = -6$

**26.** $5z - (z + 3) = 1$

**27.** $6x - (x + 4) = 6$

**28.** $4x - 2(x + 1) = -1$

**29.** $7y - (4 - 3y) = 11$

**30.** $2z - (z + 1) = 0$

**31.** $6(x + 2) - 2(2x + 1) = 0$

**32.** $4(2y - 1) - 7(y + 2) = 0$

**33.** $6z - (2z - 5) = 0$

**34.** $4(x - 2) = 12 - (x + 3)$

**35.** $(y - 2) - (y + 2) = 4y$

**36.** $9(2z - 1) = 3(z + 2)$

**37.** $7 + 3(x + 1) = 5 - (x + 1)$

**38.** $\dfrac{1}{2}(2y - 4) = \dfrac{2}{3}(9y + 3)$

**39.** $6(z - 1) - 3z = 3z + 8$

**40.** $3 - (4x + 1) = x + 2 - 5x$

**41.** $3(x - 2) = 2(x - 1) + x$

**42.** $2(1 - 3x) = 5(2x + 1) - 3(1 + x)$

**43.** $4 + y = 8 - 3[2 - (y + 2)]$

## FOR REVIEW

*Solve.*

**44.** $-\dfrac{3}{5}x = 6$

**45.** $7y = -\dfrac{14}{3}$

**46.** $\dfrac{x}{-6} = -1$

## 10.3 EXERCISES C

*Solve.*

**1.** $\dfrac{3}{4}x + \dfrac{1}{4} = 1$

**2.** $12 = -1.2y + 24$

**3.** $-\dfrac{1}{7}z + z = 18$

[Answer: 21]

**4.** $-2(z - 1) - 3(2z + 1) = -9$

**5.** $-4(-x + 1) + 3(2x + 3) - 7x = -10$

[Answer: $-5$]

## 10.4 USING ALGEBRA TO SOLVE WORD PROBLEMS

### STUDENT GUIDEPOSTS

1. Translating Words into Symbols
2. Method of Solving Word Problems
3. Number Problems
4. Consecutive Integers
5. Measurement Problems
6. Age Problems

### ① TRANSLATING WORDS INTO SYMBOLS

We have been solving arithmetic word problems since Chapter 1, but the ability to solve equations gives us a more powerful tool. Many word problems (problems that are described in words) can be solved by translating them into equations.

To solve a word problem, we need to do two things:

**1.** Change the words of the problem to an algebraic equation.

**2.** Solve the resulting equation.

We have already learned how to solve many simple equations. Now we concentrate on translating problems into equations. Some common terms and their symbolic translations are given in the following table.

| *Symbol* | + | − | · | ÷ | = |
|----------|---|---|---|---|---|
| *Terms* | sum | minus | times | divided by | equals |
| | sum of | less | of | quotient of | is |
| | and | less than | product | ratio | is equal to |
| | added to | diminished by | product of | | is as much as |
| | increased by | difference | multiplied by | | is the same as |
| | more than | difference | | | |
| | | between | | | |
| | | subtracted | | | |
| | | from | | | |
| | | decreased by | | | |

The unknown or desired quantity is represented by a letter ($x$ is often used). Some examples of translations of sentences follow.

A number increased by 3 is 8.
$$x + 3 = 8$$

Four times a number, diminished by 7, is the same as 13.
$$4 \times x - 7 = 13$$

My age five years ago was seventeen.
$$x - 5 = 17$$

8% of a number is 6.4.
$$0.08 \cdot x = 6.4$$

| **EXAMPLE 1**  **TRANSLATING SENTENCES INTO EQUATIONS** |
| --- |

Translate the word sentences into equations.

| | *Word Sentence* | *Symbolic Translation* |
| --- | --- | --- |
| **(a)** | Six times a number is 30. | $6x = 30$ |
| **(b)** | The product of a number and 8 is 32. | $x \cdot 8 = 32$ |
| **(c)** | Twice a number, increased by 7, is 28. | $2x + 7 = 28$ |
| **(d)** | Twice a number increased by 7 is 28. | $2(x + 7) = 28$ |
| **(e)** | Seven is 5 less than a number. | $7 = x - 5$ |
| **(f)** | Seven is 5 less a number. | $7 = 5 - x$ |
| **(g)** | One eighth of a number is 9. | $\frac{1}{8}x = 9$ |
| **(h)** | Four times a number, decreased by twice the number, is the same as 48. | $4x - 2x = 48$ |
| **(i)** | A number subtracted from 14 is 3. | $14 - x = 3$ |
| **(j)** | 14 subtracted from a number is 3. | $x - 14 = 3$ |
| **(k)** | A number divided by 10 is equivalent to $\frac{2}{5}$. | $\frac{x}{10} = \frac{2}{5}$ |
| **(l)** | Four more than a number is eleven. | $x + 4 = 11$ |

*Handwritten annotations:*

(e) 7 less than 5 → 7 = x - 5, 1st # written down

(f) 7 less 5 → 7 = 5 - x, 1st # written down

| **PRACTICE EXERCISE 1** |
| --- |

Use $x$ for the variable and translate each word sentence into symbols.

**(a)** Five times a number is 20.  *(handwritten: 5x = 20)*

**(b)** The product of a number and 7 is 35.

**(c)** Twice a number, increased by 3, is 11.

**(d)** Twice a number increased by 3 is 11.

**(e)** Six is 4 less than twice a number.

**(f)** Six is 4 less twice a number.

**(g)** One-tenth of a number is 13.

**(h)** Five times a number, minus twice the number, equals 10.

**(i)** A number subtracted from 5 is 12.

**(j)** 5 subtracted from a number is 12.

**(k)** A number divided by 9 is equivalent to $\frac{2}{3}$.

**(l)** Eight less than a number is 20.

Answers: **(a)** $5x = 20$  **(b)** $7x = 35$
**(c)** $2x + 3 = 11$  **(d)** $2(x + 3) = 11$
**(e)** $6 = 2x - 4$  **(f)** $6 = 4 - 2x$
**(g)** $\frac{1}{10}x = 13$  **(h)** $5x - 2x = 10$
**(i)** $5 - x = 12$  **(j)** $x - 5 = 12$
**(k)** $\frac{x}{9} = \frac{2}{3}$  **(l)** $x - 8 = 20$

///////////// **CAUTION** ⟩/////////////

Punctuation often plays an important role in translating sentences to equations. For example, notice the difference the commas make in parts **(c)** and **(d)**. In **(c)**, 7 is added to twice the number; in **(d)**, 7 is added to the number and then the sum is multiplied by two. The results are different since $2x + 7$ is not the same as $2(x + 7)$, which is $2x + 14$.

⟩///////////⟨

## ❷ METHOD OF SOLVING WORD PROBLEMS

Having practiced translating words into equations, we next consider the steps to follow when solving word problems.

| To Solve a Word Problem |
| --- |
| 1. Read the problem (perhaps several times) and determine what quantity you are asked to find. |
| 2. Make a sketch if a picture is helpful. |
| 3. Use a letter to represent the unknown quantity. *Set up a heading* |
| 4. Determine what expressions are equal and write an equation. |
| 5. Solve the equation. |
| 6. Check your answer with the original word problem. |

## ❸ NUMBER PROBLEMS

We illustrate the method of problem solving by first considering several examples that involve finding an unknown number.

| EXAMPLE 2   NUMBER PROBLEM | PRACTICE EXERCISE 2 |
| --- | --- |

Twice a number, increased by 3, is 21. What is the number?

*Heading* Let

$x$ = the desired number,　　　　Step 3

$2x$ = twice the number,

$2x + 3$ = twice the number, increased by 3.

Twice　a number　increased by　3　is　21.　　Step 4
　↓　　　↓　　　　　↓　　　↓　↓　↓
　2　·　　 $x$　　　 +　　　 3　=　21

$2x + 3 = 21$

$2x + 3 - 3 = 21 - 3$　　Subtract 3

$2x = 18$　　Simplify

$\dfrac{1}{2} \cdot 2x = \dfrac{1}{2} \cdot 18$　　Multiply by $\frac{1}{2}$　⎫
　　　　　　　　　　　　　　　　　　　　⎬ Step 5
$x = 9$　　$\frac{1}{2} \cdot 2 = 1$ and $\frac{1}{2} \cdot 18 = \frac{1}{2} \cdot \frac{2 \cdot 9}{1} = 9$　⎭

Thus, the desired number is 9.

Check:  Twice 9 is 18, and 18 increased by 3 is 21.　　Step 6

The sum of two numbers is 30. If one number is one half the other, find the two numbers.

**Answer: 10 and 20**

| **EXAMPLE 3** NUMBER PROBLEM | **PRACTICE EXERCISE 3** |
|---|---|

If seven is added to four times a number, the sum is three less than six times the number. What is the number?

Let $x$ = the desired number,

$4x + 7$ = seven added to four times the number,

$6x - 3$ = three less than six times the number.

$$4x + 7 = 6x - 3$$
$$4x - 6x + 7 = 6x - 6x - 3 \qquad \text{Subtract } 6x$$
$$-2x + 7 = -3$$
$$-2x + 7 - 7 = -3 - 7 \qquad \text{Subtract } 7$$
$$-2x = -10$$
$$-\frac{1}{2} \cdot (-2x) = -\frac{1}{2} \cdot (-10) \qquad \text{Multiply by } -\tfrac{1}{2}$$
$$x = 5$$

Thus, the number is 5.

Check:  $4 \cdot 5 + 7$ is 27 as is $6 \cdot 5 - 3$.

If five is subtracted from three times a number, the result is eleven less the number.

Let $x$ = the desired number,

$3x - 5$ = five subtracted from three times the number,

$11 - x$ = eleven less the number.

Answer: 4

## ④ CONSECUTIVE INTEGERS

Two integers are **consecutive integers** if one of them is 1 more than the other. For example, some pairs of consecutive integers are

|   |   |   |
|---|---|---|
| 6 and 7 | 21 and 22 | 78 and 79. |
| 7 is 6 + **1** | 22 is 21 + **1** | 79 is 78 + **1**. |

In general, if one integer is $x$, the next consecutive integer is $x + 1$.

| **EXAMPLE 4** CONSECUTIVE INTEGER PROBLEM | **PRACTICE EXERCISE 4** |
|---|---|

The sum of two consecutive integers is 23. Find the integers.

Let $x$ = the first integer,

$x + 1$ = the next consecutive integer.

$$x + (x + 1) = 23 \qquad \text{Their sum is 23}$$
$$2x + 1 = 23 \qquad \text{Combine like terms}$$
$$2x + 1 - 1 = 23 - 1 \qquad \text{Subtract 1}$$
$$2x = 22$$
$$\frac{1}{2} \cdot 2x = \frac{1}{2} \cdot 22 \qquad \text{Multiply by } \tfrac{1}{2}$$
$$x = 11 \qquad \text{First integer}$$
$$x + 1 = 12 \qquad \text{Second integer}$$

Thus, the numbers are 11 and 12.

Check:  11 and 12 are consecutive integers, and $11 + 12 = 23$.

Three times the smaller of two consecutive integers is 27 more than the larger. Find the integers.

Answer: 14, 15

Consecutive even or odd integers such as 2, 4, 6 or 3, 5, 7 can also be considered. In these cases, if $x$ is an even (or odd) integer, the next consecutive even (odd) integer is $x + 2$, and the next consecutive even (odd) integer after that is $x + 4$.

| EXAMPLE 5   CONSECUTIVE EVEN INTEGER PROBLEM | PRACTICE EXERCISE 5 |
|---|---|

The sum of two consecutive even integers is 2 less than twice the larger. Find the numbers.

$$\text{Let} \quad x = \text{the first even integer,}$$
$$x + 2 = \text{the second (larger) even integer.}$$

| | |
|---|---|
| $x + (x + 2) = 2(x + 2) - 2$ | |
| $2x + 2 = 2x + 4 - 2$ | Clear parentheses |
| $2x + 2 = 2x + 2$ | Combine like terms |
| $2 = 2$ | Subtract $2x$ |

Since we obtained an identity, any value of $x$ will make the equation true. Therefore, *any* two consecutive even integers solve the problem.

The sum of three consecutive odd integers is 135. Find the numbers.

Answer: 43, 45, 47

## ⑤ MEASUREMENT PROBLEMS

Problems using measurement units can be solved in a manner similar to solving number problems.

| EXAMPLE 6   MEASUREMENT PROBLEM IN CARPENTRY | PRACTICE EXERCISE 6 |
|---|---|

A board is 8 ft long. It is cut into two pieces such that one piece is three times as long as the other. Find the length of each piece.

$$\text{Let} \quad x = \text{the length of one piece,}$$
$$3x = \text{the length of the second piece.}$$

| | |
|---|---|
| $x + 3x = 8$ | The two pieces came from an 8-ft board |
| $4x = 8$ | |
| $\dfrac{1}{4} \cdot 4x = \dfrac{1}{4} \cdot 8$ | Multiply both sides by $\frac{1}{4}$ |
| $x = 2$ | Length of one piece |
| $3x = 3 \cdot 2 = 6$ | Length of second piece |

Thus, the two boards measure 2 ft and 6 ft.

Check:   6 is 3 times 2 and $6 + 2 = 8$.

A wire which is 35 cm long must be cut into four pieces of equal length and have 5 cm left over. What is the length of each of the four pieces?

Answer: 7.5 cm

## ⑥ AGE PROBLEMS

Problems involving the ages of people provide additional practice in setting up and solving word problems.

| EXAMPLE 7    AGE PROBLEM | PRACTICE EXERCISE 7 |

Sandy is 4 years older than Jose. The sum of their ages is 66 years. How old is each?

How old is Burford if you get 100 years when you multiply his age by 7 and then add 9?

Let $\quad x =$ Jose's age,

$\quad x + 4 =$ Sandy's age.    Sandy is 4 yrs older

$$x + (x + 4) = 66$$
$$2x + 4 = 66 \qquad \text{Combine like terms}$$
$$2x + 4 - 4 = 66 - 4 \qquad \text{Subtract 4}$$
$$2x = 62$$
$$\frac{1}{2} \cdot 2x = \frac{1}{2} \cdot 62 \qquad \text{Multiply by } \tfrac{1}{2}$$
$$x = 31 \qquad \text{Jose's age}$$
$$x + 4 = 35 \qquad \text{Sandy's age}$$

Thus, Sandy is 35 and Jose is 31.

Check:   35 is 4 more than 31 and 35 + 31 is 66.

Answer: 13 years old

---

## HINT

Although the problems in this section are somewhat artificial, practice at setting up and solving these simple problems will be of great benefit when solving more ''real world'' problems in the next section.

---

## 10.4  EXERCISES A

*Let x represent the unknown number and change the following into symbols. Do not solve.*

**1.** 9 times a number is 36.

**2.** 9 less than a number is 36.

**3.** 9 more than a number is 36.

**4.** 9 divided by a number is 36.

**5.** A number increased by 12 equals 20.

**6.** A number decreased by 12 equals 20.

**7.** Twice a number, decreased by 4, is 15.

**8.** Twice a number decreased by 4 is 15.

**9.** The product of a number and 5 is 40.

**10.** Three times a number, increased by 2, equals 41.

**11.** A number less 4 is 11.

**12.** A number is 4 less than 11.

**13.** 9 is 3 less than twice a number.

**14.** A number diminished by 6 is the same as 22.

**15.** The product of a number and 3, decreased by 1, equals 52.

**16.** When 8 is added to three times a number, the result is the same as twice the number, plus 12.

**17.** 3% of a number is 15.

**18.** The difference of two numbers is 12 and one of them is three times the other.

**19.** My age 3 years ago was 32.

**20.** A number divided by 4 is the same as $\frac{1}{2}$.

*Solve. (Some problems have been started.)*

**21.** Twice a number, increased by 7, is 23. What is the number?

Let  $x$ = the desired number

$2x$ = twice the number

$2x + 7$ = twice the number, increased by 7

**22.** The sum of two numbers is 20. If one number is one third of the other, find the two numbers.

Let  $x$ = one number  $x$

$\frac{1}{3} \cdot x$ = second number  $20 - x$

$y = 1/3 (20 - x)$

or  $20 - x = 1/3 x$

$20 - x + x = 1/3 x + x$  $(3/4) = 4/3 x (3/4)$

$x = 15$   $1/3 x = 5$

**23.** If 9 is subtracted from five times a number, the difference is 3 more than three times the number. What is the number?

Let  $x$ = the desired number

$5x - 9$ = 9 subtracted from five times the number

$3x + 3$ = 3 more than three times the number

**24.** The sum of two consecutive integers is 33. Find the integers.

Let  $x$ = the first integer

$x + 1$ = the next consecutive integer

**25.** The sum of three consecutive even integers is 42. Find the integers.

**26.** Twice the sum of two consecutive odd integers is 40. Find the integers.

**27** Two integers are consecutive. If twice the smaller is the same as 18 more than the larger, find the integers.

**28.** A piece of string is 12 in long. It is cut into two pieces such that one piece is twice as long as the other. Find the length of each piece.

Let $x$ = the length of one piece

**29.** If a number is added to three-fourths of itself, the result is 63. Find the number.

**30.** The sum of three consecutive integers is 66. Find the three integers.

Let     $x$ = the first integer
$x + 1$ = the second integer
$x + 2$ = the third integer
$$x + (x + 1) + (x + 2) = 66$$

**31** A wire 56 m long is cut into three pieces. The second piece is twice as long as the first. The third is 4 m more than the first. How long is each piece?

**32.** Two-thirds of a number is 108. What is the number?

Let     $x$ = the desired number
$\frac{2}{3} \cdot x$ = two thirds of the number

**33.** Two-thirds of the human body is water. If your body contains 108 lb of water, how much do you weigh? [*Hint:* Compare with Exercise 32.]

**34.** The first of two numbers is three times the second. If the first is 9300, what is the second?

Let   $x$ = the second number
$3x$ = the first number

**35.** Jodie has two savings accounts. The amount in one is three times the amount in the second. If the first totals $9300, what is the total in the second? [*Hint:* Compare with Exercise 34.]

**36.** The sum of two numbers is 21, and one is twice the other. What are the numbers?

Let   $x$ = first number
$2x$ = second number
$$x + 2x = 21$$

**37** If you double a number and then add 14, you get the same as 4 times 1 less than the number. What is the number?

**38.** Max is 8 years older than Maria. The sum of their ages is 54 years. How old is each?

Let     $x$ = Maria's age
$x + 8$ = Max's age

**39.** How old is Dyane if you get 85 years when you multiply her age by 4 and then add 1?

**40** Lucy is twice as old as Claude. Toni is 11 years older than Claude. If the sum of all three ages is 87 years, how old is each?

## FOR REVIEW

*Solve.*

**41.** $2x + 5 = 19$

**42.** $-\dfrac{1}{4}y + y = 27$

**43.** $15 - 3z = 5 + 7z$

---

ANSWERS: 1. $9x = 36$  2. $x - 9 = 36$  3. $9 + x = 36$  4. $\frac{9}{x} = 36$ or $9 \div x = 36$  5. $x + 12 = 20$  6. $x - 12 = 20$
7. $2x - 4 = 15$  8. $2(x - 4) = 15$  9. $x \cdot 5 = 40$  10. $3x + 2 = 41$  11. $x - 4 = 11$  12. $x = 11 - 4$  13. $9 = 2x - 3$
14. $x - 6 = 22$  15. $x \cdot 3 - 1 = 52$  16. $3x + 8 = 2x + 12$  17. $0.03x = 15$  18. $3x - x = 12$  19. $x - 3 = 32$
20. $\frac{x}{4} = \frac{1}{2}$  21. 8  22. 5 and 15  23. 6  24. 16 and 17  25. 12, 14, 16  26. 9, 11  27. 19 and 20  28. 4 in and 8 in
29. 36  30. 21, 22, 23  31. 13 m; 26 m; 17 m  32. 162  33. 162 lb  34. 3100  35. $3100  36. 14 and 7  37. 9
38. Max is 31, Maria is 23  39. Dyane is 21  40. Lucy is 38, Claude is 19, Toni is 30  41. 7  42. 36  43. 1

## 10.4 EXERCISES B

*Let x represent the unknown number and change the following into symbols. Do not solve.*

**1.** 4 times a number is 32.

**2.** A number increased by 3 equals 14.

**3.** A number decreased by 13 is equal to 49.

**4.** The product of a number and 8 is 96.

**5.** Twice a number, increased by 4, is 86.

**6.** Three times a number, diminished by 17, equals 42.

**7.** A number less 4 is 21.

**8.** 9 is 3 less than five times a number.

**9.** A number diminished by 2 is 18.

**10.** Twelve times a number, less 17, is 39.

**11.** The product of a number and 7, decreased by 8, equals 29.

**12.** When 15 is added to three times a number the result is the same as twice the number, plus 40.

**13.** 9% of a number is 15.

**14.** The sum of two numbers is 86, and one of them is three times the other.

**15.** Eleven times 4 less than a number is 25.

**16.** Twice the sum of a number and 8 is four times the number.

**17.** Six more than a number is 29.

**18.** Five times my age in 2 years is 200.

**19.** The quotient of 7 and some number is the same as $\frac{4}{9}$.

**20.** One third a number, less three times the reciprocal of the number is $\frac{4}{3}$.

*Solve.*

**21.** If 9 is added to six times a number, the result is 39. Find the number.

$x = a \text{ number}$

$9 - 9 + 6x = 39 - 9$

$(\frac{1}{6})6x = 30(\frac{1}{6})$

$x = 5$

**22.** The sum of two numbers is 72. The larger number is five times the smaller number. Find the two numbers.

*[handwritten: X = number   5x = 2ⁿᵈ #]*

**23.** The sum of two numbers is 48. If one is five times the other, find the two numbers.

*[handwritten: 5x + x = 48(⅙)6x = 48(⅙)   x = 8]*

**25.** Three-fourths of a filled container is composed of water. If the container holds 288 gallons of water, how much fluid is in it?

**27.** The number of acres in the Henderson farm is five times the number of acres in the Carlson farm. If the Henderson farm has 1250 acres, how many acres are there in the Carlson farm?

**29.** Two integers are consecutive. Twice the larger is 58 more than the smaller. Find the integers.

**31.** Twice the sum of two consecutive even integers is 92. Find the integers. *[handwritten: 2 (x + x + 2)]*

*[handwritten: 4x + 4 = 92 − 4   88   22 & 24]*

**33.** Barb is 9 years older than Larry. Twice the sum of their ages is 126. How old is each?

**35.** How old is Burford if you get 99 years when you multiply his age by 10 and subtract 31 years?

**37.** The sum of two numbers is 168, and one is three times the other. What are the numbers?

**39.** A steel rod is 22 m long. It is cut into two pieces in such a way that one piece is 6 m longer than the other. How long is each piece?

**24.** Three-fourths of a number is 288; what is the number?

**26.** The first of two numbers is five times the second. If the first is 1250, what is the second?

**28.** The sum of two consecutive integers is 203. Find the integers. *[handwritten: x = #   x+1 = 2ⁿᵈ #]*

*[handwritten: x + x+1 = 203 −1]*

*[handwritten: 2x = 202   101 & 102]*

**30.** The sum of three consecutive integers is 195. Find the integers.

**32.** The sum of three consecutive odd integers is 111. Find the integers.

**34.** Hector is 7 years younger than Pedro. If the sum of their ages is 39, how old is each?

**36.** Two-thirds of a number is 62; what is the number?

**38.** A rope is 168 feet long. It is to be cut into two pieces in such a way that one piece is three times as long as the other. How long is each piece?

**40.** Hortense is five times as old as Wally and half as old as Lew. The sum of their ages is 80 years. How old is each?

## FOR REVIEW

*Solve.*

**41.** $3x + 7 = 28$

**42.** $-\dfrac{1}{6}y + y = 75$

**43.** $12 - 4z = 2z + 12$

## 10.4  EXERCISES C

**1.** Two numbers are consecutive odd integers. If twice the smaller is the same as 3 more than the larger, find the integers.

**2.** International Car Rental charges $24 a day and 20¢ a mile to rent a car. If Annette rented a car for 2 days and was charged a total of $85, how many miles did she drive?

[Answer: 185 miles]

## 10.5 MORE APPLICATIONS USING ALGEBRA

### **1** PROBLEMS INVOLVING AVERAGES

This section deals with a variety of applied problems which can be solved using algebra. In Chapter 7 the average of several numbers was calculated. With algebra, if the average is known, a missing number can be calculated. Remember that

$$\text{average} = \frac{\text{sum of quantities}}{\text{number of quantities}}.$$

---

### EXAMPLE 1 AN AVERAGE PROBLEM

Russell made scores of 72, 65, and 88 on his first three tests in math. What is his average on these tests? What is the lowest score that he can make on the fourth test in order to pull his grade up to a B, which in his case is an average of 80?

His current average is

$$\frac{72 + 65 + 88}{3} = \frac{225}{3} = 75.$$

In order to pull his grade up to 80, the sum of the 4 tests must be

$$4 \times 80 = 320 \quad \text{since} \quad \frac{320}{4} = 80.$$

Let $x$ be the score on the fourth test.

| | |
|---|---|
| $72 + 65 + 88 + x = 320$ | Sum of 4 scores is 320 |
| $225 + x = 320$ | $72 + 65 + 85 = 225$ |
| $225 - 225 + x = 320 - 225$ | Subtract 225 from both sides |
| $x = 95$ | |

Russell must score 95 to bring his average up to 80.

---

**PRACTICE EXERCISE 1**

Linda sold 22, 31, 23, and 28 cars during the first four months of the year. What was the average number of cars sold per month? How many cars must she sell during May to make her average 30 for the five months?

Answer: 26; 46

---

### **2** PERCENT PROBLEMS

A new type of percent problem is the following:

Maria got a 10% raise and now makes $35,200. What was her old salary?

If we had been told that her old salary was $32,000 and that she got a 10% raise, we would have found her new salary as follows.

$$\underbrace{\text{Old salary}}_{\$32,000} + \underbrace{\text{increase}}_{(0.10)(\$32,000)} = \underbrace{\text{new salary}}_{\$35,200}.$$

The raise is 0.10($32,000) since it is 10% of $32,000. In the original problem in which we do not know the old salary, we let $x$ represent the old salary.

Old salary + increase = new salary .

$$x \quad + \quad 0.10x \quad = \quad \$35,200$$

To solve this equation, we need to combine the like terms, $x$ and $0.10x$.

$$x + 0.10x = \$35,200$$
$$x \cdot 1 + 0.10x = \$35,200 \qquad x = x \cdot 1$$
$$x(1 + 0.10) = \$35,200 \qquad \text{Factor out } x$$
$$x(1.1) = \$35,200 \qquad 1 + 0.10 = 1.10 = 1.1$$
$$\frac{x(\cancel{1.1})}{\cancel{1.1}} = \frac{\$35,200}{1.1} \qquad \text{Divide by 1.1}$$
$$x = \frac{\$35,200}{1.1} = \$32,000$$

This procedure gives the right answer, $32,000.

---

| **EXAMPLE 2    AN APPLICATION OF PERCENT** | **PRACTICE EXERCISE 2** |

The population of Pleasant, Maine, was 3645 in 1980. This was a 35% increase over the population in 1970. What was the population in 1970?

Since we need to know the population in 1970, we let $x$ be that population. The equation for population is like the one we had for salary:

Old population + increase = new population .

$$x \quad + \quad 0.35x \quad = \quad 3645$$

$$x + 0.35x = 3645$$
$$x(1 + 0.35) = 3645 \qquad \text{Factor out } x$$
$$x(1.35) = 3645 \qquad 1 + 0.35 = 1.35$$
$$\frac{x(\cancel{1.35})}{\cancel{1.35}} = \frac{3645}{1.35} \qquad \text{Divide by 1.35}$$
$$x = \frac{3645}{1.35} = 2700$$

The population was 2700 in 1970.

Wanda needs to have $3060.72 in her savings account at the end of the year in order to pay a debt. How much must she deposit in her account on the first of the year if she is paid 9% simple interest?

Answer: $2808

---

| **EXAMPLE 3    A DISCOUNT PROBLEM** | **PRACTICE EXERCISE 3** |

A color TV is on sale for $550. The dealer advertised a 20% discount for the sale. What was the old price of the TV?

Let $x$ = price before the discount (old price).
This time the equation is

old price − discount = new price .

$$x \quad - \quad (0.20)x \quad = \quad \$550$$

$$x \cdot 1 - 0.2x = 550$$
$$x(1 - 0.2) = 550$$
$$x(0.8) = 550$$
$$\frac{x(\cancel{0.8})}{\cancel{0.8}} = \frac{550}{0.8}$$

A college had a drop in enrollment of 5%. How many students did it have last year if the current enrollment is 8360?

$$x = \frac{550}{0.8}$$

$$= \$687.50$$

The price before the sale was $687.50.

Answer: 8800

## ③ GEOMETRY PROBLEMS

A summary of geometric figures and formulas is given inside the front cover. With these figures and formulas at hand, we can solve geometry word problems. Always draw the figure when working a geometry problem.

| EXAMPLE 4   A GEOMETRY PROBLEM |
|---|

| PRACTICE EXERCISE 4 |
|---|

The circumference of a circular flower bed is 182 ft. How many feet of pipe are required to reach from the edge of the bed to a fountain in the center of the bed? [Use 3.14 for $\pi$.]

The perimeter of a rectangle is 46 inches. What is the length of the rectangle if the width is 9 inches?

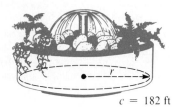

$c = 182$ ft

**Figure 10.1**

The circumference $c = 182$ ft is given. Find $r$ in

$$c = 2\pi r.$$

$$182 \text{ ft} \approx 2(3.14)r$$

$$\frac{182 \text{ ft}}{2(3.14)} = \frac{2(3.14)r}{2(3.14)} \qquad \text{Divide by } 2(3.14)$$

$$28.98 \text{ ft} \approx r \qquad \text{To the nearest hundredth}$$

About 29 ft of pipe is needed.

Answer: 14 in

| EXAMPLE 5   SOLID GEOMETRY PROBLEM |
|---|

| PRACTICE EXERCISE 5 |
|---|

The surface area of a rectangular solid is 1622 m². If the base of the solid is 22 m by 15 m, find the height.

A cylindrical storage bin with radius 10 ft is to be built. How high will it need to be in order to store 10,200 ft³ of grain?

$A = 1622$ m²

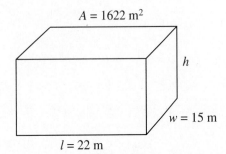

$h$

$w = 15$ m

$l = 22$ m

**Figure 10.2**

The equation to use is

$$A = 2lh + 2wh + 2lw.$$

$$1622 \text{ m}^2 = 2(22 \text{ m})h + 2(15 \text{ m})h + 2(22 \text{ m})(15 \text{ m})$$
$$1622 \text{ m}^2 = (44 \text{ m})h + (30 \text{ m})h + 660 \text{ m}^2$$
$$1622 \text{ m}^2 = (44 \text{ m} + 30 \text{ m})h + 660 \text{ m}^2$$
$$1622 \text{ m}^2 = (74 \text{ m})h + 660 \text{ m}^2$$
$$1622 \text{ m}^2 - 660 \text{ m}^2 = (74 \text{ m})h + 660 \text{ m}^2 - 660 \text{ m}^2$$
$$962 \text{ m}^2 = (74 \text{ m})h$$
$$\frac{962 \text{ m}^2}{74 \text{ m}} = \frac{(74 \text{ m})h}{74 \text{ m}}$$
$$13 \text{ m} = h$$

Answer:  approximately 32.5 ft

## 10.5  EXERCISES A

*Solve.*

1. Beth had grades of 92 and 85 on two math tests. What is the lowest score that she can make on the third test to bring her average up to 90?

2. Larry needs an average of 90 on four tests in history to get an A. What is the lowest score he can make on the fourth test if his first three scores are 96, 82, 95?

3. Rene ran 22 mi, 27 mi, and 21 mi during three weeks. How many miles must she run the fourth week to give her an average of 25 miles per week?

4. Robert's monthly sales totals were $2550, $3320, $3170, $2820, and $2110 for the first five months of the year. What amount must he sell in June to have an average of $3000 for the six months.

5. Susan makes $18,920 per year. What was her salary before she received a 10% raise?

6. A dress is discounted 30% and the sale price is $51.45. What was the original price?

7. In order to buy a new car, Walter needs to have $8856 in his account by the end of the year. How much must he invest the first of the year if the simple interest rate is 8%?

8. The student population at a college decreased by 12% over a ten-year period. If the present enrollment is 15,400, what was the enrollment ten years ago?

9. If the sales-tax rate is 5% and the marked price plus tax on a toaster is $37.59, what is the marked price?

10. The current value of a lot is $10,625. If the value decreased 15% in a year, what was the value last year?

**11.** Find the base of a parallelogram if the perimeter is 88 cm and the side is 20 cm.

**12.** The area of a triangle is 16.4 ft². What is the height if the base is 2.2 ft?

**13** Find the height of a trapezoid if the area is 247 m² and the bases are 16 m and 22 m.

**14.** A circular patio has circumference 87.9 ft. What is its diameter?

**15.** The volume of a rectangular solid is 4725 m³. If both the width and the height are 15 m, what is the length?

**16** Find the height of a cylindrical storage bin with volume 3200 m³ and radius 8 m.

$V = 3200 \text{ m}^3$    $h$

8 m

## FOR REVIEW

**17.** The sum of three consecutive integers is 135. What are the integers?

**18.** Twice a number, less 7, is the same as three times the number, less 30. What is the number?

**19.** Gus is 7 years older than Jan. If twice the sum of their ages is 66, how old is each?

**20.** A wire is 42 inches long. It is to be cut into three pieces in such a way that the second is twice the first and the third is equal to the sum of the lengths of the first two. How long is each piece?

---

ANSWERS:  1. 93   2. 87   3. 30 mi   4. $4030   5. $17,200   6. $73.50   7. $8200   8. 17,500   9. $35.80
10. $12,500   11. 24 cm   12. 14.9 ft   13. 13 m   14. 28.0 ft   15. 21 m   16. 15.9 m   17. 44, 45, 46   18. 23   19. Gus is 20, Jan is 13   20. 7 in, 14 in, 21 in

## 10.5 EXERCISES B

*Solve.*

**1.** Juan had 316 sales points one year and 324 the next. How many points must he earn the third year to have an average of 350?

**2.** Alfred needs an average of 80 on four tests to get a B. What is the minimum score that he can make on the fourth test if his first three grades are 91, 83, and 65?

3. On a backpacking trip 10 km was covered on the first day. The next two days the distances traveled were 12 km and 16 km. What distance must be covered on the fourth day if the average per day is to be 14 km?

4. Nina made $320, $285, $415, $360, and $295 the first five weeks she worked. What would she have to make the sixth week to average $350 per week?

5. After an 8% increase in salary, Henry makes $24,300. What was his salary before the raise?

6. The price of a pet mouse dropped 30% last year. If the present price is $1.05, what was the price last year?

7. The population of Lost, Nevada, increased 18% last year. If the present population is 649, what was it a year ago?

8. A sportcoat was discounted 20% and the sale price is $68. What was the original price?

9. The restaurant bill plus 15% tip was $64.63. What was the cost of the food without the tip?

10. An executive makes $114,000 per year after he took a 5% pay cut. What was his salary before the cut?

11. Find the base of a parallelogram if the perimeter is 100 inches and a side is 15 inches.

12. The perimeter of a triangle is 104 ft and two sides are 22 ft and 46 ft. What is the length of the third side?

13. What is the height of a trapezoid with area 24 yd$^2$ and bases of length 4.5 yd and 3.5 yd?

14. A circular flower bed has circumference 50 m. How many meters of water pipe would be needed to go across it?

15. Find the height of a cylindrical storage tank with volume 1020.5 ft$^3$ and radius 5 ft.

16. The volume of a rectangular solid is 735 m$^3$. If the width is 7 m and the height is 12.5 m, what is the length?

## FOR REVIEW

17. The sum of three consecutive integers is 222. What are the integers?

18. Five times a number, decreased by 8, is the same as four times 1 more than the number. Find the number.

19. Lucy is seven years older than her sister. In 3 years she will be twice as old as her sister is now. How old is each now?

20. A 36 m steel rod is cut into three pieces. The second piece is twice as long as the first and the third is three times as long as the second. How long is each piece?

## 10.5 EXERCISES C

*Solve.*

1. Wanda needs to have $3060.72 in her account at the end of the year. How much must she deposit in her account on the first of the year if the simple interest rate is 10% per year for the first six months, but will fall to 8% per year for the last six months?
[Answer: $2808.00]

2. The length of a rectangle is 24.32 ft more than its width. If the perimeter is 90.44 ft, find the width of the rectangle.

## CHAPTER 10 REVIEW

### KEY WORDS

**10.1**  An **equation** is a statement that two quantities or expressions are equal.

A **solution** to an equation is a number which, when substituted for the variable, makes the equation true.

A **linear equation** is an equation in which the variable occurs to the first power only.

A **conditional equation** is an equation that is true for some replacements of the variable and false for others.

Two equations are **equivalent** when they have exactly the same solution.

An **identity** is an equation that has every number as a solution.

A **contradiction** is an equation that has no solution.

### KEY CONCEPTS

**10.2**  When using the multiplication-division rule to solve an equation, multiply both sides by a number which makes the coefficient of the variable equal to 1.

**10.3**  **1.** Use the addition-subtraction rule before using the multiplication-division rule when a combination of rules is necessary to solve an equation.

**2.** When solving an equation involving parentheses, first use the distributive law to clear all parentheses.

**10.4**  **1.** To solve a word problem, change the words into an algebraic equation and solve.

**2.** Write out complete descriptions of the variables when solving a word problem.

**3.** Two consecutive integers can be named $x$ and $x + 1$.

**4.** Two consecutive even integers can be named $x$ and $x + 2$. Also, two consecutive odd integers can be named $x$ and $x + 2$.

**10.5**  **1.** When solving percent-increase problems, add the increase to the original amount. For example, an amount of money $x$ plus 7% interest can be represented as

$$x + 0.07x = (1 + 0.07)x = 1.07x.$$

**2.** An accurate sketch is helpful for solving a geometry problem.

### REVIEW EXERCISES

*Part I*

**10.1–** *Solve.*
**10.3**

**1.** $\dfrac{1}{3} + x = 1$

**2.** $6.7 - y = 4.2$

**3.** $3z - 5z = 6$

**4.** $\dfrac{1}{3}x = 10$

**5.** $-3y = \dfrac{6}{5}$

**6.** $-0.8z = 5.6$

**7.** $\dfrac{2}{3}x - 5 = 15$         **8.** $2y - 5 + 3y = 10$         **9.** $7z - 4 = -3z + 6$

**10.** $\dfrac{3}{2}x - \dfrac{7}{2}x = 16$         **11.** $\dfrac{1}{2}y - \dfrac{3}{2} + 2y = \dfrac{3}{2}y$         **12.** $z - (2z + 1) = 5$

**13.** $5x - 4(x - 2) = 6$         **14.** $2(x - 5) - 7(x + 5) = 0$         **15.** $9z - (z + 1) = 8z$

**16.** $9x - 3(x - 2) = 6$         **17.** $\dfrac{1}{2}(2x - 4) = \dfrac{1}{3}(6x + 9)$         **18.** $5x - (2x + 3) = 6x - (x - 1)$

**10.4** *Let x represent the unknown number and change the following into equations. Do not solve.*

**19.** 8 more than a number is 14.

**20.** 6 times a number is 72.

**21.** 5 times a number, decreased by 7, is 18.

**22.** 5 times a number decreased by 7 is 18.

**23.** A number increased by 16 is 42.

**24.** When 9 is added to 3 times a number, the result is the same as twice the number, plus 10.

*Solve.*

**25.** Twice a number, decreased by 5, is 19. What is the number?

**26.** Two-thirds of a number is 16. What is the number?

**27.** If 8 is subtracted from 3 times a number, the difference is the number plus 4.

**28.** Two odd integers are consecutive. If 3 times the first is the same as 13 more than twice the second, find the integers.

**10.5 29.** Brenda rode a bike 6 miles one day and 8 miles the next. How far would she have to ride the next day to make her average for the three days 9 miles?

**30.** On her first four tests, Anne scored 72, 85, 68, and 75. What score must she make on the fifth test in order to have an average of 80?

**31.** Frank now makes $28,558 after receiving a 9% raise. How much did he make before the raise?

**32.** A suit is discounted 20% and the sale price is $144. What was the original price?

**33.** The area of a triangle is 28.2 m$^2$ and the base is 6.2 m. Find the height.

**34.** The perimeter of a rectangle is 75.2 inches. Find the length if the width is 16.8 inches.

*Part II*

*Solve.*

**35.** A number is 3 more than another and their sum is 12. Find the numbers.

**36.** One-half my age is 16. How old am I?

**37.** $3x - 9x = 42$

**38.** $2x - (x + 7) = 12$

**39.** $4y - 3 = 2y + 7$

**40.** $2(x - 3) - 7 = 5 - (x - 3)$

**41.** A wire 28 cm long is cut into three pieces. The second piece is $\frac{1}{2}$ the first, and the third piece is 8 cm more than the first. Find the length of each piece.

**42.** Angie is 7 years older than Pete. If the sum of their ages is 51 years, find the age of each.

**43.** To make a profit, a dealer marks up his cost of an item by 60%. If a toaster has a marked price of $48, what was the cost to the dealer?

**44.** The average height of the 5 starting players on the NAU basketball team is 78 inches. If four of the players are 74 in, 76 in, 78 in, and 82 in, how tall is the fifth player?

**45.** Find the height of a storage cylinder which holds 25,600 ft$^3$ of grain and has radius 12 ft.

**46.** A rectangular solid has a base 10 m by 12 m. What is its height if the volume is 840 m$^3$?

---

ANSWERS:   1. $\frac{2}{3}$   2. 2.5   3. $-3$   4. 30   5. $-\frac{2}{5}$   6. $-7$   7. 30   8. 3   9. 1   10. $-8$   11. $\frac{3}{2}$   12. $-6$   13. $-2$   14. $-9$   15. no solution   16. 0   17. $-5$   18. $-2$   19. $8 + x = 14$   20. $6x = 72$   21. $5x - 7 = 18$   22. $5(x - 7) = 18$   23. $x + 16 = 42$   24. $3x + 9 = 2x + 10$   25. 12   26. 24   27. 6   28. 17, 19   29. 13 mi   30. 100   31. $26,200   32. $180   33. 9.1 m   34. 20.8 in   35. $\frac{9}{2}, \frac{15}{2}$   36. 32 yr   37. $-7$   38. 19   39. 5   40. 7   41. 8 cm, 4 cm, 16 cm   42. Angie is 29, Pete is 22   43. $30   44. 80 in   45. 56.6 ft   46. 7 m

*Solve.*

**1.** $x + 19 = 4$

1. _____

**2.** $-7y = 84$

2. _____

**3.** $\frac{1}{2}z = 3$

3. _____

**4.** $-2x = \frac{6}{5}$

4. _____

**5.** $3x - 2 = 4$

5. _____

**6.** $2y + 11 = 5 - 4y$

6. _____

**7.** $3 - (x - 5) = 2x - 7$

7. _____

*Let x represent the unknown number and change the following into symbols. Do not solve.*

**8.** Twice a number, increased by 11, is 8.

8. _____

**9.** Her age in 5 years will be 26.

9. _____

*Solve.*

**10.** The sum of two consecutive integers is 69. Find the integers.

10. _____

**11.** Wilma had test scores of 83 and 72. What would she need to make on the next test to give her an average of 80 on the three tests?

11. _____

**12.** The population of Heritage, Wyoming, was 5100 in 1985. This was a 20% increase over the population in 1975. What was the population in 1975?

12. _____

**13.** The area of a triangle is 4 ft$^2$. What is the height if the base is 2.5 ft?

13. _____

**14.** A circular garden has circumference 18.84 yd. What is its diameter? [Use 3.14 for $\pi$.]

14. _____

## CHAPTER 1

**1.** Consider the number line.

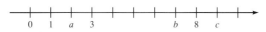

(a) What number is paired with the point *a*?

(b) What number is paired with the point *b*?

Place the correct symbol (< or >) between the numbers in each pair.

**(c)** 1    *a*          **(d)** *a*    5          **(e)** *a*    b          **(f)** *c*    4

**2.** Write 27,143,206 in expanded notation.

**3.** Write 2,000,000 + 40,000 + 300 + 20 + 5 in standard notation.

**4.** Write a word name for 86,247,159.

*Find the following sums and check your work.*

| | | | | | | | |
|---|---|---|---|---|---|---|---|
| **5.** | 279 | **6.** | 305 | **7.** | 2187 | **8.** | 4020 |
| | + 48 | | + 299 | | 6532 | | 37 |
| | | | | | + 27 | | 408 |
| | | | | | | | + 90 |

**9.** Round 27,485 to the nearest:

**(a)** ten;          **(b)** hundred;          **(c)** thousand.

**10.** Estimate the given sum by rounding to the nearest **(a)** ten **(b)** hundred.

    1953          **(a)**          **(b)**
    2247
+ 738

*Solve.*

**11.** An electronics shop has in stock 38 radios, 127 televisions, 72 calculators, and 12 watches. How many items does the owner have in his inventory?

**12.** On Monday, Sue jogged 4 mi, on Tuesday she jogged 7 mi, on Wednesday 6 mi, on Thursday 8 mi, and on Friday 3 mi. How many miles did she jog during the week?

*Find the following differences and check your work.*

| | | | | | | | |
|---|---|---|---|---|---|---|---|
| **13.** | 54 | **14.** | 607 | **15.** | 3208 | **16.** | 2007 |
| | − 28 | | − 421 | | − 1157 | | − 1348 |

*Solve.*

**17.** On the first of November, Mr. Schwacker had a balance of $427 in his checking account. During the month he made deposits of $370 and $1845. He wrote checks for $238, $475, $87, $5, and $614, and paid a service charge of $6. What was his balance at the end of the month?

**18.** Pete bought an antique car for $750. He spent $1230 restoring the car, then sold it for $4750. How much money did he make?

## CHAPTER 2

**19.** $3 \div 1 =$ _____.

**20.** $3 \div 3 =$ _____.

**21.** In the expression $7^3$, 3 is called what?

**22.** **(a)** What is the square of 4?
**(b)** What is the square root of 4?

*Find the following products and check your work by estimating the product.*

**23.**   79
  $\times\ 41$

**24.**   819
  $\times\ 78$

**25.**   3015
  $\times\ 602$

**26.**   6499
  $\times\ 4000$

*Find the following quotients and check your work.*

**27.** $7\overline{)3367}$

**28.** $34\overline{)2341}$

**29.** $615\overline{)140,006}$

*Perform the indicated operations.*

**30.** $3 - 4 \div 2 + 6$

**31.** $(2 + 3) \cdot (2 - 1) - \sqrt{25}$

**32.** $(1 + 3) \div 2 + 3 \cdot 4$

**33.** Find all divisors of 150.

**34.** Express 546 as a product of primes.

**35.** Is 90 divisible by

**(a)** 2?    **(b)** 3?    **(c)** 4?    **(d)** 5?    **(e)** 10?

**36.** Simplify each radical expression.

**(a)** $\sqrt{9}$    **(b)** $\sqrt{225}$    **(c)** $\sqrt{4}$    **(d)** $\sqrt{144}$    **(e)** $\sqrt{1}$

**37.** Sharon wants to carpet her living room with carpet selling for $21 per sq yd. A sketch showing the floor plan of the room is shown. How much will the carpet cost?

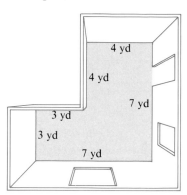

**38.** Assuming equal quality, which is the better buy, 32 ounces of soda for $1.60 (160¢) or 28 ounces for $1.12 (112¢)?

## CHAPTER 3

**39.** What fractional part of the figure has been shaded?

**40.** Let the bar be one whole unit.

| $\frac{1}{4}$ | $\frac{1}{4}$ | $\frac{1}{4}$ | $\frac{1}{4}$ |
|---|---|---|---|

What fraction of the unit is shaded below? [Consider the whole figure. There is one answer, and it is an improper fraction.]

*Express as a whole number.*

**41.** $\dfrac{11}{11}$

**42.** $\dfrac{0}{6}$

**43.** Tell whether $\frac{7}{4}$ and $\frac{19}{11}$ are equal by using the cross-product rule.

**44.** Find the missing term of the fraction. $\frac{6}{27} = \frac{2}{?}$

**45.** Reduce $\frac{70}{385}$ to lowest terms.

*Find the product.*

**46.** $\dfrac{27}{6} \cdot \dfrac{2}{9}$

**47.** $\dfrac{42}{15} \cdot \dfrac{3}{7}$

*Find the value of each power or root.*

**48.** $\left(\dfrac{3}{4}\right)^2$

**49.** $\sqrt{\dfrac{4}{81}}$

*Find the quotient.*

**50.** $\dfrac{5}{8} \div \dfrac{3}{16}$

**51.** $\dfrac{15}{4} \div \dfrac{20}{6}$

*Find the missing number.*

**52.** $\dfrac{2}{3} \cdot \square = 12$

**53.** $\dfrac{11}{4}$ of $\square$ is 22

*Change to a mixed number.*

**54.** $\dfrac{41}{9}$

**55.** $\dfrac{122}{11}$

*Change to an improper fraction.*

**56.** $13\dfrac{2}{5}$

**57.** $32\dfrac{1}{4}$

*Perform the indicated operations.*

**58.** $1\dfrac{3}{4} \cdot 3\dfrac{1}{8}$

**59.** $6\dfrac{2}{3} \div 2\dfrac{2}{5}$

*Solve.*

**60.** Laura got $\dfrac{3}{5}$ of the votes for president of her club. If the club has 75 members, how many members voted for her?

**61.** A container holds 162 gallons of water when it is $\dfrac{2}{3}$ full. What is the capacity of the container?

**62.** A container holds 162 gallons of water when it is full. If $\dfrac{1}{3}$ of a full container is drained off, how much water remains?

**63.** A child's ride makes $7\dfrac{2}{3}$ revolutions per minute. How many revolutions will be made during a $3\dfrac{3}{5}$-minute ride?

## CHAPTER 4

**64.** Use the listing method to find the LCM of 6 and 9.

**65.** Use prime factors to find the LCM of 220 and 390.

**66.** Use the special algorithm to find the LCM of 840 and 1260.

*Perform the indicated operations.*

**67.** $\dfrac{7}{12} + \dfrac{1}{12}$

**68.** $\dfrac{8}{15} - \dfrac{2}{15}$

**69.** $\dfrac{7}{12} + \dfrac{1}{18}$

**70.** $\dfrac{4}{15} + \dfrac{11}{35}$

**71.** $\dfrac{3}{5} - \dfrac{1}{7}$

**72.** $\dfrac{9}{20} - \dfrac{5}{12}$

**73.** $4\dfrac{3}{5} + 2\dfrac{7}{8}$

**74.** $9\dfrac{1}{4} - 3\dfrac{5}{6}$

**75.** $\dfrac{7}{4} - 1 + \dfrac{1}{5}$

*Evaluate each expression.*

**76.** $\left(\dfrac{3}{8} - \dfrac{1}{8}\right)^2 + \sqrt{\dfrac{1}{16}}$

**77.** $\dfrac{2}{3} \div \dfrac{4}{9} - \dfrac{3}{11} \cdot \dfrac{22}{9}$

*Answer true or false.*

**78.** $\dfrac{11}{32} < \dfrac{21}{61}$

**79.** $8\dfrac{15}{19} > 8\dfrac{7}{11}$

*Solve.*

**80.** Ann worked $7\dfrac{1}{2}$ hours, $6\dfrac{3}{4}$ hours, and $4\dfrac{3}{8}$ hours during a three-day period. How many hours did she work?

**81.** It takes Graydon $60\dfrac{1}{2}$ minutes to run a particular course. If he has been running for $38\dfrac{3}{4}$ minutes, how much longer will it be before he completes the course?

## CHAPTER 5

*Give a formal word name for each decimal.*

**82.** 42.5

**83.** 3.0007

*Write each number in decimal notation.*

**84.** two and nine thousandths

**85.** six hundred twenty-seven and nine hundredths

*Change each decimal to a fraction (or mixed number).*

**86.** 0.47

**87.** 8.131

*Change each fraction to a decimal.*

**88.** $\dfrac{4}{3}$

**89.** $\dfrac{5}{8}$

**90.** $\dfrac{4}{9}$

*Find the sum.*

**91.** 2.7 + 8.32 + 105.006

**92.** 5.2 + 5.02 + 0.502 + 50.02

**93.** Find the total deposit on the given portion of a deposit slip.

|        | Dollars | Cents |
|--------|---------|-------|
| Cash   | $  493  | 07    |
| Checks | 647     | 51    |
|        | 18      | 30    |
|        | 6047    | 28    |
|        |         |       |
| **Total** | $    |       |

**94.** Find the perimeter of the figure.

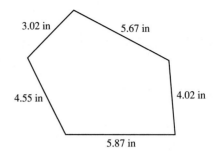

*Find the difference.*

**95.** 37.2 − 5.09

**96.** 7 − 1.006

**97.** Cindy received a check for $185.70. She spent $72.43 on a coat, $24.17 on a pair of shoes, and $13.98 for a pair of pajamas. How much did she have left?

**98.** Round 4.355 to two decimal places.

**99.** Round $47.51 to the nearest dollar.

**100.** Round $0.\overline{92}$ to the nearest hundredth.

**101.** Round $0.\overline{92}$ to three decimal places.

*Find the product.*

**102.**  24.07
        × 2.8

**103.**  6.1109
        × 4.02

**104.**  0.90001
        × 1.9

**105.** Find the area of a rectangle which is 2.31 ft long and 6.5 ft wide.

*Find the quotient.*

**106.** 3.7)‾22.385‾

**107.** 0.025)‾0.1175‾

**108.** 52)‾607.36‾

**109.** What is the approximate cost of 11 pounds of nuts selling for $3.05 per pound? The exact cost?

*Find the sum.*

**110.** $4.35 + \dfrac{4}{5}$

**111.** $5\dfrac{3}{4} + 7.84$

**112.** If Robert earns \$17.13 an hour, how much will he be paid for working $6\frac{1}{2}$ hours?

**113.** If $16\frac{3}{4}$ oz of breakfast drink sell for \$1.89, to the nearest tenth of a cent, what is the price per ounce (the unit price)?

**114.** Arrange 0.26, 2.6, 2.06, 0.026, and 20.06 in order of size with the smallest on the left.

*Perform the indicated operations.*

**115.** $5.2 \div \dfrac{1}{2} - (1.2)^2$

**116.** $\dfrac{5}{3} \cdot \dfrac{9}{20} - (0.5)^2$

## CHAPTER 6

*Are the given pairs of numbers proportional?*

**117.** 6, 5 and 7, 6

**118.** 9, 12 and 30, 40

*Solve the proportion.*

**119.** $\dfrac{a}{12} = \dfrac{5}{3}$

**120.** $\dfrac{14}{a} = \dfrac{35}{8}$

**121.** If 48 m of wire weighs 16 kg, what will 360 m of the same wire weigh?

**122.** A boat can travel 90 miles on 40 gallons of gas. How far can it travel on 56 gallons of gas?

*Change from percents to fractions.*

**123.** 37%

**124.** 0.06%

**125.** $9\dfrac{3}{8}\%$

*Change from percents to decimals.*

**126.** 58%

**127.** 220%

**128.** $10\dfrac{1}{2}\%$

*Change to percents.*

**129.** 0.81

**130.** 0.005

**131.** 20

**132.** $\dfrac{1}{6}$

**133.** $\dfrac{9}{25}$

**134.** $\dfrac{4}{9}$

*Solve the following percent problems.*

**135.** What is 38% of 40?

**136.** What percent of 96 is 12?

**137.** 72 is 120% of what number?

**138.** 2.7 is what percent of 900?

**139.** There were 202,000 votes cast for the two people in an election. If Helen received 52% of the votes, how many votes did John receive?

**140.** If there are 25 g of acid in a solution of total weight 40 g, what is the percent acid in the solution?

**141.** What is the sales tax on a refrigerator priced at $720 if the tax rate is 5%? What is the total price, including tax?

**142.** If a tax of $1.86 is charged on a dress selling for $62.00, what is the sales-tax rate?

**143.** If the tax rate for Social Security was 7.51% in 1989, how much tax did Walter pay on an income of $20,800?

**144.** Birgreta receives a 1.5% commission each month on all sales of $30,000 or less. For sales over $30,000, she is paid a 2% commission. What is her commission on sales of $58,400 during the month of June?

**145.** Considering the total commission that Birgreta received in Exercise 144, what is her overall commission rate to the nearest tenth of a percent?

**146.** Shirts are advertised for $4.50 off the original price. If the discount rate is 30%, what was the original price?

**147.** Manuel borrowed $1500 from his mother and agreed to pay her 8% per year simple interest. How much must he pay at the end of 18 months (a year and a half)?

**148.** Karen has $5200 to invest. One account pays 10% compounded annually and another pays 10.5% simple interest. Which account would give her the most interest over a 2-year period?

**149.** The price of steak increased from $1.95 per pound to $2.35 per pound. To the nearest tenth of a percent, what was the percent increase?

**150.** A dealer bought a toaster for $23.50. The markup was 30%. She was later forced to sell the toaster at a markdown of 28%. What was the percent of profit or loss based on the original cost?

## CHAPTER 7

*Exercises 151–54 refer to the bar graph showing the number of students in each class who applied for a scholarship.*

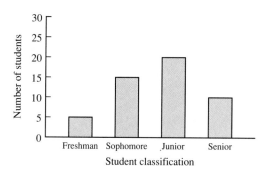

**151.** How many students applied?

**152.** What is the ratio of juniors to sophomores who applied?

**153.** What percent of those applying were freshmen?

**154.** What is the ratio of freshmen and sophomores to juniors and seniors?

*Exercises 155–59 refer to the broken-line graph showing the number of vehicles passing through an intersection (in hundreds) during an 8-hour interval.*

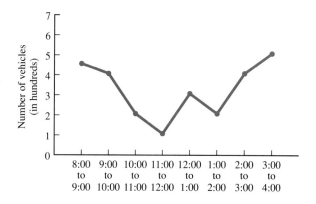

**155.** How many vehicles passed through the intersection from 8:00 to 9:00?

**156.** How many vehicles passed through the intersection from 11:00 to 12:00?

**157.** How many vehicles passed through the intersection from 8:00 to 12:00?

**158.** How many vehicles passed through the intersection during the period from 8:00 to 4:00?

**159.** What percent (to the nearest tenth) of the total number of vehicles passed through the intersection from 8:00 to 12:00?

*Find the mean and median. Give answers to nearest tenth.*

**160.** 9, 10, 17, 6

**161.** 6.2, 8.5, 4.7, 3.3, 9.4

**162.** A truck was driven 826 miles on 203 gallons of fuel. To the nearest tenth, what was the average miles per gallon?

**163.** Find the mode. 42, 35, 42, 50, 35, 70, 60, 35

## CHAPTER 8

*Complete.*

**164.** 42 in = _____ ft

**165.** 5400 cm = _____ km

**166.** 2.5 g = _____ mg

**167.** 5 lb = _____ g

**168.** $100\dfrac{km}{hr}$ = _____ $\dfrac{mi}{hr}$

**169.** 60°C = _____ °F

**170.** 3,097,600 yd² = _____ mi²

**171.** 162 ft² = _____ yd²

**172.** 2840 dm³ = _____ m³

**173.** If sugar costs 36¢ per lb, how much will the sugar cost for a cake which requires 24 oz of sugar?

**174.** If a car's gasoline tank holds 25 gal, how many liters does it hold?

*Perform the indicated operations and simplify.*

**175.**  21 hr 15 min
     +  6 hr 48 min

**176.**  13 yd 1 ft 7 in
     −   2 yd 2 ft 4 in

**177.**  25.2 cm
     +  7.9 cm

**178.**  421.6 ml
     − 382.7 ml

**179.** 4 × (2 gal 2 qt 1 pt)

**180.** 62 km ÷ 8 km

*Find the perimeter and area of the rectangle or square.*

**181.** 5 ft by 9 ft

**182.** $12\dfrac{1}{2}$ m by 6.2 m

**183.** The area shown is to be carpeted. How much will it cost if carpet is $22.50 per square yard?

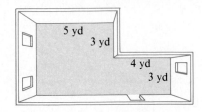

*Find the perimeter and area of each figure.*

**184.**

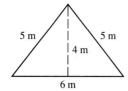

**185.**

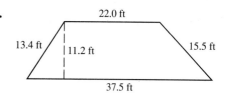

*Find the circumference and area of each circle. Use 3.14 for π and give answers to nearest tenth.*

**186.** $r = 2.6$ ft

**187.** $d = 9.2$ m

**188.** A 14-in diameter pizza costs \$7.50 and a 16-in pizza costs \$8.50. Which costs less per square inch?

*Find the volume and surface area of each solid.*

**189.** Rectangular solid: 5 m by 2 m by 8 m

**190.** Cube: 3.6 ft edge

**191.** Cylinder: $d = 5.2$ dm, $h = 7.8$ dm

**192.** Sphere: $r = 15$ in

**193.** A spherical tank has radius 42 cm. How many liters of gasoline will it hold (to the nearest liter)?

## CHAPTER 9

*Is the given statement true or false?*

**194.** $-8 < 0$

**195.** $-3 > 2$

**196.** $-3 > -2$

**197.** $|-8| = -8$

*Perform the indicated operations.*

**198.** $(-8) + (-2)$

**199.** $(-6) + 4$

**200.** $8 - (-2)$

**201.** $16 - 12$

**202.** $(-3) - (-8)$

**203.** $(-8) + 2 - 0 + 4$

**204.** $(8)(-2)$

**205.** $(-4)(-10)$

**206.** $0 \cdot (-7)$

**207.** $0 \div (-7)$

**208.** $(-24) \div (-4)$

**209.** $\dfrac{45}{-3}$

*Add.*

**210.**  $\begin{array}{r} 92 \\ -16 \\ 12 \\ \underline{-48} \end{array}$

**211.**  $\begin{array}{r} -16 \\ -24 \\ -98 \\ \underline{52} \end{array}$

**212.**  $\begin{array}{r} 416 \\ 282 \\ -705 \\ \underline{111} \end{array}$

*Is the given statement true or false?*

**213.** $\dfrac{7}{8} > -\dfrac{7}{8}$

**214.** $-8.2 = -8\dfrac{1}{5}$

**215.** $\left|-\dfrac{1}{2}\right| = -\dfrac{1}{2}$

**216.** $-\dfrac{1}{2} > -6.5$

*Perform the indicated operations.*

**217.** $\dfrac{2}{7} + \left(-\dfrac{4}{7}\right)$

**218.** $(-5.2) - (-4.3)$

**219.** $\left(-\dfrac{3}{5}\right) + \left(-\dfrac{1}{10}\right)$

**220.** $\left(-\dfrac{6}{7}\right)\left(\dfrac{14}{9}\right)$

**221.** $\left(-\dfrac{4}{5}\right) \div \left(-\dfrac{12}{35}\right)$

**222.** $\dfrac{8.8}{-4}$

**223.** $9 \div 3 - 5 \cdot 4$

**224.** $-2[3(4 - 1) - 8]$

**225.** $\left|-\dfrac{11}{3} + \dfrac{2}{3}\right|$

*Find the negative and the reciprocal.*

**226.** 6

**227.** $-8$

**228.** $\dfrac{5}{8}$

**229.** $-\dfrac{12}{7}$

**230.** 0

*Evaluate for $a = -2$, $b = 6$, and $c = -4$.*

**231.** $(a - b) + (c - a)$

**232.** $a[b - (a - c)]$

**233.** $|a - b - c|$

*Use the distributive laws to factor.*

**234.** $7x + 14y$

**235.** $3a - 6b + 9$

**236.** $-xy + xz$

*Remove parentheses and collect or combine like terms.*

**237.** $-2(a + b) + a$

**238.** $-4(2x - 3y + z + x)$

**239.** $-5[4x - (3y - x)]$

*Simplify and write without negative exponents.*

**240.** $2x^{-2}x^5$

**241.** $\dfrac{2^0 b^{-3}}{b^5}$

**242.** $(3x^{-2})^3$

*Evaluate for x = −4 and y = 3.*

**243.** $-2x^2$

**244.** $(-x)(-y)^2$

**245.** $(3x)^{-2}$

*Perform the indicated operations using scientific notation.*

**246.** $(2 \times 10^7)(3 \times 10^{-2})$

**247.** $\dfrac{0.0000008}{4000}$

## CHAPTER 10

*Solve and check.*

**248.** $x + 2 = -8$

**249.** $y + 5 = y - 3$

**250.** $2z = z + 5$

**251.** $2x + 10 = 2(x + 5)$

**252.** $\dfrac{3}{5} + y = -1$

**253.** $5z - 4 = 7z + 8$

**254.** $-2.7x = 5.4$

**255.** $\dfrac{2}{3}y = 10$

**256.** $\dfrac{1}{2}z - \dfrac{1}{5} + \dfrac{7}{2}z = \dfrac{2}{5}$

**257.** $5x - (-2x + 7) = 7$

**258.** $3(y - 8) - 4(2y + 1) = 2$

**259.** $4x - (2x + 1) = x - (5x - 2)$

*Let x represent the unknown number and change the following into equations.*

**260.** 9 times a number, decreased by 4, is 12.

**261.** When 6 is subtracted from twice a number, the result is the same as 3 times the number, plus 4.

*Solve.*

**262.** Two integers are consecutive. If 4 times the first is the same as 3 times the second, plus 8, find the integers.

**263.** Liz made grades of 92, 86, 82, and 95 on four math tests. What must she make on the fifth test in order to have an average grade of 90?

**264.** The sales-tax rate in Beach, Florida, is 4%. If a tax of $2.11 is paid on a dress, what was the price of the dress before tax?

**265.** Alice now makes $32,560 after receiving a 10% raise. How much did she make before the raise?

**266.** The area of a triangle is 23.2 cm$^2$ and the height is 2.7 cm. Find the base.

**267.** A trapezoid has area 420 ft$^2$. Find the height if the bases are 24 ft and 32 ft.

---

ANSWERS:   1. (a) 2 (b) 7 (c) $<$ (d) $<$ (e) $<$ (f) $>$   2. $20,000,000 + 7,000,000 + 100,000 + 40,000 + 3000 + 200 + 6$   3. 2,040,325   4. eighty-six million, two hundred forty-seven thousand, one hundred fifty-nine   5. 327   6. 604   7. 8746   8. 4555   9. (a) 27,490 (b) 27,500 (c) 27,000   10. (a) 4940 (b) 4900   11. 249   12. 28 mi   13. 26   14. 186   15. 2051   16. 659   17. $1217   18. $2770   19. 3   20. 1   21. exponent   22. (a) 16 (b) 2   23. 3239   24. 63,882   25. 1,815,030   26. 25,996,000   27. Q: 481, R: 0   28. Q: 68, R: 29   29. Q: 227, R: 401   30. 7   31. 0   32. 14   33. 1, 2, 3, 5, 6, 10, 15, 25, 30, 50, 75, 150   34. $546 = 2 \cdot 3 \cdot 7 \cdot 13$   35. (a) yes (b) yes (c) no (d) yes (e) yes   36. (a) 3 (b) 15 (c) 2 (d) 12 (e) 1   37. $777   38. 28 ounces of soda at 4¢ an ounce is the better buy than the 32 ounces at 5¢ an ounce   39. $\frac{2}{3}$   40. $\frac{5}{4}$   41. 1   42. 0   43. $\frac{7}{4} \neq \frac{19}{11}$   44. 9   45. $\frac{2}{11}$   46. 1   47. $\frac{6}{5}$   48. $\frac{9}{16}$   49. $\frac{2}{9}$   50. $\frac{10}{3}$   51. $\frac{9}{8}$   52. 18   53. 8   54. $4\frac{5}{9}$   55. $11\frac{1}{11}$   56. $\frac{67}{5}$   57. $\frac{129}{4}$   58. $5\frac{15}{32}$   59. $2\frac{7}{9}$   60. 45   61. 243 gallons   62. 108 gallons   63. $27\frac{3}{5}$ revolutions   64. 18   65. 8580   66. 2520   67. $\frac{2}{3}$   68. $\frac{2}{5}$   69. $\frac{23}{36}$   70. $\frac{61}{105}$   71. $\frac{16}{35}$   72. $\frac{1}{30}$   73. $7\frac{19}{40}$   74. $5\frac{5}{12}$   75. $\frac{19}{20}$   76. $\frac{5}{16}$   77. $\frac{5}{6}$   78. true   79. true   80. $18\frac{5}{8}$ hours   81. $21\frac{3}{4}$ minutes   82. forty-two and five tenths   83. three and seven ten thousandths   84. 2.009   85. 627.09   86. $\frac{47}{100}$   87. $8\frac{131}{1000}$   88. $1.\overline{3}$   89. 0.625   90. $0.\overline{4}$   91. 116.026   92. 60.742   93. $7206.16   94. 23.13 in   95. 32.11   96. 5.994   97. $75.12   98. 4.36   99. $48   100. 0.93   101. 0.929   102. 67.396   103. 24.565818   104. 1.710019   105. 15.015 sq ft   106. 6.05   107. 4.7   108. 11.68   109. $30 (10 × $3), $33.55   110. 5.15   111. 13.59   112. $111.35   113. 11.3¢   114. 0.026, 0.26, 2.06, 2.6, 20.06   115. 8.96   116. $\frac{1}{2}$   117. no   118. yes   119. 20   120. $\frac{16}{5}$   121. 120 kg   122. 126 miles   123. $\frac{37}{100}$   124. $\frac{3}{5000}$   125. $\frac{3}{32}$   126. 0.58   127. 2.2   128. 0.105   129. 81%   130. 0.5%   131. 2000%   132. $16.\overline{6}$% or $16\frac{2}{3}$%   133. 36%   134. $44.\overline{4}$%   135. 15.2   136. 12.5%   137. 60   138. 0.3%   139. 96,960   140. 62.5%   141. $36, $756   142. 3%   143. $1562.08   144. $1018   145. 1.7%   146. $15   147. $1680   148. each account pays $1092 interest over 2 years   149. 20.5%   150. 6.4% loss   151. 50   152. $\frac{4}{3}$   153. 10%   154. $\frac{2}{3}$   155. 450   156. 100   157. 1150   158. 2550   159. 45.1%   160. 10.5, 9.5   161. 6.4, 6.2   162. 4.1 mpg   163. 35   164. 3.5   165. 0.054   166. 2500   167. 2270   168. 62.1   169. 140   170. 1   171. 18   172. 2.84   173. 54¢   174. 94.6   175. 28 hr 3 min   176. 10 yd 2 ft 3 in   177. 33.1 cm   178. 38.9 ml   179. 10 gal 2 qt   180. 7.75   181. 28 ft, 45 ft$^2$   182. 37.4 m, 77.5 m$^2$   183. $945   184. 16 m, 12 m$^2$   185. 88.4 ft, 333.2 ft$^2$   186. 16.3 ft, 21.2 ft$^2$   187. 28.9 m, 66.4 m$^2$   188. 14 in: $0.049 per square inch, 16 in: $0.042 per square inch, 16 in costs less   189. 80 m$^3$, 132 m$^2$   190. 46.7 ft$^3$, 77.8 ft$^2$   191. 165.6 dm$^3$, 169.8 dm$^2$   192. 14,130 in$^3$, 2826 in$^2$   193. 310 L   194. true   195. false   196. false   197. false   198. $-10$   199. $-2$   200. 10   201. 4   202. 5   203. $-2$   204. $-16$   205. 40   206. 0   207. 0   208. 6   209. $-15$   210. 40   211. $-86$   212. 104   213. true   214. true   215. false   216. true   217. $-\frac{2}{7}$   218. $-0.9$   219. $-\frac{7}{10}$   220. $-\frac{4}{3}$   221. $\frac{7}{3}$   222. $-2.2$   223. $-17$   224. $-2$   225. 3   226. $-6, \frac{1}{6}$   227. $8, -\frac{1}{8}$   228. $-\frac{5}{8}, \frac{8}{5}$   229. $\frac{12}{7}, -\frac{7}{12}$   230. 0, no reciprocal   231. $-10$   232. $-8$   233. 4   234. $7(x + 2y)$   235. $3(a - 2b + 3)$   236. $-x(y + z)$   237. $-a - 2b$   238. $-12x + 12y - 4z$   239. $-25x + 15y$   240. $2x^3$   241. $\frac{1}{b^8}$   242. $\frac{27}{x^6}$   243. $-32$   244. 36   245. $\frac{1}{144}$   246. $6 \times 10^5$   247. $2 \times 10^{-10}$   248. $-10$   249. contradiction   250. 5   251. identity   252. $-\frac{8}{5}$   253. $-6$   254. $-2$   255. 15   256. $\frac{3}{20}$   257. 2   258. $-6$   259. $\frac{1}{2}$   260. $9x - 4 = 12$   261. $2x - 6 = 3x + 4$   262. 11, 12   263. 95   264. $52.75   265. $29,600   266. 17.2 cm   267. 15 ft

**CHAPTER 1 TEST** 1. 1, 2, 3, 4  2. "is less than"  3. associative law  4. (a) 2 (b) 8 (c) $a < b$  5. 30,000,000 + 5,000,000 + 400,000 + 8000 + 400 + 30 + 8  6. 380,407  7. one million, six hundred fifty-nine thousand, two hundred eighty-eight  8. 120  9. 515  10. 9247  11. 13,217  12. (a) 8890 (b) 8900  13. 37  14. 349  15. 207  16. 6069  17. $72  18. estimated savings: $500; exact savings: $484  19. 1255 gallons  20. 46 feet  21. $458

**CHAPTER 2 TEST** 1. commutative law  2. composite  3. when its ones digit is 0, 2, 4, 6, or 8  4. 1316  5. 843,733  6. 3,878,000  7. 1,800,000  8. 1  9. 0  10. 15  11. undefined  12. Q: 236; R: 0  13. Q: 154; R: 14  14. Q: 458; R: 12  15. $350 = 2 \cdot 5 \cdot 5 \cdot 7$ or $2 \cdot 5^2 \cdot 7$  16. $622  17. $600  18. 5  19. 625  20. 8  21. 5  22. 5  23. 9  24. 54  25. 216  26. 3  27. $550,000

**CHAPTER 3 TEST** 1. proper  2. associative law  3. $\frac{1}{6}$  4. $\frac{9}{2} = \frac{36}{8}$  5. 21  6. 33  7. $\frac{7}{3}$  8. $\frac{12}{13}$  9. $\frac{1}{28}$  10. $\frac{3}{11}$  11. 15  12. $\frac{9}{64}$  13. $\frac{7}{12}$  14. $\frac{55}{2}$ or $27\frac{1}{2}$  15. $2\frac{2}{5}$  16. $\frac{79}{11}$  17. $22\frac{1}{10}$  18. $\frac{66}{83}$  19. 315  20. $2\frac{1}{7}$  21. $20\frac{1}{4}$ gallons  22. 550 revolutions

**CHAPTER 4 TEST** 1. 210  2. 120  3. $\frac{1}{3}$  4. $\frac{2}{3}$  5. $\frac{23}{24}$  6. $\frac{17}{60}$  7. $\frac{9}{7}$ or $1\frac{2}{7}$  8. $\frac{7}{24}$  9. $9\frac{8}{9}$  10. $8\frac{13}{15}$  11. $17\frac{2}{3}$  12. $8\frac{5}{8}$  13. $8\frac{1}{6}$  14. $\frac{1}{90}$  15. true  16. false  17. $13\frac{5}{8}$ yd  18. $3\frac{11}{12}$ hours

**CHAPTER 5 TEST** 1. seventy-three and five thousandths  2. 92.06  3. $\frac{4}{25}$  4. 0.4375  5. 6.36  6. 45.1  7. 0.75  8. 110.02  9. 5.997  10. $24.60  11. 50¢; 48¢  12. 127.89  13. 10.38  14. 633.7  15. 0.6357  16. 0.006357  17. 63.57  18. $24.04  19. 24.8 mpg  20. $36.00 [using $400 \times 9$¢]; $38.60  21. 14.1  22. $8.03 \div 2.5 = 3.212$

**CHAPTER 6 TEST** 1. $\frac{13}{7}$  2. 3  3. 279 pounds  4. $\frac{19}{20}$  5. 0.008  6. 506%  7. 180%  8. 20%  9. 21.6  10. 40  11. $1156.54  12. $20.65  13. $610.20  14. 18.3%  15. 2% profit  16. $1.60; $2.00  17. $20.93  18. 12%

**CHAPTER 7 TEST** 1. 180  2. 150  3. $\frac{11}{3}$ or 11 to 3  4. 40  5. 90  6. $\frac{2}{1}$ or 2 to 1  7. $26.\overline{6}$%  8. 60%  9. 25  10. 80  11. $\frac{7}{16}$ or 7 to 16  12. 25%  13. 20  14. 18  15. 8  16. (a) 254.5 pounds (b) 255 pounds (c) no mode

**CHAPTER 8 TEST** 1. 2.5  2. 0.028  3. 14.3  4. 149  5. 432  6. 8,000,000  7. 8 yd 2 ft 7 in  8. 10,000 g  9. 12 ft$^2$  10. 31 m  11. 2009.6 cm$^3$  12. 5024 in$^2$  13. $78.75  14. 29.72 cm$^2$

**CHAPTER 9 TEST** 1. false  2. true  3. 1  4. $-3$  5. $-24$  6. $-8$  7. 0  8. $-\frac{1}{12}$  9. $\frac{5}{2}$  10. $-4.8$  11. 2  12. $\frac{12}{7}$  13. $4(3x - 2y)$  14. $7a - 21b$  15. $4x + 3$  16. $\frac{1}{a^2}$  17. $\frac{1}{3x^3}$  18. $\frac{b^4}{a^2}$  19. 4  20. 5  21. $-1$  22. $1.42 \times 10^6$  23. $7.93 \times 10^{-5}$  24. $74.25

**CHAPTER 10 TEST** 1. $-15$  2. 12  3. 6  4. $-\frac{3}{5}$  5. 2  6. $-1$  7. 5  8. $2x + 11 = 8$  9. $x + 5 = 26$  10. 34 and 35  11. 85  12. 4250  13. 3.2 ft  14. 6 yd

*All exercises are selected from the A sets of exercises.*

## 1.3

**32.** To find the total miles driven during two days, add the miles driven each day.

| | |
|---|---|
| 135 | Miles driven first day |
| + 467 | Miles driven second day |
| 602 | Total miles driven |

Thus, Todd drove 602 miles on the trip.

## 1.4

**16.**

| 23 | → | 20 | Round 23 to 20 |
|---|---|---|---|
| 16 | → | 20 | Round 16 to 20 |
| 54 | → | 50 | Round 54 to 50 |
| + 79 | → | + 80 | Round 79 to 80 |
| | | 170 | The estimated sum |

**20.**

| 329 | → | 300 | Round 329 to 300 |
|---|---|---|---|
| 556 | → | 600 | Round 556 to 600 |
| 113 | → | 100 | Round 113 to 100 |
| + 925 | → | + 900 | Round 925 to 900 |
| | | 1900 | The estimated sum |

**26.**

| 19501 | → | 20000 | Round 19501 to 20000 |
|---|---|---|---|
| 19499 | → | 19000 | Round 19499 to 19000 |
| 38201 | → | 38000 | Round 38201 to 38000 |
| + 21005 | → | + 21000 | Round 21005 to 21000 |
| | | 98000 | The estimated sum |

**34.** Round 7105 to 7100 and round 6898 to 6900. Then add to estimate the total number of books.

$$
\begin{array}{r}
7100 \\
+ \ 6900 \\
\hline
14000
\end{array}
$$

There are about 14,000 books on the two floors.

## 1.6

**29.** Round 389 to 400 and 605 to 600. Subtract to obtain an estimate for the increase in payment.

| | |
|---|---|
| 600 | |
| − 400 | |
| 200 | Estimated increase |

Subtract to find the exact increase.

| | |
|---|---|
| 605 | New payment |
| − 389 | Former payment |
| 216 | Exact increase in payment |

The estimated increase is $200 and the exact increase is $216.

## 1.7

**13.** First add the two deposits to the starting balance.

| | |
|---|---|
| 625 | Starting balance |
| 300 | One deposit |
| + 235 | Other deposit |
| 1160 | |

Next add the five amounts of the checks written.

$$
\begin{array}{r}
23 \\
18 \\
113 \\
410 \\
+ \ 32 \\
\hline
596
\end{array}
$$

To find the balance at the end of the month, subtract the value of the checks written from the amount in the account.

$$
\begin{array}{r}
1160 \\
- \ 596 \\
\hline
564
\end{array}
$$

Thus, the balance at the end of the month was $564.

**15.** First find the three distances from Salt Lake City to Seattle.

| | |
|---|---|
| 435 | Salt Lake City to Cheyenne |
| 735 | Cheyenne to Butte |
| + 597 | Butte to Seattle |
| 1787 | |

| | |
|---|---|
| 419 | Salt Lake City to Butte |
| + 597 | Butte to Seattle |
| 1016 | |

| | |
|---|---|
| 528 | Salt Lake City to Reno |
| 564 | Reno to Portland |
| + 175 | Portland to Seattle |
| 1267 | |

Thus, the shortest distance is 1016 miles (through Butte) which is 251 miles shorter than the next shortest distance (through Reno and Portland).

**18.** To find the sales on Thursday, subtract $305 from the sales on Friday.

| | |
|---|---|
| 1245 | Friday sales |
| − 305 | |
| 940 | Thursday sales |

To find the sales on Saturday, add $785 to the sales on Friday.

| | |
|---|---|
| 1245 | Friday sales |
| + 785 | |
| 2030 | Saturday sales |

**22.** Since Larry had $12 more than Jim, add $12 to Jim's total to find Larry's total.

$$
\begin{array}{rl}
43 & \text{Jim's total} \\
+\ 12 & \\
\hline
55 & \text{Larry's total}
\end{array}
$$

Mike and Jim together have:

$$
\begin{array}{r}
25 \\
+\ 43 \\
\hline
68
\end{array}
$$

Since Henri has $9 less than this total, subtract.

$$
\begin{array}{rl}
68 & \\
-\ 9 & \\
\hline
59 & \text{Henri's total}
\end{array}
$$

Add the four amounts to find the total together.

$$
\begin{array}{rl}
25 & \text{Mike's total} \\
43 & \text{Jim's total} \\
55 & \text{Larry's total} \\
+\ 59 & \text{Henri's total} \\
\hline
182 &
\end{array}
$$

Thus, the four together have $182.

## 2.2

**34.** Since 230 calories are burned in 1 hour, in 4 hours there will be 4 times 230 calories.

$$
\begin{array}{r}
230 \\
\times\ 4 \\
\hline
920
\end{array}
$$

Thus, 920 calories are burned in 4 hours.

## 2.3

**25.** Round 103 to 100 and 289 to 300.

$$
\begin{array}{rl}
300 & \\
\times\ 100 & \\
\hline
30000 & \text{Estimated total}
\end{array}
$$

Thus, about $30,000 was taken in on Saturday.

## 2.5

**16.** Don't forget this zero

$$
\begin{array}{r}
3027 \\
2\overline{)6054} \\
\underline{6}\phantom{054} \\
05 \\
\underline{4}\phantom{0} \\
14 \\
\underline{14} \\
0
\end{array}
$$

**21.** $223 \div 23$ ; $161 \div 23$

$$
\begin{array}{r}
97 \\
23\overline{)2231} \\
\underline{207}\phantom{0} \\
161 \\
\underline{161} \\
0
\end{array}
$$

**23.** $1324 \div 473$ ; $3784 \div 473$

$$
\begin{array}{r}
28 \\
473\overline{)13244} \\
\underline{946}\phantom{0} \\
3784 \\
\underline{3784} \\
0
\end{array}
$$

**29.** $755 \div 83$ ; $83 \div 83$

$$
\begin{array}{r}
91 \\
83\overline{)7553} \\
\underline{747}\phantom{0} \\
83 \\
\underline{83} \\
0
\end{array}
$$

Check:
$$
\begin{array}{r}
83 \\
\times\ 91 \\
\hline
83 \\
747\phantom{0} \\
\hline
7553
\end{array}
$$

Thus, the quotient is 91 and the remainder is 0.

**32.** To find the approximate number of calories in each bar, divide 4000 by 20, which is 200. Thus, there are about 200 calories in each bar. To find the exact number, divide 4011 by 21.

$$
\begin{array}{r}
191 \\
21\overline{)4011} \\
\underline{21}\phantom{00} \\
191 \\
\underline{189} \\
21 \\
\underline{21} \\
0
\end{array}
$$

Thus, there are exactly 191 calories in each bar.

## 2.6

**4. (a)** Multiply 35 by the value of each.

$$
\begin{array}{rl}
329 & \text{Price of 1 snow blower} \\
\times\ 35 & \text{Number of snow blowers} \\
\hline
1645 & \\
987\phantom{0} & \\
\hline
11515 &
\end{array}
$$

Thus, $11,515 was taken in on the sale of snow blowers.

**(b)** Multiply 40 by the value of each.

$$
\begin{array}{rl}
289 & \text{Price of 1 chain saw} \\
\times\ 40 & \text{Number of chain saws} \\
\hline
11560 &
\end{array}
$$

Thus, $11,560 was taken in on the sale of chain saws.

**(c)** Add to find the total taken in on both items.

$$
\begin{array}{r}
11515 \\
+\ 11560 \\
\hline
23075
\end{array}
$$

Thus, $23,075 was taken in on both.

**8.** First find the number of square feet of mirror.

$$
\begin{array}{rl}
8 & \text{Length of mirror} \\
\times\ 5 & \text{Width of mirror} \\
\hline
40 & \text{Area of mirror in square feet}
\end{array}
$$

Since each square foot costs \$3, multiply to find the total cost.

$$\begin{array}{r} 40 \\ \times\ 3 \\ \hline 120 \end{array}$$

It will cost \$120 to put a mirror on the wall.

**13. (a)** 5 sections, each seating 400, results in

$$5 \times 400 = 2000 \text{ seats.}$$

8 sections, each seating 1200, results in

$$8 \times 1200 = 9600 \text{ seats.}$$

Since there are 20 sections in total, and 13 have been considered $(5 + 8 = 13)$, there are 7 sections left $(20 - 13 = 7)$, each seating 950.

$$7 \times 950 = 6650$$

Add to find the total number of seats.

$$\begin{array}{r} 2000 \\ 9600 \\ +\ 6650 \\ \hline 18250 \end{array} \quad \text{Total seats in the dome}$$

**(b)** Multiply 18,250 by the value of each seat, \$5, to find the total amount of money taken in.

$$\begin{array}{r} 18250 \\ \times\ 5 \\ \hline 91250 \end{array}$$

Thus, \$91,250 was taken in.

**17.** $\dfrac{372¢}{12} = 31¢$     Price per pound of the 12-pound bag

$\dfrac{288¢}{9} = 32¢$     Price per pound of the 9-pound bag

Thus, the 12-pound bag is the better buy.

**20.** To find the student-teacher ratio, divide the number of students (12,100) by the number of faculty members (550).

$$\begin{array}{r} 22 \\ 550\overline{)12100} \\ \underline{1100\phantom{0}} \\ 1100 \\ \underline{1100} \\ 0 \end{array}$$

Thus, the student-teacher ratio is 22 students per teacher.

**2.7**

**13.**

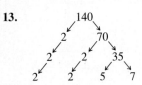

Thus, $140 = 2 \cdot 2 \cdot 5 \cdot 7$.

**2.8**

**12.**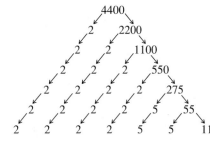

Thus, $4400 = 2 \cdot 2 \cdot 2 \cdot 2 \cdot 5 \cdot 5 \cdot 11 = 2^4 \cdot 5^2 \cdot 11$.

**32.** $5^2 - \sqrt{49} = 25 - 7$    Square and take root first

          $= 18$              Then subtract

**34.** $\sqrt{25} + 15 - 2^2 \cdot 5$

    $= 5 + 15 - 4 \cdot 5$     Take root and square first

    $= 5 + 15 - 20$      Multiply next

    $= 20 - 20$         Add

    $= 0$                Then subtract

**41.** $\sqrt{4}(9 \div 3) + 2^3$

    $= \sqrt{4}(3) + 2^3$     Divide inside parentheses first

    $= 2(3) + 8$        Take root and cube next

    $= 6 + 8$          Multiply next

    $= 14$            Then add

**43.** $(3 \cdot 7)^2 = (21)^2$    Multiply first inside parentheses

         $= 441$       Then square

**44.** $3 \cdot 7^2 = 3 \cdot 49$     Square first

       $= 147$       Then multiply

Note the difference between Exercises 43 and 44.

**46.** $(3 + 7)^2 = 10^2$     Add first inside parentheses

         $= 100$       Then square

**47.** $3^2 + 7^2 = 9 + 49$    Square first

       $= 58$        Then add

Note the difference between Exercises 46 and 47.

**3.1**

**14.** Since each shaded part of the figure corresponds to $\dfrac{1}{4}$ of a unit, there are five $\dfrac{1}{4}$'s shaded, or $\dfrac{5}{4}$ of the unit is shaded.

**3.2**

**8.** $\dfrac{8}{9}$ and $\dfrac{96}{108}$

$$8 \cdot 108 = 864 \text{ and } 9 \cdot 96 = 864$$

Since the cross products are equal, the fractions are equal.

SOLUTIONS TO SELECTED EXERCISES **613**

**13.** $\dfrac{9}{18} = \dfrac{?}{6}$

To obtain 6 from 18, divide 18 by 3. Thus 9 is divided by 3, giving 3.

$$\dfrac{9}{18} = \dfrac{9 \div 3}{18 \div 3} = \dfrac{3}{6} \longleftarrow \text{The desired number}$$

**23.** $\dfrac{36}{14} = \dfrac{18}{?}$

To get 18 from 36, divide by 2.

$$\dfrac{36}{14} = \dfrac{36 \div 2}{14 \div 2} = \dfrac{18}{7} \longleftarrow \text{The desired number}$$

**31.** $\dfrac{210}{105} = \dfrac{5 \cdot 42}{5 \cdot 21} = \dfrac{\cancel{5} \cdot 2 \cdot \cancel{21}}{\cancel{5} \cdot \cancel{21}} = \dfrac{2}{1} = 2$

**3.3**

**20.** $\dfrac{18}{84} \cdot \dfrac{36}{27} = \dfrac{\cancel{18}^{2}}{\cancel{84}_{7}} \cdot \dfrac{\cancel{36}^{3}}{\cancel{27}_{3}} = \dfrac{2}{7} \cdot \dfrac{3}{3}$

$$= \dfrac{2}{7} \cdot 1 = \dfrac{2}{7}$$

**31.** To find the area multiply the length, $\dfrac{7}{8}$, by the width, $\dfrac{4}{7}$.

$$\dfrac{7}{8} \cdot \dfrac{4}{7} = \dfrac{\cancel{7}^{1}}{\cancel{8}_{2}} \cdot \dfrac{\cancel{4}^{1}}{\cancel{7}_{1}} = \dfrac{1}{2}$$

The area is $\dfrac{1}{2}$ sq km.

**3.4**

**14.** $\dfrac{13}{15} \div \dfrac{39}{5} = \dfrac{13}{15} \cdot \dfrac{5}{39}$  Multiply by reciprocal

$$= \dfrac{\cancel{13}^{1}}{\cancel{15}_{3}} \cdot \dfrac{\cancel{5}^{1}}{\cancel{39}_{3}} = \dfrac{1}{9}$$

**22.** $\dfrac{10}{9}$ of $\square$ is 25

$\downarrow \quad \downarrow \downarrow \downarrow \downarrow$

$\dfrac{10}{9} \cdot \square = 25$

$\square = 25 \div \dfrac{10}{9}$  Related division sentence

$$= \dfrac{25}{1} \cdot \dfrac{9}{10} = \dfrac{5 \cdot \cancel{5} \cdot 9}{2 \cdot \cancel{5}} = \dfrac{45}{2}$$

**28.** The problem translates to:

$\dfrac{5}{6}$ of trip is 365

$\downarrow \quad \downarrow \quad \downarrow \quad \downarrow \quad \downarrow$

$\dfrac{5}{6} \cdot \square = 365$

$\square = 365 \div \dfrac{5}{6}$  Related division sentence

$$= \dfrac{365}{1} \cdot \dfrac{6}{5}$$

$$= \dfrac{\cancel{5} \cdot 73 \cdot 6}{1 \cdot \cancel{5}} = 438$$

The total length of the trip was 438 miles.

**3.5**

**8.** $\dfrac{257}{9}$ $\quad 9)\overline{257}$  $\dfrac{\phantom{0}28}{\phantom{0}}$
$\phantom{9)}\underline{18}$
$\phantom{9)}77$
$\phantom{9)}\underline{72}$
$\phantom{9)}\phantom{0}5$

Thus, $\dfrac{257}{9} = 28\dfrac{5}{9}$.

**17.** $32\dfrac{7}{10}$ $\quad \dfrac{10 \cdot 32 + 7}{10} = \dfrac{320 + 7}{10}$

$$= \dfrac{327}{10}$$

**3.6**

**8.** $2\dfrac{3}{5} \cdot 6\dfrac{1}{4} \cdot 12$

$$= \dfrac{13}{5} \cdot \dfrac{25}{4} \cdot \dfrac{12}{1} \quad \text{Change to improper fractions}$$

$$= \dfrac{13 \cdot \cancel{5} \cdot 5 \cdot \cancel{4} \cdot 3}{\cancel{5} \cdot \cancel{4}}$$

$$= \dfrac{13 \cdot 5 \cdot 3}{1} = 195$$

**22.** If we knew how many uniforms could be made, $\square$, we would multiply that number by $3\dfrac{1}{8}$ to obtain 25. That is,

$$3\dfrac{1}{8} \cdot \square = 25.$$

$\square = 25 \div 3\dfrac{1}{8}$  Related division sentence

$$= \dfrac{25}{1} \div \dfrac{25}{8}$$

$$= \dfrac{25}{1} \cdot \dfrac{8}{25}$$

$$= \dfrac{\cancel{25} \cdot 8}{1 \cdot \cancel{25}} = \dfrac{8}{1} = 8$$

Thus, 8 uniforms could be made.

**3.7**

**10.** Amount spent for food $= \dfrac{1}{5}$ of 18000

$$= \dfrac{1}{5} \cdot 18000$$

$$= \dfrac{18000}{5} = \$3600$$

Amount spent for housing $= \dfrac{1}{4}$ of 18000

$$= \dfrac{1}{4} \cdot 18000$$

$$= \dfrac{18000}{4} = \$4500$$

Amount spent for clothing $= \dfrac{1}{10}$ of 18000

$$= \dfrac{1}{10} \cdot 18000$$

$$= \dfrac{18000}{10} = \$1800$$

Amount spent for taxes $= \dfrac{3}{8}$ of 18000

$$= \dfrac{3}{8} \cdot 18000$$

$$= \dfrac{3 \cdot 18000}{8} = \$6750$$

**13.** If we knew the number of minutes, $\square$, we would multiply by 45 to obtain $321\dfrac{1}{2}$.

$$45 \cdot \square = 321\dfrac{1}{2}$$

$$\square = 321\dfrac{1}{2} \div 45 \qquad \text{Related division sentence}$$

$$= \dfrac{643}{2} \cdot \dfrac{1}{45}$$

$$= \dfrac{643}{90} = 7\dfrac{13}{90}$$

It will take $7\dfrac{13}{90}$ minutes to revolve $321\dfrac{1}{2}$ times.

**4.1**

**8.** $\dfrac{2}{35} + \dfrac{18}{35} + \dfrac{25}{35} = \dfrac{2 + 18 + 25}{35} \qquad$ Add numerators

$$= \dfrac{45}{35}$$

$$= \dfrac{\cancel{5} \cdot 9}{\cancel{5} \cdot 7} = \dfrac{9}{7} \qquad \text{Reduce to lowest terms}$$

**14.** $\dfrac{28}{35} - \dfrac{18}{35} = \dfrac{28 - 18}{35} \qquad$ Subtract numerators

$$= \dfrac{10}{35}$$

$$= \dfrac{\cancel{5} \cdot 2}{\cancel{5} \cdot 7} = \dfrac{2}{7} \qquad \text{Reduce to lowest terms}$$

**20.** $\dfrac{7}{10} - \dfrac{1}{10} + \dfrac{3}{10} = \dfrac{7 - 1 + 3}{10}$

$$= \dfrac{6 + 3}{10}$$

$$= \dfrac{9}{10}$$

**23.** First add the amounts done the first two days.

$$\dfrac{5}{12} + \dfrac{3}{12} = \dfrac{8}{12} = \dfrac{\cancel{4} \cdot 2}{\cancel{4} \cdot 3} = \dfrac{2}{3}$$

Subtract this total from 1 to find the amount to be done the third day.

$$1 - \dfrac{2}{3} = \dfrac{3}{3} - \dfrac{2}{3} = \dfrac{3 - 2}{3} = \dfrac{1}{3}$$

Thus, $\dfrac{1}{3}$ of the job must be done the third day.

**4.2**

**14.** Factor each number into a product of primes

$$18 = 2 \cdot 3 \cdot 3$$
$$24 = 2 \cdot 2 \cdot 2 \cdot 3$$
$$30 = 2 \cdot 3 \cdot 5$$

The LCM must consist of three 2's, two 3's, and one 5. Thus, the LCM $= 2 \cdot 2 \cdot 2 \cdot 3 \cdot 3 \cdot 5 = 360$.

**29.** The shortest length of wall will be the LCM of 8, 9, and 14. Factor each number into a product of primes.

$$8 = 2 \cdot 2 \cdot 2$$
$$9 = 3 \cdot 3$$
$$14 = 2 \cdot 7$$

The LCM must consist of three 2's, two 3's, and one 7. The LCM $= 2 \cdot 2 \cdot 2 \cdot 3 \cdot 3 \cdot 7 = 504$. Thus, the shortest length is 504 inches.

**4.3**

**8.** $4 + \dfrac{4}{5} = \dfrac{4 \cdot 5}{5} + \dfrac{4}{5} \qquad$ The LCD is 5

$$= \dfrac{20}{5} + \dfrac{4}{5}$$

$$= \dfrac{20 + 4}{5} = \dfrac{24}{5}$$

**11.**

$$\frac{1}{6} = \frac{5}{5 \cdot 6} = \frac{5}{30}$$

$$+\frac{3}{5} = +\frac{3 \cdot 6}{5 \cdot 6} = +\frac{18}{30}$$

$$\frac{23}{30}$$

The LCD is 30

**17.** $4 - \frac{4}{5} = \frac{4 \cdot 5}{5} - \frac{4}{5}$   The LCD is 5

$$= \frac{20}{5} - \frac{4}{5}$$

$$= \frac{20 - 4}{5} = \frac{16}{5}$$

**23.**

$$\frac{8}{15} = \frac{8}{3 \cdot 5} = \frac{8 \cdot 4}{3 \cdot 4 \cdot 5} = \frac{32}{60}$$

$$-\frac{3}{20} = -\frac{3}{4 \cdot 5} = -\frac{3 \cdot 3}{3 \cdot 4 \cdot 5} = -\frac{9}{60}$$

$$\frac{23}{60}$$

**29.** $\frac{8}{15} + \frac{1}{12} - \frac{5}{20} = \frac{8}{3 \cdot 5} + \frac{1}{2 \cdot 2 \cdot 3} - \frac{5}{2 \cdot 2 \cdot 5}$

$$= \frac{8 \cdot 2 \cdot 2}{2 \cdot 2 \cdot 3 \cdot 5} + \frac{1 \cdot 5}{2 \cdot 2 \cdot 3 \cdot 5} - \frac{5 \cdot 3}{2 \cdot 2 \cdot 3 \cdot 5}$$

$$= \frac{32}{60} + \frac{5}{60} - \frac{15}{60}$$

$$= \frac{32 + 5 - 15}{60}$$

$$= \frac{22}{60} = \frac{\cancel{2} \cdot 11}{\cancel{2} \cdot 30} = \frac{11}{30}$$

**4.4**

**8.**

$$15\frac{7}{8}$$
$$+\ 4$$
$$19\frac{7}{8}$$   Add the whole numbers

**10.**

$$215\frac{7}{8} = 215\frac{7}{8}$$   The LCD is 8

$$+147\frac{1}{2} = +147\frac{4}{8}$$

$$= 362\frac{11}{8} = 362 + \frac{8}{8} + \frac{3}{8}$$

$$= 362 + 1 + \frac{3}{8}$$

$$= 363\frac{3}{8}$$

**20.**

$$17\frac{6}{11}$$
$$-\ 8$$
$$9\frac{6}{11}$$   Subtract the whole numbers

**22.**

$$485\frac{9}{10} = 485\frac{9}{10}$$

$$-316\frac{2}{5} = -316\frac{4}{10}$$   The LCD is 10

$$169\frac{5}{10} = 169\frac{1}{2}$$

**26.**

$$7\frac{1}{5} = 7\frac{1}{5} = 7\frac{3}{15}$$

$$3\frac{2}{3} = 3\frac{2}{3} = 3\frac{10}{15}$$

$$+1\frac{1}{15} = 1\frac{1}{3 \cdot 5} = 1\frac{1}{15}$$

$$= 11\frac{14}{15}$$

The LCD $= 3 \cdot 5 = 15$

**30.** Subtract $1\frac{2}{3}$ from $7\frac{1}{8}$.

$$7\frac{1}{8} = 7\frac{3}{24} = 6\frac{27}{24}$$

$$-1\frac{2}{3} = 1\frac{16}{24} = 1\frac{16}{24}$$

$$= 5\frac{11}{24}$$

The LCD is 24 and we had to borrow $\frac{24}{24}$ from 7 to be able to subtract $\frac{16}{24}$.

**4.5**

**8.** $\left(\frac{3}{5}\right)^2 - \frac{1}{5} + \frac{6}{25}$

$$= \frac{9}{25} - \frac{1}{5} + \frac{6}{25}$$   Square first

$$= \frac{9}{25} - \frac{5}{25} + \frac{6}{25}$$   LCD $= 25$

$$= \frac{9 - 5 + 6}{25} = \frac{10}{25} = \frac{2}{5}$$

**10.** $\sqrt{\frac{1}{9}} + \frac{6}{7} \div \frac{3}{14} - 3 \cdot \frac{1}{9}$

$$= \frac{1}{3} + \frac{6}{7} \div \frac{3}{14} - 3 \cdot \frac{1}{9}$$   Take root first

$$= \frac{1}{3} + \frac{6}{7} \cdot \frac{14}{3} - 3 \cdot \frac{1}{9}$$   Divide by multiplying by the reciprocal

$$= \frac{1}{3} + 4 - 3 \cdot \frac{1}{9}$$   $\frac{6}{7} \cdot \frac{14}{3} = \frac{2 \cdot \cancel{3} \cdot 2 \cdot \cancel{7}}{\cancel{7} \cdot \cancel{3}} = 4$

$$= \frac{1}{3} + 4 - \frac{1}{3}$$   $3 \cdot \frac{1}{9} = \frac{3 \cdot 1}{9} = \frac{1}{3}$

$$= 4$$   $\frac{1}{3} - \frac{1}{3} = 0$

**17.** $\dfrac{17}{30} = \dfrac{17 \cdot 4}{30 \cdot 4} = \dfrac{68}{120}$

$\dfrac{21}{40} = \dfrac{21 \cdot 3}{40 \cdot 3} = \dfrac{63}{120}$

Since $68 > 63$, $\dfrac{17}{30} > \dfrac{21}{40}$.

**26.** $\dfrac{1}{4} + \dfrac{7}{40}$ is the amount done Monday.

$\dfrac{1}{4} + \dfrac{1}{5}$ is the amount done Tuesday.

Add these pairs.

$\dfrac{1}{4} + \dfrac{7}{40} = \dfrac{1 \cdot 10}{4 \cdot 10} + \dfrac{7}{4 \cdot 10} = \dfrac{10 + 7}{40} = \dfrac{17}{40}$

$\dfrac{1}{4} + \dfrac{1}{5} = \dfrac{1 \cdot 5}{4 \cdot 5} + \dfrac{1 \cdot 4}{4 \cdot 5} = \dfrac{5 + 4}{20} = \dfrac{9}{20}$

To compare $\dfrac{17}{40}$ and $\dfrac{9}{20}$, write each with the LCD of 40. Since $\dfrac{9}{20} = \dfrac{9 \cdot 2}{20 \cdot 2} = \dfrac{18}{40}$ and $\dfrac{18}{40} > \dfrac{17}{40}$, Carl did more on Tuesday. The amount is more by $\dfrac{18}{40} - \dfrac{17}{40} = \dfrac{1}{40}$.

## 4.6

**9.** Add the number of hours worked each day.

$7\dfrac{3}{8} + 6\dfrac{2}{5} + 5\dfrac{4}{5} + 8\dfrac{1}{4} + 7\dfrac{1}{2}$

$= (7 + 6 + 5 + 8 + 7) + \left( \dfrac{3}{8} + \dfrac{2}{5} + \dfrac{4}{5} + \dfrac{1}{4} + \dfrac{1}{2} \right)$

$= 33 + \left( \dfrac{15}{40} + \dfrac{16}{40} + \dfrac{32}{40} + \dfrac{10}{40} + \dfrac{20}{40} \right)$   LCD $= 40$

$= 33 + \dfrac{15 + 16 + 32 + 10 + 20}{40}$

$= 33 + \dfrac{93}{40} = 33 + 2\dfrac{13}{40} = 35\dfrac{13}{40}$

Thus, Lon worked $35\dfrac{13}{40}$ hr.

**10.** To find the amount Lon was paid, multiply the number of hours worked, $35\dfrac{13}{40}$, by the pay per hour, $\$6\dfrac{1}{4}$.

$33\dfrac{13}{40} \cdot 6\dfrac{1}{4} = \dfrac{1413}{40} \cdot \dfrac{25}{4}$

$= \dfrac{1413 \cdot \cancel{5} \cdot 5}{\cancel{5} \cdot 8 \cdot 4}$

$= \dfrac{7065}{32} = 220\dfrac{25}{32}$

Thus, Lon was paid $\$220\dfrac{25}{32}$.

**12.** The closing value was

$$121\dfrac{3}{8} + 1\dfrac{3}{4} - 3\dfrac{1}{8} + 2\dfrac{7}{8}.$$

First add $121\dfrac{3}{8} + 1\dfrac{3}{4} + 2\dfrac{7}{8}$.

$121\dfrac{3}{8} + 1\dfrac{3}{4} + 2\dfrac{7}{8}$

$= (121 + 1 + 2) + \left( \dfrac{3}{8} + \dfrac{3}{4} + \dfrac{7}{8} \right)$

$= 124 + \left( \dfrac{3}{8} + \dfrac{6}{8} + \dfrac{7}{8} \right)$   LCD $= 8$

$= 124 + \dfrac{16}{8} = 124 + 2 = 126.$

Now subtract $3\dfrac{1}{8}$ from the sum.

$$\begin{array}{rcr} 126 & = & 125\dfrac{8}{8} \\ - \quad 3\dfrac{1}{8} & = & - \quad 3\dfrac{1}{8} \\ \hline & & 122\dfrac{7}{8} \end{array}$$

Thus, the closing value was $\$122\dfrac{7}{8}$.

## 5.2

**7.** To change $0.1302$ to a fraction, note that $0.1302$ is "one thousand three hundred two **ten thousandths**." Thus,

$$0.1302 = \dfrac{1302}{10,000} = \dfrac{\cancel{2} \cdot 651}{\cancel{2} \cdot 5000} = \dfrac{651}{5000}$$

**22.** To change $\dfrac{19}{16}$ to a decimal, divide 19 by 16.

$$\begin{array}{r} 1.1875 \\ 16\overline{)19.0000} \\ \underline{16} \\ 3\,0 \\ \underline{1\,6} \\ 1\,40 \\ \underline{1\,28} \\ 120 \\ \underline{112} \\ 80 \\ \underline{80} \\ 0 \end{array}$$

Thus, $\dfrac{19}{16} = 1.1875.$

## 5.3

**34. (a)** Round $0.\overline{35}$ to the nearest tenth.

$$0.\overline{35} = 0.3535\ldots$$

Since this digit is 5, change 3 to 4 and drop remaining digits

Round to here

Thus, $0.\overline{35} \approx 0.4.$   Rounded to nearest tenth

**(b)** Round $0.\overline{35}$ to the nearest hundredth.

$$0.\overline{35} = 0.3535\ldots$$

Since this digit is 3, drop remaining digits

Round to here

Thus, $0.\overline{35} \approx 0.35.$    Rounded to nearest hundredth

**(c)** Round $0.\overline{35}$ to the nearest thousandth.

$$0.\overline{35} = 0.3535\ldots$$

Since this digit is 5, change 3 to 4 and drop remaining digits

Round to here

Thus, $0.\overline{35} \approx 0.354.$    Rounded to nearest thousandth

**47.** To approximate the total cost, use 10¢ for the price of a brick and 300 for the number of bricks. Then the approximate cost is

$$300 \cdot 10¢ = 3000¢ = \$30.00.$$

## 5.4

**29.** First find the total of all deductions.

$$\begin{array}{r} \$1237.40 \\ 725.36 \\ 2347.01 \\ +\quad 444.83 \\ \hline \$4754.60 \end{array}$$    Total deductions

Now subtract the total deductions from the total income.

$$\begin{array}{r} \$31255.75 \\ -\quad 4754.60 \\ \hline \$26501.15 \end{array}$$

The taxable income was $26,501.15.

**36. (a)** To find the total of the four injections, add

$$2.65 + 2.75 + 3.5 + 4.0 = 12.9 \text{ milligrams.}$$

**(b)** The amount left in the bottle was

$$30.5 - 12.9 = 17.6 \text{ milligrams.}$$

**37.** From the figure,

$$x = 2.3 + 4.75 + 2.3 = 9.35 \text{ ft.}$$

Also,

$$y = 11.15 - 3.25 = 7.9 \text{ ft.}$$

## 5.5

**31.** First find the area of the wall.

Area = length · width

$$= (13.5) \cdot (7.6) = 102.6 \text{ square feet}$$

Then multiply the area by the cost per square foot.

$$(102.6) \cdot (0.67) = 68.742$$

To the nearest cent, the cost of the project is $68.74.

**35.** Since each tire costs $38.75, the approximate cost of a tire is $40.00. Thus, the approximate cost of 4 tires is

$$(4)(\$40.00) = \$160.00.$$

To find the exact cost, multiply $38.75 by 4.

$$(4)(\$38.75) = \$155.00$$

Four tires will cost $155.00.

## 5.6

**8.**
$$\begin{array}{r} 7.2 \\ 0.31\overline{)2.23.2} \\ 2\ 17 \\ \hline 6\ 2 \\ 6\ 2 \\ \hline 0 \end{array}$$

**17.**
$$\begin{array}{r} 6.562 \\ 48\overline{)315.000} \\ 288 \\ \hline 27\ 0 \\ 24\ 0 \\ \hline 3\ 00 \\ 2\ 88 \\ \hline 120 \\ 96 \\ \hline 24 \end{array}$$
$\longleftarrow 2 < 5$ so drop remaining digits

The quotient, to the nearest hundredth, is 6.56.

**32.**
$$\begin{array}{r} 11.77 \\ 1.8\overline{)21.2.00} \\ 18 \\ \hline 3\ 2 \\ 1\ 8 \\ \hline 1\ 40 \\ 1\ 26 \\ \hline 1\ 40 \\ 1\ 26 \\ \hline 14 \end{array}$$

Since it is clear that 7 will continue to repeat, the quotient is $11.\overline{7}$.

**37.** Divide $382.40 by 0.147.

$$\begin{array}{r} 2\ 601.360 \\ 0.147\overline{)382.400.000} \\ 294 \\ \hline 88\ 4 \\ 88\ 2 \\ \hline 200 \\ 147 \\ \hline 53\ 0 \\ 44\ 1 \\ \hline 8\ 90 \\ 8\ 82 \\ \hline 80 \end{array}$$

To the nearest cent, the assessed value is $2601.36.

**5.7**

**2.** $3.25 - \dfrac{3}{8} = 3\dfrac{1}{4} - \dfrac{3}{8}$

$= \dfrac{13}{4} - \dfrac{3}{8}$

$= \dfrac{26}{8} - \dfrac{3}{8}$   LCD = 8

$= \dfrac{23}{8} = 2\dfrac{7}{8}$

**8.** $3.25 - \dfrac{3}{8} = 3.25 - 0.375$   Change $\dfrac{3}{8}$ to a decimal

$\begin{array}{r} 3.250 \\ -\ 0.375 \\ \hline 2.875 \end{array}$

Thus, $3.25 - \dfrac{3}{8} \approx 2.88$ to the nearest hundredth.

**20.** $(6.2)^2 - 6.6 \div 1.1$

$= 38.44 - 6.6 \div 1.1$   Square first

$= 38.44 - 6$   Divide next

$= 32.44$   Then subtract

**28.** Note that $\dfrac{2}{3} = 0.666\ldots = 0.\overline{6}$ and $\dfrac{7}{10} = 0.7$. Thus, we are arranging

$$0.\overline{6},\ 0.65,\ 6.05,\ 0.7,\ 6.5.$$

Write each number correct to two decimal places; round $0.\overline{6}$ to 0.67.

$$0.67,\ 0.65,\ 6.05,\ 0.70,\ 6.50$$

Ignoring decimal points, the order is

$$65,\ 67,\ 70,\ 605,\ 650,$$

or

$$0.65,\ 0.67,\ 0.70,\ 6.05,\ 6.50.$$

Thus, the order is $0.65,\ \dfrac{2}{3},\ \dfrac{7}{10},\ 6.05,\ 6.5.$

**5.8**

**34.** $5 \div 1.2 + 6.5$

$= 4.1\overline{6} + 6.5$   Divide first

$= 10.\overline{6}$   Then add

Don't round after dividing 5 by 1.2; let your calculator hold the quotient. Then add to that value 6.5. The result, $10.\overline{6}$ will be displayed.

**6.1**

**33.** Let $a$ be the weight of 110 ft of the wire. The proportion described by the situation is

$$\dfrac{70}{84} = \dfrac{110}{a}.$$

$70a = 84 \cdot 110$   Cross-product equation

$$a = \dfrac{84 \cdot 110}{70} = 132$$

Thus, 110 ft of wire weighs 132 pounds.

**36.** Let $a$ be the number of antelope in the preserve. The proportion described by the situation is

$$\dfrac{58}{a} = \dfrac{7}{29}.$$

$58 \cdot 29 = 7a$   Cross-product equation

$$\dfrac{58 \cdot 29}{7} = a$$

$$240 \approx a$$

There are about 240 antelope in the preserve.

**38.** Let $a$ be the number of pounds of garbage produced by 7 families of four. The proportion is

$$\dfrac{1}{115} = \dfrac{7}{a}.$$

$1 \cdot a = 7 \cdot 115$

$a = 805$

Thus, 7 families will produce 805 pounds of garbage. Notice that the number ''four'' in ''family of four'' is not used in the calculation.

**6.2**

**46.** $\dfrac{100}{3} = 33.\overline{3}$   Divide

$= 33.33\overline{3}$

$= 3333.\overline{3}\%$   Move decimal point right two places and attach %

$= 3333\dfrac{1}{3}\%$

**49.** $\dfrac{4}{7} = 0.\overline{571428} \approx 0.571$   To nearest thousandth

Move decimal point two places to the right and attach %.

$$\dfrac{4}{7} \approx 57.1\%$$

**6.3**

**15.** The question is: What is 62% of 50?

$A$ is the unknown, $P = 62$, and $B = 50$.

$A = \quad P\% \ \cdot \ B$

$A = (62)(0.01)(50)$

$\quad = 31$

Alternatively, we could use the percent proportion.

$$\dfrac{A}{50} = \dfrac{62}{100}$$

$$100(A) = (50)(62)$$

$$A = \frac{(50)(62)}{100} = 31$$

The team won 31 games.

**18.** The question is: 22 is what percent of 30?

$P$ is the unknown, $a = 22$, and $B = 30$.

$$A = P\% \cdot B$$
$$22 = P(0.01)(30)$$
$$22 = P(0.3)$$
$$\frac{22}{0.3} = P$$
$$73.\overline{3} = P$$

Alternatively, using the percent proportion we have:

$$\frac{22}{30} = \frac{P}{100}$$
$$(22)(100) = 30(P)$$
$$\frac{(22)(100)}{30} = P$$
$$73.\overline{3} = P$$

Thus, to the nearest percent, 73% of her answers were correct.

**22.** The actual decrease in weight was $220 - 187 = 33$ pounds. The question is:

33 is what percent of 220?

$P$ is the unknown. $A = 33$, and $B = 220$.

$$A = P\% \cdot B$$
$$33 = P(0.01)(220)$$
$$33 = P(2.2)$$
$$\frac{33}{2.2} = P$$
$$15 = P$$

Alternatively, using the percent proportion we have:

$$\frac{33}{220} = \frac{P}{100}$$
$$(33)(100) = 220(P)$$
$$\frac{(33)(100)}{220} = P$$
$$15 = P$$

The percent decrease was 15%.

## 6.4

**5.** The question is: $3.05 is 2.5% of what number?

The unknown is $B$, $A = 3.05$, and $P = 2.5$.

$$A = P\% \cdot B$$
$$3.05 = (2.5)(0.01)B$$
$$3.05 = (0.025)B$$
$$\frac{3.05}{0.025} = B$$
$$122 = B$$

Alternatively, using the percent proportion we have:

$$\frac{3.05}{B} = \frac{2.5}{100}$$
$$(3.05)(100) = 2.5(B)$$
$$\frac{(3.05)(100)}{2.5} = B$$
$$122 = B$$

The price before tax is $122.

**10.** The Social Security tax rate in 1989 was 7.51%. Since the maximum amount on which the tax must be paid was $48,000, and since Toni made more than this, she paid 7.51% of $48,000.

$$(0.0751)(48,000) = 3604.80$$

Thus, Toni paid $3604.80 in Social Security tax.

## 6.5

**4.** The question is: $639.12 is 12% of what number?

The unknown is $B$, $A = 639.12$, and $P = 12$.

$$639.12 = (12)(0.01)B$$
$$639.12 = (0.12)B$$
$$\frac{639.12}{0.12} = B$$
$$5326 = B$$

Using the percent proportion we would have:

$$\frac{639.12}{B} = \frac{12}{100}$$
$$(639.12)(100) = 12(B)$$
$$\frac{(639.12)(100)}{12} = B$$
$$5326 = B$$

Thus, Angela's total sales were $5326.

**10.** The discount is $40 - $34 = $6. Thus, the question is: $6 is what percent of $40? The unknown is $P$, $A = 6$, and $B = 40$.

$$A = P\% \cdot B$$
$$6 = P(0.01)(40)$$
$$6 = P(0.4)$$
$$\frac{6}{0.4} = P$$
$$15 = P$$

Alternatively, using the percent proportion we have:

$$\frac{6}{40} = \frac{P}{100}$$
$$(6)(100) = 40(P)$$
$$\frac{(6)(100)}{40} = P$$
$$15 = P$$

Thus, the discount rate is 15%.

**14.** First take 25% of $142.80.

$$(0.25)(142.80) = 35.70$$

The first discount was $35.70 and the first sale price was $142.80 − $35.70 = $107.10.

Next take 20% of $107.10.

$$(0.20)(107.10) = 21.42$$

The second discount was $21.42 making the second sale price $107.10 − $21.42 = $85.68.

## 6.6

**4.** Since 9 months is $\frac{9}{12}$ or $\frac{3}{4}$ of a year, we have $P =$ 800, $R = 12\% = 0.12$, and $T = \frac{3}{4}$.

$$
\begin{aligned}
I &= P \cdot R \cdot T \\
&= (800)(0.12)\left(\frac{3}{4}\right) \\
&= 72
\end{aligned}
$$

To pay off the loan, Lisa must pay back the principal plus the interest which is

$$A = P + I = \$800 + \$72 = \$872.$$

**8.** Since 18 months is $\frac{18}{12}$ or $\frac{3}{2}$ of a year, we have $P =$ 1213.40, $R = 16\% = 0.16$, and $T = \frac{3}{2}$.

$$
\begin{aligned}
I &= P \cdot R \cdot T \\
&= (1213.40)(0.16)\left(\frac{3}{2}\right) \\
&= 291.22 \quad \text{Rounded to nearest cent}
\end{aligned}
$$

To pay off the debt, Robert must pay $1213.40 + $291.22 = $1504.62. To find the amount paid each month, divide by 18.

$$\frac{\$1504.62}{18} = \$83.59$$

Thus, Robert pays $83.59 each month.

**11.** Since the interest is 12% compounded semiannually, the rate is 6% (one half of 12%). Since the time is 18 months, the number of periods is 3.

$$(\$625.50)(0.06) = \$37.53$$

The interest the first period is $37.53 and the principal for the second period is $625.50 + $37.53 = $663.03.

$$(\$663.03)(0.06) = \$39.78 \quad \text{Nearest cent}$$

The interest the second period is $39.78 and the principal for the third period is $663.03 + $39.78 = $702.81.

$$(\$702.81)(0.06) = \$42.17 \quad \text{Nearest cent}$$

The amount to be paid back is $702.81 + $42.17 = $744.98.

## 6.7

**5.** The actual increase in population was $415 − 329 = 86$. The question is:

$$86 \text{ is what percent of } 329?$$

The unknown is $P$, $A = 86$, and $B = 329$.

$$
\begin{aligned}
A &= P\% \cdot B \\
86 &= P(0.01)(329) \\
86 &= P(3.29) \\
\frac{86}{3.29} &= P \\
26 &\approx P \quad \text{To the nearest unit}
\end{aligned}
$$

Alternatively, using the percent proportion we have:

$$
\begin{aligned}
\frac{86}{329} &= \frac{P}{100} \\
(100)(86) &= 329(P) \\
\frac{(100)(86)}{329} &= P \\
26 &\approx P \quad \text{To the nearest unit}
\end{aligned}
$$

Thus, to the nearest percent, the percent increase was 26%.

**8.** The question is: $1824 is 15% of what number?

$B$ is the unknown, $A = 1824$, and $P = 15$.

$$
\begin{aligned}
A &= P\% \cdot B \\
1824 &= 15(0.01)B \\
1824 &= (0.15)B \\
\frac{1824}{0.15} &= B \\
12{,}160 &= B
\end{aligned}
$$

Alternatively, using the percent proportion we have:

$$
\begin{aligned}
\frac{1824}{B} &= \frac{15}{100} \\
(1824)(100) &= 15(B) \\
12{,}160 &= B
\end{aligned}
$$

Thus, Forrest's original salary was $12,160.

**16.** The total percent increase is $13\% + 9\% = 22\%$. The question is: 22% of 30 is what number? The unknown is $A$, $P = 22$, and $B = 30$.

$$
\begin{aligned}
A &= P\% \cdot B \\
A &= (22)(0.01)(30) = 6.6
\end{aligned}
$$

Alternatively,

$$
\begin{aligned}
\frac{A}{30} &= \frac{22}{100} \\
(100)A &= (30)(22) \\
A &= \frac{(30)(22)}{100} = 6.6
\end{aligned}
$$

Thus, the new mileage per gallon would be $30 + 6.6 = 36.6$ mpg.

## 6.8

**8.** The actual markdown is $125 - $95 = $30. The question is: $30 is what percent of $125? The unknown is $P$, $A = 30$, and $B = 125$.

$$A = P\% \cdot B$$
$$30 = P(0.01)(125)$$
$$30 = P(1.25)$$
$$\frac{30}{1.25} = P$$
$$24 = P$$

Alternatively,

$$\frac{30}{125} = \frac{P}{100}$$
$$(30)(100) = 125(P)$$
$$\frac{(30)(100)}{125} = P$$
$$24 = P$$

The markdown rate is 24%.

**10.** The markup was

20% of $120
↓  ↓  ↓
0.20 · $120 = $24.

The first selling price of the tools was $120 + $24 = $144. The markdown was

15% of $144.
↓  ↓  ↓
0.15 · $144 = $21.60.

The second selling price was $144 - $21.60 = $122.40.

**11.** Since the tools were bought for $120.00 and sold for $122.40, the profit was

$$122.40 - $120.00 = $2.40.$$

To find the percent profit, ask:

$2.40 is what percent of $120?

The unknown is $P$, $A = 2.40$, and $B = 120$.

$$A = P\% \cdot B$$
$$2.40 = P(0.01)(120)$$
$$2.40 = P(1.2)$$
$$\frac{2.40}{1.2} = P$$
$$2 = P$$

Alternatively,

$$\frac{2.40}{120} = \frac{P}{100}$$
$$(2.40)(100) = 120(P)$$
$$\frac{(2.40)(100)}{120} = P$$
$$2 = P$$

Thus, the percent profit was 2%.

## 7.1

**15.** With 54% of the students in K–8 and 22% of the students in high school, 76% of the students are in K–8 and high school combined. Find

76% of 42,500.
↓  ↓  ↓
0.76 · 42,500 = 32,300

There are 32,300 students in these two groups.

**22.** The ratio of math majors to the total number of students is

$$\frac{300}{1300} = \frac{3}{13}.$$

**25.** The ratio of science majors to the total number of students is

$$\frac{400}{1300} = \frac{4}{13}.$$

Change $\frac{4}{13}$ to a percent to obtain 30.8%.

## 7.2

**22.** The number of employees who earn $9 or less an hour is the total of the numbers who earn $3 to $5, $5 to $7, and $7 to $9. This total is

$$5 + 10 + 18 = 33.$$

**25.** The number of employees who earn more than $9 an hour is 18. Thus, the ratio of these employees to the total number of employees is $\frac{18}{51} = \frac{6}{17}$. Changing $\frac{6}{17}$ to a percent gives 35.3%.

## 7.3

**6.** To find the mean, add the seven numbers and divide the result by 7.

$$\frac{7.2 + 6.1 + 8.6 + 9.2 + 5.5 + 7.2 + 8.8}{7} = \frac{52.6}{7} \approx 7.5$$

**11.** To find the median, first arrange the seven numbers by size.

5.5, 6.1, 7.2, 7.2, 8.6, 8.8, 9.2

The median is the middle number, 7.2, which is the fourth number from the left.

**23.** The right column in the table is:

$$(16,200)(8) = 129,600$$
$$(20,800)(10) = 208,000$$
$$(28,600)(7) = 200,200$$
$$(32,000)(2) = 64,000$$
$$(46,000)(1) = \underline{46,000}$$
$$647,800$$

The total is 647,800. To find the average salary divide 647,800 by

$$8 + 10 + 7 + 2 + 1 = 28.$$

$$\frac{647{,}800}{28} = 23{,}135.714$$

Thus, to the nearest cent, the average salary is $23,135.71.

Since there are 28 salaries, when ranked by size, the median is the average of the middle two, both of which are $20,800. Thus, the median salary is $20,800.

The mode of the salaries is the one that occurs most frequently. Since $20,800 occurs 10 times, more frequently than any other salary, the mode is $20,800.

## 8.1

**14.** Use two unit fractions, $\dfrac{1 \text{ min}}{60 \text{ sec}}$ and $\dfrac{1 \text{ hr}}{60 \text{ min}}$

$$5400 \text{ sec} = \frac{5400 \text{ sec}}{1} \times \frac{1 \text{ min}}{60 \text{ sec}} \times \frac{1 \text{ hr}}{60 \text{ min}}$$

$$= 5400 \times \frac{1}{60} \times \frac{1}{60} \text{ hr} = 1.5 \text{ hr}$$

**23.** $88\dfrac{\text{ft}}{\text{sec}} = \dfrac{88 \text{ ft}}{1 \text{ sec}} \times \dfrac{60 \text{ sec}}{1 \text{ min}} \times \dfrac{60 \text{ min}}{1 \text{ hr}} \times \dfrac{1 \text{ mi}}{5280 \text{ ft}}$

$$= \frac{88 \times 60 \times 60}{1 \times 5280} \frac{\text{mi}}{\text{hr}} = 60 \frac{\text{mi}}{\text{hr}}$$

**31.** Since the price is given in cents per foot, change 3 mi to feet.

$$3 \text{ mi} = \frac{3 \text{ mi}}{1} \times \frac{5280 \text{ ft}}{1 \text{ mi}}$$

$$= 15{,}840 \text{ ft}$$

$$\frac{15{,}840 \text{ ft}}{1} \times \frac{20 \text{ cents}}{1 \text{ ft}} = 316800¢ = \$3168.00$$

## 8.2

**10.**

$$\begin{array}{r} 6 \text{ yd} \quad 2 \text{ ft} \quad 11 \text{ in} \\ + \ 7 \text{ yd} \quad 1 \text{ ft} \quad \ 9 \text{ in} \\ \hline 13 \text{ yd} \quad 3 \text{ ft} \quad 20 \text{ in} \end{array} = 13 \text{ yd} + 3 \text{ ft} + 1 \text{ ft} + 8 \text{ in}$$

$$= 13 \text{ yd} + 1 \text{ yd} + 1 \text{ ft} + 8 \text{ in}$$

$$= 14 \text{ yd } 1 \text{ ft } 8 \text{ in}$$

**14.** 5 da 3 hr 8 min = 4 da + 1 da + 2 hr + 1 hr + 8 min

$$= 4 \text{ da} + 24 \text{ hr} + 2 \text{ hr} + 60 \text{ min} + 8 \text{ min}$$

$$= 4 \text{ da } 26 \text{ hr } 68 \text{ min}$$

$$\begin{array}{r} 4 \text{ da} \quad 26 \text{ hr} \quad 68 \text{ min} \\ - \ 4 \text{ da} \quad \ 6 \text{ hr} \quad 40 \text{ min} \\ \hline 0 \text{ da} \quad 20 \text{ hr} \quad 28 \text{ min} \end{array} = 20 \text{ hr } 28 \text{ min}$$

**17.** 5 × (4 gal 3 qt) = 20 gal 15 qt

$$= 20 \text{ gal} + 12 \text{ qt} + 3 \text{ qt}$$

$$= 20 \text{ gal} + 3 \text{ gal} + 3 \text{ qt}$$

$$= 23 \text{ gal } 3 \text{ qt}$$

**23.** (7 hr) ÷ 3 = 2 hr with remainder 1 hr

1 hr = 60 min

60 min + 33 min = 93 min

93 min ÷ 3 = 31 min

Thus, (7 hr 33 min) ÷ 3 = 2 hr 31 min.

**28.** To make 7 loaves it would take 7 times the amount necessary for one loaf.

$$(1 \text{ lb } 12 \text{ oz}) \times 7 = 7 \text{ lb} + 84 \text{ oz}$$

$$= 7 \text{ lb} + 80 \text{ oz} + 4 \text{ oz}$$

$$= 7 \text{ lb} + 5 \text{ lb} + 4 \text{ oz}$$

$$= 12 \text{ lb } 4 \text{ oz}$$

## 8.3

**17.**

$$\underset{\text{(kg)}}{1000 \text{ g}} \quad \underset{\text{(hg)}}{100 \text{ g}} \quad \underset{\text{(dag)}}{10 \text{ g}}$$

$$0.00721 \text{ kg} = 000.721 \text{ dag}$$

$$= 0.721 \text{ dag}$$

**23.**

$$\underset{\text{(hg)}}{100 \text{ g}} \quad \underset{\text{(dag)}}{10 \text{ g}} \quad \underset{\text{(g)}}{1 \text{ g}} \quad \underset{\text{(dg)}}{0.1 \text{ g}}$$

$$4728 \text{ dg} = 4.728 \text{ hg}$$

$$= 4.728 \text{ hg}$$

**32.** $20\dfrac{\text{cg}}{\text{da}} = \dfrac{20 \text{ cg}}{1 \text{ da}} \times \dfrac{7 \text{ da}}{1 \text{ wk}} \times \dfrac{1 \text{ g}}{100 \text{ cg}}$

$$= 1.4 \frac{\text{g}}{\text{wk}}$$

**41.** 600 g = 0.6 kg

$$\frac{0.6 \text{ kg}}{1} \times \frac{6 \text{ dollars}}{1 \text{ kg}} = \$3.60$$

## 8.4

**14.** $40 \text{ L} = \dfrac{40 \text{ L}}{1} \times \dfrac{1.06 \text{ qt}}{1 \text{ L}} \times \dfrac{1 \text{ gal}}{4 \text{ qt}}$

$$= \frac{(40)(1.06)}{4} \text{ gal}$$

$$= 10.6 \text{ gal}$$

**19.** $F = \dfrac{9}{5}C + 32$

$$= \frac{9}{5}(40) + 32$$

$$= 72 + 32 + 104$$

Thus 40°C = 104°F.

**22.** $C = \dfrac{5}{9}(F - 32)$

$$= \frac{5}{9}(200 - 32)$$

$$= \frac{5}{9}(168) \approx 93$$

Thus, 200°F ≈ 93°C.

**8.5**

**14.** 1 m   0.1 m   0.01 m
(m)   (dm)   (cm)

For square units move twice the number of places. Thus, move 4 places.

$$1020 \text{ cm}^2 = .1020 \text{ m}^2 = 0.102 \text{ m}^2$$

**23.** 100 m   10 m   1 m   0.1 m   0.01 m   0.001 m
(hm)   (dam)   (m)   (dm)   (cm)   (mm)

For square units move 2(5) = 10 places.

$$4{,}352{,}000 \text{ mm}^2 = 0.0004352000 \text{ hm}^2$$
$$= 0.0004352 \text{ hm}^2$$

**28.** Divide the area into two rectangles. One is 7 yd long and 3 yd wide and the other is 3 yd long and 2 yd wide

7 yd $\times$ 3 yd = 21 yd$^2$

3 yd $\times$ 2 yd = 6 yd$^2$

21 yd$^2$ + 6 yd$^2$ = 27 yd$^2$

$$\frac{\$16.50}{1 \text{ yd}^2} \times \frac{27 \text{ yd}^2}{1} = (\$16.50)(27)$$
$$= \$445.50$$

**8.6**

**11.** $A = bh$

$= (22.2 \text{ yd})(22.2 \text{ yd})$

$\approx 492.8 \text{ yd}^2$

**16.** First find the area of the rectangle which is 28 cm long and 14 cm wide.

$$A = (28 \text{ cm})(14 \text{ cm}) = 392 \text{ cm}^2$$

Now find the area of the two triangles and subtract from the area of the rectangle. The base of each triangle is 10 cm and the height is 14 cm − 7 cm = 7 cm.

$$A = \frac{1}{2}(10 \text{ cm})(7 \text{ cm}) = 35 \text{ cm}^2$$
$$2A = 70 \text{ cm}^2$$

Thus, the area is 392 cm$^2$ − 70 cm$^2$ = 322 cm$^2$.

**8.7**

**11.** $c = 2\pi r$

$= 2\pi(7.7 \text{ mi})$

$\approx 2\left(\dfrac{22}{7}\right)(7.7 \text{ mi})$

$= 48.4 \text{ mi}$

$A = \pi r^2$

$= \pi(7.7 \text{ mi})^2$

$\approx \dfrac{22}{7}(59.29)\text{mi}^2$

$= 186.34 \text{ mi}^2$

**17.** $c = 2\pi r$

$= 2\pi(0.05 \text{ cm})$

$\approx 2(3.14)(0.05) \text{ cm}$

$= 0.314 \text{ cm}$

$A = \pi r^2$

$= \pi(0.05 \text{ cm})^2$

$\approx (3.14)(0.0025) \text{ cm}^2$

$= 0.00785 \text{ cm}^2$

**19.** Calculate the area for each hose and subtract. The radius of the first hose is $\dfrac{3}{8}$ in and the radius of the second hose is $\dfrac{1}{4}$ in.

$$A = \pi r^2$$
$$\approx 3.14\left(\frac{3}{8} \text{ in}\right)^2$$
$$\approx 0.442 \text{ in}^2$$
$$A = \pi r^2$$
$$\approx 3.14\left(\frac{1}{4} \text{ in}\right)^2$$
$$\approx 0.196 \text{ in}^2$$
$$0.442 \text{ in}^2 - 0.196 \text{ in}^2 = 0.246 \text{ in}^2$$

**22.** Find the area of the triangle and subtract 3 times the area of one hole.

$$A = \frac{1}{2}bh$$
$$= \frac{1}{2}(9 \text{ in})(7.8 \text{ in})$$
$$= 35.1 \text{ in}^2$$
$$A = \pi r^2$$
$$= \pi(1 \text{ in})^2 \qquad \text{Radius is } \frac{1}{2}(2 \text{ in}) = 1 \text{ in}$$
$$\approx 3.14 \text{ in}^2$$

The area of the three circles is

$$3(3.14 \text{ in}^2) = 9.42 \text{ in}^2$$
$$35.1 \text{ in}^2 - 9.42 \text{ in}^2 = 25.68 \text{ in}^2$$

**8.8**

**9.** Convert each number to decimals:
9.5 in by 3.4 in by 8.75 in

$$V = (9.5 \text{ in})(3.4 \text{ in})(8.75 \text{ in})$$
$$= 282.625 \text{ in}^3$$

**16.** Use the unit fraction $\dfrac{1 \text{ yd}^3}{27 \text{ ft}^3}$.

$$324 \text{ ft}^3 = \frac{324 \text{ ft}^3}{1} \times \frac{1 \text{ yd}^3}{27 \text{ ft}^3}$$
$$= \frac{324}{27} \text{ yd}^3 = 12 \text{ yd}^3$$

**17.** From hm to m is 2 places to the right. For cubic units move $3(2) = 6$ places to the right.

$$0.0085 \text{ hm}^3 = 0008500. \text{ m}^3$$
$$= 8500 \text{ m}^3$$

**20.** $525\text{L} = 525{,}000 \text{ ml}$

$\qquad = 525{,}000 \text{ cm}^3 \qquad 1 \text{ ml} = 1 \text{ cc} = 1 \text{ cm}^3$

$\qquad = 0.525 \text{ m}^3 \qquad$ Move $3(2) = 6$ places to left

**25.** $A = 2lh + 2wh + 2lw$

$\qquad = 2(15.5 \text{ cm})(20.2 \text{ cm}) + 2(6.4 \text{ cm})(20.2 \text{ cm}) +$
$\qquad \quad 2(15.5 \text{ cm})(6.4 \text{ cm})$

$\qquad = 1083.16 \text{ cm}^2$

**30.** $V = e^3$

$\qquad = (22 \text{ cm})^3$

$\qquad = 10{,}648 \text{ cm}^3$

Since water weighs $1\dfrac{\text{gm}}{\text{cm}^3}$ the answer is 10,648 g.

**31.** Area of bottom $= (22 \text{ cm})^2 = 484 \text{ cm}^2$

Area of 4 sides $= 4(22 \text{ cm})^2 = 4(484)\text{cm}^2$

$\qquad\qquad\qquad\qquad = 1936 \text{ cm}^2$

Total required $= 484 \text{ cm}^2 + 1936 \text{ cm}^2$
$\qquad\qquad\qquad\quad = 2420 \text{ cm}^2$

## 8.9

**2.** $V = \pi r^2 h$

$\qquad = \pi (12.2 \text{ in})^2 (30 \text{ in}) \qquad$ Radius is $\dfrac{1}{2}$ the diameter

$\qquad \approx (3.14)(12.2)^2 (30) \text{in}^2$

$\qquad \approx 14{,}020.7 \text{ in}^3$

**5.** $A = 2\pi rh + 2\pi r^2$

$\qquad \approx 2(3.14)(12.2 \text{ in})(30 \text{ in}) + 2(3.14)(12.2 \text{ in})^2$

$\qquad \approx 3233.2 \text{ in}^2$

**8.** $V = \dfrac{4}{3}\pi r^3$

$\qquad = \dfrac{4}{3}\pi \left(\dfrac{3}{4} \text{ in}\right)^3$

$\qquad \approx \left(\dfrac{4}{3}\right)(3.14)\left(\dfrac{3}{4}\right)^3 \text{ in}^3$

$\qquad \approx 1.766 \text{ in}^3$

**12.** $A = 4\pi r^2$

$\qquad = 4\pi (2.8 \text{ yd})^2$

$\qquad \approx 4(3.14)(2.8)^2 \text{ yd}^2$

$\qquad \approx 98.47 \text{ yd}^2$

**13.** Calculate the volume of the can and the volume of the pie pan and compare the two.

$$V = \pi r^2 h$$
$$= \pi (4 \text{ cm})^2 (12 \text{ cm})$$

$$\approx (3.14)(4)^2 (12) \text{ cm}^3$$
$$= 602.88 \text{ cm}^3$$
$$V = \pi r^2 h$$
$$= \pi (10 \text{ cm})^2 (3 \text{ cm})$$
$$\approx (3.14)(10)^2 (3) \text{ cm}^3$$
$$= 942 \text{ cm}^3$$

$$\frac{942 \text{ cm}^3}{1} \times \frac{1 \text{ can}}{602.88 \text{ cm}^3} = 1.6 \text{ cans}$$

## 9.2

**27.** $-42° + 57° = 15° \qquad$ Temperature at 2:00 P.M.

$\quad\; 15° + (-41°) = -26° \qquad$ Temperature at 5:00 P.M.

## 9.3

**31.** To find the difference between the temperatures, subtract.

$$35 - (-13) = 35 + 13 = 48$$

The temperature changed by 48°.

## 9.4

**37.** Each check could be considered as $-12$ dollars. Thus, $4(-12 \text{ dollars}) = -48 \text{ dollars}$.

## 9.5

**55.** $217.93 + 493.16 = 711.09$

$\qquad\qquad\qquad\qquad\qquad$ Balance plus deposit

$16.25 + 92.87 + 388.14 = 497.26$

$\qquad\qquad\qquad\qquad\qquad$ Sum of checks

$711.09 - 497.26 = 213.83$

$\qquad\qquad\qquad\qquad\qquad$ Balance at end of month

## 9.6

**19.** $a - (b - c) = (-2) - (3 - (-5))$

$\qquad\qquad\qquad\qquad\qquad\qquad a = -2, \, b = 3, \, c = -5$

$\qquad\qquad = -2 - (3 + 5)$

$\qquad\qquad\qquad\qquad\qquad$ Inside parentheses first

$\qquad\qquad = -2 - 8 = -10$

**22.** $\dfrac{a}{b} - \dfrac{b}{c} = \dfrac{(-2)}{3} - \dfrac{3}{(-5)} \qquad a = -2, \, b = 3, \, c = -5$

$\qquad\qquad = -\dfrac{2}{3} + \dfrac{3}{5} \qquad -\dfrac{3}{(-5)} = \dfrac{3}{5}$

$\qquad\qquad = -\dfrac{10}{15} + \dfrac{9}{15} = -\dfrac{1}{15}$

**23.** Since any number plus its negative is zero,

$$\frac{a}{b} + \left(-\frac{a}{b}\right) = 0$$

for any values of $a$ and $b$.

**46.** $\frac{1}{2}x - 2y - 2x + \frac{1}{4}y = \frac{1}{2}x - 2x - 2y + \frac{1}{4}y$

$$= \left(\frac{1}{2} - 2\right)x - \left(2 - \frac{1}{4}\right)y$$

$$= \left(\frac{1}{2} - \frac{4}{2}\right)x - \left(\frac{8}{4} - \frac{1}{4}\right)y$$

$$= -\frac{3}{2}x - \frac{7}{4}y$$

**56.** $x - (-x + y - 1) = x + x - y + 1$

Change all signs in parentheses

$$= 2x - y + 1$$

**59.** $2[x - (4x - 3)] = 2[x - 4x + 3]$

Innermost parentheses first

$$= 2[-3x + 3] \qquad \text{Collect like terms}$$

$$= -6x + 6 \qquad \text{Distributive law}$$

**9.7**

**21.** $(2x)^{-1} = \frac{1}{(2x)^1} = \frac{1}{2x}$

**22.** $2x^{-1} = 2\frac{1}{x^1} = \frac{2}{x} \qquad$ Compare Exercises 21 and 22

**30.** $3a^2 = 3(-2)^2 \qquad a = -2$

$$= 3 \cdot 4 \qquad \text{Square then multiply}$$

$$= 12$$

**31.** $(3a)^2 = (3(-2))^2 \qquad a = -2$

$$= (-6)^2 \qquad \text{Multiply inside parentheses first}$$

$$= 36 \qquad \text{Compare Exercises 30 and 31}$$

**37.** $a^2 - b^2 = (-2)^2 - (3)^2 \qquad a = -2 \text{ and } b = 3$

$$= 4 - 9 = -5 \qquad \text{Square then subtract}$$

**38.** $(a - b)^2 = ((-2) - 3)^2 \qquad a = -2 \text{ and } b = 3$

$$= (-5)^2 \qquad \text{Subtract inside parentheses first}$$

$$= 25 \qquad \text{Compare Exercises 37 and 38}$$

**44.** $(40{,}000{,}000)(20{,}000) = (4 \times 10^7)(2 \times 10^4)$

$$= (4 \times 2)(10^7 \times 10^4)$$

$$= 8 \times 10^{11}$$

**10.1**

**26.** Add $\frac{1}{4}$ to both sides of the equation.

$$-\frac{1}{4} + z = \frac{3}{4}$$

$$\left(-\frac{1}{4} + \frac{1}{4}\right) + z = \frac{3}{4} + \frac{1}{4}$$

$$0 + z = \frac{3 + 1}{4}$$

$$z = \frac{4}{4} = 1$$

**29.** Add 3.9 to both sides of the equation.

$$-2.6 = z - 3.9$$

$$-2.6 + 3.9 = z - 3.9 + 3.9$$

$$1.3 = z + 0$$

$$1.3 = z$$

**10.2**

**11.** $\frac{3}{4}y = -6$

$$\frac{4}{3} \cdot \frac{3}{4}y = \frac{4}{3} \cdot (-6) \qquad \text{Multiply by } \frac{4}{3}, \text{ the reciprocal of } \frac{3}{4}$$

$$y = -8$$

**17.** $-2.5y = 7.5$

$$\frac{-2.5y}{-2.5} = \frac{7.5}{-2.5} \qquad \text{Divide by } -2.5$$

$$y = -3$$

**10.3**

**18.** $x + 9 + 6x = 2 + x + 1$

$$x + 6x + 9 = 2 + 1 + x$$

$$7x + 9 = 3 + x \qquad \text{Collect like terms on each side}$$

$$7x - x + 9 = 3 \qquad \text{Subtract } x \text{ from both sides}$$

$$6x + 9 = 3$$

$$6x + 9 - 9 = 3 - 9 \qquad \text{Subtract 9 from both sides}$$

$$6x = -6$$

$$\frac{6x}{6} = \frac{-6}{6}$$

$$x = -1$$

**21.** $3y + \frac{5}{2}y + \frac{3}{2} = \frac{1}{2}y + \frac{5}{2}y$

$$\left(\frac{6}{2} + \frac{5}{2}\right)y + \frac{3}{2} = \left(\frac{1}{2} + \frac{5}{2}\right)y \qquad \text{Collect like terms}$$

$$\frac{11}{2}y + \frac{3}{2} = \frac{6}{2}y$$

$$\frac{11}{2}y + \frac{6}{2}y + \frac{3}{2} = \frac{6}{2}y - \frac{6}{2}y \qquad \text{Subtract } \frac{6}{2}y$$

$$\frac{5}{2}y + \frac{3}{2} = 0$$

$$\frac{5}{2}y + \frac{3}{2} - \frac{3}{2} = 0 - \frac{3}{2} \qquad \text{Subtract } \frac{3}{2}$$

$$\frac{5}{2}y = -\frac{3}{2}$$

$$\frac{2}{5} \cdot \frac{5}{2}y = \frac{2}{5} \cdot \left(-\frac{3}{2}\right)$$

$$y = -\frac{3}{5}$$

**27.** $6z - (3z + 8) = 10$

$6z - 3z - 8 = 10$     Clear parentheses

$3z - 8 = 10$

$3z - 8 + 8 = 10 + 8$     Add 8

$3z = 18$

$\dfrac{3z}{3} = \dfrac{18}{3}$     Divide by 3

$z = 6$

**33.** $9z - (3z - 18) = 36$

$9z - 3z + 18 = 36$     Clear parentheses

$6z + 18 = 36$

$6z + 18 - 18 = 36 - 18$     Subtract 18

$6z = 18$

$\dfrac{6z}{6} = \dfrac{18}{6}$     Divide by 6

$z = 3$

**34.** $7(x - 5) = 10 - (x + 1)$

$7x - 35 = 10 - x - 1$     Clear parentheses

$7x - 35 = 9 - x$     Collect like terms

$7x + x - 35 = 9 - x + x$     Add $x$

$8x - 35 = 9$

$8x - 35 + 35 = 9 + 35$     Add 35

$8x = 44$

$\dfrac{8x}{8} = \dfrac{44}{8}$     Divide by 8

$x = \dfrac{11}{2}$

**42.** $3(2 - 4x) = 4(2x - 1) - 2(1 + x)$

$6 - 12x = 8x - 4 - 2 - 2x$     Clear parentheses

$6 - 12x = 6x - 6$     Collect like terms

$6 - 12x - 6x = 6x - 6x - 6$     Subtract $6x$

$6 - 18x = -6$

$6 - 6 - 18x = -6 - 6$     Subtract 6

$-18x = -12$

$\dfrac{-18x}{-18} = \dfrac{-12}{-18}$     Divide by $-18$

$x = \dfrac{2}{3}$

**10.4**

**27.** Let     $x =$ the first integer

$x + 1 =$ the next consecutive integer

$2x =$ twice the smaller

$x + 1 + 18 = 18$ more than the larger.

$2x = x + 1 + 18$

$2x = x + 19$

$2x - x = x - x + 19$

$x = 19$

$x + 1 = 20$

The integers are 19 and 20.

**31.** Let     $x =$ the first length

$2x =$ the second length

$x + 4 =$ the third length.

Since the wire is 56 m long the pieces must add to give 56.

$x + 2x + x + 4 = 56$

$4x + 4 = 56$

$4x = 52$

$x = 13$

$2x = 26$

$x + 4 = 17$

The lengths are 13 m, 26 m, and 17 m.

**37.** Let     $x =$ the number

$2x + 14 =$ double the number plus 14

$4(x - 1) = 4$ times 1 less than the number.

$2x + 14 = 4(x - 1)$

$2x + 14 = 4x - 4$

$-2x + 14 = -4$

$-2x = -18$

$x = 9$

**40.** Let     $x =$ Claude's age

$2x =$ Lucy's age

$x + 11 =$ Toni's age.

Since the sum of their ages is 87,

$x + 2x + x + 11 = 87.$

$4x + 11 = 87$

$4x = 76$

$x = 19$

$2x = 38$

$x + 11 = 30$

Lucy is 38, Claude is 19, and Toni is 30.

**10.5**

**4.** Let $x$ = amount to be sold in June.

To have an average of $3000 for 6 months, the total must be $6(3000) = 18,000$.

$$2550 + 3320 + 3170 + 2820 + 2110 + x = 18000$$

$$13970 + x = 18000$$

$$x = 18000 - 13970$$

$$x = 4030$$

Robert's sales must total $4030 in June to average $3000.

**8.** Let $x$ = enrollment ten years ago.

$$x - 0.12x = 15400$$

$$(1 - 0.12)x = 15400$$

$$0.88x = 15400$$

$$\frac{0.88x}{0.88} = \frac{15400}{0.88}$$

$$x = 17,500$$

The enrollment ten years ago was 17,500.

**13.** Let $h$ = height of the trapezoid.

16 m

$A = 247 \text{ m}^2$    $h$

22 m

$$A = \frac{1}{2}(b_1 + b_2)h$$

$$247 = \frac{1}{2}(16 + 22)h$$

$$247 = \frac{1}{2}(38)h$$

$$247 = 19h$$

$$\frac{247}{19} = h$$

$$13 = h$$

The height is 13 m.

**16.** Let $h$ = height of the storage bin.

$V = 3200 \text{ m}^3$    $h$

8 m

$$3200 = \pi(8)^2 h$$

$$3200 \approx (3.14)(64)h$$

$$\frac{3200}{(3.14)(64)} \approx h$$

$$15.9 \approx h$$

The height is approximately 15.9 m.

**Absolute value of a number**   The number of units that the number is from zero on a number line. **(9.1)**

**Addends**   Two or more numbers to be added. **(1.2)**

**Algebraic expression**   An expression containing variables and numbers combined by the basic operations of addition, subtraction, multiplication, and division. **(9.6)**

**Algorithm**   A procedure or method used to perform a certain task. **(2.5)**

**Area**   The number of square units in a two-dimensional figure. **(8.5)**

**Associative law**   The law which states that regrouping sums or products does not change the results. **(1.3)**

**Average (arithmetic mean)**   The sum of the quantities divided by the number of quantities. **(7.3)**

**Bar graph**   A method of presenting numerical data by using a bar to show quantity. **(7.1)**

**Binary operation**   An operation on two numbers at a time. **(1.3)**

**Borrowing in a subtraction problem**   The reverse of carrying in an addition problem. **(1.6)**

**Canceling factors**   The process of dividing out common factors by crossing them out. **(3.2)**

**Carry number**   A number which must be added to the next column to the left in an addition problem. **(1.3)**

**Circle**   A plane figure which has the property that all of its points are the same distance from a point called the center. **(8.7)**

**Circle graph**   A method of presenting numerical data in pie-shaped sectors of a circle. **(7.1)**

**Circumference**   The distance around a circle. **(8.7)**

**Coefficient (numerical coefficient)**   The numerical factor of a term of an algebraic expression. **(9.6)**

**Commission**   Income that is a percent of sales. **(6.5)**

**Common multiple**   A number which is a multiple of two or more numbers. **(4.2)**

**Commutative law**   The law which states that changing the order of sums or products does not change the results. **(1.2)**

**Composite number**   A natural number greater than 1 which is not prime. **(2.7)**

**Compound interest**   A type of interest that is paid on interest as well as principal. **(6.6)**

**Conditional equation**   An equation which is true for some replacements of the variable and false for others. **(10.1)**

**Consecutive integers**   Integers with the property that one immediately follows the other. **(10.4)**

**Contradiction**   An equation which has no solutions. **(10.1)**

**Counting number**   *See* natural number. **(1.1)**

**Cross products**   Products formed by multiplying the numerator of one fraction by the denominator of another, and vice-versa. **(3.2)**

**Cube**   A rectangular solid which has $l = w = h$. **(8.8)**

**Cube of a number**   A number raised to the third power. **(2.8)**

**Cylinder**   A solid with circular ends of the same radius. **(8.9)**

**Decimal fractions (decimals)**   A fraction which has denominator equal to a power of 10. **(5.1)**

**Decimal point**   The period which separates the ones digit from the tenths digit in a decimal fraction. **(5.1)**

**Denominator**   The bottom number in a fraction. **(3.1)**

**Diameter**   The number which is twice the radius of a circle. **(8.7)**

**Difference**   The result of a subtraction problem. **(1.5)**

**Digit**   One of the first ten whole numbers, 0, 1, 2, 3, 4, 5, 6, 7, 8, or 9. **(1.1)**

**Discount**   A reduction in the regular price of an item when it is put on sale. **(6.5)**

**Distributive law**   The law which states that quantities like the following are equal.

$$5 \times (7 + 2) = (5 \times 7) + (5 \times 2)$$
$$5 \times (7 - 2) = (5 \times 7) - (5 \times 2) \qquad \textbf{(2.3)}$$

**Dividend**   The number being divided in a division problem. **(2.4)**

**Divisor**   The dividing number in a division problem. **(2.4)**

**English system of measurement**   The measurement system used in the United States. **(8.1)**

**Equation**   A statement that two quantities or expressions are equal. **(6.1)**

**Equivalent equations**   Equations which have the same solutions. **(10.1)**

**Exponent**   The number which tells how many times a number is multiplied by itself. **(2.8)**

**Factor**   Numbers or letters which are multiplied in a term. **(9.6)**

**Fraction**   A quotient of two whole number. **(3.1)**

**Gram**   The unit for weight in the metric system. **(8.3)**

**Histogram**   A type of bar graph on which the width of the bar corresponds to the class interval and the length of the bar to the class frequency. **(7.2)**

**Identities of addition and multiplication**   The number 0 is the identity of addition and 1 is the identity of multiplication. **(9.6)**

**Identity**   An equation which has every number as a solution. **(10.1)**

**Improper fraction**   A fraction which has numerator greater than or equal to the denominator. **(3.1)**

**Integers**   The numbers which consist of the natural numbers and their negatives together with zero. **(9.1)**

**Interest**   Money paid to borrow money or money paid on an investment. **(6.6)**

**Irrational numbers**   All numbers on the number line which are not rational numbers. **(9.5)**

**Least common denominator (LCD)**   The least common multiple (LCM) of all denominators in an addition or subtraction problem. **(4.3)**

**Least common multiple (LCM)**   The smallest of all common multiples of two or more counting numbers. **(4.2)**

**Like fractions**   Fractions which have the same denominator. **(4.1)**

**Like terms**   *See* Similar terms. **(9.6)**

**Linear equation**   An equation in which the variable occurs to the first power only. **(10.1)**

**Liter**   The unit for volume in the metric system **(8.3)**

**Markdown**   The amount of a price reduction. **(6.8)**

**Markup**   The amount by which the selling price is increased over the cost. **(6.8)**

**Mean (average)**   The sum of values divided by the number of values. **(7.3)**

**Median**   The middle value of several values or the average of the two middle values. **(7.3)**

**Meter**   The unit for length in the metric system. **(8.3)**

**Metric system of measurement**   The system of measurement used in most of the countries of the world. **(8.3)**

**Minuend**   The number before the minus sign or the top number in a subtraction problem. **(1.5)**

**Mixed number**   The sum of a whole number and a proper fraction. **(3.5)**

**Mode**   The value that occurs most frequently in a set of values. **(7.3)**

**Multiple of a number**   The product of the number and a whole number. **(2.1)**

**Natural number**   The numbers used in counting: 1, 2, 3, 4, 5, 6, . . . . **(1.1)**

**Negatives**   The negative of $a$ is $-a$. **(9.6)**

**Number line**   A line used to "picture" numbers. **(1.1)**

**Numeral**   A symbol used to represent a number. **(1.1)**

**Numerator**   The top number in a fraction. **(3.1)**

**Parallelogram**   A four-sided figure whose opposite sides are parallel. **(8.6)**

**Percent**   The number of parts in one hundred parts. **(6.1)**

**Perfect square**   The square of a whole number. **(2.8)**

**Perimeter**   The distance around a figure. **(8.5)**

**Prime number**   A natural number which has exactly two different divisors, 1 and itself. **(2.7)**

**Principal**   Money borrowed or invested. **(6.6)**

**Product**   The number which results in a multiplication problem **(2.1)**

**Proper fraction**   A fraction which has numerator less than the denominator. **(3.1)**

**Proportion**   A statement that two ratios are equal. **(6.1)**

**Quotient**   The number which is the result of a division problem. **(2.4)**

**Radical**   The symbol $\sqrt{\phantom{x}}$ used to indicate the square root. **(2.8)**

**Radius of a circle**   The distance from the center to each point on a circle. **(8.7)**

**Rate of interest**   The percent interest. **(6.6)**

**Ratio**   A comparison of two numbers using division. **(6.1)**

**Rational numbers**   The fractions formed by dividing pairs of integers. **(9.5)**

**Real numbers**   The rational numbers together with the irrational numbers. **(9.5)**

**Reciprocal**   The fraction formed by interchanging numerator and denominator. **(3.4)**

**Remainder**   The final difference in a division problem. **(2.5)**

**Sale price**   Regular price minus the discount. **(6.5)**

**Scientific notation** A number written as a number between 1 and 10 times a power of 10. **(9.7)**

**Similar terms** Terms which contain the same variables. **(9.6)**

**Simple interest** Interest that is charged only on the principal. **(6.6)**

**Solution of an equation** A number which makes the equation true. **(10.1)**

**Sphere** A solid in the shape of a ball. **(8.9)**

**Square** A rectangle with $l = w$. **(8.5)**

**Square of a number** A number raised to the second power. **(2.8)**

**Square root of a perfect square** The number which is squared to give the perfect square. **(2.8)**

**Statistics** An area of mathematics dealing with the collection, presentation, and interpretation of numerical data. **(7.1)**

**Subtrahend** The number following the minus sign or the bottom number in a subtraction problem. **(1.5)**

**Sum** The result of an addition problem. **(1.2)**

**Surface area** The area of all the surfaces of an object. **(8.8)**

**Term** A part of an algebraic expression which is a product of numbers and variables and which is separated from the rest of the expression by plus (or minus) signs. **(9.6)**

**Trapezoid** A four-sided figure with two parallel sides. **(8.6)**

**Triangle** A three-sided figure. **(8.6)**

**Unit fraction** A fraction which has value 1 and is used in measurement conversions. **(8.1)**

**Unit price** The total price divided by the number of units. **(5.7)**

**Unlike fractions** Fractions with different denominators. **(4.1)**

**Variables** Letters used to represent numbers. **(9.6)**

**Volume** The number of cubic units in a solid. **(8.8)**

**Whole number** The natural numbers together with zero: 0, 1, 2, 3, 4, . . . . **(1.1)**

# APPLICATIONS INDEX

Numbers in parentheses are problem numbers.

## AGRICULTURE

Farming, 24, 125(**90**), 224–25, 226(**3, 4**), 433(**31**), 458, 482(**24**), 583(**27**), 588(**16**), 589(**15**), 592(**45**)

Gardening, 57(**61**), 466(**27**), 467(**27**), 481(**19**), 490(**29**), 586, 589(**14**), 594(**14**)

Ranching, 30, 43, 447(**43**), 464, 466(**26**), 467(**26**), 468(**2**), 475(**19**), 502(**92**)

## CONSTRUCTION

Glass, 487–88, 490(**31**), 500(**60**)

House construction, 224, 257(**4**), 292(**36**), 294(**36**), 399(**1–8**), 439

Insulation, 475(**18**), 480(**25**), 488, 490(**32**), 491(**32**), 496, 497(**15**), 501(**70**)

Masonry, 78(**27**), 103(**14**), 196, 199(**29**), 201(**29**), 227(**5**), 255(**47**)

Metal, 487, 491(**31**), 497(**14**)

Painting, 441(**29**), 470–71, 475(**2**), 482(**25**), 488, 494, 496(**14**), 497(**15**)

Plumbing, 196

Road construction, 15(**57–60**), 227(**3, 4**), 230(**37**), 484, 491(**29**), 504(**13**)

Wood, 164(**30**), 498(**1, 2**), 578

## CONSUMER AFFAIRS

Bills, 259, 264(**12, 13**), 267(**12, 13**)

Checking account, 25(**34**), 43, 45(**31**), 51(**13**), 53(**12**), 57(**64**), 60(**21**), 237–38, 239(**23–25**), 240(**2, 23–25**), 248(**40**), 249(**40**), 259, 260, 263, 264(**16, 17**), 265(**31**), 268(**16, 17**), 269(**31**), 318(**96**), 513, 514(**26**), 515(**26**), 521, 522, 523(**37**), 527, 528, 530(**55, 56**), 532(**55, 56**), 596(**17**), 600(**93**)

Housing, 44(**29**), 57(**65**), 422(**5–10**)

Purchasing, 25(**31**), 32(**35**), 36(**54**), 38(**54**), 46, 47, 51(**14**), 53(**13, 19, 21**), 60(**17**), 73(**35**), 94(**31**), 97, 98, 99, 100(**7–11**), 101(**12, 13, 16–18, 22, 23**), 102(**4, 7–10**), 103(**11, 12, 16–19, 23**), 111(**33–35**), 112(**33, 35**), 123(**31, 34, 35**), 127(**16, 17**), 134(**22**), 136(**22**), 175, 178(**9, 10**), 180(**9, 10**), 183(**54**), 254, 256(**48**), 257(**2, 5**), 265(**30, 34**), 267(**44–47**), 269(**34**), 270(**44–47**), 276, 278(**25, 29, 31**), 279(**35**), 280(**25**), 281(**29, 31, 35**), 282(**2**), 293(**49**), 295(**49**), 297, 300, 302(**29–32**), 303(**15, 29–32**), 304(**34**), 304(**1, 2**), 311(**47**), 315(**24, 33, 34**), 316, 317(**63, 64, 66**), 318(**99, 100, 101, 103**), 319(**11**), 320(**18, 20**), 326(**35**), 328(**35**), 350(**25**), 440(**25**), 448(**43**), 460–61, 466(**28**), 467(**28**), 473(**17, 19**), 475(**17**), 475(**1**), 480(**24**), 501(**85, 86**), 530(**58**), 597(**37, 38**), 600(**97, 109**), 601(**113**), 604(**173, 183**), 605(**188**)

Rentals, 278(**33**), 281(**33**), 293(**50**), 295(**50**), 302(**35**), 304(**35**), 522, 528, 583(**2**)

Savings account, 30–31, 46, 50(**10**), 303(**14**), 313(**4**), 349(**25**), 521, 581(**35**), 585, 598(**1**)

## ECONOMICS

Benefits, 58(**81**), 103(**21**), 123(**32**), 163(**32**), 420(**23**)

Commission, 349(**19**), 350(**19**), 358–61, 363(**1–6**), 364(**7**), 365(**1–7**), 366(**1**), 372(**13**), 373(**13**), 391(**30, 31**), 392(**44**), 602(**144, 145**)

Discount, 361–63, 364(**8–16**), 365(**8–16**), 366(**2**), 372(**14**), 373(**14**), 391(**32, 33**), 394(**12, 17**), 585, 587(**6**), 589(**8**), 591(**32**), 602(**146**)

Expenses, 26(**33**), 46(**32**), 313(**2**), 395, 402(**9–20**), 522

Fees, 77, 156(**33**)

Gross National Product, 8(**48**)

## Interest

Interest, 349(**20**), 366–71, 371(**1–4**), 372(**5–12**), 372(**1, 2**), 373(**3–12**), 373(**1–5**), 380(**17, 18**), 381(**17, 18**), 391(**34, 35**), 392(**48**), 394(**13, 18**), 587(**6**)

Investment, 33(**34**), 34(**1–4**), 195, 530(**57**), 532(**57**), 602(**148**)

Loans, 154(**34**), 254, 257(**3**), 290, 292(**35, 38**), 294(**35**), 311(**45**), 313(**45**), 315(**25**), 392(**43**), 602(**147**)

Markup/markdown, 382–86, 387(**1–12**), 388(**1–12**), 388(**1, 2**), 391(**38, 39**), 392(**41, 46**), 394(**15, 16**), 404(**33–40**), 405–6, 419(**24–30**), 592(**43**)

Money, 236–37, 274–76

Profit, 33(**32**), 51(**20**), 596(**18**), 602(**150**)

Retirement, 340(**3**)

Revenue, 33(**33**), 52(**18, 20, 22**), 78(**25**), 95, 99(**4**), 100(**5**), 102(**2**), 315(**27**)

Sales, 10(**50**), 46(**31**), 54(**1**), 102(**5**), 103(**13**), 104(**1, 2**), 123(**33**), 127(**27**), 227(**11**), 228(**1, 2**), 232(**17**), 338(**30**), 339(**1**), 350(**22**), 403(**21–32**), 408(**1–10**), 414, 418(**21**), 422(**22**), 532(**58**), 584, 587(**4**), 588(**1**)

Stock, 101(**21**), 226(**12**), 227(**12**)

Taxes, 54(**20**), 262, 265(**29**), 268(**34**), 282(**3**), 292(**37**), 294(**37**), 311(**42, 43**), 313(**42, 43**), 317(**79**), 318(**98**), 340(**2**), 351–56, 356(**1–8**), 357(**9–12**), 357(**1–12**), 358(**1, 2**), 364(**17, 18**), 365(**19, 20**), 366(**17–20**), 391(**26–29**), 392(**45, 49**), 393(**11**), 396, 410(**30–32**), 587(**9**), 602(**141–143**), 608(**264**)

Wages, 58(**79**), 153, 154(**33**), 156(**34**), 173(**20**), 174(**20**), 179(**11**), 180(**11**), 225, 226(**10**), 227(**10**), 276, 278(**26, 32, 34**), 280(**26**), 281(**32**), 293(**51, 52**), 295(**51, 52**), 297, 301(**14**), 313(**47**), 316(**39**), 317(**62**), 349(**21**), 350(**21**), 358(**17**), 409(**19–28**), 410(**29**), 416, 418(**23**), 422(**23**), 587(**5**), 589(**4**), 601(**112**)

## EDUCATION

Classes and teachers, 339(**66, 67**), 398, 400(**21–32**), 420(**24–30**)

Exams, 143, 335(**29**), 349(**18**), 350(**17**), 357(**17**), 391(**23**), 413, 415, 420(**22**), 421(**1, 2**), 584, 587(**1, 2**), 588(**2**), 591(**30**), 594(**11**), 607(**263**)

Grade point average, 290, 292(**39**), 294(**39**), 302(**33**), 304(**33**), 316(**52**)

Students, 50(**6**), 52(**1**), 54(**2**), 71, 101(**20**), 103(**20**), 125(**89**), 134(**19**), 164(**31**), 178(**3**), 216(**30**), 325, 336(**64, 65**), 398, 399(**9–14**), 400(**15–20**), 401(**33–40**), 424(**32–40**), 603(**151–54**)

## ENGINEERING

Electrical engineering, 226(**5**), 338(**28**), 434(**31**), 442(**2**), 501(**84**), 595(**11**)

Fluid flow, 432, 433(**30**), 434(**30**), 480(**19**)

Fuel, 10(**49**), 51(**17**), 60(**19**), 164(**28**), 168(**19**), 176, 178(**1, 2**), 184(**55, 56**), 269(**38**), 303(**16**), 448(**42**), 452, 605(**193**)

Machine shop, 478, 480(**21–23**), 481(**21–23**), 482(**1, 2**), 504(**14**)

Manufacturing, 73(**36**), 281(**34**), 414, 492(**1**)

Printing, 45(**30**)

Rotations, 79(**27**), 174(**19**), 177, 179(**13, 14**), 180(**13, 14**), 183(**51, 52**), 184(**57**), 186(**22**), 598(**63**)

## GEOMETRY

Area and perimeter, 9(**50**), 50(**4**), 53(**5**), 96, 100(**6**), 102(**6**), 151, 154(**31**), 156(**31**), 175(**23**), 214(**31**), 226(**8**), 279(**37**), 281(**37**), 313(**44**), 402(**1–8**), 438, 458, 460, 461, 464, 467(**25**), 478, 487

Lengths, 160, 163(**27, 29**), 169(**19**), 174(**22**), 179(**8**), 224, 442, 578, 580(**28**), 581(**31**), 583(**38, 39**), 588(**14, 20**), 589(**20**), 592(**41**)

# INDEX